Juvenile Delinquency
An Integrated Approach

James W. Burfeind
Professor
Department of Sociology
University of Montana
Missoula, Montana

Dawn Jeglum Bartusch
Assistant Professor
Department of Sociology and Criminology
Valparaiso University
Valparaiso, Indiana

JONES AND BARTLETT PUBLISHERS
Sudbury, Massachusetts
BOSTON   TORONTO   LONDON   SINGAPORE

World Headquarters
Jones and Bartlett Publishers
40 Tall Pine Drive
Sudbury, MA 01776
978-443-5000
info@jbpub.com
www.jbpub.com

Jones and Bartlett Publishers Canada
6339 Ormindale Way
Mississauga, Ontario L5V 1J2
Canada

Jones and Bartlett Publishers International
Barb House, Barb Mews
London W6 7PA
United Kingdom

Jones and Bartlett's books and products are available through most bookstores and online booksellers. To contact Jones and Bartlett Publishers directly, call 800-832-0034, fax 978-443-8000, or visit our website, www.jbpub.com.

Substantial discounts on bulk quantities of Jones and Bartlett's publications are available to corporations, professional associations, and other qualified organizations. For details and specific discount information, contact the special sales department at Jones and Bartlett via the above contact information or send an email to specialsales@jbpub.com.

Production Credits
Publisher—Public Safety Group: Kimberly Brophy
Acquisitions Editor: Stefanie Boucher
Associate Editor: Janet Morris
Production Director: Amy Rose
Associate Production Editor: Carolyn F. Rogers
Director of Marketing: Alisha Weisman
Marketing Associate: David Weliver
Manufacturing and Inventory Coordinator: Amy Bacus
Composition: Carlisle Publishers Services
Cover Design: Anne Spencer
Senior Photo Researcher: Kimberly Potvin
Photo Researcher: Christine McKeen
Cover Image: Photo © Robert Deal/ShutterStock, Inc.;
    column © Ron Chapple/Thinkstock/Alamy Images
Chapter Opener Image: © Masterfile
Printing and Binding: Malloy, Inc.
Cover Printing: Malloy, Inc.

Library of Congress Cataloging-in-Publication Data
Burfeind, James W., 1953-
    Juvenile delinquency : an integrated approach / James W. Burfeind,
Dawn Jeglum Bartusch.
        p. cm.
    Includes bibliographical references and index.
    ISBN 0-7637-3628-7 (hardcover : alk. paper)
    1. Juvenile delinquency. 2. Juvenile justice, Administration of.
I. Bartusch, Dawn Jeglum. II. Title.
HV9069.B79 2006
364.36—dc22

                                        2005024693

Printed in the United States of America
09 08 07 06 05   10 9 8 7 6 5 4 3 2 1

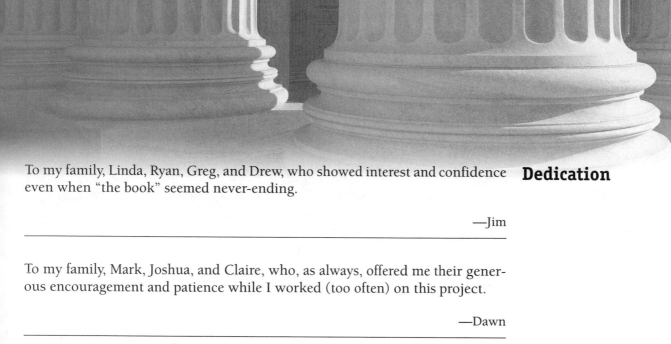

To my family, Linda, Ryan, Greg, and Drew, who showed interest and confidence even when "the book" seemed never-ending.

Dedication

—Jim

To my family, Mark, Joshua, and Claire, who, as always, offered me their generous encouragement and patience while I worked (too often) on this project.

—Dawn

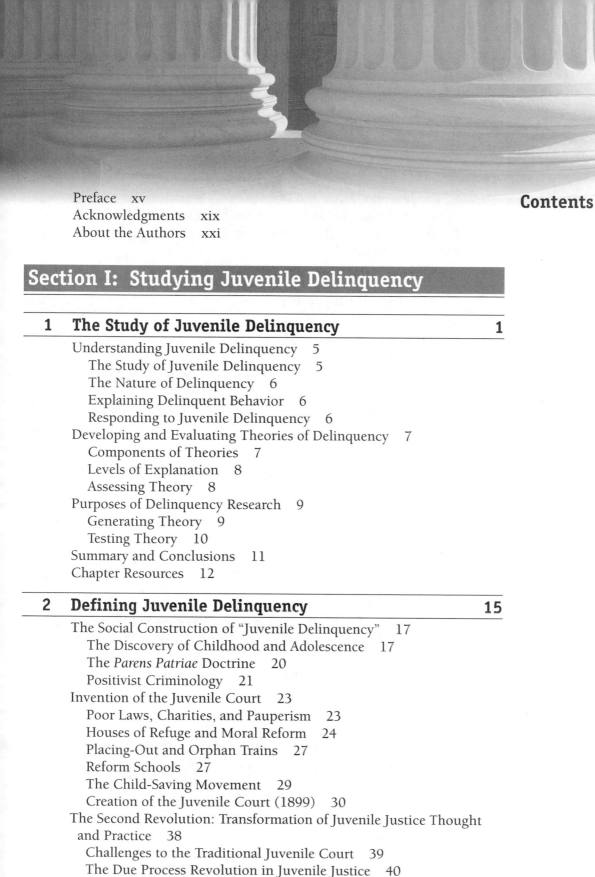

Contents

Section I: Studying Juvenile Delinquency

Section II: The Nature of Delinquency

Section III: Explaining Delinquent Behavior

8   Biological and Psychological Approaches   289

9   Situational and Routine Dimensions of Delinquency 337

10 Social Control Theories: Family Relations   373

11 Social Learning Theories: Peer-Group Influences   423

12 Social Structure Theories: Community, Strain, and Subcultures   467

13 Labeling and Critical Criminologies   515

14 Understanding Delinquency   567

Section IV: Responding to Juvenile Delinquency

Welcome to *Juvenile Delinquency: An Integrated Approach*. We believe that most students have a natural curiosity about juvenile delinquency. Perhaps they were once involved in delinquency, but were never caught. Regardless of their personal histories, students come to class with an almost endless list of questions. Why do some kids engage in delinquent acts? Is delinquent behavior a normal part of adolescence? Is the delinquency problem growing worse? Are adolescents becoming more violent? Why is delinquency rampant in some areas? What causes delinquent behavior? How can we best respond to the problem of juvenile delinquency? Can delinquency be prevented and controlled? As criminologists and authors of this book, we share this interest and desire to understand juvenile delinquency—that's why we wrote this text.

As we describe in Chapter 1, the scientific study of delinquent behavior uses two basic tools: theory and research. Theories of delinquency provide a systematic presentation of key causal factors, and offer insight into the causes of juvenile delinquency. Delinquency research seeks either to test theory or to provide sufficient information to develop theory. Thus, theory and research go hand-in-hand. The primary purpose of this book is to cultivate an understanding of juvenile delinquency by integrating theory and research.

■ Organization of the Text

Juvenile Delinquency: An Integrated Approach is divided into four main sections, containing 15 chapters. The first two sections focus on *defining* and *describing juvenile delinquency*. The third section of the book concentrates on *explaining delinquent behavior,* and the fourth section considers *responding to juvenile delinquency* through contemporary juvenile justice systems.

In Section I, *Studying Delinquency,* we describe the basic components of theory and conceptual tools for assessing theory, as well as research methods and sources of data for studying delinquent behavior. In this section we also discuss the invention and historical transformation of the concept of "juvenile delinquency" and the juvenile justice system.

In Section II, *The Nature of Delinquency,* we present a trilogy of chapters in which we consider the nature of delinquent offenses, offenders, and patterns of offending. We begin by examining the extent of delinquent behavior and the types of offenses in which young people are involved. We then attempt to answer the question "Who are the offenders?" by exploring the social correlates of delinquency: age, gender, race, and social class. Next, we consider patterns of offending, presenting the key elements of the developmental perspective and influential developmental models.

In Section III, *Explaining Delinquent Behavior,* we present the primary sociological, psychological, and biological theories that criminologists have offered to explain delinquency and social responses to it. First, we describe classical and positivist schools of thought in criminology, and consider the question "Is delinquent behavior chosen or determined?" We then examine biological and psychological approaches to delinquency, as well as social control, social learning, and social structure theories. In this section, we also explore characteristics of situations and the routine activities of adolescents that create opportunities for delinquent behavior. Finally, we consider social responses to delinquency from the perspectives of labeling theory and critical criminologies.

Section IV, *Responding to Juvenile Delinquency,* considers contemporary juvenile justice. There is no single juvenile justice system in the United States; rather, each state has its own juvenile justice system, made up of various components at both the state and local levels and operated by both public and private agencies. Thus, contemporary juvenile justice is hardly a "system." Juvenile justice practices and procedures are examined with regard to law enforcement, courts, and corrections, providing a full understanding of contemporary juvenile justice in action.

■ Distinctive Features

In addition to the pedagogical features that we describe in the following section, this text includes these distinctive features:

Chapter Opening Case Studies

We begin each chapter with a case study or excerpt that illustrates key ideas addressed in the chapter. These case studies provide students with a compelling introduction to the topic at hand. The cases are not sensationalized accounts of recent high-profile cases, that fail to illustrate the characteristics and true nature of the vast majority of delinquent offenders and acts. Instead, most of the cases we use are drawn from classic works in the field of criminology. At the end of each chapter, a critical thinking question revisits the case study and invites students to consider the case study again, in light of what they have read in the chapter.

Applying Criminological Theory and Research

Application boxes appear throughout this book in an effort to enhance understanding of juvenile delinquency. These boxes provide students with practical, relevant, and engaging applications of criminological theory and research. We offer five types of application boxes, which serve distinct purposes.

- *Case in Point* boxes provide real-life case examples, appellate court cases, or statutory law to illustrate points of discussion.
- *Research in Action* boxes describe delinquency research, highlighting particular programs of research, and offering insight into how researchers actually carry out their work. For example, these boxes often describe the measurement of variables used to test theories.
- *Theory into Practice* boxes illustrate how theory is translated into policy or practice. These boxes describe specific programs or strategies, derived from the theories we discuss, for preventing or reducing delinquency.
- *Expanding Ideas* boxes elaborate on key points or highlight theoretical issues presented in the text. For example, *Expanding Ideas* boxes sometimes list and discuss the basic elements or propositions of theories.
- *Links* boxes provide descriptions of Internet sites relevant to the theories and research we discuss. The actual Internet addresses for these sites will be listed and continually updated at http://criminaljustice.jbpub.com.

Emphasis on Integration of Theory and Research

As the title of the book indicates, we emphasize the integration of theory and research in understanding juvenile delinquency. We discuss research within its theoretical context. For example, we consider families, peers, and gangs within the context of the theoretical traditions that most actively address these arenas. This integrated approach helps students understand how social scientists actually "do" criminology by developing theory and conducting research.

Throughout the book, we present the key theories of delinquent behavior, along with the most relevant research used to test these theories. However, our coverage of research is not encyclopedic. We do not attempt to cite every study in every area of research, but instead we discuss selected research studies thoroughly. This approach provides *depth* of understanding, rather than sheer *breadth* of coverage. Neither do we try to oversimplify our presentation of theory and research. Much of what is interesting and insightful about delinquency theory and research is lost when it is offered in an abbreviated fashion, and we do not want to lose the richness of this field of study.

■ Resources to Aid Learning

Juvenile Delinquency: An Integrated Approach includes several pedagogical features that will assist students in mastering the material we present.

Beginning of Chapter

- **Chapter outlines:** Identify the major themes and topics discussed in each chapter.
- **Chapter objectives:** Alert students to the issues and concepts they should understand after reading each chapter. The list of objectives also includes key terms and theories students should watch for as they read.

In Text

- **Running glossary:** Supplies students with definitions of key terms.
- **Application boxes:** Enhance understanding of juvenile delinquency by providing students with practical, relevant, and engaging applications of theory and research. (See our descriptions on the preceding page of the different types of boxes we use throughout this text.)

End of Chapter

- **Chapter summaries and conclusions:** Draw attention to important points from the chapter, and provide conclusions to chapter materials.
- **Lists of theories:** Highlight the theories presented in the chapter.
- **Critical thinking questions:** Invite readers to apply knowledge acquired through reading the chapter, and to consider the chapter opening case study in light of the materials presented in the chapter.
- **Suggested readings:** Offer references to primary sources discussed in the chapter, and to interesting applications of key concepts.

■ Supplements

Juvenile Delinquency: An Integrated Approach is accompanied by several supplements to aid instructors. These supplements, available from the book's Web site at http://criminaljustice.jbpub.com, include:

- A detailed outline of each chapter
- A test bank of questions for exam preparation
- PowerPoint presentations to accompany each chapter

Acknowledgments

We are grateful to the many individuals who have provided thoughtful reviews of our chapters. Numerous reviewers, most of whom were anonymous to us, gave helpful comments. These reviewers included Rob Balch (The University of Montana), Scott Decker (University of Missouri–St. Louis), Dusten Hollist (The University of Montana); Donald Shoemaker (Virginia Polytechnic Institute and State University), and Randy Blazak (Portland State University). While we sometimes stubbornly declined reviewers' suggestions, many of their comments helped us greatly improve the text. Karen Heimer and Terrie Moffitt also read chapters and offered their encouragement and support of this project. Two students at The University of Montana were of tremendous help: Katie Murphy and Katherine Georger. We appreciate their work.

We are especially grateful for the commitment and dedication of the Jones and Bartlett team, which helped us in every way to bring this project to life. We especially appreciate the enthusiasm, creativity, and patience of Stefanie Boucher, Janet Morris, and Carolyn Rogers. Carolyn gently kept us focused on the tasks at hand, and with great competence moved the project forward at a remarkable pace. Early in the project, Chambers Moore offered encouragement and expertise. Laura Passin carefully copyedited each chapter and offered us valuable suggestions for improving the text.

James W. Burfeind is Chair and Professor of Sociology at The University of Montana. He earned a Ph.D. in criminology and urban sociology from Portland State University. He has extensive experience in juvenile probation and parole, and adolescent residential care. Dr. Burfeind's teaching and research interests are in criminological theory, juvenile delinquency, juvenile justice, delinquency prevention, and program evaluation. He is coauthor, with Ted D. Westermann, of *Crime and Justice in Two Societies: Japan and the United States* (1991, Brooks/Cole).

Dawn Jeglum Bartusch is Assistant Professor of Sociology and Criminology at Valparaiso University in Indiana. In 1998, she earned her Ph.D. in Sociology from the University of Wisconsin–Madison, where her work was supported by a Jacob K. Javits Fellowship from the US Department of Education. Dr. Bartusch's research interests include interactionist explanations of delinquency, gender differences in involvement in delinquent behavior, and the social context in which crime and delinquency occur. Her research has appeared in *Criminology, Social Forces, Law and Society Review,* and the *Journal of Abnormal Psychology.*

About the Authors

Studying Juvenile Delinquency

I

The Study
of Juvenile
Delinquency

Chapter Outline

- Understanding juvenile delinquency
- Developing and evaluating theories of delinquency
- Purposes of delinquency research

Chapter Objectives

After completing this chapter, students should be able to:

- Understand the approach and structure of this book.
- Describe the key components of theory.
- Describe the relationship between theory and research.
- Identify the purposes of research.
- Understand key terms:
  juvenile delinquency
  theory
  concepts
  propositions
  theory of delinquency
  level of explanation
  inductive theorizing
  deductive theorizing

CASE IN POINT

Rick: A "Delinquent Youth"

The youth court adjudicated or judged Rick, a 14-year-old, a "delinquent youth," for motor vehicle theft and placed him on formal probation for six months. He and a good friend took without permission a car that belonged to Rick's father. They were pulled over by the police for driving erratically—a classic case of joyriding.

Rick was already a familiar figure in the juvenile court. When Rick was 12, he was referred to the court for "deviant sex" for an incident in which he was caught engaging in sexual activity with a 14-year-old girl. The juvenile court dealt with this offense "informally." A probation officer met with Rick and his parents to work out an agreement of informal probation that included "conditions" or rules, but no petition into court. Not long after this first offense, Rick was taken into custody by the police for curfew violation and, on a separate occasion, vandalism—he and his good friend had gotten drunk and knocked down numerous mailboxes along a rural road. In both of these instances, Rick was taken to the police station and released to his parents. Even though Rick's first formal appearance in juvenile court was for the auto theft charge, he was already well-known to the police and probation departments.

Rick was a very likable kid; he was pleasant and personable. He expressed a great deal of remorse for his delinquent acts and seemed to genuinely desire to change. He had a lot going for him; he was goal-directed, intelligent, and athletic. He interacted well with others, including his parents, teachers, and peers. His best friend, an American Indian boy who lived on a nearby reservation, was the same age as Rick and had many similar personal and social characteristics. Not surprisingly, the boy also had a very similar offense record. In fact, Rick and his friend were often "companions in crime," committing many of their delinquent acts together.

Rick was the adopted son of older parents who loved him greatly and saw much ability and potential in him. They were truly perplexed by the trouble he was in, and they struggled to understand why Rick engaged in delinquent acts and what needed to be done about it. Rick, too, seemed to really care about his parents. He spent a good deal of time with them and apparently enjoyed their company. Because Rick was adopted as an infant, these parents were the people he considered family.

Rick attended school regularly and earned good grades. He was not disruptive in the classroom or elsewhere in the school. In fact, teachers reported that he was a very positive student both in and out of class and that he was academically motivated. He did his homework and handed in assignments on time. He was also actively involved in sports—football, wrestling, and track and field.

Rick's six months of formal probation for auto theft turned into a two-year period as he continued to get involved in delinquent acts. Through regular meetings and enforcement of probation conditions, his probation officer tried to work with Rick to break his pattern of delinquency. Such efforts were to no avail. Rick continued to offend, resulting in an almost routine series of court hearings that led to the extension of his probation supervision period.

The continuing pattern of delinquency included a long list of property and status offenses: minor in possession of alcohol, numerous curfew violations, continued vandalism, minor theft (primarily shoplifting), and continued auto theft, usually involving joyrides in his father's car.

Rick's "final" offense was criminal mischief, and it involved extensive destruction of property. Once again, Rick and his best friend "borrowed" his father's car, got drunk, and drove to Edina, an affluent suburb of Minneapolis. For no apparent reason, they parked the car and began to walk along France Avenue, a major road with office buildings along each side. After walking a while, they started throwing small rocks toward buildings, seeing how close they could get. Their range increased quickly and the rocks soon reached their targets, breaking numerous windows. The "fun" turned into thousands of dollars worth of window breakage in a large number of office buildings.

Because of the scale of damage, Rick faced the possibility of being placed in a state training school. As a potential "loss of liberty case," Rick was provided with representation by an attorney. This time, the juvenile court's adjudication process followed formal procedures, including involvement of a prosecutor and a defense attorney. In the preliminary hearing, Rick admitted to the petition (statement of charges against him), and the case was continued to a later date for disposition (sentencing). In the meantime, the judge ordered a predisposition report.

The predisposition report is designed to individualize the court's disposition to "fit the offender." The investigation for the report uses multiple sources of information, including information from the arresting officer, parents, school personnel, coaches, employers, friends, relatives, and, most importantly, the offending youth. The predisposition report tries to describe and explain the pattern of delinquency and then offer recommendations for disposition based on the investigation. In Rick's case, the predisposition report attempted to accurately describe and explain his persistent pattern of property and status offending, and it offered a recommendation for disposition. Finding no information to justify otherwise, the probation officer recommended that Rick be committed to the Department of Corrections for placement at the Red Wing State Training School. Depending on one's viewpoint, the state training school represented either a last ditch effort for rehabilitation or a means of punishment through restricted freedom. Either way, Rick was viewed as a chronic juvenile offender, with little hope for reform.

It was one of those formative experiences. I [coauthor Jim Burfeind] was fresh out of college and newly hired as a probation officer. I was meeting with two experienced attorneys—one the defense, the other the prosecutor. Almost in unison, it seemed, they turned to me and asked, "Why did Rick do this? Why did he develop such a persistent pattern of delinquency?" They wanted to make sense of Rick's delinquency, and they wondered how the juvenile court could best respond to his case.

I had become familiar with Rick only in the previous few weeks when his case was reassigned to me as part of my growing caseload as a new probation officer. Now, meeting with the attorneys to gather information for the predisposition report, I was being asked to explain Rick's pattern of delinquent behavior to two legal experts who had far more experience in the juvenile justice system than I did. I was, after all, new to the job. How could I possibly know enough to offer an explanation? I also had the daunting responsibility of making a recommendation for disposition that the judge would most likely follow completely. Rick's future was at stake, and my recommendation would determine the disposition of the juvenile court.

As I attempted to respond to the attorneys sitting in front of me, my mind was flooded with questions. The answers to these questions became the basis for my predisposition report—an attempt to explain Rick's delinquent behavior and, based on this understanding, to recommend what should be done through court disposition. The questions with which I wrestled included the following:

- Is involvement in delinquency common among adolescents—that is, are most youths delinquent? Maybe Rick was just an unfortunate kid who got caught.

- Are Rick's offenses fairly typical of the types of offenses in which youths are involved?

- Will Rick "grow out" of delinquent behavior?

- Is Rick's pattern of offending much the same as those of other delinquent youths?

- Do most delinquent youths begin with status offenses and then persist and escalate into serious, repetitive offending? (*Status offenses* are acts, such as truancy and running away, that are considered offenses when committed by juveniles but are not considered crimes if committed by adults.)

- Is there a rational component to Rick's delinquency so that punishment by the juvenile court would deter further delinquency?

- Did the fact that Rick was adopted have anything to do with his involvement in delinquency? Might something about Rick's genetic makeup and his biological family lend some insight into his behavior?

- What role did Rick's use of alcohol play in his delinquency?

- Are there family factors that might relate to Rick's involvement in delinquency?

- Were there aspects of Rick's school experiences that might be related to his delinquency?

- What role did Rick's friend play in his delinquent behavior?

- Did the youth court's formal adjudication of Rick as a "delinquent youth" two years earlier label him and make him more likely to continue in delinquent behavior?

- Should the juvenile court retain jurisdiction for serious, repeat offenders like Rick?

- What should the juvenile court try to do with Rick: punish, deter, or rehabilitate him?

- Should the juvenile court hold Rick less responsible for his acts than an adult because he has not fully matured?

Perhaps this list of questions seems a little overwhelming to you now. We don't present them here with the expectation that you will be able to answer them. Instead, we present them to prompt you to think about what causes juvenile delinquency and to give you an idea of the types of questions that drive the scientific study of delinquent behavior. Throughout this book, we address these types of questions as we define delinquency; consider the nature of delinquent offenses, offenders, and offending; and present a variety of theories to explain delinquent behavior. We return to Rick's story and these questions in Chapter 14. After reading the next 12 chapters, you should have the tools necessary to think about and respond to these questions in a whole new light.

■ Understanding Juvenile Delinquency

The questions that shape the scientific study of juvenile delinquency constitute attempts to *define, describe, explain,* and *respond to* delinquent behavior. Rather than being asked with regard to a particular case like Rick's, the questions that inspire the study of juvenile delinquency are cast more broadly in order to understand delinquent behavior as it occurs among adolescents.

An understanding of delinquent behavior builds upon explanations that have been offered in theories and findings that have been revealed in research. The primary purpose of this book is to cultivate an understanding of juvenile delinquency by integrating theory and research. Throughout the book, we focus on the central roles that theory and research play in the study of delinquency, because these two components form the core of any scientific inquiry.

Before we go any further, we must define what we mean by "juvenile delinquency." This definition is far more complicated than you might think. In the next chapter, we offer a thorough discussion of the social construction and transformation of the concept of juvenile delinquency. Here we offer a brief working definition of <u>juvenile delinquency</u> as actions that violate the law, committed by a person who is under the legal age of majority.

juvenile delinquency
Actions that violate the law, committed by a person who is under the legal age of majority.

Our exploration of juvenile delinquency reflects the four basic tasks of the scientific study of delinquency—to *define, describe, explain,* and *respond to* delinquent behavior. The first two major sections of this book are devoted to defining and describing juvenile delinquency, the third section to explaining delinquent behavior, and the final section to contemporary ways of responding to juvenile delinquency. Responses to delinquent behavior, however, should be based on a thorough understanding of delinquency. Thus, an understanding of juvenile delinquency must come first.

The Study of Juvenile Delinquency

The first section of this book describes the historical transformation of the concept of juvenile delinquency and the methods and data sources researchers use to study involvement in delinquent behavior. We begin by developing a working understanding of what we commonly call "juvenile delinquency" (Chapter 2). This includes not only the social, political, and economic changes that led to the social construction of juvenile delinquency as a legal term, but also the contemporary transformations that have dramatically altered how we as a society view, define, and respond to juvenile delinquency. We then explore how researchers "measure" delinquency (Chapter 3). We describe the research process, various methods of gathering data and doing research on juvenile delinquency, and sources of data on crime and delinquency.

The Nature of Delinquency

The second section of this book presents a trilogy of chapters in which we describe the nature of delinquent offenses, offenders, and patterns of offending. Any attempt to explain juvenile delinquency must first be able to accurately describe the problem in terms of these three dimensions. Chapters 4 through 6 report research findings that describe the extent of delinquent *offenses* (Chapter 4), the social characteristics of delinquent *offenders* (Chapter 5), and the developmental patterns of delinquent *offending* (Chapter 6).

Explaining Delinquent Behavior

The third section of this book examines a variety of explanations of delinquency that criminologists have proposed in theories and examined in research related to those theories. These chapters are organized in terms of the major themes that run through seven different groups of theories. One group of theories, for example, emphasizes the importance of peer group influences on delinquency. These theories, called social learning theories, address how delinquent behavior is learned in the context of peer group relations (Chapter 11). Six other themes are also considered: the question of whether delinquency is chosen or determined (Chapter 7); the role of individual factors, including biological characteristics and personality, in explaining delinquent behavior (Chapter 8); situational and routine dimensions of delinquency (Chapter 9); the importance of social relationships, especially family relations and school experiences, in controlling delinquency (Chapter 10); the structure of society, and how societal characteristics motivate individual behavior (Chapter 12); and social and societal responses to delinquency (Chapter 13). We also apply these various explanations to Rick's case, which opened this chapter, and examine integrated theoretical approaches (Chapter 14).

Throughout the book, as we present theoretical explanations for delinquency, we weave together theories and the most relevant research that criminologists have conducted to test those theories.

Responding to Juvenile Delinquency

The final section of this book comprises a single chapter that describes contemporary juvenile justice (Chapter 15). We have deliberately chosen to keep the discussion of juvenile justice in one chapter, in order to provide an undivided view of its structure and process. The formal system of juvenile justice includes police, courts, and corrections. Yet a substantial amount of juvenile delinquency is dealt with informally, sometimes by agencies outside the "system." Juvenile justice encompasses efforts at prevention, together with informal and formal action taken by the traditional juvenile justice system. Formal procedures, such as taking youths into custody and adjudicating them as delinquent youths, are central to the task of responding to juvenile delinquency. But informal procedures designed to prevent delinquency and divert youths from the juvenile justice system are far more common.

■ Developing and Evaluating Theories of Delinquency

In 1967, two noted sociologists, Travis Hirschi and Hanan Selvin, observed that theories of delinquency suggest a "sequence of steps through which a person moves from law abiding behavior to . . . delinquency."[1] Criminological theories try to identify and describe the key causal factors that make up this "sequence of steps" leading to delinquent behavior. In doing so, theories of delinquency emphasize certain factors as being causally important and then describe how these factors are interrelated in producing delinquent behavior. Stated simply: "a <u>theory</u> is an explanation."[2]

Components of Theories

Like other scientific theories, theories of delinquency are composed of two basic parts: concepts and propositions. <u>Concepts</u> isolate and categorize features of the world that are thought to be causally important.[3] Different theories of juvenile delinquency incorporate and emphasize different concepts. For example, the theories of delinquency we consider in later chapters include concepts such as personality traits, intelligence, routine activities of adolescents, relationship ties (called attachments), associations with delinquent friends, and social disorganization of neighborhoods. Concepts require definition.[4] Definitions serve two functions: they clarify concepts and provide common understanding, and they describe how concepts will be measured for the purpose of research.

<u>Propositions</u> tell how concepts are related. Scientific theories use propositions to make statements about the relationships between concepts.[5] Some propositions imply a *positive linear relationship* in which the "concepts increase or decrease together in a relatively straight-line fashion."[6] For example, some theories offer the proposition that the number of delinquent friends is positively related to delinquent behavior: as the number of delinquent friends increases, so does the likelihood of delinquency. In a *negative linear relationship,* the concepts vary in opposite directions. For instance, one theory offers the proposition that level of attachment and delinquency are negatively related: as attachment increases, delinquent behavior decreases. Relationships between concepts may also be *curvilinear.* Here, too, the concepts vary together, either positively or negatively, but after reaching a certain level, the relationship moves in the opposite direction. For example, researchers have found that parental discipline is related to delinquency in a curvilinear fashion.[7] Delinquent behavior is most frequent when parental discipline is either lacking or excessive, but it is least common when levels of discipline are moderate. If you think of parental discipline as a continuum, delinquency is highest on the two ends of the discipline continuum, when discipline is lax or excessive, and lowest in the middle, when discipline is moderate.

Different theories may offer competing propositions. One theory may propose that two concepts are related in a particular way, whereas another theory may claim that they are unrelated. For example, one of the major issues in delinquency theory is the role of the family in explaining delinquent behavior. One major theory contends that the family is essentially unrelated to delinquent behavior and that delinquent peers are an important factor in explaining delinquency. Another

theory An explanation that makes a systematic and logical argument regarding what is important and why.

concepts Isolated features of the world that are thought to be causally important.

propositions Theoretical statements that tell how concepts are related.

theory of delinquency
A set of logically related propositions that explain why and how selected concepts are related to delinquent behavior.

level of explanation The realm of explanation—individual, microsocial, or macrosocial—that corresponds to the types of concepts incorporated into theories.

influential theory proposes the opposite relationship, arguing that family relations are strongly related to delinquency, whereas peer relations are less important in explaining delinquency.[8]

To summarize, a **theory of delinquency** is a set of logically related propositions that explain why and how selected concepts are related to delinquent behavior.[9] A theory offers a logically developed argument that certain concepts are important in causing delinquent behavior. The purpose of theory, then, is to explain juvenile delinquency.

Levels of Explanation

Theories of delinquency operate at three different **levels of explanation**: individual, microsocial, and macrosocial.[10] On the *individual* level, theories focus on traits and characteristics of individuals, either innate or learned, that make some people more likely than others to engage in delinquent behavior. The *microsocial* level of explanation considers the social processes by which individuals become the "kinds of people" who commit delinquent acts.[11] Criminologists have emphasized family relations and delinquent peer group influences at this level. Some microsocial theories also point to the importance of the structural context of social interaction.[12] Race, gender, and social class, for example, influence social interaction not only within families and peer groups, but in virtually all social contexts. As a result, the distinction between social process and social structure is not always clear, nor is it always useful as a means of categorizing theoretical explanations.[13] At the *macrosocial* level, societal characteristics such as social class and social cohesiveness are used to explain group variation in rates of delinquency.[14] For example, poverty, together with the absence of community social control, is central to several explanations of why gang delinquency is more common in lower-class areas.[15]

The level of explanation—individual, microsocial, or macrosocial—corresponds to the types of concepts incorporated into a theory.[16] Individual-level explanations tend to incorporate biological and psychological concepts. Microsocial explanations most often use social psychological concepts, but may incorporate structural concepts that influence social interaction. Macrosocial explanations draw extensively on sociological concepts. Theories can be combined to form "integrated theories" (see Chapter 14), which sometimes merge different levels of explanation into a single theoretical framework.

Assessing Theory

We have proposed that concepts and propositions are the bare essentials of theory.[17] These components, however, do not automatically produce a valid explanation of delinquency. We can begin to assess the *validity* of theory—the degree to which it accurately and adequately explains delinquent behavior—by paying attention to several key dimensions of theory.[18] We highlight these dimensions (e.g., clarity, consistency, testability, applicability) in the following list of questions. We invite you to ask yourself these questions as you evaluate the theories of delinquency we present in later chapters and consider how well they explain delinquent behavior.

> **1. Conceptual clarity:** *How clearly are the theoretical concepts identified and defined?[19] How well do the concepts and propositions fit together—how compatible, complementary, and congruent are they?[20]*

2. **Logical consistency**: *Does the theoretical argument develop logically and consistently? Do the concepts and propositions depict a causal process leading to delinquency?*

3. **Parsimony**: *How concise is the theory in terms of its concepts and propositions?* This question concerns economy of explanation. Generally, simpler is better. So if two theories explain delinquency equally well, we should favor the theory that offers the more concise explanation with the smaller number of concepts.

4. **Scope**: *What is the theory attempting to explain?*[21] Some theories try to explain a wide variety of criminal acts and criminal offenders. Others focus on particular types of offenses or offenders. *What question is the theory designed to answer?* Theories of delinquency usually address one of two basic questions: (1) How and why are laws made and enforced? and (2) Why do some youths violate the law?[22] Far more theories try to answer the second question than the first.[23]

5. **Level of explanation**: *At what level (individual, microsocial, or macrosocial) does the theory attempt to explain delinquency?*

6. **Testability**: *To what extent can the theory be tested—verified or disproved by research evidence?* It is not enough for a theory simply to "make sense" by identifying key concepts and then offering propositions that explain how these concepts are related to delinquency.[24] Rather, theories must be constructed in such a way that they can be subjected to research verification.[25]

7. **Research validity**: *To what extent has the theory been supported by research evidence?*

8. **Applicability and usefulness**: *To what extent can the theory be applied practically? In other words, to what extent is the theory useful in policy and practice?*

These questions reflect key concerns in assessing theory. In the end, theory is the foundation for the accumulation of knowledge, and it is indispensable for an understanding of juvenile delinquency. However, theory must be tested through research. Together, theory and research constitute the two basic components of a scientific approach to juvenile delinquency.

■ Purposes of Delinquency Research

Delinquency research serves two vital purposes: to *generate or develop theory*, and to *test theory*.[26] In Chapter 3, we discuss research methods and sources of data used in the study of delinquency. Here we briefly describe the two purposes of research as it relates to theory.

Generating Theory

Research is sometimes used to gain sufficient information about juvenile delinquency to theorize about it.[27] Despite the old adage, "the data speak for themselves," research findings about delinquency require interpretation, and it is this interpretation that yields theory. As a result, the development of theoretical

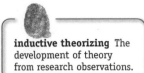

inductive theorizing The development of theory from research observations.

explanations of delinquency requires a long, hard look at the "facts" of delinquency (repeated and consistent findings), in order to isolate and identify key concepts and then explain how these concepts are related to delinquent behavior. Along this line, Donald Shoemaker defines theory as "an attempt to make sense out of observations."[28] The difficult task of making theoretical sense of research observations is sometimes referred to as "grounded theory" or **inductive theorizing**.[29] In the process of inductive theorizing, research involves collecting data and making empirical observations, which are then used to develop theory.

For example, Sheldon and Eleanor Glueck, whose work we discuss more fully in later chapters, spent their entire careers attempting to uncover the most important empirical findings about juvenile delinquency. They referred to their task as *Unraveling Juvenile Delinquency*—the title of their most important book.[30] The Gluecks' work was heavily criticized for being atheoretical, or without theory.[31] Their research, however, was clearly directed at providing empirical observations that would allow for the development of a theoretical explanation of delinquency, even though they never developed such a theory.[32] In recent years, their data and findings have become the basis for an important new theory called "life-course theory," which we describe in Chapter 10.

Testing Theory

deductive theorizing The evaluation of theoretical statements through research.

Research also provides the means to evaluate theory and to choose among alternative theories.[33] In contrast to inductive theorizing, **deductive theorizing** begins with theoretical statements and then attempts to test the validity of theoretical predictions.[34]

As we already discussed, theories advance explanations of delinquency in which propositions identify certain concepts and describe how they are related to delinquent behavior. These theoretically predicted relationships can be tested through research and either verified or disproved. For example, one simple proposition of differential association theory (presented in Chapter 11) is that attitudes favoring delinquency are learned in the context of "intimate personal groups."[35] The predicted relationship portrayed here is that youths develop attitudes from peer group relations, and delinquent behavior is then an expression of these attitudes:

| peer group relations | → | delinquent attitudes | → | delinquent behavior |

If research findings support the theoretical propositions tested, then the theory is verified or confirmed. If research findings are not consistent with the predicted relationships, then the theory is disproved.

Different theories often offer different predictions. To continue with the previous example, differential association theory and social bond theory (presented in Chapter 10) provide competing predictions about the relationships between peers, attitudes, and delinquent behavior. In contrast to differential association theory, social bond theory contends that attitudes are largely a product of family relationships.[36] Delinquent attitudes result in delinquent behavior. Associations with delinquent peers then follow from delinquent behavior as youths seek out friendships with others like themselves. The relationships predicted by social bond theory are as follows:

delinquent attitudes $\longrightarrow$ delinquent behavior $\longrightarrow$ delinquent peer group

As this brief example illustrates, theories have empirical implications, and one purpose of research is to enable scholars to choose among competing theories.[37]

The preceding discussion of the two purposes of delinquency research implies that the processes of inductive theorizing and deductive theorizing are completely distinct. The former is used to generate or develop theory; the latter is used to test theory. We must acknowledge, however, the complexity of the relationship between theory and research, and note that the distinction between the two purposes of research is not necessarily clear-cut. Even within the process of deductive theorizing, for example, an element of inductive theorizing exists. In deductive theorizing, researchers begin with theoretical predictions and then use empirical observations to test those propositions. The research results may lead to modification or refinement of the theory being tested. The latter part of this process, in which observations are interpreted and may result in a revised statement of theory, is consistent with the process of inductive theorizing. Although the relationship between theory and research is complex, it is clear that the development of theory and the performance of research go hand in hand.[38]

■ Summary and Conclusions

The scientific study of juvenile delinquency attempts to describe and explain delinquent behavior through theory and research. Theory seeks to provide a systematic and logical argument that specifies what is important in causing delinquency and why. Like other scientific theories, theories of delinquency are composed of *concepts* and *propositions*. It is necessary to assess the validity of theories, including those we apply to explain delinquency. We provided a series of questions that you can use to evaluate the theories of delinquency we present in later chapters.

The second basic component of the scientific method is research. In relation to theory, research serves two purposes: to generate theory and to test theory. Research is sometimes used to gain sufficient information about juvenile delinquency so that it becomes possible to theorize about it. The development of theory from research observations is called inductive theorizing.[39] Research is also used to evaluate or test theory in a process called deductive theorizing.

As we noted earlier, the primary purpose of this book is to cultivate an understanding of juvenile delinquency by integrating theory and research. This chapter has offered an overview of the key elements of a scientific approach to juvenile delinquency, focusing especially on theory. We describe research methods in Chapter 3. With this basic understanding of theory and its relationship to research, we can begin our study of juvenile delinquency on solid ground. The first two sections of this book present criminologists' efforts to define and describe juvenile delinquency, the third major section presents explanations of juvenile delinquency that have been offered in theory and tested in research, and the fourth section considers contemporary responses to delinquency. Throughout the book, we present theoretical explanations of delinquency together with the most relevant research that has tested those theories.

CRITICAL THINKING QUESTIONS

1. Define theory without using the words "concept" or "proposition."
2. Why does a scientific approach to juvenile delinquency depend on theory?
3. Develop your own example of inductive theorizing. Develop your own example of deductive theorizing.
4. As you read Rick's story at the beginning of this chapter, what factors seemed most significant to you in considering why Rick engaged in delinquency? Why?

SUGGESTED READING

Gibbons, Don C. *Talking About Crime and Criminals: Problems and Issues in Theory Development in Criminology.* Englewood Cliffs, NJ: Prentice Hall, 1994.

GLOSSARY

concepts: Isolated features of the world that are thought to be causally important.

deductive theorizing: The evaluation of theoretical statements through research.

inductive theorizing: The development of theory from research observations.

juvenile delinquency: Actions that violate the law, committed by a person who is under the legal age of majority.

level of explanation: The realm of explanation—individual, microsocial, or macrosocial—that corresponds to the types of concepts incorporated into theories.

propositions: Theoretical statements that tell how concepts are related.

theory: An explanation that makes a systematic and logical argument regarding what is important and why.

theory of delinquency: A set of logically related propositions that explain why and how selected concepts are related to delinquent behavior.

REFERENCES

Akers, Ronald L. *Criminological Theories: Introduction, Evaluation, and Application.* 4th ed. Los Angeles, CA: Roxbury, 2004.

Babbie, Earl. *The Practice of Social Research.* 8th ed. Belmont, CA: Wadsworth, 1998.

Bohm, Robert M. *A Primer on Crime and Delinquency Theory.* 2nd ed. Belmont, CA: Wadsworth, 2001.

Cloward, Richard A., and Lloyd E. Ohlin. *Delinquency and Opportunity: A Theory of Delinquent Gangs.* New York: Free Press, 1960.

Cohen, Albert K. *Delinquent Boys: The Culture of the Gang.* New York: Free Press, 1955.

———. *Deviance and Control.* Englewood Cliffs, NJ: Prentice Hall, 1966.

Cohen, Bernard P. *Developing Sociological Knowledge: Theory and Method.* Englewood Cliffs, NJ: Prentice Hall, 1980.

Curran, Daniel J., and Claire M. Renzetti. *Theories of Crime.* 2nd ed. Boston, MA: Allyn and Bacon, 2001.

Gibbons, Don C. *The Criminological Enterprise: Theories and Perspective.* Englewood Cliffs, NJ: Prentice Hall, 1979.

————. *Talking About Crime and Criminals: Problems and Issues in Theory Development in Criminology.* Englewood Cliffs, NJ: Prentice Hall, 1994.

Gibbons, Don C., and Marvin D. Krohn. *Delinquent Behavior.* 5th ed. Englewood Cliffs, NJ: Prentice Hall, 1991.

Gibbs, Jack P. "The State of Criminological Theory." *Criminology* 25 (1987):821–840.

Glaser, Barney, and Anselm L. Straus. *The Discovery of Grounded Theory.* Chicago, IL: Aldine, 1967.

Glueck, Sheldon, and Eleanor Glueck. *Unraveling Delinquency.* Cambridge, MA: Harvard University Press, 1950.

Hepburn, John R. "Testing Alternative Models of Delinquency Causation." *Journal of Criminal Law and Criminology* 67 (1976):450–460.

Hirschi, Travis. *Causes of Delinquency.* Berkeley, CA: University of California Press, 1969.

Hirschi, Travis, and Hanan C. Selvin. *Delinquency Research: An Appraisal of Analytic Methods.* New York: Free Press, 1967.

Jensen, Gary F. "Parents, Peers, and Delinquent Action: A Test of the Differential Association Perspective." *American Sociological Review* 78 (1972):562–575.

Laub, John H., and Robert J. Sampson. "The Sutherland–Glueck Debate: On the Sociology of Criminological Knowledge." *American Journal of Sociology* 96 (1991):1402–1440.

Sampson, Robert J., and John H. Laub. *Crime in the Making: Pathways and Turning Points Through Life.* Cambridge, MA: Harvard University Press, 1993.

Shaw, Clifford R., and Henry D. McKay. *Juvenile Delinquency and Urban Areas: A Study of Rates of Delinquency in Relation to Differential Characteristics of Local Communities in American Cities.* Rev. ed. Chicago: University of Chicago Press, 1969.

Shoemaker, Donald J. *Theories of Delinquency: An Examination of Explanations of Delinquent Behavior.* 4th ed. New York: Oxford University Press, 2000.

Short, James F., Jr. "The Level of Explanation Problem Revisited." *Criminology* 36 (1998):3–36.

Stark, Rodney. *Sociology.* 7th ed. Belmont, CA: Wadsworth, 1998.

Stinchcombe, Arthur L. *Constructing Social Theories.* New York: Harcourt, Brace, and World, 1968.

Sutherland, Edwin H., Donald R. Cressey, and David F. Luckenbill. *Principles of Criminology.* 11th ed. Dix Hills, NY: General Hall, 1992.

Turner, Jonathan. *The Structure of Sociological Theory.* Rev. ed. Homewood, IL: Dorsey Press, 1978.

Vold, George B., Thomas J. Bernard, and Jeffrey B. Snipes. *Theoretical Criminology.* 5th ed. New York: Oxford University Press, 2002.

ENDNOTES

1. Hirschi and Selvin, *Delinquency Research,* 66.
2. Bohm, *Primer,* 1.
3. Turner, *Structure of Sociological Theory,* 2–3.
4. Bohm, *Primer,* 2. See Bernard P. Cohen, *Developing Sociological Knowledge,* 140–148, for a full discussion of concept definition.
5. Vold, Bernard, and Snipes, *Theoretical Criminology,* 4.
6. Bohm, *Primer,* 2.
7. Glueck and Glueck, *Unraveling Delinquency.*
8. Sutherland, Cressey, and Luckenbill, *Principles of Criminology,* 211–214; and Hirschi, *Causes of Delinquency,* 140–146.
9. Stark, *Sociology,* 2; and Curran and Renzetti, *Theories of Crime,* 2.
10. Short, "Level of Explanation."
11. Albert K. Cohen, *Deviance and Control,* 43; and Gibbons, *Criminological Enterprise,* 9.
12. Sampson and Laub, *Crime in the Making*; Sutherland, Cressey, and Luckenbill, *Principles of Criminology*; and Short, "Level of Explanation."
13. Akers, *Criminological Theories,* 4–5.

14. Albert K. Cohen, *Deviance and Control*, 43; Gibbons, *Criminological Enterprise*, 9; and Akers, *Criminological Theories*, 4.
15. Shaw and McKay, *Juvenile Delinquency*; Albert K. Cohen, *Delinquent Boys*; and Cloward and Ohlin, *Delinquency and Opportunity*.
16. Short points out, in "The Level of Explanation Problem Revisited" (3), that the level of explanation corresponds to the unit of observation and the unit of analysis.
17. Our discussion of delinquency theory comprising concepts and propositions makes theory seem simple and straightforward. But we must admit that, among social scientists, "there is still no agreed-upon view of what theory is" (Bernard P. Cohen, *Developing Sociological Knowledge*, 170). See also Gibbs, "State of Criminological Theory."
18. Drawn from Bernard P. Cohen, *Developing Sociological Knowledge*, 191–192.
19. Shoemaker, *Theories of Delinquency*, 9.
20. Akers, *Criminological Theories*, 6–7; and Shoemaker, *Theories of Delinquency*, 9.
21. Akers, *Criminological Theories*, 6–7; and Curran and Renzetti, *Theories of Crime*, 3.
22. Akers, *Criminological Theories*, 2–6. Renowned criminologist Edwin Sutherland defined criminology as the study of law making, law breaking, and law enforcement (Sutherland, Cressey, and Luckenbill, *Principles of Criminology*, 3).
23. Akers, *Criminological Theories*, 4. Gibbons (*Talking About Crime*, 9–11, 73–76) describes two key criminological questions: "Why do they do it?" and "the rates question." The first question addresses "the origins and development of criminal acts and careers," and the second question addresses "organizations, social systems, social structures, and cultures that produce different rates of behaviors of interest" (9). See also Gibbons, *Criminological Enterprise*, 9; Gibbons and Krohn, *Delinquent Behavior*, 85–86; and Short, "Level of Explanation," 7.
24. Akers, *Criminological Theories*, 7.
25. Stinchcombe, *Constructing Social Theories*.
26. Bernard P. Cohen, *Developing Sociological Knowledge*, vii, 10; and Stark, *Sociology*, 3.
27. Stark, *Sociology*, 3.
28. Shoemaker, *Theories of Delinquency*, 7.
29. Glaser and Straus, *Discovery of Grounded Theory*; and Babbie, *Practice of Social Research*, 4, 60–64.
30. Glueck and Glueck, *Unraveling Delinquency*, 1950.
31. Gibbons and Krohn, *Delinquent Behavior*, 83–84.
32. Laub and Sampson, "Sutherland–Glueck Debate;" and Sampson and Laub, *Crime in the Making*.
33. Bernard P. Cohen, *Developing Sociological Knowledge*, 10.
34. Babbie, *Practice of Social Research*, 4.
35. Sutherland, Cressey, and Luckenbill, *Principles of Criminology*, 88–89.
36. Jensen, "Parents;" Hepburn, "Testing Alternative Models;" and Hirschi, *Causes of Delinquency*.
37. Stark, *Sociology*, 2; and Bernard P. Cohen, *Developing Sociological Knowledge*, 10.
38. Gibbons, *Talking About Crime*, 7.
39. Stark, *Sociology*, 3.

Defining Juvenile Delinquency

2

Chapter Objectives

After completing this chapter, students should be able to:

- Identify the major historical developments that led to the social construction of "juvenile delinquency" as a social and legal concept.
- Describe the roots of the juvenile court in nineteenth-century developments such as poor laws, houses of refuge, placing-out, reform schools, and the child-saving movement.
- Describe the character of the original juvenile court—its philosophy, jurisdiction, and procedures.
- Distinguish four legal categories of juvenile delinquency.
- Understand key terms:
  social constructionist perspective
  parens patriae
  poor laws
  pauperism
  houses of refuge

placing-out
reform schools
child-saving movement
rehabilitative ideal
"best interests of the child"
due process of law
status offender
balanced and restorative justice
status offense

Juvenile delinquency, as we know it today, is a relatively recent concept. This does not mean, however, that young people in the past were more compliant than they are today. In fact, Socrates (470–399 BCE) offered a critique that

CASE IN POINT

The "Stubborn Child Law"

Deuteronomy 21:18–21

> [18] If a man has a stubborn and rebellious son who does not obey his father and mother and will not listen to them when they discipline him, [19] his father and mother shall take hold of him and bring him to the elders at the gate of his town. [20] They shall say to the elders, "This son of ours is stubborn and rebellious. He will not obey us. He is a profligate and a drunkard." [21] Then all the men of his town shall stone him to death. You must purge the evil from among you. All Israel will hear of it and be afraid.

In November 1646, the governing body of Massachusetts Bay Colony took these verses almost verbatim and made them into law. The colonial codes of Connecticut, Rhode Island, and New Hampshire followed suit. Though substantially amended, the Massachusetts law remained in effect until 1973.

In his book, *Stubborn Children: Controlling Delinquency in the United States, 1640–1981,* John Sutton points out that "the 'stubborn child law' was legally distinctive in three ways: (1) It defined a special legal obligation that pertained to children, but not to adults; (2) it defined the child's parents as the focus of that obligation; and (3) it established rules to govern when public officials could intervene in the family and what actions they could take." While the "bare words" of these stubborn child laws made it a capital offense for a child to disobey parents, they also "established rules to govern when public officials could intervene in the family and what actions they could take." We will explore this new approach to child misconduct in this chapter.

Source: Sutton, *Stubborn Children,* 11–12, numbers added.

sounds amazingly contemporary: "The children now love luxury. They have bad manners, contempt for authority, they show disrespect for adults and love to talk rather than work or exercise. They no longer rise when adults enter the room. They contradict their parents, chatter in front of company, gobble down food at the table, and intimidate teachers."[1] So it seems that a concern over the "next generation" is perennial.[2]

Scholars say that juvenile delinquency was "invented" or socially constructed in order to indicate that the concept is a product of a great many social, political, economic, and religious changes. With regard to juvenile delinquency, these changes began in the Renaissance (roughly 1300–1600), but were most pronounced during the Enlightenment (mid-1600s–late 1700s) and the Industrial Revolution (1760–mid 1900s).[3] This transformation of thought and practice eventually led to a series of legal changes at the end of the nineteenth century that created the legal status of "juvenile delinquent" and a separate legal system that included juvenile courts and reformatories.[4] The use of a separate legal status and legal system for juveniles spread rapidly throughout the United States in the early twentieth century. It was not long, however, before the legal philosophy of the juvenile court began to be seriously questioned. Beginning in the 1960s, a "second revolution" significantly altered contemporary definitions of juvenile delinquency and practices of juvenile justice.[5]

Using a **social constructionist perspective**, this chapter traces the historical origins and recent transformations of juvenile delinquency as a legal concept. We also consider the associated changes in juvenile justice practices. As our perspectives toward juvenile delinquency have changed, so too has our legal response to it. Focusing on this correspondence, this chapter covers four areas: (1) the social construction of "juvenile delinquency," (2) the invention of the juvenile court, (3) the transformation of juvenile justice systems, and (4) the legal definitions of "juvenile delinquency."

social constructionist perspective An attempt to understand the many social, political, and economic factors that lead to the development of an idea, concept, or view.

■ The Social Construction of "Juvenile Delinquency"

At least three historical developments led to the social construction of "juvenile delinquency": the "discovery" of childhood and adolescence, the English common law doctrine of *parens patriae,* and the rise of positivist criminology. As a result, the concept of juvenile delinquency came to signify a separate and distinct status for young people, both socially and legally. Sociologists use the concept *status* to refer to the position or rank of a person or group within society, with the position being determined by certain individual or group traits.[6] Juvenile delinquency is a status determined both by age (less than the legal age of majority) and behavior (actions that violate the law).

The Discovery of Childhood and Adolescence

Today, we take for granted that childhood and adolescence are separate stages in life, unique from other stages. As Socrates' critique of youth indicated, the Greeks also considered this age group distinctive. Both Plato (about 427–347 BCE) and

Aristotle (384–322 BCE) offered commentary that distinguished and categorized the behaviors, attitudes, and emotions of young people as different from those of people at other ages. While their characterizations suggest that the Greeks considered adolescence a separate stage of life, they advance a rather negative view toward youth, suggesting that, by modern standards, the Greeks were not very understanding of the trials and tribulations of adolescence.[7]

Throughout the Middle Ages (476–1450 CE), European literature made frequent reference to the "ages of life." This literary theme included seven stages, beginning with infancy and ending with old age; the third stage was adolescence. Shakespeare, for example, refers to the seven stages of life in his famous passage:

> *All the world's a stage*
> *And all the men and women merely players;*
> *They have their exits and their entrances;*
> *And one man in his time plays many parts;*
> *His acts being seven ages. . . .*
> *(As You Like It, II, vii, 139).*[8]

However, these age categories lacked distinction and were not used outside of literature.[9]

In his widely cited book, *Centuries of Childhood,* Philippe Aries argues that the idea of childhood did not exist in medieval society.[10] Basing his argument on a variety of historical records, he contends that during the Middle Ages, few distinctions were made between young people and adults. In fact, the only culturally prescribed age distinction was that of infancy. Young people beyond the age of five were viewed as adults "on a smaller scale."[11]

Aries contends that during the late sixteenth and seventeenth centuries noticeable social distinctions associated with childhood began to emerge. "For instance, a special type of dress for children appeared. Moreover, it was not until the sixteenth and seventeenth centuries that fables, fairy stories, and nursery rhymes became the property of children."[12] Renaissance culture, however, failed to differentiate adolescence from childhood.[13] Thus, while childhood was established as a separate age category by the seventeenth century, it was not distinguished from adolescence until the late eighteenth century.[14]

The slowness with which the companion concepts of childhood and adolescence emerged during the Renaissance was most likely a result of the harsh living conditions of medieval Europe. As LaMar Empey and Mark Stafford write, "under conditions like these, the cultural prescriptions we consider important today were lacking, especially those provisions for close ties between parent and child, that stress the importance of the nuclear family, that take delight in the innocence and beauty of children, and that provide long years of total economic support for a phase of the life cycle known as childhood."[15]

Aries claims that the slow development of childhood and adolescence does *not* suggest that "children were neglected, forsaken, or despised."[16] Rather, it simply signifies a lack of awareness of the special needs of childhood. He acknowledges, however, that attitudes toward children during the Middle Ages were largely "indifferent" and treatment of children was often harsh and punitive.[17] Empey and Stafford expand on this, referring to the pre-Renaissance period as the "history of indifference to children," characterized by infanticide, abandonment, poor care of infants, disease and death, and harsh and punitive discipline.[18] Their portrayal of the history of childhood is consistent with the observations of historian Lloyd deMause, who writes, "The history of childhood is a nightmare from which we are only beginning to awaken. The further back one goes, the lower the level of child care, and the more likely children are to be killed, abandoned, beaten, terrorized, and sexually abused."[19]

Family life and child-rearing emerged as matters of great importance in Renaissance society.[20] Numerous treatises and manuals were written in the sixteenth and seventeenth centuries, and many remained popular until the eighteenth century.[21] These manuals instructed parents on how to train their children in the "new morality," involving etiquette, obedience, respect for others, self-control, and modesty.[22] The training tools emphasized in these manuals included supervision, strict discipline, and insistence on decency and modesty. Parents were instructed that they must control what their children read, hear, and do.

These manuals were clear expressions of a new view of children, childhood, and child-rearing: children began to be viewed as innocent and vulnerable. It was the responsibility of parents to protect their children from the evils of a corrupt world and to train them so that they developed moral character and religious faith and devotion.[23] Aries summarizes, "The family ceased to be simply an institution for the transmission of a name and an estate—it assumed a moral and spiritual function, it moulded bodies and souls. The care expended on children inspired new feelings, a new emotional attitude. . . ."[24]

While the Renaissance is often described as an awakening from the darkness of the Middle Ages, the Enlightenment of the eighteenth century constitutes another sharp departure from traditional thinking. Building on the importance of training and education stressed in treatises and manuals popular at the close of the Renaissance, Enlightenment philosophers elaborated on the newly "discovered" stages of childhood and adolescence.

In 1693, John Locke (1632–1704), a physician, wrote *Some Thoughts Concerning Education,* a treatise on child-rearing that went through twenty-six editions before 1800.[25] Because he believed that children are born morally neutral (blank slates) and mature into a product of the influences of experience, Locke stressed the importance of parents in a child's life, especially for purposes of supervision, discipline, and moral training. In many ways, Locke's work represents a culmination of Renaissance views toward childhood and youth as periods of vulnerability and dependency, requiring protection and training. He used the term "education" not to refer to the importance of schools, but to stress the importance of training by parents in the form of child-rearing practices.

Another Enlightenment philosopher, Jean-Jacques Rousseau (1712–1778), provided a clear departure from Renaissance views. He argued that people were inherently good, but that social and political institutions corrupted this good and caused all sorts of trouble for them: "The Author of Nature makes all things good; man meddles with them and they become evil."[26] As a result, he argued for permissiveness in child-rearing, so that parents and schools would not corrupt the innocence of children.

Rousseau also pointed to the "distinctive human plight" confronted by adolescents during the transition from childhood to adulthood—a plight characterized by moral and sexual tensions.[27] Adolescence is actually one of five stages of development that Rousseau advanced in his controversial book, *Emile* (1762). In this work, Rousseau provided the first systematic consideration of the stages of development, emphasizing how these stages differ and how these differences influence learning and necessitate appropriate educational methods.[28]

The slow discovery of childhood and adolescence was not complete until the Enlightenment, when Rousseau's idea of developmental stages capped a growing awareness of age distinctions across the life course. The awareness of developmental stages and differentiation based on age had important implications for the structure of family life, for child-rearing, and for education. Ideas about the innocence, vulnerability, and dependence of childhood, as well as the moral and sexual tensions of adolescence, resulted in an increasing emphasis on the family and school as institutions of socialization. Gradually, the view developed that young people require protection, nurture, supervision, discipline, training, and education in order to grow and mature into healthy and productive adults.[29] This point of view culminated in the late 1800s, when it was widely held that youngsters needed to be "excused from participation in the larger society while they concentrated on personal growth."[30] This was a significant departure from the harsh lives of children in Medieval Europe and colonial America and their often forced involvement in the labor force during most of the early Industrial Revolution (beginning about 1760).[31] The slow discovery of childhood and adolescence is depicted in a timeline in **Figure 2-1.**

The *Parens Patriae* Doctrine

The development of the English legal doctrine of *parens patriae* coincides with the discovery of childhood and adolescence. This far-reaching legal doctrine emerged slowly in the late fourteenth and early fifteenth centuries in response to a series of cases heard before the English chancery courts.[32] *Parens patriae* became a central tenet of the equity law that was derived from chancery court decisions. Adopted in the United States as a part of the Anglo-Saxon legal tradition

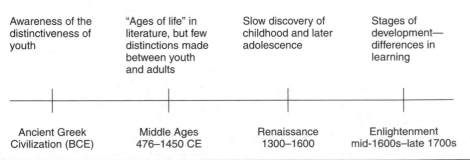

Awareness of the distinctiveness of youth	"Ages of life" in literature, but few distinctions made between youth and adults	Slow discovery of childhood and later adolescence	Stages of development— differences in learning
Ancient Greek Civilization (BCE)	Middle Ages 476–1450 CE	Renaissance 1300–1600	Enlightenment mid-1600s–late 1700s

Figure 2-1 The Discovery of Childhood and Adolescence. While the ancient Greeks were aware that the age of youth was distinctive, characterized by distinctive behaviors, attitudes, and emotions, the conception of developmental stages and social differentiation based upon age was "discovered" very slowly beginning in the Renaissance. Childhood and adolescence, as we know the companion concepts today, were not given full expression as developmental stages in the life course until the late nineteenth century.

of England, *parens patriae* provided the fundamental legal authority for the idea of juvenile delinquency and the early juvenile court.

The Latin phrase <u>*parens patriae*</u> means literally "parent of the country."[33] As a Renaissance legal doctrine, *parens patriae* vested far-reaching power in the king as sovereign and supreme guardian over his land and people—the "king's prerogative."[34] Chancery courts were established to provide just settlements to disputes but to do so in a way that maintained the king's prerogative. These disputes arose mainly with regard to property rights and inheritance. Chancery court decisions sought the orderly transfer of property interests and feudal duties, especially when the "crown's interests" were at stake economically and politically.[35]

Parens patriae provided the king with considerable legal authority to support and protect the social and political order of English feudal society.[36] Attached to this authority, however, was a duty that the king had to his subjects in return for the allegiance paid to him. In practice, this duty was concerned primarily with the social welfare of certain dependent groups. Chancery court cases centered on three dependent groups: children, those who were mentally incompetent, and those in need of charity.[37] Under *parens patriae,* the king was established as protector and guardian of these dependent classes.

With regard to children, *parens patriae* was applied most extensively to cases in which the wardship or guardianship of young children was at issue.[38] In medieval England, though all infants were considered wards of the king because of the king's prerogative, cases involving the wardship of children of the landed gentry were of primary legal concern.[39] These children were heirs to an estate or had already inherited an estate from a deceased father. Gaining custody and control of these children and their estates through wardship could prove profitable for relatives.

Gradually, the chancery courts extended the doctrine of *parens patriae* to include the general welfare of children; the proper care, custody, and control of children was to the "crown's interests."[40] This included the ability of the courts to assume and exercise parental duties—to act *in loco parentis*—when parents failed to provide for the child's welfare.[41] Implicit in this doctrine are the developmental concepts of childhood and adolescence in which it is the parents' responsibility to protect, nurture, supervise, discipline, train, and educate children. Insuring the general welfare of children was a means to maintain the power of the monarchy and the feudal structure of English society.[42] It is important to note that the chancery courts did not have jurisdiction over children charged with criminal offenses. Juvenile offenders were handled within the framework of the regular court system. As a result, the *parens patriae* doctrine of equity law "embraced the dependent and not the delinquent child."[43]

parens patriae Literally means "parent of the country." The legal authority of courts to assume parental responsibilities when the natural parents fail to fulfill their duties.

Positivist Criminology

To say that juvenile delinquency is socially constructed means that it is a product of prevailing thoughts and perspectives. The two historical developments we have considered so far correspond closely in time and perspective. A third historical development took root somewhat later but had an equally strong influence on the idea of juvenile delinquency.

Positivist criminology is an approach or school of thought that emerged in the last half of the nineteenth century and flourished to such a degree that it came to dominate the field of study for most of the twentieth century. While positivist criminology will be discussed more fully in Chapter 7, we want to point out here some of its basic tenets, which have influenced conceptions of juvenile delinquency.

First and foremost, positivist criminology is based on *positivism*—the use of scientific methods to study crime and delinquency.[44] The scientific method involves systematic observation, measurement, description, and analysis so that scientists can look for, uncover, and draw conclusions about the regularities and patterns of crime and delinquency. The prospect of a scientific approach to crime and delinquency quickly became popular because of the hope it offered to better understand and respond to the problem of crime through individual and social improvement.

The scientific approach advanced by positivism assumes that crime and delinquency are caused or determined by identifiable factors. This cause-and-effect relationship is referred to as *determinism*. Determinism holds that, given the presence or occurrence of certain causal factors, crime and delinquency invariably follow. According to positivism, causal factors can be systematically observed and measured, and causal processes can be analyzed and described.

The patterns and regularities of crime that early positivist criminologists observed led them to conclude that some people are more likely to commit crime than others. A number of individual and social characteristics were explored to differentiate criminals from noncriminals. Public fear of crime in rapidly growing cities also centered on differentiation, and public opinion was often driven by the belief that immigrant groups who were poor, uneducated, unemployed, and lived according to a foreign culture constituted a "dangerous class."[45]

Early positivist criminologists often contended that the degree of difference between criminals and noncriminals was so great that it was most accurately portrayed as a pathology.[46] Early versions of positivist thought emphasized biological and psychological differences between criminals and noncriminals, claiming that criminals suffered from individual pathologies such as physiological defects, mental inferiority, insanity, and a tendency to give in to passion. These individual pathologies were usually thought to have biological roots.

The rise of sociology as a discipline in the late 1800s was associated with differentiation that emphasized social pathologies related to rapid urbanization and industrialization. Rapid social change was thought to dissolve the organization of social life, resulting in lack of social control. Accordingly, young people who experienced social pathologies (such as divorce and family disruption, lack of parental supervision, poverty, cultural heterogeneity, and residential instability) were considered more likely to engage in delinquent acts and become adult criminals, destined for a life of crime.

Armed with scientific methods to discover the causes of crime and delinquency, positivist criminologists seek to use this understanding to bring about change in individual criminals and their social environment. Using a medical analogy, criminologists in the early 1900s sought effective treatment and rehabilitation, directed at the individual and the social pathologies that caused crime. While the pathological emphasis of positivist criminology gives the impression that criminality is determined and therefore unchangeable, the use of the scientific method to uncover the causes of crime gave hope that these pathologies could be understood and treated. The *rehabilitative ideal* emerged and prospered, especially with regard to children and adolescents.[47] Charles Cooley, a well-known sociologist, writing in 1896, declared: "when an individual actually enters upon a criminal career, let us try to catch him at a tender age, and subject him to rational social discipline, such as is already successful in enough cases to show that it might be greatly extended."[48]

■ Invention of the Juvenile Court

The concept of "juvenile delinquency" is a clear expression of the three historical developments we described: the discovery of childhood and adolescence, the English equity law doctrine of *parens patriae,* and the growing dominance of positivist thought in criminology. The legal designation of a "delinquent child" is an important part of the legislation that created the first juvenile court in the United States in 1899. In this section, we trace the evolution of thought and practice that led to the origins of the juvenile court.

Poor Laws, Charities, and Pauperism

Colonial America widely accepted the Christian theology of original sin, and many people believed that poverty and crime were expressions of this natural depraved state. Based on this view, three social institutions were emphasized in colonial times: family, church, and community. David Rothman writes, "Families were to raise their children to respect law and authority, the church was to oversee not only family discipline but adult behavior, and the members of the community were to supervise one another, to detect and correct first signs of deviancy."[49] When these institutions functioned well, towns were "spared the turbulence of vice and crime" and "enjoyed a high degree of order and stability."[50] However, when these institutions were unable to change the "sin nature" of children, a strong, uncompromising response was thought to be necessary. As we have already seen, "stubborn child laws" were developed in colonial America to "establish rules to govern when public officials could intervene in the family and what actions they could take."

In accordance with the Enlightenment concepts of childhood and adolescence, colonial Americans took family life and parenting very seriously. Benjamin Wadsworth, a minister and author of the popular child-rearing booklet, *The Well-Ordered Family* (1712), charged parents to "train up a child in the way wherein he should go" (Proverbs 22:6). Parents were to "govern their children well, restrain, reprove, correct them as there is occasion," and in so doing, instill moral integrity, self-reliance, and civic responsibility.[51]

Civic responsibility was a strong obligation in colonial America. It generated individual and social obligation to the poor, first informally, and then through formal provisions of law. In fact, the first colonial poor laws, legislated in the latter part of the seventeenth century, stipulated a community obligation to support

poor laws Laws enacted in colonial America that established a civic duty of private citizens to "relieve" the poor. These laws usually provided a definition of residence so that outsiders could not benefit from private relief. Legal authority was also granted for governmental agencies or private relief societies to separate poor children from their "undeserving" parents.

pauperism The view, popularly held throughout the nineteenth century, that children growing up in poverty, surrounded by depravity in their neighborhood and family, are destined to lives of crime and degradation.

and "relieve" the poor. However, these local statutes merely stated the obligation, without specifying who should be considered "poor" or what provisions were to be provided to the poor.[52]

Colonial **poor laws** mirrored, but did not merely duplicate, those that had developed earlier in sixteenth-century England.[53] Sharing similar philosophy and purpose with the *parens patriae* doctrine, colonial communities (and later cities and states) developed a system for protecting poor children and, if necessary, separating them from their "undeserving parents."[54] This system grew to include laws, passed by local legislative bodies, regulating the poor; the creation of charitable organizations and relief societies; and, especially in urban places, government-sponsored institutions. Poor laws provided legal authority for governmental entities and authorized private philanthropic agencies to separate poor children from their parents and to apprentice these children to local residents. The apprenticeship system kept relief costs down because the child's labor paid for care, education, and training. However, the overall quality of the care and training was questionable, and in many cases, apprenticeship was merely a "business proposition" in which the child provided slave labor for a term.[55]

Gradually, the view developed that poverty, if left unchecked, will lead children to "a future of crime and degradation," a process known throughout much of the nineteenth century as **pauperism**.[56] This point of view contrasted sharply from the doctrine of original sin, widely accepted in earlier times. Instead of focusing on the sinful nature of individuals, pauperism emphasizes a breakdown in social order.[57]

The working assumption of pauperism led to the creation of various institutions to help the poor. Institutional charity was thought to be superior to physical and nutritional care offered in the home, or financial support paid to the poor, because these noninstitutional forms of relief were thought to have " 'pauperized' the poor by creating habits of idleness and dependency."[58] Institutional relief efforts included the almshouse, workhouse, and poor house, which sought to motivate the poor out of poverty by hard work and strict discipline.[59] However, reformers soon realized that these institutional settings could similarly pauperize children by exposing them to adults "addicted to idleness and intemperance."[60]

Houses of Refuge and Moral Reform

In the first quarter of the nineteenth century, the state's parental authority derived from poor laws, and institutional efforts to respond to pauperism became increasingly focused on the plight of urban poor children. Because of the fear of pauperism, reformers were most concerned about the placement of poor children into pauper institutions, such as the municipal almshouse, "where they are liable to acquire bad habits and principles, and lay the foundation for the career of worthlessness and improvidence."[61] In fact, these reformers were more concerned about the placement of children into pauper institutions than about the mixing of delinquent children with adult criminals in jails and prisons. It was believed that children placed in adult penal facilities were already beyond the potential for reform.[62]

As an expression of this perspective, the New York House of Refuge was established in 1824 by the Society for the Reformation of Juvenile Delinquents, the successor to the Society for the Prevention of Pauperism. The House of Refuge dealt both with children who were convicted of crimes and those who were vagrant, but in practice almost all of its children were vagrants from pauper families.[63] State legislation gave the Society authority to manage the institution and the children under its custody.[64] __Houses of refuge__ followed soon in Boston (1825) and Philadelphia (1828), and "for a quarter of a century the activities of these three institutions defined institutional treatment of juvenile delinquents."[65]

David Rothman characterizes house of refuge reformers as conservative Protestants who desired to prevent pauperism and to protect and reform children, thereby sustaining order and stability in society.[66] Through legislation, "they widened the scope of permissible state intercession," and their emphasis on prevention, reform, and protection proved to be the "seeds of what came to be called the juvenile court."[67]

Not every vagrant or delinquent child was committed to a house of refuge, however; only those that could still be "rescued" and were not too far down the road of crime were admitted.[68] The focus of houses of refuge was to protect the "predelinquent." Little distinction was made between "pauper, vagrant, or criminal children"—all required protection and reform.[69] Reformers were convinced that these children were victims rather than offenders and that they needed to be removed from evil influences of urban poverty: "Pauperism then was the enemy; juvenile delinquency, like intemperance, ignorance, and gambling, was the symptom."[70] Reformers intended the house of refuge to be a sanctuary or haven, where children could be isolated from the wickedness of the world and where moral reform could take place.[71]

Assuming parental responsibility in an institutional setting proved to be a difficult task for houses of refuge. Though reformers firmly believed that parental neglect was the single greatest cause of delinquency, houses of refuge were not intended literally to replace parents by providing a surrogate home environment. Instead, houses of refuge were intended to provide moral reform.[72] John Sutton refers to houses of refuge as "moral institutions."[73] Moral reform involved four basic elements: a daily regimen, strict discipline, education, and work. David Rothman's provocative book, *The Discovery of the Asylum,* speaks of the house of refuge as a "well-ordered asylum."[74] Discipline was strictly enforced, based largely on solitary confinement and corporal punishment. Despite humanitarian and religious intentions, houses of refuge frequently became sites of physical abuse.[75]

As Case in Point, "A Typical Day at the New York House of Refuge" reveals, education and work consumed the daily life of children in houses of refuge. School in the early mornings and late at night, both before and after work, was intended not only to provide academic skills and achievement, but also to promote self-discipline and to instill morality and religion. Children in houses of refuge were also expected to work long and hard, doing physical labor and repetitive tasks such as making brass nails, finishing shoes, and wicker work.[76] The labor of children was sometimes contracted to manufacturers to provide revenue for houses of refuge. The apprenticeship system was also used, justified as a

houses of refuge The first institutional facilities in the United States for poor, vagrant children. Both private and public refuges sought to protect and reform the "predelinquent."

A Typical Day at the New York House of Refuge

At sunrise, the children are warned, by the ringing of a bell, to rise from their beds. Each child makes his own bed, and steps forth, on a signal, into the Hall. They then proceed, in perfect order, to the Wash Room. Thence they are marched to parade in the yard, and undergo an examination as to their dress and cleanliness; after which, they attend morning prayer. The morning school then commences, where they are occupied in summer, until 7 o'clock. A short intermission is allowed, when the bell rings for breakfast; after which, they proceed to their respective workshops, where they labor until 12 o'clock, when they are called from work, and one hour allowed them for washing and eating dinner. At one, they again commence work and continue at it until five in the afternoon, when the labor of the day terminated. Half an hour is allowed for washing and eating their supper, and at half-past five, they are conducted to the school room where they continue at their studies until 8 o'clock. Evening Prayer is performed by the Superintendent; after which, the children are conducted to their dormitories, which they enter, and are locked up for the night, when perfect silence reigns throughout the establishment. The foregoing is the history of a single day, and will answer for every day in the year, except Sundays, with slight variation during stormy weather, and the short days in winter.

Source: Quoted in Mennel, *Thorns & Thistles,* 18–19.

means for children to develop occupational skills. Apprenticeships accounted for about 90% of the children released each year from houses of refuge.[77] The most common apprenticeship placement for boys was with farmers, whereas for girls, maid service was the only socially acceptable form of indenture.[78]

The enthusiasm of house of refuge reformers was contagious, and numerous institutions of similar design opened across the United States during the 1840s and 1850s.[79] The philosophy and authority for placing children in houses of refuge was derived from the English equity law doctrine of *parens patriae*. In fact, the *parens patriae* doctrine was introduced into American law in an 1838 Pennsylvania Supreme Court case called *Ex parte Crouse*, in which the commitment of a young girl to a house of refuge was contested. While this case makes only brief mention of the doctrine, the intent and meaning appears deliberate: under the philosophy and purpose of the *parens patriae* doctrine, the government is granted legal authority to assume custody (guardianship) and parental responsibility.[80]

Based on her mother's petition, Mary Ann Crouse was committed by a justice of the peace to the Philadelphia House of Refuge for being incorrigible. The girl's father sought her release through a writ of habeas corpus, arguing that she had been deprived of the right to trial by jury guaranteed under the state constitution. The Pennsylvania Supreme Court denied the motion, holding that parental custody and control of children is a natural, but not inalienable right, and if the parents fail to properly supervise, train, and educate their children, their rights as parents can be taken over by the state.[81] As one legal scholar noted: the judicial reasoning in *Ex Parte Crouse* authorized state government to "invade the home, replace the parents, and take custody of the child."[82]

Placing-Out and Orphan Trains

Even though houses of refuge continued to open during the 1850s, critics began to argue that "not only the discipline, but every detail of the routine made the houses of refuge indistinguishable from prisons."[83] Rather than becoming models of care, houses of refuge had become juvenile prisons, unable to nurture and reform children through an institutional approach.[84]

Beginning in the 1850s, reformers returned to the traditional belief that family homes, not institutions, were the best places for reform.[85] After leaving training for the ministry at Yale University, Charles Loring Brace (1826–1890) founded the New York Children's Aid Society in 1853. With the fervor of an evangelist, he argued that urban poverty bred a "dangerous class," prone to crime and violence, and the "poison" of society. Fueled by this fear, the New York Children's Aid Society sought to "drain the city" of poor and delinquent children through a practice called **placing-out**.[86] Placing-out involved taking groups of vagrant children, sometimes referred to as "urban waifs" or "street urchins," west by railroad, on "orphan trains," for placement with farming families. Brace believed that "the best of all Asylums for the outcast child is the farmer's home . . . the cultivators of the soil are in America our most solid and intelligent class."[87]

Placing-out was apparently well received in many communities. Reports indicate that community members were excited and willing to take these youths into their homes, whether because of the prospect of free farm labor or a sense of civic obligation.[88] Placing-out programs were soon implemented by other organizations, but they were not without critics. Some saw Brace's unabashed faith in the reforming powers of rural family life as naive. Critics contended that it is next to impossible to take a poor, vagrant child off the streets and expect him or her to adjust to rural family life. The solution, according to these critics, was not to do away with the placing-out program, but to use the institutional setting beforehand for discipline and reform.[89] The New York Juvenile Asylum and the Pine Farm facility operated by the Boston Children's Aid Society were based upon this more formal placing-out model.

Reform Schools

The development of **reform schools**, beginning in the mid-nineteenth century, represents another way in which institutions were used to respond to the problem of dependency and juvenile crime. As the name implies, reform schools emphasized formal schooling. Instead of sandwiching school around a full day of work as houses of refuge did, reform school operated on a traditional school schedule.[90]

Many reform schools used a cottage or family system in which children were divided into small "families" of forty or fewer. Each family had its own

placing-out The practice, begun in the mid-1800s, in which philanthropic groups took vagrant and wayward urban children west by railroad to be placed in farm families.

reform schools In the mid-1800s a new form of institution began to replace houses of refuge. These institutions emphasized education and operated with traditional school schedules. Many reform schools also used a cottage or family system in which children were grouped into "families" of forty or fewer.

cottage, matron and/or patron (mother or father), and schedule. Cottages were used to make the facility more like a family and less like a prison. Affectional discipline, rather than physical discipline, was used in an effort to generate conformity and instill good citizenship.[91] Some reform schools embraced Brace's conviction of reform through rural family and farm life and were therefore located in rural areas.[92]

Cottage reform schools spread widely across the United States in the latter half of the 1800s, but the degree of emphasis on the family ideal and the roles of schooling and work varied greatly. Some reform schools provided only large congregate housing of children, with little resemblance to family units.[93] Contract labor of children to manufacturers was a part of most reform schools, but after the Civil War, child labor became more exploitative in some schools.[94] In addition, as farming opportunities diminished, training in agriculture provided in some rural reform schools became less marketable as a learned trade. Similarly, changes in the nature of work brought on by the Industrial Revolution meant a significant reduction in apprenticeship opportunities—the means by which most children were released from reform schools. In response to this, some reform schools, especially those in the west, began to emphasize vocational education and deemphasize a family environment. These programs were oriented toward vocational education and often included military drill and organization.[95]

Placement or commitment to reform schools was based on the legal authority of the state, under *parens patriae,* to take over parental custody and control. The reform school, however, provided the context for the first significant legal challenges to the *parens patriae* doctrine, in the Illinois Supreme Court case *O'Connell v. Turner* (1870).[96]

Michael O'Connell was sent to the Illinois State Reform School under a state statute that authorized youths between the ages of six and seventeen to be committed based upon the charge of being "a vagrant, or destitute of proper parental care, wandering about the streets, or committing mischief or growing up in mendicancy, ignorance, idleness and vice."[97] A writ of habeas corpus challenged this commitment power when the decision was based solely on the dependency and neglect of the child, without criminal conviction. The Illinois Supreme Court agreed, ruling that a general consideration of the child's "moral welfare and the good of society" are not sufficient reasons for commitment to a reform school. The state law allowing this was ruled unconstitutional, and officials were ordered to discharge Michael O'Connell.[98]

The state Supreme Court's opinion questioned the state's *parens patriae* authority when protective custody was ordered based on the subjective consideration of parental care and supervision. The court's opinion stressed that parents have a right and responsibility to rear and educate their children that cannot be preempted by the government except under "gross misconduct [by the child] or almost total unfitness on the part of the parents."[99]

While the O'Connell decision focused on Illinois law and procedure and had considerable impact in that state, reformers continued their efforts to allow and encourage governmental intervention in the lives of dependent, neglected, and delinquent children. The Chicago Reform School closed in 1872, two years after the court's decision, and legislation repealed jurisdiction over "misfortune" (dependency and neglect) cases.[100]

The Child-Saving Movement

The O'Connell decision and the closure of the Chicago Reform School were signs of changing societal attitudes. By the end of the Civil War, juvenile institutions had become increasingly custodial and repressive: "Those who sought to reform juvenile delinquents in mid-19th century America spoke the lofty language of nurture and environmentalism. Reform schools, they claimed, were not prisons but home-like institutions, veritable founts of generous sentiment. In fact, they were prisons, often brutal and disorderly ones."[101] By the late nineteenth century, little enthusiasm and hope remained for the institutions that had once been heralded as places of protection and reform for vagrant and delinquent children.[102]

Despite these concerns, the problems of urban poverty and delinquency persisted and, in fact, grew worse. Forces of industrialization, urbanization, and immigration weakened the cohesiveness of communities and the abilities of communities and families to socialize and control children effectively.[103] The Illinois Board of State Commissioners of Public Charities expressed grave concern in an 1898 report: "There are at the present moment in the State of Illinois, especially in the city of Chicago, thousands of children in need of active intervention for their preservation from physical, mental and moral destruction. Such intervention is demanded, not only by sympathetic consideration for their well-being, but also in the name of the commonwealth, for the preservation of the State. If the child is the material out of which men and women are made, the neglected child is the material out of which paupers and criminals are made."[104]

The Board of Charity lamented the lack of resources for juveniles. Dependent children were left to charity institutions that housed both adults and children. Delinquent children were incarcerated in adult jails, with adult criminals, and they were adjudicated in adult courts, without special consideration. The few juvenile institutions that did exist were largely private and often sponsored by religious denominations. Reformers felt that the state had shirked its responsibility for the care and protection of children.[105]

Despite these problems, the late nineteenth century was a time of optimism, not pessimism. The period between 1880 and 1920 is known as the Progressive Era in US history. The government's role in reform was reassessed, especially in terms of program and policy delivery, administration, and government structure.[106] In the late 1800s, these progressive principles were applied to the problem of juvenile delinquency by the **child-saving movement**. The child-saving movement was a loose collection of women across the United States from middle- and upper-class backgrounds who exercised considerable influence in mobilizing change in how government dealt with dependent, neglected, and delinquent children.

Scholars have long debated the motives behind the child savers' reform efforts. Traditional explanations of the child-saving movement emphasize the "noble sentiments and tireless energy of middle class philanthropists."[107] Another point of view holds that child savers were progressive reformers seeking

child-saving movement
A loose collection of women from middle- and upper-class backgrounds who exercised considerable influence in mobilizing change in how government dealt with dependent, neglected, and delinquent children. One particular child-saving group, the Chicago Women's Club, is largely responsible for the creation of the first juvenile court in Chicago.

to "alleviate the miseries of urban life and to solve social problems by rational enlightenment and scientific methods."[108] Still others argue that the child-saving movement was an effort by the ruling class to repress newly arriving immigrants and the urban poor and to preserve its own way of life.[109] Regardless of which explanation is correct, leading child savers were prominent, influential, philanthropic women, who were "generally well educated, widely traveled, and had access to political and financial resources."[110] Additionally, child savers viewed their work as a humanitarian "moral enterprise," seeking to "strengthen and rebuild the moral fabric of society."[111]

Child-saving was largely women's work. Middle-class women "extended their housewifely roles into public service and used their extensive political contacts and economic resources to advance the cause of child welfare."[112] The child savers argued that women were uniquely suited to work with dependent and delinquent children.[113] Women involved in the child-saving movement proclaimed that the domestic role of women made them better equipped to take on the task of child-saving: "the child savers... vigorously defended the virtue of traditional family life and emphasized the dependence of the social order on the proper socialization of children. They promoted the view that women were more ethical and genteel than men, better equipped to protect the innocence of children, and more capable of regulating their education and recreations."[114]

Creation of the Juvenile Court (1899)

The child savers realized that "child-welfare reform could only be accomplished with the support of political and professional organizations."[115] At the 1898 annual meeting of the Chicago Bar Association, a resolution was introduced calling for the appointment of a committee to "investigate existing conditions relative to delinquent and dependent children" and to develop legislation for reform.[116] One month later, the Illinois Conference of Charities devoted most of its program to child-saving issues. Frederick Wine's closing speech expressed the sentiment of the conference (see the Case in Point feature on this topic). His words closely agreed with those of the Chicago Bar Association's resolution calling for "a separate system" for juveniles. The Conference of Charities, like the Bar Association, concluded its meeting by appointing a committee to cooperate with other child-saving agencies to draft a juvenile court bill.[117]

A collaborative effort between the Chicago Women's Club, the Chicago Bar Association, and the Illinois Conference of Charities resulted in a bill that was introduced to the House of Representatives in February 1899 and shortly thereafter to the Senate. The bill was passed on April 14th, the last day of the session.[118] See Case in Point, "Excerpts from *An Act to Regulate the Treatment and Control of Dependent, Neglected, and Delinquent Children* (The Illinois Juvenile Court Act of 1899)," for particularly relevant sections of the act.

The Juvenile Court Act did not represent radical reform; rather, it consolidated existing practices.[119] In fact, in the years before the bill was drafted, a number of other states already practiced some of the innovations advanced in the Juvenile Court Act. For example, Massachusetts, in 1874, and New York, in 1892, passed laws that provided for separate trials of minors, apart from adults, and Massachusetts developed a system of probation in 1846.[120] Anthony Platt, author of *The Child Savers: The Invention of Delinquency*, observes that "special

CASE IN POINT

Excerpt from Frederick Wine's Closing Speech to the Illinois Conference of Charities, 1898

We make criminals out of children who are not criminals by treating them as if they were criminals. That ought to be stopped. What we should have, in our system of criminal jurisprudence, is an entirely separate system of courts for children, in large cities, who commit offenses which would be criminal in adults. We ought to have a "children's court" in Chicago, and we ought to have a "children's judge," who should attend to no other business. We want some place of detention for those children other than prison. . . . No child ought to be tried unless he has a friend in court to look after his real interest. There should be someone there who has the confidence of the judge, and who can say to the court, "Will you allow me to make an investigation of this case? Will you allow me to make a suggestion to the court?"

Source: Platt, *Child Savers,* 132.

provisions for the protection and custody of 'delinquent' children apart from adult offenders existed in the United States long before the enactment of the juvenile court in 1899."[121]

Nonetheless, the creation of the juvenile court culminated a century-long evolution of thought and practice by which juveniles were differentiated from adults both in terms of development and control.[122] The new juvenile court established a separate system that is noteworthy in terms of (1) structure and jurisdiction, (2) legal authority under the expansion of *parens patriae,* and (3) legal philosophy and process.

Structure and Jurisdiction of the Juvenile Court

The Illinois Juvenile Court Act was the "first legislation in the United States to specifically provide for a separate system of juvenile justice."[123] As described in Section 3 of the act, the juvenile court was made up of a designated judge of the circuit court, a "special courtroom," and separate records ("Juvenile Record"). The act specifically refers to this court as the "Juvenile Court." The personnel that made up the first juvenile court were separate and distinct from the personnel that made up the adult court (see Case in Point, "Personnel of the Original Chicago Juvenile Court," page 34).

This new court structure was designed to remove children from the adult criminal justice system, and create special programs for delinquent, dependent, and neglected children.[124] As the US Supreme Court observed in the groundbreaking case, *In re Gault* (1967): "The early reformers were appalled by the adult procedures and penalties and by the fact that children could be given long prison sentences and mixed in jails with hardened criminals."[125] The child savers' goal was first to create a separate system of justice for juveniles, and then to distinguish that system from the adult criminal justice system in terms of legal philosophy and process.

CASE IN POINT

Excerpts from *An Act to Regulate the Treatment and Control of Dependent, Neglected, and Delinquent Children* (The Illinois Juvenile Court Act of 1899)

Section 1. Definitions. This act shall apply only to children under the age of sixteen (16) years. . . . For the purposes of this act the words **dependent child** and **neglected child** shall mean any child who for any reason is destitute or homeless or abandoned; or dependent upon the public for support; or has not proper parental care or guardianship; or who habitually begs or receives alms; or who is found living in any house of ill fame or with a vicious or disreputable person; or whose home, by reason of neglect, or cruelty or depravity on the part of parents, guardian or other person in whose care it may be, is an unfit place for such a child; and any child under the age of eight (8) years who is found peddling or selling any article or singing or playing any musical instrument upon the streets or giving any public entertainment. The words **delinquent child** shall include any child under the age of sixteen (16) who violates any law of the State or any city or village ordinance. . . .

Section 3. Juvenile Court. In counties having over 500,000 population the **judges** of the circuit court shall, at such times as they shall determine, designate one or more of their number whose duty it shall be to hear all cases coming under this act. A special courtroom, to be designated as the **juvenile courtroom,** shall be provided for the hearing of such cases, and the findings of the court shall be entered in a book or books to be kept for that purpose and known as the **"Juvenile Record,"** and the court may, for convenience, be called the "Juvenile Court. . . ."

Section 6. Probation Officers. The court shall have authority to appoint or designate one or more discreet persons of good character to serve as probation officers during the pleasure of the court; said probation officers to receive no compensation from the public treasury. In case a probation officer shall be appointed by any court, it shall be the duty of the clerk of the court, if practical, to notify the said probation officer in advance when any child is to be brought before the said court; it shall be the duty of the said probation officer to make such investigation as may be required by the court; to be present in court in order to represent the interests of the child when the case is heard; to furnish the court such information and assistance as the judge may require; and to take such charge of any child before and after trial as may be directed by the court.

Section 7. Dependent and Neglected Children. When any child under the age of sixteen (16) years shall be found to be dependent or neglected within the meaning of this act, the court may make an order **committing the child** to the care of some suitable State institution, or to the care of some reputable citizen of good moral character, or to the care of some training school or an industrial school, as provided by law, or to the care of some association willing to receive it embracing in its objects the purpose of caring or obtaining homes for dependent or neglected children, which association shall have been accredited as hereinafter provided. . . .

Section 9. Disposition of Delinquent Children. In the case of a delinquent child the court may continue the hearing from time to time and may **commit the child** to the care and guardianship of a probation officer duly appointed by the court and may allow said child to remain in its own home, subject to the visitation of the probation officer; such child to report to the probation officer as often as may be required and subject to be returned to the court for further proceeding, whenever such action may appear to be necessary, or the court may commit the child to the care and guardianship of the probation officer, to be placed in a suitable family home, subject to the friendly supervision of such probation officer, or it may authorize the said probation officer to board out the said child in some suitable family home, in case provision is made by voluntary contribution or otherwise for the payment of the board of such child, until a suitable provision may be made for the child in a home without such payment; or the court may commit the child, if a boy, to a training school for boys, or if a girl, to an industrial school for girls. Or, if the child is found guilty of any criminal offense, and the judge is of the opinion that the best interest requires it, the court may commit the child to any institution within said county incorporated under the laws of this State for the care of delinquent children, or provided by a city for the care of such offenders, or may commit the child, if a boy over the age of ten (10) years, to the State reformatory, or if a girl over the age of ten (10) years, to the State Home for Juvenile Female Offenders. In no case shall a child be committed beyond his or her minority. A child committed to such institution shall be subject to the control of the board of managers thereof, and the said board shall have power to parole such child on such conditions as it may prescribe, and the court shall, on the recommendation of the board, have power to discharge such child from custody whenever in the judgement of the court his or her reformation shall be complete; or the court may commit the child to the care and custody of some association that will receive it, embracing in its objects the care of neglected and dependent children and that has been duly accredited as hereinafter provided. . . .

Source: Illinois Statute 1899, Section 131 (emphasis added).

Personnel of the Original Chicago Juvenile Court

Descriptions of the procedures of the early juvenile court indicate that it was informal and family-like. The image conjured up is of a kindly judge, sitting in a big chair at a table next to a frightened youth (in a much smaller chair). The judge dispenses fatherly wisdom to a repentant and receptive lad. When one looks at the positions that made up the original juvenile court, however, a very different picture emerges. The early juvenile court was composed primarily of law enforcement personnel, not judicial personnel. In his study of the first juvenile court in Chicago, Anthony Platt provides a roster of the original court's personnel.

The personnel of Cook County juvenile court consisted of

1. **six probation officers** paid from private sources, particularly the Chicago Woman's Club,
2. **"one colored woman** who devoted her entire time to the work, free of charge, and whose services are invaluable to the court as she takes charge of all the colored children,"
3. **twenty-one truant officers** paid by and responsible to the Board of Education,
4. **sixteen police officers,** paid by the Chicago police department, assigned to "assist the general probation officers in their visitation work," and
5. **thirty-six private citizens** who were occasionally responsible for supervising children on probation.

In effect, the court staff was primarily composed of police and truant officers, thus facilitating the arrest and disposition of delinquent youth. The juvenile court provided its own policing machinery and removed many distinctions between the enforcement and adjudication of laws.

Source: Platt, *Child Savers,* 139–140.

The full title of the legislation that created the juvenile court indicates that the new court was granted jurisdiction over both juvenile delinquents and dependent and neglected children. As such, the juvenile court was deliberately created to have broad jurisdiction over almost all juvenile matters. The act defined a *delinquent child* as "any child under the age of sixteen (16) who violates any law of the State or any city or village ordinance" (see Case in Point, "Excerpts").[126] The definition of a *dependent* and *neglected child* was much longer and far more sweeping, covering a wide range of conditions from which children must be protected, including homelessness, lack of parental care or guardianship, and parental neglect and abuse. Notice too that dependency and neglect includes a child who "habitually begs or receives alms" or "any child under eight

(8) years who is found peddling or selling any article or singing or playing any musical instrument upon the streets or giving any public entertainment." Taken together, the newly established juvenile court was given broad jurisdiction in all matters of dependency, neglect, and delinquency.

Such broad jurisdiction, however, blurred the distinctions among dependent, neglected, and delinquent children. Regardless of the reason for referral, the early juvenile court was ready and willing to step in if natural parents failed to fulfill their proper function.[127]

Legal Authority: *Parens Patriae*

Robert Mennel observes that "the creation of the juvenile court represented both a restatement and an expansion of the *parens patriae* doctrine."[128] A 1905 Pennsylvania Supreme Court case, *Commonwealth v. Fisher,* expressed the legal authority of the new juvenile court under *parens patriae*: "To save a child from becoming a criminal, or from continuing in a career of crime, to end in maturer years in public punishment and disgrace, the legislatures surely may provide for the salvation of such a child, if its parents or guardians be unable or unwilling to do so, by bringing it into one of the courts of the state without any process at all, for the purpose of subjecting it to the state's guardianship and protection."[129]

Sections 7 and 9 of the Illinois Juvenile Court Act of 1899 prescribe the authority of the juvenile court in dealing with dependent and neglected children, and providing "disposition of delinquent children" (see Case in Point, "Excerpts"). For dependent and neglected children, the court was granted the authority to "make an order committing the child to the care . . . of some reputable citizen of good moral character," or to a public or private agency or institution. For delinquent children, the act stated that the juvenile court "may commit the child to the care and guardianship of the probation officer." This "care and guardianship" authorized the probation officer to allow the child to remain at home under probation supervision, or to place the child in "a suitable family home" or institution. In either case, the duty of probation officers, as specified in this law (Sections 6 and 9), was to conduct investigations into the social background of youths, "represent the interests of the child" in court, and provide "friendly supervision" when the child was committed to them by the court. Section 6 of the act points out that probation officers were "to receive no compensation from the public treasury." They worked as volunteers or were paid by private philanthropic organizations, such as the Chicago Women's Club.

The act also identified a variety of institutional commitments, including training and industrial schools, the State Home for Juvenile Female Offenders, and the boys' State Reformatory. As an indication of the blurred distinction between delinquent and dependent children, the act stated that a juvenile found guilty of a criminal offense could be placed with a philanthropic association that provided "care of neglected and dependent children." It is important to recognize that the original juvenile court statute allowed for commitment to a wide variety of private and public institutions and agencies.[130]

Although the spirit of the Illinois Juvenile Court Act affirmed the value of home and family life, the variety of commitment options stipulated in the act allowed the juvenile court to assume parental responsibility, either through a probation officer, an institution, or a philanthropic association. This authority clearly reflects the *parens patriae* doctrine.[131] The closing passage of the act reaffirms this:

> *This act shall be liberally construed, to the end that its purpose may be carried out, to wit: That the care, custody, and discipline of a child shall approximate... that which should be given by its parents and in all cases where it can properly be done the child placed in an improved family home and become a member of the family by legal adoption or otherwise.*[132]

Legal Philosophy and Process: The "Rehabilitative Ideal"

In advocating for the juvenile court, the child savers sought not only a separate legal system for juveniles, but also a legal philosophy and process that distinguished juvenile courts from adult criminal courts. In a 1909 article in the *Harvard Law Review*, Judge Julian Mack, the second judge of the Chicago Juvenile Court, declared:

> *Why is it not just and proper to treat these juvenile offenders as we deal with the neglected children, as a wise and merciful father handles his own child whose errors are not discovered by the authorities? Why is it not the duty of the State, instead of asking merely whether a boy or a girl has committed the specific offense, to find out what he is, physically, mentally, morally, and then if it learns that he is treading the path that leads to criminality, to take him in charge, not so much to punish as to reform, not to degrade but to uplift, not to crush but to develop, not to make him a criminal but a worthy citizen.*[133]

rehabilitative ideal The traditional legal philosophy of the juvenile court, which emphasizes assessment of the youth and individualized treatment, rather than determination of guilt and punishment.

This distinctive legal philosophy of the original juvenile court has been called the **rehabilitative ideal** because of its emphasis on assessment and reform, rather than the determination of guilt and punishment as in criminal courts.[134] The rehabilitative ideal is clearly founded on the *parens patriae* doctrine, giving the state, through the juvenile courts, the authority and obligation to assume parental responsibility. Rehabilitation became the focus of the new juvenile court, and procedures were developed to reflect and facilitate this ideal. In the same article, Judge Mack characterized the judicial procedures that should accompany this distinctive legal philosophy:

> *The child who must be brought into court should, of course, be made to know that he is face to face with the power of the state, but he should at the same time, and more emphatically, be made to feel that he is the object of its care and solicitude. The ordinary trappings of the courtroom are out of place in such hearings. The judge on a bench, looking down upon the boy standing at the bar, can never evoke a proper sympathetic spirit. Seated at a desk, with the child at his side, where he can on occasion put his arm around his shoulder and draw the lad to him, the judge, while losing none of his judicial dignity, will gain immensely in the effectiveness of the work.*[135]

The *parens patriae* doctrine was adopted from the English chancery courts, which did not have criminal jurisdiction. Thus, the juvenile court was created as a civil court, not as a criminal court.[136] The civil law tradition of *parens patriae,* together with the rehabilitative ideal, resulted in at least three important implications for the legal procedures of the early juvenile court: (1) diminished criminal responsibility of juveniles, (2) a child welfare approach operating on the concept of the "best interests of the child" and (3) informal and family-like procedures.

1. **Diminished criminal responsibility of juveniles:** Drawn from the developmental concepts of childhood and adolescence, the early juvenile court held that children and adolescents less than sixteen years of age lacked the capacity to commit crime. This presumption of incapacity acknowledged that young people could not be held legally responsible for their offenses because they lacked physical and mental maturity.[137] Viewed in this way, juveniles were not charged with or convicted of criminal offenses, and rehabilitation, not punishment, was the appropriate outcome of the juvenile court process. Judge Mack states, "children were no longer to be dealt with as criminals, but rather through the *parens patriae* power of the state were to be treated as wards of the state, not fully responsible for their conduct and capable of being rehabilitated."[138]

2. **A child welfare approach—the "best interests of the child":** The child savers envisioned the juvenile court as a welfare system, rather than a judicial system.[139] As a result, the prevailing goal of the juvenile court was to protect, nurture, reform, and regulate the dependent, neglected, and delinquent child. The role of the juvenile court was not to determine guilt or innocence, but to ascertain the character and needs of an offender by analyzing his or her social background so that the court could make a full determination of what was in the "<u>best interests of the child</u>."[140]

   The early juvenile court's intense focus on the individual juvenile offender, rather than the offense, coincides with the rise of positivist criminology in the late nineteenth and early twentieth centuries. Using detailed social histories, the juvenile court sought to uncover the causes of a youth's delinquent behavior and thereby provide a "proper diagnosis."[141] The identification of pathological traits and conditions was then used to develop a treatment program that was individualized to the child.[142]

   Because the juvenile court was primarily interested in determining the "best interests of the child," based upon a scientific assessment of the "total child," it gave little consideration to the reason for referral—dependency, neglect, or delinquency.[143] The referral offense was merely a symptom that the juvenile court had to assess more thoroughly in order to uncover the "real needs" of the child.[144] According to Harvey Baker, judge of the early Boston juvenile court, "The court does not confine its attention to just the particular offense which brought the child to its notice. For example, a boy who comes to court for some trifle as failing to wear a badge when selling papers may be held on probation for months because of difficulties at school; and a boy who comes in for playing ball on the street may . . . be committed to a reform school because he is found to have habits of loafing, stealing or gambling which can not be corrected outside."[145]

"best interests of the child" The overarching interest of the traditional juvenile court to assess the needs of the youth and then to seek physical, emotional, mental, and social well-being for that youth through court intervention.

3. **Informal and family-like procedures:** In an effort to bring about the re-habilitative ideal, the original juvenile court discarded the rules of criminal procedure. Anthony Platt observes that "the administration of juvenile justice differed in many important respects from the criminal court process. A child was not accused of a crime but offered assistance and guidance; intervention in his life was not supposed to carry the stigma of a criminal record; judicial records were not generally available to the press or public, and hearings were conducted in relative privacy; proceedings were informal and due process safeguards were not applicable due to the court's civil jurisdiction."[146]

In place of due process of law, the juvenile court developed an informal process in which the judge, much like a parent, tried to find out all about the child.[147] Judge Tuthill, the first judge of the Chicago Juvenile Court described his approach as follows: "I have always felt, and endeavored to act in each case, as I would were it my own son who was before me in the library at home, charged with some misconduct."[148]

The physical setting of the juvenile court was intended to facilitate the rehabilitative ideal. The new juvenile court building that opened in Chicago in 1907 was designed to provide an informal, family-like setting for juvenile court hearings: "The hearings will be held in a room fitted up as a parlor rather than a court, around a table instead of a bench.... The hearing will be in the nature of a family conference, in which the endeavor will be to impress the child with the fact that his own good is sought alone."[149]

Juvenile court reformers also introduced euphemistic legal terminology in order to avoid reference to the harsh, adversarial process of adult criminal courts.[150] To initiate the juvenile court process, a *petition* is filed *"in the welfare of the child,"* whereas the formal legal document that initiates the adult criminal process is an *indictment* or an *information.* The proceedings of juvenile courts are referred to as *hearings,* instead of *trials,* as in adult courts. Juvenile courts find youths to be *delinquent,* rather than *guilty of an offense.* Finally, juvenile delinquents are given a *disposition,* instead of a *sentence,* as in adult criminal courts.

The juvenile court was clearly an idea ripe for its time. The Illinois Juvenile Court Act of 1899 was "a prototype for legislation in other states and juvenile courts were quickly established in Wisconsin (1901), New York (1901), Ohio (1902), Maryland (1902), and Colorado (1903)."[151] By 1925, all but two states, Maine and Wyoming, had juvenile court laws.[152] The juvenile justice systems that emerged from legislation were composed of newly created juvenile courts together with a collection of private and public institutions and community programs, all embracing the rehabilitative ideal and empowered by *parens patriae.*[153]

■ The Second Revolution: Transformation of Juvenile Justice Thought and Practice

Although the new juvenile court system proved wildly popular and spread rapidly, it was not without critics. Some scholars questioned the intent of child-saving reformers, arguing that they were self-serving, middle- and upper-class women whose sole goal was to maintain the status quo by implementing new and powerful control strategies through the creation of the juvenile court.[154] Other scholars claimed that, while reformers may have been well-intended, the rhetoric of reform was never really achieved in the newly created juvenile justice systems. From their point of view the resulting systems were even more punitive and authoritarian than the earlier child welfare systems that were used in combination with adult criminal courts. In addition, the new juvenile justice systems were given extensive, almost unbridled, authority under the rehabilitative ideal and expanded *parens patriae.*[155]

Despite occasional criticism, juvenile courts across the United States achieved high regard in the decades following their creation. The few reports of problems were viewed as "minor imperfections soon to be corrected by a continually improving system."[156] The confidential records and closed hearings of juvenile courts made them inaccessible and effectively above reproof.

Challenges to the Traditional Juvenile Court

Shortly after World War II, criticism of the juvenile court began to mount. In 1946, criminologist Paul Tappan wrote a widely read and influential article entitled "Treatment without Trial?"[157] In this article, Tappan criticized the juvenile court's failure to provide due process of law. **Due process of law** refers to the procedural rights established in the Constitution (especially the Bill of Rights) and extended through appellate court decisions. Under the guise of the rehabilitative ideal, and empowered by *parens patriae,* the procedures of the original juvenile court were informal and family-like, making the rules of criminal procedure inapplicable. In a 1966 ruling, the Supreme Court observed, "There is evidence, in fact, that there may be grounds for concern that the child receives the worst of both worlds: that he gets neither the protections accorded to adults nor the solicitous care and regenerative treatment postulated for children."[158]

due process of law
Procedural rights established in the Constitution (including the Bill of Rights) and extended through appellate court decisions that are based upon individual freedoms and limitation of governmental powers.

Beginning in the 1960s and persisting until the 1980s, legal challenges were mounted against the informalities of the juvenile justice system.[159] This movement involved a series of Supreme Court cases that radically altered juvenile justice procedures.

The 1960s also ushered in empirical challenges to the juvenile justice system.[160] Most actively in the 1970s, evaluation research seriously questioned the effectiveness of individualized treatment, rehabilitation, and community control.[161] While this research considered both juvenile and adult correctional methods, it directed significant attention at the rehabilitative ideal of the juvenile court. In his book, *Radical Non-Intervention,* Edwin Schur argued for a drastic reduction in the juvenile justice system's reliance on treatment and rehabilitation. Instead, he advocated a "return to the rule of law," involving the reduction of discretionary powers of the juvenile court, diversion of less serious offenders, and intervention for only the most serious crimes.[162]

In the years following these challenges, the juvenile justice system was altered dramatically. Prevailing views of juvenile delinquency and the proper approach of the juvenile justice system changed significantly. A number of criminologists have argued that these changes were so consequential as to constitute a revolution comparable to the one that first created the juvenile court—

a "second revolution."[163] Three areas of change have been most pronounced: (1) the due process revolution, (2) enactment of the Juvenile Justice and Delinquency Prevention Act of 1974, and (3) contemporary initiatives for punishment and accountability.

The Due Process Revolution in Juvenile Justice

Since its inception, the juvenile court has been described as civil, rather than criminal. With the express purpose of protection and reform, the rehabilitative ideal of the juvenile court made the due process protections afforded criminal defendants unnecessary.[164] In the ten-year period from 1966 to 1975, however, the US Supreme Court took an activist stand in establishing due process requirements for the juvenile justice system. Five cases handed down during this period dramatically altered the procedures of the traditional juvenile justice system. The following case summaries are drawn from an overview of significant Supreme Court cases offered by Howard Snyder and Melissa Sickmund from the National Center for Juvenile Justice.[165] The case details provide insight into the dynamics of legal change occurring during the juvenile due process revolution.

Kent v. United States (1966)[166]

While on probation for an earlier offense, Morris Kent, age 16, was charged with rape and robbery for an incident in which he broke into a woman's apartment, raped her, and stole her wallet. Kent confessed to these offenses, as well as to several similar incidents. The judge of the juvenile court waived the case to adult court after making a "full investigation." Without a waiver hearing, the juvenile court judge did not describe the investigation or the grounds for the waiver. Kent was subsequently found guilty in criminal court on 6 counts of breaking and entering and robbery and sentenced to 30 to 90 years in prison. Kent's attorney appealed the waiver, arguing that it was invalid. He also filed a writ of habeas corpus asking the State to justify Kent's detention. Appellate courts rejected both the appeal and the writ. In appealing to the US Supreme Court, Kent's attorney argued that the judge had not made a complete investigation and that Kent was denied constitutional rights simply because he was a minor. The Court ruled the waiver invalid, stating that Kent was entitled to a hearing that measured up to "the essentials of due process and fair treatment." The opinion of the Court stated that Kent should have been granted a formal hearing on the motion of waiver, with representation, and that his attorney must be given access to all records involved in the waiver. The ruling also stated that the juvenile court must provide a written statement of the reasons for waiver.

In re Gault (1967)[167]

Gerald Gault, age 15, was on probation in Arizona for a minor property offense when he and a friend made a crank telephone call to an adult neighbor, asking her, "Are your cherries ripe today?" and "Do you have big bombers?" Identified by the neighbor, the youths were arrested and detained. The victim did not appear at the adjudication hearing, and the court never resolved the issue of whether Gault made the "obscene" remarks. Nonetheless, Gault was committed to a training school. The maximum sentence for an adult would have been a $50 fine or 2 months in jail. An attorney obtained for Gault after the trial filed a writ of habeas corpus that was eventually heard by the US Supreme Court. The issue presented in the case was the denial of Gault's constitutional rights for due process of law, including notice of charges, counsel, questioning of witnesses, protection against self-incrimination, transcript of the proceedings, and appellate review. The Court ruled that, in hearings that could result in commitment to an institution, juveniles have the right to notice and counsel, the right to question witnesses, and the right to protection against self-incrimination. The Court did not rule on a juvenile's right to appellate review or transcripts, but it encouraged the States to provide those rights. The Court based its ruling on the fact that Gault was being punished rather than helped by the juvenile court. The Court explicitly rejected the doctrine of *parens patriae* as the founding principle of juvenile justice, describing the concept as "murky" and of "dubious historical relevance." The Court concluded that the handling of Gault's case violated the due process clause of the Fourteenth Amendment: "Juvenile court history has again demonstrated that unbridled discretion, however benevolently motivated, is frequently a poor substitute for principle and procedure."

In re Winship (1970)[168]

Samuel Winship, age 12, was charged with stealing $112 from a woman's purse in a store. A store employee claimed to have seen Winship running from the scene just before the woman noticed the money was missing; others in the store stated that the employee was not in a position to see the money being taken. Winship was adjudicated delinquent and committed to a training school. New York juvenile courts operated under the civil court standard of "preponderance of evidence." The court agreed with Winship's attorney that there was "reasonable doubt" of Winship's guilt, but based its ruling on the "preponderance of evidence." Upon appeal to the Supreme Court, the central issue in the case was whether "proof beyond a reasonable doubt" should be considered among the "essentials of due process and fair treatment" required during the adjudicatory stage of the juvenile court process. The Court rejected lower court arguments that juvenile courts were not required to operate on the same standards as adult courts because juvenile courts were designed to "save" rather than to "punish" children. The Court ruled that the "reasonable doubt" standard should be required in all delinquency adjudications.

McKeiver v. Pennsylvania (1971)[169]

Joseph McKeiver, age 16, was charged with robbery, larceny, and receiving stolen goods. He and 20 to 30 other youths allegedly chased 3 youths and took 25 cents from them. McKeiver met with his attorney for only a few minutes before his adjudicatory hearing. At the hearing, the attorney's request for a jury trial was denied by the court. McKeiver was subsequently adjudicated and placed on probation. His case was appealed to the Pennsylvania State Supreme Court, which cited recent decisions of the US Supreme Court that attempted to include more due process in juvenile court proceedings without eroding the essential benefits of the juvenile court. The State Supreme Court affirmed the lower court, arguing that of all due process rights, trial by jury is most likely to "destroy the traditional character of juvenile proceedings." The US Supreme Court found that the due process clause of the Fourteenth Amendment did not require jury trials in juvenile court. The impact of the Court's *Gault* and *Winship* decisions was to enhance the accuracy of the juvenile court process in the fact-finding stage. In *McKeiver,* the Court argued that juries are not known to be more accurate than judges in the adjudication stage and could be disruptive to the informal atmosphere of the juvenile court by tending to make it more adversarial.

Breed v. Jones (1975)[170]

In 1970, Gary Jones, age 17, was charged with armed robbery. Jones appeared in Los Angeles juvenile court and was adjudicated delinquent on the original charge and two other robberies. At the dispositional hearing, the judge waived jurisdiction over the case to criminal court. Counsel for Jones filed a writ of habeas corpus, arguing that the waiver to criminal court violated the double jeopardy clause of the Fifth Amendment. The court denied this petition, saying that Jones had not been tried twice because juvenile adjudication is not a "trial" and does not place a youth in jeopardy. Upon appeal, the US Supreme Court ruled that an adjudication in juvenile court, in which a juvenile is found to have violated a criminal statute, is equivalent to a trial in criminal court. Thus, Jones *had* been placed in double jeopardy. The Court also specified that jeopardy applies at the adjudication hearing when evidence is first presented. Waiver cannot occur after jeopardy attaches.[171]

In the years that followed these cases, the US Supreme Court continued to hear cases that had impact on the proceedings of the juvenile justice system, but the number and scope of these cases diminished. Taken together, however, these five cases dramatically changed the character and procedures of the juvenile justice system. The rehabilitative ideal of the traditional juvenile court, together with its *parens patriae* authority, was diminished drastically, making juvenile courts more like criminal courts.

The Juvenile Justice and Delinquency Prevention Act of 1974

The Juvenile Justice and Delinquency Prevention (JJDP) Act of 1974 embodied a series of reforms to redefine juvenile delinquency and redirect the legal philosophy, authority, and procedures of the juvenile justice system. Three groups directly influenced these reform efforts: the President's Commission on Law Enforcement and Administration of Justice, the National Council on Crime and Delinquency, and the National Advisory Commission on Criminal Justice Standards and Goals.[172]

President's Commission on Law Enforcement and Administration of Justice

Established in 1965 by executive order of President Johnson, the President's Commission on Law Enforcement and Administration of Justice was charged with the task of examining the extent and nature of crime and what could be done about it. As part of this task, the Commission was directed to conduct an analysis of juvenile crime and the workings of the juvenile justice system and then to make recommendations based on the analysis. The Commission drew attention to the widening gap between the rehabilitative ideal of the original juvenile court and the actual juvenile justice practices: "The great hopes originally held for the juvenile court have not been fulfilled. It has not succeeded significantly in rehabilitating delinquent youth, in reducing or even stemming the tide of delinquency, or in bringing justice and compassion to the child offender."[173] In particular, the Commission expressed grave concern about the juvenile court's power over children with regard to noncriminal conduct, suggesting that perhaps this facet of the juvenile court's authority over children should be eliminated.[174]

Based on its inquiry, the Commission recommended handling minor offenders in the community instead of juvenile court (diversion). The Commission advocated the development of community resources (especially neighborhood centers, called "youth service bureaus") that would provide a wide variety of services to youth and families and make referrals to community-based programs.[175] These community-based efforts were justified in terms of delinquency prevention: "What research is making increasingly clear is that delinquency is not so much an act of individual deviancy as a pattern of behavior produced by a multitude of pervasive societal influences well beyond the reach of the actions of any judge, probation officer, correctional counselor, or psychiatrist."[176]

Several recommendations of the Commission were noteworthy: narrowing the jurisdiction of the juvenile court to youth who violated the criminal law, limiting the use of detention and incarceration, expanding use of informal dispositions at prejudicial screenings, and developing community alternatives to formal court proceedings.[177] For serious offenders, however, the Commission envisioned a more formal, punitive system: "Court adjudication and disposition of those offenders should no longer be viewed solely as a diagnosis and prescription for cure, but should be frankly recognized as an authoritative court judgment expressing society's claim to protection. . . . Accordingly, the adjudicatory hearing should be consistent with basic principles of due process."[178]

National Council on Crime and Delinquency

In 1966, the President's Commission requested that the National Council on Crime and Delinquency (NCCD) survey state and local correctional agencies and institutions across the United States. The survey showed widespread use of detention facilities for juveniles accused of noncriminal conduct (status offenses, dependency, and neglect), often without court petitions. As a result of the survey, the NCCD recommended that "no child be placed in any detention facility unless he is a delinquent or alleged delinquent and there is substantial probability that he will commit an offense dangerous to himself or the community."[179] The NCCD additionally recommended that noncriminal youths, including dependent and neglected children, should not be placed in detention facilities or committed to institutions with delinquent offenders.

National Advisory Commission on Criminal Justice Standards and Goals

The third group to marshal reform of the juvenile justice system through the JJDP Act was the National Advisory Commission on Criminal Justice Standards and Goals. Established in 1971, the National Advisory Commission was created to formulate model criminal and juvenile justice standards, goals, and practices. After extensive study, the National Advisory Commission concluded that "first priority should be given to preventing juvenile delinquency, to minimizing the involvement of young offenders in the juvenile and criminal justice system, and to reintegrating delinquent and young offenders into the community."[180] The National Advisory Commission also recommended that "the delinquency jurisdiction of the court should be limited to those juveniles who commit acts that if committed by an adult would be criminal, and that juveniles accused of delinquent conduct would not under any circumstances be detained in facilities for housing adults accused or convicted of crime."[181]

The message coming from these three groups was consistent: the juvenile justice system is not the panacea the child savers hoped it would be. The findings and recommendations of these three groups formed the basis for the JJDP Act of 1974. This act was the first major federal initiative to address juvenile delinquency in a comprehensive manner.[182] Primary responsibility for juvenile justice had historically existed at the state and local levels; the JJDP Act established a leadership role for the federal government through the creation of the Office of Juvenile Justice and Delinquency Prevention (OJJDP). The JJDP Act established juvenile justice goals and policies and committed ongoing financial assistance to aid their implementation at the state and local levels.[183] The most assertive parts of the JJDP Act were four "system reform mandates," which attorney Kathleen Kositzky Crank summarizes as follows.

status offender A juvenile who has committed an act that would *not* be a crime if committed by an adult.

1. **Deinstitutionalization of status offenders.** The deinstitutionalization of status offenders mandate provides, as a general rule, that no **status offender** (a juvenile who has committed an act that would not be a crime if committed by an adult) or non-offender [a dependent and neglected child] may be held in secure detention or confinement. . . .

2. **Sight and sound separation.** The separation mandate provides that juveniles shall not be detained or confined in a secure institution in which they may have contact with incarcerated adults, including inmate trustees. This requires complete separation such that there is no sight or sound contact with adult offenders in the facility. . . .

3. **Jail and lockup removal.** The jail and lockup removal mandate establishes as a general rule that all juveniles who may be subject to the original jurisdiction of the juvenile court based on age and offense limitations established by State law cannot be held in jails and law enforcement lockups in which adults may be detained or confined. [Several exceptions to this mandate are specified in law.] . . .

4. **Disproportionate minority confinement.** The disproportionate minority confinement (DMC) mandate requires States to address efforts to reduce the number of minority youth in secure facilities where the proportion of minority youth in confinement exceeds the proportion such groups represent in the general population. In order to meet the DMC mandate, States go through stages of data gathering, analysis and problem identification, assessment, program development, and systems improvement initiatives.[184]

In addition to these mandates, the JJDP Act called for a preventive approach to the problem of delinquency. OJJDP established a formula grant program for states and communities to develop policies, practices, and programs directed at crime prevention in local areas. In addition, communities were encouraged to develop alternatives to the juvenile justice system. "Community-based programs, diversion, and deinstitutionalization became the banners of juvenile justice policy in the 1970s."[185]

The juvenile justice reforms initiated by the JJDP Act of 1974 and carried out by the OJJDP are, in many ways, consistent with the juvenile due process revolution that occurred between 1966 and 1975.[186] Both reform efforts questioned the rehabilitative ideal and the *parens patriae* authority of the traditional juvenile justice system. The due process changes made the juvenile justice system more like the adult justice system while still acknowledging the need for a separate system. The JJDP Act refined this separation by establishing certain mandates to deinstitutionalize status offenders and non-offenders and to keep juveniles out of the adult criminal justice system, especially jails. In addition, the Act authorized federal initiatives for a preventive, community-based approach to juvenile delinquency.

Getting Tough: Initiatives for Punishment and Accountability

The 1980s saw a dramatic shift in juvenile justice law and practice at both the federal and state levels. A 1984 report by the National Advisory Committee for Juvenile Justice and Delinquency Prevention states, "federal effort in the area of juvenile delinquency should focus primarily on the serious, violent, or chronic offender."[187] Soon after, OJJDP began to devote attention to the identification and control of serious juvenile offenders.[188] It liberally sponsored research on chronic, violent offenders and funded state and local programs designed to prevent and control violence and the use of drugs.[189]

At the state level, legislatures passed laws to crack down on juvenile crime, reflecting a widespread reconsideration of juvenile justice philosophy, jurisdiction, and authority and a more punitive approach to juvenile delinquency.[190] Four areas of legal change have been most pronounced: (1) transfer provisions in state law, (2) enhanced sentencing authority for juvenile courts, (3) reduction in juvenile court confidentiality, and (4) balanced and restorative justice efforts.

Transfer Provisions

All states have enacted laws that allow juveniles to be tried in adult criminal courts. While these laws vary from state to state, transfer provisions fall into three main categories: judicial waiver, concurrent jurisdiction, and statutory exclusion.[191]

1. **Judicial waiver:** Juvenile court judges are granted statutory authority to waive juvenile court jurisdiction and transfer cases to criminal court. Waiver decisions are discretionary and therefore subject to due process review. While judicial waiver statutes vary from state to state, the basic idea is that certain types of offenses and offenders, especially violent ones, are beyond the scope of the juvenile court. States may use terms other than judicial waiver, including *certification, remand,* or *bind over* for criminal prosecution. States may also *transfer* or *decline,* rather than waive, jurisdiction.

2. **Concurrent jurisdiction:** Under this statutory provision, prosecutors are granted authority to file certain types of cases in either juvenile or criminal court. Some states, for example, allow prosecutors, at their discretion, to file felony offenses directly in adult criminal courts. State appellate courts have usually taken the view that prosecutor discretion is equivalent to a charging decision made in criminal cases; therefore, it is not subject to judicial review for due process. Transfer under concurrent jurisdiction provisions is also known as *prosecutor waiver, prosecutor discretion,* or *direct file.*

3. **Statutory exclusion:** With statutory exclusion, state statutes exclude certain juvenile offenders and offenses from juvenile court jurisdiction. Under statutory exclusion provisions, cases originate in criminal rather than juvenile court. State statutes usually set age and offense limits for excluded offenses.

While judicial waiver is the oldest and most common transfer provision, almost all states have expanded their statutory provisions for transferring juvenile cases to adult court.[192] In an effort to reduce discretion in the transfer decision, many state legislatures have since the 1970s moved toward mechanisms that are based on age or seriousness of offense or both, without case-specific considerations. According to Melissa Sickmund, "Although not typically thought of as transfers, large numbers of youth younger than 18 are tried in criminal court in the 13 states where the upper age of juvenile court is set at 15 or 16."[193]

Sentencing Authority

A second area of transformation in the 1980s was the enactment of state laws that give both criminal and juvenile courts expanded sentencing options in juvenile cases. This change resulted in a more punitive approach to juvenile delinquency. Traditionally, juvenile court dispositions were individualized and based on the background characteristics of the offender. Indeterminate sentencing laws allowed the juvenile court judge to customize the disposition to fit the offender's needs and situation, with rehabilitation as the clear primary goal. As states shifted the purpose of their juvenile justice systems away from rehabilitation and toward punishment, accountability, and public safety, juvenile case dispositions began to be based more on the offense than the offender. "Offense-based dispositions tend to be determinate and proportional to the offense; retribution and deterrence replace rehabilitation as the primary goal."[194]

Beginning in the mid-1970s, a number of states changed their statutes to allow for punishment in juvenile court disposition. New York's Juvenile Justice Reform Act of 1976 provided for secure confinement and mandatory treatment of serious juvenile offenders, followed by strict parole standards and intensive supervision upon release. By 1997, at least 16 states had followed New York's lead by adding or modifying laws to require minimum periods of incarceration for certain violent or serious offenders.[195]

A number of states have also raised the maximum age of the juvenile court's continuing jurisdiction over juvenile offenders. In these states, the dispositional order may extend the juvenile court's jurisdiction beyond the upper age of original jurisdiction (usually 18 years old).[196] Illinois' habitual juvenile offender law, for example, allows the juvenile court to commit juvenile offenders who meet the law's criteria of habitual offender to the Department of Corrections until these offenders are 21 years old.[197]

Confidentiality

A third area of juvenile justice transformation concerns the traditional confidentiality of juvenile justice proceedings and records. In almost every state, "legislatures have recently made significant changes in how information about juvenile offenders is treated by the justice system."[198] Laws allowing for the release of court records to other justice agencies, schools, victims, and the public have been enacted in most states. These laws also establish the circumstances under which media access is allowed. A number of states also permit or even require the juvenile court to notify school districts about juveniles charged with or convicted of serious or violent crimes.[199]

Balanced and Restorative Justice

These initiatives for punishment and accountability have replaced the rehabilitative ideal and *parens patriae* authority of the original juvenile court. Sections of state law that declare the purpose of the juvenile court now speak of holding juveniles accountable to victims and communities, having juveniles accept responsibility for their criminal actions, promoting public safety, and deterring potential offenders. Instead of a primary emphasis on rehabilitation, these declarations of purpose now promote competency development so that delinquent youth can become responsible citizens. This new orientation of the juvenile court is referred to as **balanced and restorative justice**, and it emphasizes offender accountability, community safety, and offender competency development.[200]

balanced and restorative justice A contemporary orientation in juvenile justice that emphasizes offender accountability, community safety, and offender competency development.

In the mid-1990s, one leg of this "balanced" approach began to receive more emphasis than the other two: offender accountability. This increasing emphasis on accountability was spearheaded by federal legislation that endorsed a more punitive approach to juvenile justice. In 1996, Congress passed the Balanced Juvenile Justice and Crime Prevention Act of 1996, which announced the serious need for change in the operating philosophy and procedures of state juvenile justice systems. In late 1997, OJJDP began to administer a large federal grant incentive program enacted in law: the Juvenile Accountability Incentive Block Grants Program (JAIBG). The Act appropriated $250 million for the program, including $232 million for state block grants. States requesting funding were required to demonstrate that their laws, policies, and procedures fulfilled several expectations: (1) juveniles, age 15 years or over, who are alleged to have committed a "serious violent crime" must be criminally prosecuted in adult court; (2) sanctions should be imposed for every delinquency act, including probation violations, and sanctions must escalate for each subsequent, more serious offense or probation violation; (3) records of juvenile felony offenders who have a prior delinquency adjudication are to be treated in a manner equivalent to that of adult records, including submission of such records to the FBI; and (4) state law must not prohibit juvenile court judges from issuing court orders that require parental supervision of juvenile offenders.[201]

Grant funds from the JAIBG were used by states in twelve "program purpose areas" stipulated by OJJDP. These twelve program purpose areas have

CASE IN POINT

Purpose Clause of the Montana Youth Court Act

41-5-102. Declaration of purpose. The Montana Youth Court Act must be interpreted and construed to effectuate the following express legislative purposes:

(1) to preserve the unity and welfare of the family whenever possible and to provide for the care, protection, and wholesome mental and physical development of a youth coming within the provisions of the Montana Youth Court Act;

(2) to prevent and reduce youth delinquency through a system that does not seek retribution but that provides:

(a) immediate, consistent, enforceable, and avoidable consequences of youth's actions;

(b) a program of supervision, care, rehabilitation, detention, competency development, and community protection for youth before they become adult offenders;

(c) in appropriate cases, restitution as ordered by the youth court; and

(d) that whenever removal from the home is necessary, the youth is entitled to maintain ethnic, cultural, or religious heritage whenever appropriate.

(3) to achieve the purposes of subsections (1) and (2) in a family environment whenever possible, separating the youth from the parents only when necessary for the welfare of the youth or for the safety and protection of the community;

(4) to provide judicial procedures in which the parties are ensured a fair, accurate hearing and recognition and enforcement of their constitutional and statutory rights.

Source: Montana Code Annotated 2005.

since been expanded to sixteen.[202] Among the program areas are the construction and staffing of juvenile detention or correctional facilities; the hiring of judges, prosecutors, defense attorneys, and probation officers; accountability-based sanctions programs; graduated sanctions; restorative justice programs; gun courts; and drug courts.[203]

The wording of "purpose clauses" in many state juvenile court acts was changed to adopt balanced and restorative language, in order to demonstrate an operating philosophy consistent with the requirements of JAIBG, thus making these states eligible for block grants.[204] The wording of the "Declaration of Purpose" of the Montana Youth Court Act reflects this balanced and restorative justice orientation (see Case in Point, "Purpose Clause of the Montana Youth Court Act").

■ Legal Definitions of "Juvenile Delinquency"

In its zeal to save children, the original juvenile court showed little interest in distinguishing the different types of children that came under its jurisdiction.[205] Based on the state law that created the first juvenile court—*An Act to Regulate the Treatment and Control of Dependent, Neglected, and Delinquent Children*—the original juvenile court assumed broad jurisdiction over not only delinquent offenders, but also dependent and neglected children. While the Juvenile Court Act provided definitions of delinquent and dependent and neglected children, the primary interest of the early court was to act in the "best interests of the child," regardless of the reason the child came before the court. An early English reformer, Mary Carpenter, captured this sentiment when she said, "there is no distinction between pauper, vagrant, and criminal children, which would require a different system of treatment."[206] In fact, the original juvenile court was founded on the premise that the poor, neglected child, as a "predelinquent," must be protected from criminal influences.[207] Dependency, neglect, and delinquency are all stops on the road to crime.

Reformers soon realized, however, that restricting the new juvenile court's definition of delinquency to violations of criminal law might make it function much as a criminal court does.[208] Within two years, amendments to the original act broadened the definition of delinquent to include a youth "who is *incorrigible*; or who knowingly associates with thieves, vicious or immoral persons; or who is growing up in idleness and crime; or who knowingly frequents a house of ill-fame; or who knowingly patronizes any policy shop or place where any gaming device is, or shall be operated."[209] Barry Feld observes that this undefined term, *incorrigible*, "introduced a major element of vagueness, imprecision, and subjectivity into the court's inquiry into a youth's 'condition of delinquency.' "[210] This broadened definition of juvenile delinquency now included behavior that was defined by law as illegal only for juveniles—those individuals under the legal age. Such noncriminal but illegal acts by juveniles are commonly called **status offenses**. Status offenses include acts like running away, truancy, ungovernability, and liquor law violations.

Traditionally, juvenile courts have had two primary areas of jurisdiction: (1) juvenile delinquency, which includes violations of criminal law and status offenses; and (2) dependency, neglect, and child abuse.[211] The working assumption behind both areas of jurisdiction has been the rehabilitative ideal and the *parens patriae* authority of the juvenile court: if parents are unable or have failed to provide proper care for their children, then the court can assume parental responsibility in the best interests of the child.

In the early 1960s, criticism began to mount over the juvenile court's broad jurisdiction and its failure to distinguish different types of cases. This criticism was most pronounced with regard to the court's failure to distinguish status and criminal law offenders. Don Gibbons and Marvin Krohn summarize the argument:

> *Critics have suggested that these categories of behavior (status offenses) are so vaguely defined that nearly all youngsters could be made the subject of court attention. What is a "vicious or immoral person"?*

status offense An act that is illegal for a juvenile but would *not* be a crime if committed by an adult.

> *What is "incorrigibility"? These are highly subjective characterizations of persons or situations. Further, considerable doubt has been expressed about whether these kinds of acts are predictive of serious antisocial conduct. Perhaps courts would be better off not to concern themselves with these relatively benign activities and conditions, and, at the very least, youngsters who have not been charged with criminal offenses should not be processed in the juvenile court with those who have engaged in criminal law violations.*[212]

In 1961, the California legislature created a separate section of the juvenile code to specify three different areas of jurisdiction for the juvenile court: (1) dependent and neglected children (nondelinquents), (2) juveniles who violate the state criminal code, and (3) juveniles who are beyond parental control or who engage in conduct harmful to themselves (i.e., status offenders).[213] The following year, New York passed legislation that established a family court system with jurisdiction over all areas of family life. In addition, the legislation established a person in need of supervision (PINS) classification to provide a separate designation for status offenders. This legal separation of status and criminal law offenders allowed the juvenile justice system to approach these two groups differently. Following the lead of New York, statutory law in many states soon provided a separate legal category for status offenders. Various acronyms were used: YINS (youth in need of supervision), MINS (minor in need of supervision), CHINS (children in need of supervision), and JINS (juveniles in need of supervision).

This differentiation between status offenders and juvenile criminal law offenders plays an important role in contemporary trends in juvenile justice. The juvenile due process revolution has been applied most extensively to juvenile criminal offenders. Status offenses are normally handled like dependency and neglect cases in terms of both civil-like procedures and dispositional provisions for social services. Alternatives to the juvenile justice system, in the form of diversion and deinstitutionalization, have been developed most extensively for use with status offenders. The more punitive approach to juvenile justice has been applied most frequently to serious, violent juvenile offenders.

The state statutes that define juvenile delinquency are similar in form throughout the United States. Contemporary statutes typically use four legal categories, often with varying names but with similar legal conceptualization: serious delinquent youth, delinquent youth, youth in need of supervision, and dependent and neglected youth. The statutory definitions for these categories under the Montana Youth Court Act illustrate this legal categorization (see Case in Point, "Adjudication Classifications in the Montana Youth Court Act").

Adjudication Classifications in the Montana Youth Court Act

Delinquent youth means a youth who is adjudicated under formal proceedings under the Montana Youth Court Act as a youth:

(a) who has committed an offense that, if committed by an adult, would constitute a criminal offense; or

(b) who has been placed on probation as a delinquent youth or a youth in need of intervention and who has violated any condition of probation.

Serious juvenile offender means a youth who has committed an offense that would be considered a felony offense if committed by an adult and that is an offense against a person, an offense against property, or an offense involving dangerous drugs.

Youth in need of intervention means a youth who is adjudicated as a youth and who:

(a) commits an offense prohibited by law that if committed by an adult would not constitute a criminal offense, including but not limited to a youth who: (1) violates any Montana municipal or state law regarding alcoholic beverages; (2) continues to exhibit behavior, including running away from home or habitual truancy, beyond the control of the youth's parents, foster parents, physical custodian, or guardian despite the attempt of the youth's parents, foster parents, physical custodian, or guardian to exert all reasonable efforts to mediate, resolve, or control the youth's behavior; or

(b) has committed any of the acts of a delinquent youth but whom the youth court, in its discretion, chooses to regard as a youth in need of intervention.

Youth in need of care means a youth who has been adjudicated or determined, after a hearing, to be or to have been abused, neglected, or abandoned.

Source: Montana Code Annotated 2005. 41-5-103. Definitions. (11), (38), and (51). 41-3-102. Definitions (34).

■ Summary and Conclusions

Juvenile delinquency is a concept that was *socially constructed*—a product of myriad social, political, economic, and religious changes. Three historical developments laid the foundation for the idea of juvenile delinquency: (1) the discovery of childhood and adolescence as separate and distinct stages of life; (2) the emergence of *parens patriae* in English equity law, which gave legal authority to the state for protective control of children when parents failed to fulfill child-rearing responsibilities; and (3) the rise of positivist criminology, which introduced scientific methods to the study and control of crime and delinquency.

The creation of the juvenile court in Chicago in 1899 clearly reflected these historical developments. Reformers envisioned a child welfare system, rather than a judicial system. As a result, the prevailing goal of the early juvenile court was to protect, nurture, reform, and regulate the dependent, neglected, and delinquent child. Since the court was pursuing the "best interests of the child," little distinction was made between types of offenders. The civil law tradition of *parens patriae,* together with the *rehabilitative ideal,* provided the new juvenile court with a distinctive legal philosophy, structure, and process.

The traditional juvenile court came under attack shortly after World War II. Criticism centered on its disregard of due process and its failure to provide effective rehabilitation through individualized treatment. Beginning in the 1960s, at least three forces dramatically changed the character of juvenile justice systems across the United States: (1) the juvenile due process revolution from 1966 to 1975, (2) the Juvenile Justice and Delinquency Prevention Act of 1974, and (3) a growing emphasis on punishment and accountability in the 1980s and 1990s. Instead of an intense focus on the individual offender, contemporary juvenile justice systems emphasize offender accountability, public safety, and offender competency development, an approach called *balanced and restorative justice.*

Contemporary legal definitions of juvenile delinquency continue to distinguish juvenile offenders from adult criminals and to provide for a separate system and process of justice. Legal definitions continue to emphasize the dependency of children and adolescents and their need for protection and nurture. In addition, the family unit is affirmed as the key institution of socialization, providing "care, protection, and wholesome mental and physical development of a youth."[214] However, contemporary legal definitions of juvenile delinquency commonly specify at least four different legal classifications of juveniles over which the juvenile court maintains jurisdiction: (1) dependent and neglected children; (2) status offenders, sometimes called "youth in need of intervention" or some variant of that term; (3) delinquent youth who violate the criminal code; and (4) serious delinquent offenders who have committed felony offenses.

CRITICAL THINKING QUESTIONS

1. How does the "invention" of juvenile delinquency and the juvenile court reflect legal innovations suggested in the stubborn child laws of colonial America? (Refer back to the case that opened the chapter.)

2. What does it mean to say that juvenile delinquency was *socially constructed?*

3. How was the creation of the juvenile court a culmination of earlier reforms, such as houses of refuge, placing-out, and reform schools?

4. How did the due process revolution change the character of the juvenile justice system?

5. In what ways does contemporary juvenile justice emphasize accountability?

6. Distinguish the legal categories: "youth in need of care," "youth in need of intervention," "delinquent youth," and "serious delinquent offender."

SUGGESTED READINGS

Fox, Sanford J. "Juvenile Justice Reform: An Historical Perspective." *Stanford Law Review* 22 (1970):1187–1239.

Joseph, Hawes. *Children in Urban Society: Juvenile Delinquency in Nineteenth-Century America.* New York: Oxford University Press, 1971.

Mennel, Robert M. *Thorns & Thistles: Juvenile Delinquents in the United States 1825–1940.* Hanover, NH: University Press of New England, 1973.

Platt, Anthony. *The Child Savers: The Invention of Delinquency.* 2nd ed. Chicago: University of Chicago Press, 1977.

Rothman, David J. *The Discovery of the Asylum: Social Order and Disorder in the New Republic.* Boston: Little, Brown, and Company, 1971.

Schlossman, Steven. *Love and the American Delinquency: The Theory and Practice of "Progressive" Juvenile Justice.* Chicago: University of Chicago Press, 1977.

GLOSSARY

balanced and restorative justice: A contemporary orientation in juvenile justice that emphasizes offender accountability, community safety, and offender competency development.

"best interests of the child": The overarching interest of the traditional juvenile court to assess the needs of the youth and then to seek physical, emotional, mental, and social well-being for that youth through court intervention.

child-saving movement: A loose collection of women from middle- and upper-class backgrounds who exercised considerable influence in mobilizing change in how governments dealt with dependent, neglected, and delinquent children. One particular child-saving group, the Chicago Women's Club, is largely responsible for the creation of the first juvenile court in Chicago.

due process of law: Procedural rights established in the Constitution (including the Bill of Rights) and extended through appellate court decisions that are based upon individual freedoms and limitation of governmental powers.

houses of refuge: The first institutional facilities in the United States for poor, vagrant children. Both private and public refuges sought to protect and reform the "predelinquent."

parens patriae: Literally means "parent of the country." The legal authority of courts to assume parental responsibilities when the natural parents fail to fulfill their duties.

pauperism: The view, popularly held throughout the nineteenth century, that children growing up in poverty, surrounded by depravity in their neighborhood and family, are destined to lives of crime and degradation.

placing-out: The practice, begun in the mid-1800s, in which philanthropic groups took vagrant and wayward urban children west by railroad to be placed in farm families.

poor laws: Laws enacted in colonial America that established a civic duty of private citizens to "relieve" the poor. These laws usually provided a definition of residence so that outsiders could not benefit from private relief. Legal authority was also granted for governmental agencies or private relief societies to separate poor children from their "undeserving" parents.

reform schools: In the mid-1800s a new form of institution began to replace houses of refuge. These institutions emphasized education and operated with traditional school schedules. Many reform schools also used a cottage or family system in which children were grouped into "families" of forty or fewer.

rehabilitative ideal: The traditional legal philosophy of the juvenile court, which emphasizes assessment of the youth and individualized treatment, rather than determination of guilt and punishment.

social constructionist perspective: An attempt to understand the many social, political, and economic factors that lead to the development of an idea, concept, or view.

status offender: A juvenile who has committed an act that would *not* be a crime if committed by an adult. See **status offense.**

status offense: An act that is illegal for a juvenile but would *not* be a crime if committed by an adult.

REFERENCES

Albert, Rodney L. "Juvenile Accountability Incentive Grants Program." Washington, DC: Office of Juvenile Justice and Delinquency Prevention, 1998.

Allen, Francis A. *The Borderland of Criminal Justice: Essays in Law and Criminology.* Chicago: University of Chicago Press, 1964.

Andrew, Chyrl, and Lynn Marble. "Changes to OJJDP's Juvenile Accountability Program." Washington, DC: Office of Juvenile Justice and Delinquency Prevention, 2003.

Aries, Philippe. *Centuries of Childhood: A Social History of Family Life.* Translated by Robert Baldick. New York: Random House, 1962.

Bazemore, Gordon. *Balanced and Restorative Justice for Juveniles: A Framework for Juvenile Justice in the 21st Century.* Washington, DC: Office of Juvenile Justice and Delinquency Prevention, 1997.

Bazemore, Gordon, and Mark Umbreit. "Balanced and Restorative Justice: Program Summary." Washington, DC: Office of Juvenile Justice and Delinquency Prevention, 1994.

Beirne, Piers. "Adolphe Quetelet and the Origins of Positivist Criminology." *American Journal of Sociology* 92 (1987):1140–1169.

Beirne, Piers, and James Messerschmidt. *Criminology.* 3rd ed. Boulder, CO: Westview Press, 2000.

Berger, Peter L., and Thomas Luckman. *The Social Construction of Reality.* New York: Anchor Books, 1967.

Binder, Arnold, Gilbert Geis, and Dickson D. Bruce, Jr. *Juvenile Delinquency: Historical, Cultural, and Legal Perspectives.* Cincinnati, OH: Anderson, 1997.

Brummer, Chauncey E. "Extended Juvenile Jurisdiction: The Best of Both Worlds?" *Arkansas Law Review* 54 (2002):777–822.

Cogan, Neil Howard. "Juvenile Law, Before and After the Entrance of 'Parens Patriae.' " *South Carolina Law Review* 22 (1970):147–181.

Cooley, Charles H. " 'Nature v. Nurture' in the Making of Social Careers." *Proceedings of the National Conference of Charities and Corrections (PNCCC).* (1896): 399–405.

Crank, Kathleen Kositzky. "The JJDP Mandates: Rationale and Summary." *Fact Sheet #22.* Washington, DC: Office of Juvenile Justice and Delinquency Prevention, 1995.

Curtis, George B. "The Checkered Career of *Parens Patriae:* The State as Parent or Tyrant." *DePaul Law Review* 25 (1976):895–915.

Danegger, Anna E., Carole E. Cohen, Cheryl D. Hayes, and Gwen A. Holden. *Juvenile Accountability Incentive Block Grants: Strategic Planning Guide.* Washington, DC: Office of Juvenile Justice and Delinquency Prevention, 1999.

DeFrances, Carole J., and Kevin J. Strom. "Juveniles Prosecuted in State Criminal Courts." Washington, DC: Bureau of Justice Statistics, 1997.

Degler, Carl. *At Odds: Women and the Family in America from the Revolution to the Present.* New York: Oxford University Press, 1980.

deMause, Lloyd, ed. *The History of Childhood.* New York: Psychohistory Press, 1974.

deMause, Lloyd. "The Evolution of Childhood." In deMause, *The History of Childhood,* 1–73.

Empey, LaMar T., and Mark C. Stafford. *American Delinquency: Its Meaning and Construction.* 3rd ed. Belmont, CA: Wadsworth, 1991.

Empey, LaMar T., Mark C. Stafford, and Carter H. Hay. *American Delinquency: Its Meaning and Construction.* 4th ed. Belmont, CA: Wadsworth, 1999.

Feld, Barry C. *Bad Kids: Race and the Transformation of the Juvenile Court.* New York: Oxford University Press, 1999.

Ferdinand, Theodore N. "History Overtakes the Juvenile Justice System." *Crime and Delinquency* 37 (1991):204–224.

Fox, Sanford J. "Juvenile Justice Reform: An Historical Perspective." *Stanford Law Review* 22 (1970):1187–1239.

Freivalds, Peter. "Balanced and Restorative Justice Project (BARJ)." Washington, DC: Office of Juvenile Justice and Delinquency Prevention, 1996.

Gallagher, Catherine A. "Juvenile Offenders in Residential Placement." Washington, DC: Office of Juvenile Justice and Delinquency Prevention, 1999.

Gibbons, Don C., and Marvin D. Krohn. *Delinquent Behavior.* 5th ed. Englewood Cliffs, NJ: Prentice Hall, 1991.

Griffin, Patrick, Patricia Torbet, and Linda Szymanski. *Trying Juveniles as Adults in Criminal Court: An Analysis of State Transfer Provisions.* Washington, DC: Office of Juvenile Justice and Delinquency Prevention, 1998.

Hall, G. Stanley. *Adolescence: Its Psychology and its Relationship to Physiology, Anthropology, Sociology, Sex, Crime, Religions, and Education.* New York: Appelton, 1904.

Hawes, Joseph. *Children in Urban Society: Juvenile Delinquency in Nineteenth-Century America.* New York: Oxford University Press, 1971.

Hellum, Frank. "Juvenile Justice: The Second Revolution." *Crime and Delinquency* 25 (1979):299–317.

Howell, James C. *Juvenile Justice and Youth Violence.* Thousand Oaks, CA: Sage, 1997.

Illick, Joseph E. "Child-Rearing in Seventeenth-Century England and America." In deMause, *The History of Childhood,* 303–350.

Jensen, Gary F., and Dean G. Rojek. *Delinquency and Youth Crime.* Prospect Heights, IL: Waveland Press, 1998.

Kaplan, Louise J. *Adolescence: The Farewell to Childhood.* New York: Simon and Schuster, 1984.

Kett, Joseph F. *Rites of Passage: Adolescence in America: 1790–Present.* New York: Basic Books, 1977.

Krisberg, Barry, and James Austin. "History of the Control and Prevention of Juvenile Delinquency in America." In *The Children of Ishmael: Critical Perspective on Juvenile Justice,* edited by Barry Krisberg and James Austin, 7–50. Palo Alto, CA: Mayfield, 1978.

———. *Reinventing Juvenile Justice.* Newbury Park, CA: Sage, 1993.

Krisberg, Barry, Ira M. Schwartz, Paul Litsky, and James Austin. "The Watershed of Juvenile Justice Reform." *Crime and Delinquency* 32 (1986):5–38.

Lerman, Paul. *Community Treatment and Control.* Chicago: University of Chicago Press, 1975.

Lipton, Douglas, Robert Martinson, and Judith Wilks. *The Effectiveness of Correctional Treatment.* New York: Praeger, 1975.

Mack, Julian W. "The Juvenile Court." *Harvard Law Review* 23 (1909):104–122.

Martinson, Robert. "What Works?—Questions and Answers about Prison Reform." *Public Interest* 32 (1974):22–54.

Matza, David. *Delinquency and Drift.* New York: Wiley, 1964.

Mennel, Robert M. *Thorns & Thistles: Juvenile Delinquents in the United States 1825–1940.* Hanover, NH: University Press of New England, 1973.

National Advisory Commission on Criminal Justice Standards and Goals. *Task Force Report on Corrections* (Standard 22.3). Washington, DC: GPO, 1973.

National Council on Crime and Delinquency. "Corrections in the United States." Data summarized in President's Commission on Law Enforcement and Administration of Justice, *Task Force Report: Corrections.* Washington, DC: GPO, 1967.

Chapter Resources

Chapter Resources

Pickett, Robert S. *House of Refuge: Origins of Juvenile Reform in New York State, 1815–1857.* Syracuse, NY: Syracuse University Press, 1969.

Pisciotta, Alexander W. "Saving the Children: The Promise and Practice of *Parens Patriae, 1838–98.*" *Crime and Delinquency* 28 (1982):410–425.

Platt, Anthony. *The Child Savers: The Invention of Delinquency.* 2nd ed. Chicago: University of Chicago Press, 1977.

President's Commission on Law Enforcement and Administration of Justice. *The Challenge of Crime in a Free Society.* Washington, DC: GPO, 1967.

———. *Task Force Report: Juvenile Delinquency and Youth Crime.* Washington, DC: GPO, 1967.

———. *Task Force Report: Corrections.* Washington, DC: GPO, 1967.

Puzzanchera, Charles M. "Delinquency Cases Waived to Criminal Court, 1990–1999." Washington, DC: Office of Juvenile Justice and Delinquency Prevention, 2003.

Rainville, Gerard A., and Steven K. Smith. "Juvenile Felony Defendants in Criminal Courts." Washington, DC: Office of Juvenile Justice and Delinquency Prevention, 2003.

Raley, Gordon. "The JJDP Act: A Second Look." *Juvenile Justice Journal* 2 (1995):11–18.

Rendleman, Douglas R. "Parens Patriae: From Chancery to the Juvenile Court." *South Carolina Law Review* 23 (1971):205–259.

Robertson, Priscilla. "Home as a Nest: Middle Class Childhood in Nineteenth-Century Europe." In deMause, *The History of Childhood,* 407–431.

Rothman, David J. *The Discovery of the Asylum: Social Order and Disorder in the New Republic.* Boston: Little, Brown, and Company, 1971.

Rubin, H. Ted. "The Nature of the Court Today." *The Juvenile Court* 6, no. 3 (1996):40–52.

———. *Juvenile Justice: Policy, Practice, and Law.* 2nd ed. New York: Random House, 1985.

Schlossman, Steven. *Love and the American Delinquency: The Theory and Practice of "Progressive" Juvenile Justice.* Chicago: University of Chicago Press, 1977.

Schur, Edwin M. *Radical Non-Intervention: Rethinking the Delinquency Problem.* Englewood Cliffs, NJ: Prentice Hall, 1973.

Shepherd, Robert E., Jr. "The Juvenile Court at 100 Years: A Look Back." *Juvenile Justice* 6, no. 2 (1999):13–21.

Sickmund, Melissa. "Juveniles in Court." *Juvenile Offenders and Victims National Report Series.* Washington, DC: Office of Juvenile Justice and Delinquency Prevention, 2003.

Snyder, Howard N., and Melissa Sickmund. *Juvenile Offenders and Victims: 1999 National Report.* Washington, DC: Office of Juvenile Justice and Delinquency Prevention, 1999.

Snyder, Howard N., Melissa Sickmund, and Eileen Poe-Yamagata. *Juvenile Transfers to Criminal Courts in the 1990's: Lessons Learned From Four Studies.* Washington, DC: Office of Juvenile Justice and Delinquency Prevention, 2000.

Sommerville, John. *The Rise and Fall of Childhood.* Beverly Hills, CA: Sage, 1982.

Stark, Rodney. *Sociology.* 7th ed. Belmont, CA: Wadsworth, 1998.

Strom, Kevin J., Steven K. Smith, and Howard N. Snyder. "Juvenile Felony Defendants in Criminal Courts." Washington, DC: Bureau of Justice Statistics, 1998.

Sutton, John. *Stubborn Children: Controlling Delinquency in the United States, 1640–1981.* Berkeley, CA: University of California Press, 1988.

Tappan, Paul. "Treatment Without Trial?" *Social Problems* 24 (1946):306–311.

Taylor, Robert W., Eric J. Fritsch, and Tory J. Caeti. *Juvenile Justice: Policies, Programs, and Practices.* New York: Glencoe/McGraw-Hill, 2002.

Thornton, William E., Jr., Lydia Voigt, and William G. Doerner. *Delinquency and Justice.* 2nd ed. New York: Random House, 1987.

Wilson, John J. and James C. Howell. *Comprehensive Strategy for Serious, Violent, and Chronic Juvenile Offenders: Program Summary.* Washington, DC: Office of Juvenile Justice and Delinquency Prevention, 1993.

ENDNOTES

1. Quoted in Jensen and Rojek, *Delinquency and Youth Crime,* 5.
2. Sommerville, *Rise and Fall.*
3. Berger and Luckman, *Social Construction of Reality*; Empey, Stafford, and Hay, *American Delinquency*; Feld, *Bad Kids*; Platt, *Child Savers.*

4. Sutton, *Stubborn Children*.

5. Hellum, "Juvenile Justice."

6. Stark, *Sociology*, 33, 228.

7. Hall, *Adolescence*, 522–523; and Sommerville, *Rise and Fall*, 29.

8. Thornton, Voigt, and Doerner, *Delinquency and Justice*, 4.

9. Aries, *Centuries of Childhood*, 19.

10. Ibid.

11. Ibid., 10.

12. Thornton, Voigt, and Doerner, *Delinquency and Justice*, 5; these observations are based on Aries, *Centuries of Childhood*.

13. Aries, *Centuries of Childhood*, 29.

14. Kett, *Rites of Passage*.

15. Empey and Stafford, *American Delinquency*, 30.

16. Aries, *Centuries of Childhood*, 128.

17. Ibid., 130.

18. Empey and Stafford, *American Delinquency*, 30.

19. deMause, "Evolution," 1. See also Illick, "Child-Rearing."

20. Illick, "Child-Rearing."

21. Aries, *Centuries of Childhood*, 381–390, especially 389–390; Illick, "Child-Rearing," 311–313; and Sommerville, *Rise and Fall*, 109–115.

22. Aries, *Centuries of Childhood*, 114–119.

23. Aries, *Centuries of Childhood*, 119–121; deMause, "Evolution;" Empey and Stafford, *American Delinquency*, 36; Illick, "Child-Rearing;" Robertson, "Home as a Nest;" and Sommerville, *Rise and Fall*.

24. Aries, *Centuries of Childhood*, 412–413.

25. Illick, "Child-Rearing," 318–320; and Sommerville, *Rise and Fall*, 121–124.

26. Cited in Sommerville, *Rise and Fall*, 127, also 127–133. See also Thornton, Voigt, and Doerner, *Delinquency and Justice*, 6.

27. Kaplan, *Adolescence*, 51.

28. Thornton, Voigt, and Doerner, *Delinquency and Justice*, 6; Kaplan, *Adolescence*, 63.

29. Degler, *At Odds*, 66.

30. Sommerville, *Rise and Fall*, 179, also 177.

31. Ibid., 160–178.

32. Cogan, "Juvenile Law," 148; Rendleman, "Parens Patriae," 208.

33. *Oxford English Dictionary*, http://dictionary.oed.com/.

34 The sovereignty of the king and his extensive power to accomplish his interests are referred to as the "king's prerogative," or *prerogative regis* (Curtis, "Checkered Career," 896; see also Cogan, "Juvenile Law," 155–156).

35. Feld, *Bad Kids*, 52; Rendelman, "Parens Patriae," 208–209. See also Cogan, "Juvenile Law," 155–156, and Curtis, "Checkered Career," 896.

36. Curtis, "Checkered Career," 896; and Rendelman, "Parens Patriae," 209.

37. Curtis, "Checkered Career," 896–898; and Cogan, "Juvenile Law," 156.

38. Curtis, "Checkered Career," 897–898.

39. Cogan, "Juvenile Law," 174; and Curtis, "Checkered Career," 897. Curtis refers to this practical limitation of applying *parens patriae* only to cases involving property as the "property nexus" (897); see also Cogan, 148–151.

40. Cogan, "Juvenile Law," 149–152,181. See also Rendleman, "Parens Patriae," 207–209.

41. Feld, *Bad Kids*, 52.

42. Curtis, "Checkered Career," 899. See also Rendleman, "Parens Patriae," 223–224.

43. Curtis, "Checkered Career," 899. See also Fox, "Juvenile Justice Reform," 1193, and Rendleman, "Parens Patriae."

44. Beirne and Messerschmidt, *Criminology*, 72–82.

45. Beirne, "Adolphe Quetelet," 1145–1146.

46. Matza, *Delinquency and Drift*, 11.

47. Allen, *Borderland*; Platt, *Child Savers*; and Feld, *Bad Kids*.

48. Cooley, "Nature v. Nurture," 405.

49. Rothman, *Discovery*, 16.

50. Ibid.

51. Quoted in Rothman, *Discovery,* 16.
52. Rothman, *Discovery.*
53. Ibid., 20.
54. Rendleman, "Parens Patriae," 233, also 212. See also Rothman, *Discovery,* 20.
55. Rendleman, "Parens Patriae," 212.
56. Fox, "Juvenile Justice Reform," 1189.
57. Rothman, *Discovery,* 17–18.
58. Rendleman, "Parens Patriae," 213.
59. Fox, "Juvenile Justice Reform," 1200.
60. Rendleman, "Parens Patriae," 214.
61. Quoted in Mennel, *Thorns & Thistles,* 8.
62. Mennel, *Thorns & Thistles,* 9.
63. Rendleman, "Parens Patriae," 215; and Rothman, *Discovery,* 207.
64. Fox, "Juvenile Justice Reform," 1190; and Rendleman, "Parens Patriae," 216.
65. Mennel, *Thorns & Thistles,* 4.
66. Ibid., 5.
67. Rendleman, "Parens Patriae," 217. See also Feld, *Bad Kids,* 48.
68. Fox, "Juvenile Justice Reform," 1190–1191; and Pickett, *House of Refuge,* 56.
69. Quoting Mary Carpenter, an English penal reformer (Fox, "Juvenile Justice Reform," 1193).
70. Mennel, *Thorns & Thistles,* 10. See also Fox, "Juvenile Justice Reform," 1191.
71. Mennel, *Thorns & Thistles,* 18. See also Feld, *Bad Kids,* 51.
72. Pickett, *House of Refuge,* 52, 55–57.
73. Sutton, *Stubborn Children.*
74. Rothman, *Discovery,* Chapter 9, especially 210–216.
75. Rothman, *Discovery,* 231–234; and Mennel, *Thorns & Thistles,* 19–20.
76. Mennel, *Thorns & Thistles,* 20–21.
77. Ibid., 22–23.
78. Ibid., 21–23.
79. Rothman, *Discovery,* 209.
80. Rendleman, "Parens Patriae," 219, 237; Curtis, 901–902; and Fox, "Juvenile Justice Reform," 1193.
81. *Ex parte Crouse,* 4, Wharton (PA) 9 (1838) at 11. Fox, "Juvenile Justice Reform," 1205–1206; Krisberg and Austin, "America," 16; Krisberg and Austin, "United States," 18; Pisciotta, "Saving the Children," 410–412; and Rendleman, "Parens Patriae," 218–219.
82. Curtis, "Checkered Career," 258–259.
83. Rothman, *Discovery,* 258–259.
84. Empey, Stafford, and Hay, *American Delinquency,* 40–41.
85. Mennel, *Thorns & Thistles,* 35.
86. Ibid., 37.
87. Ibid.
88. Ibid., 39.
89. Ibid., 43–48.
90. Ibid., 49.
91. Schlossman, *Love.*
92. Mennel, *Thorns & Thistles,* 52–55.
93. Ibid., 52.
94. Ibid., 59.
95. Ibid., 74.
96. *People ex rel. O'Connell v. Turner,* 55 Ill. (1870). The discussion here is drawn from Rendleman, "Parens Patriae," 233–236, and Fox, "Juvenile Justice Reform," 1216–1221.
97. The Illinois statute is quoted in Fox, "Juvenile Justice Reform," 1214.
98. Rendleman, "Parens Patriae," 234.
99. Quoted in Fox, "Juvenile Justice Reform," 1219.
100. Fox, "Juvenile Justice Reform," 1220.
101. Kett, *Rites of Passage,* 132. See also Rothman, *Discovery.*
102. Mennel, *Thorns & Thistles,* 124.
103. Feld, *Bad Kids,* 47, see also 24–28.

104. Quoted in Mennel, *Thorns & Thistles*, 129. See also Platt, *Child Savers*, 132.

105. Fox, "Juvenile Justice Reform;" Mennel, *Thorns & Thistles*; and Platt, *Child Savers*.

106. Sutton, *Stubborn Children*, 125–132. See also Feld, *Bad Kids*, 34–36; Krisberg and Austin, "America," 24; and Krisberg and Austin, "United States," 27.

107. Platt, *Child Savers*, 10. See Hawes, *Children in Urban Society*, for this point of view.

108. Platt, *Child Savers*, 10.

109. For this critical–revisionist point of view, see Platt, *Child Savers*; Rothman, *Discovery*; Schlossman, *Love*; and Sutton, *Stubborn Children*.

110. Platt, *Child Savers*, 77.

111. Ibid., 75, also 3.

112. Ibid., 83.

113. Ibid., 79.

114. Ibid., 7; see also the quotation of Christopher Lasch on page 76.

115. Platt, *Child Savers*, 130–131.

116. Ibid., 131.

117. Platt, *Child Savers*, 131–132; and Mennel, *Thorns & Thistles*, 130.

118. Platt, *Child Savers*, 133–134; and Krisberg and Austin, "United States," 29.

119. Feld, *Bad Kids*, 55–56, 75; Hellum, "Juvenile Justice," 299; Platt, *Child Savers*, 135; and Sutton, *Stubborn Children*, 132.

120. Mennel, *Thorns & Thistles*, 43–44, 131; and Platt, *Child Savers*, 9.

121. Platt, *Child Savers*, 101.

122. Feld, *Bad Kids*, 56, 75.

123. Taylor, Fritsch, and Caeti, *Juvenile Justice*, 89.

124. Feld, *Bad Kids*, 56; Platt, *Child Savers*, 10; and Sutton, *Stubborn Children*, 121.

125. *In re Gault*, 387 U.S. 1, 87 S. Ct. 1428 (1967).

126. Illinois Statute 1899, Section 131.

127. Platt, *Child Savers*, 135; and Feld, *Bad Kids*, 62.

128. Mennel, *Thorns & Thistles*, 132. See also Platt, *Child Savers*, 137.

129. *Commonwealth v. Fisher*, 213 Pennsylvania 48 (1905).

130. Ferdinand, "History," 207; and Fox, "Juvenile Justice Reform," 1229.

131. Feld, *Bad Kids*, 62; and Platt, *Child Savers*, 135.

132. *Revised Statutes of Illinois*, 1899, Sec. 21. Quoted in Hawes, *Children in Urban Society*, 170.

133. Mack, "Juvenile Court," in Feld, *Bad Kids*, 4.

134. Allen, *Borderland*; Feld, *Bad Kids*; and Platt, *Child Savers*, 43, 45.

135. Mack, "Juvenile Court," in Feld, *Bad Kids*, 7.

136. Feld, *Bad Kids*, 68.

137. Brummer, "Extended Juvenile Jurisdiction," 777.

138. Mack, "Juvenile Court," 109.

139. Feld, *Bad Kids*, 66.

140. Feld, *Bad Kids*, 65–69; and Platt, *Child Savers*, 141.

141. Feld, *Bad Kids*, 66.

142. Feld, *Bad Kids*, 60; and Rothman, *Discovery*, 43.

143. Feld, *Bad Kids*, 66.

144. Ibid.

145. Quoted in Platt, *Child Savers*, 142.

146. Platt, *Child Savers*, 137–138.

147. Feld, *Bad Kids*, 66.

148. Quoted in Platt, *Child Savers*, 144.

149. Quoted in Platt, *Child Savers*, 143.

150. Feld, *Bad Kids*, 68; and Schlossman, *Love*.

151. Platt, *Child Savers*, 139.

152. Mennel, *Thorns & Thistles*, 132.

153. Ferdinand, "History," 207.

154. Platt, *Child Savers*; Krisberg and Austin, "America;" and Krisberg and Austin, "United States."

155. Feld, *Bad Kids*; Fox, "Juvenile Justice Reform;" and Schlossman, *Love*.

156. Hellum, "Juvenile Justice," 301.

157. Tappan, "Treatment Without Trial?"; and Ferdinand, "History," 210.

Chapter Resources

158. *Kent v. United States,* 383 U.S. 541, 86 S. Ct. 1045 (1966).
159. Hellum, "Juvenile Justice," 301–302.
160. Ibid., 302–303.
161. Feld, *Bad Kids,* 92–94; Ferdinand, "History," 212–213; Hellum, "Juvenile Justice," 302–303; Lerman, *Community Treatment and Control;* Lipton, Martinson, and Wilks, *Effectiveness;* and Martinson, "What Works?"
162. Schur, *Radical Non-Intervention.*
163. Ferdinand, "History;" Hellum, "Juvenile Justice;" and Howell, *Juvenile Justice.*
164. *In re Gault,* 387 U.S. 1, 87 S.Ct. 1428 (1967); Feld, *Bad Kids;* Ferdinand, "History;" Snyder and Sickmund, *Juvenile Offenders,* 87.
165. These case summaries are taken verbatim or paraphrased from the work of Snyder and Sickmund, *Juvenile Offenders,* 90–92.
166. *Kent v. United States,* 383 U.S. 541, 86 S.Ct. 1045 (1966).
167. *In re Gault,* 387 U.S. 1, 87 S.Ct. 1428 (1967).
168. *In re Winship,* 397 U.S. 358, 90 S.Ct. 1068 (1970).
169. *McKeiver v. Pennsylvania,* 403 U.S. 528, 91 S.Ct. 1976 (1971).
170. *Breed v. Jones,* 421 U.S. 519, 95 S.Ct. 1779 (1975).
171. End of excerpted case summaries from Snyder and Sickmund, *Juvenile Offenders,* 90–92.
172. Crank, "JJDP Mandates;" and Howell, *Juvenile Justice,* 15–19.
173. President's Commission on Law Enforcement and Administration of Justice, *Challenge,* 80.
174. President's Commission on Law Enforcement and Administration of Justice, *Juvenile Delinquency,* 27; and Crank, "JJDP Mandates," 1.
175. President's Commission on Law Enforcement and Administration of Justice, *Challenge,* 83.
176. Ibid., 80.
177. Ibid., 81.
178. Ibid.
179. National Council on Crime and Delinquency, "Corrections," 211, as cited in Howell *Juvenile Justice,* 17.
180. National Advisory Commission on Criminal Justice Standards and Goals, *Task Force,* 23, as cited in Howell, *Juvenile Justice,* 18.
181. National Advisory Commission on Criminal Justice Standards and Goals, *Task Force,* 259, as cited in Howell, *Juvenile Justice,* 18.
182. Raley, "JJDP Act."
183. Shepherd, "Look Back," 20.
184. Excerpted from Crank, "JJDP Mandates," 2–4. See also Snyder and Sickmund, *Juvenile Offenders,* 88.
185. Snyder and Sickmund, *Juvenile Offenders,* 88. See also Krisberg et al., "Watershed."
186. Feld, *Bad Kids,* 97–106.
187. National Advisory Committee for Juvenile Justice and Delinquency Prevention (1984:9), cited in Krisberg et al., "Watershed," 7.
188. Wilson and Howell, *Comprehensive Strategy.*
189. Krisberg et al., "Watershed," 7–9.
190. Snyder and Sickmund, *Juvenile Offenders,* 88–89.
191. The following is drawn extensively from Sickmund, "Juveniles in Court," 6. See also DeFrances and Strom, "Juveniles Prosecuted;" Griffin, Torbet, and Szymanski, *Trying Juveniles;* Puzzanchera, "Cases Waived;" Rainville and Smith, "Juvenile Felony Defendants;" Strom, Smith, and Snyder, "Juvenile Felony Defendants;" and Snyder, Sickmund, and Poe-Yamagata, *Juvenile Transfers.*
192. Sickmund, "Juveniles in Court," 7.
193. Ibid., 10.
194. Snyder and Sickmund, *Juvenile Offenders,* 108.
195. Rubin, *Juvenile Justice,* 35.
196. Snyder and Sickmund, *Juvenile Offenders,* 108.
197. 705 ILCS 405/5-815.
198. Snyder and Sickmund, *Juvenile Offenders,* 101.
199. Ibid.
200. Snyder and Sickmund, *Juvenile Offenders,* 87, 89; Bazemore, *Balanced;* Freivalds, "BARJ;" and Bazemore and Umbreit, *Balanced.*
201. Albert, "Juvenile Accountability," 1.

202. Andrew and Marble, "Changes."
203. Albert, "Juvenile Accountability," 1. See also Danegger et al., *Juvenile Accountability,* series published by OJJDP called *JAIBG Bulletin.*
204. Snyder and Sickmund, *Juvenile Offenders,* 89.
205. Platt, *Child Savers,* 135.
206. Quoted in Binder, Geis, and Bruce, *Juvenile Delinquency,* 201.
207. Platt, *Child Savers,* Chapter 5.
208. Feld, *Bad Kids,* 64; and Sutton, *Stubborn Children.*
209. Hawes, *Children in Urban Society,* 186, emphasis added; and Feld, *Bad Kids,* 64.
210. Feld, *Bad Kids,* 64.
211. Rubin, "Nature," 40–41.
212. Gibbons and Krohn, *Delinquent Behavior.*
213. Ibid., 15–16.
214. *Montana Code Annotated 2005,* 41-5-102.

Measuring Delinquency

3

Chapter Outline

- Studying causes of delinquency
- Research methods for studying crime and delinquency
- Sources of data on crime and delinquency
- Methods, theory, and policy

Chapter Objectives

After completing this chapter, students should be able to:

- Describe the relationship between theory and research methodology.
- Recognize a variety of research methodologies, including case studies, ethnography, ecological analysis, and analysis of data from various sources.
- Describe what types of information various research methodologies are designed to reveal.
- Identify the strengths and weaknesses of each of the three major sources of data on crime and delinquency: official records, victimization surveys, and self-report surveys.
- Understand key terms:
  analytic induction
  ethnography
  ecological analysis
  survey research
  Uniform Crime Reporting program

crime rates

crime trends

National Incident-Based Reporting System

cross-sectional survey

longitudinal survey

validity

reliability

criminal career

career criminals

selective incapacitation

CASE IN POINT

Some Thoughts on the Relationship Between Measurement and Theory

The following excerpt from the "Foreword" of *Measuring Delinquency*, by Michael Hindelang, Travis Hirschi, and Joseph Weis, demonstrates the sometimes conflicted relationship between adequate measurement of the phenomenon of interest and the development and testing of theory.

When social scientists focus their attention on their measurement procedures, what was formerly solid ground for confident statements about causes and consequences often appears incapable of supporting anything more substantial than proposals for additional research. As a consequence, researchers concerned with using data to construct or test sociological theories have a clear tendency to grow impatient with what they see as excessive concern for the accuracy and precision of measurement. At some point, they say, we have to get on with the tasks of social research. It is true that our measures are not perfect, that they might be refined and refined again, but the fact is that these refinements probably wouldn't make that much difference anyway—and social scientists should take data over groundless speculation any time.

There is another problem with concentration on measurement that is of concern to researchers: The idea that social science should somehow concentrate on the improvement of measuring devices before proceeding with further research on important problems . . . seems to many to overlook the place of theory in deciding what should be measured and how it should be measured.

Source: Hindelang, Hirschi, and Weis, *Measuring Delinquency,* 9.

The scientific study of delinquency is based on the ability to gather accurate and valid data. However, gathering data that accurately represent the occurrence of delinquency is a persistent problem in criminology. Over time, various research methods have been used to obtain data from a variety of sources, including agencies that constitute the juvenile justice system, victims, and offenders themselves. In this chapter, we explore these research methods and sources of data.

We begin with a brief description of the research process and the criteria for establishing cause and effect in delinquency research. Next, we offer examples of various research methods used to study delinquency, including comparison of offenders and non-offenders, case studies in the method of analytic induction, ethnographic work on violence, ecological analysis of delinquent behavior in Chicago, and survey analysis. We then examine various sources of data used in the statistical analysis of crime and delinquency, including "official data" and data from victimization and self-report surveys. We compare these three data sources and discuss the strengths and weaknesses of each, focusing on what each source can really tell us about delinquent behavior. Finally, we explore the relationships among research methodology, theory, and public policy, using the example of the recent criminal careers debate in criminology.

Studying Causes of Delinquency

Before we consider specific research methods and sources of data used to study delinquency, we offer a brief description of the research process to facilitate understanding of the research findings we present in later chapters. We also summarize the criteria for establishing cause and effect in research on crime and delinquency.

The Research Process

In Chapter 1, we discussed two purposes of research: to generate theory (the inductive research process) and to test theory (the deductive research process). Both inductive and deductive research processes involve identification of key concepts, hypothesis development, empirical observation, and data collection and analysis, although the order in which these activities occur varies for the two research processes. The approach taken throughout this book integrates theory and research—we present theories of delinquency along with the research used to test them. Thus, an awareness of the deductive research process will help you better understand the discussions in later chapters. Research in Action, "The Theoretically Driven Research Process," summarizes the five basic activities of this process.[1]

Causal Analysis

Theories incorporate concepts and propositions to depict causal processes leading to delinquency. As Research in Action, "The Theoretically Driven Research Process," points out, researchers must construct concrete measures of concepts, called *variables*. Some variables cause or determine others in a cause-and-effect relationship.[2] The proposed cause is the *independent variable*, and the proposed effect is the *dependent variable*. Consider this simple model:

$$A \longrightarrow B$$

In this model, "A" is the independent variable, which causes or leads to "B," the dependent variable. "A" might be number of hours spent studying, which causes "B," success in school as measured by grade point average. Rarely, however, does a single variable alone cause a dependent variable.[3]

Relationships among variables and causal sequences related to delinquency are tested in the research process. Causal sequences described in theory are often difficult to test through research for at least three reasons. First, most theories propose multiple causes of delinquent behavior. Rarely is a single variable sufficient to explain delinquency. Instead, multiple independent variables are measured in relation to delinquent behavior. Second, independent variables are

RESEARCH IN **ACTION** **The Theoretically Driven Research Process**

1. CONCEPTUALIZE
Identify and clarify key theoretical concepts.

2. OPERATIONALIZE
Construct techniques for measurement of variables.
"Operationalization is the construction of actual, concrete measurement techniques" (Babbie, *Practice of Social Research,* 5). In research, *variables* are the measurable equivalent of concepts. In *Research in Action* boxes throughout this book, we offer examples of variables that researchers have used to measure a variety of concepts.

3. HYPOTHESIZE
Develop hypotheses or statements about the expected relationships between variables (or measures of concepts) (Stark, *Sociology,* 23).
Hypotheses are the testable counterpart of propositions. In the deductive research process, hypotheses are derived from theory.

4. COLLECT DATA
Observe and measure variables using one or more research strategies, including survey research, field studies, experiments, or unobtrusive research.
- **Survey research** involves asking questions through written questionnaires on which respondents record their own answers, or through interviews in which interviewers ask questions and record answers. In delinquency research, survey questions cover areas such as family background, school experiences and grades, friendships, leisure activities, attitudes, aspirations, and most importantly, involvement in delinquent behavior.

Surveys used to study delinquency are typically administered to a sample of juveniles who may or may not represent the entire adolescent population.

- **Field studies** involve first-hand observation, often supplemented with intensive interviews (Agnew, *Juvenile Delinquency*, 76). As Earl Babbie writes, "perhaps the most natural technique for doing social research involves simply going where the action is and observing it" (*Practice of Social Research*, 8). Studies of street gangs, for example, have long been based on observation of gang activities and interaction and on interviews with gang members. (See, for example, a classic study of gangs by Short and Strodtbeck, and more recent research by Venkatesh and his colleagues). When field studies involve active participation by the researcher, they are referred to as *participant observation*.

- **Experiments** test the effect of an experimental stimulus (Babbie, *Practice of Social Research*, 252). In experiments, researchers can control the level of a stimulus or variable, and then observe the effect of this stimulus on individuals. Experiments are not often used in the study of delinquency. Ethical considerations prevent experimental exposure to many of the social causes of delinquency identified in theory (e.g., lack of parental supervision or association with delinquent peers). Also, questions exist about whether responses to a controlled stimulus in an artificial environment are the same as in the "real world."

- **Unobtrusive research** enables social scientists to gather data without influencing what they are studying. (Such influence is a danger in the previous three methods of data collection). One form of unobtrusive research is *analysis of existing data*. Agencies that deal with delinquent youths, such as police departments, courts, and correctional institutions, keep records that document their activities. These records are the basis for much delinquency research. *Historical analysis* is a second form of unobtrusive research that provides valuable information for understanding delinquency. For example, in Chapter 2, we discussed Platt's analysis of historical changes that led to the concept of "delinquency" and the "invention" of the juvenile court (*Child Savers*). Platt's research dramatically altered prevailing views about the origins of juvenile delinquency and the juvenile court.

5. ANALYZE DATA AND TEST HYPOTHESES
Assess theoretically predicted relationships between variables through hypothesis testing.

After data are collected, researchers must analyze and interpret them. Data analysis allows researchers to test hypotheses about the relationships between variables. Put simply, "We must compare what we observe with what the hypothesis said we would see" (Stark, *Sociology*, 23).

themselves often interrelated in complex ways. In fact, some independent variables "cause" other independent variables, which in turn are related to delinquency. Thus, the name "independent" is a bit of a misnomer. Third, it is difficult to observe, measure, and establish cause-and-effect relationships using social science data. Though there have been tremendous advances in data collection and analytic techniques, establishing causation has been a difficult and controversial task for social scientists.

Because the causes of delinquency are many and varied, the causal sequences leading to delinquent behavior are hard to untangle. In Research in Action, "Establishing Cause and Effect," we summarize the criteria for establishing cause and effect. Keep these criteria in mind as you consider the causal processes described in various theories in later chapters.

■ Research Methods for Studying Crime and Delinquency

Social scientists have used a variety of research methods to address questions about crime and delinquency. We examine several of these methods to offer a sense of the richness and complexity of the research enterprise, the extent to which theory determines methodology, and the ways in which answers to questions of interest depend on one's choice of research method. The methods we explore are not exhaustive. Instead, they illustrate the great diversity of research tools available to criminologists.

Comparing Offenders and Non-Offenders

Sheldon and Eleanor Glueck conducted research on crime and delinquency for forty years (1930–1970) at Harvard University. "Their primary interests were in discovering the causes of juvenile delinquency and adult criminality and in assessing the overall effectiveness of correctional treatment in restraining criminal careers."[4] In their research, the Gluecks took an interdisciplinary approach, exploring sociological, psychological, and biological factors that might contribute to delinquency.

The Gluecks are best known for *Unraveling Juvenile Delinquency,* a study of the causes of delinquency that was conducted in the 1940s and published in 1950.[5] In this study, the Gluecks matched 500 officially defined "delinquent" boys from two correctional facilities in Massachusetts with a control group of 500 nondelinquent boys from Boston public schools. "Nondelinquent status was determined on the basis of official record checks and interviews with parents, teachers, local police, social workers, and recreational leaders as well as the boys themselves."[6] The goal of this control group design was to maximize differences in delinquency between "persistent" delinquents and nondelinquents. Both groups contained only white males between the ages of 10 and 17 who had grown up in lower-class Boston neighborhoods. The two samples were matched case by case on age, ethnicity (birthplace of parents), intelligence, and residence in low-income neighborhoods. Thus, as a result of the study design, involvement in delinquency could not be attributed to gender, age, race, ethnicity, IQ, or residence in low-income areas—factors that were taken into account or "con-

Establishing Cause and Effect

In *Delinquency Research: An Appraisal of Analytic Methods,* Hirschi and Selvin provide an extensive discussion of the methods of scientific analysis used to study delinquency. They offer three criteria for establishing cause and effect.

1. **Association:** The cause and effect are empirically associated. This means that change in one variable is related to change in another. The statistical measure of this associated change is *correlation.* Though correlation does not prove causation, it is an important first step in establishing a causal connection between an independent variable and a dependent variable.

2. **Temporal order:** The cause precedes the effect in time. As simple as this sounds, it is sometimes difficult to establish which variable occurs first. The independent variable must precede the dependent variable. It is often proposed, for example, that drug use causes delinquent behavior. However, research has established that delinquent acts usually occur before drug use, failing to support the temporal order of the proposition (Elliott, Huizinga, and Ageton, *Explaining Delinquency;* Elliott, Huizinga, and Menard, *Multiple Problem Youth*).

3. **Lack of spuriousness:** The relationship between the proposed cause and effect cannot be spurious, or explained away by the influence of some other factor that causes both. Babbie (*Practice of Social Research,* 74–75) offers this illustration: "There is a positive correlation between ice cream sales and deaths due to drowning: the more ice cream sold, the more drowning, and vice versa. The third variable here is *season* or *temperature.* Most drowning deaths occur during summer—the peak period for ice cream sales. There is no direct link between ice cream and drowning, however." The correlation between ice cream sales and drowning deaths is spurious because both are related to temperature. Following the same logic, some have proposed that the relationship between drug use and delinquency is spurious because other causal factors, such as having deviant peers, influence both drug use and delinquency (Elliott, Huizinga, and Ageton, *Explaining Delinquency;* Elliott, Huizinga, and Menard, *Multiple Problem Youth*).

Source: Hirschi and Selvin, *Delinquency Research.*

trolled for" in the matching procedure. Research in Action, "Gluecks' Matching of Delinquents and Nondelinquents," provides an excerpt from *Unraveling Juvenile Delinquency,* which shows the list of matched pairs of delinquents and nondelinquents. This excerpt demonstrates how closely individuals were matched on national origin, age, and IQ.

Gluecks' Matching of Delinquents and Nondelinquents

Following is an excerpt from Appendix B of *Unraveling Juvenile Delinquency* by Sheldon and Eleanor Glueck. In this Appendix, the Gluecks list all pairs of 500 delinquents and 500 nondelinquents in their classic study, matched on national origin, age, and IQ.

Case Number		National Origin		Age*		Total I.Q.	
Delinquents	Non-Del.	Delinquents	Non-Del.	Delinquents	Non-Del.	Delinquents	Non-Del.
1	907	Italian	Italian	15 - 3	15 - 4	77	87
2	831	Irish	Irish	14 - 3	13 - 4	117	120
3	581	Italian	Italian	11 - 9	12 - 11	88	93
4	974	English	Scotch	13 - 6	13 - 10	53	58
5	658	Portuguese	Portuguese	14 - 1	13 - 5	75	87
6	845	Irish	Irish	14 - 9	14 - 4	107	107
7	649	Eng. Can.	Eng. Can.	15 - 3	15 - 11	69	73
8	533	Italian	Italian	12 - 8	12 - 11	91	97
9	700	Italian	Italian	14 - 0	13 - 6	93	100
10	524	Italian	Italian	12 - 7	12 - 11	88	98

*Age is represented by year and month. For example, "15 - 5" is 15 years 5 months old.

Source: Glueck and Glueck, *Unraveling Juvenile Delinquency,* page 297. Copyright © 1950 The Commonwealth Fund. Reproduced with permission of the publisher.

The Gluecks' study is remarkable for the thoroughness of its matching design, among other reasons. This design and the breadth of the Gluecks' scope of inquiry would be difficult to replicate today, even with the advances that have been made in methodology and analytic techniques. Comparing delinquents with a matched control group of nondelinquents offers a unique way to isolate the factors leading to delinquency.

From 1939–1948, the Gluecks and their research team conducted extensive interviews with the boys in both the delinquent and nondelinquent samples, as well as their parents, teachers, neighbors, and employers. They also gathered information from the records of public and private social service agencies, juvenile courts, probation and parole departments, and correctional institutions.[7] The collection of data from multiple sources was one of the strengths of this research project. A second strength was the collection of data about a wide variety of factors thought to be related to delinquency, including the following:

- family life: disciplinary practices, supervision, attachment between parent and child, family structure
- school performance: grades, behavior in school, attendance and truancy records, educational aspirations

- peer relationships
- recreational activities and use of leisure time
- temperament and personality development
- history of criminal justice system contacts and sanctions
- family background: economic status, parental criminality and alcohol use
- body structures[8]

The Gluecks also conducted follow-up research on these samples from 1949 to 1963, gathering data at two time points, when the individuals were 25 and 32 years old.[9] With these extensive data, the Gluecks found that family life was the most important factor distinguishing delinquents from nondelinquents. Based on these findings, they developed a scale to predict involvement in delinquency that included parent–child attachment, parental supervision, and disciplinary practices.[10] The Gluecks also discovered the now readily accepted relationship between age and crime (e.g., the decline in offending with age), and the stability of delinquency over the life course of some offenders.[11]

In the early 1990s, Robert Sampson and John Laub resurrected the Gluecks' original data and reanalyzed them with the benefit of modern statistical techniques. They published their results in an award-winning book, *Crime in the Making: Pathways and Turning Points Through Life*.[12] Sampson and Laub's findings are consistent with those of the Gluecks, confirming the research conducted decades earlier.

Over the years, the Gluecks' research has been sharply criticized on both methodological and substantive grounds.[13] Some critics point out errors in the Gluecks' analytic strategy and in their interpretation of results. Others criticize them for what became an unpopular focus on the family as a predictor of delinquency and for their openness to consider biological factors as predictors of delinquent behavior.[14] Despite the methodological inadequacies of their work, the Gluecks were interested in the same research questions that shape current criminological debates, particularly those concerning the relationship between age and crime, the stability of crime over the life course, and the value of longitudinal research (in which information is gathered from the same individuals at more than one point in time). In many respects, the Gluecks' work foreshadowed the central concerns of the discipline of criminology today.[15]

Sutherland's Use of Case Studies and Analytic Induction

Edwin Sutherland developed differential association theory, one of the most influential theories in twentieth-century American criminology. (See Chapter 11 for a discussion of the theory.) In this chapter, we are concerned not with the theory itself, but with the methodology used to arrive at the theory.

Before Sutherland's first statement of his theory in 1939,[16] the "multiple-factor approach" was the prevailing paradigm for explaining crime. This was essentially an "everything but the kitchen sink" approach. It incorporated a wide range of factors thought to be related to crime, but it lacked a theoretical foundation. Sutherland was profoundly dissatisfied with this atheoretical approach and sought to develop a scientific explanation of crime that identified "conditions which are always present when the phenomenon being explained [crime] is present but which are never present when the phenomenon is absent."[17]

To arrive at a scientific generalization for explaining crime, Sutherland used a method called **analytic induction**, which was developed, applied, and introduced to Sutherland by his colleague Alfred Lindesmith.[18] Analytic induction consists of seven steps, beginning with the definition of the behavior to be explained and ending with the examination of cases used to test the hypothesis.[19] The seven steps of analytic induction are listed in Expanding Ideas, "Steps of Analytic Induction." While Sutherland fruitfully applied the method of analytic induction to study the causes of crime and develop his theory, his colleagues used the technique to study drug addiction[20] and embezzlement.[21]

Sutherland and his associate, Donald Cressey, explain how this use of case studies in analytic induction is different from the individual case study method.[22] In the latter, the individual is the unit of analysis, and all traits or conditions of the individual are studied together. When case studies are used in the process of analytic induction, the hypothesis being tested is of primary concern, and the researcher is interested only in individual traits or conditions that have some bearing on that hypothesis.

EXPANDING IDEAS Steps of Analytic Induction

The method of **analytic induction** that Edwin Sutherland used in his research and the development of *differential association theory* consists of these seven steps:

1. Roughly define the behavior to be explained (in this case, crime).
2. Formulate a tentative hypothesis to explain the behavior.
3. Study one case and determine "whether the hypothesis fits the facts in that case."
4. If the hypothesis does not fit the facts, either more precisely redefine the behavior to be explained or reformulate the hypothesis.
5. "Practical certainty" is achieved when one has examined a small number of cases and found none that disproves the hypothesis. "[A] negative case disproves the explanation and requires a reformulation."
6. Continue examining cases, redefining the behavior, and reformulating the hypothesis until "a universal relationship is established."
7. Examine cases outside the area of defined behavior "to make certain that the final hypothesis does not apply to them."

Source: Sutherland and Cressey, *Principles of Criminology*, 71–72.

Ethnography

<u>Ethnography</u> is a form of field study, which we mentioned in Research in Action, "The Theoretically Driven Research Process." Ethnography involves "describing social or cultural life based on direct, systematic observation, such as becoming a participant in a social system."[23] Ethnographers often immerse themselves as participants in the social system they are studying; take extensive field notes about interactions they observe or participate in; sometimes conduct in-depth interviews with key participants in the social system; and then offer a detailed, descriptive analysis of that system. Elijah Anderson, a well-known ethnographer, describes how ethnographers try to illuminate the social and cultural dynamics of the settings they study by answering questions such as how participants in a particular setting perceive the situation, what assumptions they make, and what consequences result from their choices and behaviors in the setting.[24]

ethnography A research method that involves "describing social or cultural life based on direct, systematic observation, such as becoming a participant in a social system" (Vogt, *Dictionary*, 83).

Ethnographic studies are typically based on sustained interactions between the researcher and those whom he or she is studying—interactions that last several months or possibly even years. Participant observation, the primary ethnographic research tool, "implies that the researcher is closely associated with the daily activities of the subjects under study and is in a position to observe the events involving these subjects as they naturally occur. If enough time is spent in the field there should be sufficient data to establish a pattern that provides an understanding of how, why, and under what conditions certain events (including criminal events) have taken place."[25]

The great value of this type of research lies in its description of a particular phenomenon as it truly exists. The artificial qualities of experiments or surveys, for example, are stripped away in ethnographic research. Environments and "subjects" are not controlled or manipulated by the researcher; the ethnographer simply provides a detailed account of naturally occurring interactions and social processes. Ethnography also allows researchers, through their direct participation in particular settings, to understand more fully the meanings and definitions shared by participants in those settings that motivate their behavior. The researcher's vantage point *from within* the setting being studied is the hallmark and primary strength of ethnographic research. Discussing the use of ethnography to study crime and deviance, Jeff Ferrell and Mark Hamm write, "research methods which stand outside the lived experience of deviance or criminality can perhaps sketch a faint outline of it, but they can never fill that outline with essential dimensions of meaningful understanding."[26]

The major drawback to ethnographic research is that everything the ethnographer describes is viewed through the lens of his or her perceptions and subjectivity. When conducting their research, ethnographers must try to set aside their own values and assumptions, and be as objective as possible in describing and analyzing what they see. Although it is often difficult to recognize the influence of one's own assumptions, the quality of ethnographic research depends on the researcher's ability to identify and override those assumptions and to offer a bias-free account of the situation.[27]

Another drawback of ethnographic research (and of field studies more generally) is that, because of the amount of time this type of research requires, it is difficult for researchers to observe or interview more than a small number of individuals.[28] In addition, an observer or interviewer may affect the behavior and ideas of research subjects or even the events that transpire. For example, a researcher studying the functions of violence in gang life might find gang members disinclined to carry out violent acts in his or her presence.

Social scientists have used ethnographic techniques to address diverse questions in the field of criminology. Scott Decker, for example, used interviews and direct observations over a three-year period to explore the functions and normative character of gang violence.[29] (See Chapter 11 for a discussion of Decker's research.) Thomas Schmid and Richard Jones used participant observation and focused interviews with prisoners to examine the strategies that first-time, short-term inmates use to adapt to life in a maximum security prison.[30] Thomas Vander Ven used observational research to study fear of victimization in a Latino neighborhood.[31]

Elijah Anderson used participant observation and in-depth interviews over a four-year period to explore youth violence in the inner city.[32] Alienation from social institutions, including the criminal justice system, is common in the context of persistent poverty and deprivation. Examining patterns of behavior for four years, Anderson discovered that, for some people living in the most impoverished, crime-ridden areas of the inner city, the "code of the street" has replaced the authority of civil law as the standard for acceptable behavior. Particularly for young people, this code or set of informal rules governs behavior, especially concerning the use of violence. Anderson writes, "The code of the street emerges where the influence of the police ends and personal responsibility for one's safety is felt to begin, resulting in a kind of 'people's law,' based on 'street justice.' "[33] In this system of street justice, one gains respect through the display of "nerve," which represents the threat of vengeance for aggression or disrespect. Possession of respect, in turn, "shields" an individual from violent victimization. Thus the code of the street both fosters and requires violence in some inner city neighborhoods.

Ecological Analysis

ecological analysis A research method used to explore the geographic distribution of crime and delinquency and the social conditions that characterize areas with high rates of crime and delinquency. In ecological analysis, geographic areas, such as neighborhoods or cities, rather than the individuals who reside in them, are the units of analysis.

Clifford Shaw and Henry McKay used police and juvenile court records to study the ecological distribution of delinquent behavior—the way delinquency rates varied across a city.[34] Their work grew out of their perception of the inadequacies of the individual-level psychological and biological explanations of delinquency that were popular at the time they began their work. Shaw and McKay were instead interested in how juvenile delinquents (officially defined) were geographically distributed across the city of Chicago and in the social conditions that characterized areas with high delinquency rates.[35] To study these issues, they conducted **ecological analysis** of the distribution of delinquency rates in Chicago. We discuss Shaw and McKay's social disorganization theory in Chapter 12. Here we briefly describe the method of ecological analysis that led them to the statement of their theory.

To explore the geographic distribution of delinquency, Shaw and McKay gathered data on approximately 60,000 male delinquents from juvenile court and police records for several time periods from 1900 to 1933.[36] They then plotted the home address of each delinquent on a map of the city of Chicago and observed that delinquency was concentrated in or near areas zoned industrial or commercial.[37]

Shaw and McKay divided Chicago into 140 "square-mile areas."[38] For each of these areas, they gathered information about community characteristics, including population change, percentage of families receiving government aid, median rent, home ownership, and percentage of immigrant and African American residents. Shaw and McKay then examined the relationship between delinquency rates and community characteristics in each of the 140 square-mile areas. By focusing on this aggregate-level relationship, they discovered that delinquency rates were highest in areas characterized by decreasing populations, low rents, high percentages of families receiving government aid, and high percentages of "foreign born" residents.[39] Shaw and McKay also discovered that these areas were patterned in concentric zones in which delinquency rates were stable over time, despite population turnover.

The strength of Shaw and McKay's work is that it focused attention on previously neglected community-level factors in predicting delinquency. By exploring the social characteristics of communities associated with high rates of delinquency, Shaw and McKay literally brought a different level of understanding to the issue of delinquent behavior. They were not uninterested, however, in the individual-level factors leading to delinquency. In fact, their efforts to understand delinquent behavior included the life histories or case studies of individual delinquents.[40] Still, their enduring legacy is tied to their interest in community characteristics that contribute to delinquency.

The primary weakness of Shaw and McKay's work is that it rests on data from police and juvenile court records. Shaw and McKay defined a male juvenile delinquent as a boy under the age of 17 who was brought before the juvenile court (or another court having jurisdiction) on a delinquency petition, or whose case was disposed of without a court appearance.[41] But as we discuss later in this chapter, not all delinquents are known to police, so official records do not represent all youths who have violated the law. Furthermore, official responses to delinquency vary by class and race—a fact that may have contributed to Shaw and McKay's findings of higher delinquency rates in impoverished, immigrant and African American neighborhoods. Despite the drawbacks of official data, they represented the best information available for addressing the questions that interested Shaw and McKay.

Contemporary Ecological Analysis

Criminologist Robert Sampson has led a resurgence of interest in ecological analysis and the study of community-level factors that contribute to high crime rates in some areas.[42] According to Sampson, a basic assumption of ecological analysis is that social systems, such as neighborhoods, have qualities that exist apart from the characteristics of individuals who constitute those social systems.[43] In several articles, Sampson has used neighborhoods, rather than individuals, as the unit of analysis, and has explored the structural characteristics of communities that are associated with high crime rates.[44]

For example, Sampson used victimization survey data to explore the aggregate-level effect of neighborhood family structure (the percentage of households in the neighborhood that are affected by divorce or separation, headed by a woman, or occupied by a single individual) on victimization rates.[45] The reasoning is that family dissolution undermines informal social control, and single-individual households impair guardianship and create increased opportunities for crime. Results indicated that neighborhood family structure strongly affected the risk of victimization.[46] In a separate study, Sampson examined the relationship between a variety of neighborhood characteristics—unemployment, income inequality, racial composition, residential mobility, population density, and family structure—and rates of criminal victimization.[47] He found that residential mobility, population density, and family structure had strong effects on victimization rates.

In more recent work, Sampson and his colleagues examined the community characteristic of collective efficacy (a combination of neighborhood cohesion and informal social control) and its association with violent crime and victimization rates.[48] They found that collective efficacy, measured at the neighborhood level, was negatively related to violence and mediated many of the effects of neighborhood structural characteristics (economic disadvantage and residential instability) on crime and victimization rates. (See Chapter 12 for more details about this study.)

Sampson's influential studies, like those of Shaw and McKay, illustrate the value of aggregate-level research in which communities, rather than individuals, are the unit of analysis. The structural characteristics and dynamics of communities exert powerful influences on the individuals who reside in them. Ecological analysis captures those influences, missed in individual-level research.

Survey Research

survey research A research method involving the administration of a survey instrument (questionnaire or interview) to a sample of respondents, coding of data gathered, statement of hypotheses, and analysis of data in order to make inferences from the sample to the population from which it was drawn.

<u>Survey research</u> is the most common strategy for collecting data in the study of delinquency. It is based on the simple idea that if you want to know what people do or think, you should ask them. Survey research involves asking, through questionnaires or interviews, a series of questions, sometimes on a single topic, but usually on a wide range of issues. This is complex and involves several stages: selection of a sample of respondents, writing and testing of questions to be asked, administration of the survey instrument, coding of data gathered, statement of hypotheses, analysis of data gathered, and inferences from the sample to the population from which it was drawn.

For most questions that social scientists address, it is not possible to survey *everyone* in the population of interest. This approach is time-consuming, costly, and unnecessary. Instead, researchers can survey a carefully selected *representative sample*—a sample that is similar in terms of social characteristics to the population from which it was drawn.[49] For example, suppose we select a sample of adults in the United States that is 80 percent female and 20 percent male. This sample would not be representative of the US population, which is roughly 52 percent female and 48 percent male. If researchers select representative samples, they can then make inferences about the behaviors or thoughts of the population of interest based on the survey responses of those in the sample.

For example, suppose we want to know about the relationship between associations with delinquent peers and youths' own involvement in delinquency. We certainly cannot survey all adolescents in America. But we can select a representative sample of adolescents from across the country, ask them about their own involvement in delinquency and that of their friends, analyze the relationship between the two sets of variables, and then make inferences about the relationship between having delinquent peers and youths' own delinquency for the entire population of adolescents in the United States. The key to this process is the selection of a sample that is truly representative of the population of interest. The quality of survey research rests on both accurate sample selection and rigorous data analysis.[50]

Given the complexity of survey research, many sources of error exist in survey analysis. These sources include sample selection, survey questions, interviewers, the coding and entry of data, and data analysis.[51] For example, the way in which a sample is selected, its size, and the refusal of some individuals included in the sample to participate in the survey or to answer specific questions are all sources of error. Respondents' faulty recollection, misunderstanding of survey questions, and unwillingness to provide accurate answers to some questions also produce error in surveys. For surveys administered through interviews, the interviewers are a potential source of error. For example, they might misread questions or inaccurately record answers. Finally, error can take place in the research process when survey data are incorrectly coded or entered into computer-readable files and when researchers incorrectly analyze the data or inaccurately interpret their findings.[52]

Researchers use a variety of techniques to analyze survey data. Designing the analysis includes three steps: "the specification of the hypotheses to be tested, the choice of the specific variables to use in those tests, and the selection of appropriate statistical procedures to analyze the data."[53] In most cases, social scientists who conduct large-scale surveys are interested in explaining people's behaviors or attitudes. This typically involves exploring complex relationships among several variables using techniques that can take into account multiple variables simultaneously. For example, suppose we want to examine the relationship between parenting practices (or youths' perceptions of them) and youths' involvement in delinquency. In a survey of adolescents, we might ask questions about attachment to parents, parental supervision, and parents' discipline strategies, as well as questions about youths' delinquent behaviors. In statistical analysis of the data gathered, we could then examine these elements of parenting and the extent to which they are associated with or predict youths' delinquent involvement.

For decades, criminologists have used surveys to gather a wealth of information about the frequency and distribution of delinquency and the causes and correlates of offending. We offer three examples of survey research by noted criminologists to provide a sense of how survey analysis is conducted and what it can tell us about crime and delinquency.

The Richmond Youth Study

Travis Hirschi is best known for his book *Causes of Delinquency*, based on data from the Richmond Youth Study, begun in 1964. In this study, Hirschi selected an original sample of 5,545 students from the approximately 17,500 youths entering junior and senior high schools in the Richmond, California area. Complete survey data were obtained for 4,077 students, or 73.5% of the original sample.[54]

Researchers administered written questionnaires to students in schools.[55] The survey instrument included questions about attitudes toward school and teachers, involvement in school activities, leisure time activities, peer group associations, behavior problems at school, involvement in delinquency, neighborhood characteristics, employment and earnings, aspirations and expectations, attachments to parents, and parents' background.[56] Many of the questions asked in the survey were derived from and used to test Hirschi's social control theory, described in Chapter 10.

In addition to self-report survey data, Hirschi also gathered information on study participants from school and police records. School records provided achievement test scores and grade point averages, as well as demographic information, such as gender and race.[57] Police records included information such as total number and types of offenses committed by study participants, age at first offense, and date of most recent offense.[58]

Hirschi analyzed this wealth of data using various statistical techniques. He examined how respondents' involvement in delinquency (self-reported or as indicated in police records) varied by individual characteristics, such as social class and race.[59] Hirschi also explored sets of variables measuring attachment to parents, peers, and school; commitment to conventional lines of action; involvement in conventional activities; and belief in the moral order. (See Chapter 10.) He statistically analyzed the strength of the relationships between these variables and involvement in delinquency.

The National Youth Survey

Delbert Elliott and his colleagues began the National Youth Survey (NYS) in 1976. This survey was "designed to meet three primary objectives: (1) to provide a comprehensive description of the prevalence and incidence of delinquent behavior and drug use in the American youth population [prevalence is the proportion of the population involved in offending, and incidence is the number of offenses committed]; (2) to examine the causal relationship between delinquent behavior and drug use; and (3) to test an integrated theoretical model of delinquent behavior."[60]

The NYS involved a nationally representative sample of youths who were 11 to 17 years old in 1976, when the study began. Of the original sample of 2,360 youths, 1,725 (or 73%) agreed to participate in the study and completed interviews in the first "wave" of data collection in 1977.[61] Researchers used face-to-face interviews to gather data. In addition to demographic information about respondents, the survey gathered extensive information about respondents' alcohol and drug use, involvement in delinquent behavior (both minor and serious), and social psychological characteristics. Research in Action, "The National

RESEARCH IN ACTION

The National Youth Survey

Supported by the National Institute of Mental Health, the National Youth Survey (NYS) was designed to offer researchers a better understanding of youth involvement in both conventional and delinquent behaviors. From respondents, the survey gathered seven "waves" of data from 1977 to 1987. For a detailed description of the NYS design and sample, see Elliott, Huizinga, and Ageton (*Explaining Delinquency*, Chapter 5). The NYS contains data on self-reported delinquency and victimization, drug and alcohol use, exposure to delinquent peers, attitudes toward deviance, labeling, family background, parental discipline, parental aspirations for youth, disruptive events in the home, school and work status, community involvement, and perceptions of neighborhood problems, including crime.

The following questions were asked in the third "wave" of data collection of the National Youth Survey.

"Normlessness":
Responses on a five-point scale ranging from "strongly agree" to "strongly disagree."
1. *It's important to be honest with your parents, even if they become upset or you get punished.*
2. *To stay out of trouble, it is sometimes necessary to lie to teachers.*
3. *Making a good impression is more important than telling the truth to friends.*
4. *It's okay to lie if it keeps your friends out of trouble.*
5. *You have to be willing to break some rules if you want to be popular with your friends.*

Attitudes toward Deviance:
Responses on a four-point scale ranging from "very wrong" to "not wrong at all." How wrong is it for someone your age to . . .
1. *Cheat on school tests?*
2. *Purposely damage or destroy property that does not belong to him or her?*
3. *Break into a vehicle or building to steal something?*
4. *Get drunk once in awhile?*

Exposure to Delinquent Peers:
Responses on a five-point scale ranging from "all of them" to "none of them." Think of your friends. During the last year how many of them have . . .
1. *Cheated on school tests?*
2. *Used marijuana or hashish?*
3. *Stolen something worth more than $50?*
4. *Suggested you do something that was against the law?*
5. *Gotten drunk once in awhile?*

Self-Reported Delinquency:

Respondents were asked to provide their "best estimate of the ex-act number of times" they had done each of several behaviors during the past year. If a respondent reported engaging in a par-ticular behavior 10 or more times in the past year, he or she was asked to select which of the following six responses "best de-scribes how often you were involved in this behavior": Once a month, Once every 2–3 weeks, Once a week, 2–3 times a week, Once a day, 2–3 times a day. The list about which respondents were asked included 47 behaviors, not all of which were delin-quent. **How many times in the last year have you . . .**

1. *Purposely damaged or destroyed property belonging to a school?*
2. *Stolen or tried to steal a motor vehicle, such as a car or mo-torcycle?*
3. *Stolen or tried to steal something worth more than $50?*
4. *Run away from home?*
5. *Lied about your age to gain entrance or to purchase something, for example, lying about your age to buy liquor or get into a movie?*
6. *Carried a hidden weapon other than a plain pocket knife?*
7. *Attacked someone with the idea of seriously hurting or killing him or her?*
8. *Had sexual intercourse with a person of the opposite sex?*
9. *Been involved in gang fights?*
10. *Hit or threatened to hit other students?*
11. *Sold hard drugs such as heroin, cocaine, and LSD?*
12. *Used force or strong-arm methods to get money or things from a teacher or other adult at school?*

Source: Elliott, *National Youth Survey.*

Youth Survey," describes the NYS in greater detail and provides examples of questions from the NYS (see also Links, "The National Youth Survey").

To analyze the data gathered in the NYS, Elliott and his colleagues used sta-tistical techniques designed to examine relationships among several sets of vari-ables simultaneously. In addition, given the availability of data from the same respondents at multiple points in time, Elliott and his colleagues (and many other criminologists who have analyzed the NYS data) have been able to exam-ine issues of causal ordering and the ability of theoretical variables of interest to predict later drug use and delinquency.[62]

The Community Survey of the Project on Human Development in Chicago Neighborhoods

The Community Survey, begun in 1994, is part of the ambitious Project on Hu-man Development in Chicago Neighborhoods (PHDCN).[63] In the Community Survey, researchers divided Chicago into 343 neighborhoods and interviewed 8,782 residents of those neighborhoods.[64] The goal of the survey was to gather

The National Youth Survey

The data collected in the NYS are stored in the Substance Abuse and Mental Health Data Archive, which is maintained by the Inter-University Consortium for Political and Social Research at the University of Michigan. These data can be accessed via a link provided at:

http://criminaljustice.jbpub.com/burfeind

information about the social, economic, organizational, and political structures of Chicago neighborhoods. Respondents were interviewed about a variety of neighborhood factors, including neighborhood cohesion, social capital, informal social control, social disorder, availability of programs and services, organizational involvement, and criminal victimization of residents. (See Chapter 12 for a detailed description of the PHDCN and research findings derived from it.)

To analyze data from the Community Survey, researchers have used sophisticated statistical techniques that enable them to examine simultaneously multiple levels of analysis. Many questions of interest to them concern differences across neighborhoods, rather than differences among individuals who reside in a particular neighborhood. Because of the way the survey sample was selected and data were gathered, Community Survey researchers have been able to address these complex questions. For example, Robert Sampson, Stephen Raudenbush, and Felton Earls asked: To what extent do informal social control and social cohesion vary across neighborhoods, and how are these characteristics related to neighborhood crime rates?[65] In their multi-level analysis, they examined variation in perceptions of social control and cohesion across individuals *within* neighborhoods and also *across* neighborhoods.

■ Sources of Data on Crime and Delinquency

Data on crime and delinquency come primarily from three sources: "official" records maintained by law enforcement agencies and courts, surveys of individuals who have been victims of crime, and surveys of individuals who self-report involvement in offending. We discuss each of these, comparing the data gathered from these sources, and considering the strengths and weaknesses of each type of data.

"Official Data"

"Official data" on juvenile delinquency are gathered by governmental agencies within the criminal justice system. These data reveal the extent of delinquency with which these agencies deal and the characteristics of offenses and offenders they encounter. The two primary sources of official data on delinquency are the Uniform Crime Reporting program and *Juvenile Court Statistics*.

Uniform Crime Reporting Program

Uniform Crime Reporting (UCR) program Provides "official data" on crime and delinquency, voluntarily reported by over 17,000 law enforcement agencies across the United States and compiled by the Federal Bureau of Investigation. These data reveal the extent of crime and delinquency with which the reporting agencies deal and the characteristics of offenses and offenders whom they encounter.

The **Uniform Crime Reporting (UCR) program** is a nationwide effort that was begun by the Federal Bureau of Investigation (FBI) in 1930.[66] The UCR program includes "more than 17,000 city, county, and state law enforcement agencies voluntarily reporting data on crimes brought to their attention."[67] Although participation is voluntary, the vast majority of law enforcement agencies report crime data to the FBI as part of the UCR program. In 2003, participating agencies represented 93% of the total population of the United States.[68] As the program title implies, the UCR provides *uniformity* in crime reporting by requiring law enforcement agencies to use standardized offense definitions. In this way, variation in local statutes does not affect the nature or extent of offenses reported. Links, "Official Data—Uniform Crime Reporting Program," provides Internet connections to sites containing UCR data and information based on official data about the nature and extent of delinquency.

Offenses included in the UCR are divided into two categories: Part I Index offenses and Part II offenses. Part I offenses include the violent crimes of murder and nonnegligent manslaughter, forcible rape, robbery, and aggravated assault, and the property crimes of burglary, larceny-theft, motor vehicle theft, and arson. Part II offenses include all other criminal and delinquent acts. See **Table 3-1** for a list of Part II offenses. Data on both Part I and Part II offenses are gathered and submitted to the FBI on a monthly basis.

The UCR includes primarily three types of information: offenses known to police, crimes cleared, and persons arrested. Offenses known to police are those reported by victims, witnesses, or other sources or discovered by police officers. The category of "crimes cleared" is distinct from "offenses known to police."

Official Data—Uniform Crime Reporting Program

The Bureau of Justice Statistics site contains links to the Uniform Crime Reports and the *Sourcebook of Criminal Justice Statistics*.

Additional information regarding the criminal justice system and Web sites containing official statistics about crime is available through the National Criminal Justice Reference Service, a clearinghouse for the exchange of criminal justice information.

The links to these sites can be accessed at:

http://criminaljustice.jbpub.com/burfeind

Table 3-1 Uniform Crime Reporting Program Part II Offenses

- Simple assaults
- Forgery and counterfeiting
- Fraud
- Embezzlement
- Stolen property offenses (buying, receiving, possessing)
- Vandalism
- Weapons offenses (carrying, possessing, etc.)
- Prostitution and commercialized vice
- Sex offenses (except forcible rape, prostitution, and commercialized vice)
- Drug abuse violations
- Gambling
- Offenses against the family and children (nonsupport, neglect, desertion, or abuse of family and children)
- Driving under the influence
- Liquor law violations
- Drunkenness offenses
- Disorderly conduct
- Vagrancy
- All other offenses ("All violations of state and/or local laws except those listed above and traffic offenses.")
- Suspicion
- Curfew and loitering laws (persons under age 18)
- Runaways (persons under age 18)

Source: Federal Bureau of Investigation, *Crime in the United States 2003,* 497–498.

Crimes are cleared in one of two ways: (1) by arrest of at least one person, who is charged with committing an offense and turned over to the court for prosecution; or (2) by exceptional means when a factor beyond law enforcement control prevents the agency from formally charging an offender (e.g., death of the offender, or refusal of the victim to cooperate with prosecution).[69] A law enforcement agency may clear multiple crimes with the arrest of one individual, or it may clear one crime with the arrest of many individuals. In 2003, the nationwide clearance rate was 46.5% for Part I violent crimes and 16.4% for Part I property crimes (excluding arson).[70] Clearance rates are higher for violent crimes than for property crimes because violent crimes are often more "vigorously" investigated than property crimes, and because victims or witnesses often identify the offenders.[71]

For juvenile offenders, a clearance by arrest is recorded "when an offender under the age of 18 is cited to appear in juvenile court or before other juvenile authorities," even though a physical arrest may not have occurred.[72] The UCR reveals that, in 2003, juvenile offenders accounted for 12.2% of clearances for violent crimes and 19.3% of those for property crimes.[73]

The third type of data in the UCR is information about persons arrested, such as age, gender, and race. The UCR provides extensive data on crimes committed by various subgroups based on these offender characteristics. For example, the UCR presents data separately on offenses committed by juveniles under the age of 18 and on those committed by adults age 18 or older. It also presents data separately for males and females and for different racial and ethnic groups. The total number of persons arrested, recorded in the UCR, does not equal the total number of persons who have committed crimes, but rather only the number apprehended by police. In addition, the number of persons arrested does not equal the number of arrests, because an individual offender may be arrested for multiple crimes. For arrested offenders who have committed multiple crimes, only the most serious offense is recorded in the UCR. This is another reason that the actual amount of crime committed in the United States is higher than the amount revealed by the UCR.

UCR data are presented in various ways. **Crime rates** show the number of offenses per a specific portion of the population, such as the number of violent Index offenses per 100,000 individuals. **Crime trends** show rates over time and the percent change in rates from one time point to another. For example, the UCR shows the percent change in crime rates from the previous year and the percent change in rates over the previous ten years.

In addition to national crime statistics, the UCR also presents data separately by various geographic areas: by state; by region of the country; and by urban, suburban, and rural areas. The presentation of data by geographic area reveals the ecological distribution of crime in the United States and permits comparisons among neighboring jurisdictions and among areas with common characteristics such as population size.[74]

Redesign of the UCR Program

In the early 1980s, law enforcement agencies called for evaluation and modernization of the UCR program. A comprehensive, three-phase redesign effort followed, resulting in a new UCR program called the **National Incident-Based Reporting System (NIBRS)**.[75] "The goals of NIBRS are to enhance the quantity, quality, and timeliness of crime data collection . . . and to improve the methodology used in compiling, analyzing, auditing, and publishing the collected crime statistics."[76] The NIBRS offers detailed information about criminal incidents, including when and where an incident occurred, the type and value of property stolen, the relationship between victim and offender, and the age, gender, and race of both offender and victim. The differences between the original UCR program and the NIBRS are outlined in **Table 3-2.**[77] The major differences are the greater detail and larger number of offenses included in the NIBRS and the inclusion in the NIBRS of all offenses occurring in a single incident (compared to the UCR, which records only the most serious offense per incident). Despite this change in the recording of multiple-offense incidents, differences in crime rates based on the two reporting formats are small.[78]

Juvenile Court Statistics

In addition to the UCR, juvenile court records provide another official source of data about juvenile delinquency. *Juvenile Court Statistics* is compiled from data

crime rates Show the number of offenses per a specific portion of the population, such as the number of violent Index offenses per 100,000 individuals.

crime trends Show crime rates over time and the percent change in rates from one point in time to another.

National Incident-Based Reporting System (NIBRS) Provides more detailed official data on criminal incidents than the Uniform Crime Reporting program does. NIBRS resulted from revisions to the UCR program and, like the UCR, gathers information from law enforcement agencies across the United States.

Table 3-2	Differences between Uniform Crime Reporting Program and National Incident-Based Reporting System

Ramona Rantala and Thomas Edwards outline the following differences between the original Uniform Crime Reporting program and the National Incident-Based Reporting System.

Uniform Crime Reporting Program

- Consists of monthly aggregate crime counts for eight Index crimes.
- Records one offense per incident as determined by "hierarchy rule," which suppresses counts of lesser offenses in multiple-offense incidents.
- Does not distinguish between attempted and completed crimes.
- Applies "hotel rule" to burglary.
- Records rape of females only.
- Collects weapon information for murder, robbery, and aggravated assault.
- Provides counts on arrests for the 8 Part I Index crimes and 21 other Part II offenses (listed in Table 3-1).

National Incident-Based Reporting System

- Consists of individual incident records for the 8 Index crimes and 38 other offenses with details on:
  - Offense
  - Offender
  - Victim
  - Property
- Records each offense occurring in incident.
- Distinguishes between attempted and completed crimes.
- Expands burglary "hotel rule" to include rental storage facilities.
- Records rape of males and females.
- Restructures definition of assault.
- Collects weapon information for all violent offenses.
- Provides details on arrests for the 8 Index crimes and 49 other offenses.

Source: Rantala and Edwards, *Effects of NIBRS,* 1.

that state and county agencies provide to the National Juvenile Court Data Archive.[79] The most recent juvenile court statistics from 2000 are based on individual case-level data from 1,675 jurisdictions and court-level aggregate data from 313 jurisdictions.[80] Together, these jurisdictions contained 70% of the nation's juvenile population in 2000.[81] *Juvenile Court Statistics* provides information about offenses charged; age, gender, and race of offenders; referral sources; detention practices; petitioning decisions; and dispositions ordered. Reporting agencies use their own definitions and coding categories when providing juvenile court data, so the data are not uniform across jurisdictions. However, the National Juvenile Court Data Archive "restructures contributed data into standardized coding categories."[82] (See Links, "Official Data—*Juvenile Court Statistics*," for the Internet address of this archive.)

Links — **Official Data—Juvenile Court Statistics**

Data used to compile *Juvenile Court Statistics* are stored in the National Juvenile Court Data Archive at the National Center for Juvenile Justice in Pittsburgh. "The Archive contains the most detailed information available on juveniles involved in the juvenile justice system and on the activities of U.S. juvenile courts" (Puzzanchera et al., *Juvenile Court Statistics 2000*, x). The Archive's Web site, which contains a summary of Archive holdings and procedures for accessing data, can be found via a link provided at:

http://criminaljustice.jbpub.com/burfeind

The "unit of count" in *Juvenile Court Statistics* is the number of "cases disposed."[83] "A 'case' represents a juvenile processed by a juvenile court on a new referral, regardless of the number of law violations contained in the referral."[84] For example, a youth charged in a single referral with three offenses represents one case. "The fact that a case is 'disposed' means that a definite action was taken as a result of the referral—i.e., a plan of treatment was selected or initiated."[85] The treatment plan does not have to be completed for a case to be considered disposed. "For example, a case is considered to be disposed when the court orders probation, not when a term of probation supervision is completed."[86]

Because not all juveniles arrested are referred to juvenile court, the number of cases processed by juvenile courts is smaller than the number of juveniles arrested. UCR data from 2000 show 1,560,289 arrests of juvenile offenders.[87] In that same year, *Juvenile Court Statistics* data are based on only 1,148,341 delinquency and status offense cases processed (1,040,843 delinquency cases and 107,498 status offense cases).[88] Recall, however, that these juvenile court data represent jurisdictions containing only 70% of the nation's juvenile population.

The discrepancy between juvenile arrests and referrals to juvenile court points to the primary weakness of juvenile court data: it is not available for all juvenile offenders. Only a portion of juveniles who commit offenses come to the attention of law enforcement personnel, and only a portion of known offenders are referred to juvenile court. Thus, the offenders included in juvenile court statistics are not representative of all offenders. Other drawbacks of juvenile court data include their inconsistent publication schedule and the relatively limited information they provide about juvenile offenders.

Strengths of Official Data

The UCR program provides uniform, nationwide data about crime and delinquency. The FBI goes to great lengths to ensure that law enforcement agencies across the country use consistent definitions and procedures for reporting crimes as part of the UCR program. This uniformity, combined with the nationwide scope of the program, enables researchers to compare crime statistics across jurisdictions and examine the nature and extent of crime for the nation as a whole. Because the UCR program has been in existence for more than seventy years, researchers are also able to use UCR data to explore trends in crime rates over long periods of time. In addition, factors that affect the validity of UCR data (e.g., police discretion in arrest decisions) might be expected to remain relatively constant over time, so that data from successive years are truly comparable, and trends based on UCR data reveal genuine fluctuations in crime rates, rather than incidental factors that influence crime reporting.[89] Another strength of the UCR program is that it offers a good deal of information about the demographic characteristics of offenders (e.g., age, gender, and race).

Weaknesses of Official Data

Official data are characterized by several weaknesses. Foremost among them is the vast number of offenses not included in official statistics. Victimization survey data indicate that more than half of the criminal victimizations that occur annually in the United States are *not* reported to police. According to the National Crime Victimization Survey, in 2003, victims reported to police only 48% of violent victimizations and 38% of property crimes.[90] The majority of offenses are not known to police and therefore not included in official statistics. Moreover, a strong relationship exists between the seriousness of an offense and the likelihood that it will be cleared through an arrest. So, the less serious the offense, the more likely that it will be excluded from official data. In addition, certain types of crimes, including so-called victimless offenses (e.g., drug abuse violations and prostitution), are particularly likely to go unreported to police and thus be excluded from official data.

A second weakness of official data concerns the effects of policing practices on official statistics. The criteria that govern arrest decisions may vary among law enforcement agencies and within agencies over time. For example, an agency might decide to "crack down" on prostitution, and conduct operations that result in an increased number of vice arrests. Yet this increase does not mean that prostitution has increased in that jurisdiction, or that it is necessarily a greater problem there than in other jurisdictions that show fewer arrests for prostitution. Rather, the increased number of arrests reflects changes in policing practices.

A related criticism is that the criteria governing arrest decisions appear to vary for different segments of the population. For example, race discrimination in the arrest process is well documented and indicates that, for committing the same crimes, blacks are more likely than whites to be arrested.[91] Such discrimination distorts official data, particularly about the characteristics of offenders.

Law enforcement agencies may also manipulate crime data for political purposes and thus provide an inaccurate picture of the nature and extent of crime. For example, an agency may underreport crime in its jurisdiction in an attempt to show that it has curbed crime. Another agency might overreport crime in an effort to acquire additional resources for its crime-fighting activities.

Finally, police officers, like everyone else, are fallible and sometimes simply make errors in recording crime data. In addition, although the FBI works hard to achieve uniformity of data across jurisdictions, some variation in crime coding and recording practices is inevitable, given the huge number of agencies participating in the UCR program.

Victimization Surveys

Crime data are gathered not only from police agencies and courts, but also from those who have been victimized. Begun in 1972, the National Crime Victimization Survey (NCVS) was developed to overcome problems with data from official sources.[92] Criminologists expected the NCVS crime rate and crime trend data to be more reliable than UCR data. The NCVS provided a systematic way to gather information about offenses unreported to police and thus excluded from UCR data. In 1989, the NCVS was redesigned to incorporate improved methods of helping respondents recall victimizations and more direct questions about sexual victimizations.

The primary objectives of the NCVS included estimation of the number and types of crimes unreported to police and the gathering of detailed information about the victims and consequences of crime—information not gathered in the UCR. Victimization surveys were designed to provide data regarding the following:

- situational factors, such as where the crime occurred, time of day at which it occurred, how many victims were involved, whether or not a weapon was used, and self-protective actions by the victim and the results of those actions
- victim characteristics, such as gender, age, race, educational attainment, income, marital status, and relationship to the offender
- consequences of the victimization, such as whether or not the victim was injured, and cost of medical expenses incurred

Data gathered in the NCVS enable researchers to estimate the likelihood of victimization for various types of crime for the US population as a whole, as well as for specific demographic subgroups. For example, with data on victim characteristics, researchers can estimate how the likelihood of becoming a robbery victim differs for males and females, or how racial status influences one's chances of becoming a rape victim.

The NCVS is conducted by the US Census Bureau, which selects the national sample and interviews respondents. Most interviews are conducted in person. In 2003, the sample consisted of 83,660 households, and 149,040 persons age 12 or older who resided in them.[93] The NCVS is based on a panel design, meaning that the households and individuals included in the survey provide data at multiple points in time. Households selected as part of the NCVS remain in the sample for 3 years, and individuals are interviewed twice a year.[94] The NCVS has achieved a high response rate: in 2003, 92% of eligible households and 86% of individuals selected to participate in the NCVS actually completed the survey.[95] The NCVS asks respondents about personal crime victimizations (including rape and sexual assault, robbery, aggravated assault, simple assault, and purse-snatching/pocket-picking) and property crime victimizations (including burglary, theft, and motor vehicle theft).

RESEARCH IN ACTION

National Crime Victimization Survey

The following victimization questions are asked in the National Crime Victimization Survey.

I'm going to read some examples that will give you an idea of the kinds of crimes this study covers.

As I go through them, tell me if any of these happened to you in the last 6 months, that is since _____ __, 20__.

Was something belonging to YOU stolen, such as —

(a) Things that you carry, like luggage, a wallet, purse, briefcase, book —
(b) Clothing, jewelry, or calculator —
(c) Bicycle or sports equipment —
(d) Things in your home — like a TV, stereo, or tools —
(e) Things outside your home such as a garden hose or lawn furniture —
(f) Things belonging to children in the household —
(g) Things from a vehicle, such as a package, groceries, camera, or cassette tapes —

OR

(h) Did anyone ATTEMPT to steal anything belonging to you?

(Other than any incidents already mentioned,) has anyone —

(a) Broken in or ATTEMPTED to break into your home by forcing a door or window, pushing past someone, jimmying a lock, cutting a screen, or entering through an open door or window?
(b) Has anyone illegally gotten in or tried to get into a garage, shed or storage room?

OR

(c) Illegally gotten in or tried to get into a hotel or motel room or vacation home where you were staying?

(Other than any incidents already mentioned,) since _____ __, 20__, were you attacked or threatened OR did you have something stolen from you —

(a) At home including the porch or yard —
(b) At or near a friend's, relative's, or neighbor's home —
(c) At work or school —
(d) In places such as a storage shed or laundry room, a shopping mall, restaurant, bank, or airport —
(e) While riding in any vehicle —
(f) On the street or in a parking lot —
(g) At such places as a party, theater, gym, picnic area, bowling lanes, or while fishing or hunting —

OR

(h) Did anyone ATTEMPT to attack or ATTEMPT to steal anything belonging to you from any of these places?

Incidents involving forced or unwanted sexual acts are often difficult to talk about. (Other than any incidents already mentioned,) have you been forced or coerced to engage in unwanted sexual activity by —

(a) Someone you didn't know before —
(b) A casual acquaintance —
OR
(c) Someone you know well?

If respondents answer "yes" to any of these questions, they are then asked (1) how many times the incident occurred, and (2) to describe what happened.

Source: Bureau of Justice Statistics Web site: http://www.ojp.usdoj.gov/bjs/pub/pdf/ncvs104.pdf

During the NCVS interview process, one adult answers background questions about the household, such as family income, number of household members, and whether the family owns or rents the home. That same respondent also answers questions about household victimizations during the previous six months, such as burglary, motor vehicle theft, and larceny from the premises.[96] Each member of the household age twelve or older is then interviewed about personal victimizations during the previous six months. The NCVS also gathers background information, such as age, sex, and education, for all members of the household. Research in Ac-

Links   **National Crime Victimization Survey**

Information about the National Crime Victimization Survey (NCVS), along with statistics derived from it, can be found in the "Crime and Victims Statistics" section of the Bureau of Justice Statistics Web site. A copy of the survey questionnaire and information about the redesign of the NCVS in the early 1990s are also available via a link provided at:

http://criminaljustice.jbpub.com/burfeind

The data collected in the National Crime Victimization Survey are part of the National Archive of Criminal Justice Data, which is maintained by the Inter-University Consortium for Political and Social Research at the University of Michigan. These data, along with other data regarding criminal justice, can be accessed via a link provided at:

http://criminaljustice.jbpub.com/burfeind

tion, "National Crime Victimization Survey," includes sample questions from the NCVS, and Links, "National Crime Victimization Survey," includes connections to Web sites containing information about the NCVS.

Comparing NCVS and UCR Data

NCVS estimates suggest that "the actual amount of crime is roughly two to four times higher than official statistics would indicate, depending on the type of crime being considered."[97] In addition, the crime trends revealed by NCVS and UCR data are often in different directions, with NCVS data showing downward trends in crime rates and UCR data showing upward trends.

In a recent study, Robert O'Brien compared violent crime trends based on data from the NCVS and the UCR program.[98] He focused on the crimes of rape, robbery, and aggravated assault. From 1973–1992, NCVS data reveal trends in violent crime rates that are either flat or downward, yet UCR data for the same time period show an upward trend. Previous studies found similar divergence between the two data sources.[99] When researchers take into consideration the effects of time itself, however, the trends revealed by NCVS and UCR data are more similar.[100] In a sophisticated analysis, O'Brien found that, though violent crime trends based on NCVS and UCR data are not strongly related, they also are not independent of each other.[101] He concluded that the flat or downward trend portrayed by NCVS data is the more plausible one, and attributed the upward trend in violent crime rates based on UCR data to increases in police productivity in recording crimes.[102]

Strengths of Victimization Surveys

The primary strength of the NCVS, and of victimization surveys more generally, is their ability to provide information about crimes that victims do not report to police. Recall that NCVS data reveal that, in 2003, victims reported to police only 48% of violent crimes and 38% of property crimes.[103] By surveying victims directly, the NCVS provides a systematic way to capture offenses unreported to police that would otherwise go unnoticed in crime statistics. O'Brien's research also suggests that, by providing a measure of crime unfiltered by the policing process, the NCVS may offer the more accurate picture of crime rates over time.[104]

Because the NCVS also gathers information about demographic characteristics of individuals and households that have been victimized, it offers insights into the risks of victimization. Using victimization survey data, researchers can determine how the risks of becoming the victim of various types of crime differ by social characteristics, such as age, gender, race, and social class. (See Chapter 5 for a discussion of the social correlates of victimization.) The NCVS highlights the dynamic nature of crime as a "social transaction" that involves not just an offender, but also a victim.[105]

Weaknesses of Victimization Surveys

The primary weakness of victimization surveys concerns the offenses that they omit or are unable to capture, including homicide, most crimes classified as Part II offenses in the UCR, "victimless" crimes, and status offenses.[106] The NCVS includes questions about all UCR Part I offenses except homicide and arson, but the only Part II offense it includes is simple assault. Thus, by design, it excludes crimes such as vandalism, weapons offenses, fraud, forgery, and embezzlement. Victimless crimes, such as alcohol and drug use violations, prostitution, and gambling, are also omitted from the NCVS, as are juvenile status offenses, including curfew violations and running away. Some types of crime, such as employee theft, income tax violations, fraud, embezzlement, and violations of antitrust laws, are simply difficult to assess from a victim's perspective.[107] Thus the scope of offenses included in victimization surveys is more limited than that of UCR data.

In addition, some victimizations are omitted from NCVS data not by design, but because survey respondents do not report them. Some offenses may go unreported because the adult answering questions about household victimizations is unaware of all victimizations during the previous six months. Other offenses may go unreported because respondents choose not to reveal them. Despite revisions to the survey instrument, designed to encourage respondents to report personal crimes of violence, offenses such as sexual assault may be particularly likely to be underreported. Respondents may also systematically underreport offenses because of the relationship between victim and offender. To explore this possibility, one researcher first obtained from police records information about reported crimes and then interviewed the victims of those crimes about recent victimizations. He found that victims reported to the interviewer 76% of known incidents when the offender was a stranger to the victim, 57% of known incidents when the offender was known to the victim, and only 22% of known incidents when the offender was a relative of the victim.[108]

A second limitation of the NCVS is that it underestimates juvenile victimizations. This is true in part because the NCVS excludes victims under 12 years of age, and in part because young respondents appear reluctant to provide interviewers with information about crimes committed against them. In a recent study, Edward Wells and Joseph Rankin explained why the NCVS may not provide an accurate picture of juvenile victimization: "Young respondents may be hesitant to talk about victimizations to an adult stranger who is also an official Bureau of the Census interviewer, particularly when most of the offenses are by either their peers . . . or a family member. Being less conversant with legal labels and categories, young respondents also may interpret and label events differently than those intended in the NCVS questionnaire. Last, adolescents may be reluctant to reveal victimizations in which they themselves are involved in illegal activities."[109] Wells and Rankin's research, described later in this chapter, indicates that the NCVS underestimates juvenile victimization, and that caution is required when using NCVS data to estimate the level of crime against juveniles.[110]

A third weakness of victimization surveys is that they offer limited information about offenders. Obviously, victimization surveys can provide information about offender characteristics, such as gender, race, and estimated age, only for those offenses involving personal contact between offender and victim.

Finally, victimization surveys carry with them the shortcomings inherent in the survey method. Perhaps most important among these is the problem of recall. Respondents, like everyone else, have faulty memories and may not accurately recall past victimizations and the details related to them. This problem intensifies as the length of time between the victimization event and the survey interview increases.[111] The NCVS tries to minimize this problem by "bounding" all interviews after the first one to a six-month time frame. This means that, after the initial interview, the respondent is asked only about victimizations that have occurred during the previous six months since the last interview. Research has shown that this is an effective way to minimize recall errors.[112]

Another problem of survey methodology is that respondents may not provide truthful responses to survey questions, or they may provide what they consider to be "socially desirable" responses. Respondents may also simply misunderstand the questions and thus inadvertently give incorrect answers. In interview situations, language barriers may also prevent respondents from providing accurate responses.

Self-Report Surveys

Self-report surveys, like UCR data and victimization surveys, are a major source of data on crime and delinquency. Their popularity grew in the 1960s and 1970s as researchers began to recognize problems associated with official statistics and the ability of self-reports to overcome these problems. Self-report surveys ask individuals directly, through either questionnaires or interviews, about their involvement in crime and delinquency. Thus, these surveys avoid the filter of the criminal justice system and provide information about offenders, regardless of whether or not those offenders have been arrested or officially processed.

Self-report surveys also provide demographic information about offenders, such as age, race, and gender, as well as information—unavailable through official data or victimization surveys—about personal characteristics of offenders, such as family background and social class. Importantly, self-report surveys enable researchers to explore the attitudes, beliefs, motivations, and personality characteristics of offenders.

Self-report surveys can be either cross-sectional or longitudinal. A **cross-sectional survey** is one conducted at a single point in time. A cross-sectional design offers researchers a glimpse of a cross section or "slice" of the population at a particular time. The Richmond Youth Study, described earlier in this chapter, included a cross-sectional self-report survey used to explore the relationship between informal social control and delinquency.[113]

Researchers must be extremely cautious when using cross-sectional data to try to draw conclusions about causal order or change. A longitudinal research design is better suited to questions of causal order and change. A **longitudinal survey** gathers information from the same individuals at more than one point in time. The National Youth Survey, in which researchers gathered information from respondents at seven different points in time from 1977 to 1986, is an example of a longitudinal survey.

Longitudinal surveys are relatively expensive and complex to implement compared to cross-sectional surveys. Yet longitudinal designs have become popular among researchers interested in processes of change and in crime and delinquency over the life course. As you will see later in this chapter, however, their popularity has not gone unchallenged. Researchers have also discovered problems unique to longitudinal designs. Janet Lauritsen, for example, used data from the NYS to explore the value of longitudinal self-report data for studying the age–crime relationship.[114] She found "testing effects" indicating that the repeated interviews themselves had effects on respondents' answers to questions at later "waves" of data collection.

cross-sectional survey A self-report survey conducted at a single point in time. A cross-sectional research design provides a glimpse of a cross section of the population at a particular time.

longitudinal survey A self-report survey that gathers information from the same individuals at more than one point in time. A longitudinal research design is better suited than a cross-sectional design to address questions of causal order and change.

Comparing Self-Report Data with UCR and NCVS Data

Given that UCR data represent only crimes known to police, it is not surprising that self-report surveys reveal much higher rates of offending than those suggested by UCR data. O'Brien points out, however, that it is difficult to compare UCR and self-report data—even data from the nationally representative National Youth Survey—because of the limited age range of self-report data.[115] With that caveat in mind, O'Brien notes that the NYS in 1980 revealed an aggravated assault rate of 1,400 per 10,000 (for 15–21 year-olds), whereas the NCVS and UCR for 1980 showed rates of 90.3 and 29.9 per 10,000, respectively, for the same age group. "Clearly, the NCVS and UCR rates would be much higher if they were based only on those aged 15 to 21, but they would not be fifteen to sixteen times higher (1,400/90.3)."[116] O'Brien reports that, with regard to trends in crime rates over time, self-report data converge with those from the UCR and NCVS for some types of offenses.[117]

Though the three data sources show substantial differences in crime rates for some offenses, all three reveal similar patterns in the demographic characteristics of those who commit serious crimes. "For serious crimes, all three data sources indicate that, in relation to their proportions in the population, males offend substantially more than do females, and African-Americans offend more than do whites. All three measures indicate a far greater rate of street crime committed by youth than by middle-aged or elderly persons."[118] (See Chapter 5 for a discussion of the social characteristics of offenders.)

In a 2003 study, David Farrington and his colleagues compared conclusions about delinquency careers derived from court referrals with those derived from self-reports.[119] They used longitudinal data from the Seattle Social Development Project, which began in 1985 and included both self-report and court referral data for 808 youths from age 11 to age 17. The two data sources agreed about some aspects of delinquency careers and disagreed about others. Self-reports and court referrals were consistent in showing that the proportion of the sample involved in offending increased during the juvenile years and that the younger the age at which an individual began offending, the larger the number of offenses committed.[120] When compared to court referral data, however, self-reports showed a higher proportion of the sample involved in offending, a higher frequency of offending among those who committed offenses, offending beginning at younger ages, and less continuity in offending over time.[121] These results led Farrington and his colleagues to conclude that "criminal career research based on self-reports sometimes yields different conclusions compared with research based on official records."[122]

Edward Wells and Joseph Rankin recently compared self-report and NCVS data to test the effectiveness of the NCVS in assessing juvenile victimizations. They examined NCVS data and self-report victimization data from the National Youth Survey and the Monitoring the Future study, both national samples of young people. Wells and Rankin matched the three surveys as closely as possible in terms of time covered by the survey (1978), age ranges of respondents (13–19 years old), and wording and response format of questions. Their results showed remarkably clear patterns. First, the National Youth Survey and Monitoring the Future study revealed strikingly similar estimates of juvenile victimization for both violent and property offenses. Second, the victimization levels suggested by the NCVS diverge substantially and consistently from those revealed by the National Youth Survey and Monitoring the Future. The self-report surveys both indicated much higher victimization rates than did the NCVS.[123]

Wells and Rankin contend that the large discrepancies between victimization and self-report surveys in estimates of criminal victimization of juveniles cannot be explained by differences in the technical designs of the surveys or by measurement or sampling error. "Rather, they may reflect more substantial differences in the social dynamics of the interviews by which the survey data are elicited from adolescent respondents."[124] Whatever the reasons for the discrepancies, the results of this study indicate that the NCVS underestimates juvenile victimization.[125]

Strengths of Self-Report Data

The primary strength of self-report data is that they offer information about the delinquent acts of those who have not been arrested or officially processed, and thus they provide a broader and less biased picture of delinquency and crime in America. Self-reports also provide data about relatively minor forms of offending and drug and alcohol use. These behaviors are often omitted from official data because more serious offenses are most likely to lead to arrest. Because they capture both minor and serious forms of offending, apart from the filter of criminal justice system processing, self-reports reveal far more delinquency than do official data.

Self-report data also provide researchers with opportunities to consider questions they would be unable to address with either official or victimization data. In addition to questions about involvement in delinquency and drug use, self-report surveys may include questions regarding respondents' attitudes and beliefs about law violation; perceptions of opportunities for success in school or work; and family interactions, such as parent–child attachment, parental supervision, and disciplinary practices. Answers to these types of questions are needed by researchers interested in testing various theories about the causes of crime and delinquency. Researchers have great flexibility in designing self-report survey instruments that will provide them with the information they need to test the theoretical perspectives of interest to them.

Weaknesses of Self-Report Data

Scholars have raised several concerns about self-report data on crime and delinquency. Among the more serious concerns is a question about the expectation that offenders will candidly report their involvement in crime and delinquency—particularly in offenses that have gone undetected by police. Some respondents may intentionally underreport their involvement in offending or may simply forget some of the offenses they have committed. The problem of recall may be worse for those who have committed the most offenses. Other respondents may exaggerate or overreport their involvement in offending.

A second weakness of self-report surveys concerns sampling design and its potential effects on survey responses. In many self-report studies, samples of respondents are drawn from student populations in school settings, and surveys are often administered through the schools. Obviously, in these cases, individuals not attending school—including truants, dropouts, and institutionalized youth—are excluded from the samples. Thus, the individuals excluded are also those most likely to be delinquent. This undersampling of serious delinquents may result in underestimation of the true amount of delinquency and in an incomplete or inaccurate picture of the characteristics of offenders.[126]

The selection of samples of respondents in school settings also results in samples that are fairly homogeneous in many respects.[127] For example, a researcher who draws a sample from a single high school should not expect a great deal of variation among respondents in social class, as students who reside in the same community or neighborhood probably have relatively similar family incomes. The problem with fairly homogeneous samples is that they prevent researchers from exploring the full range of possible responses to some questions of interest. For example, suppose a researcher is interested in the relationship between social class and delinquency, and draws a sample of respondents from a school in a lower-class neighborhood. That researcher is unlikely to be able to conclude much about the offending of upper-class individuals because the responses to the survey questions about social class will likely be "truncated" or limited mostly to those indicating lower-class status.[128] In addition, samples selected in school settings allow researchers to examine involvement in juvenile delinquency but offer no information on adult crime.

A third criticism of self-report surveys is that some fail to capture the full range of delinquent behaviors and focus instead on relatively minor offenses, such as underage drinking, truancy, and petty theft. Minor offenses may be more difficult for offenders to remember than serious offenses, and thus the emphasis on minor offenses is likely to increase the problem of recall error.[129] The exclusion of serious offenses also limits researchers' abilities to address theoretically significant questions and to compare self-reports of offending to official or victimization data, which better represent serious crime and delinquency.

While the exclusion of serious offenses from self-reports is a legitimate criticism, not all self-report surveys are subject to this problem. The National Youth Survey, for example, was the first self-report survey to incorporate questions about serious offenses, such as rape, assault, and the sale of hard drugs. The volume of serious delinquency revealed by NYS respondents indicates that, even in face-to-face interviews, respondents were willing to report serious delinquent acts they had committed.

Finally, self-report surveys have been criticized for reasons of both validity and reliability. **Validity** refers to the extent to which "a measurement instrument . . . measures what it is supposed to measure."[130] For example, questions about the number of times a respondent has stolen something or damaged someone else's property would be valid measures of delinquency. Questions about lying to parents or arguing with teachers, however, would not be valid measures of delinquent behavior, because even though lying and school conflicts might be related to delinquency, they are not considered delinquent acts themselves. The validity of measures is compromised to the extent that respondents systematically underreport or overreport their delinquent acts. Validity is also threatened when respondents have limited recall or forget past behaviors or when they misunderstand survey questions. For example, if we

validity The extent to which "a measurement instrument . . . measures what it is supposed to measure" (Vogt, *Dictionary*, 240).

ask survey respondents to report the number of times they engaged in delinquency in the past two years, the time frame is likely to be too long for them to be able to offer accurate responses, and so the measure will not be valid.

To assess the validity of self-reports of delinquency, researchers have compared responses from self-report surveys to official police records. Michael Hindelang and his colleagues, in their extensive study of the measurement of delinquency, examined the level of consistency between self-reports of delinquency and official records of involvement in delinquent behavior.[131] They found relatively strong relationships between official records and self-reported delinquency, indicating that those most likely to report involvement in delinquent behavior were also those most likely to have official records of offending. Hindelang and his colleagues concluded that most self-report surveys provide valid measures of delinquent behavior.[132] They also discovered, however, that self-reports of delinquency are less valid for black males, who tend to underreport their involvement in serious offenses.[133]

Reliability refers to the extent to which repeated measurements of a variable produce the same or similar responses over time. Suppose, for example, an individual takes an IQ test once each year for three years. She scores 132 on the first test, 102 on the second test, and 153 on the third. These three tests would not be considered reliable measures of IQ because of the great variation in scores over time.

reliability The extent to which repeated measurements of a variable produce the same or similar responses over time.

Reliability is a concern with self-report surveys for two reasons. First, in longitudinal surveys, repeated measurement itself can affect responses, leading to a decline in the reporting of delinquency at later "waves" of data collection and therefore to unreliable estimates of the true amount of delinquent behavior.[134] Recall Janet Lauritsen's work, cited earlier in this chapter, on testing effects of repeated measurement in longitudinal self-report surveys.[135]

Second, reliability is related to sample size and random error in responses (or differences due to chance). The smaller one's sample size, the greater the random error and the less reliable the measures. Most self-report surveys are conducted with relatively small samples, at least compared to official data or victimization surveys. In addition, serious offenses are relatively rare events, so respondents in a typical self-report survey are unlikely to report involvement in many serious delinquent acts. Thus, small sample sizes, combined with the rarity of serious crime, increase the random error associated with measures of serious offending and decrease the reliability of such measures in self-report surveys.

Although researchers must be mindful of reliability issues when using self-report data, Michael Hindelang and his colleagues found that reliability poses less cause for concern than does validity. They found "impressive" measures of reliability for the self-report survey items they explored.[136]

■ Methods, Theory, and Policy

In this chapter, we have introduced several sources of data and research methods. Which methodology is best suited for addressing questions about crime and delinquency? The answer depends on the question at hand and the theory underlying it. A researcher's theoretical perspective suggests a particular methodological or research strategy. One must evaluate the adequacy of any research design in terms of its relevance to questions of theory and policy.[137]

A relatively recent debate in criminology illustrates the integral relationships among methodology, theory, and policy. The debate concerns the value of the "criminal career" paradigm and of longitudinal data. It was sparked in 1986 by a report in which Alfred Blumstein and his colleagues expressed the value of studying criminal careers and the necessity of longitudinal data to do so.[138] (We discuss criminal careers in detail in Chapter 6.) A **criminal career** is defined as "the longitudinal sequence of offenses committed by an offender who has a detectable rate of offending during some period."[139] Several aspects of criminal careers are important:

- ages at which offending begins and ends ("age of onset" and "age of desistance")
- duration of "career"
- frequency of offending ("incidence" and "lambda")
- seriousness of offending
- proportion of the population involved in offending at any given time ("prevalence")

In the study of criminal careers, researchers recognize the possibility that different causal factors may account for different elements of a criminal career.[140] As Blumstein and his colleagues write, "different sets of 'causes' may influence individuals' initiation of criminal activity, the frequency with which they commit crimes, the types of crimes they commit, and their termination of criminal activity."[141] This perspective suggests that, to understand all aspects of involvement in crime, researchers must gather data throughout offenders' criminal careers. In other words, exploration of the distinct aspects of criminal careers "requires" longitudinal data that provide information about individuals over a period of time.

The study of criminal careers enables researchers to determine the existence of **career criminals**, characterized by "some combination of a high frequency of offending, a long duration of the criminal career, and high seriousness of offenses committed."[142] It is important to note that *criminal careers* and *career criminals* are distinct concepts. Blumstein and his colleagues clearly see value in using longitudinal data to try to identify career criminals early in their criminal careers and in developing incarceration policies based on this belief in the existence of career criminals or chronic offenders.[143]

On the other side of the debate stand Michael Gottfredson and Travis Hirschi, who argue that the concept of criminal careers is without merit, that the expenditure of funds to study criminal careers is unjustified, and that longitudinal data (crucial to the study of criminal careers) are unnecessary in the study of causes of crime and delinquency.[144] Gottfredson and Hirschi derive these conclusions from their views about the invariance of the age–crime curve, and from their *general theory of crime,* which argues that *stable* characteristics of individuals lead to criminal involvement, and that various aspects of criminal careers (e.g., prevalence and incidence of crime) are influenced by the same factors.[145]

criminal career "The longitudinal sequence of offenses committed by an offender who has a detectable rate of offending during some period" (Blumstein, Cohen, and Farrington, "Criminal Career Research," 2).

career criminals Individuals characterized by "some combination of a high frequency of offending, a long duration of the criminal career, and high seriousness of offenses committed" (Blumstein, Cohen, and Farrington, "Criminal Career Research," 22).

(See Chapter 5 for a discussion of the age–crime curve and Chapter 10 for coverage of this general theory of crime.)

According to Gottfredson and Hirschi's general theory, as criminal propensity increases, aspects of criminal careers such as frequency of offending, duration of criminal career, and age of termination of offending also tend to increase. If this theory is correct, then researchers need not study all aspects of criminal careers, because the factors related to all aspects are the same. In addition, Gottfredson and Hirschi argue that criminal propensity peaks in the middle to late teens and then declines rapidly.[146] They also contend that this relationship between age and crime exists for all offenders—even the most frequent and serious offenders. Thus, Gottfredson and Hirschi view the search for "career criminals" or "chronic offenders" as futile because they do not believe that any segment of the criminal population is characterized by a high rate of offending that does not decline with age. Based on their theoretical perspective, Gottfredson and Hirschi contend that longitudinal data are unnecessary in the study of crime because an understanding of an offender's criminal propensity at any point in time is sufficient to understand it at *all* points in time.

The debate between Blumstein and his colleagues and Gottfredson and Hirschi is relevant not only to the relationship between theory and method, but also to the issue of public policy on crime control. By viewing participation in offending and frequency of offending as separate elements of criminal careers, Blumstein and his colleagues see two ways to decrease crime: (1) reduce the proportion of the population committing crime (i.e., reduce participation or prevalence), or (2) reduce the rate at which active offenders commit crime (i.e., reduce frequency or incidence). "The first approach relates to a general social policy of developing prevention strategies directed at the total population; the second approach relates more narrowly to the identification of effective treatment or control alternatives for those who have begun to commit crimes."[147]

Consistent with the second approach, Blumstein and his colleagues advocate for the possibility of **selective incapacitation**—a policy consistent with their view of the existence of career criminals or chronic offenders. A policy of selective incapacitation would involve the incarceration of offenders identified as having high individual rates of offending. Ideally, such identification, based in part on prior criminal record, would occur relatively early in offenders' criminal careers, so that selective incapacitation might achieve its maximum crime reduction potential.

Gottfredson and Hirschi view the policy of selective incapacitation as both unwise and impossible to achieve. They maintain that chronic offenders cannot be identified early enough in their criminal careers for selective incapacitation to be a feasible policy. Moreover, they contend that selective incapacitation strategies "cannot simply duplicate existing criminal justice practices," which base incarceration decisions on the nature of the current offense and prior criminal record.[148] Yet, according to Gottfredson and Hirschi, one must question the legitimacy of using information beyond that presently employed in the criminal justice system (current offense and prior record) to make incarceration decisions.

It is beyond the scope of this discussion to bring any resolution to the heated debate about the value of the criminal career paradigm and of longitudinal data. We have briefly summarized the debate here only to give the reader a sense of

selective incapacitation
A policy involving the incarceration of offenders identified as having high individual rates of offending. A selective incapacitation policy is consistent with a belief in the existence of "career criminals" or "chronic offenders."

the vital role that theory plays in determining appropriate research strategy. One's choice of research method cannot stand apart from considerations of theory. The debate between Blumstein and his colleagues and Gottfredson and Hirschi illustrates how tightly theory, method, and policy are interwoven.

■ Summary and Conclusions

Delinquency research is used both to develop theory and to test it. We began this chapter with a summary of the deductive research process (for testing theory) and of the criteria for establishing cause and effect. Awareness of both of these issues will help in understanding the theories and research we present in later chapters. We then examined a variety of research methods and data sources used to study crime and delinquency. These methods include comparison of offenders and non-offenders, case studies, ethnography, ecological analysis, and survey analysis.

The Gluecks used longitudinal data from multiple sources to compare 500 delinquent boys with 500 nondelinquent boys. They matched these two groups on a case-by-case basis and discovered that family life was a crucial factor distinguishing delinquents from nondelinquents. Sutherland used case studies and the method of analytic induction in his attempt to arrive at a scientific generalization for explaining crime. In this seven-step method, one begins with the definition of the behavior to be explained and ends with examination of cases that either support or disprove the hypothesis.

Social scientists have also used ethnography to study deviance and crime. Ethnographic studies involve sustained interaction between the researcher—often acting as a participant–observer—and those whom he or she is studying. Ethnographers attempt to provide a detailed account of interactions and social processes as they naturally occur. In well-known research, Anderson uses ethnography to explore the "code of the street" governing youth violence in the inner city.

Ecological analysis helps researchers understand the geographic distribution of crime and delinquency. In classic research, Shaw and McKay used official data to plot the home addresses of Chicago delinquents and explore the characteristics of communities with the highest delinquency rates. In contemporary research, Sampson and his colleagues have revived interest in community-level factors that contribute to high crime rates in some geographic areas.

Survey analysis is perhaps the most widely used method in research on crime and delinquency. Surveys often provide detailed information about respondents' behaviors, attitudes, and beliefs. Many criminologists have used surveys to gather data on the frequency and distribution of delinquent behavior and the causes and correlates of offending. We presented the Richmond Youth Study, the National Youth Survey, and the Community Survey of the Project on Human Development in Chicago Neighborhoods as examples of survey analysis by noted criminologists.

In this chapter, we also examined the three primary sources of data on crime and delinquency: agencies within the criminal justice system that provide official data, victimization surveys, and self-report surveys of offenders. We discussed the strengths and weaknesses of each of these data sources and tried to convey how each is best suited to address particular types of questions about crime and offenders. When evaluating data on crime, delinquency, and offenders, it is important to keep in mind the strengths and weaknesses of the data source and to consider carefully what that source is designed to tell us about crime and delinquency.

We closed the chapter with a discussion of the integral relationships among theory, methodology, and policy. One's theoretical perspective prompts one to choose particular research strategies. We can evaluate the adequacy of a given strategy only in terms of its capacity to test a particular theory. We illustrated the relationships among theory, methodology, and policy by exploring the recent criminological debate about the value of the criminal career paradigm.

CRITICAL THINKING QUESTIONS

1. Why is it necessary to demonstrate association, temporal order, and lack of spuriousness to establish a cause-and-effect relationship?

2. How would you determine which research methodology (for example, case study, ethnography, statistical analysis of UCR or victimization data) is best suited for a particular research question?

3. Identify the strengths and weaknesses of each of the three major sources of data on crime and delinquency: official data, victimization survey data, and self-report survey data. What is each source designed to tell us?

4. What is the problem with relying on UCR data to examine the crime rate in America?

5. Suppose you want to understand the nature and extent of the drug trade in the Robert Taylor housing project in Chicago. What research methodology would you use to study this problem and why?

6. Explain the relationship between methodology and theory. How does theory influence one's choice of research method? How might methodology influence the development of theory?

SUGGESTED READINGS

Anderson, Elijah. "Violence and the Inner-City Street Code." In *Childhood and Violence in the Inner City*, edited by Joan McCord, 1–30. New York: Cambridge University Press, 1997.

Hindelang, Michael J., Travis Hirschi, and Joseph G. Weis. *Measuring Delinquency*. Beverly Hills, CA: Sage, 1981.

Mosher, Clayton J., Terance D. Miethe, and Dretha M. Phillips. *The Mismeasure of Crime*. Thousand Oaks, CA: Sage, 2002.

O'Brien, Robert M. "Crime Facts: Victim and Offender Data." In *Criminology: A Contemporary Handbook*, 3rd ed., edited by Joseph F. Sheley, 59–83. Belmont, CA: Wadsworth, 2000.

GLOSSARY

analytic induction: A research method, developed by Alfred Lindesmith, used to arrive at a scientific generalization for explaining behavior. Analytic induction consists of seven steps, beginning with the definition of the behavior to be explained and ending with the examination of cases used to test the hypothesis.

career criminals: Individuals characterized by "some combination of a high frequency of offending, a long duration of the criminal career, and high seriousness of offenses committed" (Blumstein, Cohen, and Farrington, "Criminal Career Research," 22).

crime rates: Show the number of offenses per a specific portion of the population, such as the number of violent Index offenses per 100,000 individuals.

crime trends: Show crime rates over time, and the percent change in rates from one point in time to another.

criminal career: "The longitudinal sequence of offenses committed by an offender who has a detectable rate of offending during some period" (Blumstein, Cohen, and Farrington, "Criminal Career Research," 2).

cross-sectional survey: A self-report survey conducted at a single point in time. A cross-sectional research design provides a glimpse of a cross section of the population at a particular time.

ecological analysis: A research method used to explore the geographic distribution of crime and delinquency, and the social conditions that characterize areas with high rates of crime and delinquency. In ecological analysis, geographic areas, such as neighborhoods or cities, rather than the individuals who reside in them, are the units of analysis.

ethnography: A research method that involves "describing social or cultural life based on direct, systematic observation, such as becoming a participant in a social system" (Vogt, *Dictionary*, 83).

longitudinal survey: A self-report survey that gathers information from the same individuals at more than one point in time. A longitudinal research design is better suited than a cross-sectional design to address questions of causal order and change.

National Incident-Based Reporting System: Provides more detailed "official data" on criminal incidents than the Uniform Crime Reporting program does. NIBRS resulted from revisions to the UCR program, and, like the UCR, gathers information from law enforcement agencies across the United States.

reliability: The extent to which repeated measurements of a variable produce the same or similar responses over time.

selective incapacitation: A policy involving the incarceration of offenders identified as having high individual rates of offending. A selective incapacitation policy is consistent with a belief in the existence of "career criminals" or "chronic offenders."

survey research: A research method involving the administration of a survey instrument (questionnaire or interview) to a sample of respondents, coding of data gathered, statement of hypotheses, and analysis of data in order to make inferences from the sample to the population from which it was drawn.

Uniform Crime Reporting program: Provides "official data" on crime and delinquency, voluntarily reported by over 17,000 law enforcement agencies across the United States, and compiled by the Federal Bureau of Investigation. These data reveal the extent of crime and delinquency with which the reporting agencies deal, and the characteristics of offenses and offenders whom they encounter.

validity: The extent to which "a measurement instrument . . . measures what it is supposed to measure" (Vogt, *Dictionary*, 240).

REFERENCES

Agnew, Robert. *Juvenile Delinquency: Causes and Control.* Los Angeles: Roxbury, 2002.

Anderson, Elijah. *Code of the Street: Decency, Violence, and the Moral Life of the Inner City.* New York: Norton, 1999.

———. "Going Straight: The Story of a Young Inner-City Ex-Convict." *Punishment and Society* 3 (2001):135–152.

———. "Jelly's Place: An Ethnographic Memoir." *Symbolic Interaction* 26 (2003):217–237.

———. *A Place on the Corner.* Chicago: University of Chicago Press, 1978.

———. "The Social Ecology of Youth Violence." *Crime and Justice* 24 (1998):65–104.

———. *Streetwise: Race, Class, and Change in an Urban Community.* Chicago: University of Chicago Press, 1990.

———. "Violence and the Inner-City Street Code." In *Childhood and Violence in the Inner City,* edited by Joan McCord, 1–30. New York: Cambridge University Press, 1997.

Babbie, Earl. *The Practice of Social Research,* 8th ed. Belmont, CA: Wadsworth, 1998.

Becker, Howard S. *Outsiders: Studies in the Sociology of Deviance.* New York: Free Press, 1963.

Blumstein, Alfred, Jacqueline Cohen, and David P. Farrington. "Criminal Career Research: Its Value for Criminology." *Criminology* 26 (1988):1–35.

———. "Longitudinal and Criminal Career Research: Further Clarifications." *Criminology* 26 (1988):57–74.

Blumstein, Alfred, Jacqueline Cohen, and Richard Rosenfeld. "Trend and Deviation in Crime Rates: A Comparison of UCR and NCS Data for Burglary and Robbery." *Criminology* 29 (1991):237–263.

Blumstein, Alfred, Jacqueline Cohen, Jeffrey A. Roth, and Christy A. Visher. *Criminal Careers and "Career Criminals."* Vol. I. Washington, DC: National Academy Press, 1986.

Bureau of Justice Statistics. *Criminal Victimization in the United States, 1995: A National Crime Victimization Survey Report.* Washington, DC: US Department of Justice, 2000.

Byrne, James M., and Robert J. Sampson. *The Social Ecology of Crime.* New York: Springer-Verlag, 1986.

Catalano, Shannan M. *Criminal Victimization, 2003.* Washington, DC: US Department of Justice, Bureau of Justice Statistics, 2004.

Coughlin, Brenda C., and Sudhir Alladi Venkatesh. "The Urban Street Gang after 1970." *Annual Review of Sociology* 29 (2003):41–64.

Cressey, Donald R. *Other People's Money: A Study of the Social Psychology of Embezzlement.* Glencoe, IL: Free Press, 1953.

Decker, Scott H. "Collective and Normative Features of Gang Violence." In Pope, Lovell, and Brandl, *Voices from the Field,* 160–181.

Elliott, Delbert S. *National Youth Survey (United States): Wave III, 1978.* 2nd ICPSR Edition Codebook. Ann Arbor, MI: Inter-University Consortium for Political and Social Science Research, 1988.

Elliott, Delbert S., David Huizinga, and Suzanne S. Ageton. *Explaining Delinquency and Drug Use.* Beverly Hills, CA: Sage, 1985.

Elliott, Delbert, David Huizinga, and Scott Menard. *Multiple Problem Youth: Delinquency, Substance Abuse and Mental Health.* New York: Springer-Verlag, 1989.

Empey, LaMar T., Mark C. Stafford, and Carter H. Hay. *American Delinquency: Its Meaning and Construction.* 4th ed. Belmont, CA: Wadsworth, 1999.

Farrington, David P., Darrick Jolliffe, David J. Hawkins, Richard F. Catalano, Karl G. Hill, and Rick Kosterman. "Comparing Delinquency Careers in Court Records and Self-Reports." *Criminology* 41 (2003):933–958.

Farrington, David P., Rolf Loeber, Magda Stouthamer-Loeber, Welmoet B. Van Kammen, and Laura Schmidt. "Self-Reported Delinquency and a Combined Delinquency Seriousness Scale Based on Boys, Mothers, and Teachers: Concurrent and Predictive Validity for African-Americans and Caucasians." *Criminology* 34 (1996):493–517.

Federal Bureau of Investigation. *Crime in the United States 1999: Uniform Crime Reports.* US Department of Justice. Washington, DC: GPO, 2000.

———. *Crime in the United States 2003: Uniform Crime Reports.* US Department of Justice. Washington, DC: GPO, 2004.

Ferrell, Jeff, and Mark S. Hamm, eds. *Ethnography at the Edge: Crime, Deviance, and Field Research.* Boston: Northeastern University Press, 1998.

Fowler, Floyd J., Jr., and Thomas W. Mangione. *Standardized Survey Interviewing: Minimizing Interviewer-Related Error.* Newbury Park, CA: Sage, 1990.

Gaylord, Mark S., and John F. Galliher. *The Criminology of Edwin Sutherland.* New Brunswick, NJ: Transaction Books, 1988.

Glueck, Sheldon, and Eleanor Glueck. *Delinquents and Nondelinquents in Perspective.* Cambridge, MA: Harvard University Press, 1968.

———. *Predicting Delinquency and Crime.* Cambridge, MA: Harvard University Press, 1967.

———. *Unraveling Juvenile Delinquency.* Cambridge, MA: Harvard University Press, 1950.

Gottfredson, Michael R., and Travis Hirschi. *A General Theory of Crime.* Stanford, CA: Stanford University Press, 1990.

———. "The Methodological Adequacy of Longitudinal Research on Crime." *Criminology* 25 (1987):581–614.

———. "Science, Public Policy, and the Career Paradigm." *Criminology* 26 (1988):37–55.

———. "Self-Control and Opportunity." In *Control Theories of Crime and Delinquency,* edited by Chester L. Britt and Michael R. Gottfredson, 5–19. New Brunswick, NJ: Transaction, 2003.

———. "The True Value of Lambda Would Appear To Be Zero: An Essay on Career Criminals, Criminal Careers, Selective Incapacitation, Cohort Studies, and Related Topics." *Criminology* 24 (1986):213–234.

Hagedorn, John. "Back in the Field Again: Gang Research in the Nineties." In *Gangs in America,* edited by C. Ronald Huff, 240–259. Newbury Park, CA: Sage, 1991.

———. *People and Folks.* Chicago: Lakeview Press, 1988.

Harris, Anthony R., and James W. Shaw. "Looking for Patterns: Race, Class, and Crime." In *Criminology,* edited by Joseph F. Sheley, 129–161. Belmont, CA: Wadsworth, 2000.

Hashima, Patricia Y., and David Finkelhor. "Violent Victimization of Youth Versus Adults in the National Crime Victimization Survey." *Journal of Interpersonal Violence* 14 (1999):799–820.

Hindelang, Michael J., Travis Hirschi, and Joseph G. Weis. "Correlates of Delinquency: The Illusion of Discrepancy Between Self-Report and Official Measures." *American Sociological Review* 44 (1979):995–1014.

———. *Measuring Delinquency.* Beverly Hills, CA: Sage, 1981.

Hirschi, Travis. *Causes of Delinquency.* Berkeley, CA: University of California Press, 1969.

Hirschi, Travis, and Michael R. Gottfredson. "Age and the Explanation of Crime." *American Journal of Sociology* 89 (1983):552–584.

———. "In Defense of Self-Control." *Theoretical Criminology* 4 (2000):55–69.

Hirschi, Travis, and Hanan C. Selvin. *Delinquency Research: An Appraisal of Analytic Methods.* New York: Free Press, 1967.

Jankowski, Martin Sanchez. "Ethnography, Inequality, and Crime in the Low-Income Community." In *Crime and Inequality,* edited by John Hagan and Ruth D. Peterson, 80–94. Stanford, CA: Stanford University Press, 1995.

Kleck, Gary. "On the Use of Self-Report Data to Determine the Class Distribution of Criminal and Delinquent Behavior." *American Sociological Review* 47 (1982):427–438.

Laub, John, and Robert Sampson. "The Sutherland-Glueck Debate: On the Sociology of Criminological Knowledge." *American Journal of Sociology* 96 (1991):1402–1440.

———. "Unraveling Families and Delinquency: A Reanalysis of the Gluecks' Data." *Criminology* 26 (1988):355–380.

Lauritsen, Janet L. "The Age–Crime Debate: Assessing the Limits of Longitudinal Self-Report Data." *Social Forces* 77 (1998):127–155.

Lindesmith, Alfred R. *Opiate Addiction.* Bloomington, IL: Principia Press, 1947.

Menard, Scott, and Herbert C. Covey. "UCR and NCS: Comparisons Over Space and Time." *Journal of Criminal Justice* 16 (1988):371–384.

Messner, Steven F. "The 'Dark Figure' and Composite Indices of Crime: Some Empirical Explorations of Alternative Data Sources." *Journal of Criminal Justice* 12 (1984):435–444.

Moore, Joan. *Homeboys.* Philadelphia: Temple University Press, 1978.

O'Brien, Robert M. "Comparing Detrended UCR and NCS Crime Rates Over Time: 1973–1986." *Journal of Criminal Justice* 18 (1990):229–238.

———. *Crime and Victimization Data.* Beverly Hills, CA: Sage, 1985.

———. "Crime Facts: Victim and Offender Data." In *Criminology: A Contemporary Handbook,* 3rd ed., edited by Joseph F. Sheley, 59–83. Belmont, CA: Wadsworth, 2000.

———. "Police Productivity and Crime Rates: 1973 1992." *Criminology* 34 (1996):183–207.

Padilla, Felix. *The Gang as an American Enterprise.* New Brunswick, NJ: Rutgers University Press, 1992.

Pastore, Ann L., and Kathleen Maguire, eds. *Sourcebook of Criminal Justice Statistics* 1999. US Department of Justice, Bureau of Justice Statistics. Washington, DC: GPO, 2000.

Piquero, Alex R., David P. Farrington, and Alfred Blumstein. "The Criminal Career Paradigm." *Crime and Justice* 30 (2003):359–506.

Platt, Anthony. *The Child Savers: The Invention of Delinquency.* 2nd ed. Chicago: University of Chicago Press, 1977.

Pope, Carl E., Rick Lovell, and Steven G. Brandl, eds. *Voices from the Field: Readings in Criminal Justice Research.* Belmont, CA: Wadsworth, 2001.

Puzzanchera, Charles, Anne L. Stahl, Terrence A. Finnegan, Nancy Tierney, and Howard N. Snyder. *Juvenile Court Statistics 2000.* US Department of Justice. Washington, DC: Office of Juvenile Justice and Delinquency Prevention, 2004.

Rantala, Ramona R., and Thomas J. Edwards. *Effects of NIBRS on Crime Statistics.* Bureau of Justice Statistics Special Report. US Department of Justice. Washington, DC: GPO, 2000.

Sampson, Robert J. "Crime and Public Safety: Insights from Community-Level Perspectives on Social Capital." In *Social Capital and Poor Communities,* edited by Susan Saegert, J. Phillip Thompson, and Mark R. Warren, 89–114. New York: Russell Sage Foundation, 2001.

———. "Local Friendship Ties and Community Attachment in Mass Society: A Multilevel Systemic Model." *American Sociological Review* 53 (1988):766–779.

———. "Neighborhood and Crime: The Structural Determinants of Personal Victimization." *Journal of Research in Crime and Delinquency* 22 (1985):7–40.

———. "Neighborhood Family Structure and the Risk of Personal Victimization." In *The Social Ecology of Crime,* edited by James M. Byrne and Robert J. Sampson, 25–46. New York: Springer-Verlag, 1986.

————. "Organized for What? Recasting Theories of Social Dis(Organization)." In *Crime and Social Organization,* edited by Elin Waring and David Weisburd, 95–110. New Brunswick, NJ: Transaction, 2002.

————. "Whither the Sociological Study of Crime?" *Annual Review of Sociology* 26 (2000):711–714.

Sampson, Robert J., and W. Byron Groves. "Community Structure and Crime: Testing Social-Disorganization Theory." *American Journal of Sociology* 94 (1989):774–802.

Sampson, Robert J., and John H. Laub. *Crime in the Making: Pathways and Turning Points Through Life.* Cambridge, MA: Harvard University Press, 1993.

Sampson, Robert J., Jeffrey D. Morenoff, and Felton Earls. "Beyond Social Capital: Spatial Dynamics of Collective Efficacy for Children." *American Sociological Review* 64 (1999):633–660.

Sampson, Robert J., Jeffrey D. Morenoff, and Thomas Gannon-Rowley. "Assessing 'Neighborhood Effects': Social Processes and New Directions in Research." *Annual Review of Sociology* 28 (2002):443–478.

Sampson, Robert J., Stephen W. Raudenbush, and Felton Earls. "Neighborhoods and Violent Crime: A Multilevel Study of Collective Efficacy." *Science* 277 (1997):918–924.

Schmid, Thomas J., and Richard S. Jones. "Ambivalent Actions: Prison Adaptation Strategies of First-Time, Short-Term Inmates." In Pope, Lovell, and Brandl, *Voices from the Field,* 182–201.

Sealock, Miriam, and Sally Simpson. "Unraveling Bias in Arrest Decisions: The Role of Juvenile Offender Typescripts." *Justice Quarterly* 15 (1998):427–457.

Shaw, Clifford R. *The Jack-Roller: A Delinquent Boy's Own Story.* Chicago: University of Chicago Press, 1930.

————. *The Natural History of a Delinquent Career.* Philadelphia, PA: Saifer, 1931.

Shaw, Clifford R., and Henry D. McKay. *Juvenile Delinquency and Urban Areas: A Study of Rates of Delinquency in Relation to Differential Characteristics of Local Communities in American Cities.* Rev. ed. Chicago: University of Chicago Press, 1969.

————. *Social Factors in Juvenile Delinquency: A Study of the Community, the Family, and the Gang in Relation to Delinquent Behavior.* Report of the National Commission on Law Observance and Enforcement, Causes of Crime, Volume II. Washington, DC: GPO, 1931.

Shaw, Clifford R., Frederick M. Zorbaugh, Henry D. McKay, and Leonard S. Cottrell. *Delinquency Areas: A Study of the Geographic Distribution of School Truants, Juvenile Delinquents, and Adult Offenders in Chicago.* Chicago: University of Chicago Press, 1929.

Short, James F., and Fred L. Strodtbeck. *Group Process and Gang Delinquency.* Chicago: University of Chicago Press, 1965.

Siegel, Larry J., and Joseph J. Senna. *Juvenile Delinquency: Theory, Practice, and Law.* 7th ed. Belmont, CA: Wadsworth, 2000.

Stark, Rodney. *Sociology.* 7th ed. Belmont, CA: Wadsworth, 1998.

Sutherland, Edwin H. *Principles of Criminology.* 3rd ed. Philadelphia: Lippincott, 1939.

Sutherland, Edwin H., and Donald R. Cressey. *Principles of Criminology.* 7th ed. Philadelphia: Lippincott, 1966.

Turner, A. G. *The San Jose Methods Test of Known Crime Victims.* National Criminal Justice Information and Statistics Service, Law Enforcement Assistance Administration. Washington, DC: GPO, 1972.

Vander Ven, Thomas M. "Fear of Victimization and the Interactional Construction of Harassment in a Latino Neighborhood." In Pope, Lovell, and Brandl, *Voices from the Field,* 141–159.

Venkatesh, Sudhir Alladi. "The Gang in the Community." In *Gangs in America,* 2nd ed., edited by C. Ronald Huff, 241–255. Thousand Oaks, CA: Sage, 1996.

————. "Gender and Outlaw Capitalism: A Historical Account of the Black Sisters United 'Girl Gang.' " *Signs* 23 (1998):683–709.

————. "The Social Organization of Street Gang Activity in an Urban Ghetto." *American Journal of Sociology* 103 (1997):82–111.

Venkatesh, Sudhir Alladi, and Steven D. Levitt. "'Are We a Family or a Business?' History and Disjuncture in the Urban American Street Gang." *Theory and Society* 29 (2000):427–462.

Vigil, Diego. *Barrio Gangs.* Austin: University of Texas Press, 1988.

Vogt, W. Paul. *Dictionary of Statistics and Methodology: A Nontechnical Guide for the Social Sciences.* Newbury Park, CA: Sage, 1993.

Weisberg, Herbert F., and Bruce D. Bowen. *An Introduction to Survey Research and Data Analysis.* San Francisco: W. H. Freeman and Co, 1977.

Chapter Resources

Wells, L. Edward, and Joseph Rankin. "Juvenile Victimization: Convergent Validation of Alternative Measurements." In Pope, Lovell, and Brandl, *Voices from the Field,* 267–287.

Woltman, H. F., and J. M. Bushery. *A Panel Bias Study in the National Crime Survey.* Proceedings of the Social Statistics Section of the American Statistical Association, 1975.

ENDNOTES

1. Research in Action, "The Theoretically Driven Research Process," draws heavily from Babbie, *Practice of Social Research,* 80–81. He summarizes these activities when he says, "Social research (in fact, all science) is organized around two activities: measurement and interpretation" (2).

2. Babbie, *Practice of Social Research,* 4.

3. Ibid.

4. Laub and Sampson, "Unraveling," 355.

5. Glueck and Glueck, *Unraveling Juvenile Delinquency.*

6. Sampson and Laub, *Crime in the Making,* 26. See Glueck and Glueck (*Unraveling Juvenile Delinquency,* Chapter 4) for a detailed description of the selection and matching of delinquents and nondelinquents.

7. Glueck and Glueck, *Unraveling Juvenile Delinquency,* Chapter 5, 41–53.

8. See Glueck and Glueck, *Unraveling Juvenile Delinquency,* Chapters 5 and 6, for a detailed description of the information gathered.

9. Glueck and Glueck, *Delinquents,* 1968.

10. Glueck and Glueck, *Unraveling Juvenile Delinquency,* Chapter 20, 257–271.

11. Glueck and Glueck, *Unraveling Juvenile Delinquency and Delinquents.* See also Glueck and Glueck, *Predicting Delinquency and Crime.*

12. Sampson and Laub, *Crime in the Making.* See also Laub and Sampson, "Unraveling."

13. Laub and Sampson, "Unraveling," and Sampson and Laub, *Crime in the Making,* (Chapter 2) provide thorough discussions of the criticisms of *Unraveling Juvenile Delinquency,* the Gluecks' best-known work. See also Laub and Sampson, "Debate," for an analysis of the contentious debate between the Gluecks and Edwin Sutherland, the author of differential association theory.

14. See Laub and Sampson, "Unraveling," 357–361, for a summary of criticisms of the Gluecks' research.

15. See Laub and Sampson, "Debate," for an assessment of the Gluecks' contributions to criminology.

16. Sutherland, *Principles of Criminology.*

17. Sutherland and Cressey, *Principles of Criminology,* 72.

18. Gaylord and Galliher, *Criminology,* 124–126.

19. Sutherland and Cressey, *Principles of Criminology,* 71–72.

20. Lindesmith, *Opiate Addiction.*

21. Cressey, *Other People's Money.*

22. Sutherland and Cressey, *Principles of Criminology,* 69–72.

23. Vogt, *Dictionary,* 83.

24. Anderson, "Violence," 4. See references list for Anderson's other works.

25. Jankowski, "Ethnography," 81.

26. Ferrell and Hamm, *Ethnography at the Edge,* 10. See also Becker, *Outsiders,* 165–176; Becker calls for ethnographic research as a means of achieving a clearer understanding of deviance.

27. Anderson, "Violence," 4.

28. Agnew, *Juvenile Delinquency,* 77.

29. Decker, "Collective." See Padilla, *American Enterprise;* Hagedorn, "Back" and *People and Folks;* Vigil, *Barrio Gangs;* and Moore, *Homeboys,* who also use ethnographic methods to examine gangs.

30. Schmid and Jones, "Ambivalent Actions."

31. Vander Ven, "Fear," 143.

32. Anderson, *Code of the Street,* "Social Ecology," and "Violence." See also Jankowski, "Ethnography;" Jankowski uses ethnography to study the relationship between crime and inequality in a low-income community.

33. Anderson, *Code of the Street,* 10.

34. Shaw and McKay, *Social Factors* and *Juvenile Delinquency;* and Shaw et al., *Delinquency Areas.*
35. Empey, Stafford, and Hay, *American Delinquency,* 142.
36. Shaw and McKay, *Juvenile Delinquency,* Chapter 3.
37. Ibid.
38. Ibid., 53.
39. Ibid., Chapter 6.
40. Shaw, *Jack-Roller* and *Natural History.*
41. Shaw and McKay, *Juvenile Delinquency,* 43.
42. See, for example, Sampson, "Neighborhood and Crime," "Neighborhood Family Structure," "Friendship Ties," "Crime and Public Safety," and "Organized for What?"; Byrne and Sampson, *Social Ecology of Crime;* Sampson and Groves, "Community Structure and Crime;" Sampson, Raudenbush, and Earls, "Neighborhoods and Violent Crime;" and Sampson, Morenoff, and Earls, "Beyond Social Capital." For reviews of research on community-level factors and their theoretical importance, see Sampson, "Whither;" and Sampson, Morenoff, and Gannon-Rowley, "Assessing."
43. Sampson, "Friendship Ties," 766.
44. For example, Sampson, "Neighborhood and Crime," and "Neighborhood Family Structure;" Sampson and Groves, "Community Structure and Crime;" and Sampson, Raudenbush, and Earls, "Neighborhoods and Violent Crime."
45. Sampson, "Neighborhood Family Structure."
46. Ibid.
47. Sampson, "Neighborhood and Crime."
48. Sampson, Raudenbush, and Earls, "Neighborhoods and Violent Crime." See also Sampson, Morenoff, and Earls, "Beyond Social Capital," and Sampson, "Crime and Public Safety."
49. Vogt, *Dictionary,* 196.
50. See Hirschi and Selvin, who in *Delinquency Research* contend that the quality of research depends more on rigorous data analysis than on study design.
51. Fowler and Mangione, *Standardized Survey Interviewing,* 12–14.
52. See Fowler and Mangione, *Standardized Survey Interviewing,* 12–14, for a more detailed discussion of sources of error in survey research.
53. Weisberg and Bowen, *Introduction,* 103.
54. Hirschi, *Causes of Delinquency,* 35–36.
55. Ibid., 37.
56. Appendix C of Hirschi's *Causes of Delinquency* provides the complete questionnaire.
57. Hirschi, *Causes of Delinquency,* 39.
58. Ibid., 41.
59. Ibid., Chapter 5.
60. Elliott, Huizinga, and Ageton, *Explaining Delinquency,* 91.
61. Ibid., 92.
62. See, for example, Elliott, Huizinga, and Ageton, *Explaining Delinquency.*
63. See the PHDCN internet site for information about the project and publications based on data gathered through it: http://www.hms.harvard.edu/chase/projects/chicago/
64. Sampson, Raudenbush, and Earls, "Neighborhoods."
65. Ibid.
66. Federal Bureau of Investigation, *2003,* 3.
67. Ibid.
68. Ibid.
69. Ibid., 255.
70. Ibid.
71. Ibid.
72. Ibid.
73. Ibid.
74. Ibid., 499.
75. See Federal Bureau of Investigation, 2003, 4–5, for a description of the phases of the redesign.
76. Rantala and Edwards, *Effects,* 1, quoting from the Federal Bureau of Investigation, CJIS Newsletter, NIBRS edition, volume 4.
77. Rantala and Edwards, *Effects,* 1.
78. Ibid.

79. Puzzanchera et al., *Juvenile Court Statistics 2000,* 73.
80. Puzzanchera et al., *Juvenile Court Statistics 2000,* 74. Individual case-level data include detailed information about "the characteristics of each delinquency and status offense case handled by courts, generally including the age, gender, and race of the youth referred; the date and source of referral; the offenses charged; detention; petitioning; and the date and type of disposition" (73). Court-level aggregate data "typically provide counts of the delinquency and status offense cases handled by courts in a defined time period (calendar or fiscal year)" (73). These aggregate data are sometimes abstracted from courts' annual reports.
81. Puzzanchera et al., *Juvenile Court Statistics 2000,* 74.
82. Ibid., 3.
83. Ibid., 1.
84. Ibid.
85. Ibid.
86. Ibid.
87. Federal Bureau of Investigation, *2000,* 226.
88. Puzzanchera et al., *Juvenile Court Statistics 2000,* 74. Given the sometimes long lag between arrest and referral to juvenile court, and the differences in case handling across jurisdictions, one must be cautious when making the difficult and imperfect comparison of arrests to court referrals.
89. Siegel and Senna, *Juvenile Delinquency,* 41.
90. Catalano, *Criminal Victimization, 2003,* 10.
91. Sealock and Simpson, "Unraveling Bias."
92. O'Brien, "Crime Facts." From 1972 to 1990, the NCVS was called the National Crime Survey.
93. Catalano, *Criminal Victimization, 2003,* 11. In addition to the national sample, from 1972 until 1975, the National Crime Survey also included surveys of 26 large cities (O'Brien, "Crime Facts," 65).
94. Bureau of Justice Statistics, *Criminal Victimization,* 165.
95. Catalano, *Criminal Victimization, 2003,* 11.
96. O'Brien, *Crime and Victimization Data,* 39–61, describes in detail the NCVS sampling and data collection processes.
97. Wells and Rankin, "Juvenile Victimization," 270.
98. O'Brien, "Police Productivity."
99. Messner, "Dark Figure;" O'Brien, *Crime and Victimization Data*; and Menard and Covey, "UCR and NCS."
100. O'Brien, "Comparing;" and Blumstein, Cohen, and Rosenfeld, "Criminal Career Research."
101. O'Brien, "Police Productivity," 204.
102. Ibid.
103. Catalano, *Criminal Victimization, 2003,* 10.
104. O'Brien, "Police Productivity."
105. Wells and Rankin, "Juvenile Victimization," 270.
106. O'Brien, *Crime and Victimization Data,* 66.
107. Empey, Stafford, and Hay, *American Delinquency,* 98–99.
108. Turner, *San Jose,* cited in O'Brien, "Crime Facts," 68.
109. Wells and Rankin, "Juvenile Victimization," 272.
110. Wells and Rankin, "Juvenile Victimization," 283. See also Hashima and Finkelhor, who in "Violent Victimization" caution that the NCVS may underestimate juvenile victimization.
111. Turner, *San Jose,* cited in O'Brien, "Crime Facts," 68.
112. Woltman and Bushery, *Panel Bias Study,* cited in O'Brien, "Crime Facts," 68.
113. Hirschi, *Causes of Delinquency.*
114. Lauritsen, "Age–Crime Debate."
115. O'Brien, "Crime Facts," 75.
116. Ibid.
117. Ibid., 79–80.
118. O'Brien, "Crime Facts," 80. See also Hindelang, Hirschi, and Weis, "Correlates," and *Measuring Delinquency.*
119. Farrington, Jolliffe, Hawkins, Catalano, Hill, and Kosterman, "Comparing Delinquency Careers."
120. Ibid., 954–955.
121. Ibid., 952–955.

122. Ibid., 933–934.
123. Wells and Rankin, "Juvenile Victimization."
124. Ibid., 283.
125. Ibid.
126. Harris and Shaw, "Looking for Patterns," 137.
127. Kleck, "Self-Report Data," 431.
128. See Kleck, "Self-Report Data," 431–432, for a discussion of class-homogeneous research sites and conclusions about the relationship between social class and crime.
129. Kleck, "Self-Report Data," 431.
130. Vogt, *Dictionary,* 240.
131. Hindelang, Hirschi, and Weis, "Correlates," and *Measuring Delinquency.*
132. Hindelang, Hirschi, and Weis, *Measuring Delinquency.*
133. Hindelang, Hirschi, and Weis, *Measuring Delinquency,* 213. See also Farrington, Loeber, Stouthamer-Loeber, Van Kammen, and Schmidt, "Self-Reported," for an exploration of the validity of self-report data and differences in validity by race.
134. See Lauritsen, "Age–Crime Debate," 140–144.
135. Lauritsen, "Age–Crime Debate."
136. Hindelang, Hirschi, and Weis, *Measuring Delinquency,* Chapter 4.
137. Gottfredson and Hirschi, "Methodological Adequacy," 582.
138. Blumstein, Cohen, Roth, and Visher, *Criminal Careers.* See also Blumstein, Cohen, and Farrington, "Criminal Career Research," and "Longitudinal;" and Piquero, Farrington, and Blumstein, "Criminal Career Paradigm."
139. Blumstein, Cohen, and Farrington, "Criminal Career Research," 2. See also Blumstein, Cohen, Roth, and Visher, *Criminal Careers,* 12–15.
140. Blumstein, Cohen, and Farrington, "Criminal Career Research," 4.
141. Ibid.
142. Ibid., 22.
143. Blumstein, Cohen, Roth, and Visher, *Criminal Careers*; Blumstein, Cohen, and Farrington, "Criminal Career Research," and "Longitudinal."
144. Gottfredson and Hirschi, "True Value," "Methodological Adequacy," and "Science."
145. Hirschi and Gottfredson, "Age;" Gottfredson and Hirschi, "True Value," and *General Theory of Crime.* See also Gottfredson and Hirschi, "Self-Control and Opportunity," for a discussion of the role of opportunity in their general theory of crime, and Hirschi and Gottfredson, "Defense," for a response to critiques of the theory.
146. Gottfredson and Hirschi, "True Value," 219.
147. Blumstein, Cohen, and Farrington, "Criminal Career Research," 6–7.
148. Gottfredson and Hirschi, "True Value," 217.

The Nature
of Delinquency

II

The Extent of
Delinquent Offenses

4

Chapter Objectives

After completing this chapter, students should be able to:

- Describe fully the extent to which juveniles are involved in crime.
- Identify the types of delinquent offenses that occur most frequently.
- Provide an accurate portrayal of the trends in juvenile crime.
- Describe where and when delinquent offenses occur most frequently.
- Increase interpretive skills by being exposed to delinquency data presented in tables and figures.
- Understand key terms:
  prevalence
  incidence
  relative frequency
  ecology of delinquency
  spatial distribution
  temporal distribution

Jack-Rollers: Specializing in Gang Robbery

Throughout the 1930s, Clifford Shaw produced three different biographical accounts of delinquent careers that he called the "delinquent boy's own story." According to Shaw, these stories show how the delinquent youth "conceives his role in relation to other persons and the interpretations which he makes of the situations in which he lives. It is in the personal document that the child reveals his feelings of inferiority and superiority, his fears and worries, his ideals and philosophy of life, his antagonisms and mental conflicts, his prejudices and rationalizations."[1] Shaw also offered a numbered list of the youth's arrest, school, and work records; a description of the boy's social and cultural background; a summary of treatment; and a discussion of the case by a well-known sociologist or psychologist.

In the first "own story," the "delinquent boy," Stanley, offers an account of a particular incident involving his two preferred types of crime: jack-rolling and burglary. Jack-rolling refers to the strong-arm robbery of "live ones" (drunks with money). While Stanley was involved in a variety of crimes, ranging from minor to serious and violent, he showed a definite preference for jack-rolling and burglary. He was successful at it, and he knew it. His partners in crime were a group of youths who called themselves "The United Quartet Corporation." Shaw offers an insightful footnote to this incident: "This is an interesting illustration of an organized delinquent gang. It not only functioned according to the code of the criminal world, but developed definite techniques which were adapted to the particular delinquency in which the group specialized. These techniques usually vary with different types of offenses. For example, the techniques employed in 'jack-rolling' will be quite different from those used in burglary. It should be noted that these techniques can be transmitted from one person to another or from one group to another, in much the same manner that any cultural element is disseminated through society."[2]

My fellow-workers were fast guys and good pals. We were like brothers and would stick by each other through thick and thin. We cheered each other in our troubles and loaned each other dough. A mutual understanding developed, and nothing could break our confidence in each other. "Patty" was a short, sawed-off Irish lad—big, strong, and heavy. He had served two terms in St. Charles [a state reform school for delinquent boys]. "Maloney" was another Irish lad, big and strong, with a sunny disposition and a happy outlook on life. He had done one term in St. Charles and had already been in County Jail. Tony was an Italian lad, fine-looking and daring. He had been arrested several times, served one term in St. Charles, and was now away from home, because of a hard-boiled stepfather. We might have been young, but we sure did pull off our game in a slick way. . . .

One day we were strolling along West Madison Street "taking in the sights," or, in other words, looking for "live ones." At the corner of Madison and Desplaines we saw a drunk who was talking volubly about how rich he was and that the suitcase he had in his hand was full of money. We were too wise to believe that, but we thought he might have a little money, so we would try. We tried to lure him into an alley to rob him, but he was sagacious even if he was drunk. He wanted to take me up to a room for an immoral purpose, but we decided that was too dangerous, so we let him go his way and then shadowed him.

He went on his aimless way for a long time, and we followed him, wherever he went. He finally went into a hotel and registered for a room. I saw his room number and then registered for a room on the same floor. Then we went up and worked our plans. It was not safe for all of us to go to his room, for that would arouse suspicion. One man could do the job and the others would stand by because they might be needed. But who would do the job? We always decided such things by a deck of cards. The cards were dealt, we drew, and it fell to my lot to do the deed. I was a little nervous inwardly, but did not dare to show it outwardly. A coward was not tolerated in our racket. Woe betide the one who shows it outwardly.

Putting on a bold front, I stepped into the hall and surveyed the field. Then I went to the drunk's door, my spinal nerves cold as ice. I tried the door and it was open, and that saved me a lot of work and nerve. The occupant was snoring, dead drunk, so the way was clear, I had a "sop" (blackjack) with me to take care of him if he woke up. I rifled the room, picked his pockets, and took the suitcase to our room. With great impatience we ripped it open, only to stare at a bachelor's wardrobe. That was quite a blow to our expectations, but we dragged everything out, and at the bottom our labor was rewarded by finding a twenty-dollar bill. With the thirteen dollars I had found on his person and the twenty-dollar bill, we had thirty-three dollars—eight dollars and twenty-five cents apiece. We debated what to do. Since the job would be found out and suspicion would be directed toward us, it was decided to separate for a day or two, then we would not be caught in a bunch. We divvied up the clothes. I got a pair of pants and some other small articles. Then we separated.

Source: Shaw, *Jack-Roller*, 96–98. Copyright © 1930, University of Chicago Press. All rights reserved. Reprinted with permission of the publisher.

In Chapter 3 we saw that most offenses committed by juveniles are never reported to authorities and that many juveniles involved in illegal acts are never arrested or officially processed by the juvenile justice system. Criminologists are also quick to point out that those who do come to the attention of the police or juvenile courts may not be representative of all delinquent offenders.[3] One of the primary purposes of self-report surveys is to uncover the full extent of delinquency, regardless of whether or not the illegal acts are reported or whether offenders are arrested or officially processed. Self-report surveys have their limitations, however, including problems of memory and the question of openness and honesty of self-reporting.

This chapter explores the extent of delinquent offenses. Are most juveniles involved in delinquent offenses? If so, how frequently are they involved? What types of delinquent offenses occur most often? Is delinquency increasing, decreasing, or staying the same? Are violent delinquent acts increasing? When and where do most delinquent offenses occur? To address these questions, we rely on data from self-report surveys, but we also turn to juvenile arrest and juvenile court data. The focus here is *not* on individual offenders or offending, but on juveniles as a group, so that we can obtain a general understanding of delinquent behavior. Therefore, we provide group or *aggregate data* on the extent of delinquent offenses. The extent of delinquent offenses includes four primary considerations: prevalence and incidence; relative frequency; trends; and spatial (place) and temporal (time) distribution.

■ Prevalence and Incidence

prevalence The proportion of youth involved in delinquent acts.

One of the most basic areas of study regarding delinquency is the extent to which youth report being involved in delinquent behavior. This consideration begins with the prevalence and incidence of delinquent behavior among adolescents. <u>Prevalence</u> refers to the proportion of youth involved in delinquent acts. Prevalence is often stated in terms of a percentage—the percent of youth who said they were involved in a delinquent act. <u>Incidence</u> is the average number of offenses committed per adolescent, both delinquent and nondelinquent, or the average number of offenses committed by each delinquent youth.

incidence The number of offenses committed by adolescents in general, or by each delinquent youth.

When considering the prevalence and incidence of delinquency, remember that "juvenile delinquency is an umbrella term for acts as widely different as petty theft or teenager experimentation with alcohol, at one extreme, and aggravated assault or homicide at the other. Researchers are likely to produce differing results when they employ measuring instruments that zero in on different collections of delinquent acts."[4] If only minor forms of delinquent behavior are measured through a self-report instrument, delinquency may be found to be incredibly widespread. On the other hand, if only serious delinquent acts are included, or if only arrest data are used, it may appear that relatively few youth are "delinquent."[5]

Self-Report Data

Self-report studies indicate consistently that the number of youth who commit crimes is far greater than official statistics lead us to believe. In a nationwide self-report survey by Jay Williams and Martin Gold, interviews were conducted with 847 boys and girls, ages 13–16. Eighty-eight percent of the juveniles admitted committing at least one delinquent act in the previous 3 years. Only 20% had contact with the police, and only 4% showed up in police records. Sixteen youth were sent to juvenile court, and only 11 teenagers were declared "delinquent" by the court.[6]

The National Youth Survey

As we saw in Chapter 3, the National Youth Survey (NYS) involves a comprehensive set of behavioral items that measures the full range of delinquent offenses. Because of the size and representativeness of the sample, the NYS provides some of the most valid and reliable estimates of the prevalence and incidence of delinquency. **Table 4-1** shows some, but not all, of the questionnaire items used to measure delinquent behavior, as administered to 1,719 youth between the ages of 11 and 17. To sample the findings: 16% of the youth reported

Table 4-1	Prevalence and Incidence of Delinquent Behavior in the National Youth Survey	

Respondents were asked to provide their "best estimate of the *exact number* of times" they had done each of several behaviors during the past year. The list included 47 behaviors, not all of which were delinquent.

In the last year have you . . .	Prevalence Percent reporting one or more offenses	Incidence Mean number of offenses per youth
Purposely damaged or destroyed school property?	16	.96
Stolen or tried to steal a motor vehicle, such as a car or motorcycle?	1	.02
Stolen or tried to steal something under $5?	18	1.28
Stolen something worth $5 to $50?	6	.27
Stolen or tried to steal something worth more than $50?	2	.06
Run away from home?	6	.09
Lied about your age to gain entrance or to purchase something?	27	2.80
Been loud, rowdy, or unruly in a public place?	32	3.13
Carried a hidden weapon other than a plain pocket knife?	6	.99
Attacked someone with the idea of seriously hurting or killing him or her?	6	.17
Been involved in gang fights?	13	.30
Hit or threatened to hit parents?	6	1.04
Hit or threatened to hit a teacher?	8	.50
Hit or threatened to hit other students?	48	6.01

Source: Ageton and Elliott, "Incidence," 10. Published in McGarrell and Flanagan, *Sourcebook 1984,* 363–364, 366.

purposely damaging or destroying school property in the past year, and 48% reported hitting or threatening to hit another student. Twenty-seven percent said that they lied about their age to gain entrance or to purchase something, 6% reported that they ran away from home, and 6% disclosed that they hit or threatened to hit their parents. Six percent reported carrying a weapon other than a plain pocket knife, and 6% said that they attacked someone with the idea of seriously hurting or killing that person. These findings clearly indicate that a sizable portion of youth reported involvement in minor acts of misconduct, but relatively few reported involvement in more serious forms of lawbreaking. When prevalence is considered across a variety of offenses, almost two-thirds of the youths reported involvement in less serious offenses like minor theft, minor assault, and property damage, whereas about one-fifth reported involvement in serious forms of assault including aggravated assault, sexual assault, and gang fights.[7]

To gauge the amount of crime committed by young people, it is useful to consider not only the proportion of youth involved in crime (prevalence), but also the frequency of their involvement. Frequency is usually reported in terms of the incidence of delinquency—the average number of crimes committed per youth. Incidence rates from the NYS are also reported in Table 4-1, indicating the mean number of offenses per youth. There is wide variation in the frequency of particular delinquent acts, but generally, the lower the prevalence, the lower the incidence. For example, only 1% of the youth reported involvement in motor vehicle theft, and among all youth in the survey, the mean frequency of involvement for this offense was .02 per youth. In contrast, 48% of the youth reported that they hit or threatened to hit other students, and the average number of these incidents reported was 6. Generally, the incidence, or frequency, of delinquency is less than one per youth for all but the least serious types of offenses.

Monitoring the Future

Monitoring the Future (MTF) is an annual survey of high school seniors that asks questions about delinquent acts and drug use in the previous 12-month period. Conducted by the Institute of Social Research at the University of Michigan, the MTF survey asks about a broad range of activities, many of which can be classified as delinquent behavior. **Table 4-2** shows that a considerable portion of high school seniors report involvement in delinquent offenses (prevalence). While most self-reported offenses are minor, a surprising percentage of high school seniors report involvement in fairly serious offenses. Thirteen percent reported damaging school property on purpose, 10% said they took something worth over $50, almost 20% indicated they took part in a group fight, and about 12% reported hurting someone so badly that that person needed medical attention. Only 8%, however, said they were arrested or taken to the police station.

The incidence of self-reported delinquency revealed in Table 4-2 indicates clearly that delinquent acts are a one- or two-time occurrence. The various behaviors which constitute delinquency are infrequent and sporadic—seldom do youth continue in a repetitive pattern of delinquency.[8] For example, while 20% of the youth reported that they had participated in a group fight, only 9.4% had been involved two or more times.

Table 4-2	Prevalence and Incidence of Self-Reported Delinquency in the Monitoring the Future Survey, 2003				
In the last 12 months have you . . .	Not at all (%)	Once (%)	Twice (%)	3 or 4 times (%)	≥5 times (%)
Argued or had a fight with either of your parents?	10.3	10.5	13.8	22.9	42.5
Damaged school property on purpose?	86.8	6.3	3.9	1.4	1.6
Gone into some house or building when you weren't supposed to be there?	77.0	10.5	6.8	3.2	2.5
Taken a car that didn't belong to someone in your family without permission of the owner?	94.7	2.3	1.1	0.7	1.2
Taken something from a store without paying for it?	73.2	12.1	5.4	4.0	5.2
Taken something not belonging to you worth under $50?	72.3	13.4	5.8	4.0	4.4
Taken something not belonging to you worth over $50?	90.4	4.3	1.6	1.3	2.4
Used a knife or gun or some other thing (like a club) to get something from a person?	96.1	1.5	0.9	0.8	0.7
Hurt someone badly enough to need bandages or a doctor?	88.0	5.6	3.1	1.7	1.6
Taken part in a fight where a group of your friends were against another group?	80.2	10.3	4.7	3.0	1.7
Gotten into a serious fight in school or at work?	85.7	7.2	4.0	1.5	1.6
Been arrested and taken to a police station?	92.0	4.5	1.9	0.7	0.9
Received a ticket (or been stopped and warned) for moving [traffic] violation?	71.0	17.4	6.8	2.7	2.1
Received a traffic ticket after:					
. . . drinking alcohol?	94.0	4.8	0.8	0.1	0.3
. . . smoking marijuana/hashish?	94.5	4.2	0.8	0.2	0.3
Involved in a driving accident after:					
. . . drinking alcohol?	96.4	2.8	0.5	0.1	0.2
. . . smoking marijuana/hashish?	97.4	1.9	0.3	0.1	0.2

Source: Pastore and Maguire, *Sourcebook of Criminal Justice Statistics Online 2004,* Tables 3.46, 3.49, 3.52, 3.56.

Table 4-3 provides prevalence and incidence findings from MTF for student drug use. This table reports on not only high school seniors, but also 8th and 10th graders, who were added to the survey in 1991. The prevalence of drug use is indicated by the percent of youth reporting that they had "ever used" (*lifetime prevalence*). Similar to the prevalence of delinquent offenses, the percentage of students reporting that they used different types of drugs is strikingly high, especially for alcohol and marijuana.

The incidence of drug use is represented by the percent reporting use during the "last 12 months" (*annual prevalence*) and "last 30 days" (*30-day prevalence*).[9] The 30-day and annual prevalence figures drop considerably from those reporting "ever used" (lifetime prevalence), indicating that for most forms of drug use, frequent use is not common. The glaring exceptions to this observation are alcohol and marijuana use: 47.5% of the high school seniors report

Table 4-3	Prevalence and Incidence of Drug Use in the Monitoring the Future Survey, 2004		
		Ever used (%)	Used in last 12 months (%)	Used in last 30 days (%)
Alcohol*	8th Grade	45.6	37.2	19.7
	10th Grade	66.0	59.3	35.4
	12th Grade	76.6	70.1	47.5
Marijuana/ Hashish	8th Grade	16.3	11.8	6.4
	10th Grade	35.1	27.5	15.9
	12th Grade	45.7	34.3	19.9
Inhalants	8th Grade	17.3	9.6	4.5
	10th Grade	12.4	5.9	2.4
	12th Grade	10.9	4.2	1.5
Hallucinogens	8th Grade	3.5	2.2	1.0
	10th Grade	6.4	4.1	1.6
	12th Grade	9.7	6.2	1.9
Cocaine	8th Grade	3.4	2.0	0.9
	10th Grade	5.4	3.7	1.7
	12th Grade	8.1	5.3	2.3
Amphetamines	8th Grade	7.5	4.9	2.3
	10th Grade	11.9	8.5	4.0
	12th Grade	15.0	10.0	4.6

*Alcohol figures are from 2003—Pastore and Maguire *Sourcebook of Criminal Justice Statistics Online 2004*, Table 3.66.

Source: Johnston, O'Malley, Bachman, and Schulenberg, *Monitoring the Future.*

alcohol consumption in the last 30 days and 19.9% report marijuana use in the last 30 days. One other startling observation is the high incidence and prevalence of inhalant use by 8th-grade students: 17.3% report that they have used inhalants, with 9.6% indicating that they have used inhalants in the last 12 months and about 4.5% indicating inhalant use in the last 30 days.

Self-report data allow us to see that, while delinquent acts are common among adolescents, relatively few youths report frequent or repetitive involvement. Nonetheless, a substantial proportion of adolescents reports involvement in crime, and some of these offenses are quite serious. Furthermore, of those youths who commit delinquent acts, only a small portion come to the attention of the police, relatively few are arrested, and even fewer are processed by juvenile courts.

Official Data

Uniform Crime Reporting Program

When juvenile offenders come to the attention of the police and are arrested, they are counted in the Uniform Crime Reporting (UCR) program. As we described in Chapter 3, the UCR program primarily gathers three types of information: offenses known to police, crimes cleared, and persons arrested. Of the three types, offenses known to the police provide the clearest picture of the volume and trends of crime; however, the age of the offender is not recorded and specific information on juvenile crimes is therefore unavailable.[10] The extent of juvenile crime, as represented by the UCR, is therefore based on juvenile arrest and clearance data. Keep in mind, however, just what these data report. Because these data are collected only when juvenile offenders come to the attention of the police, much juvenile crime goes unrecorded by the UCR. As a result, arrest and clearance data are poor indicators of the extent of juvenile delinquency. Rather, these data reflect police action taken in response to juvenile crime, as well as the proportion of crime known to the police that is attributed to juveniles.[11] Nonetheless, these data provide insight into the delinquency problem encountered by the police and the rest of the juvenile justice system.

In the 2000 census, more than 72 million persons in the United States were below age 18, commonly referred to as *juveniles*.[12] This number represents about 25% of the US population. Those in the adolescent years (ages 13 through 17) numbered 20.1 million, or about 7% of the US population. **Table 4-4** shows that almost 1.6 million of the arrests recorded in 2003 were of juveniles, whereas 8 million arrests were of adults. While the number of juvenile arrests is much smaller than the number of adult arrests, juveniles are disproportionately involved in crime, accounting for 16.3% of the arrests for all UCR crimes and 25.3% of the arrests for Index crimes. The proportion of juveniles arrested for property Index crimes is far greater than it is for violent Index crimes.

Arrest rates are another measure of the extent of juvenile crime. These rates take into account the size of a population group by dividing the number of arrests within a given group by the population of that group and then multiplying by a standardized population unit. For juveniles, the arrest rate is stated in terms of the number of arrests per 100,000 juveniles. The juvenile arrest rate of 2,140 for all UCR crimes is significantly lower than the adult arrest rate of 3,682 per 100,000 adults. However, the juvenile arrest rate for property Index crimes (450) is far greater than that for adults (372). In contrast, the juvenile arrest rate for violent Index crimes (89) is substantially lower than the adult violent crime arrest rate (163) (see Table 4-4).

Table 4-4	Juvenile and Adult Arrests and Arrest Rates, 2003*					
	Number of arrests		**Arrest rate per 100,000 in age group**		**Percent of arrests**	
Type of Crime	**Juveniles**	**Adults**	**Juveniles***	**Adults**	**Juveniles**	**Adults**
All Crimes	1,563,149	8,018,274	2,140	3,682	16.3	83.7
Violent Crime Index	64,799	354,165	89	163	15.5	84.5
Property Crime Index	328,823	810,537	450	372	28.9	71.1
Total Crime Index	393,622	1,164,702	539	535	25.3	74.7

*Juvenile arrest rates are calculated using ages 0–17.

Source: Federal Bureau of Investigation, *Crime in the United States 2003*, Table 38, 280. Population estimates: U.S. Bureau of the Census.

Even though the UCR data addressing juvenile offenses are based on the number and rate of juvenile arrests, these data provide a sense of the extent of juvenile delinquency in terms of the volume confronted by the juvenile justice system at its entry point: law enforcement. As such, UCR data report *arrest* prevalence, rather than *offense* prevalence. The proportion of youth actually involved in crime is not represented by arrest data.

Juvenile Court Statistics

As we described in Chapter 3, *Juvenile Court Statistics* provides data on cases processed by juvenile courts across the United States. If most crime goes unreported and never comes to the attention of the police, an even smaller portion of juvenile crime is referred to juvenile courts. As a result, neither arrest nor juvenile court data provide accurate estimates of the extent of juvenile delinquency. Nonetheless, data from *Juvenile Court Statistics* provide valuable information on the extent to which cases are referred to and processed by juvenile courts. As such, these data represent society's "official" response to the problem of delinquency.

In 2000, the most recent year for which data is available, juvenile courts handled 1,633,300 delinquency cases. The 1997 *Juvenile Court Statistics* reveals 158,500 petitioned status cases. This latter number does not include status cases dealt with informally by social welfare agencies or juvenile courts (without petition). The case rate expresses the number of cases per 1,000 juveniles, for the ages of 10 through 17. For every 1,000 juveniles in the population, there were 53.2 delinquency cases and 5.5 petitioned status cases.[13]

Prevalence and Incidence in Brief

Our analysis of the prevalence and incidence of juvenile delinquency leads to the unambiguous conclusion that a substantial portion of adolescents is involved in crime, and that some juveniles are involved in quite serious offenses. For most youth, however, delinquency is not a frequent or regular activity. Moreover, relatively few youths are dealt with officially by the police or the juvenile courts. The limited frequency of offending suggests that most youths do not develop a persistent pattern of delinquent behavior—a topic addressed more fully in Chapter 6. These data also raise the question of whether delinquency involvement varies by age, gender, race, and class—a consideration taken up in Chapter 5.

Why Are Young People Disproportionately Involved in Crime?

The disproportionate involvement of young people in crime is a clear indication that there is something distinctive about the adolescent and young adult years that generates high rates of crime for this particular age range. Yet not all adolescents are involved in criminal acts, and most do not develop serious, repetitive patterns of delinquent behavior. So what is it about the adolescent and young adult years that is conducive to criminal involvement for some, but not all youth?

This is the fundamental question addressed in a research initiative, begun in 1986 by the Office of Juvenile Justice and Delinquency Prevention (OJJDP), called the Program of Research on Causes and Correlates of Delinquency.[14] Drawing from three coordinated research projects in Denver, Pittsburgh, and Rochester (NY), this program of research sought to improve "understanding of

RESEARCH IN ACTION

Risk Factors for Adolescent Behavioral Problems

Risk Factors	Adolescent Problem Behaviors				
	Substance Abuse	Delinquency	Teenage Pregnancy	School Dropout	Violence
Community					
Availability of drugs	√				
Availability of firearms		√			√
Community laws and norms favorable toward drug use, firearms, and crime	√	√			√
Media portrayals of violence					√
Transition and mobility	√	√		√	
Low neighborhood attachment and community disorganization	√	√			√
Extreme economic deprivation	√	√	√	√	√
Family					
Family history of the problem behavior	√	√	√	√	√
Family management problems	√	√	√	√	√
Family conflict	√	√	√	√	√
Favorable parental attitudes and involvement in the problem behavior	√	√			√
School					
Academic failure beginning in late elementary school	√	√	√	√	√
Lack of commitment to school	√	√	√	√	
Peer and Individual					
Early and persistent antisocial behavior	√	√	√	√	√
Rebelliousness	√	√		√	
Friends who engage in the problem behavior	√	√	√	√	√
Gang involvement	√	√			√
Favorable attitudes toward the problem behavior	√	√	√	√	
Early initiation of the problem behavior	√	√	√	√	√
Constitutional factors	√	√			√

THEORY INTO PRACTICE **Comprehensive Strategy for Serious, Violent, and Chronic Juvenile Offenders**

Six Principles for Preventing and Reducing Juvenile Delinquency

1. **Strengthening the family** in its primary responsibility to instill moral values and provide guidance and support to children.
2. **Supporting core social institutions** (schools, churches, youth service organizations, community organizations) in their roles to develop capable, mature, and responsible youth.
3. **Promoting delinquency prevention** as the most cost-effective approach to reducing juvenile delinquency.
4. **Intervening immediately and effectively when delinquent behavior first occurs** to prevent delinquent offenders from becoming chronic offenders or from committing progressively more serious and violent crimes.
5. **Establishing a system of graduated sanctions** that holds each juvenile offender accountable, protects public safety, and provides programs and services that meet identified treatment needs.
6. **Identifying and controlling the small percent of serious, violent, and chronic juvenile offenders** who commit the majority of juvenile felony-level offenses.

Source: Coolbaugh and Hansel, "The Comprehensive Strategy," 2.

serious delinquency, violence, and drug use by examining how youth develop within the context of family, school, peers, and community."[15]

This and other research identifies conclusively five major "categories of causes and correlates of juvenile delinquency."[16] These causes and correlates can be thought of as *risk factors*—traits, conditions, and experiences that make delinquency and other problem behaviors more likely. Stated simply, these categories of factors place youth "at risk" for problem behaviors, including substance abuse, delinquency, teen pregnancy, dropping out of school, and violence.[17] The five categories of risk include "(1) individual characteristics such as alienation, rebelliousness, and lack of bonding to society; (2) family influences such as parental conflict, child abuse, and family history of problem behavior (substance abuse, criminality, teen pregnancy, and school dropouts); (3) school experiences such as early academic failure and lack of commitment to school; (4) peer group influences such as friends who engage in problem behaviors (minor criminality, gangs, and violence); (5) neighborhood and community factors such as economic deprivation, high rates of substance abuse and crime, and low neighborhood attachment."[18]

The extent to which these risk factors are experienced by youth determines the degree to which delinquent behavior occurs. Therefore, when we ask, "What is it about the adolescent and young adult years that is conducive to criminal involvement for some, but not all youth?" the answer lies in the degree to which these different categories of risk are present in a youth's life. Research in Action, "Risk Factors for Adolescent Behavioral Problems," page 121, provides a graphic summary of these risk factors with respect to five main problem behaviors.[19]

Building on the Causes and Correlates research, OJJDP has also sponsored and promoted a "research-based framework" of delinquency prevention and reduction called Comprehensive Strategy for Serious, Violent, and Chronic Juvenile Offenders.[20] Comprehensive Strategy consists of two principal components: (1) "preventing youth from becoming delinquent through prevention strategies for all youth, with a focus on those at greatest risk," and (2) "improving the juvenile justice system response to delinquent offenders through a system of graduated sanctions and a continuum of treatment alternatives that include immediate intervention, intermediate sanctions, community-based corrections, and aftercare services."[21]

Six principles, identified in Theory Into Practice, "Comprehensive Strategy for Serious, Violent, and Chronic Juvenile Offenders," guide the Comprehensive Strategy. The delinquency prevention component of the Strategy is based on the premise that risk factors must be first identified through research and then addressed through a broad range of prevention programs directed at the family, school, and community.

■ Relative Frequency of Different Types of Offenses

The prevalence and incidence of juvenile delinquency provide a general understanding of the extent to which young people are involved in crime. It is also useful to consider the types of crime in which juveniles are most involved. Relative to each other, what types of juvenile offenses are most common? We refer to this as the **relative frequency** of juvenile offenses. This too can be studied by looking at different sources of data to see if identifiable patterns can be observed.

Self-Report Data

Both the National Youth Survey and the Monitoring the Future survey reveal that a sizable portion of youth reports involvement in a wide variety of minor misconduct, but relatively few youths are involved in serious forms of lawbreaking. In addition, only the less serious forms of delinquency are committed with any frequency. As a general rule, the more serious the offense, the lower the prevalence and incidence. More specifically, these self-report data indicate that the most common forms of adolescent misconduct are using alcohol and marijuana, lying about age, skipping school, committing minor theft, damaging the property of others, committing disorderly conduct, making threats of physical harm, and fighting.[22] This is readily apparent in Tables 4-1, 4-2, and 4-3.

Thus the prevalence and incidence of different types of delinquency reveal a clear picture: various types of minor delinquent behavior are far more common among adolescents than are more serious forms of delinquency, and delinquent

relative frequency In comparison to each other, the types of delinquent offenses that occur most often.

youth commit these minor illegal acts with far greater frequency. This is not to say that serious violent crime by juveniles is nonexistent or unimportant. Self-report data indicate clearly that youth are involved in serious crime, but in terms of relative frequency, minor delinquency significantly outpaces serious delinquency. As a result, the nature of delinquency is most accurately characterized as involving minor offenses.

The Monitoring the Future survey reports that while alcohol and marijuana use is remarkably common among high school seniors, other forms of drug use are not nearly so common (Table 4-3). Among high school seniors, 47.5% reported alcohol use and 19.7% reported marijuana use in the last 30 days. On the other hand, use in the last 30 days for amphetamines was only 4.6%; for cocaine, 2.3%; for hallucinogens, 1.9%; and for inhalants, 1.5%.

The use of drugs other than alcohol and marijuana is associated with greater likelihood of involvement in all types of delinquent offenses. **Figure 4-1** shows a consistent pattern of drug use and involvement in a variety of delinquent offenses. Clearly, youths who use drugs other than marijuana and alcohol are more likely to engage in a wide variety of offenses than are youths who do not report serious drug use. However, criminologists question whether the relationship between drugs and delinquency is a causal one.[23] In Chapter 11, we examine the

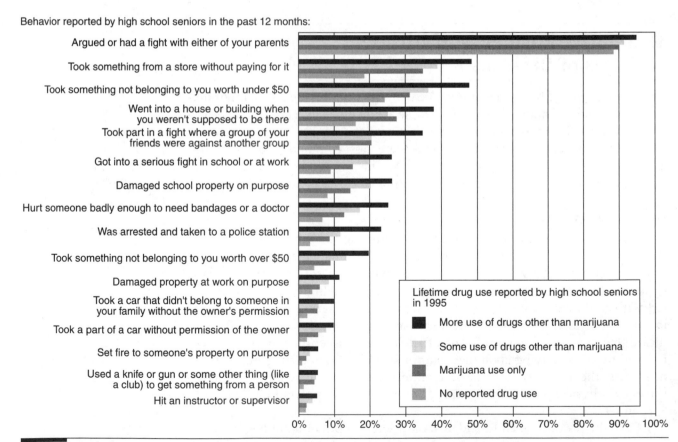

Figure 4-1 Drug use and self-reported delinquency. Law-violating behavior is more common for youth who use drugs.
Source: Snyder and Sickmund, *Juvenile Offenders,* 76.

issues of causal order and causal effect that make up this debate. For now, it is important to note that drug use and other types of delinquency are associated and tend to occur together.[24]

Official Data

Uniform Crime Reporting Program

Juveniles are arrested for some offenses more frequently than others. Juveniles account for 29% of all arrests for property Index crimes but only 15% of arrests for all violent Index crimes. This difference in involvement is revealed more clearly with regard to particular offenses. **Figure 4-2** shows that juveniles account for 51% of arrests for arson; 39% of arrests for vandalism; 29% of arrests for burglary and motor vehicle theft, and 28% of arrests for larceny–theft—all property offenses. For violent crimes, the largest portions of arrests involving juveniles are for robbery (24%), other assaults (19%), and forcible rape (16%). Juveniles also account for a disproportionate number of arrests for several crimes that can be classified as public order crimes, including disorderly conduct (30%) and liquor laws (22%). Thus, juvenile crime, as indicated by the proportion of arrests involving juveniles, is predominantly property and public order crime.[25]

Arrest rates also demonstrate that juvenile crime is predominantly property crime. The juvenile arrest rate for property Index crimes is 450 per 100,000

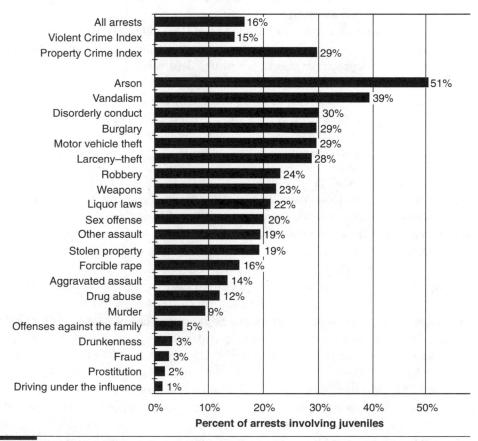

Figure 4-2 Percent of arrests involving juveniles. In 2003, juveniles were involved in fewer than 1 in 6 arrests for violent crime and 1 in 3 for property crime.
Source: Snyder, "Juvenile Arrests 2003," 4.

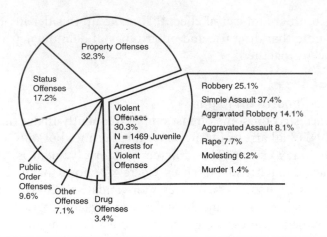

Figure 4-3 Juvenile arrests of violent offenders by type of offense. Donna Hamparian and her colleagues' study of "The Violent Few" found that even violent offenders were involved in a variety of offenses and that property offenses were most common.
Source: Hamparian et al., *Violent Few,* 12.

adolescents, compared to an arrest rate for violent Index crimes of 89 per 100,000 adolescents (see Table 4-4). Property offenses were most common even among youths who had been arrested for violent crime. In a study of juveniles who had been arrested for violent crime, Donna Hamparian and her colleagues found that arrest records of 1,222 violent offenders revealed "diverse patterns of criminal activity, ranging from murder and other violent offenses to a broad variety of property and other nonviolent offenses."[26] They concluded that "violent juvenile offenders, as a group, do not specialize in the types of crimes they commit."[27] Arrest records revealed that, even among these violent juvenile offenders, property offenses were the most common type of arrest (see **Figure 4-3**). Property offenses were followed by violent offenses; however, the majority of violent offenses were simple assault.

Juvenile Court Statistics

Reporting on delinquency cases in juvenile courts across the United States, *Juvenile Court Statistics* provides a representation of the relative frequency of different types of juvenile offenses that is very consistent with self-report and UCR data. **Table 4-5** indicates that the largest portion of all juvenile court cases are property offenses (42%), with larceny–theft accounting for almost 20% of all delinquency cases in juvenile court. Person offenses, over half of which are simple assault, are 23% of all cases. Public order offenses are 23% of all juvenile court cases, with the largest component being obstruction of justice (10% of all cases). Thus, juvenile court cases are largely property offenses, but cases are concentrated in a few specific offense types: larceny–theft (19%), simple assault (15%), drug violations (11%), obstruction of justice (10%), burglary (7%), vandalism (7%), disorderly conduct (5%), and trespassing (4%).[28]

Conspicuously absent from the juvenile court data reported so far are status offense cases. *Juvenile Court Statistics* reports only status offense cases formally petitioned into the juvenile court, because not all states provide the number and

Table 4-5	Juvenile Court Delinquency Caseload, 1999		
Most serious offense	Number of cases	10 year change	% of total cases
Total delinquency	1,673,000	27%	100%
Person offenses	387,100	55	23
Criminal homicide	1,800	−21	0.1
Forcible rape	4,200	−19	0.2
Robbery	25,100	−9	1.5
Aggravated assault	55,800	−5	3.3
Simple assault	255,900	95	15.3
Other violent sex offenses	11,600	52	0.7
Other person offenses	32,700	95	23
Property offenses	706,200	−9	42
Burglary	113,900	−22	6.8
Larceny–theft	322,100	−6	19.3
Motor vehicle theft	38,500	−45	2.3
Arson	8,600	28	0.5
Vandalism	111,400	12	6.7
Trespassing	58,700	12	3.5
Stolen property offenses	26,300	−11	1.6
Other property offenses	26,800	−4	1.6
Drug law violations	191,200	169	11
Public order offenses	388,600	74	23
Obstruction of justice	171,800	115	10.3
Disorderly conduct	90,600	67	5.4
Weapons offenses	39,800	32	2.4
Liquor law violations	19,900	21	1.2
Nonviolent sex offenses	13,700	10	0.8
Other public order offenses	52,700	75	3.2

Source: Puzzanchera et al., *Juvenile Court Statistics, 1999,* 6 and 7.

processing of nonpetitioned status offense cases.[29] In those states that report both petitioned and nonpetitioned status offense cases, it is typical for nonpetitioned cases to at least equal the number of petitioned cases, and in a number of states, nonpetitioned cases outnumber petitioned cases 2 to 1.[30] As a result, *Juvenile Court Statistics* does not provide an accurate representation of the number of status offense cases. If we conservatively estimate that nonpetitioned status offense cases are at least equal in number to petitioned cases, then the juvenile courts confront more than 300,000 cases a year, making status offense cases about 15% of the total case load per year. Status offenses are then one of the more

Table 4-6	Number and Percent of Petitioned Status Offense Cases in Juvenile Court, 1997	
Petitioned status offenses	Number of cases	Percent of total petitioned status cases
TOTAL CASES	158,500	100
Runaway	24,000	15
Truancy	40,500	26
Ungovernability	21,300	13
Liquor	40,700	26
Miscellaneous	32,100	20

Source: Charles Puzzanchera et al., *Juvenile Court Statistics 1997*, 37.

frequent types of juvenile court cases. **Table 4-6** shows that among status offense cases, truancy (26%) and liquor violations (26%) are most common.

Relative Frequency in Brief

All forms of delinquency data confirm that minor forms of delinquent behavior are far more common than serious, violent offenses. Property offenses far exceed violent offenses. Various self-report studies, arrest data (UCR), and juvenile court cases (*Juvenile Court Statistics*) indicate that certain types of delinquency are most common and occur with greater frequency than other offenses:

        alcohol and marijuana violations

        minor theft

        property damage and vandalism

        disorderly conduct

        fighting and simple assault

        truancy

Even though these types of delinquent offenses predominate, delinquent youths rarely specialize in the type of crime they commit.[31] In addition, their involvement is infrequent, sporadic, and nonrepetitive.[32] Even if juveniles are involved in multiple offenses, they typically commit a variety of relatively minor offenses.[33]

Why Are Some Types of Delinquent Offenses More Common Than Others?

Minor forms of delinquency are far more common than serious violent offenses simply because of the sheer number of youths that engage in these acts. As we have seen, the vast majority of youths who commit delinquent acts are involved infrequently, sporadically, and non-repetitively in minor offenses. Many youths engage in delinquency, but it is of a minor variety. Only a small portion of youth exhibits serious, repetitive patterns of violent offending. It is estimated that serious violent offenders account for more than half of all serious crimes committed by juveniles.[34]

Serious, violent offenders have been studied extensively in recent years, most notably by a panel of experts—the Study Group on Serious and Violent Juvenile Offenders. By reviewing research in this area, the Study Group concluded that the developmental patterns of serious violent offenders are significantly different from the more typical non-serious offenders. Serious and violent juvenile offenders (SVJ) start offending early and continue much longer than non-serious offenders. Their involvement begins with a variety of problem behaviors, such as stubborn behavior, defiance, bullying, lying, and disobedience, and progresses to a variety of delinquent acts and then to serious violent delinquency. SVJ offenders tend to have weak social ties, antisocial peers, and low academic achievement. Poor, socially disorganized neighborhoods also seem to play a role in the development of serious, violent offending.[35]

Psychologist Terrie Moffitt refers to this SVJ offender group as *life-course persistent offenders*.[36] In contrast to this small group of offenders, a much larger group is involved in relatively minor offenses during the adolescent years only. Moffitt refers to this group as *adolescence-limited offenders*. Adolescence-limited offenders are involved in a variety of minor offenses, but their involvement is sporadic and experimental. Moffitt's developmental theory is explained more fully in Chapter 6.

■ Trends in Delinquent Offenses

If we were to ask the general public how juvenile delinquency has changed in recent years, most would respond that juvenile crime is growing increasingly worse.[37] Not only are delinquent offenses growing in number, the general public imagines, but juveniles are also becoming more violent. Is this true? To answer this question, we will examine trends in delinquent offenses.

Self-Report Data

Monitoring the Future (MTF), the annual survey of high school seniors, was designed to chart changes in attitudes, values, and behavior from one class of seniors to the next, beginning with the class of 1975. Earlier we considered the incidence and prevalence of delinquent activity in the class of 2003. But how do the different classes of high school seniors over the years compare to one another in terms of delinquent acts? The activities included in the MTF survey tend to be relatively minor. Nonetheless, these data provide the only annual depiction of self-reported delinquency that is representative of a sizeable segment of American youth.[38]

Figure 4-4 shows the trends in annual prevalence from 1988 through 2003. Represented are the percent of high school seniors reporting that they had gotten into a serious fight, taken something worth under $50, taken something worth over $50, and been arrested and taken to a police station. The survey question about being arrested has been asked since 1993. The serious fight and theft questions were chosen because they represent both the most common and serious forms of property and violent crime.

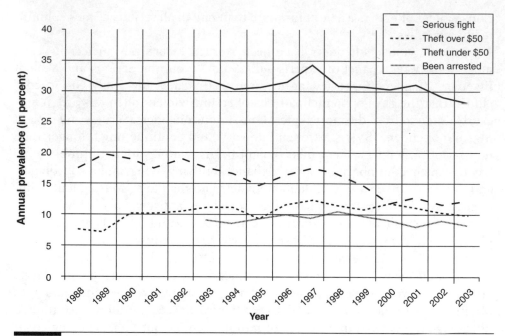

Figure 4-4 Prevalence trends in self-reported delinquency among high school seniors: fighting, theft, and arrest, 1988–2003.

Source: Pastore and Maguire, *Sourcebook of Criminal Justice Statistics Online 2004*, Table 3.46.

While the data show fluctuations from year to year, the general trend for involvement in serious fighting has declined, while the trend for serious theft has increased, but only slightly. Minor theft has declined somewhat. In 1989, almost 20% of the high school seniors reported being involved in a serious fight, but by 2003, the percentage had dropped to about 12%. The level of change for serious theft is not as great, but it is in the opposite direction. In 1989, just over 8% of the high school seniors surveyed reported taking something worth over $50. By 2003, the prevalence of serious theft increased to almost 10% of those surveyed. The prevalence of minor theft declined from 32% to 28% from 1988 to 2003. Arrest fluctuated between 9% and 10% over the course of the 11-year period in which it was reported, indicating no measurable change in the percentage of high school seniors reporting that they were arrested and taken to a police station.

We also looked previously at self-reported drug use of high school seniors and 8th and 10th graders, who were added to the MTF survey in 1991. Alcohol, marijuana, and cigarette use in the last 30 days, as reported by high school seniors, reveal a consistent trend across the 16-year period reported in **Figure 4-5.** In the late 1980s and very early 1990s, use of alcohol, marijuana, and cigarettes declined. In 1992, drug use swung upward and 30-day prevalence increased until 1997, when a slight downturn began. While alcohol and cigarette use in the last 30 days was considerably lower in 2003 than it was in 1988, marijuana and cocaine use was higher in 2003 than it was in 1988. Trends in cocaine use in the last 30 days appear flat in the graph, but that is a reflection of the scale masking change when the percentage of use is low. Cocaine use fell the first half of the 1990s, increased slightly in the second half of the decade, and declined again in the early 2000s.

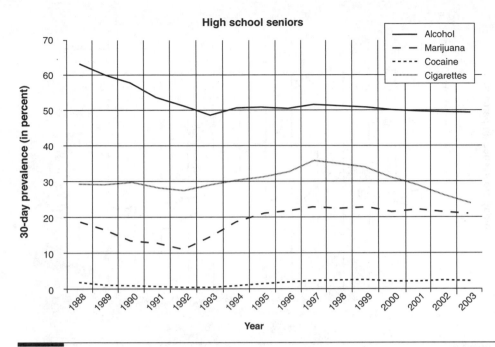

Figure 4-5 Thirty-day prevalence trends in self-reported alcohol, marijuana, cocaine, and cigarette use among high school seniors, 1988–2003.

Source: Pastore and Maguire, *Sourcebook of Criminal Justice Statistics Online 2004,* Table 3.69.

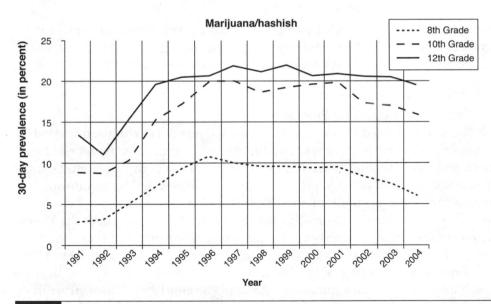

Figure 4-6 Thirty-day prevalence trends in self-reported marijuana/hashish use among 8th, 10th, and 12th graders, 1991–2004.

Source: Johnston, O'Malley, Bachman, and Schulenberg, *Monitoring the Future.*

Figures **4-6** and **4-7** show self-reported marijuana and inhalant use in the last 30 days for 8th, 10th, and 12th graders. The trend in marijuana/hashish use is similar across the three grades. After 1992, use of marijuana in the last 30 days increased, peaking in 1996/1997, and then leveling off in the late 1990s, followed by a decline. Across the years, high school seniors report higher involvement in

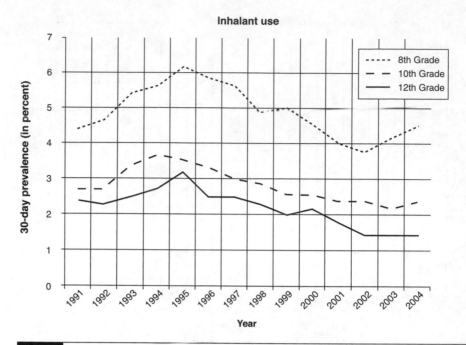

Figure 4-7 Thirty-day prevalence trends in self-reported inhalant use among 8th, 10th, and 12th graders, 1991–2004.
Source: Johnston, O'Malley, Bachman, and Schulenberg, *Monitoring the Future*.

marijuana use than the other two grades. Inhalant use is just the opposite: 8th graders report the highest percentage of use and 12th graders the lowest. Use of inhalants in the last 30 days peaked in 1995 and has since declined.

Official Data

Uniform Crime Reporting Program

UCR data can be used to assess trends in the volume of violent juvenile crime by monitoring four crimes that make up the Violent Crime Index: murder and non-negligent manslaughter, forcible rape, robbery, and aggravated assault. In the same way, trends in juvenile property crime can be assessed by monitoring four crimes that make up the Property Crime Index: burglary, larceny–theft, motor vehicle theft, and arson. To analyze trends, it is best to express both the Violent Crime Index and the Property Crime Index in terms of an arrest rate—the number of juvenile arrests for Index crimes per 100,000 juveniles, ages 10 to 17.[39]

Figure 4-8 shows that, after a stable pattern for much of the 1980s, the juvenile Violent Crime Index increased between 1988 and 1994. This increase in juvenile arrest rate drew national attention to the problem of juvenile violence. However, after peaking in 1994, the juvenile arrest rate for violent Index crime dropped each year from 1994 through 2003. Howard Snyder observes: "For all Violent Crime Index offenses combined, the number of juvenile arrests in 2003 was the lowest since 1987."[40]

The Property Crime Index is another story. **Figure 4-9** reveals that, despite a notable dip in 1983 and 1984, the juvenile arrest rate for property Index crimes was quite stable for the 15-year period from 1980 to 1994. The following nine years saw a significant drop in the arrest rate for juvenile property crime. The 2003 rate was the lowest since the 1960s.[41]

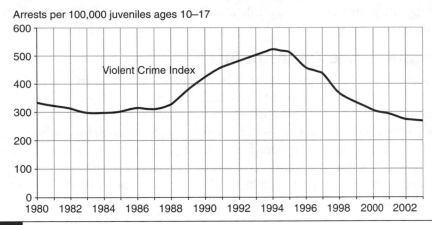

Figure 4-8 Juvenile arrest rate for Violent Crime Index offenses. In 2003, the juvenile arrest rate for the Violent Crime Index was at its lowest level since 1988—41 below the peak year of 1994.
Source: Snyder, "Juvenile Arrests 2003," 5.

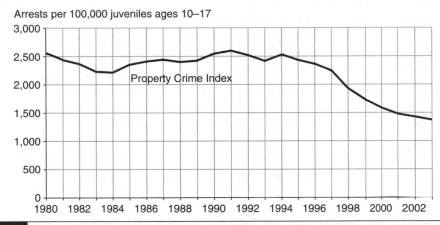

Figure 4-9 Juvenile arrest rate for Property Crime Index offenses. In 2003, the juvenile arrest rate for the Property Crime Index was at its lowest level since the 1960s.
Source: Snyder, "Juvenile Arrests 2003," 5.

Juvenile Court Statistics

In recent years, juvenile courts in the United States have dealt with more cases than ever. In 2000, juvenile courts handled 53 delinquency cases for every 1,000 juveniles in the population. This rate represents a 25% increase in case rates from 1989 to 1996, and then a decline of 14% through 2000.[42] Case rates for petitioned status offense cases also showed a substantial increase over a 10-year period, with runaway increasing by 71% and truancy by 74%.[43]

In conjunction with the case rate increase, there is a growing tendency for police to make referrals to the juvenile court, rather than to deal with juveniles taken into custody within the police department and then release them. In 1972, 51% of all juveniles taken into police custody were referred to juvenile court, while 45% were handled within the department. By 2003, police were far more likely to refer cases to juvenile court (71%) than to deal with juvenile matters within the police department (20%).

Thus, the growing number of juvenile court cases must be tempered by the realization that these trends depict changes in how juvenile offenses are dealt with by the police and the juvenile court more than they indicate trends in juvenile delinquency.

Juvenile Crime Trends in Brief

A nice, clean conclusion about trends in juvenile crime simply is not possible because different sources of data give different pictures. However, if we keep in mind what the data are measuring, more accurate conclusions are possible. Monitoring the Future provides a self-reported indication of high school seniors' involvement in crime. Questions from the survey tend to concentrate on less serious forms of crime. In addition, less serious offenses have low rates of reporting and clearance, making them more difficult to track with official sources of data, like the UCR. For this reason, self-report surveys are good indicators of trends in less serious juvenile crime, such as less serious forms of violent and property crime, and public order, drug use, and status offenses. Reporting is less problematic for violent crime, making arrest data more valid and reliable as a measure of violent juvenile crime trends. With these qualifications, we can draw several conclusions about trends.

Juvenile property crimes occur at a consistently higher rate than juvenile violent crimes. Throughout much of the 1990s, juvenile property crimes fluctuated some, but a substantial increase was not evident. In fact, juvenile arrest data for property crime showed a decline after 1994. Violent crime by juveniles increased in the late 1980s and into the 1990s, but since 1994, the Violent Crime Index for juveniles has declined. Drug use by adolescents, including the use of alcohol, declined throughout the 1980s and early 1990s, then increased somewhat in the mid-1990s, then declined again in the latter part of the decade. With the exception of an increase in violent juvenile crime in the late 1980s and early 1990s, the incidence of juvenile crime has not changed dramatically in the last 25 years.

Is There an Epidemic of Youth Violence?

The increase in the Violent Crime Index from 1988 to 1994 focused national attention on the problem of juvenile violence.[44] In addition, projections of a growing juvenile population, in conjunction with increasing arrest rates, warned of continued growth in juvenile crime—growth to unprecedented levels.[45] Criminologists, policy analysts, and governmental leaders feared a new generation of juvenile offenders, one that was incredibly violent. Many believed that these trends, taken together, provided just the right ingredients for an epidemic of youth violence.

In 1995, renowned criminologist Alfred Blumstein advanced the view, supported by research, that the major share of this growth in youth violence was due to a dramatic rise in youth homicide.[46] He noted three major changes that occurred in the short period from 1985 to 1992: "(1) homicide rates by youth eighteen and under have more than doubled, while there has been no growth in homicide rates by adults twenty-four and older, (2) the number of homicides juveniles commit with guns has more than doubled, while there has been no growth in non-gun homicides, and (3) the arrest rate of non-white juveniles on drug charges has more than doubled, while there has been no growth in the drug arrest rate for white juveniles."[47]

Blumstein's explanation centered on the crack cocaine epidemic in inner-city areas that produced a flourishing illicit drug market. Young blacks were recruited to maintain and extend market shares of drug dealers. Blumstein observed, "Because these markets are illegal, the participants must arm themselves for self-protection, and the resulting 'arms race' among young people results in a more frequent resorting to guns as a major escalation of the violence that has often characterized encounters among teenage males."[48]

The use of handguns by young males involved in the illicit drug market is key to Blumstein's argument: "In view of both the recklessness and bravado that is often characteristic of teenagers, and their low level of skill in settling disputes other than through the use of physical force, many of the fights that would otherwise have taken place and resulted in nothing more serious than a bloody nose can now turn into shooting as a result of the presence of guns."[49] Thus, the "deadly nexus" of youthfulness, crack cocaine markets, and handguns fueled the rise in violent crime, especially in terms of homicide rates for young, black males.[50]

At the same time as Blumstein was drawing attention to the role that crack cocaine and hand guns played in increasing youth violence, others were exploring the effect that changing age composition has on crime rates. In 1995, political scientist John DiIulio declared that the United States was "sitting on a demographic crime bomb."[51] It was projected that, between 1995 and 2007, the population between age 15 to 17 ("the age group responsible for two-thirds of all juvenile arrests") would increase by 21%.[52] In addition, minority populations and the number of children in poverty were also predicted to grow.[53] DiIulio provided the following summary on the connection of these population trends with crime: "the large population of seven- to 10-year-old boys now growing up fatherless, Godless, and jobless—and surrounded by deviant, delinquent, and criminal adults—will give rise to a new and more vicious group of predatory street criminals than the nation has known."[54] Writing with William Bennett and John Walters a year later, DiIulio referred to this new generation of violent juvenile offenders as "superpredators." While demographic trends play a role in the emergence and the numbers of superpredators, the underlying cause, according to the authors, is "moral poverty—children growing up without love, care, and guidance from responsible adults."[55]

Economist Steven Levitt has disputed the view that population age structure contributes heavily to crime rates and trends.[56] His research revealed that a projected increase in the number of crime-prone teens, even if it contributes to the crime rate, will be offset by a growing number of senior citizens—a group with a relatively low crime rate. The effects of growth in these two population groups on crime rates is counterbalancing, producing a stable crime rate.

Even though growth in the size of the adolescent population has been used to explain the boom in juvenile violence that began in the mid-1980s and persisted until the mid-1990s, it has not stood up well in research. While demographic effects undoubtedly play an important role, age structure is not the only influence, nor is its influence constant.[57] Other changes in society, such as level of drug use, availability of hand guns, economic conditions, and tougher laws and enforcement practices are also necessary to explain crime trends.[58]

■ The Ecology of Juvenile Offenses: Spatial and Temporal Distribution

One of the oldest observations about crime and delinquency is that offenses are not evenly distributed in place and time. In fact, the first attempt to collect national crime data, in France in 1825, contained information about *when* and *where* crime occurred. These data included the location of the crime; the time of year; and the age, sex, residence, occupation, and educational status of both the accused and the convicted.[59] The resulting data were published by the French Ministry of Justice as the *Compte*. It was hoped that the "hard facts" of crime could be used to understand the causes of crime and solve this pervasive problem.[60]

One hundred years later, American criminology was taking root. Two pioneering sociologists at the University of Chicago, Robert Park and Ernst Burgess, observed that a variety of social problems, including crime and delinquency, were distributed *ecologically*—in a geographic pattern associated with the growth of cities.[61] Based on this observation, Clifford Shaw and Henry McKay began their extensive study of juvenile delinquency in Chicago. They found that juvenile delinquency varied considerably across the city. Offenses were concentrated in the central part of the city and diminished as distance from the city center increased. Shaw and McKay argued that this "geographic distribution" of delinquency was closely connected to the social characteristics of neighborhoods.[62] We describe Shaw and McKay's work more thoroughly in Chapter 12.

Criminologists have maintained a steady interest in the **ecology of delinquency** by examining the **spatial** and **temporal distribution** of offenses.[63] Here we focus on the spatial distribution of delinquent offenses across three areas of different size—urban, suburban, and rural. We examine the commonly accepted notion that delinquent offenses occur at a much higher rate in urban areas than in rural areas. We also consider the temporal dimension of time of day—when delinquency occurs.

ecology of delinquency The spatial and temporal distribution of delinquent offenses—how offenses vary across place and time.

spatial distribution The geographic occurrence of delinquency—where delinquency occurs in terms of place.

temporal distribution The time aspect of delinquency—when delinquency occurs.

Self-Report Data

Data from the National Youth Survey show some, but not many, urban–rural differences in the prevalence and incidence of delinquent offenses (**Table 4-7**). A greater percentage of urban youth, as compared to suburban or rural youth, report that they have been involved in family property damage and various forms of fighting. Incidence figures reveal that urban youth are involved more frequently in property damage, various forms of theft, fighting, and sexual intercourse. Rural youth report greater frequency of hitting or threatening to hit.

Official Data: Uniform Crime Reporting Program

Official data reflect not only the behavior of offenders, but also the behavior of victims and justice agencies and personnel.[64] As such, we can anticipate spatial and temporal variation in juvenile delinquency according to official data. However, official sources report only a limited amount of data in terms of ecological variation.

Table 4-7	Prevalence and Incidence of Delinquent Behavior by Place of Residence in the National Youth Survey, 1976					
	Prevalence Percent reporting one or more offenses			Incidence Mean number of offenses per youth		
In the last year have you . . .	Urban	Suburban	Rural	Urban	Suburban	Rural
Purposely damaged or destroyed school property?	17	17	14	2.08	.75	.29
Stolen or tried to steal a motor vehicle, such as a car or motorcycle?	2	1	0	.02	.02	.00
Stolen or tried to steal something under $5?	19	18	16	2.81	.95	.42
Stolen something worth $5 to $50?	8	5	3	.51	.13	.27
Stolen or tried to steal something worth more than $50?	4	2	1	.15	.04	.01
Run away from home?	6	5	7	.08	.08	.10
Lied about your age to gain entrance or to purchase something?	31	29	19	5.39	2.53	.92
Been loud, rowdy, or unruly in a public place?	34	33	29	4.41	2.25	3.37
Carried a hidden weapon other than a plain pocket knife?	11	5	4	1.26	.86	.94
Attacked someone with the idea of seriously hurting or killing him or her?	10	4	6	.40	.08	.10
Been involved in gang fights?	17	13	8	.47	.29	.16
Hit or threatened to hit parents?	5	7	6	.11	.43	2.81
Hit or threatened to hit a teacher?	12	7	6	.40	.26	.94
Hit or threatened to hit another student?	51	46	48	9.69	4.31	5.44
Used alcoholic beverages (beer, wine, and hard liquor)?	43	50	43	2.02	2.26	2.00
Used marijuana–hashish ("grass," "pot," "hash")?	21	19	12	1.79	1.62	1.3
Sold hard drugs such as heroin, cocaine, and LSD?	2	0	0	.03	.01	.64

Source: Ageton and Elliott, "Incidence." Published in McGarrell and Flanagan, Sourcebook 1984, 373–375.

UCR data provide juvenile arrest statistics that are broken down by urban, suburban, and nonmetropolitan population groups. **Table 4-8** shows that juveniles account for a larger portion of the arrests in cities, as compared to suburban and nonmetropolitan areas. In nonmetropolitan areas, 10% of the arrests involved juveniles, compared to 16.5% in suburban areas, and 17.8% in cities.

Juvenile arrest rates across these population categories would provide a more direct measure of ecological variation than do clearance or percent arrest data. However, the UCR does not provide such information. Some time ago, Howard Snyder and Ellen Nimick, of the National Center for Juvenile Justice, calculated juvenile arrest rates and found that the juvenile property crimes arrest rate was 2,216 (per 100,000 youth ages 10 through 17) for youths in cities, compared to 1,297 for suburban youth, and 625 for youth in rural areas.[65] These findings indicate that juvenile delinquency, as measured by arrest rates, is a much greater problem in urban than in rural areas.

Table 4-8	Percent of Arrests Involving Juveniles, by Place of Residence, 2003		
Type of crime	City %	Suburban %	Nonmetropolitan %
Total Crime Index	17.8	16.5	10.0
Violent Crime Index	16.3	15.9	10.2
Property Crime Index	29.7	29.2	23.9

Source: Federal Bureau of Investigation, *Crime in the United States 2003*, Tables 46, 64, and 58, pages 293, 311, and 320.

Extensive variation also exists in juvenile arrest rates across different states. Using UCR data, Howard Snyder calculated state juvenile arrest rates for the Violent and Property Crime Indexes, drug abuse violations, and alcohol violations. **Table 4-9** reveals that the juvenile arrest rate for the Violent Crime Index varies from a low of 47 per 100,000 juveniles in Vermont, to a high of 897 in Illinois.[66] Similar variation is apparent for the other reported categories of crime.

The new generation of UCR data, called the National Incident-Based Reporting System (NIBRS), collects information on each crime reported to law enforcement, including the time the crime was committed.[67] Analysis of this temporal dimension for juvenile and adult crime from 1991 to 1996 appears in **Figure 4-10.** The different graphs reveal that adult robbery and aggravated assault follow similar temporal patterns, while juvenile patterns for these two offenses differ considerably. The percent of serious violent incidents for adults increases steadily from 6 AM until its peak at 11 PM. The juvenile pattern rises sharply after school and drops steadily after 9 PM. Further analysis reveals that juvenile violence on non-school days follows a temporal pattern similar to that of adult violence.[68] These findings on the temporal patterns of juvenile violence have spurred a host of after-school programs that provide supervised activity for youth after the school day, such as programs sponsored by Boys and Girls Clubs.

Ecology of Delinquency in Brief

The taken-for-granted urban–rural difference in juvenile delinquency does not receive strong support in the data. Self-report data show little variation in the proportion of youth who report involvement in crime across three different groups of population size: urban, suburban, and rural. At most, a slightly greater number of urban youth report involvement in property damage and fighting. When involved in delinquent acts, urban youth report more frequent involvement in property damage, theft, and fighting. Official data, however, show greater ecological variation in involvement in delinquency. Juveniles are involved in a larger portion of the serious crimes in cities, compared to suburban and rural areas. Juvenile arrest rates in cities are greater than in suburban or rural areas.

The new generation of UCR data provides a temporal dimension to crime, allowing us to see when crime occurs most frequently. The temporal pattern for juvenile crime is distinct from adult crime. Most importantly, juvenile violence peaks in the after-school hours.

Table 4-9 State Juvenile Arrest Rates per 100,000 Persons, Ages 10–17, 2002

State	Reporting Coverage	2002 Juvenile Arrest Rate Violent Crime Index	Property Crime Index	Drug Abuse	Weapons
United States	77%	295	1,511	571	105
Alabama	84	138	773	219	30
Alaska	91	257	2,375	547	69
Arizona	95	259	1,938	740	69
Arkansas	52	169	1,393	308	61
California	99	365	1,225	549	162
Colorado	81	231	2,215	729	144
Connecticut	70	197	1,147	471	77
Delaware	85	330	1,405	447	244
District of Columbia	0	NA	NA	NA	NA
Florida	99	517	2,170	718	100
Georgia	49	263	1,320	440	103
Hawaii	77	286	1,669	429	35
Idaho	97	157	2,254	498	111
Illinois	23	898	2,323	2,541	384
Indiana	69	337	1,352	454	34
Iowa	91	244	1,957	417	37
Kansas	49	168	1,211	458	35
Kentucky	23	291	1,646	668	60
Louisiana	71	398	1,949	533	71
Maine	100	99	2,004	541	33
Maryland	59	299	1,630	797	133
Massachusetts	73	428	709	399	33
Michigan	96	172	964	324	54
Minnesota	83	184	2,046	614	86
Mississippi	54	120	1,735	551	78
Missouri	84	298	1,685	585	97
Montana	66	157	2,182	270	32
Nebraska	91	107	2,266	669	68
Nevada	71	221	2,083	313	63
New Hampshire	64	118	1,062	725	15
New Jersey	97	354	1,039	763	178
New Mexico	64	307	1,144	545	142
New York	33	314	1,485	706	96
North Carolina	83	310	1,563	417	154
North Dakota	90	61	2,146	381	42
Ohio	57	185	1,105	345	66
Oklahoma	98	243	1,476	439	83
Oregon	84	133	1,826	520	58
Pennsylvania	85	398	1,257	564	98
Rhode Island	98	257	1,464	577	138
South Carolina	54	407	1,548	636	142
South Dakota	69	80	1,686	597	76
Tennessee	85	195	899	372	83
Texas	98	194	1,383	538	61
Utah	95	175	2,490	565	105
Vermont	86	47	750	323	15
Virginia	86	128	862	370	73
Washington	84	230	2,031	496	92
West Virginia	51	54	541	122	22
Wisconsin	91	349	3,207	884	231
Wyoming	98	106	1,649	824	76

Source: Snyder, "Juvenile Arrests 2000," 11.

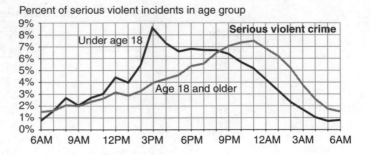

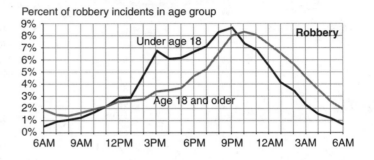

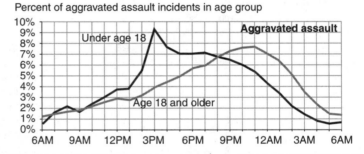

Figure 4-10 Temporal patterns of juvenile and adult violence. Juvenile violence peaks in the after-school hours and is distinct from the temporal patterns of adult violence.
Source: Snyder and Sickmund, *Juvenile Offenders*, 64.

Why Are Delinquent Offenses Distributed in Place and Time?

Delinquent offenses are not evenly distributed in place and time. This ecological variation is a major area of study in criminology and has spawned much research and theory. Research like that of Shaw and McKay in Chicago tries to document ecological variation, showing that delinquent offenses are distributed geographically. In addition, a number of theories have been developed to explain such ecological variation.

This research often concentrates on structural characteristics of the social environment that influence rates of delinquency. Sociologists refer to these organizational features of the social environment as *social structures*. Characteristics such as age, race, gender, and social class, together with cultural norms and traditions and institutional controls, are the center of attention in research and theory on social structures.

Criminological research attempts to measure empirically these structural characteristics and evaluate the strength of their influence on delinquency rates. Shaw and McKay, for example, identified and measured three key structural characteristics that disrupt community social organization and, in turn, are related to high rates of crime and delinquency: low economic status, ethnic heterogeneity, and residential mobility.[69] In Chapter 12, we examine these structural factors more fully.

■ Summary and Conclusions

Using aggregate data, this chapter has addressed the extent of delinquent offenses by exploring the proportion of juveniles that are involved in delinquent offenses and the frequency of their involvement. We have also examined the relative frequency of different types of offenses and the ecology of offenses in terms of spatial and temporal distribution.

To understand the extent of delinquency we examined prevalence and incidence. Self-report data lead to the clear conclusion that a substantial portion of adolescents report involvement in delinquent acts, but relatively few indicate that their involvement is frequent or repetitive. Furthermore, among those youth who commit delinquent acts, only a small portion come to the attention of the police, relatively few are arrested, and even fewer are processed by juvenile courts.

Virtually all sources of data indicate that minor forms of delinquent behavior are far more common than serious, violent offenses. Property offenses, especially property destruction and theft, are among the most common types of offenses. Alcohol and marijuana use also exceeds serious, violent offenses in prevalence and incidence. Among violent offenses, minor forms are most common, particularly simple assault. Truancy and liquor law violations are relatively common status offenses.

Despite data limitations, we can draw certain conclusions about delinquency trends. Because much property crime goes unreported, self-report data provide the most accurate indication of trends. Across a variety of offense types, self-report delinquency data generally show a fairly stable pattern since the late 1980s. Self-report drug use shows a decline since the mid-1990s. Even though juvenile property arrests fluctuated in the first half of the 1990s, a notable drop followed in the second half of the decade. Violent juvenile crime arrest trends indicate an increase in the late 1980s and into the early 1990s, but a steady decline since then. According to UCR data, drug use declined through the 1980s and early 1990s, then increased some, then declined again in the latter part of the decade. In general, with the exception of an increase in violent juvenile crime in the late 1980s and early 1990s, the incidence of juvenile crime has not changed dramatically in the last 25 years.

Urban–rural differences in juvenile delinquency are most extensive for arrest rates, but small when measured with self-report data. However, there is more variation in arrest rates between different cities than between urban and rural population groups. Perhaps the most important ecological dimension of delinquent offenses is time of day. Juvenile violence peaks dramatically in the after-school hours.

CRITICAL THINKING QUESTIONS

1. With regard to juvenile delinquency, to what do the terms *prevalence* and *incidence* refer? What conclusions can be drawn about the prevalence and incidence of delinquent offenses?

2. What types of delinquent offenses are most common? Why do these forms of delinquent behavior occur more frequently than others?

3. In recent years, has there been a juvenile crime wave? Has there been an epidemic of youth violence?

4. What is the *ecology* of delinquency?

5. Do most delinquent youth develop a persistent and specialized pattern of offending, as did Stanley in the case that opened this chapter?

SUGGESTED READINGS

Snyder, Howard N. "Juvenile Arrests 2003." Washington, DC: Office of Juvenile Justice and Delinquency Prevention, 2005.

Snyder, Howard N., and Melissa Sickmund. *Juvenile Offenders and Victims: 1999 National Report.* Washington, DC: Office of Juvenile Justice and Delinquency Prevention, 1999.

GLOSSARY

ecology of delinquency: The spatial and temporal distribution of delinquent offenses—how offenses vary across place and time.

incidence: The number of offenses committed by adolescents in general, or by each delinquent youth.

prevalence: The proportion of youths involved in delinquent acts.

relative frequency: In comparison to each other, the types of delinquent offenses that occur most often.

spatial distribution: The geographic occurrence of delinquency—where delinquency occurs in terms of place.

temporal distribution: The time aspect of delinquency—when delinquency occurs.

REFERENCES

Ageton, Suzanne S., and Delbert S. Elliott. "The Incidence of Delinquent Behavior in a National Probability Sample of Adolescents." Boulder, CO: Behavioral Research Institute, 1978.

Beirne, Piers. "Adolphe Quetelet and the Origins of Positivist Criminology." *American Journal of Sociology* 92 (1987):1140–1169.

Beirne, Piers, and James Messerschmidt. *Criminology.* 3rd ed. Boulder, CO: Westview Press, 2000.

Bennett, William J., John J. DiIulio, and John P. Walters. *Body Count: Moral Poverty and How to Win America's War Against Crime and Drugs.* New York: Simon and Schuster, 1996.

Bernard, Thomas. "Juvenile Crime and the Transformation of Juvenile Justice: Is There a Juvenile Crime Wave?" *Justice Quarterly* 16 (1999):336–356.

Blumstein, Alfred. "Disaggregating the Violence Trends." In *The Crime Drop in America,* edited by Alfred Blumstein and Joel Wallman, 13–44. Boston: Cambridge University Press, 2000.

———. "Violence by Young People: Why the Deadly Nexus?" *National Institute of Justice Journal,* 229 (August 1995):2–9.

———. "Youth Violence, Guns, and the Illicit-Drug Industry." *Journal of Criminal Law and Criminology* 86 (1995):10–36.

Browning, Katherine, David Huizinga, Rolf Loeber, and Terence P. Thornberry. "Causes and Correlates of Delinquency Program." In *OJJDP Fact Sheet #100.* Washington, DC: Office of Juvenile Justice and Delinquency Prevention, 1999.

Bursik, Robert J., Jr. "Social Disorganization and Theories of Crime and Delinquency: Problems and Prospects." *Criminology* 26 (1988):519–551.

Byrne, James M., and Robert J. Sampson. *The Social Ecology of Crime.* New York: Springer-Verlag, 1986.

Catalano, Richard F., and David J. Hawkins. *Risk Focused Prevention: Using the Social Development Strategy.* Seattle: Developmental Research and Programs, Inc., 1995.

Cook, Philip J., and John Laub. "The Unprecedented Epidemic in Youth Violence." In *Crime and Justice,* edited by Mark H. Moor and Michael Tonry, 101–138. Chicago, IL: University of Chicago Press, 1998.

Coolbaugh, Kathleen, and Cynthia J. Hansel. *The Comprehensive Strategy: Lessons Learned From Pilot Sites.* Washington, DC: Office of Juvenile Justice and Delinquency Prevention, 2000.

Coordinating Council of Juvenile Justice and Delinquency Prevention. *Combating Violence and Delinquency: The National Juvenile Justice Action Plan.* Washington, DC: Office of Juvenile Justice and Delinquency Prevention, 1996.

Davis, Nanette. *Youth Crisis: Growing Up in the High-Risk Society.* Westport, CT: Praeger, 1999.

DiIulio, John J., Jr. "Arresting Ideas: Tougher Law Enforcement is Driving Down Urban Crime." *Policy Review* 74 (1995):12–16.

Elliott, Delbert S., and Suzanne S. Ageton. "Reconciling Differences in Estimates of Delinquency." *American Sociological Review* 45 (1980):95–110.

Elliott, Delbert S., David Huizinga, and Suzanne S. Ageton. *Explaining Delinquency and Drug Use.* Beverly Hills, CA: Sage, 1985.

Elliott, Delbert S., David Huizinga, and Scott Menard. *Multiple Problem Youth: Delinquency, Substance Use, and Mental Health Problems.* New York: Springer-Verlag, 1989.

Farrington, David P. "Predictors, Causes, and Correlates of Male Youth Violence." In *Youth Violence,* edited by Michael Tonry and Mark H. Moore, 421–475. Chicago: University of Chicago Press, 1988.

Federal Bureau of Investigation. 2004. *Crime in the United States 2003: Uniform Crime Reports.* US Department of Justice. Washington, DC: GPO, 2004. Also available at http://www.fbi.gov/ucr/ucr.htm#cius.

Fox, James Alan. "Demographics and U.S. Homicide." In *The Crime Drop in America,* edited by Alfred Blumstein and Joel Wallman, 288–318. Boston: Cambridge University Press, 2000.

Gibbons, Don C., and Marvin D. Krohn. *Delinquent Behavior.* 5th ed. Englewood Cliffs, NJ: Prentice Hall, 1991.

Hamparian, Donna Martin, Joseph M. Davis, Judith M. Jacobson, and Robert E. McGraw. *The Young Criminal Years of the Violent Few.* Washington, DC: Office of Juvenile Justice and Delinquency Prevention, 1985.

Hawkins, David J., Richard F. Catalano, and J. Y. Miller. "Risk and Protective Factors for Alcohol and Other Drug Problems in Adolescence and Early Adulthood: Implications for Substance Abuse Programs." *Psychological Bulletin* 112 (1992):64–105.

Hawkins, J. David, Todd I. Herrenkohl, David P. Farrington, Devon Brewer, Richard F. Catalano, Tracy W. Harachi, and Lynn Cothern. "Predictors of Youth Violence." Washington, DC: Office of Juvenile Justice and Delinquency Prevention, 2000.

Hawkins, David, and Joseph G. Weis. *The Social Development Model: An Integrated Approach to Delinquency Prevention.* Washington, DC: Office of Juvenile Justice and Delinquency Prevention, 1980.

Hindelang, Michael J., Travis Hirsch, and Joseph G. Weis. "Correlates of Delinquency: The Illusion of Discrepancy Between Self-Report and Official Measures." *American Sociological Review* 44 (1979):95–1014.

Howell, James C., ed. *Guide for Implementing the Comprehensive Strategy for Serious, Violent, and Chronic Juvenile Offenders.* Washington, DC: Office of Juvenile Justice and Delinquency Prevention, 1995.

———. *Preventing and Reducing Juvenile Delinquency: A Comprehensive Framework.* Thousand Oaks, CA: Sage, 2003.

Huizinga, David, and Delbert S. Elliott. "Juvenile Offenders: Prevalence and Incidence, and Arrest Rates by Race." *Crime and Delinquency* 33 (1987):206–223.

Chapter Resources

Huizinga, David, Rolf Loeber, Terence P. Thornberry, and Lynn Couthern. "Co-Occurrence of Delinquency and Other Problem Behaviors." Washington, DC: Office of Juvenile Justice and Delinquency Prevention, 2000.

Jensen, Gary F., and Dean G. Rojek. *Delinquency and Youth Crime.* 3rd ed. Prospect Heights, IL: Waveland Press, 1998.

Johnston, Lloyd D., Patrick M. O'Malley, Jerald G. Bachman, and John E. Schulenberg. *Monitoring the Future.* Available at http://www.monitoringthefuture.org (accessed September 12, 2005).

Kelley, Barbara Tatem, David Huizinga, Terence P. Thornberry, and Rolf Loeber. *Epidemiology of Serious Violence.* Washington, DC: Office of Juvenile Justice and Delinquency Prevention, 1997.

Kornhauser, Ruth Rosner. *Social Sources of Delinquency: An Appraisal of Analytic Models.* Chicago: University of Chicago Press, 1978.

Levitt, Steven. "The Limited Role of Changing Age Structure in Explaining Aggregate Crime Rates." *Criminology* 37 (1999):581–599.

Loeber, Rolf, and David P. Farrington, eds. *Serious and Violent Juvenile Offenders: Risk Factors and Successful Intervention.* Thousand Oaks, CA: Sage, 1998.

Maguire, Kathleen, and Ann L. Pastore, eds. *Sourcebook of Criminal Justice Statistics 2000.* Washington, DC: Bureau of Justice Statistics, 2001.

McGarrell, Edmund F., and Timothy J. Flanagan, eds. *Sourcebook of Criminal Justice Statistics 1984.* Washington, DC: Bureau of Justice Statistics, 1985.

Moffitt, Terrie E. "Adolescence-Limited and Life-Course-Persistent Antisocial Behavior: A Developmental Taxonomy." *Psychological Review* 100 (1993):674–701.

Park, Robert E., Ernst W. Burgess, and R. D. McKenzie. *The City.* Chicago: University of Chicago Press, 1967.

Pastore, Ann L., and Kathleen Maguire, eds. *Sourcebook of Criminal Justice Statistics Online 2004.* Available at http://www.albany.edu/sourcebook (accessed September 12, 2005).

Puzzanchera, Charles M. "Self-Reported Delinquency by 12-Year-Olds, 1997." In *OJJDP Fact Sheet* #03. Washington, DC: Office of Juvenile Justice and Delinquency Prevention, 2000.

Puzzanchera, Charles, Anne L. Stahl, Terrence A. Finnegan, Howard N. Snyder, Rowen S. Poole, and Nancy Tierney. *Juvenile Court Statistics 1997.* US Department of Justice. Washington, DC: Office of Juvenile Justice and Delinquency Prevention, 2000.

Puzzanchera, Charles, Anne L. Stahl, Terrence A. Finnegan, Nancy Tierney, and Howard N. Snyder. *Juvenile Court Statistics 1998.* US Department of Justice. Washington, DC: Office of Juvenile Justice and Delinquency Prevention, 2003.

Puzzanchera, Charles, Anne L. Stahl, Terrence A. Finnegan, Nancy Tierney, and Howard N. Snyder. *Juvenile Court Statistics 2000.* US Department of Justice. Washington, DC: Office of Juvenile Justice and Delinquency Prevention, 2004.

Rumsey, Elissa, Charlotte A. Kerr, and Barbara Allen-Hagen. *Serious and Violent Juvenile Offenders.* Washington, DC: Office of Juvenile Justice and Delinquency Prevention, 1998.

Sampson, Robert J., and W. Byron Groves. "Community Structure and Crime: Testing Social Disorganization Theory." *American Journal of Sociology* 94 (1989):774–802.

Shaw, Clifford R. *The Jack-Roller: A Delinquent Boy's Own Story.* Chicago: University of Chicago Press, 1930.

Shaw, Clifford R., and Henry D. McKay. *Juvenile Delinquency and Urban Areas: A Study of Rates of Delinquency in Relation to Differential Characteristics of Local Communities in American Cities.* Rev. ed. Chicago: University of Chicago Press, 1969.

Shaw, Clifford R., Frederick M. Zorbaugh, Henry D. McKay, and Leonard S. Cottrell. *Delinquency Areas: A Study of the Geographic Distribution of School Truants, Juvenile Delinquents, and Adult Offenders in Chicago.* Chicago: University of Chicago Press, 1929.

Sickmund, Melissa. "Juveniles in Court." In *Juvenile Offenders and Victims National Report Series.* Washington, DC: Office of Juvenile Justice and Delinquency Prevention, 2003.

———. *Offenders in Juvenile Court, 1997.* Washington, DC: Office of Juvenile Justice and Delinquency Prevention, 2000.

Siegel, Larry J., Brandon C. Welsh, and Joseph J. Senna. *Juvenile Delinquency: Theory, Practice, and Law.* 8th ed. Belmont, CA: Wadsworth, 2003.

Snyder, Howard N. "Juvenile Arrests 2002." Washington, DC: Office of Juvenile Justice and Delinquency Prevention, 2004.

Snyder, Howard N. "Juvenile Arrests 2003." Washington, DC: Office of Juvenile Justice and Delinquency Prevention, 2005.

Snyder, Howard N., and Melissa Sickmund. "Challenging the Myths." *1999 National Report Series: Juvenile Justice Bulletin.* Washington, DC: Office of Juvenile Justice and Delinquency Prevention, 2000.

———. *Juvenile Offenders and Victims: 1999 National Report.* Washington, DC: Office of Juvenile Justice and Delinquency Prevention, 1999.

Steffensmeier, Darrell, and Miles D. Harer. "Making Sense of Recent U.S. Crime Trends, 1980 to 1996/1998: Age Composition Effects and Other Explanations." *Journal of Research in Crime and Delinquency* 36 (1999):235–274.

Sutherland, Edwin, Donald R. Cressey, and David F. Luckenbill. *Principles of Criminology.* 11th ed. Dix Hills, NY: General Hall, 1992.

Thornberry, Terence P., David Huizinga, and Rolf Loeber. "The Causes and Correlates Studies: Findings and Policy Implications." *Juvenile Justice* 9, no. 1 (2004):3–19.

U.S. Bureau of the Census. Available at http://www.census.gov.

Weis, Joseph G., and David Hawkins. *Preventing Delinquency.* Washington, DC: GPO, 1981.

Williams, Jay R., and Martin Gold. "From Delinquent Behavior to Official Delinquency." *Social Problems* 20 (1972):209–229.

Wilson, John J., and James C. Howell. *Comprehensive Strategy for Serious, Violent, and Chronic Juvenile Offenders: Program Summary.* Washington, DC: Office of Juvenile Justice and Delinquency Prevention, 1993.

ENDNOTES

1. Shaw, *Jack-Roller,* 3–4.
2. Ibid., 97, fn 7.
3. Gibbons and Krohn, *Delinquent Behavior,* 21; and Huizinga and Elliott, "Juvenile Offenders," 210.
4. Gibbons and Krohn, *Delinquent Behavior,* 45; and Hindelang, Hirschi and Weis, "Correlates of Delinquency."
5. Gibbons and Krohn, *Delinquent Behavior,* 45.
6. Williams and Gold, "From Delinquent Behavior," 213; and Gibbons and Krohn, *Delinquent Behavior,* 45.
7. Huizinga and Elliott, "Juvenile Offenders."
8. Huizinga et al., "Co-Occurrence," 2.
9. The use of the term "prevalence" in "annual prevalence" and "30-day prevalence" to indicate incidence may be a bit confusing. These units of analysis report the percent of youth reporting use in the last year and in the last 30 days. In this regard, they can be viewed as indicating prevalence of drug use in a given time period. However, the recency of drug use is also represented and implied in this frequency. In this way, annual and 30-day prevalence provide evidence of incidence.
10. Jensen and Rojek, *Delinquency and Youth Crime,* 82.
11. Snyder, "Juvenile Arrests 2003," 2.
12. US Bureau of the Census, Census 2000 Summary File 1, Matrix PCT12; and Snyder and Sickmund, *Juvenile Offenders and Victims,* 2.
13. Puzzanchera et al., *2000,* 7–8. The number of petitioned status offenses and case rates are from 1997 (Puzzanchera et al., *1997,* 37). The 1998 reporting of *Juvenile Court Statistics* changed with regard to petitioned status offense cases because of extensive variation among communities in handling and counting status offense cases. As discussed later, the number of petitioned status offense cases is a problematic estimate of volume of status offenses dealt with by juvenile courts (Puzzanchera et al., *1998,* 4).
14. Thornberry, Huizinga, and Loeber, "Causes and Correlates."
15. Browning, Huizinga, Loeber, and Thornberry, "Causes and Correlates," 1; and Thornberry, Huizinga, and Loeber, "Causes and Correlates."
16. Wilson and Howell, *Comprehensive Strategy,* 13.
17. Coolbaugh and Hansel, *Comprehensive Strategy,* 3; Hawkins, Catalano, and Miller, "Risk;" Weis and Hawkins, *Preventing Delinquency*; and Hawkins and Weis, *Social Development Model.*
18. Wilson and Howell, *Comprehensive Strategy,* 13.
19. In *Youth Crisis: Growing Up in the High-Risk Society,* Nanette Davis emphasizes the larger social context of risk. She argues that our contemporary youth crisis is composed of a *high social crisis* where social and moral order is broken down, combined with *high risk* for some youth, together with a *lack of social justice* (a shared sense of obligation to each other).
20. Coolbaugh and Hansel, *Comprehensive Strategy,* 1; Loeber and Farrington, *Serious and Violent*; Howell, *Guide*; and Wilson and Howell, *Comprehensive Strategy.*

Chapter Resources

21. Coolbaugh and Hansel, *Comprehensive Strategy,* 1–2.
22. McGarrell and Flanagan, *Sourcebook,* 363–364, 366; and Maguire and Pastore, *Sourcebook,* 222–223, 232, 234, 237.
23. Elliott, Huizinga, and Ageton, *Explaining Delinquency.*
24. Elliott and Ageton, "Reconciling Differences," 101.
25. Snyder, "Juvenile Arrests 2003."
26. Hamparian et al., *Violent Few,* 7.
27. Ibid.
28. Puzzanchera et al., *1999,* 7; and Sickmund, "Juveniles in Court," 12.
29. Puzzanchera et al., *1997,* 1.
30. Ibid., 70–92.
31. Hamparian et al., *Violent Few*; Farrington, "Predictors;" and Elliott, Huizinga, and Ageton, *Explaining Delinquency.*
32. Huizinga et al., "Co-occurrence," 2.
33. Elliott, Huizinga, and Menard, *Multiple Problem Youth*; Hamparian et al., *Violent Few*; and Huizinga et al., "Co-occurrence."
34. Rumsey, Kerr, and Allen-Hage, *Serious and Violent,* 2.
35. Rumsey, Kerr, and Allen-Hage, *Serious and Violent,* 2–3. See also Hawkins et al., "Predictors of Youth Violence," and Huizinga et al., "Co-occurrence."
36. Moffitt, "Developmental Taxonomy."
37. Blumstein, "Youth Violence," 10.
38. Jensen and Rojek, *Delinquency and Youth Crime,* 132.
39. Snyder, "Juvenile Arrests 2003," 4–9.
40. Ibid., 4.
41. Ibid., 5.
42. Puzzanchera et al., *1999,* 8.
43. The 10-year period reported here is 1988–1997 (Puzzanchera et al., *1997,* 37).
44. Snyder, "Juvenile Arrests 2003," 5.
45. Coordinating Council on Juvenile Justice and Delinquency Prevention, *Combating Violence and Delinquency,* 1.
46. Blumstein, "Youth Violence," 10.
47. Ibid., 29.
48. Ibid., 10.
49. Ibid., 30–31.
50. Blumstein, "Violence by Young People."
51. DiIulio, "Arresting Ideas," 15.
52. Snyder and Sickmund, *Juvenile Offenders and Victims,* 2.
53. Ibid., 2–8.
54. DiIulio, "Arresting Ideas," 15.
55. Bennett, DiIulio, and Walters, *Body Count.*
56. Levitt, "Limited Role."
57. Fox, "Demographics and U.S. Homicide;" and Steffensmeier and Harer, "Making Sense."
58. Bernard, "Juvenile Crime;" Blumstein, "Disaggregating the Violence Trends;" Cook and Laub, "Unprecedented Epidemic;" and Steffensmeier and Harer, "Making Sense," 262–266.
59. Beirne, "Adolphe Quetelet," 1148.
60. Beirne and Messerschmidt, *Criminology,* 75.
61. Park, Burgess, and McKenzie, *The City.*
62. Kornhauser, *Social Sources of Delinquency*; Sampson and Groves, "Community Structure and Crime;" Shaw and McKay, *Juvenile Delinquency*; and Shaw et al., *Delinquency Areas.*
63. Byrne and Sampson, *Social Ecology of Crime.*
64. Jensen and Rojek, *Delinquency and Youth Crime,* 91.
65. Ibid., 91.
66. Snyder, "Juvenile Arrests 2003," 11.
67. Snyder and Sickmund, *Juvenile Offenders and Victims,* 64.
68. Ibid., 65.
69. Sampson and Groves, "Community Structure and Crime," 774–775. See also Kornhauser, *Social Sources of Delinquency,* 63–64, and Robert Bursik, "Social Disorganization," 520.

5

Age, Gender, Race, and Class of Offenders

Chapter Objectives

After completing this chapter, students should be able to:

- Describe how involvement in delinquency is related to the social characteristics of age, gender, race, and social class.
- Explain why the relationships between these social characteristics and delinquency differ for different types of data (Uniform Crime Reports, self-reports, victimization data).
- Describe how the likelihood of victimization is related to age, gender, race, and social class.
- Understand key terms:
  social correlates
  age–crime curve
  crime-prone years
  aging out of crime

age effect
age composition effect
convergence hypothesis
ecological correlation
ecological fallacy

CASE IN **POINT**

Perceptions of Criminals in Chestnut Hill: The Social Correlates of Crime and Delinquency

Elijah Anderson is an ethnographer whose recent work describes the "code of the street" and the ways in which poor inner-city youth use violence or the threat of violence to acquire respect and to shield themselves from further violence. In impoverished neighborhoods, the code of the street often replaces legitimate forms of social control. Anderson begins his work with a description of the neighborhood in which he conducted his research.

Chestnut Hill . . . is a predominantly residential community consisting mostly of affluent and educated white people, but it is increasingly becoming racially and ethnically mixed. . . . The business and shopping district along Germantown Avenue draws shoppers from all over the city. . . . You see many different kinds of people—old and young, black and white, affluent, middle class, and working class. . . .

Once in a while . . . a violent incident does occur in Chestnut Hill. A holdup occurred at the bank in the middle of the day not long ago, ending in a shoot-out on the sidewalk. The perpetrators were black, and two black men recently robbed and shot up a tavern on the avenue. Such incidents give the residents here the simplistic yet persistent view that blacks commit crime and white people do not. That does not mean that the white people here think that the black people they ordinarily see on the streets are bound to rob them: many of these people are too sophisticated to believe that all blacks are inclined to criminality. But the fact that black people robbed the bank and that blacks commit a large number of crimes in the area does give a peculiar edge to race relations, and the racial reality of street crime affects the relations between blacks and whites. Because everybody knows that the simplistic view does exist, even middle-class blacks have to work consciously against that stereotype—although the whites do as well. Both groups know the reality that crime is likely to be perpetrated by young black males.

Source: Anderson, *Code of the Street,* 16–17.

To understand theoretical explanations of delinquency, you must first grasp the nature of delinquent behavior itself, including an understanding of offenders. Who are the offenders? What are the social characteristics that tend to distinguish them from non-offenders? When you think of delinquent offenders, what picture comes to mind? Young or old? Male or female? Rich or poor? Black, white, or another color? What, if anything, do age, gender, social class, and race or ethnicity have to do with the likelihood of offending?

The social characteristics that tend to distinguish offenders from non-offenders are often called the **social correlates** of delinquency. Correlation is "the extent to which two or more things are related ('co-related') to one another."[1] So, by social correlates, we mean social characteristics that are statistically associated with each other. It is important to distinguish correlation from causation. As we show later in this chapter, gender and delinquency are correlated, with males being more likely than females to be involved in offending. But gender does not directly *cause* delinquency. Instead, perhaps parents supervise daughters more closely than sons, leading to greater opportunities for delinquent behavior for sons than for daughters. Thus, lack of parental supervision might be a cause of delinquency and might mediate between gender and delinquency.

In discussing social correlates of delinquency, we place the emphasis on the word "social." In this chapter we examine age, gender, race, and social class as social correlates of offending. An individual can be defined according to these characteristics—for example, a young female, or a wealthy African American person. But each of these characteristics is fundamentally *social*, rather than individual, in nature. To be female in contemporary American culture, for example, has meaning far beyond the biological sex of an individual. Gender is socially defined, and it determines and defines positions of power—or lack of it—within society. The same is true for race, social class, and age.

Once we have established that age, gender, race, and social class are social correlates of delinquency, the next logical question is, "*Why* are these characteristics related to offending?" In this chapter, we focus primarily on the social correlates of delinquency, providing evidence that helps us answer the question, "Who are delinquent offenders?" We also briefly address the question of why these social characteristics are related to involvement in delinquency. We address this question in greater detail in later chapters.

social correlates Social characteristics (such as age, gender, race, and social class) that are statistically related to involvement in delinquent behavior and that tend to distinguish offenders from non-offenders.

■ Age

There are few factors that criminologists agree are undeniably related to crime. Age is one of those rare factors. An indisputable link exists between age and involvement in crime. Crime is predominantly a pursuit of the young. Involvement in crime tends to increase with age during the teenage years, peaking in mid-adolescence to early adulthood, and then declines rapidly with age. When we graph the relationship between age and crime, we generally see a bell-shaped curve called the **age–crime curve** (see **Figure 5-1**). The age–crime relationship varies somewhat depending on the type of offense and other factors we discuss later in this chapter.

age–crime curve The bell-shaped curve generally observed when one plots or graphs the relationship between age and crime. The age–crime curve typically shows an increase in delinquent involvement during the teenage years, a peak in mid-adolescence to early adulthood, and then a rapid decline.

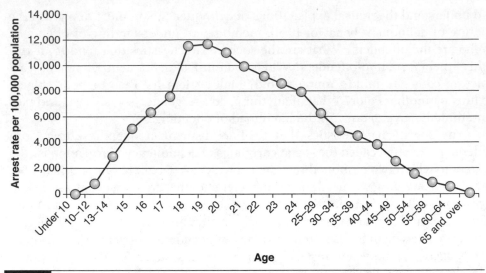

Figure 5-1 The "age–crime curve." This figure shows the bell-shaped curve one typically sees when graphing the relationship between age and crime. This curve illustrates the consistent finding that crime tends to increase with age during the teenage years, peak in mid-adolescence to early adulthood, and then decline rapidly with age.

Note: Arrest figures based on 10,843 reporting agencies, with 2003 estimated population 204,034,545.

Sources: Federal Bureau of Investigation, *Crime in the United States 2003,* 280–281. US Bureau of the Census, "Monthly Postcensal Resident Population." *National Population Estimates for the 2000s.* Estimates for July 1, 2003.

Table 5-1 shows, for each type of offense, the proportion of offenders arrested who are juveniles (under the age of 18) and adults (ages 18 and over). In 2003, juveniles accounted for 16.3% of all arrests and 25.3% of arrests for Index offenses. These numbers are remarkable when one considers that juveniles in the 13–17 age range (those most likely to be involved in delinquency) constituted only 7.2% of the US population in 2003.[2]

Table 5-1 shows that juveniles are more likely to be arrested for property than for violent Index crimes. Juveniles accounted for 28.9% of all persons arrested for property Index crimes, but only 15.5% of those arrested for violent Index crimes. But, compared to other violent crimes, robbery shows a higher percentage of juvenile arrests. For robbery, 23.7% of those arrested were juveniles. Juveniles are disproportionately represented in arrests for arson, for which they account for 50.8% of arrests, and for vandalism, for which they account for 39.4% of arrests.

We should note that statistics on the proportion of arrests attributable to juveniles can be somewhat misleading if a small number of juveniles is responsible for a large percentage of arrests. It may be, for example, that one juvenile is arrested six times for burglary. The percentages in Table 5-1 correctly state the proportions of juvenile and adult arrests. But they mask the reality that a relatively small number of "chronic offenders" account for a disproportionate share of crimes and arrests. Conversely, it is also true that more than one offender may be arrested for a single crime.

Arrest rates per age-eligible population offer another way to look at juvenile involvement in delinquency. These rates are adjusted to take into account the size of the juvenile and adult populations of potential offenders. **Table 5-2** shows arrest rates per 100,000 population for selected offenses, for juveniles ages

Table 5-1	Juvenile and Adult Arrests, by Type of Crime, 2003			
	Juveniles		Adults	
Type of crime	Number	Percentage	Number	Percentage
All crimes	1,563,149	16.3	8,018,274	83.7
Index crimes	393,622	25.3	1,164,702	74.7
Violent crimes	64,799	15.5	354,165	84.5
Homicide	783	8.6	8,336	91.4
Forcible rape	2,966	16.1	15,480	83.9
Robbery	17,900	23.7	57,767	76.3
Aggravated assault	43,150	13.7	272,582	86.3
Property crimes	328,823	28.9	810,537	71.1
Burglary	59,870	29.2	144,891	70.8
Larceny–theft	232,322	28.4	584,726	71.6
Motor vehicle theft	30,874	29.1	75,347	70.9
Arson	5,757	50.8	5,573	49.2
Non-Index crimes	1,069,527	13.3	6,953,572	86.7
Other assaults	170,168	19.4	706,937	80.6
Weapons offenses	27,492	23.3	90,352	76.7
Sex offenses (except forcible rape, prostitution)	12,747	20.0	51,012	80.0
Fraud	5,642	2.7	202,827	97.3
Embezzlement	826	6.9	11,160	93.1
Stolen property offenses (Buying, receiving, possessing)	17,184	19.2	72,376	80.8
Vandalism	76,042	39.4	117,041	60.6
Gambling	1,151	15.5	6,263	84.5
Disorderly conduct	136,970	30.2	316,675	69.8
Drug abuse violations	137,658	11.7	1,034,564	88.3
Liquor laws	96,592	22.4	335,320	77.6

Note: Arrest figures based on 10,843 reporting agencies, with 2003 estimated population 204,034,545.

Source: Federal Bureau of Investigation, *Crime in the United States 2003*, 280–281.

10–17, juveniles ages 0–17, and adults. When we calculated arrest rates for juveniles, we first considered only the juvenile population at ages 10–17 because relatively few juveniles engage in delinquency at ages younger than 10. Column 1 shows the number of offenses committed by juveniles at ages 10–17, divided by the juvenile population at ages 10–17 (and then multiplied by 100,000). We then considered the entire juvenile population (ages 0–17). Column 2 shows the number of offenses committed by all juveniles, divided by the total juvenile population in 2003.

Table 5-2	Arrest Rates per 100,000 Population, for Juveniles and Adults, 2003		
Type of crime	Juveniles (ages 10–17)	Juveniles (ages 0–17)	Adults
All crimes	4,612.0	2,140.0	3,682.1
Index crimes	1,159.7	538.9	534.8
Violent crimes	191.3	88.7	162.6
Homicide	2.3	1.1	3.8
Forcible rape	8.8	4.1	7.1
Robbery	53.2	24.5	26.5
Aggravated assault	127.1	59.1	125.2
Property crimes	968.4	450.2	372.2
Burglary	175.6	82.0	66.5
Larceny–theft	684.7	318.1	268.5
Motor vehicle theft	92.0	42.3	34.6
Arson	16.1	7.9	2.6
Non-Index crimes	3,452.3	1,601.1	3,147.2
Other assaults	499.6	233.0	324.6
Weapons offenses	80.9	37.6	41.5
Sex offenses (except forcible rape, prostitution)	36.8	17.5	23.4
Fraud	16.6	7.7	93.1
Embezzlement	2.5	1.1	5.1
Stolen property offenses (Buying, receiving, possessing)	51.0	23.5	33.2
Vandalism	220.0	104.1	53.7
Gambling	3.4	1.6	2.9
Disorderly conduct	404.6	187.5	145.4
Drug abuse violations	410.0	188.5	475.1
Liquor laws	287.9	132.2	154.0

Notes: Arrest figures based on 10,843 reporting agencies, with 2003 estimated population 204,034,545.

Population figures from census data for 2003: juvenile (ages 10–17) population = 33,498,951; juvenile (ages 0–17) population = 73,043,506; adult population = 217,766,271. Arrest rates were calculated for column 1 using arrests and population estimates only for juveniles ages 10–17, and for column 2 using arrests and population estimates for juveniles ages 0–17.

Sources: Federal Bureau of Investigation, Crime in the United States 2003, 280–281.

US Bureau of the Census, "Monthly Postcensal Resident Population." In National Population Estimates for the 2000s. Estimates for July 1, 2003.

Column 1 of Table 5-2 reveals, for juveniles at ages 10–17, much higher arrest rates for property Index offenses (968.4) than for violent Index offenses (191.3). The high rate for property Index offenses is driven primarily by the high rate for larceny–theft (684.7). When we compare arrest rates for the total juvenile population (column 2) and total adult population (column 3), we see similar rates for all Index offenses combined (538.9 for juveniles and 534.8 for adults). But this similarity is due to offsetting differences for juveniles and adults for violent and property offenses. For violent Index crimes, arrest rates are higher for adults (162.6) than for juveniles (88.7). But for property Index crimes, arrest rates are higher for juveniles (450.2) than for adults (372.2). For all non-Index offenses combined, the arrest rate is almost twice as high for adults (3,147.2) as for juveniles (1,601.1), although the juvenile arrest rate is substantially higher if we focus on those ages 10–17 (3,452.3).

Crime-Prone Years

So far, we have examined crimes by juveniles versus crimes by adults. This distinction, however, is somewhat artificial and ignores the continuity of crime from adolescence into young adulthood. The age–crime curve is smooth, not disjointed at age 18. The <u>**crime-prone years**</u>, defined as the age period when people are most likely to be involved in crime, extend into young adult years, with some variation depending on the type of offense.

Table 5-3 shows the age distribution of individuals arrested for all types of crime combined, and for violent and property Index offenses separately. This table clearly shows the continuity of crime from adolescence into young adulthood. While individuals in the 13–24 age range constitute only 17.1% of the US population, they account for 44.9% of arrests for all crimes and 51.3% of arrests for Index crimes.

Table 5-3 reveals that the crime-prone years are somewhat older for violent crimes than for serious property offenses. Serious property crime is concentrated in the 13–21 age range. Individuals in this age group constituted 12.8% of the population in 2003, yet they accounted for 45.5% of all property Index crime arrests. Violent Index crime, on the other hand, is concentrated in the 16–29 age range. While individuals in this age group constituted 19.3% of the 2003 population, they accounted for 50.0% of all violent Index crime arrests. Compared to violent crime arrests, property crime arrests show higher percentages of offenders in all age categories through age 20. Then a shift occurs, and violent crime arrests show higher percentages of offenders in all age categories after age 20.

Aging Out of Crime

Relatively few people who are involved in crime during adolescence and young adulthood continue offending into later adulthood (see Chapter 6). Most stop committing crime due to sociological, psychological, and biological forces in their lives. Criminologists often refer to this process as <u>**aging out of crime**</u>. Even given the continuity of offending from adolescence into young adulthood, the process of aging out of crime is fairly rapid. Arrests for serious property crime, for example, peak at ages 17–18, but then drop off quickly (see Table 5-3). By age 22, arrest figures for property crime are about half of what they were at ages

crime-prone years The age period when people are most likely to be involved in crime. Generally, the crime-prone years range from mid-adolescence into early adulthood.

aging out of crime Termination of involvement in crime following adolescence and young adulthood, due to sociological, psychological, and biological forces.

| Table 5-3 | Percentage of Offenders at Various Ages, by Type of Crime, 2003 |

Age	Percent of US Population	All Crimes	Index Crimes	Violent Index Crimes	Property Index Crimes
12 and under	18.0%	1.4%	2.6%	1.4%	3.1%
13–14	2.9	3.8	6.5	3.7	7.5
15	1.4	3.1	4.9	2.9	5.6
16	1.4	3.8	5.5	3.5	6.3
17	1.4	4.2	5.8	4.0	6.4
18	1.4	4.9	5.9	4.6	6.4
19	1.4	5.0	5.1	4.4	5.3
20	1.4	4.7	4.3	4.3	4.4
21	1.4	4.3	3.8	4.3	3.7
22	1.4	4.0	3.5	4.1	3.3
23	1.4	3.7	3.2	3.8	2.9
24	1.4	3.3	2.8	3.5	2.6
25–29	6.6	12.5	10.7	13.6	9.6
30–34	7.1	10.7	9.6	11.6	8.8
35–39	7.4	10.0	9.0	10.6	8.4
40 and over	43.9	20.5	16.8	19.9	15.7

Notes: Arrest figures based on 10,843 reporting agencies, with 2003 estimated population 204,034,545. Due to rounding, columns may not sum precisely to 100 percent.

Sources: Federal Bureau of Investigation, *Crime in the United States 2003,* 280–281. US Bureau of the Census, "Monthly Postcensal Resident Population." *National Population Estimates for the 2000s.* Estimates for July 1, 2003.

17–18. The decline in arrests for violent crime is more gradual. Violent crime arrests peak at age 18, and then taper off steadily through the 20s. They do not decline to half their peak rate until the early 30s.

Victimization Surveys and Self-Reports

So far we have presented official data on age and crime from the Uniform Crime Reporting (UCR) program. Because these data are available only for *arrested* offenders, they may not accurately represent involvement in crime. Studies based on victimization and self-report data, however, do tend to paint a similar picture of age and crime.

Michael Hindelang used data from the National Crime Victimization Survey (NCVS), 1973–1977, to examine victims' perceptions of offender age for crimes in which the victim saw the offender.[3] Victimization data were fairly consistent with UCR data. NCVS data revealed that rates of offending were highest for young adults (ages 18–20), followed by adolescents (ages 12–17), and finally older adults (over age 20). Hindelang found that juveniles were less involved than adults in serious crimes—a finding also consistent with UCR data.

Researchers who have used self-reports of delinquency and crime have also found a peak in offending from late adolescence to early adulthood, followed by a decline in criminal involvement.[4] Thus, the relationship between age and crime holds up across all three sources of data.

Age Composition Effect

Several studies explore the age composition effect on crime.[5] In these studies, researchers are not interested in the ages of individual offenders, but rather in the age composition of the total population and its effect on overall crime rates. The **age effect** refers to the fact that, while juveniles constitute a small portion of the US population, they commit a disproportionate share of crime. The **age composition effect** attributes changes in crime rates to changes in population demographics, such that crime rates rise as the number of people in the age group most likely to commit crimes (mid-adolescence to early adulthood) increases, and crime rates fall as the number of people in their crime-prone years decreases. For example, some have attributed increases in crime rates in the 1960s and 1970s to the fact that "baby boomers" reached adolescence and early adulthood during those years, and thus the proportion of the population in the crime-prone years increased.

Darrell Steffensmeier and his colleagues examined UCR data from more than 30 years (1953–1984) and found a large age composition effect on Index crime rates. Age composition of the population was a more significant predictor of crime rate than any other factor in their analysis.[6] Lawrence Cohen and Kenneth Land used data from 1947–1984 and found that crime rate trends for both murder and motor vehicle theft corresponded to changes in age composition of the population.[7] These crimes increased as the proportion of the population at crime-prone ages increased with the aging of the baby boomers and then declined as baby boomers matured beyond crime-prone ages. In a similar study, Darrell Steffensmeier and Miles Harer found large effects of age composition on crime rates during the 1980s. They also found smaller age composition effects during the early 1990s, however, and they concluded that recent declines in crime rates cannot be attributed solely to the age composition of the population.[8]

Debate about the Age–Crime Relationship

Criminologists agree that involvement in crime diminishes with age, but they have been vigorously debating the strength and universality of the age–crime relationship for over two decades.[9] Travis Hirschi and Michael Gottfredson sparked this debate in 1983 with their influential paper entitled "Age and the Explanation of Crime."[10] In it, they call the age–crime distribution "one of the brute facts of criminology" and argue several controversial points.[11] We focus on two of those points.

First, Hirschi and Gottfredson maintain that the relationship between age and crime does not vary across time, place, or demographic subgroups.[12] In other words, the shape of the age–crime curve is the same for different historical time periods, for different cultures, for different racial groups, and for males and females. Hirschi and Gottfredson also argue that less variation exists in the age–crime curve for different types of offenses than UCR data lead us to believe. As we will discuss shortly, much evidence contradicts this proposition of invariance.

age effect The disproportionate involvement of young people in crime.

age composition effect Concerns the age composition of the total population and its effect on overall crime rates. This effect traces changes in crime rates to changes in population demographics, such that crime rates rise as the number of people in their crime-prone years increases and fall as the number of people in their crime-prone years decreases. For example, crime rates rose as "baby boomers" reached mid-adolescence and fell as they aged into adulthood.

Second, Hirschi and Gottfredson maintain that criminological and sociological theories are inadequate for explaining the age–crime relationship—precisely because the relationship is so universal.[13] They write, "If the age effect cannot be even partially explained by historical trends or cross-cultural comparisons, if it is unaffected by introduction of such gross correlates of crime as sex and race, if it appears when other known causes of crime . . . are held constant. . . , then there is reason to believe that efforts to explain the age effect with the theoretical and empirical variables currently available to criminology are doomed to failure."[14] Studies of this issue have produced mixed results. Some researchers have found that sociological variables are unable to explain the age–crime relationship.[15] But other research has found that variables derived from criminological theory are able to at least partially explain the relationship between age and crime.[16]

Hirschi and Gottfredson's first proposition—that the age distribution of crime does not vary across social and cultural conditions—has motivated many researchers, including Darrell Steffensmeier and his colleagues, to test that proposition empirically. Like many criminologists, Steffensmeier does not dispute the *general* shape of the age–crime curve, but he does argue that the *exact* properties of that curve might vary for a variety of reasons, including cultural, historical, racial, and offense-type factors.[17] For example, the peak age of offending might be earlier for property crimes than for violent crimes, as UCR data suggest. Or the peak age of offending might have declined over time, so that offenders today tend to be younger than those of 50 years ago. Or the rate of decline in criminal involvement might be more gradual for crimes that are less risky than for those that involve more risk to the offender. All of these possibilities, if true, provide evidence contrary to Hirschi and Gottfredson's proposition of invariance. We turn now to studies that have tested the invariance proposition.

Variation Over Time

Darrell Steffensmeier and Emilie Allan observe that, with the industrialization of society, the once relatively smooth transition from childhood to adulthood was replaced by a more stressful process. They argue that this change is responsible for higher levels of offending by younger individuals today than in the past.[18] In preindustrial societies, formal "rites of passage" provided visible transition points at which young people assumed more responsible, more "adult" roles as productive members of society. But in postindustrial American society, the stresses of adolescence have been aggravated by a lack of formal processes "for moving people smoothly from protected childhood to autonomous adulthood."[19] Thus, the structure of contemporary society encourages delinquency by increasing status anxiety for adolescents. This increased anxiety should result in a decline in the peak age of offending, as offending becomes more heavily concentrated among the young.

Several studies have found support for this hypothesis.[20] Steffensmeier and his colleagues, for example, used UCR data for three time points (1940, 1960, and 1980), and for over 25 types of offenses to examine the uniformity of the age–crime curve over time and across crime type.[21] They found substantial differences in the age–crime curve over time. All offenses except gambling showed an earlier age peak in offending and a more skewed curve in 1980 than in 1940 or 1960.[22] David Greenberg also explored the question of variation over time in the age–crime curve using UCR data from 1952 to 1987. He found "dramatic historical variability" in the peak age of involvement for homicide and aggravated assault. "Between 1952 and 1987 the peak age declined from 24 to 18 for homicide and from 23 to 21 for aggravated assault. . . . In 1952, 8.4% of the homicide arrests and 10.5% of the aggravated assault arrests were of offenders under the age of 18; by 1987 these figures had become 16.1 and 19.5%, just about double."[23]

Variation Across Cultures and Social Groups

The rationale for expecting variation in the age–crime curve across cultures is basically the same as that for expecting variation over time. While adolescents in contemporary American society tend to be economically dependent on adults and excluded from productive roles, young people in some other cultural settings are more fully integrated into the structures of economic productivity. Steffensmeier and Allan write, "In small preindustrial societies, the passage to adult status is relatively simple and continuous. Formal 'rites of passage' at relatively early ages . . . avoid much of the status ambiguity and role conflict that torment American adolescents. Youths often begin to assume responsible and economically productive roles well before they reach full physical maturity. It is not surprising, therefore, to find that such societies have significantly flatter and less skewed age–crime patterns."[24] David Greenberg, writing from a Marxist perspective, agrees with this view.[25] He explains delinquency primarily in terms of status problems associated with adolescence in capitalist societies that exclude young people from productive work roles. (See Chapter 13 for a discussion of Greenberg's perspective.)

Researchers have not conducted a thorough statistical comparison of age–crime curves across societies.[26] But some have compared the age–crime relationship for different social groups within societies. One study of Israeli men found no differences in the age–crime curve across social groups defined according to religious orthodoxy, ethnicity, and socioeconomic status.[27] A study of young British males found differences in the age–crime relationship across groups defined in terms of educational attainment.[28]

Variation by Race

Evidence suggests that the age–crime relationship is not the same for blacks and whites in the United States.[29] Steffensmeier and Allan predicted racial variation in the age–crime curve on the basis of persistent racial discrimination.[30] Because of discrimination, legitimate opportunities for blacks to participate fully in the adult labor market are more limited than opportunities for whites. For this reason, Steffensmeier and Allan predicted that levels of adult offending will be higher among African Americans than among whites and that "the proportion of total African-American crime that is committed by African-American adults will be greater than the proportion of total white crime that is committed by white adults."[31] They found support for this hypothesis when they compared the percentages of black and white adults arrested for property crimes (after taking into account differences in population size). For all ten types of property offenses they examined, "the adult percentage of arrests is greater for blacks than for whites, in some cases strongly so."[32]

Variation by Crime Type

Crimes differ substantially in the rewards they offer. Young people often receive reinforcement and high social status from their peers for involvement in mischievous or "hell-raising" offenses, such as vandalism, burglary, petty theft, auto theft, and violations of drug and liquor laws. But the rewards for committing these types of offenses change with age. Mischievous offending becomes less gratifying as individuals mature into adulthood and acquire legitimate roles that increase the social costs of offending.

Opportunities to commit delinquent or criminal acts also change with age. Compared to juveniles, adults spend more time participating in conventional roles, such as working, spending time with a spouse or children, or attending college courses; furthermore, they spend less time with peers who encourage and reinforce offending. Adults, therefore, presumably have fewer opportunities to participate in the mischievous forms of offending that characterize adolescence. But adults hold positions that provide them with more opportunities than juveniles have to participate in crimes like fraud and embezzlement. Because both the rewards and opportunities associated with different types of offenses vary with age, we can reasonably expect variation in the age–crime relationship across offenses.

Steffensmeier and his colleagues examined twenty-seven UCR offenses and found significant variation in the age distribution of crime across offense types.[33] For example, relatively lucrative property crimes, such as fraud and forgery, peaked at later ages and declined more slowly than more mischievous, "low-yield, high-risk" property offenses, such as vandalism and auto theft.

Age and Delinquency in Brief

The conclusion we can draw from the vast body of literature on age and crime is complicated. Numerous studies have demonstrated that variation exists in the age–crime curve across time, social groups, and offenses. The *exact* shape of the age–crime curve varies by several factors, including time periods, cultural settings, racial groups, and offense types. However, and more importantly, the *general* assertion that crime is a pursuit primarily of the young is undisputed. Multiple sources of data (UCR and self-report and victimization surveys) all point to the same fact: for most types of offenses, the period of adolescence and young adulthood is the peak time of offending for most individuals. The peak age of offending for violent crimes, though, is somewhat older than for most types of property offenses.

There are exceptions to this general age–crime curve—individuals who begin offending early and continue to commit crimes long after most people their age have stopped. We discuss these offenders in Chapter 6, in which we explore developmental patterns of offending.

Why Are Young People So Involved in Crime?

In our discussion of the debate about invariance of the age–crime relationship, we suggested reasons for the general relationship between age and crime. Here, we briefly summarize several explanations.

First, as individuals mature into young adulthood, they assume more responsible roles (for example, employment, college attendance, marriage, military service), and thus they have "more to lose" if caught committing crime. Though adolescents may be willing to participate in risky offenses, adults who have "greater stakes in conformity" are unlikely to want to take the same risks. Robert Sampson and John Laub have written about the social control effects of job stability and marriage.[34] They used the Gluecks' classic dataset containing information about 500 delinquents and 500 nondelinquents, from childhood to adulthood, and found that strong social ties achieved through employment and marriage inhibit criminal behavior.[35] Adult entry into the labor market also means that individuals can acquire the financial resources they need in legitimate ways and so have less need to resort to crime.[36]

Second, the legal costs of offending increase with the transition from adolescence to adulthood. The penalties for most offenses are more severe for adults than for juveniles. This should serve as a greater deterrent for adults, who may be more likely than juveniles to consider the consequences of their criminal actions. In addition, because of the legal costs of offending, many "chronic offenders" are incarcerated by their mid-20s and unable to commit further crimes.

Third, the structure of opportunities to engage in delinquency and crime changes with the transition from adolescence to adulthood. While opportunities to participate in offenses like fraud and embezzlement increase with age, opportunities to engage in more mischievous forms of offending tend to decrease with age. The demands of time that accompany adulthood (e.g., time spent working or attending college classes) mean that less time is available for delinquent activities. Compared to adolescents, adults spend less time just "hanging out" with peers. In the transition from adolescence to adulthood, lifestyle routines and peer associations change in ways that offer fewer opportunities for the types of offenses more common in adolescence.

Fourth, adolescence is often accompanied by a "party" culture that at least tolerates and often values and reinforces offending.[37] Status among adolescent peers is sometimes achieved through bravado displayed in offending. Criminal activities are less likely to be rewarded or reinforced in adulthood, when expectations for responsible and productive behavior are greater than in adolescence.

Finally, some researchers have argued that young people simply have greater physical abilities to commit crimes than do older people. Walter Gove focuses primarily on "physically demanding crimes," and suggests that "physical strength, energy, psychological drive, and the reinforcement effect of the adrenaline high" can account for the decline in offending with age.[38] He argues that these physical attributes peak at the same ages as offending and that their rapid decline contributes to the rapid decline in offending with age. Steffensmeier and Allan tested Gove's hypothesis and found that biological factors such as physical strength cannot account for the association between age and crime.[39] They compared the age–crime curve for burglary (an offense that requires strength and endurance) with the age curve for physical strength. While burglary peaked at about age 18, the peak for physical strength was fairly constant from early 20s through late 30s. Furthermore, the decline from the peak age was abrupt for burglary but gradual for physical strength.[40]

■ Gender

Gender and crime are certainly related, but the strength of the relationship varies depending on data source (official vs. self-report data) and type of offense. Crime and delinquency are committed primarily by males.

Official Data

UCR data indicate that males are disproportionately involved in crime and delinquency. Males and females constitute roughly equal shares of the US population, yet 76.8% of persons arrested in 2003 were male.[41] In 2003, males accounted for 82.2% of persons arrested for violent Index crimes and 69.2% of those arrested for property Index crimes. The gender difference in arrests was particularly great for murder, rape, robbery, burglary, weapons offenses, and sex offenses. For all of these offenses, more than 85% of those arrested were male. Only for prostitution and running away were arrests higher for females than for males.[42]

Table 5-4 presents arrest percentages by gender and age for various offenses. This table illustrates that the proportions of males and females arrested for the offenses shown are similar for juveniles and adults. For property Index crimes, for example, males account for 68.2% of juveniles arrested and 69.7% of adults arrested.

When we focus only on juveniles (the first three columns of Table 5-4), we see that males account for 81.6% of juvenile arrests for violent Index crimes and 68.2% of juvenile arrests for property Index crimes. Among juveniles, the gender difference in arrests is largest for murder, rape, robbery, burglary, arson, weapons offenses, and sex offenses. For these crimes, more than 87% of juveniles arrested in 2003 were male. Crime in general is a male phenomenon, especially violent crime.

The third column of Table 5-4 shows that the proportions of juvenile females arrested are highest for larceny–theft, simple assault ("other assaults"), forgery and counterfeiting, fraud, prostitution, curfew violations, and running away. Males and females are more equally represented in arrests for status offenses (runaways, curfew violations) than for most other types of offenses. Females are more likely than males to attract the scrutiny of law enforcement and be arrested for status offenses,[43] even though males are more likely to report engaging in such offenses.[44] Meda Chesney-Lind, a feminist criminologist who has written extensively about why this is the case, argues that juvenile justice officials respond more harshly to status offenses by females than males in an attempt to control female sexual activity.[45]

Official vs. Victimization Data

Michael Hindelang examined the level of agreement between official (UCR) and victimization (National Crime Survey) data from 1972 to 1976.[46] He found close agreement regarding female involvement in crime (primarily Index offenses other than murder).[47] Hindelang concluded that gender is a true correlate of crime and delinquency. In other words, the relationship between gender and crime is a consequence of females being less involved in crime, rather than of biases in the criminal justice system that might lead to fewer female arrests.

Table 5-4	Arrests, by Gender, Age, and Type of Crime, 2003					
	Under age 18			Age 18 and over		
Type of crime	Total number	Male percentage	Female percentage	Total number	Male percentage	Female percentage
All crimes	1,563,149	71.0	29.0	8,018,274	77.9	22.1
Index crimes	393,622	70.4	29.6	1,164,702	73.5	26.5
Violent crimes	64,799	81.6	18.4	354,165	82.3	17.7
Homicide	783	90.7	9.3	8,336	89.6	10.4
Forcible rape	2,966	98.0	2.0	15,480	98.8	1.2
Robbery	17,900	91.1	8.9	57,767	89.2	10.8
Aggravated assault	43,150	76.4	23.6	272,582	79.7	20.3
Property crimes	328,823	68.2	31.8	810,537	69.7	30.3
Burglary	59,870	88.2	11.8	144,891	85.5	14.5
Larceny–theft	232,322	60.6	39.4	584,726	63.8	36.2
Motor vehicle theft	30,874	83.1	16.9	75,347	83.5	16.5
Arson	5,757	87.6	12.4	5,573	81.1	18.9
Non-Index crimes	1,169,527	71.2	28.8	6,853,572	77.2	21.4
Other assaults	170,168	67.5	32.5	706,937	77.8	22.2
Vandalism	76,042	86.2	13.8	117,041	81.9	18.1
Weapons offenses	27,492	88.9	11.1	90,352	92.7	7.3
Forgery and counterfeiting	3,328	64.5	35.5	75,860	59.5	40.5
Fraud	5,642	66.9	33.1	202,827	55.1	44.9
Prostitution	972	31.3	68.7	50,714	35.9	64.1
Sex offenses (except forcible rape, prostitution)	12,747	90.7	9.3	51,012	91.7	8.3
Drug abuse violations	137,658	83.5	16.5	1,034,564	81.4	18.6
Curfew violations	95,052	69.7	30.3	—	—	—
Runaways	87,396	41.3	58.7	—	—	—

Note: Arrest figures based on 10,843 reporting agencies, with 2003 estimated population 204,034,545.

Source: Federal Bureau of Investigation, *Crime in the United States 2003*, adapted from Tables 39 and 40, pages 282–285.

Self-Report Data

Self-report studies confirm that crime is a male phenomenon, but the gender difference in offending is generally smaller in self-report studies than in official data.[48] In a widely cited study, Rachelle Canter notes that the male to female ratios of involvement in delinquency based on UCR data generally range from 3:1 to 7:1, but the ratios based on self-report data range from 1.2:1 to 2.5:1.[49]

Canter uses NYS data to examine the size and pattern of gender differences in self-report delinquency. She finds gender differences in overall delinquency, with boys reporting roughly twice as many delinquent acts as girls. She also finds small, but consistent, gender differences in most specific types of delinquent behavior. These gender differences are due to a higher number of male offenders and a higher frequency of offending among males.[50] While Canter finds consistent gender differences in involvement in delinquency, these differences are far less dramatic than are those indicated by official data.

Canter also finds that males and females report involvement in the same types of offenses.[51] The exception to this similarity is for serious offenses, particularly violent ones. Males are substantially more likely than females to report involvement in serious offenses. This finding is consistent with both prior research and UCR data.[52] Contrary to UCR data, however, the NYS provides no evidence of greater female involvement in "traditionally female crimes," such as prostitution and running away.[53] Canter attributes the discrepancy between her findings and those based on UCR data to harsher responses by law enforcement officials when females commit these acts than when males do.[54] Thus, girls are more heavily represented in official data for these offenses than self-reports indicate they should be.

Gender Differences in Gang Participation

Researchers have used self-report data to study gender differences in gang participation. Two surveys—the Denver Youth Survey and the Rochester Youth Development Study—provide data from samples of "high-risk" youth (those at high risk of involvement in delinquency based on factors such as gender and residence in economically disadvantaged, high-crime neighborhoods). Because these surveys overrepresent serious and chronic offenders, they are well suited to addressing questions about gang participation.

Research based on these surveys reveals gender differences in gang participation, in the proportion of gang members involved in serious delinquency (but not moderate or minor delinquency), and in the frequency of offending. First, research indicates that female involvement in gangs, although less than male involvement, is greater than prior research based on official data suggests.[55] One study, using data from the Denver Youth Survey, showed that females constituted 20–25% of gang members during 3 years of the survey.[56] These percentages, while perhaps initially surprising, are consistent with other research.[57] Second, another study, using data from the Rochester Youth Development Study, showed that male gang members are more likely than are female gang members to engage in serious delinquency and alcohol use. The same study found no gender differences in rates of involvement in more minor offenses.[58] Finally, studies reveal that male gang members report a higher frequency of offending than do female gang members for all types of offenses.[59]

In sum, gender differences exist among gang members in both rate of involvement in delinquency and frequency of offending. Still, it is worth noting that, when gang members are compared to nonmembers, "gang members of *both* sexes are significantly more likely to have participated in delinquency, including serious delinquency and substance abuse, and to have committed these acts at much higher frequencies than nonmembers."[60]

Narrowing of the Gender Gap in Delinquency?

Feminist criminologists have advanced a controversial proposition about the relationship between gender and crime. In the mid-1970s, Freda Adler and Rita Simon argued that, as a result of the feminist movement, gender equality would emerge as male and female roles became more similar. This equality across gender would, in turn, lead to more similar crime rates and patterns for males and females, as opportunities for female involvement in crime expanded along with female roles in society.[61] In other words, according to this **convergence hypothesis**, the gender gap in crime and delinquency would narrow, and male and female rates of offending would converge as a function of gender equality.

Support for the convergence hypothesis has been mixed.[62] Roy Austin examined UCR data from 1965 through 1986 and self-report data and found some evidence of convergence. He analyzed trends (separately for juveniles and adults) in UCR data (for murder, aggravated assault, robbery, larceny–theft, burglary, motor vehicle theft, and arson) for the entire 22-year time frame and for two separate time periods (1965–1975 and 1975–1986).[63]

For juveniles, the ratios of male to female arrest rates for the entire 22-year period showed significant downward trends for all offenses except murder and robbery, indicating a narrowing of the gender gap in arrests for aggravated assault, larceny–theft, burglary, motor vehicle theft, and arson. Evidence of convergence was greater in the earlier period (1965–1975) than in the later (1975–1986). During the earlier period, the ratios of arrest rates showed downward trends for all seven offenses. During the later period, downward trends in the ratios existed for only three offenses: aggravated assault, burglary, and auto theft. For two offenses, murder and larceny–theft, the trend in male to female arrest ratios was upward, indicating divergence rather than convergence.[64] For adults, significant findings of convergence were less consistent in the earlier period than they were for juveniles, but were similar to those for juveniles in the later period. Austin interpreted his results as support for the convergence hypothesis. But he also acknowledged that, for some offenses, the trends toward convergence were inconsistent, and for others, the data provided evidence of divergence.[65]

Darrell Steffensmeier and his colleagues have provided a wealth of empirical evidence indicating that the gender gap has not narrowed in the way or to the extent feminist theorists predicted.[66] They have shown that females are not "catching up" with males in the commission of serious or violent offenses—the types of crime typically thought of as "masculine."[67] Exploring change over a 30-year period from 1960 to 1990, Steffensmeier shows that, among juvenile offenders, the female percentage of arrests decreased 3.1% for homicide and increased only 3.6% for aggravated assault, 2.8% for weapons offenses, and 3.1% for robbery. Among adult offenders, the female percentage of arrests decreased 1.4% for homicide and 1.8% for aggravated assault and increased only 1.9% for weapons offenses and 4.0% for robbery.[68]

Steffensmeier finds, instead, that female arrests have increased most dramatically in recent decades for "minor property offenses"—larceny–theft, fraud, and forgery. Offenses in these categories include shoplifting, credit card fraud, and writing bad checks. From 1960 to 1990, among juvenile offenders, the

convergence hypothesis
The prediction that the gender gap in crime and delinquency would narrow and that male and female rates of offending would converge as a function of increasing gender equality.

female percentage of arrests increased 13.5% for larceny–theft, 17.2% for fraud, and 9.1% for forgery. The figures for adult offenders are 13.2% for larceny–theft, 31.1% for fraud, and 19.3% for forgery.[69] Steffensmeier and his colleagues argue that these increases among females in minor property crimes do not reflect equality of gender roles or increased opportunities for women in the economic sphere. In fact, "the largest increases . . . in the female share of arrests for [minor property crimes] occurred between 1960 and 1975, before the women's movement had gained much momentum."[70] Instead, increases in the female share of arrests for minor property crimes reflect the increasing economic marginalization of women. Steffensmeier and Streifel write, "a large segment of the female population faces greater economic insecurity today than 25 years ago. The economic pressures on women have been aggravated by rising rates of divorce, illegitimacy, and female-headed households, coupled with continued segregation of women into low-paying and traditionally female occupations. . . . Growing economic insecurity increases the pressures to commit traditional female consumer-based crimes such as shoplifting, check fraud, theft of services, and welfare fraud."[71]

Steffensmeier and Streifel found that changes in the female share of arrests for property crimes were also due to more formal policing, which "tends to increase the visibility of female offending (especially its less serious forms)" and "which has contributed to more 'official' counting of female offending."[72]

Steffensmeier's many studies suggest that a narrowing of the gender gap in crime has not occurred for most offenses. For the few crimes that show convergence in recent decades in male and female participation, the gender equality hypothesis is not the most plausible explanation for that convergence.

Gender and Delinquency in Brief

Anthony Harris calls gender "the single most powerful predictor of officially and unofficially known criminal deviance in this society."[73] All three data sources—official data, victimization surveys, and self-report surveys—show that males are disproportionately involved in crime and delinquency. The gender gap in crime revealed in self-report data, however, is generally smaller than the gap suggested by official data. The extent of gender difference also varies by type of offense. The gender gap is particularly wide for violent offenses, which are committed overwhelmingly by males. The gap is smaller for status offenses, for which boys and girls are more equally likely to be arrested. Self-report data indicate that female involvement in gangs is greater than UCR data suggest, but that male gang members are more heavily involved in serious delinquency than are female gang members.

Finally, the convergence hypothesis predicts that male and female rates of offending will converge as a function of increased gender equality. This hypothesis has received mixed support. The best evidence suggests that the gender gap in crime has not narrowed in the way or to the extent that the convergence hypothesis predicts. Instead, female arrest rates have increased for minor property offenses like larceny–theft (shoplifting). Steffensmeier and his colleagues argue that this trend reflects the increasing economic marginalization of women, not liberation or equality of opportunity across gender.

Why Are Males So Involved in Crime?[74]

In the last three decades, there has been an explosion of interest in gender and crime. Many criminologists have tried to determine the extent of the gender gap in crime and delinquency, and others have worked to develop adequate theories for explaining gender differences in offending.

Traditional theories of crime and delinquency were developed primarily to explain male offending. Criminologists have debated the ability of traditional theories to explain female offending as well. Some have called for separate "gender-specific" theories of female delinquency.[75] Yet others have demonstrated that the processes leading to delinquent behavior are similar across gender, and that traditional "gender-neutral" theories can explain male and female delinquency equally well.[76] Steffensmeier and Allan argue in favor of "gender-neutral" theories, noting that "measures of bonds, associations, learning, parental controls, perceptions of risk, and so forth have comparable effects across genders."[77] But they caution that this finding of similar effects applies primarily to minor offending. Traditional theories may be less capable of explaining the substantial gender differences in serious offending.

If the processes leading to less serious delinquency are similar across gender, then how do we explain the gender gap in offending? Some suggest that, while the factors leading to delinquency are similar across gender, the *level of exposure* to these factors might differ for males and females, and that this difference might explain males' greater involvement in delinquency. For example, parental supervision might have the same inhibitory effect on delinquency for males and females. But parents might supervise daughters more closely than sons, and this difference in level of supervision might account for higher levels of delinquency among males. This "differential exposure" explanation has received empirical support.[78]

So what are the factors that lead to delinquency to which males and females might be differentially exposed or that might have different effects on delinquent behavior across gender? A complete review of the vast literature on gender differences in delinquency is beyond the scope of this chapter, but we briefly identify some of the factors researchers have considered.

Gender Differences in the Effects of Parenting

A large body of research, grounded mainly in control theories (see Chapter 10), concerns gender differences in the effects of parent–child relationships on delinquency. These studies have focused primarily on supervision and attachment between parent and child. Although the results are somewhat mixed, researchers have typically found that parent–child attachment has a stronger inhibitory effect on delinquency for females than for males and that daughters are subject to higher levels of supervision than are sons.[79] Stephen Cernkovich and Peggy Giordano, for example, found that parents supervise and control daughters to a greater extent than they do sons and that such control prevents delinquency.[80] They also found that attachment to parents differs significantly for daughters and sons, with daughters reporting both more open communication and more conflict with parents and sons reporting a higher level of caring and trust. In this study, most measures of attachment were moderately related to delinquency.

Other researchers, however, have not found gender differences in attachment or the effect of attachment on delinquency. Canter, for example, looked for gender differences in family bonds to see if they could account for the gender gap in self-reported delinquency. She found that, although girls reported significantly less delinquency than boys, girls did not report significantly stronger family bonds. In fact, family bonds had a stronger inhibitory effect on serious delinquency for males than for females.[81]

Interactionist Perspective on Gender and Delinquency

In the early 1990s, Ross Matsueda and Karen Heimer developed differential social control theory, an interactionist perspective consistent with social learning theory (see Chapter 11).[82] According to differential social control theory, role-taking is the causal factor most closely related to delinquency.[83] Role-taking involves seeing and evaluating one's own behavior from the standpoint of others.

Heimer and Matsueda and their colleagues explored gender differences in the processes of social interaction that lead to delinquency.[84] Heimer defined "gender definitions" as "societal definitions of femininity and masculinity."[85] She hypothesized that, for girls, the role-taking process results in internalized feminine gender definitions that reduce the likelihood of delinquency because the societal view of femininity is inconsistent with law violation in a way that the societal view of masculinity is not. She also hypothesized that, due to gender differences in socialization experiences, "anticipating disapproval of delinquency from parents and peers should have a larger deterrent effect on girls' than on boys' delinquency because girls are more likely than boys to be affected by others' reactions."[86] Heimer found support for her hypothesis about gender definitions but not for her hypothesis about anticipated disapproval of delinquency. In a separate study, Bartusch and Matsueda found that an interactionist model derived from differential social control theory accounted for a substantial portion of the gender gap in delinquency.[87]

Gender Bias in the Criminal Justice System

If we use official data to address the question of why the gender gap in delinquency exists, we might conclude that gender bias in the criminal justice system is the cause. Research has shown that females are more likely than males to be arrested for status offenses,[88] even though males are more likely to report involvement in such offenses.[89]

For more serious forms of delinquency, though, gender bias might operate in the opposite direction. Girls may be less likely than boys to be arrested for serious offenses because of chivalry on the part of police officers, or because of the "invisibility" of female involvement in serious offenses. Some have argued that police officers, subscribing to traditional notions of appropriate gender role behavior, treat female delinquents more leniently than males. Merry Morash found support for this hypothesis and showed that males were more likely than females to acquire police records for involvement in "typical male" delinquency.[90] Others have found that female involvement in serious delinquency is less "visible" to police because it contradicts gender stereotypes that portray serious delinquent behavior as a male phenomenon.[91]

Despite evidence of gender bias in the decision to arrest (and in other parts of the criminal justice system), such bias alone cannot account for the gender gap in delinquency. Though the gender gap is smaller when we examine self-report data, it still exists. Thus it appears that bias in the criminal justice system accounts for part of the difference between male and female arrest rates for delinquency, but not all of it.

■ Race

The relationship between race and involvement in delinquency is by no means straightforward. To understand this relationship, it is crucial to consider the source of data (official data, victimization survey, or self-report survey). The conclusions we draw about the relationship between race and delinquency depend on which source of data we use.

In our discussion of race and delinquency, we try to be as racially and ethnically inclusive as possible. Only limited data on delinquency exist for racial and ethnic groups other than African Americans and Caucasians, however. For example, official data include only four race categories (see **Table 5-5**) and typically do not indicate Hispanic or non-Hispanic origin.

Official Data

The UCR includes data only for individuals who have been arrested. But offenses committed by members of some racial or ethnic groups may be more likely to result in arrest, and members of some racial or ethnic groups may be more likely to report offenses to the police.[92] These factors can produce biased estimates of racial differences in rates of offending. Because many factors (such as nature of the offense and demeanor of the suspect) come into play in the arrest process, those who are arrested do not constitute a sample that is representative of all offenders. In other words, we cannot look at those who have been arrested and accurately assume that others who have offended but avoided arrest are similar in terms of age, gender, race, or other social characteristics. This is an important point to keep in mind when using official data to explore the relationship between delinquency and social characteristics, including race.

UCR data indicate that African Americans are strongly overrepresented in involvement in delinquency. In 2003, African Americans constituted 12.8% of the total US population and 15.5% of the juvenile population (under age 18). Whites constituted 76.6% of the juvenile population.[93] Yet, of juveniles arrested in 2003, 26.6% were black and 70.6% were white (see Table 5-5).[94] An additional 1.3% of juveniles arrested were American Indian and 1.6% were Asian. Racial differences in arrest rates become more stark when we look at specific types of offenses. Table 5-5 compares across race the percentages of juveniles arrested for various offenses.

Table 5-5 reveals that, given their percentage of the juvenile population, black youth are consistently overrepresented in arrests for both Index and non-Index crimes. The disproportionate likelihood of arrest for blacks is most glaring for violent offenses, particularly robbery and homicide. When we look only at arrests for violent crime, we find that African American youth account for 45.0% of arrests, while white youth account for 52.7%. Black youth constitute 62.3% of those arrested for robbery and 48.1% of those arrested for homicide.

Table 5-5 Juvenile Arrests, by Race and Type of Crime, 2003

Type of crime	Whites Number	Whites Percentage	Blacks Number	Blacks Percentage	American Indians Number	American Indians Percentage	Asians Number	Asians Percentage
All crimes	1,098,012	70.6	413,236	26.6	19,917	1.3	24,636	1.6
Index crimes	259,624	66.3	119,694	30.6	5,200	1.3	7,017	1.8
Violent crimes	34,012	52.7	29,012	45.0	582	0.9	877	1.4
Homicide	381	48.9	375	48.1	10	1.3	13	1.7
Forcible Rape	1,895	64.1	988	33.4	48	1.6	27	0.9
Robbery	6,278	35.2	11,208	62.3	86	0.5	277	1.6
Aggravated assault	25,458	59.3	16,441	38.3	438	1.0	560	1.3
Property crimes	225,612	69.0	90,682	27.7	4,618	1.4	6,140	1.9
Burglary	42,586	71.4	15,581	26.1	709	1.2	781	1.3
Larceny–theft	161,082	69.8	61,715	26.7	3,424	1.5	4,655	2.0
Motor vehicle theft	17,281	56.1	12,417	40.3	431	1.4	656	2.1
Arson	4,663	81.3	969	16.9	54	0.9	48	0.8
Non-Index crimes	838,388	72.0	293,542	25.2	14,717	1.3	17,619	1.5

Note: Arrest figures based on 10,839 reporting agencies, with 2003 estimated population 203,489,015.

Source: Federal Bureau of Investigation, *Crime in the United States 2003*, 289.

The numbers are less striking for property crimes, but here, too, black juveniles are overrepresented in arrests.

To know whether the overrepresentation of blacks that we find in UCR data is due to greater involvement of blacks in offending or to biases in the processes that lead to arrest, we must compare official statistics to data from other sources, including self-reports and victimization surveys.

Self-Report Data

Because self-report surveys ask individuals directly about their involvement in delinquency, they avoid the potential biases of official data. Individuals responding to the surveys are asked to report the offenses they have committed, even if those offenses have gone undetected by police. The primary limitation to using self-reports to assess delinquency concerns sample size. Because the sample size for most self-report surveys is rather small, and because serious or violent delinquency is a relatively rare event, it is difficult to detect many serious or violent offenses with self-reports.[95] One would have to survey a large number of people to find even a few who report having committed serious delinquent offenses.

A second limitation of self-report surveys—particularly important when examining racial differences in offending—is that the data they provide may not be equally valid for blacks and whites. Some researchers have found that blacks are more likely than whites to underreport serious or violent offenses.[96] Though a more recent study found no difference across race in the validity of self-reports, this issue is one that must be kept in mind when interpreting self-report data.[97]

An influential, early study using self-report data found negligible differences between blacks and whites in rates of offending. Travis Hirschi found that 49% of black boys and 44% of white boys in his sample reported committing one or more delinquent offenses during the previous year. This 5 percentage point difference in self-report data is inconsistent with the 24 percentage point difference between the two groups in official data: 42% of the black boys and 18% of the white boys in the sample had police records.[98]

More recently, a series of studies using data from the NYS have supported the early finding that whites and blacks are basically equally likely to engage in delinquency.[99] David Huizinga and Delbert Elliott used NYS data from five time points (annual data from 1976 to 1980) and for various types of offenses and found almost no significant differences between racial groups in the proportions of persons involved in delinquency.[100] Similarly, Delbert Elliott and Suzanne Ageton compared the proportions of black and white youth reporting that they had committed one or more offenses and found no significant differences by race in the prevalence of delinquency.[101] (*Prevalence* refers to the proportion of a group involved in offending.)

Yet when Elliott and Ageton turned their attention from prevalence to *incidence,* defined as the number of offenses committed by each delinquent youth, they found differences by race. As **Table 5-6** shows, however, these racial differences in frequency of offending were statistically significant only for the combined scale of 47 delinquent offenses and for serious property offenses.

The combined scale of offenses ("total self-reported delinquency") included serious crimes against persons, serious crimes against property, illegal service crimes, public disorder crimes, status offenses, and hard drug use. For this combined delinquency scale, Elliott and Ageton found that "blacks report three offenses for every two reported by whites."[102] For serious crimes against property, "blacks report more than two offenses for every offense reported by whites."[103] It was primarily this large difference across race in the frequency of serious property offenses that accounted for the significant racial difference in the combined scale. The difference across race in the frequency of serious crimes against persons was substantial, but not statistically significant.[104] We must use caution when interpreting this finding because of the small sample of blacks in the NYS and the relative rarity of serious crimes against persons.[105]

Elliott and Ageton found similar percentages of blacks and whites reporting offenses toward the low end of the frequency distribution. But racial differences exist at the high end of the frequency distribution (with high frequency defined as 200 or more offenses on the combined delinquency scale, and 55 or more offenses on the serious property crime scale). For the combined delinquency scale, 9.8% of black respondents reported engaging in 200 or more offenses, whereas only 4.1% of white respondents reported this many offenses. Similarly, for the serious property crime scale, 4.2% of black respondents reported engaging in 55 or more offenses, whereas only 1.9% of white respondents reported this many offenses.[106] In interpreting the effect of these racial differences at the high end of the frequency distribution on the overall relationship between race and delinquency, however, one must consider that the group of high frequency or "chronic" offenders is quite small and constitutes a small portion of the total number of offenders.[107]

Table 5-6	Average Frequency of Self-Reported Delinquency, by Race and Type of Offense, 1976		
	Race		Status of difference across race
Type of offense	White	Black	
Total self-reported delinquency	46.79	79.20	Statistically significant
Serious crimes against persons (Sexual assault, aggravated assault, simple assault, robbery)	7.84	12.96	Not significant
Serious crimes against property (Vandalism, burglary, auto theft, larceny, stolen goods, fraud, joyriding)	8.93	20.57	Statistically significant
Illegal service crimes (Prostitution, selling drugs, buying/providing liquor for minors)	1.85	1.71	Not significant
Public disorder crimes (Carrying a concealed weapon, hitchhiking, disorderly conduct, drunkenness, panhandling, making obscene phone calls, marijuana use)	14.98	16.50	Not significant
Status offenses (Runaway, sexual intercourse, alcohol use, truancy)	14.84	16.19	Not significant
Hard drug use (Amphetamines, barbiturates, hallucinogens, heroin, cocaine)	1.26	0.18	Not significant

Note: Sample sizes = 1,357 white respondents, 259 black respondents.
Source: Elliott and Ageton, "Reconciling," 101–102.

In summary, self-report surveys tend to show that blacks and whites are equally likely to be involved in delinquency (prevalence) and are equal in frequency of offending (incidence) for most types of delinquency. Racial differences exist only in frequency of serious offenses. These findings are contrary to the conclusion we reached when we relied on official data. How do we explain the inconsistency?

First, higher arrest rates for blacks than for whites might indicate discrimination in the criminal justice system. Many opportunities exist to exercise discretion within the justice system, and history has shown us that this discretion often disadvantages blacks.[108]

Second, official data and self-report data assess different kinds of offenses. Official data are available only for individuals who have been arrested. The likelihood of being arrested increases with the number of offenses an individual commits and with the seriousness of those offenses. Thus, official data tell us a good deal about individuals who are arrested for serious offenses, but offer us little information about individuals who commit relatively minor offenses, which are less likely to be detected by police. Conversely, because the sample size of most self-report surveys is fairly small, and because serious offenses occur relatively infrequently, self-report surveys are unlikely to offer us much information about serious delinquency. Also, individuals surveyed may be unlikely to provide information about serious offenses they have committed, regardless of assurances of confidentiality. Recall, too, that studies suggest that blacks may be more likely than whites to underreport serious or violent offenses. Thus, self-report data tell us more about individuals who commit minor offenses than about those who commit serious delinquency.

Third, we noted that blacks are more likely than whites to be among the small group of high frequency offenders. Because the likelihood of arrest increases with frequency of offending, blacks may be more likely than whites to be represented in official arrest data. Because the number of chronic offenders is relatively small, though, this explanation alone cannot account for the fairly wide gap between official and self-report data.

Each of these explanations could account, in part, for the discrepancy between official and self-report data regarding the relationship between race and delinquency. In truth, probably all of these factors are operating to some extent. To try to understand the true relationship between race and delinquency, in light of the inconsistent pictures painted by official and self-report data, we turn to a third source of data: victimization surveys.

Victimization Survey Data

The National Crime Victimization Survey (NCVS), conducted since 1973, provides data based on victims' perceptions of the characteristics of offenders, regardless of whether or not the offenses were reported to the police. Thus, information about the race of offenders is available only for crimes involving personal contact between offender and victim, such as rape, robbery, and aggravated assault. One must keep in mind this limitation when using NCVS data.

Darnell Hawkins and his colleagues report that "analyses of racial differences in victimization survey data show patterns that are generally consistent with those of official records."[109] Other researchers examined NCVS data and found that personal crime victims described 51% of juvenile offenders as white and 41% as black.[110] Like official data, victimization survey data indicate that, compared to the percentage of blacks in the general population, blacks are overrepresented among offenders.

We may reach this same conclusion when we examine either official or victimization data because both sources provide data primarily on serious types of crime. Self-report data, however, are a better measure of minor forms of offending and thus lead to a different conclusion about the relationship between race and delinquency.

Race and Delinquency in Brief

Race is a factor that should be included in any analysis of the social correlates of delinquency. The bulk of evidence suggests that blacks are disproportionately represented among the population of offenders—at least among those who commit serious offenses. For decades, some social scientists have argued that official data on offenders are suspect because of the biases inherent in the criminal justice system. Yet victimization data tend to confirm the conclusions drawn from official data. Both official and victimization data are weighted more heavily toward serious offenses. These data show that blacks are overrepresented among offenders, given their proportion of the general population. Thus, it appears that, if we are interested in *serious* offenses, official data can be used to reliably assess racial differences in offending rates. Self-report data, on the other hand, best capture more minor offenses, and show that blacks and whites are equally likely to engage in delinquency—except at the high end of the frequency continuum, where blacks are overrepresented.

Why Are African Americans Disproportionately Involved in Crime?

Some social scientists have used the concept of subcultural values and norms to explain racial differences in rates of offending.[111] Results from empirical research on the question of subcultural values that tolerate violence and deviance are mixed. Some studies show that blacks are more likely than whites to condone deviance or the use of violence in some situations.[112] But a larger body of research provides strong evidence to the contrary.[113]

Given the strong relationship in the United States between race and socioeconomic status, some researchers have turned to strain theory (see Chapter 12) for an explanation of racial differences in offending. These researchers have asked whether racial differences in access to legitimate opportunities can account for racial differences in crime and delinquency. Gary LaFree and his colleagues, for example, examined the effects of opportunities provided by education, employment, income, and family stability on involvement in crime (robbery, burglary, and homicide), and how these effects differ for blacks and whites.[114] They used census and UCR data gathered annually for a period of more than 30 years, and found that, for whites, crime rates declined as legitimate opportunities increased—as predicted by strain theory. For blacks, however, their findings contradicted common assumptions about opportunities and crime. Crime rates for blacks increased as educational attainment and family income increased, and they declined as the percentage of families headed by females increased.

More recently, Stephen Cernkovich and his colleagues examined the ability of strain and social control theories to explain the relationship between involvement in crime and adherence to the ideology of the "American dream," which they defined as "the promise of economic and material success."[115] They found that, relative to whites, blacks are more strongly committed to the American dream (as measured in economic terms). The effect of this commitment on crime, however, is consistent with strain theory only for whites. For blacks, commitment to the American dream has no significant effect on crime. In addition, the relationship between commitment to the American dream and offending is inconsistent with social control theory for both blacks and whites. These studies suggest that the causal mechanisms central to some of the leading theories of crime and delinquency are inadequate to account for racial disparities in rates of offending.[116]

Much of the most promising research on racial differences in offending has focused on structural, rather than individual-level, variables. Some criminologists advocate a community-level approach that considers community structures and cultures and how they contribute to different crime rates across groups, including racial groups.[117] Several influential social scientists, including William Julius Wilson and Robert Sampson, have examined community-level factors and have written extensively about differences in the types of communities in which blacks and whites tend to reside. In *The Truly Disadvantaged*, Wilson details the plight of the American "ghetto underclass," which occupies inner-city neighborhoods characterized by high rates of violent crime, unemployment, out-of-wedlock births, and female-headed households.[118] This underclass consists overwhelmingly of African Americans, yet Wilson looks beyond racism to "changes in the urban economy" and the "class transformation of the inner city" to locate the origins of this underclass. In *When Work Disappears*, Wilson discusses the devastating effects on inner-city blacks of diminished opportunities for legitimate employment, brought about by the decline in jobs for low-skilled workers in the mass production system and the transition to a service economy in which jobs require more training and education.[119] The adverse effects of these labor market changes tend to be concentrated in inner-city neighborhoods where poor blacks reside. This concentration of economic disadvantage contributes to the disintegration of neighborhood institutions that would otherwise serve many functions, including crime control.[120]

In their theory of "race, crime, and urban inequality," Sampson and Wilson stress the importance of examining communities as the unit of analysis. They argue that differences between the types of communities in which blacks and whites tend to reside may be the real cause of the relationship we see between race and crime. In other words, blacks may be more likely than whites to commit crime because blacks are more likely to live in communities where they are exposed to structural conditions that lead to crime, such as concentrated poverty, family disruption, and residential instability.[121]

Empirical research has supported this explanation of racial differences in offending. For example, one study found that racial differences in offending disappeared once the researchers took into account neighborhood factors (percentage of neighborhood families or households characterized by poverty, female head of household, public assistance, no employed family member, male joblessness, out-of-wedlock births). "When African American youths and white youths were compared without regard to neighborhood context, African American youths were more frequently and more seriously delinquent than white youths. When African American youths did *not* live in underclass neighborhoods, their delinquent behavior was similar to that of the white youths. . . . Once individually measured factors were accounted for, residence in underclass neighborhoods was significantly related to delinquent behavior while ethnicity was not."[122] These findings suggest that what has often appeared as racial difference in rates of offending might actually be an artifact of the reality that African Americans are more likely than whites to live in neighborhoods of concentrated disadvantage.

Recent studies of racial differences in offending highlight the important point that race and socioeconomic status are closely intertwined in American society. The disturbing reality is that African Americans are more likely than whites to live in conditions of extreme poverty. Any thorough analysis of race and crime must consider this context of disadvantage to avoid attributing differences in rates of offending to race when they are actually due to socioeconomic status or community context.

■ Social Class[123]

UCR data do not contain information about the social class of persons arrested. Yet, for decades, criminologists acted under the assumption that crime was a "lower-class" phenomenon. Social class is an important factor in many of the theories we present in later chapters. Some of the most influential theories in criminology are based, either directly or indirectly, on the assumption that the disadvantage of a lower-class environment has far-reaching effects on individuals, their social relations, and the communities in which they live.[124] For example, lower-class individuals lack opportunities for success normally afforded others and experience strain as a result. The strain of poverty and lack of opportunity motivates involvement in crime. Additionally, those from lower-class backgrounds are exposed daily to criminal traditions in the neighborhood, and lower-class communities often lack effective means of social control.

ecological correlation A correlation between two variables based on data combined for a group, such as a geographic area or social group. The relationship between delinquency rates and median family income in a geographic area is an example of an ecological correlation between crime and social class.

The belief in an *individual-level* relationship between class and crime was derived largely from ecological research, which showed a strong and consistent correlation between class and delinquency, measured through official data. An **ecological correlation** is "a correlation between two variables based on grouped data such as averages for a geographic area or for social groups."[125] Summarizing influential early ecological research, Michael Hindelang and his colleagues note the strong relationships found between rates of delinquency in an area and factors such as median rental cost and percentage of families receiving public assistance in that area.[126] Although the ecological class–crime relationship is strong, it is incorrect to infer individual-level relationships from ecological data. This mistake is called the **ecological fallacy**, defined as "an error of reasoning committed by coming to conclusions about individuals based only on data about groups."[127] Hindelang and his colleagues explain that such inferences are incorrect because "ecological correlations generally overestimate individual-level correlations by a substantial margin."[128] Yet criminologists typically made the mistake of relying on ecological data to make inferences about class and crime at the individual level. Their assumptions remained virtually unchallenged until researchers began to gather self-report data.

ecological fallacy "An error of reasoning committed by coming to conclusions about individuals based only on data about groups" (Vogt, *Dictionary*, 78).

In the late 1970s, however, a heated and enduring debate emerged in criminology about the relationship between social class and delinquency. In the

1960s and 1970s, self-report methods became widely accepted as a means of gathering data about involvement in crime and delinquency. Self-reports challenged the conclusion that members of lower social classes were more likely than others to engage in crime and delinquency. Numerous studies relying on self-report data showed no relationship between social class and the likelihood of offending.[129] So began the debate about the class–crime relationship.

The Early Debate

In 1978, Charles Tittle and his colleagues reviewed 35 studies of social class and crime or delinquency and declared that the relationship between social class and criminality was a "myth."[130] They concluded that self-report data revealed only a *slight* negative association between class and crime, and that official data showed a decreasing relationship over time between class and crime. In a later paper, Tittle also critiqued eight major criminological theories that suggest hypotheses about an inverse class–crime relationship. He concluded that the theoretical basis for a hypothesis of a negative class–crime relationship was weak, and called the relationship "both theoretically and empirically problematic."[131]

The conclusion that the class–crime relationship was a "myth" did not go unchallenged. John Braithwaite conducted a more extensive review of the literature than Tittle and his colleagues did and came to a different conclusion. Braithwaite pointed to "neglected evidence which suggests that self-reports exaggerate the proportion of delinquency committed by the middle class." He concluded that "class is one of the very few correlates of criminality which can be taken, on balance, as persuasively supported by a large body of empirical evidence."[132]

Gary Kleck was next to critique Tittle's work. Kleck argued that findings based on self-report data of no class differences in offending were due to "class linked bias in self-report studies" and other problems associated with the self-report method.[133] Based on his review of the literature, he concluded that "lower-class respondents are generally more likely than middle-class respondents to give dishonest or incomplete responses to questions about their criminal or delinquent behavior."[134] He also criticized self-report methods for their inclusion of "trivial" offenses that may be hard to recall; their selection of respondents from relatively "class-homogeneous research sites" (such as schools or neighborhoods); and their exclusion of school dropouts, who, according to Kleck, are more likely than those in school to be lower-class boys and to be delinquent. Kleck, then, disregards the value of self-report data and therefore Tittle's conclusions based on self-report data.

Measurement of Social Class and Delinquency

The measurement of key concepts is one of the most important but problematic aspects of research on class and crime. Inadequate measures can lead researchers to reject sound theories or to accept faulty ones.[135] Some researchers have explored the effects of alternative measures of social class on the class–crime relationship.[136] Others have examined the effects of various crime and delinquency measures.[137] Still others have considered simultaneously both class and crime measures.[138]

Social Class Measures

John Hagan and Bill McCarthy observed that self-report studies, which rarely show a relationship between class and delinquency, typically assess the class of one's family of origin, using measures of parents' occupational status.[139] They argued that parental occupational status is of little relevance to youths' delinquency. Furthermore, they maintained that, if a relationship exists between one's class of origin and delinquent behavior, it is indirect, "operating through a variety of family, school and other mediating variables."[140] Thus, according to Hagan and McCarthy, we should expect only a weak relationship between class and delinquency, given the typical measurement of class in criminological research.

As an alternative, Hagan and McCarthy measured not only parents' social class, but also the "current class conditions" of youth, indicated by their inclusion in either a sample of school attendees or a sample of homeless or "street" youth.[141] Living on the street is a strong indicator of the most impoverished immediate circumstances.

Hagan and McCarthy tested the effects of youths' current class conditions (and of parents' social class) on involvement in serious theft. They found that children from families of the lowest social class, measured by parents' class category, were most likely to resort to life on the street. More importantly, they found that the experience of life on the street accounted for more of the difference in theft involvement between the school and street samples than did characteristics of the samples such as gender, age, family structure, parental control, and school involvement.[142] These results indicated that class, when measured in terms of youths' characteristics, rather than relatively far removed parental characteristics, had a strong effect on youths' involvement in serious theft. One might argue that this study is premised on a somewhat artificial circumstance, because relatively few youth are in a class situation as desperate or extreme as life on the street. Still, this study demonstrates the value of exploring youths' own status in research on class and delinquency, rather than simply measuring class of family of origin.

Delinquency Measures

In an important study of the social correlates of delinquency, Michael Hindelang and his colleagues tried to explain apparently inconsistent findings from official and self-report data regarding class and delinquency.[143] They note that official and self-report data assess different types of offenses or "domains of behavior." While official data provide information primarily about those who commit serious offenses, self-reports typically reveal more about those who commit minor offenses. Hindelang and his colleagues reviewed several studies that use both types of data and found that the conclusions we can draw from official and self-report data about the class–delinquency relationship are not at odds—if we consider the same types of offenses with both data sources. By design, then, the focus of this comparative study was on relatively trivial offenses because those are the types of delinquent acts typically assessed with self-report data. According to Hindelang and his colleagues, when one examines *comparable, relatively minor offenses* using both official and self-report data, the relationship between class and crime is similarly weak for both data sources.[144]

Both Class and Crime/Delinquency Measures

Margaret Farnworth and her colleagues measured both class and crime in several ways to determine whether inadequate measurement accounted for prior findings of no class–delinquency relationship.[145] First, they measured social class using variables based on a status attainment model (occupation and education of wage earners in the household). Second, they developed a "neo-Marxist" measure based on relation to the means of production: surplus population, working class, and managers/bourgeoisie. Third, they constructed variables to represent the "underclass," measured in terms of households below the poverty level, households receiving welfare, or unemployment of the principal wage earner. Finally, they measured "persistent economic need" in two ways: principal wage earner unemployment over time and welfare support.[146]

Farnworth and her colleagues were similarly thorough in their measurement of delinquency, constructing three scales and two measures of persistent offending. First, they created a scale of general delinquency, containing thirty-four items measuring behaviors ranging from minor infractions (skipping school, lying about age) to serious law violations (aggravated assault, gang fights). Second, they created a street crimes scale containing ten serious offenses, such as burglary, robbery, selling drugs, and car theft. Third, they created a common delinquency scale of minor offenses, such as theft of items worth $100 or less, joyriding, and property damage. To assess persistent offending, they developed a cumulative measure of delinquency, combining data from all four rounds of data collection. They also developed a cumulative measure of police reports of all delinquent acts except traffic offenses, also from all four rounds of data collection.

This study highlights the impact that alternative measures of class and delinquency can have on research results. The relationship between social class and delinquent behavior varied greatly depending on the measures used. When class was measured using variables from status attainment models—the kinds of variables used most often in prior research—no strong or consistent relationship existed between class and delinquency. When class was measured using neo-Marxist and underclass variables, however, an inverse relationship existed between class and serious street crime.[147]

The measurement of delinquency also affected research results. When Farnworth and her colleagues examined data from a single time point, none of the class measures were significantly related to the general or common delinquency scales. But a significant relationship existed between the serious street crimes scale and underclass measures of economic well-being. This study also supports the value of longitudinal research, which incorporates multiple time points. Farnworth and her colleagues found the most consistent evidence of a class–delinquency relationship when they examined persistent economic need and delinquent behavior over time. They concluded that, when class and delinquency are measured in ways most consistent with theories of delinquency, the expected negative relationship exists between class and delinquency. "This is especially the case when persistent underclass status is related to persistent delinquency and to victimizing street crimes associated with lifestyles among the urban poor."[148]

Self-Report Data

Though self-reports, like any form of data, are limited in some ways, they provide the best information available to date to explore the relationship between social class and crime. Early self-report studies in the 1960s and 1970s tended to show no relationship between class and crime or delinquency.[149] These studies, however, measured primarily trivial offenses. Not until the NYS began in 1976 did self-reports include data on the kinds of serious offenses that would allow a thorough examination of the class–crime relationship and permit a reasonable comparison between self-reports and official data.[150]

National Youth Survey

The NYS enabled researchers to overcome the problem of different "domains of behavior" in self-reports and official data.[151] Using NYS data, Elliott and Ageton developed a typology of six categories of offenses (see Table 5-6). In addition to these six subscales, they also used in their analysis a general self-report scale that included the full range of delinquent behaviors. "This measure was conceptualized as a parallel measure to official arrest records and includes 46 items selected so as to be representative of the full range of official acts for which juveniles could be arrested. The set included all but one of the UCR Part I offenses (homicide was excluded), 60% of Part II offenses, and a wide range of UCR 'other' offenses."[152]

Elliott and Ageton measured social class in terms of the occupational status of the primary wage earner, with a variable containing three categories: lower-class, working-class, or middle-class.[153] They found significant class differences in delinquent behavior for the general delinquency scale and the serious crimes against persons subscale. Lower-class youth scored higher on the general delinquency scale than did youth from the working and middle classes. The striking class difference in this study, though, was for serious crimes against persons. For this subscale, "lower-class youth report nearly four times as many offenses as do middle-class youth and one-and-one-half times as many as working-class youth."[154] Elliott and Ageton found no class differences for the other five subscales of offenses. The class differences in the general delinquency scale and the crimes against persons subscale were primarily at the high end of the frequency continuum.[155] In other words, "lower-class youth are found disproportionately among high frequency offenders."[156]

Elliott and Huizinga extended this study to include data gathered annually over a five-year period.[157] They examined both prevalence and incidence of offending for persons from different social classes. (As we noted earlier, prevalence refers to the proportion of a group involved in delinquency, and incidence refers to the frequency of offending.) They found no class differences in the prevalence of delinquency when they measured delinquent behavior using a general scale of offenses. They also found, however, that, for males only, significant class differences in prevalence existed for serious offenses: felony assault, felony theft, and robbery. Middle-class males were less likely to engage in serious offenses than lower- or working-class males.[158] Elliott and Huizinga concluded that, with the exception of serious delinquency, class differences in prevalence of offending were sometimes significant, but were neither strong nor consistent over time or across gender.[159] Regarding incidence of offending, their results are similar to Elliott and Ageton's.

Elliott and Huizinga note that class differences in delinquency appear greater with incidence than with prevalence measures.[160] Yet many previous self-report studies have relied on prevalence measures that do not include the most serious offenses. Elliott and Huizinga argue that these weaknesses of self-report measures in earlier studies might be responsible for the typical, but incorrect, finding of no relationship between class and involvement in crime.[161] When delinquency is measured thoroughly and includes serious offenses, we see a class–crime relationship more like that indicated by official data than that indicated by previous self-report studies relying on more limited delinquency measures.

Philadelphia Birth Cohort Study

Like the NYS, the Philadelphia birth cohort study includes self-reports of serious offenses.[162] In addition, the Philadelphia study includes official measures of offending for the subjects who provided self-report data and thus allows researchers to compare self-reports and official data.

Terence Thornberry and Margaret Farnworth used data from the Philadelphia study to examine the class–crime relationship.[163] They found that, when the "domain problem" was overcome, self-reports and official data provided fairly similar portrayals of the class–crime relationship. In other words, as the types of offenses represented in self-reports and official data became more consistent, the discrepancies in research results based on these two sources of data diminished.[164]

Thornberry and Farnworth also found a negative relationship between social class and adult offending, particularly among blacks. This inverse relationship, however, did not hold for juveniles. Neither self-report nor official measures of offending were strongly related to social class among juveniles as measured by father's occupation. Thornberry and Farnworth found, though, that social class is more strongly related to offending when it is measured in terms of the respondent's own class, rather than that of his family of origin. This finding is consistent with the work of Hagan and McCarthy, who argue for more thoughtful measures of juvenile social class in delinquency research, rather than the typical use of parental occupation as a substitute for juvenile social class.

Social Class and Delinquency in Brief

Ecological research shows a strong and consistent relationship between class and crime. Many criminologists incorrectly inferred individual-level relationships from ecological research. But early self-report studies challenged long-held beliefs in a negative individual-level relationship between class and crime. Critics pointed out problems with these early studies, most notably their focus on relatively trivial offenses. Later self-report surveys, such as the NYS, overcame the drawbacks of earlier studies, and provide us with a sound body of research from which to draw conclusions about the class–crime relationship.

The NYS, focusing on the full range of delinquent offenses, reveals significant class differences in delinquency, particularly in serious crimes against persons. Elliott and Ageton found that lower-class youth are significantly more likely than working- and middle-class youth to report involvement in serious crimes against persons. This result is due primarily to class differences at the high end of the frequency of offending continuum, where lower-class youth are disproportionately found.[165] Elliott and Huizinga found significant class differences in both prevalence and incidence of serious offenses, though these relationships did not hold for more minor forms of offending.[166]

The general conclusion we can draw is that, when self-report delinquency measures include serious offenses that are most comparable to offenses represented in official data, a consistent negative relationship exists between class and delinquency. Lower-class youth are more likely than their working-class and middle-class counterparts to engage in serious delinquent behavior and to commit a higher frequency of offenses. When we examine more minor forms of offending, the class–crime relationship is generally weak.

Why Are the Economically Disadvantaged So Involved in Serious Crime?

Attempts to explain why social class is related to serious crime and delinquency have occurred on several levels. Some researchers have explored the relationship at the individual level, examining social psychological processes that might explain how social class affects crime. Others have looked to the economic conditions of neighborhoods, attempting to explain higher rates of offending in more impoverished communities. Still others, exploring the class–crime relationship from a Marxist perspective, have pointed to the capitalist foundations of modern American society to explain the relationship.

Individual-Level Research

Karen Heimer explored the relationship between social class and parenting practices, using differential association and social learning theories of crime (see Chapter 11). She tried to specify "an explanation of the links between socioeconomic stratification, parenting practices, cultural definitions of violence, and violent delinquency."[167] Heimer used self-report data (NYS) to test these hypotheses about the mechanisms through which socioeconomic status influences violent delinquency:[168]

1. Parents of lower socioeconomic status (SES) will be more likely than those of higher status to use coercive discipline strategies (for example, threatening, yelling, restricting privileges, or punishing physically), which will increase the likelihood that their children will develop attitudes favoring violence.

2. Lower-status parents will be less likely than those of higher status to supervise their children closely, which will increase the likelihood of their children associating with aggressive peers and developing attitudes favoring violence.

3. Lower-status parents will be less likely than those of higher status to disapprove of aggression, which will increase the likelihood that their children will develop attitudes favoring violence.

4. Attitudes favoring violence "will increase the likelihood of subsequent violent delinquency and will mediate the effects on violent delinquency of SES, parenting variables, and association with aggressive friends."

With the exception of the third hypothesis, Heimer found strong support for these hypotheses. Her work demonstrates that socioeconomic status affects violent delinquency indirectly, through a process involving parenting practices, peer associations, and attitude formation.

Like Heimer, Bradley Entner Wright and his colleagues explored social psychological mechanisms that intervene between SES and delinquency.[169] They attempted to explain the inconsistency between theory, which proposes an inverse relationship between SES and crime, and empirical research, which tends to show little or no relationship between the two. Wright and his colleagues proposed that this inconsistency could be reconciled by recognizing that SES both positively and negatively affects delinquency, and that these opposite effects might "cancel each other out" and lead to the faulty conclusion that SES and delinquency are not related.[170]

Relying on classic criminological theories, such as anomie and subculture theory (see Chapter 12), Wright and his colleagues hypothesized that *low* socioeconomic status will increase the likelihood of involvement in delinquency. In this case, SES will have a negative effect on delinquent behavior.[171] The processes mediating between SES and delinquency, according to these theories, include strain caused by limited opportunities, alienation, diminished educational and occupational aspirations, and diminished self-control. Relying on social psychological theories, such as power-control theory (see Chapter 13), Wright and his colleagues hypothesized that *high* socioeconomic status will increase the likelihood of involvement in delinquency. Here, SES will have a positive effect on delinquent behavior.[172] The processes mediating between SES and delinquency, according to these theories, include "taste for risk" or the search for excitement and thrills, absence of parental controls, and perception of limited risk of detection or punishment (all of which are greatest among the upper class, according to power-control theory).

Wright and his colleagues found strong support for these hypotheses. Socioeconomic status affected delinquency indirectly—positively through some intervening variables, and negatively through others. Thus, in the end, the positive and negative effects of SES on delinquent behavior "cancelled each other out." These offsetting effects create the illusion that SES is unrelated to delinquency and might explain why prior studies that have not considered these opposing effects have typically found no relationship between SES and delinquent behavior.

Neighborhood-Level Research

Because race and class are intertwined in America, the community-level processes we discussed earlier in this chapter apply here also. William Julius Wilson's work on the creation of an "underclass" through the concentration of economic disadvantage sheds a great deal of light on the class–crime link.[173] From Wilson's work and the social disorganization model of crime and delinquency (see Chapter 12), Robert Bursik and Harold Grasmick derived hypotheses about the relationship between the socioeconomic composition of neighborhoods and rates of crime.[174]

Bursik and Grasmick measured community-level economic well-being in two ways. First, they measured general socioeconomic status in terms of median family income, median level of education, and percentage of the population in professional or managerial occupations. Second, they measured severe economic deprivation in terms of percentage of families with incomes below the poverty level, unemployment rate, public assistance rate per 100 residents, and percentage of the population that is black.[175] Bursik and Grasmick found that general socioeconomic status of a neighborhood had no significant effect on delinquency rates. But severe economic deprivation significantly affected delinquency rates, even after other factors were taken into account. The effects of economic deprivation on delinquency rates were both direct and indirect, through variables measuring social disorganization and the ability of residents to control the behavior of those in the neighborhood (for example, residential mobility, rates of owner occupancy, percentage of children in two-parent households).[176]

The measurement of severe economic deprivation in Bursik and Grasmick's study is consistent with Wilson's conceptualization of an urban underclass. Bursik and Grasmick's work makes a number of contributions to research on class and crime. It demonstrates (1) the importance of distinguishing measures of general socioeconomic status from those of economic deprivation, (2) the role of economic deprivation in explaining delinquency, (3) the importance of social disorganization processes as intervening mechanisms between deprivation and crime, and (4) the value of exploring the class–crime relationship at the neighborhood level, rather than at the individual level.[177]

Marxist Paradigm

Some criminologists have turned to Marxist theories to explain the class–crime relationship. We discuss the Marxist paradigm in Chapter 13. Marxist criminologists locate the origins of crime in capitalism, which creates competing economic classes within society.[178] Divisions among the classes result in attempts by the ruling class to control those who are relatively powerless. This control is accomplished in part through the legal system, which the ruling class uses to define legitimate behavior among the working class and to protect its own interests. In the Marxist view, both this process of "criminalization" of the actions of the working class, and the responses of workers to ruling class controls, account for higher rates of offending among the working class.

■ Victimization

The National Crime Victimization Survey (NCVS) reveals that 24.2 million crimes occurred in the United States in 2003.[179] As **Table 5-7** indicates, the majority of these offenses were property crimes (76.9%), rather than crimes against persons (22.3%). Victimization rates were more than seven times greater for property crimes than for personal crimes, primarily because of the relative frequency of theft.

The risks of becoming a victim of crime are not the same for all people. Victimization risk varies by several factors, including age, gender, race, and social

Table 5-7	Number, Percentage, and Rate of Victimizations, by Type of Offense, 2003		
Type of offense	Number	Percentage of all crimes	Victimization rates*
All crimes	24,212,800	100.0%	
Violent crimes	5,401,720	22.3	22.6
Rape/sexual assault	198,850	0.8	0.8
Robbery	596,130	2.5	2.5
Aggravated assault	1,101,110	4.5	4.6
Simple assault	3,505,630	14.5	14.6
Property crimes	18,626,380	76.9	163.2
Burglary	3,395,620	14.0	29.8
Motor vehicle theft	1,032,470	4.3	9.0
Theft	14,198,290	58.6	124.4

*Rates per 1,000 persons age 12 and older for personal crimes, and per 1,000 households for property crimes.

Note: Due to rounding, middle column may not sum precisely to 100 percent.

Source: Catalano, Criminal Victimization, 2003, 2.

class. The social correlates of offending are basically identical to the social correlates of victimization. Those who are most likely to commit offenses are also most likely to become crime victims.

Age

As we have shown, young people are disproportionately involved in crime. They are also more likely than older persons to be victims of crime.[180] **Figure 5-2** shows victimization rates (per 1,000 persons age 12 and older) by age for violent crimes (rape/sexual assault, robbery, and assault). With striking similarity to rates of offending for many types of crime, rates of violent victimization peak at ages 16–19.[181] The decline from this peak age range is so steep that, by ages 25–34, rates of violent victimization are half the rates in the late teen years. NCVS data do not include information about murder, but UCR data show that young people between the ages of 17 and 29, especially males, are disproportionately likely to be victims of homicide.[182]

Gender

Males are more likely than females to commit delinquent or criminal acts. NCVS data reveal that males are also more likely to be crime victims. **Table 5-8** shows violent crime victimization rates by gender. For all crimes of violence combined (excluding murder), the victimization rate for males is 1.4 times the rate for females (26.3 compared to 19.0). For robbery, victimization rates for males are double the rates for females. Only for the crime of rape are females more likely than males to be victimized. The rape victimization rate in 2003 was 7.5 times greater for females than males.

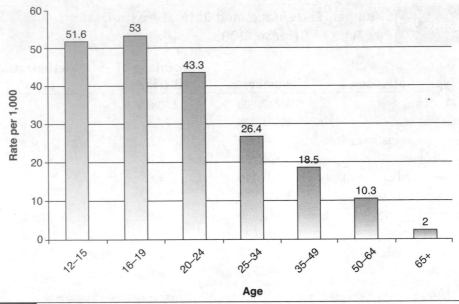

Figure 5-2 Victimization rates for violent crimes, by age, 2003. This figure shows victimization rates (per 1,000 persons age 12 and older) by age for the violent crimes of rape/sexual assault, robbery, and assault. These victimization rates are strikingly similar to arrest rates by age for many types of crime.

Source: Catalano, *Criminal Victimization, 2003, 7.*

Table 5-8 Victimization Rates for Violent Crime, by Gender, 2003*		
Type of offense	**Males**	**Females**
All crimes of violence (except homicide)	26.3	19.0
Rape/sexual assault	0.2†	1.5
Robbery	3.2	1.6
Aggravated assault	5.9	3.3
Simple assault	17.1	12.4

*Rates per 1,000 persons age 12 and older.

†Based on 10 or fewer sample cases.

Source: Catalano, *Criminal Victimization, 2003, 7.*

Race

Table 5-9 shows victimization rates by race and ethnicity for violent crimes. Blacks are more likely than people of other races to be the victims of all types of violent crime except rape/sexual assault. The race difference in victimization is particularly high for robbery, for which blacks are three times as likely as whites, and almost twice as likely as those of other races, to be victimized. Victimization rates are similar for Hispanics and non-Hispanics for all violent offenses except rape/sexual assault, for which non-Hispanics are twice as likely as those of Hispanic origin to be victimized.

UCR data reveal that blacks are also more likely than whites and those of other races to be the victims of homicide. Although blacks constitute about

Table 5-9	**Victimization Rates for Violent Crime, by Race and Ethnicity, 2003***				
Type of offense	**Black**	**White**	**Other****	**Hispanic**	**Non-Hispanic**
Violent crimes	29.1	21.5	16.0	24.2	22.3
Rape/sexual assault	0.8[†]	0.8	0.2[†]	0.4[†]	0.9
Robbery	5.9	1.9	3.4	3.1	2.4
Aggravated assault	6.0	4.2	5.4	4.6	4.6
Simple assault	16.3	14.7	7.0	16.1	14.4

*Rates per 1,000 persons age 12 and older.

**Other races include Asians, Native Hawaiians, other Pacific Islanders, Alaska Natives, and American Indians.

[†]Based on 10 or fewer sample cases.

Source: Catalano, *Criminal Victimization, 2003,* 7.

12.8% of the US population, 47.8% of homicide victims in 2003 were black and 48.0% were white. The numbers are similar for juvenile homicide victims (those under age 18): in 2003, 46.3% were black and 49.5% were white.[183]

Social Class

Economic disadvantage is associated with violent crime victimization. **Table 5-10** shows victimization rates by household income. For violent crime, victimization rates generally decline as household income increases. For all crimes of violence combined (excluding murder), the victimization rate is 49.9 for those whose household income is less than $7,500, but only 17.5 for those whose household income is $75,000 or more.

Table 5-10	**Victimization Rates for Violent and Property Crime, by Household Income, 2003***						
Type of offense	**Less than $7,500**	**$7,500– $14,999**	**$15,000– $24,999**	**$25,000– $34,999**	**$35,000– $49,999**	**$50,000– $74,999**	**$75,000 or more**
Violent crimes	49.9	30.8	26.3	24.9	21.4	22.9	17.5
Rape/sexual assault	1.6[†]	1.8[†]	0.8[†]	0.9[†]	0.9[†]	0.5[†]	0.5[†]
Robbery	9.0	4.0	4.0	2.2	2.1	2.0	1.7
Aggravated assault	10.8	7.9	4.5	5.0	4.8	5.2	2.7
Simple assault	28.2	17.0	17.0	16.9	13.5	15.2	12.6
Property crimes	204.6	167.7	179.2	180.7	177.1	168.1	176.4
Burglary	58.0	42.2	38.4	35.3	27.6	24.9	20.8
Motor vehicle theft	6.3	7.3	8.9	12.3	9.5	8.4	11.9
Theft	140.3	118.3	131.9	133.1	140.0	134.7	143.7

*Rates per 1,000 persons age 12 and older for personal crimes, and per 1,000 households for property crimes.

[†]Based on 10 or fewer sample cases.

Source: Catalano, *Criminal Victimization, 2003,* 8.

The relationship between economic disadvantage and property crime is less straightforward. Burglary victimization rates decrease as household income increases, but victimization rates for theft and motor vehicle theft show no clear pattern in relation to household income.

■ Summary and Conclusions

In this chapter, we examined age, gender, race, and social class as social correlates of offending. Age and gender, in particular, are strongly and consistently related to involvement in crime, which is primarily a pursuit of young males. For most offenses, the peak ages of offending for most individuals are during adolescence and young adulthood. The peak age of offending varies somewhat by type of offense and is older for violent crimes than for property offenses. The gender gap in crime also varies by type of offense and is greater for violent than for property crimes. While males are disproportionately involved in almost all forms of crime, this is especially true for violence. Although some theorists predicted that the gender gap in crime would narrow as a function of increasing gender role equality, it appears that this convergence generally has not occurred for offenses other than minor property crimes.

The relationships between race and social class and involvement in crime are less clear-cut than those observed for age and gender. But the bulk of evidence suggests that African Americans and those who are economically disadvantaged are more likely than others to be involved in serious crime. Self-report data suggest, however, that race and class differences in rates of offending do not hold for minor forms of crime and delinquency. The race and class differences that exist for serious crime cannot be explained solely by biases inherent in the criminal justice system.

Finally, those who are most likely to be offenders are also most likely to be crime victims. Young people, males, African Americans, and economically disadvantaged persons are more likely than others to be victims of crime.

CRITICAL THINKING QUESTIONS

1. Why are age, gender, race, and social class best viewed as social—rather than individual—correlates of offending?

2. Why do official statistics and self-report data paint different pictures of the social correlates of offending?

3. Given what you have learned about the disproportionate involvement in crime of young people, males, African Americans, and those who are economically disadvantaged, describe the kinds of programs and efforts you believe would be most successful in preventing delinquency and crime.

4. What accounts for the similarities between social correlates of offending and social correlates of victimization? In other words, why are those who are most likely to commit crime also most likely to be victims of it?

5. Consider the perceptions of criminals in the Chestnut Hill neighborhood, described at the opening of this chapter. How do the social correlates of offending (age, gender, race, and social class) shape people's perceptions of crime and criminals?

SUGGESTED READINGS

Elliott, Delbert S., and Suzanne S. Ageton. "Reconciling Race and Class Differences in Self-Reported and Official Estimates of Delinquency." *American Sociological Review* 45 (1980):95–110.

Hindelang, Michael J., Travis Hirschi, and Joseph G. Weis. "Correlates of Delinquency: The Illusion of Discrepancy Between Self-Report and Official Measures." *American Sociological Review* 44 (1979):995–1014.

Hirschi, Travis, and Michael Gottfredson. "Age and the Explanation of Crime." *American Journal of Sociology* 89 (1983):552–584.

Sampson, Robert J., and William Julius Wilson. "Toward a Theory of Race, Crime, and Urban Inequality." In *Crime and Inequality*, edited by John Hagan and Ruth D. Peterson, 37–54. Stanford, CA: Stanford University Press, 1995.

Steffensmeier, Darrell, and Emilie Allan. "Gender and Crime: Toward a Gendered Theory of Female Offending." *Annual Review of Sociology* 22 (1996):459–487.

Steffensmeier, Darrell, Emilie Anderson Allan, Miles D. Harer, and Cathy Streifel. "Age and the Distribution of Crime." *American Journal of Sociology* 94 (1989):803–831.

GLOSSARY

age composition effect: Concerns the age composition of the total population and its effect on overall crime rates. This effect traces changes in crime rates to changes in population demographics, such that crime rates rise as the number of people in their crime-prone years increases and fall as the number of people in their crime-prone years decreases. For example, crime rates rose as "baby boomers" reached mid-adolescence and fell as they aged into adulthood.

age–crime curve: The bell-shaped curve generally observed when one plots or graphs the relationship between age and crime. The age–crime curve typically shows an increase in delinquent involvement during the teenage years, a peak in mid-adolescence to early adulthood, and then a rapid decline.

age effect: The disproportionate involvement of young people in crime.

aging out of crime: Termination of involvement in crime following adolescence and young adulthood, due to sociological, psychological, and biological forces.

convergence hypothesis: The prediction that the gender gap in crime and delinquency would narrow and that male and female rates of offending would converge as a function of increasing gender equality.

crime-prone years: The age period when people are most likely to be involved in crime. Generally, the crime-prone years range from mid-adolescence into early adulthood.

ecological correlation: A correlation between two variables based on data combined for a group, such as a geographic area or social group. The relationship between delinquency rates and median family income in a geographic area is an example of an ecological correlation between crime and social class.

ecological fallacy: "An error of reasoning committed by coming to conclusions about individuals based only on data about groups" (Vogt, *Dictionary*, 78).

social correlates of delinquency: Social characteristics (such as age, gender, race, and social class) that are statistically related to involvement in delinquent behavior and that tend to distinguish offenders from non-offenders.

REFERENCES

Adler, Freda. *Sisters in Crime: The Rise of the New Female Criminal.* New York: McGraw-Hill, 1975.

Alarid, Leanne Fiftal, Velmer S. Burton, Jr., and Francis T. Cullen. "Gender and Crime Among Felony Offenders: Assessing the Generality of Social Control and Differential Association Theories." *Journal of Research in Crime and Delinquency* 37 (2000):171–199.

Anderson, Elijah. *Code of the Street: Decency, Violence, and the Moral Life of the Inner City.* New York: Norton, 1999.

Austin, Roy L. "Recent Trends in Official and Female Crime Rates: The Convergence Controversy." *Journal of Criminal Justice* 21 (1993):447–466.

Barnes, Grace M., and Michael P. Farrell. "Parental Support and Control as Predictors of Adolescent Drinking, Delinquency, and Related Problem Behaviors." *Journal of Marriage and the Family* 54 (1992):763–776.

Bartusch, Dawn Jeglum, and Ross L. Matsueda. "Gender, Reflected Appraisals, and Labeling: A Cross-Group Test of an Interactionist Theory of Delinquency." *Social Forces* 75 (1996):145–176.

Beirne, Piers, and Richard Quinney. *Marxism and Law.* New York: Wiley, 1982.

Bjerregaard, Beth, and Carolyn Smith. "Gender Differences in Gang Participation, Delinquency, and Substance Use." *Journal of Quantitative Criminology* 9 (1993):329–355.

Blumenthal, Monica D. "Predicting Attitudes toward Violence." *Science* 176 (1972):1296–1303.

Braithwaite, John. "The Myth of Social Class and Criminality Reconsidered." *American Sociological Review* 46 (1981):36–57.

Bursik, Robert J., Jr., and Harold G. Grasmick. "Economic Deprivation and Neighborhood Crime Rates, 1960–1980." *Law and Society Review* 27 (1993):263–283.

Campbell, Anne. *The Girls in the Gang.* 2nd ed. Cambridge, MA: Basil Blackwell, 1991.

Canter, Rachelle J. "Sex Differences in Self-Reported Delinquency." *Criminology* 20 (1982):373–393.

Cao, Liqun, Anthony Adams, and Vickie J. Jensen. "A Test of the Black Subculture of Violence Thesis: A Research Note." *Criminology* 35 (1997):367–379.

Catalano, Shannan M. *Criminal Victimization, 2003.* Washington, DC: US Department of Justice, Bureau of Justice Statistics, 2004.

Cernkovich, Stephen A., and Peggy C. Giordano. "Family Relationships and Delinquency." *Criminology* 25 (1987):295–321.

Cernkovich, Stephen A., Peggy C. Giordano, and Meredith D. Pugh. "Chronic Offenders: The Missing Cases in Self-Reported Delinquency Research." *Journal of Criminal Law and Criminology* 76 (1985):705–732.

Cernkovich, Stephen A., Peggy C. Giordano, and Jennifer L. Rudolph. "Race, Crime, and The American Dream." *Journal of Research in Crime and Delinquency* 37 (2000):131–170.

Chambliss, William J., and Robert B. Seidman. *Law, Order, and Power.* 2nd ed. Reading, MA: Addison-Wesley, 1982.

Chesney-Lind, Meda. "Girls' Crime and Women's Place: Toward a Feminist Model of Female Delinquency." *Crime and Delinquency* 35 (1989):5–29.

———. "Judicial Paternalism and the Female Status Offender: Training Women to Know Their Place." *Crime and Delinquency* 23 (1977):121–130.

Cohen, Lawrence E., and Kenneth C. Land. "Age Structure and Crime: Symmetry versus Asymmetry and the Projection of Crime Rates through the 1980s." *American Sociological Review* 52 (1987):170–183.

Colvin, Mark, and John Pauly. "A Critique of Criminology: Toward an Integrated Structural-Marxist Theory of Delinquency Production." *American Journal of Sociology* 89 (1983):513–551.

Curtis, Lynn A. *Violence, Race, and Culture*. Lexington, MA: D. C. Heath, 1975.

D'Alessio, Stewart J., and Lisa Stolzenberg. "Race and the Probability of Arrest." *Social Forces* 81 (2003):1381–1397.

Elliott, Delbert S., and Suzanne S. Ageton. "Reconciling Race and Class Differences in Self-Reported and Official Estimates of Delinquency." *American Sociological Review* 45 (1980):95–110.

Elliott, Delbert S., Suzanne S. Ageton, David Huizinga, B. A. Knowles, and Rachelle J. Canter. *The Prevalence and Incidence of Delinquent Behavior: 1976–1980: National Estimates of Delinquent Behavior by Sex, Race, Social Class and Other Selected Variables*. Boulder, CO: Behavioral Research Institute, 1983.

Elliott, Delbert S., and David Huizinga. "Social Class and Delinquent Behavior in a National Youth Panel." *Criminology* 21 (1983):149–177.

Elliott, Delbert S., David Huizinga, and Suzanne S. Ageton. *Explaining Delinquency and Drug Use*. Beverly Hills, CA: Sage, 1985.

Erlanger, Howard S. "The Empirical Status of the Subculture of Violence Thesis." *Social Problems* 22 (1974):280–292.

Esbensen, Finn-Aage, and David Huizinga. "Gangs, Drugs, and Delinquency in a Survey of Urban Youth." *Criminology* 31 (1993):565–585.

Fagan, Jeffrey. "Social Processes of Drug Use and Delinquency among Gang and Non-Gang Youths." In *Gangs in America,* 1st ed., edited by C. Ronald Huff, 183–222. Newbury Park, CA: Sage, 1990.

Farnworth, Margaret, Terence P. Thornberry, Marvin D. Krohn, and Alan J. Lizotte. "Measurement in the Study of Class and Delinquency: Integrating Theory and Research." *Journal of Research in Crime and Delinquency* 31 (1994):32–61.

Farrington, D. P., R. Loeber, M. Stouthamer-Loeber, W. B. VanKammen, and L. Schmidt. "Self-Reported Delinquency and a Combined Delinquency Seriousness Scale Based on Boys, Mothers, and Teachers: Concurrent and Predictive Validity for African-Americans and Caucasians." *Criminology* 34 (1996):493–514.

Federal Bureau of Investigation. *Crime in the United States 2003. Uniform Crime Reports*. Washington, DC: US Department of Justice, Federal Bureau of Investigation, 2004.

Gove, Walter. "The Effect of Age and Gender on Deviant Behavior: A Bio-Psychosocial Perspective." In *Gender and the Life Course,* edited by Alice Rossi, 115–144. Hawthorne, NY: Aldine, 1985.

Gove, Walter R., and Robert D. Crutchfield. "The Family and Juvenile Delinquency." *The Sociological Quarterly* 23 (1982):301–319.

Greenberg, David F. "Age, Crime, and Social Explanation." *American Journal of Sociology* 91 (1985):1–21.

———. "Delinquency and the Age Structure of Society." In *Crime and Capitalism,* edited by David F. Greenberg, 118–139. Palo Alto, CA: Mayfield, 1981.

———. "The Historical Variability of the Age–Crime Relationship." *Journal of Quantitative Criminology* 10 (1994):361–373.

Hagan, John. "Destiny and Drift: Subcultural Preferences, Status Attainments, and the Risks and Rewards of Youth." *American Sociological Review* 56 (1991):567–582.

Hagan, John, A. R. Gillis, and John Simpson. "Clarifying and Extending Power-Control Theory." *American Journal of Sociology* 95 (1990):1024–1037.

———. "The Class Structure of Gender and Delinquency: Toward a Power-Control Theory of Common Delinquent Behavior." *American Journal of Sociology* 90 (1985):1151–1178.

Hagan, John, and Bill McCarthy. "Streetlife and Delinquency." *British Journal of Sociology* 43 (1992):533–561.

Hagan, John, and Ruth D. Peterson, eds. *Crime and Inequality*. Stanford, CA: Stanford University, 1995.

Hagan, John, John Simpson, and A. R. Gillis. "Class in the Household: A Power-Control Theory of Gender and Delinquency." *American Journal of Sociology* 92 (1987):788–816.

Hansen, Kirstine. "Education and the Crime–Age Profile." *British Journal of Criminology* 43 (2003):141–168.

Chapter Resources

Harris, Anthony R. "Sex and Theories of Deviance: Toward a Functional Theory of Deviant Type Scripts." *American Sociological Review* 42 (1977):3–16.

Hawkins, Darnell F., John H. Laub, Janet L. Lauritsen, and Lynn Cothern. "Race, Ethnicity, and Serious and Violent Juvenile Offending." In *Juvenile Justice Bulletin*. Washington, DC: Office of Juvenile Justice and Delinquency Prevention, 2000.

Heimer, Karen. "Gender, Interaction, and Delinquency: Testing a Theory of Differential Social Control." *Social Psychology Quarterly* 59 (1996):39–61.

———. "Gender, Race, and the Pathways to Delinquency: An Interactionist Explanation." In Hagan and Peterson, *Crime and Inequality,* 140–173.

———. "Socioeconomic Status, Subcultural Definitions, and Violent Delinquency." *Social Forces* 75 (1997):799–833.

Heimer, Karen, and Stacy DeCoster. "The Gendering of Violent Delinquency." *Criminology* 37 (1999):277–312.

Heimer, Karen, and Ross L. Matsueda. "Role-Taking, Role Commitment, and Delinquency: A Theory of Differential Social Control." *American Sociological Review* 59 (1994):365–390.

Hill, Gary D., and Maxine P. Atkinson. "Gender, Familial Control, and Delinquency." *Criminology* 26 (1988):127–149.

Hill, Gary D., and Elizabeth M. Crawford. "Women, Race, and Crime." *Criminology* 28 (1990): 601–626.

Hindelang, Michael J. "Sex Differences in Criminal Activity." *Social Problems* 27 (1979):143–156.

———. "Variations in Sex-Race-Age-Specific Incidence Rates of Offending." *American Sociological Review* 46 (1981):461–474.

Hindelang, Michael J., Travis Hirschi, and Joseph G. Weis. "Correlates of Delinquency: The Illusion of Discrepancy Between Self-Report and Official Measures." *American Sociological Review* 44 (1979):995–1014.

———. *Measuring Delinquency.* Beverly Hills, CA: Sage, 1981.

Hirschi, Travis, *Causes of Delinquency.* Berkeley, CA: University of California Press, 1969.

Hirschi, Travis, and Michael Gottfredson. "Age and the Explanation of Crime." *American Journal of Sociology* 89 (1983):552–584.

Horowitz, Ruth, and Anne E. Pottieger. "Gender Bias in Juvenile Justice Handling of Seriously Crime-Involved Youths." *Journal of Research in Crime and Delinquency* 28 (1991):75–100.

Huizinga, David, and Delbert S. Elliott. "Juvenile Offenders: Prevalence, Offender Incidence, and Arrest Rates by Race." *Crime and Delinquency* 33 (1987):206–223.

———. "Reassessing the Reliability and Validity of Self-Report Delinquency Measures." *Journal of Quantitative Criminology* 2 (1986):293–327.

———. *Self-Reported Measures of Delinquency and Crime: Methodological Issues and Comparative Findings.* Boulder, CO: Behavioral Research Institute, 1984.

Jensen, Gary F., and Raymond Eve. "Sex Differences in Delinquency: An Examination of Popular Sociological Explanations." *Criminology* 13 (1976):427–448.

Johnson, Richard E. "Social Class and Delinquent Behavior: A New Test." *Criminology* 18 (1980):86–93.

Kleck, Gary. "On the Use of Self-Report Data to Determine the Class Distribution of Criminal and Delinquent Behavior." *American Sociological Review* 47 (1982):427–433.

Koita, Kiyofumi, and Ruth A. Triplett. "An Examination of Gender and Race Effects on the Parental Appraisal Process: A Reanalysis of Matsueda's Model of the Self." *Criminal Justice and Behavior* 25 (1998):382–400.

Krohn, Marvin D., Ronald L. Akers, Marcia J. Radosevich, and Lonn Lanza-Kaduce. "Social Status and Deviance." *Criminology* 18 (1980):303–318.

LaFree, Gary, Kriss A. Drass, and Patrick O'Day. "Race and Crime in Postwar America: Determinants of African-American and White Rates, 1957–1988." *Criminology* 30 (1992):157–188.

Lanctot, Nadine, and Marc LeBlanc. "Explaining Deviance by Adolescent Females." *Crime and Justice* 29 (2002):113–202.

Laub, John H. "Urbanism, Race, and Crime." *Journal of Research in Crime and Delinquency* 20 (1983):183–198.

Leonard, Eileen B. *Women, Crime and Society: A Critique of Theoretical Criminology.* New York: Longman, 1982.

Liska, Allen E., and Paul E. Bellair. "Violent-Crime Rates and Racial Composition: Convergence Over Time." *American Journal of Sociology* 101 (1995):578–610.

Liska, Allen E., John R. Logan, and Paul E. Bellair. "Race and Violent Crime in the Suburbs." *American Sociological Review* 63 (1998):27–38.

Liu, Xiaoru, and Howard B. Kaplan. "Explaining the Gender Difference in Adolescent Behavior: A Longitudinal Test of Mediating Mechanisms." *Criminology* 37 (1999):195–215.

Macmillan, Ross. "Violence and the Life Course: The Consequences of Victimization for Personal and Social Development." *Annual Review of Sociology* 27 (2001):1–22.

Matsueda, Ross L. "Reflected Appraisals, Parental Labeling, and Delinquency: Specifying a Symbolic Interactionist Theory." *American Journal of Sociology* 97 (1992):1577–1611.

Matsueda, Ross L., and Karen Heimer. "Race, Family Structure, and Delinquency: A Test of Differential Association and Social Control Theories." *American Sociological Review* 52 (1987):826–840.

McCarthy, Bill, John Hagan, and Todd S. Woodward. "In the Company of Women: Structure and Agency in a Revised Power-Control Theory of Gender and Delinquency." *Criminology* 37 (1999):761–788.

Mead, George H. *Mind, Self and Society.* Chicago: University of Chicago Press, 1934.

Morash, Merry. "Establishment of a Juvenile Record: The Influence of Individual and Peer Group Characteristics." *Criminology* 22 (1984):97–112.

Paschall, Mallie J., Robert L. Flewelling, and Susan T. Ennett. "Racial Differences in Violent Behavior Among Young Adults: Moderating and Confounding Effects." *Journal of Research in Crime and Delinquency* 35 (1998):148–165.

Peeples, Faith, and Rolf Loeber. "Do Individual Factors and Neighborhood Context Explain Ethnic Differences in Juvenile Delinquency?" *Journal of Quantitative Criminology* 10 (1994):141–157.

Piquero, Alex R., Randall MacIntosh, and Matthew Hickman. "The Validity of a Self-Reported Delinquency Scale: Comparisons across Gender, Age, Race, and Place of Residence." *Sociological Methods and Research* 30 (2002):492–529.

Poe-Yamagata, Eileen, and Jeffrey A. Butts. *Female Offenders in the Juvenile Justice System.* Washington DC: Office of Juvenile Justice and Delinquency Prevention, 1996.

Quinney, Richard. *Class, State, and Crime.* 2nd ed. New York: Longman, 1980.

Rossi, Peter H., Emily Waite, Christine E. Bose, and Richard E. Berk. "The Seriousness of Crimes: Normative Structure and Individual Differences." *American Sociological Review* 39 (1974):224–237.

Rowe, A., and Charles Tittle. "Life Cycle Changes and Criminal Propensity." *Sociological Quarterly* 18 (1977):223–236.

Rowe, David C., Alexander T. Vazsonyi, and Daniel J. Flannery. "Sex Differences in Crime: Do Means and Within-Sex Variation Have Similar Causes?" *Journal of Research in Crime and Delinquency* 32 (1995):84–100.

Sampson, Robert J., and Dawn Jeglum Bartusch. "Legal Cynicism and (Subcultural?) Tolerance of Deviance: The Neighborhood Context of Racial Differences." *Law and Society Review* 32 (1998):777–804.

Sampson, Robert J., and John H. Laub. "Crime and Deviance in the Life Course." *Annual Review of Sociology* 18 (1992):63–84.

———. "Crime and Deviance over the Life Course: The Salience of Adult Social Bonds." *American Sociological Review* 55 (1990):609–627.

———. *Crime in the Making: Pathways and Turning Points Through Life.* Cambridge, MA: Harvard University Press, 1993.

Sampson, Robert J., and William Julius Wilson. "Toward a Theory of Race, Crime, and Urban Inequality." In Hagan and Peterson, *Crime and Inequality*, 37–54.

Seydlitz, Ruth. "The Effects of Age and Gender on Parental Control and Delinquency." *Youth and Society* 23 (1991):175–201.

———. "The Effects of Gender, Age, and Parental Attachment on Delinquency: A Test for Interactions." *Sociological Spectrum* 10 (1990):209–225.

Shavit, Yossi, and Arye Rattner. "Age, Crime and the Early Life Course." *American Journal of Sociology* 93 (1988):1457–1470.

Shaw, Clifford R., and Henry D. McKay. *Juvenile Delinquency and Urban Areas.* Chicago: University of Chicago Press, 1942.

———. *Juvenile Delinquency and Urban Areas.* 2nd ed. Chicago: University of Chicago Press, 1969.

Shover, Neal, Stephen Norland, Jennifer James, and William E. Thornton. "Gender Roles and Delinquency." *Social Forces* 58 (1979):162–175.

Simon, Rita. *The Contemporary Woman and Crime.* Washington, DC: National Institute of Mental Health, 1975.

Chapter Resources

Smith, Douglas A., and Raymond Paternoster. "The Gender Gap in Theories of Deviance: Issues and Evidence." *Journal of Research in Crime and Delinquency* 24 (1987):140–172.

Smith, Douglas A., and Christy A. Visher. "Sex and Involvement in Deviance/Crime: A Quantitative Review of Empirical Literature." *American Sociological Review* 45 (1980):691–701.

Snyder, Howard N., and Melissa Sickmund. *Juvenile Offenders and Victims: A National Report.* Washington, DC: US Department of Justice, Office of Juvenile Justice and Delinquency Prevention, 1995.

Spitzer, Steven. "Toward a Marxian Theory of Deviance." *Social Problems* 22 (1975):638–651.

Steffensmeier, Darrell. "Crime and the Contemporary Woman: An Analysis of Changing Levels of Female Property Crime, 1960–75." *Social Forces* 57 (1978):566–583.

———. "National Trends in Female Arrests, 1960–1990: Assessments and Recommendations for Research." *Journal of Quantitative Criminology* 9 (1993):411–440.

———. "Sex Differences in Patterns of Adult Crime, 1965–77: A Review and Assessment." *Social Forces* 58 (1980):1080–1108.

Steffensmeier, Darrell, and Emilie Allan. "Age-Inequality and Property Crime: The Effects of Age-Linked Stratification and Status-Attainment Processes on Patterns of Criminality Across the Life Course." In Hagan and Peterson, *Crime and Inequality,* 95–115.

———. "Gender and Crime: Toward a Gendered Theory of Female Offending." *Annual Review of Sociology* 22 (1996):459–487.

———. "Looking for Patterns: Gender, Age, and Crime." In *Criminology,* 3rd ed., edited by Joseph F. Sheley, 85–127. Belmont, CA: Wadsworth, 2000.

Steffensmeier, Darrell, Emilie Anderson Allan, Miles D. Harer, and Cathy Streifel. "Age and the Distribution of Crime." *American Journal of Sociology* 94 (1989):803–831.

Steffensmeier, Darrell, and Michael J. Cobb. "Sex Differences in Urban Arrest Patterns, 1934–79." *Social Problems* 29 (1981):37–50.

Steffensmeier, Darrell, and Miles D. Harer. "Making Sense of Recent U.S. Crime Trends, 1980 to 1996/1998: Age Composition Effects and Other Explanations." *Journal of Research in Crime and Delinquency* 36 (1999):235–274.

Steffensmeier, Darrell, and Renee Hoffman Steffensmeier. "Trends in Female Delinquency: An Examination of Arrest, Juvenile Court, Self-Report, and Field Data." *Criminology* 18 (1980):62–85.

Steffensmeier, Darrell, and Cathy Streifel. "Age, Gender, and Crime Across Three Historical Periods: 1935, 1960, and 1985." *Social Forces* 69 (1991):869–894.

———. "Time-Series Analysis of the Female Percentage of Arrests for Property Crimes, 1960–1985: A Test of Alternative Explanations." *Justice Quarterly* 9 (1992):77–103.

Steffensmeier, Darrell, Cathy Streifel, and Miles D. Harer. "Relative Cohort Size and Youth Crime in the United States, 1953–1984." *American Sociological Review* 52 (1987):702–710.

Teilmann, Katherine S., and Pierre H. Landry, Jr. "Gender Bias in Juvenile Justice." *Journal of Research in Crime and Delinquency* 18 (1981):47–80.

Thornberry, Terence P., and Margaret Farnworth. "Social Correlates of Criminal Involvement: Further Evidence on the Relationship between Social Status and Criminal Behavior." *American Sociological Review* 47 (1982):505–518.

Tittle, Charles R. "Social Class and Criminal Behavior: A Critique of the Theoretical Foundation." *Social Forces* 62 (1983):334–358.

Tittle, Charles R., and Robert F. Meier. "Specifying the SES/Delinquency Relationship." *Criminology* 28 (1990):271–299.

Tittle, Charles R., Wayne J. Villemez, and Douglas A. Smith. "The Myth of Social Class and Criminality: An Empirical Assessment of the Empirical Evidence." *American Sociological Review* 43 (1978):643–656.

Uggen, Christopher. "Class, Gender, and Arrest: An Intergenerational Analysis of Workplace Power and Control." *Criminology* 38 (2000):835–862.

US Bureau of the Census. "Monthly Postcensal Resident Population, by Single Year of Age, Sex, Race, and Hispanic Origin." In *National Population Estimates for the 2000s.* Estimates for July 1, 2003. Retrieved March 2005 from Bureau of the Census website.

Vogt, W. Paul. *Dictionary of Statistics and Methodology: A Nontechnical Guide for the Social Sciences.* Newbury Park, CA: Sage, 1993.

Warr, Mark. "Age, Peers, and Delinquency." *Criminology* 31 (1993):17–40.

Williams, James Herbert, Charles D. Ayers, Wade S. Outlaw, Robert D. Abbott, and J. David Hawkins. "The Effects of Race in Juvenile Justice: Investigating Early Stage Processes." *Journal for Juvenile Justice and Detention Services* 16 (2001):77–91.

Williams, Jay R., and Martin Gold. "From Delinquent Behavior to Official Delinquency." *Social Problems* 20 (1972):209–229.

Wilson, William Julius. *The Truly Disadvantaged: The Inner City, the Underclass, and Public Policy.* Chicago: University of Chicago Press, 1987.

———. *When Work Disappears.* New York: Knopf, 1996.

Wolfgang, Marvin E., and Franco Ferracuti. *The Subculture of Violence: Toward an Integrated Theory in Criminology.* London: Tavistock, 1967.

Wolfgang, Marvin E., Robert M. Figlio, and Thorsten Sellin. *Delinquency in a Birth Cohort.* Chicago: University of Chicago Press, 1972.

Wolfgang, Marvin E., Terence P. Thornberry, and Robert M. Figlio, eds. *From Boy to Man, from Delinquency to Crime.* Chicago: University of Chicago Press, 1987.

Wright, Bradley R. Entner, Avshalom Caspi, Terrie E. Moffitt, Richard A. Miech, and Phil A. Silva. "Reconsidering the Relationship Between SES and Delinquency: Causation But Not Correlation." *Criminology* 37 (1999):175–194.

Chapter Resources

ENDNOTES

1. Vogt, *Dictionary,* 48.
2. US Bureau of the Census, *National Population Estimates.*
3. Hindelang, "Variations."
4. Elliott, Ageton, Huizinga, Knowles, and Canter, *Prevalence and Incidence*; and Rowe and Tittle, "Life Cycle Changes."
5. Steffensmeier and Harer, "Making Sense;" Steffensmeier, Streifel, and Harer, "Relative Cohort Size;" and Cohen and Land, "Age Structure and Crime."
6. Steffensmeier, Streifel, and Harer, "Relative Cohort Size."
7. Cohen and Land, "Age Structure and Crime."
8. Steffensmeier and Harer, "Making Sense," 257–258.
9. Steffensmeier, Allan, Harer, and Streifel, "Age," 803.
10. Hirschi and Gottfredson, "Age."
11. Ibid., 552.
12. Ibid.
13. Ibid., 554.
14. Ibid., 566–567.
15. Shavit and Rattner, "Early Life Course."
16. Warr, "Age, Peers, and Delinquency."
17. Steffensmeier, Allan, Harer, and Streifel, "Age."
18. Steffensmeier and Allan, "Looking for Patterns."
19. Ibid.
20. Steffensmeier, Allan, Harer, and Streifel, "Age;" Steffensmeier and Streifel, "Age, Gender, and Crime;" and Greenberg, "Historical Variability."
21. Steffensmeier, Allan, Harer, and Streifel, "Age."
22. Steffensmeier and Streifel, in "Age, Gender, and Crime," extended this analysis to include gender. They asked whether the age–crime curve was the same for males and females and whether any changes in the curve over time were the same across gender. They found that the age–crime distribution was similar for males and females (except for prostitution), and that a change had occurred over time toward a younger and more peaked curve—a change similar in magnitude for males and females.
23. Greenberg, "Historical Variability," 370.
24. Steffensmeier and Allan, "Age-Inequality and Property Crime," 105.
25. Greenberg, "Delinquency."
26. In "Delinquency" (11–14), Greenberg presents data on the age–crime relationship over time in several industrial and preindustrial countries (e.g., United States, England, France, Norway, India, Uganda). Although there appear to be substantial differences in age–crime curves across countries, Greenberg does not formally test the statistical significance of these differences.
27. Shavit and Rattner, "Early Life Course."
28. Hansen, "Education."

29. Steffensmeier and Allan, "Age-Inequality and Property Crime;" Greenberg, "Age;" Wolfgang, Thornberry, and Figlio, *From Boy to Man*; and Laub, "Urbanism, Race, and Crime."

30. Steffensmeier and Allan, "Looking for Patterns," 113; and Steffensmeier and Allan, "Age-Inequality and Property Crime," 103–105.

31. Steffensmeier and Allan, "Looking for Patterns," 113.

32. Steffensmeier and Allan, "Age-Inequality and Property Crime," 104.

33. Steffensmeier, Allan, Harer, and Streifel, "Age."

34. Sampson and Laub, *Crime in the Making*; Sampson and Laub, "Crime and Deviance in the Life Course;" and Sampson and Laub, "Salience of Adult Social Bonds."

35. Sampson and Laub, *Crime in the Making*; and Sampson and Laub, "Salience of Adult Social Bonds."

36. Steffensmeier and Allan, "Looking for Patterns."

37. Hagan, "Destiny and Drift." See also Warr, "Age, Peers, and Delinquency," 38.

38. Gove, "Effect," 138.

39. Steffensmeier and Allan, "Age-Inequality and Property Crime," 98–100.

40. Ibid., 100.

41. Federal Bureau of Investigation, *Crime in the United States 2003*, 282–285.

42. Ibid., 287.

43. Chesney-Lind, "Judicial Paternalism;" and Teilmann and Landry, "Gender Bias."

44. Canter, "Sex Differences."

45. Chesney-Lind, "Judicial Paternalism" and "Toward a Feminist Model."

46. Hindelang, "Sex Differences." The National Crime Survey is now called the National Crime Victimization Survey.

47. Hindelang, "Sex Differences."

48. Jensen and Eve, "Sex Differences in Delinquency;" Smith and Visher, "Sex and Involvement;" Steffensmeier and Steffensmeier, "Trends in Female Delinquency;" and Canter, "Sex Differences." But see also Hindelang, Hirschi, and Weis, *Measuring Delinquency* (137–155), who argue that gender differences in delinquency are similar across data sources (self-report and official data), at least for more serious offenses.

49. See Canter, "Sex Differences," 373–374, for numerous references to studies based on UCR and self-report data.

50. Canter, "Sex Differences," 388.

51. Ibid., 380.

52. See, for example, Smith and Visher, "Sex and Involvement;" and Hindelang, "Sex Differences."

53. See also Hindelang, Hirschi, and Weis, who in *Measuring Delinquency* (141–142), found no gender differences in self-reports of incorrigibility, measured as running away and defying or hitting parents.

54. Canter, "Sex Differences," 388.

55. Bjerregaard and Smith, "Gender Differences;" and Esbensen and Huizinga, "Gangs."

56. Esbensen and Huizinga, "Gangs," 571–572.

57. Campbell, *Girls in the Gang*; Fagan, "Social Processes;" and Bjerregaard and Smith, "Gender Differences."

58. Bjerregaard and Smith, "Gender Differences."

59. Esbensen and Huizinga, "Gangs;" and Bjerregaard and Smith, "Gender Differences."

60. Bjerregaard and Smith, "Gender Differences," 347.

61. Adler, *Sisters in Crime*; and Simon, *Contemporary Woman*.

62. See Austin, "Recent Trends," for a review of the research literature on the "convergence controversy."

63. Austin, "Recent Trends."

64. Ibid., 457–461.

65. See also Poe-Yamagata and Butts, who in *Female Offenders* analyzed UCR data and interpreted their findings as supporting the convergence hypothesis.

66. Steffensmeier, "Crime," "Sex Differences," and "National Trends;" Steffensmeier and Steffensmeier, "Trends in Female Delinquency;" Steffensmeier and Cobb, "Urban Arrest Patterns;" Steffensmeier and Streifel, "Time-Series Analysis;" and Steffensmeier and Allan, "Gender and Crime."

67. Steffensmeier, "Sex Differences," and "National Trends;" Steffensmeier and Steffensmeier, "Trends in Female Delinquency;" Steffensmeier and Cobb, "Urban Arrest Patterns;" and Steffensmeier and Allan, "Gender and Crime."

68. Steffensmeier, "National Trends," 420.

69. Ibid.

70. Steffensmeier and Allan, "Gender and Crime," 469.

71. Steffensmeier and Streifel, "Time-Series Analysis," 81. See also Steffensmeier, "Sex Differences," and "National Trends," 424–425; and Steffensmeier and Allan, "Gender and Crime," 469.

72. Steffensmeier and Streifel, "Time-Series Analysis," 82.

73. Harris, "Sex," 14.

74. Lanctot and LeBlanc, in "Explaining Deviance," review the literature on the gender gap in deviance and offer three perspectives for explaining the gap.

75. Adler, *Sisters in Crime*; and Leonard, *Women, Crime and Society.*

76. See, for example, Smith and Paternoster, "Gender Gap;" Bartusch and Matsueda, "Gender;" and Liu and Kaplan, "Explaining the Gender Difference."

77. Steffensmeier and Allan, "Gender and Crime," 466.

78. Smith and Paternoster, "Gender Gap;" Bartusch and Matsueda, "Gender;" and Rowe, Vazsonyi, and Flannery, "Sex Differences in Crime."

79. Alarid, Burton, and Cullen, "Gender and Crime;" Barnes and Farrell, "Parental Support;" Smith and Paternoster, "Gender Gap;" Cernkovich and Giordano, "Family Relationships and Delinquency;" Gove and Crutchfield, "Family and Juvenile Delinquency;" Shover, Norland, James, and Thornton, "Gender Roles and Delinquency;" and Jensen and Eve, "Sex Differences in Delinquency." See also Seydlitz, "Parental Attachment," and "Parental Control." Seydlitz found that the inhibitory effects of parental attachment and control on delinquency varied by *both* gender and age; and Hill and Atkinson, "Gender," found that females were not necessarily subjected to more control than males, but rather that the *type* of familial control varied by gender.

80. Cernkovich and Giordano, "Family Relationships and Delinquency." See also Shover et al., "Gender Roles and Delinquency."

81. Canter, "Sex Differences."

82. Heimer and Matsueda, "Role-Taking;" and Matsueda, "Reflected Appraisals."

83. Heimer, "Gender, Interaction, and Delinquency," 41.

84. Heimer, "Gender, Race," and "Gender, Interaction, and Delinquency;" and Bartusch and Matsueda, "Gender." See also Heimer and DeCoster, "Gendering of Violent Delinquency;" and Koita and Triplett, "Examination."

85. Heimer, "Gender, Interaction, and Delinquency."

86. Ibid., 43.

87. Bartusch and Matsueda, "Gender."

88. Chesney-Lind, "Judicial Paternalism;" and Teilmann and Landry, "Gender Bias."

89. Canter, "Sex Differences."

90. Morash, "Establishment."

91. Horowitz and Pottieger, "Gender Bias."

92. Hawkins, Laub, Lauritsen, and Cothern, "Race," 1.

93. US Bureau of the Census, *National Population Estimates.*

94. Federal Bureau of Investigation, *Crime in the United States 2003,* 289.

95. Hawkins, Laub, Lauritsen, and Cothern, "Race," 2; and Cernkovich, Giordano, and Pugh, "Chronic Offenders."

96. Hindelang, "Variations;" Hindelang, Hirschi, and Weis, *Measuring Delinquency,* 176; Huizinga and Elliott, *Self-Reported Measures*; and Huizinga and Elliott, "Reassessing."

97. Farrington, Loeber, Stouthamer-Loeber, VanKammen, and Schmidt, "Self-Reported Delinquency." See also Piquero, MacIntosh, and Hickman, "Validity."

98. Hirschi, *Causes of Delinquency,* 75–76.

99. Elliott and Ageton, "Reconciling;" Huizinga and Elliott, "Juvenile Offenders;" and Hindelang, Hirschi, and Weis, *Measuring Delinquency.*

100. Huizinga and Elliott, "Juvenile Offenders," 210–212.

101. Elliott and Ageton, "Reconciling," 103.

102. Ibid., 102.

103. Ibid.

104. The sample of black respondents is relatively small (259) compared to the sample of white respondents (1,357) in the NYS. Therefore, more random error is associated with the reports by blacks than with those by whites. As a result, even the seemingly large difference in the frequency of serious crimes against persons between blacks and whites is not statistically significant (Elliott and Ageton, "Reconciling," 102).

105. Heimer, "Gender, Race" (316, note 3), points out that "since [predatory crimes against persons] are relatively rare occurrences and the number of blacks in their sample is modest, Elliott and Ageton may be unable to detect a pattern in the sample that does exist in the population. The finding of greater involvement in violence [by blacks] by numerous studies using victimization and official data suggest that this indeed may be the case."

106. Elliott and Ageton, "Reconciling," 103–104.

107. In their classic study, *Delinquency in a Birth Cohort,* Wolfgang, Figlio, and Sellin (247–248) found that "chronic" offenders, defined as those who committed more than four offenses, constituted only 18 percent of the total number of offenders in their study. Wolfgang and his colleagues also found that nonwhites were five times more likely than whites to be included in the chronic offender category.

108. Williams, Ayers, Outlaw, Abbott, and Hawkins, "Effects of Race." But see also D'Alessio and Stolzenberg, "Race and the Probability of Arrest," who found that disproportionately high arrest rates for blacks were due to greater involvement in offending compared to whites, rather than to racially biased law enforcement.

109. Hawkins, Laub, Lauritsen, and Cothern, "Race," 3.

110. Snyder and Sickmund, *Juvenile Offenders and Victims.*

111. See, for example, Wolfgang and Ferracuti, *Subculture of Violence*; and Curtis, *Violence, Race, and Culture.*

112. See, for example, Blumenthal, "Predicting Attitudes toward Violence."

113. See, for example, Sampson and Bartusch, "Legal Cynicism;" Cao, Adams, and Jensen, "Test;" Erlanger, "Empirical Status;" and Rossi, Waite, Bose, and Berk, "Seriousness of Crimes."

114. LaFree, Drass, and O'Day, "Postwar America."

115. Cernkovich, Giordano, and Rudolph, "American Dream," 131.

116. See also Hill and Crawford, "Women, Race, and Crime." Studies by Matsueda and Heimer, "Race," and Heimer, "Gender, Race," however, demonstrated that the social psychological mechanisms central to differential association and interactionist theories of delinquency can help explain racial differences in offending. Yet, Matsueda and Heimer, in "Race," also found that racial differences in family structure, rather than social psychological mechanisms, were the driving force behind racial differences in delinquency in their empirical models.

117. Hawkins et al., "Race."

118. Wilson, *Truly Disadvantaged.*

119. Wilson, *When Work Disappears.*

120. Sampson and Wilson, "Toward a Theory."

121. Ibid.

122. Peeples and Loeber, "Individual Factors," 141.

123. We use the terms "social class" and "socioeconomic status" interchangeably, mainly because the criminologists whose work we review have used the terms as though they were interchangeable. But Terence Thornberry and Margaret Farnworth, in "Social Correlates" (507), correctly point out that the concepts of social class and socioeconomic status are distinct. Explaining the distinction, they write, "Social class refers to major social cleavages that demarcate a relatively small number of *discrete* groups within society. This property is in direct contrast to the properties of social status, which refers to the manner in which individuals are arrayed along a *continuous* status hierarchy." (Emphasis added.) Clearly then, social class and socioeconomic status must be measured in different ways. The differences between social class and socioeconomic status may be responsible for some of the discrepant findings about the relationship between economic well-being and crime. Thus, these differences should be kept in mind when reading this section. See Thornberry and Farnworth, "Social Correlates" (506–507), for a thorough discussion of the issue.

124. Tittle, "Social Class."

125. Vogt, *Dictionary,* 78.

126. Hindelang, Hirschi, and Weis, *Measuring Delinquency.*

127. Vogt, *Dictionary,* 78.

128. Hindelang, Hirschi, and Weis, *Measuring Delinquency,* 184.

129. For example, Hirschi, *Causes of Delinquency;* Johnson, "Social Class;" Krohn, Akers, Radosevich, and Lanza-Kaduce, "Social Status and Deviance;" and Williams and Gold, "Official Delinquency."

130. Tittle, Villemez, and Smith, "Myth."

131. Tittle, "Social Class," 353.

132. Braithwaite, "Reconsidered," 36.

133. Kleck, "Self-Report Data."

134. Ibid., 430.

135. Farnworth, Thornberry, Krohn, and Lizotte, "Measurement," 33.

136. Hagan and McCarthy, "Streetlife and Delinquency."

137. Hindelang, Hirschi, and Weis, "Correlates of Delinquency."

138. Farnworth, Thornberry, Krohn, and Lizotte, "Measurement," 33; and Tittle and Meier, "Specifying."

139. Hagan and McCarthy, "Streetlife and Delinquency," 533.

140. Ibid., 534.

141. Hagan and McCarthy, "Streetlife and Delinquency." Tittle and Meier, in "Specifying" (277–278), draw attention to the need for studies—like Hagan and McCarthy's work—that directly measure youths' social status, rather than simply the status of parents.

142. Hagan and McCarthy, "Streetlife and Delinquency," 556.

143. Hindelang, Hirschi, and Weis, "Correlates of Delinquency."

144. Ibid., 1011.

145. Farnworth, Thornberry, Krohn, and Lizotte, "Measurement."

146. Ibid., 46.

147. Ibid.

148. Ibid., 55.

149. For a review, see Hindelang, Hirschi, and Weis, "Correlates of Delinquency."

150. Elliott and Ageton, "Reconciling;" Elliott and Huizinga, "Social Class;" and Elliott, Huizinga, and Ageton, *Explaining Delinquency*.

151. Hindelang, Hirschi, and Weis, "Correlates of Delinquency."

152. Elliott and Huizinga, "Social Class," 151.

153. Elliott and Ageton, "Reconciling."

154. Ibid., 102.

155. Ibid.

156. Ibid., 104.

157. Elliott and Huizinga, "Social Class."

158. Ibid., 160.

159. Ibid., 159–161.

160. Ibid.

161. Ibid., 174.

162. Wolfgang, Figlio, and Sellin, *Delinquency*.

163. Thornberry and Farnworth, "Further Evidence."

164. Ibid.

165. Elliott and Ageton, "Reconciling."

166. Elliott and Huizinga, "Social Class."

167. Heimer, "Socioeconomic Status," 800.

168. Ibid., 812.

169. Wright, Caspi, Moffitt, Miech, and Silva, "Reconsidering."

170. Ibid., 175.

171. Ibid., 178.

172. Ibid., 178–179.

173. Wilson, *Truly Disadvantaged,* and *When Work Disappears*.

174. Bursik and Grasmick, "Economic Deprivation."

175. Ibid., 271–272.

176. Ibid., 276–277.

177. Ibid., 276.

178. See, for example, Colvin and Pauly, "Critique of Criminology;" Beirne and Quinney, *Marxism and Law*; Chambliss and Seidman, *Law, Order, and Power*; Quinney, *Class, State, and Crime*; and Spitzer, "Toward a Marxian Theory."

179. Catalano, *Criminal Victimization, 2003*, 1.

180. Macmillan, "Violence."

181. Catalano, *Criminal Victimization, 2003*, 7.

182. Federal Bureau of Investigation, *Crime in the United States 2003,* 17.

183. Ibid., 17.

Developmental Patterns of Offending

6

Chapter Objectives

After completing this chapter, students should be able to:

- Describe the age-determined patterning of delinquent behavior.
- Identify the defining characteristics of "chronic offenders" or "career criminals."
- Describe key elements of the developmental perspective: age of onset of problem behaviors, continuity and change in problem behaviors, progression of seriousness, and generality of deviance.
- Identify specific models that are good examples of the developmental perspective.
- Understand key theories and terms:
  Theories:
  Developmental theory
  Loeber's developmental pathways model
  Patterson's early- and late-starter models
  Moffitt's adolescence-limited and life-course-persistent offenders

Terms:
cohort
recidivists
chronic offenders
criminal career
participation
frequency
career criminals
age of onset
behavioral continuity
escalation
generality of deviance
desistance
pathway
persisters
experimenters
early starter
late starter
adolescence-limited offenders
life-course-persistent offenders

CASE IN POINT

Patterns of Offending: The Development of Delinquency for Two Brothers

This case describes the behavior of Mark, age 10, and Shawn, age 14, and their interactions with their mother, Donna, who was a single parent. They lived in low-income housing in a large metropolitan city. The family's history was "quite tumultuous." The case is presented by the therapist to whom the court referred the family. It illustrates several elements of the developmental perspective that we discuss in this chapter, including continuity in problem behaviors, progression of seriousness of offenses, and co-occurrence of problem behaviors.

Donna had been court-ordered to seek therapy as a result of Shawn's truancy and delinquent behavior. He had a history of disruptive behavior in school (fighting, swearing, and refusing to work on assignments) and had been expelled numerous times in the past year. His response was simply to stop going to school. While truant, he had been caught shoplifting on three separate occasions. His most recent offense was an arrest for burglarizing a home, which led to formal adjudication and referral for treatment.

Intake information from Donna indicated that she had received complaints from neighbors and teachers concerning Shawn's stealing since he was 8 years old. Although money was occasionally missing from her purse, she did not consider stealing a problem in the home. She did identify lying, fighting, failure to come home on time, and suspected drug and alcohol use as current problems with Shawn. She described Shawn as having been a difficult child to control and said that he had "always been a handful." She professed to have used spankings as punishment when the children were younger but had stopped this practice following an abuse complaint.

Mark was apparently following in his older brother's footsteps. He was disobedient at home, fought both at school and in the community, lied, and occasionally threw temper tantrums and wet the bed. Donna expressed concern about Mark's also getting into legal trouble and seemed more eager to discuss her difficulties with Mark than those with Shawn.

Donna described her attempts to control the children as consisting mostly of arguing, threatening, and yelling at the boys. She felt that spanking was an effective punisher but was at a loss now that she no longer spanked. It was observed that Donna used a cold, aversive tone when interacting with the boys. With adults, however, she often was pleasant and engaging. Physical affection between mother and sons was not observed, and it was noticed that the family's interactions were mostly concerned with problems and were aversive in nature.

[The therapist then describes the type of treatment provided to the family in therapy sessions. Donna married Bill, who "seemed to be a stabilizing influence on Donna's life and had agreed to participate in therapy to work on Mark's problems." Following an arrest for car theft, Shawn was sent to a residential facility for delinquent youth.]

The prognosis for Mark and his new family is good. His participation in delinquent activity had been curtailed, and he seemed to enjoy the caring attention of Bill. Donna had also been able to reduce her aversiveness when interacting with Mark and showed genuine warmth on occasion. Shawn, however, fits the classic pattern of continued antisocial behavior. His stint in juvenile corrections holds a high probability of improving his current criminal skills and adding new ones to his repertoire. It is highly likely that he will continue to commit delinquent acts throughout adolescence and into adulthood. It is also likely that he will serve additional time in adult criminal or psychiatric facilities.

Source: Moore and Arthur, "Juvenile Delinquency," 212–213. Copyright © 1989, Springer-Verlag. All rights reserved. Reproduced with permission of the publisher.

In Chapter 5, we discussed the relationship between age and crime and the consistent research finding that adolescents and young adults are disproportionately involved in crime—the age effect. We also explored the continuity of criminal behavior from adolescence into young adulthood (the crime prone years), and the process of aging out of crime. This discussion of the age–crime relationship implies a *patterning* to involvement in delinquency that is heavily *age determined.* To understand delinquency fully, we must understand not only offenses and offenders, but also the common patterns to delinquent offending.

In this chapter, we consider the nature of delinquency in terms of *developmental patterns of offending.* This area of study, which has emerged primarily since the early 1990s, is sometimes referred to as "developmental criminology"[1] or the study of "criminal careers."[2] The developmental perspective fills a gap in criminology, which has tended to focus too heavily on the behavior of adolescents alone. Robert Sampson and John Laub, whose life-course perspective on crime and delinquency has made a significant contribution to developmental criminology, write, "The age–crime curve has had a profound impact on the organization and content of sociological studies of crime by channeling research to a focus on adolescents. As a result sociological criminology has traditionally neglected the theoretical significance of childhood characteristics and the link between early childhood behaviors and later adult outcomes."[3]

We begin this chapter with research on the small portion of offenders who develop serious patterns of delinquency—"chronic offenders" or "career criminals."[4] This research shares several key elements with the developmental perspective of the last decade. The developmental perspective provides theoretical approaches with which to frame empirical findings regarding criminal careers.[5] We then discuss the logic of the developmental perspective, focusing on five major themes: age of onset of problem behaviors, continuity of problem behaviors, progression of seriousness, co-occurrence of problem behaviors, and desistance from offending. We illustrate this perspective with three developmental models of delinquent behavior: (1) the work of Rolf Loeber and his colleagues on authority conflict, covert, and overt pathways to delinquency; (2) the work of Gerald Patterson and his colleagues on early- and late-starter routes to delinquency; and (3) Terrie Moffitt's work on "adolescence-limited" and "life-course-persistent" offenders.

■ "Chronic Offenders" and "Career Criminals"

At the end of Chapter 3, we introduced the "criminal career" paradigm and the methodological and policy debates surrounding it. In this chapter, we elaborate on this paradigm and discuss its relationship to the more recent developmental perspective in criminology.

Wolfgang's "Chronic Offenders"

In 1972, Marvin Wolfgang, Robert Figlio, and Thorsten Sellin published what has become a classic study in criminology: *Delinquency in a Birth Cohort.*[6] In this groundbreaking research, Wolfgang and his colleagues studied the "cohort" of all boys born in 1945 and residing in Philadelphia from at least their 10th to 18th birthdays.[7] A <u>cohort</u> is a group of "individuals (within some population definition) who experienced the same event within the same time interval."[8] In *Delinquency in a Birth Cohort,* the cohort was defined by year of birth, and included 9,945 boys, of whom 3,475 had at least one recorded police contact.[9] School records provided extensive background information about the boys, including birth date, race, country of origin, IQ scores, achievement level, behavior problems, and highest grade completed.[10] Philadelphia police records provided information about the number and type of offenses committed by members of the cohort, as well as dispositions of cases.[11]

With this wealth of data, Wolfgang and his colleagues explored the onset and progression of delinquency and the social correlates of offending. While this study produced many important results, it is probably most noted for its findings regarding "chronic offenders." Wolfgang, Figlio, and Sellin divided the cohort into three groups: non-offenders, one-time offenders, and <u>recidivists</u> or those who committed multiple offenses resulting in police contact. Of the 3,475 boys in the cohort who had at least one police contact, 1,235 boys had two to four contacts and 627 had five or more contacts.[12] The boys with five or more police contacts were defined as **chronic offenders**.

Table 6-1 shows the classification of offenders and the distribution of their offenses, and it reveals several remarkable findings. First, while almost 35% of the cohort had at least one police contact, 65% had no police contacts. Second, almost half (46%) of those with police contacts (1,613 out of 3,475) were *one-time offenders*. They accounted for 15.8% of recorded offenses. The remaining 54% of those with police contacts were classified as *recidivists*.[13] Third, about one-third (36%) of the delinquents had 2 to 4 contacts with police. These offenders were labeled *non-chronic recidivists*, and they accounted for 32.3% of recorded offenses.[14] Finally, the most remarkable finding in *Delinquency in a Birth Cohort* concerns the 627 boys with 5 or more police contacts. While these *chronic offenders* or *chronic recidivists* represented only 6.3% of the cohort of 9,945 boys (or 18% of the 3,475 boys with police contacts), they were responsible for 51.9% of the offenses that resulted in police contact.[15] This small group of persistent offenders is often called the "chronic 6 percent."

cohort A group of "individuals (within some population definition) who experienced the same event within the same time interval" (Ryder, *Cohort,* 845). For example, all persons born in Chicago in 1985 form a cohort.

recidivists Individuals who commit multiple delinquent or criminal offenses resulting in police contact.

chronic offenders The small proportion of offenders who engage in a disproportionate share of offenses, particularly serious and violent ones. In the research of Wolfgang and his colleagues, chronic offenders were defined as those with five or more police contacts (Wolfgang, et al., *Delinquency*).

Table 6-1	**Classification of Offenders and Offenses in *Delinquency in a Birth Cohort***				
Juveniles	Number of cases	Percent of original sample	Percent of delinquent sample	Total number of offenses	Percent of total offenses
Original sample	9,945				
Delinquents	3,475	34.9%		10,214	
One official police contact	1,613	16.2	46.4%	1,613	15.8%
Two to four police contacts	1,235	12.4	35.6	3,296	32.3
Five or more police contacts	627	6.3	18.0	5,305	51.9

Source: Wolfgang, Figlio, and Sellin, *Delinquency,* 89. Copyright © 1972, University of Chicago Press. All rights reserved. Reproduced with permission of the publisher.

Chronic offenders not only committed a disproportionate share of offenses, they also committed more serious offenses than did other delinquents, as indicated by their disproportionate involvement in Index offenses. Chronic offenders committed 1,726 Index offenses out of the 2,728 committed by the entire cohort. More specifically, chronic offenders were responsible for:

- 63.3% of all Index offenses committed by the cohort
- 71.4% of all homicides
- 72.7% of all rapes
- 69.9% of all robberies
- 69.1% of all aggravated assaults[16]

The longitudinal design of this study also revealed interesting findings regarding age and its relationship to delinquency. Wolfgang and his colleagues confirmed the age distribution of crime found in prior cross-sectional research. Few offenses were committed by boys under the age of 11; from age 11 to age 13, the proportion of offenses committed increased steadily; from age 14 to age 16, the proportion of offenses committed rose sharply; and offending peaked at age 16 and then declined rapidly at age 17.[17] Wolfgang and his colleagues also found that boys who began involvement in delinquency at younger ages committed more offenses than boys who began offending later in life.

In a follow-up study, Wolfgang, Thornberry, and Figlio chose a random 10% sample of the original 1945 birth cohort (975 individuals) and traced them through adulthood to age 30.[18] This study was designed to investigate the link between juvenile delinquency and adult criminality. A relationship existed between chronic offending as a juvenile and persistent offending to age 30. "Those juvenile offenders with extensive delinquency records were more likely to have extensive adult records. Indeed, 45 percent of the juvenile chronic offenders were also classified as chronic offenders during the adult years." The proportion of crimes attributable to chronic offenders in this study is striking. "Chronic offenders accounted for 74 percent of all arrests and 82 percent of all index arrests, although they represented only 15 percent of the total sample and 32 percent of the official offenders."[19]

In 1990, Tracy, Wolfgang, and Figlio published a second birth cohort study. In this research, the cohort included 27,160 males and females who were born in 1958 and resided in Philadelphia at least from their 10th to 18th birthdays.[20] In general, the results of this study were quite similar to those of the earlier 1945 birth cohort study.[21] Police records showed that 33% of the males in the 1958 cohort had at least one police contact before age 18.[22] Among males, chronic offenders made up 7.5% of the total cohort and 23% of all delinquents in the cohort.[23] As in the earlier study, chronic offenders accounted for a disproportionate share of offenses, particularly violent ones. In the 1958 cohort, chronic male offenders were responsible for 68% of Index offenses.[24]

The work of Wolfgang and his colleagues reveals that delinquents who begin offending at a relatively young age tend to accumulate lengthy criminal careers that extend well into adulthood, and that a relatively small number of chronic offenders account for the majority of crimes. These vitally important findings paved the way for the developmental perspective in criminology.

Blumstein's "Criminal Careers" and "Career Criminals"

Criminal Careers

The criminal career approach focuses on individual offenders and their patterns of offending over time.[25] Alfred Blumstein, Jacqueline Cohen, and David Farrington define a **criminal career** as "the longitudinal sequence of offenses committed by an offender who has a detectable rate of offending during some period."[26] As we noted in Chapter 3, a criminal career is characterized by several distinct aspects including *age of onset,* or age at which delinquent behavior begins; *age at which offending ends; length or duration of criminal career;* and *frequency of offending* while an individual is an active offender.[27] According to the criminal career paradigm, it is important to distinguish these elements because different causal factors may account for different elements of a criminal career.[28] For example, age of onset may be associated with family influences such as parenting practices or family disruption, whereas age at which offending ends may be related to employment opportunities. Blumstein and his colleagues use and advocate for longitudinal data to study patterns of offending over time and to capture fully all aspects of criminal careers.[29]

It is also important to distinguish those who participate in criminal careers from those who do not. The distinction between participation in offending and frequency of offending is central to the criminal career perspective. In the terminology of this paradigm, **participation**, sometimes referred to as *prevalence,* is "the proportion of a population who are active offenders at any given time."[30] In other words, prevalence "reflects the pervasiveness of offenders in a population."[31] **Frequency** of offending, sometimes referred to as *incidence,* is "the average annual rate at which [the] subgroup of active offenders commits crimes."[32] The term *lambda* is also used in the criminal career literature to refer to an active offender's frequency of offending or individual crime rate.[33]

Research on criminal careers shows that most individuals who participate in offending commit only one or a limited number of offenses during late adolescence or early adulthood, when delinquent behavior is almost normative, and they desist in the young adult years without escalating into serious offending.[34] In other words, the participation rate is high for adolescents and young adults, but frequency of offending is low and criminal career length is short for *most* delinquents. These findings demonstrate the nonchronic nature of delinquency and support Hirschi and Gottfredson's general age–crime curve (see Chapter 5).[35]

As we mentioned earlier, a premise of the criminal career paradigm is that different aspects of criminal activity, such as participation, frequency, and duration of criminal career, may require different causal explanations. In contrast, the general theory of crime (see Chapter 10) contends that different aspects of criminal activity are all caused by a single underlying propensity toward crime, and that a general theory of criminal propensity is sufficient to explain all aspects of criminal activity, including participation, frequency, and persistence.[36] Douglas Smith and his colleagues tested these competing hypotheses.[37] Their primary research question was whether different variables, derived from various theories, were related to participation, frequency, and persistence of offending, or whether the same variables were associated with all three dimensions. Results

criminal career "The longitudinal sequence of offenses committed by an offender who has a detectable rate of offending during some period" (Blumstein et al., "Criminal Career Research," 2).

participation "The proportion of a population who are active offenders at any given time" (Blumstein et al., "Criminal Career Research," 3). Sometimes called *prevalence.*

frequency "The average annual rate at which [the] subgroup of active offenders commits crimes" (Blumstein et al., "Criminal Career Research," 3). Sometimes called *incidence.*

were somewhat mixed. Contrary to the general theory of crime, a single underlying variable did not account for both participation and frequency of offending.[38] However, the variables associated with different dimensions of offending were not entirely distinct. Rather, a "core" set of variables, including exposure to delinquent peers, was related to all three dimensions of offending. This study indicates that researchers should keep in mind the criminal career paradigm premise that different aspects of criminal careers *may* relate differently to explanatory variables.

In addition, the distinction in the criminal career approach between participation in offending and frequency of offending makes evident the very different crime prevention strategies associated with each. For example, a neighborhood crime rate may be high because a large proportion of the people who live there each commit a small number of crimes (i.e., high rate of participation) or because a relatively small proportion of the people who live there each commit a large number of crimes (i.e., high frequency of offending) or both. A high rate of participation suggests a crime prevention policy of general deterrence aimed at the entire population of the neighborhood. A high frequency of offending suggests a crime prevention policy aimed at identifying and treating or controlling those who have already become involved in crime.[39]

Career Criminals

career criminals
Characterized by "some combination of a high frequency of offending, a long duration of the criminal career, and high seriousness of offenses committed" (Blumstein et al., "Criminal Career Research," 22).

The distinction of various aspects of criminal careers enables researchers to identify <u>career criminals</u>, characterized by "some combination of a high frequency of offending, a long duration of the criminal career, and high seriousness of offenses committed."[40] Blumstein's concept of career criminals is consistent with Wolfgang's concept of chronic offenders. These high rate offenders begin delinquent involvement at a relatively young age and persist in offending after most individuals have aged out of crime in late adolescence or early adulthood. In other words, they are exceptions to the general age–crime curve discussed in Chapter 5. Career criminals constitute a relatively small proportion of all offenders but account for a disproportionate share of offenses, particularly serious ones.[41]

The notion of career criminals has been controversial because of its policy implications.[42] If we can identify offenders who are career criminals based on their patterns of offending over time, then presumably we can target them with intervention and crime prevention strategies specific to career criminals. As we discussed in Chapter 3, however, the idea of *selective incapacitation* of career criminals, or the incarceration of serious, long-term offenders for the period of their criminal careers characterized by high rates of offending, has been hotly debated.[43]

Theoretical Perspective on Criminal Careers, Career Criminals, and Chronic Offenders

In terms of framing research questions on patterns of offending and guiding discourse on crime prevention policy, the work of Wolfgang, Blumstein, and their colleagues is among the most important research in criminology.[44] In terms of theory, however, the criminal careers construct is not well defined or developed. As Wayne Osgood and David Rowe write, "research on criminal careers has been largely divorced from theoretical criminology."[45] Responding to the 1988 debate between Blumstein and his colleagues and Gottfredson and Hirschi, one criminologist asks, "Where is the actual theory?"[46]

Blumstein and his colleagues plainly acknowledge that "the construct of the criminal career is not a theory of crime."[47] While the criminal career perspective maintains that different causal factors may account for different elements of criminal careers, it does not delve into theory to attempt to determine what those causal factors might be. Some have moved toward filling this gap. Charles Tittle, for example, suggests how researchers might apply labeling and social control theories to the study of criminal careers.[48] Others discuss methodological tools that would enable researchers to incorporate theoretical variables into their analyses of criminal careers.[49]

Some have developed new theoretical perspectives that "harmonize" with the criminal career paradigm.[50] Blumstein and his colleagues write, "while not a theory itself, the construct of the criminal career should prove to be very valuable for the development of theory."[51] Indeed, it has. Raymond Paternoster and Robert Brame note how recent developmental theories of crime might explain the separate elements of criminal careers—age of onset, duration of career, frequency of offending, and age at termination of offending—identified by Blumstein and his colleagues and others in the criminal career tradition.[52] A great asset of the contemporary developmental perspective is its ability to account theoretically for aspects of offending identified by the empirically rich, but theoretically weak, criminal career perspective.

■ The Developmental Perspective

The central tenet of the developmental perspective in criminology is that the development of problem behaviors tends to occur in an "orderly, progressive fashion" that is highly age-determined.[53] Researchers working from this perspective are concerned with explaining the stability (and instability) of problem behaviors over time for individual offenders. Rolf Loeber and Marc LeBlanc describe developmental criminology as the study of (1) "the development and dynamics of problem behaviors and offending with age," and (2) the causal factors that occur before or along with the development of problem behaviors and affect this development.[54]

Many years ago, Lee Robins observed that even though "adult antisocial behavior virtually *requires* childhood antisocial behavior . . . most antisocial children do *not* become antisocial adults."[55] The developmental perspective attempts to account for this paradoxical finding—simultaneously explaining why antisocial behavior is stable across the life-course for some offenders, while it is only a brief adolescent excursion from conformity for *most* offenders whose "criminal careers" are characterized by change rather than stability.[56]

In addressing this paradox, developmental theorists speak of *heterogeneity* or variation across individuals in factors that influence involvement in delinquency and crime. They also speak of *stability* of individual differences over time in both the potential to commit crime and the social factors related to offending. "Developmental theories are friendly to the notion that both persistent individual differences and changing life circumstances are related to involvement in crime, and that these factors affect different groups of offenders in different ways."[57] For example, some who engage in delinquency during adolescence have a history dating back to early childhood of difficult and problematic behaviors, and their progression from childhood problem behaviors to adolescent delinquency is characterized by continuity of behaviors in various settings, including home and school. Others who engage in delinquency during adolescence may have no history of problem behaviors prior to their offending in adolescence and show no progression or continuity in offending, but rather may be drawn briefly to delinquent behavior by social or peer influences. Developmental criminologists try to formulate explanations for delinquency that take into account these kinds of individual differences in paths leading to delinquent behavior.

In this section, we describe five key themes of the developmental perspective: age of onset of problem behaviors, continuity and change in problem behaviors, progression of seriousness of offenses, generality of deviance or co-occurrence of problem behaviors, and desistance from offending.

Age of Onset of Problem Behaviors

As we discussed in Chapter 5, most offenders begin involvement in delinquent behavior during adolescence, remain active offenders for a relatively brief period of time, and age out of crime in late adolescence or early adulthood. Loeber and LeBlanc contend that the focus in criminology on adolescent offending is due in part to a reliance on official records, which generally provide little information about offending prior to adolescence.[58] Yet, as they point out, this focus "ignores the finding that early onset of offending is not uncommon, is predictive of frequent, persisting, varied, and serious offenses in males, and may result from causal factors that differ from those associated with later onset of delinquency."[59]

Developmental criminologists maintain that some offenders display problem behaviors early in the life-course, and that these offenders are the ones most likely to continue problem behaviors into adulthood and to develop stable patterns of offending.[60] Loeber and LeBlanc provide a thorough review of the research literature on **age of onset** of offending—the age at which an individual begins involvement in delinquent or criminal acts. They conclude that early age of onset is related to the stability and frequency of offending over time and the diversity of offenses committed. Those who begin involvement in problem behaviors early in the life-course "tend to commit crimes at a much higher rate than those with a later age at onset."[61] It is important to note that research shows that those who begin offending at a relatively young age do not commit more offenses than those who begin later in life simply because of the longer time span of their offending careers. They commit more offenses in part because of their higher rate or frequency of offending.[62]

Research results on the relationship between age of onset of offending and diversity of offenses committed are mixed. Some research shows that early-onset offenders tend to commit a wider variety of offenses than those who begin offending at later ages.[63] Other studies, however, suggest that age of onset is not related—or at least not strongly related—to variety of offenses committed or to

age of onset The age at which an individual begins involvement in delinquent or criminal acts.

"offense specialization."[64] Attempting to resolve this contradiction, Alex Piquero and his colleagues examined the relationship between age of onset and offense specialization.[65] They found an inverse relationship—the younger the age of onset, the greater the variety of offenses committed. But they also found that this relationship disappeared once they took into account offender's age. They concluded that, regardless of age of onset of offending, age itself affects offense versatility by producing a decline in the rate of offending, and therefore in versatility as well.[66]

Studies have also shown that early-onset offenders tend to engage in more serious offenses than do late-onset offenders.[67] Using data from the National Youth Survey (NYS), researchers found that early-onset offenders—defined as those who began offending before age 13—showed significantly higher rates of serious offenses than did late-onset offenders. Among males, rates of serious offending were 2.1 to 2.9 times higher for early- than for late-onset offenders.[68]

In sum, the bulk of evidence suggests that those who begin offending at a relatively young age engage in a *higher rate* of offending, a *wider variety* of offenses, and crimes of *greater seriousness* than do those who begin offending at a later age. These features of offending also characterize chronic offenders or career criminals, for whom early age of onset is typically a defining characteristic. Numerous studies demonstrate that early age of onset of offending is related to chronic offending.[69]

"How Early Can We Tell?"

In a study of the predictors of conduct disorder and delinquency, Jennifer White and her colleagues asked, "How early can we tell?"[70] Using data from a longitudinal study of approximately 1,000 subjects in Dunedin, New Zealand, they examined the ability of child characteristics assessed at ages 3 and 5 to predict conduct disorder at age 11 and delinquency at age 15. These preschool characteristics included the following:

- behavior problems at age 5, as reported by the mother
- "externalizing" behaviors (hyperactivity and aggression) at age 3, as observed and reported by the research staff
- difficulty managing the child as a baby, as reported by the mother when the child was three years old
- two motor skills variables (a measure of perceptual and visual–motor integration, and a measure of physical coordination), assessed at age 5[71]

Of these 5 characteristics, the first 3 behavioral variables were most important in predicting and classifying conduct disorder at age 11.[72] The 5 preschool characteristics were also able to predict and correctly classify delinquency status at age 15 for the majority of subjects in the study.[73] These results suggest that we can "tell" quite early in the life course—adolescent delinquency is foreshadowed in problem behaviors observed as early as ages 3 and 5. In a separate study, Loeber and his colleagues also found that, even as toddlers and preschoolers, some children distinguish themselves by their difficult temperament, which is related through the developmental process to problem behaviors and delinquency later in life.[74]

Later in this chapter, we describe the developmental models proposed by Terrie Moffitt[75] and Gerald Patterson,[76] who argue that the routes to delinquency are different for those who begin involvement in problem behaviors at an early age and those who begin later. Most individuals who engage in delinquency do not become serious or chronic offenders. But the research literature clearly suggests that those who engage in problem behaviors at a relatively young age are the ones most likely to become frequent, serious, chronic offenders. "Adolescent involvement in serious delinquency and crime appears to have deep developmental roots."[77]

Continuity and Change in Problem Behaviors

As we noted earlier in this chapter, for some offenders, antisocial behavior is stable over time, and their criminal careers are marked by continuity over the life course. For most offenders, however, antisocial behavior is not stable over time, and relatively brief involvement in delinquency represents a change in patterns of behavior.[78] The developmental perspective attempts to account for both stability and change in patterns of offending.

Although the developmental perspective is relatively recent, the notion of continuity in antisocial behavior is not new. In the 1950s and 1960s, Sheldon and Eleanor Glueck proposed the hypothesis of "longitudinal consistency" and found continuity over time in individuals' involvement in delinquency and crime.[79] Robert Sampson and John Laub more recently used the Gluecks' data and earlier research to develop and test their own life-course theory of crime.[80] This theory is discussed in Chapter 10. Like other developmental perspectives, life-course theory includes the concept of **behavioral continuity**, or continuity in behavior over time.

behavioral continuity
Patterns of behavior that are consistent and stable over time, resulting from stable individual attributes or styles of interaction.

This concept of continuity pertains to individual attributes and interactional styles, rather than to specific behaviors.[81] When we say that continuity in behavior exists over time, we are not saying that identical behaviors are displayed over time. Instead, behavioral continuity refers to consistent patterns of behavior. For example, a child may be characterized in infancy as having a difficult temperament and may be aggressive in preschool, have problems interacting with peers and teachers in elementary school, and engage in delinquency in adolescence. This sequence indicates continuity of problem behaviors, though the specific behaviors change over time.

Numerous studies, from the field of developmental psychology and other academic disciplines, provide evidence that antisocial behavior is stable or persistent across the life course.[82] These studies have shown not only continuity in offending from adolescence to adulthood, but also continuity between childhood conduct problems and later involvement in delinquency and crime. For example, in his review of sixteen studies, Daniel Olweus found a strong relationship between early aggressive behavior and involvement in crime later in life. In the studies he reviewed, the average correlation between early aggression and later crime was 0.68 (on a 0 to 1 scale, with 0 representing no relationship and 1 representing a perfect relationship).[83]

Recall, however, the paradox stated earlier in this chapter: antisocial behavior in childhood is a strong predictor of antisocial behavior in adulthood, yet most antisocial children do not become antisocial adults.[84] So while stability characterizes the problem behaviors of some antisocial individuals, change in behavior patterns characterizes even more individuals. As Sampson and Laub write, "antisocial behavior appears to be highly stable and consistent only in a relatively small number of males whose behavior problems are quite extreme."[85] These individual differences in patterns of offending point to the need to consider different developmental paths to delinquency and crime for different offenders. Later in this chapter, we describe specific developmental models that do exactly that: attempt to account for the continuity in problem behaviors for some individuals and the change in behavior patterns for others.

Explaining Continuity

What explains continuity for individuals who display problem behaviors that are "extreme" and stable over time? Several possibilities exist. First, according to the *population heterogeneity* explanation, there may be persistent differences across individuals in the "underlying potential" or "propensity" to commit antisocial acts.[86] For example, those who are more aggressive in childhood tend to be more aggressive during adolescence and adulthood as well.[87] Shawn Bushway and his colleagues summarize this explanation: "According to this proneness explanation, individuals vary in the probability with which they will commit crime at all points in time because they differ with respect to some risk factor (impulsivity, criminal propensity, or an antisocial trait, etc.) that is established early in life and remains, at least relatively, stable over time."[88] This interpretation of continuity in offending is the basic premise of Michael Gottfredson and Travis Hirschi's general theory of crime (see Chapter 10).[89]

A second explanation for continuity in offending is called the *state dependence* explanation.[90] It suggests that initial involvement in crime has a causal effect on later offending because that initial involvement changes the "personal characteristics or life chances" of the offender in some way that decreases inhibitions to or increases incentives for future crime.[91] This explanation for continuity is compatible with various criminological theories, including social control, differential association, social learning, and labeling theories.[92] (These theories are discussed in Chapters 10, 11, and 13.)

In a sense, the state dependence interpretation of continuity is more "social" than the population heterogeneity explanation because it takes into consideration the relationship between offenders and their environments. For example, the state dependence explanation is consistent with the scenario that, after committing delinquent acts, one might begin to "hang out" with like-minded delinquent peers who provide incentive for further delinquency. Offenders often belong to peer groups that reinforce delinquent behavior and make it difficult to quit offending.

Third, continuity in antisocial behavior may be a function of failure to master "prosocial" developmental tasks from preschool to adolescence. Barbara Kelley and her colleagues suggest that, to prevent disruptive and delinquent behavior, children must master developmental tasks such as being honest, respecting other people (especially authority figures), respecting others' property, and solving interpersonal problems nonaggressively.[93] Because these skills are developed over time, early childhood failure to begin acquiring prosocial skills may set in motion the development over time of patterns of disruptive behaviors in place of prosocial ones.

Finally, Rolf Loeber suggests that some behaviors, such as hyperactivity and substance use, may serve as *catalysts* for continuity of problem behaviors.[94] Loeber's argument is that problem behaviors are prone to exist or persist when a catalyst is present, but tend to become less likely over time when a catalyst is absent.[95] Loeber cites numerous studies which suggest that hyperactivity and substance use may be related to the continuity of problem behaviors in a catalytic way.[96]

Explaining Change

What explains change for individuals whose problem behaviors are less stable over time? Later in this chapter, we present specific developmental models that propose explanations for the lack of continuity for those who offend for brief periods of time. These explanations of the *beginnings* of involvement in delinquency include peer influences and "social mimicry." Sampson and Laub, in their life-course theory (discussed in Chapter 10), are interested in behavioral changes in terms of both initiation into offending and desistance from it.[97] They attribute changes in patterns of offending to changes in social bonds and thus in informal social control. For example, desistance from crime may result from strengthened social bonds related to employment, marriage, or educational pursuits.[98]

Progression of Seriousness

Few delinquents begin their offending careers with the most serious types of delinquent acts. Rather, most individuals who eventually become serious or chronic offenders begin with relatively minor problem behaviors. A key theme of the developmental perspective and a consistent finding in the research literature is that a progression or escalation in seriousness of offenses characterizes the criminal careers of most serious delinquents.[99] **Escalation** is "the tendency for offenders to move to more serious offense types as offending continues."[100] As Barbara Kelley and her colleagues write, based on their research using data from the all-male sample in the Pittsburgh Youth Study, "the development of disruptive and delinquent behavior by boys generally takes place in an orderly, progressive fashion, with less serious problem behaviors preceding more serious problem behaviors."[101] David Farrington uses the image of "stepping stones" to illustrate the progression from childhood problem behaviors to adult crime.[102]

Kelley and her colleagues describe the developmental process relating problem behaviors early in life to delinquency later:[103]

> *After birth, the earliest problem noted is generally the infant's difficult temperament. Although activity level is one dimension of temperament, hyperactivity becomes more apparent when children are able to walk. Overt conduct problems, such as aggression, are usually not recognized until age 2 or later, when the child's mobility and physical strength have increased. During the preschool years, the*

escalation Progression of seriousness of offenses, or "the tendency for offenders to move to more serious offense types as offending continues" (Blumstein et al., *Criminal Careers*, 84).

quality of the child's social contacts becomes evident, including excessive withdrawal or poor relationships with peers and/or adults. Academic problems rarely emerge clearly before the child attends first or second grade. Beginning at elementary school age and continuing through early adolescence, covert or concealing conduct problems, such as truancy, stealing, and substance use, become more apparent. . . . For youth age 12 and older, the prevalence of delinquency and associated recidivism increases.

This developmental ordering of problem behaviors is illustrated in **Figure 6-1.** The work of Kelley and her colleagues makes it clear that the precursors of delinquency in adolescence may be observed at very early ages. Obviously, because of different cognitive and behavioral capabilities at different ages, the manifestations or displays of problem behaviors change over time, from early childhood through adolescence.[104] Yet, as Kelley and her colleagues point out, "a child can exhibit considerable continuity in disruptive and antisocial behaviors, even though the behaviors are manifested differently with increasing age."[105]

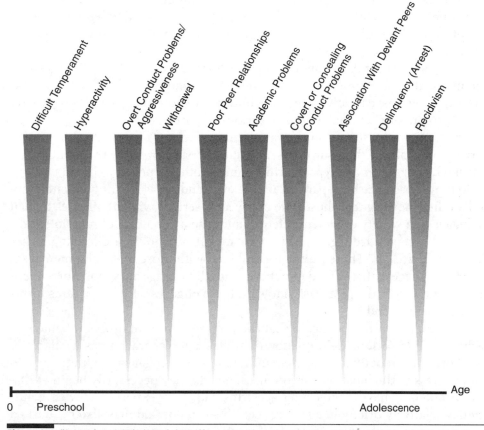

Figure 6-1 "Approximate Ordering of the Different Manifestations of Disruptive and Antisocial Behaviors in Childhood and Adolescence." This figure, by Kelley and her colleagues, shows continuity over time in problem behaviors. Although specific behavioral manifestations change as children age, a *pattern* of disruptive and antisocial behaviors in interaction with others still exists.

Source: Kelley, Loeber, Keenan, and DeLamatre, "Developmental Pathways," 4.

Rolf Loeber has conducted extensive research on the development of disruptive and delinquent behaviors. Examining median ages of onset for various behaviors, Loeber and his colleagues found this general sequence:[106]

- stubborn behavior beginning at about age 9
- minor "covert" acts (such as shoplifting and frequent lying) beginning at about age 10
- defiance and disobedience beginning at about age 11
- minor aggression (such as bullying and annoying others) and property damage (such as vandalism and fire setting) beginning at about age 12
- serious aggression involving physical fighting and violence, serious property offenses (such as fraud, burglary, and theft), and "authority avoidance" (such as truancy, running away, and staying out late) beginning at about age 13

Of course, not all delinquents progress through the entire sequence, or even through the sequence in this order. But, by examining the median ages of onset for various behaviors, researchers begin to understand the typical ages at which those behaviors emerge, and the general progression from minor to serious offending. Loeber and his colleagues later separated this sequence of behaviors into three conceptually distinct pathways, which we discuss later in this chapter.

Using self-report data from the NYS, Delbert Elliott and his colleagues also documented the progression of seriousness in delinquency.[107] They found that those who increased involvement in delinquency showed a general progression from "exploratory" to non-serious to serious offenses.

In a review of research on the causes and correlates of violent crime, David Farrington explored the issue of *specialization* versus *versatility* in offending: do youth who engage in one type of crime commit other types as well (versatility), or do they "specialize" in a particular type of offending?[108] Farrington cited several studies showing that those who engaged in serious, violent offenses tended to engage in a variety of violent offenses and also a good deal of nonviolent offending. He concluded, "generally, youth are versatile in their offending rather than specialized."[109] These findings, combined with those of developmental researchers, suggest that offenders who eventually engage in serious or violent delinquency begin the progression toward these offenses with less serious forms of nonviolent offending.

It is important to note again that the progression from problem behaviors early in life to serious delinquency in adolescence characterizes a relatively small proportion of offenders. Most individuals who engage in delinquency do not progress to the most serious offenses, nor do they necessarily begin to display problem behaviors at young ages. Instead, they experiment with delinquency relatively briefly in adolescence. This brief period may also be marked by a progression of seriousness of offenses, but this progression is characterized neither by "deep developmental roots" of problem behaviors nor by involvement in serious offenses—as the criminal careers of serious, chronic offenders are.

Generality of Deviance or Co-Occurrence of Problem Behaviors

The progression of seriousness is ultimately concerned with patterns of problem behavior that move from less to more serious, culminating in serious delinquency. Another developmental question for empirical study concerns the <u>generality of deviance</u>, or the extent to which juvenile delinquency is a component of a larger group of problem behaviors, such as drug and alcohol use, mental health problems, behavior problems and underachievement in school, and precocious and risky sexual behavior, that tend to occur together or "co-occur." Seriously delinquent youth often appear to suffer difficulties in many areas of life, and their delinquency seems to be part of a more general pattern of problem behaviors.[110]

A good deal of empirical research has explored this issue. Some studies relying on official data show "substantial" co-occurrence of delinquency and other problem behaviors.[111] In other studies, researchers have used self-report data to explore the co-occurrence of problem behaviors. Elliott and his colleagues, for example, used data from the NYS to consider "multiple problem youth."[112] They examined the co-occurrence of delinquency, substance use, and mental health problems and found a progression from delinquent behavior to drug use and mental health problems. Minor delinquency tended to precede more serious forms of offending. Involvement in minor and serious delinquency tended to precede both drug use and mental health problems. However, many individuals characterized by one problem did not progress to the next stage of problem behavior.[113]

Other studies have also shown that serious delinquency often occurs along with drug use and mental health problems, as well as promiscuous sexual behavior, alcohol use, and school failure or dropout.[114] Helene White, for example, studied the stability of problem behaviors over time and their ability to predict later drug use.[115] She found that several problem behaviors—delinquency, substance use, school misconduct and underachievement, and precocious sexual behavior—tend to "cluster" together, supporting the notion of "multiple problem youth."[116] But she also found that the clusters of problem behaviors vary somewhat for males and females, and that the relationships among specific problem behaviors are not stable over time—a finding contrary to the generality of deviance argument.[117]

In a more recent study of the co-occurrence of problem behaviors, David Huizinga and several colleagues used self-report data gathered at three research sites (Rochester, NY; Denver, CO; and Pittsburgh, PA) as part of the Program of Research on the Causes and Correlates of Delinquency, funded by the Office of Juvenile Justice and Delinquency Prevention.[118] Research in Action, "The Program of Research on the Causes and Correlates of Delinquency," provides a description of this program of research (see also Links, "Causes and Correlates of Delinquency"). Huizinga and his associates examined the co-occurrence of persistent serious delinquency with persistent drug use, school problems (subject earned below average grades or dropped out of school), and mental health problems. ("Persistent" problems were defined as those displayed in at least two of the three years examined.)

generality of deviance
The extent to which problem behaviors—such as juvenile delinquency, drug and alcohol use, mental health problems, behavior problems and underachievement in school, and risky sexual behavior—tend to occur together or "co-occur."

The Program of Research on the Causes and Correlates of Delinquency

The Program of Research on the Causes and Correlates of Delinquency was initiated in 1986 by the Office of Juvenile Justice and Delinquency Prevention. This program consists of three coordinated longitudinal projects:

- Denver Youth Survey, directed by David Huizinga at the University of Colorado;
- Pittsburgh Youth Study, directed by Rolf Loeber, Magda Stouthamer-Loeber, and David Farrington at the University of Pittsburgh;
- Rochester Youth Development Study, directed by Terence Thornberry at the University at Albany, State University of New York.

These projects are "designed to improve the understanding of serious delinquency, violence, and drug use by examining how youth develop within the context of family, school, peers, and community" (Web site).

All three projects use a similar research design and are longitudinal, "involving repeated contacts with youth during a substantial portion of their developmental years" (internet site). The participants in these studies are from inner cities and are considered to be at high risk for involvement in delinquency and drug use. To gather data, researchers use face-to-face interviews with the youth, their primary caregivers, and in two research sites, their teachers. Researchers also collect data from official agencies, including police, courts, schools, and social services. At all three sites, researchers use identical core measures, including "self-reported delinquency and drug use; community and neighborhood characteristics; youth, family, and peer variables; and arrest and judicial processing histories" (Web site).

The Denver Youth Survey sample consists of 1,527 children (806 boys and 721 girls), who were 7, 9, 11, 13, or 15 years old in 1987 and lived in a disadvantaged, high-crime neighborhood.

The Pittsburgh Youth Study sample consists of 1,517 boys, who were in the first, fourth, or seventh grade of the Pittsburgh public school system when the study began. This sample includes "the top 30 percent of boys with the most disruptive behavior" and "a random sample of the remaining 70 percent who showed less disruptive behavior" (internet site).

The Rochester Youth Development Study sample consists of 1,000 children (729 boys and 271 girls), who were in the seventh or eighth grade of the Rochester public school system in 1988. Boys were "oversampled," as were students from high-crime neighborhoods.

Source: Web site http://ojjdp.ncjrs.org/programs/ProgSummary.asp?pi=19

Links

Causes and Correlates of Delinquency

Information about the Program of Research on the Causes and Correlates of Delinquency—consisting of the Denver Youth Survey, the Pittsburgh Youth Study, and the Rochester Youth Development Study—is available via a link provided at:

http://criminaljustice.jbpub.com/burfeind

The site contains an overview of the program and its research design, descriptions of the three longitudinal projects comprising the program of research, and a list of publications.

The primary finding of this study was that "a large proportion of persistent serious delinquents are not involved in persistent drug use, nor do they have persistent school or mental health problems."[119] **Table 6-2** shows the percentages of persistent serious delinquents who were also characterized by other persistent problems. In Denver and Pittsburgh, about 55–56% of the males who were persistent serious delinquents showed no persistent drug use, school problems, or mental health problems. In Rochester, the comparable number was 38.8%. The percentages were quite similar for females. Among those for whom delinquency was combined with one other persistent problem behavior, drug use was the most common problem. When researchers examined persistent drug use alone or in combination with other problem behaviors, they found that "among

Table 6-2	Co-Occurrence of Persistent Serious Delinquency and Other Persistent Problem Behaviors				
	Males			**Females**	
Problem Behavior	Denver	Pittsburgh	Rochester	Denver	Rochester
None	55.2%	56.4%	38.8%	54.4%	39.9%
Drug use only	21.4	24.3	17.7	34.4	3.6
School problems only	4.9	2.9	7.2	0.0	3.6
Mental health problems only	4.6	5.0	5.6	0.0	0.0
Drug use and school problems	6.4	4.3	17.2	11.3	21.7
Drug use and mental health problems	4.9	5.7	3.2	0.0	7.8
School and mental health problems	1.8	0.0	4.7	0.0	8.3
Drug use, school, and mental health problems	0.9	1.4	5.6	0.0	15.1

Notes: Percentages in this table represent persistent serious delinquents who also have other persistent problems. The Pittsburgh Youth Study sample consists of males only. Due to rounding, columns may not sum to 100.

Source: Huizinga, Loeber, Thornberry, and Cothern, "Co-Occurrence," 5, 6.

males who were serious delinquents, 34–44 percent were also drug users; 46–48 percent of female serious delinquents were also drug users."[120] In addition, "for males, as the number of persistent problems other than delinquency increases, so does the likelihood that an individual will be a persistent serious delinquent."[121] In other words, the combination of other problem behaviors constitutes a "risk factor" for serious delinquency.

From this study, Huizinga and his colleagues concluded that a large proportion of serious delinquents were not characterized by other persistent problem behaviors. They also noted, however, that the degree of overlap between persistent serious delinquency and other persistent problem behaviors "suggests that a large number of persistent serious delinquents face additional problems that need to be addressed."[122]

The general conclusion we can draw from the many studies of the co-occurrence of problem behaviors is that various forms of deviance often occur together. Delinquency is often part of a more general pattern of problem behaviors. However, these problem behaviors appear to be less closely related than some theorists would have us believe. The relationships among problem behaviors are sometimes weak and vary over time and across groups (for example, differences for males and females). Importantly, Helene White also points out that "various problem behaviors follow different developmental paths." As an example, she notes, "delinquency peaks between ages 15 and 17 and then declines, whereas polydrug use increases through adolescence into young adulthood."[123] The developmental perspective also holds open the possibility that some adolescents are "multiple problem youth" who participate in several forms of deviance, while others are more limited in their deviant involvement. As White writes, "It is possible that some adolescents are 'generalists' and others are 'specialists.' "[124] In other words, stable differences may exist across individuals in the clustering of deviant behaviors.

Explaining the Generality of Deviance

How do we explain the co-occurrence of problem behaviors revealed in many studies? Wayne Osgood and his colleagues explore two possibilities.[125] First, causal links may exist between forms of deviance such that involvement in one form leads to involvement in another. For example, drug use may lead to delinquency, which in turn may lead to mental health problems. Second, deviant behaviors may be related because of an underlying common cause or "shared influence."[126] This hypothesis is perhaps most clearly articulated in Gottfredson and Hirschi's general theory of crime (see Chapter 10).[127]

Osgood and his colleagues tested these competing explanations of the generality of deviance using longitudinal self-reports of "heavy alcohol use, marijuana use, use of other illicit drugs, dangerous driving, and other criminal behavior."[128] Their results were somewhat mixed. Consistent with Gottfredson and Hirschi's theory, a common underlying cause or shared influence accounted for almost all relationships among the various types of deviance. But this "general tendency" or common cause did not fully explain the stability of separate deviant behaviors over time.[129] The hypothesis of causal links between different forms of deviance, though, received little support. Only the effect of marijuana use on later use of other illicit drugs was significant, and it was significant for one time period in the study, but not for another.[130] These findings led Osgood and his colleagues to conclude: "A theory that addresses only the general construct [or common cause] can never fully account for the separate [deviant] behaviors, though it might account for much of each of them. Each behavior is, in part, a manifestation of a more general tendency and, in part, a unique phenomenon."[131]

Other research supports that conclusion and shows *partial* support for the hypothesis that a common cause underlies different forms of deviance. A study by White and her colleagues, for example, showed that serious delinquency and substance use shared some common causes, but were also affected by unique predictors.[132]

The notion that various problem behaviors tend to co-occur is appealing in a sense because it suggests that these problem behaviors can be dealt with collectively. But the sometimes weak associations among problem behaviors suggest that prevention and intervention efforts should focus on specific problems, rather than attempting to address the full range of problem behaviors that *sometimes* occur together.

Desistance from Offending[133]

At some point, most offenders cease involvement in crime and delinquency. But criminologists have typically been more interested in the question of why individuals begin offending than why they stop.[134] Recent theory and research, however, consider <u>desistance</u> from offending as a key component of the developmental perspective. John Laub and Robert Sampson point out that both the definition and measurement of desistance are problematic.[135] They distinguish desistance from termination of offending. "Termination is the time at which criminal activity stops. Desistance, by contrast, is the causal process that supports the termination of offending."[136] Thus, desistance is a "social transition," rather than an event in one's criminal career.[137]

desistance "The causal process that supports the termination of offending" (Laub and Sampson, "Understanding Desistance," 2001:11).

Although involvement in offending tends to decline with age (see Chapter 5), desistance does not occur simply as a function of age.[138] Desistance may occur at any age, and it is linked to different factors at different ages.[139] In addition, factors leading to desistance cannot be viewed simply as the opposite of factors leading to involvement in offending. Christopher Uggen and Irving Piliavin use the term "asymmetrical causation" to refer to the idea that predictors of desistance are often distinct from predictors of initiation into offending.[140]

What, then, accounts for the process of desistance from offending? Laub and Sampson conducted a thorough review of the literature on desistance and concluded: "Desistance stems from a variety of complex processes—developmental, psychological, and sociological—and thus there are several factors associated with it. The key elements seem to be aging; a good marriage; securing legal, stable work; and deciding to 'go straight,' including a reorientation of the costs and benefits of crime."[141]

In their book, *Crime in the Making,* Sampson and Laub presented an age-graded theory of informal social control to explain crime and deviance over the life course (see Chapter 10).[142] In this theory, they proposed that adult involvement in offending is influenced not only by early life experiences, including delinquency, but also by social ties in adulthood that facilitate informal social control (e.g., family, work, and military service). In other words, they proposed that life-course changes that strengthen social bonds to society in adulthood will lead to desistance from offending.[143] To test their hypotheses, Sampson and Laub used data from the Gluecks' classic study of delinquency and crime (see Chapter 3). They found that "adult social bonds to work and family had similar consequences for the life trajectories of the 500 delinquents and 500 nondelinquent controls. That is, job stability and marital attachment in adulthood were significantly related to changes in adult crime—the stronger the adult ties to work and family, the less crime and deviance among both delinquents and controls. . . . The major turning points in the life course for men who refrained from crime and deviance in adulthood were stable employment and good marriages."[144] Other criminologists have also found that marriage is strongly related to desistance from offending.[145]

In addition, Laub, Sampson, and their colleague Daniel Nagin contend that, because the development of strong social relationships (like marriages) is gradual and cumulative, the influence of strong social ties on desistance from offending will also be gradual and cumulative.[146] In their research, they found support for this idea.

Sampson and Laub's interpretation of the link between marriage and desistance rests on the assumption that strong ties to conventional institutions create stakes in conformity and thus inhibit crime and delinquency. (This assumption is derived from control theory, which we discuss in Chapter 10). Mark Warr interpreted the link between marriage and desistance in a different way. He used data from the NYS to explore the effects of marriage and exposure to delinquent peers on desistance. Like Sampson and Laub, Warr found that marriage contributed to desistance from offending. But Warr found that this effect occurred through the altering of relationships with delinquent peers.[147] Warr's research "shows that the transition to marriage is followed by a dramatic decline in time spent with friends as well as reduced exposure to delinquent peers, and that these factors largely explain the association between marital status and delinquent behavior."[148] Reduced interaction with delinquent peers as a result of marriage limits both opportunities and motivation to engage in offending. (This interpretation is consistent with differential association and social learning theories, which we discuss in Chapter 11.)

In addition to the influence of marriage on desistance, researchers have also examined the effects of work on desistance. Christopher Uggen used data from a national work experiment to compare control group offenders with treatment group offenders who were given minimum wage jobs.[149] He found that the influence of work on desistance differed depending on the age of offenders. Those age 27 or older were less likely to report crime and arrest when they were provided with jobs. For those under the age of 27, jobs had no effect on desistance.[150]

Research has also shown a connection between earnings (both legal and illegal) and desistance. One study found that current and expected future *criminal* earnings influenced the likelihood of continued offending, and that individuals with higher *legal* earnings were more likely than others to desist from offending.[151]

As we noted earlier, desistance does not occur simply as a function of age. But in a study of age, expectations, and desistance, some researchers found that desistance is linked to age.[152] Together with past experiences, age influences desistance by changing the assessment of risks and rewards of offending. In other words, perceptions of the risks and rewards of crime change with age, and some individuals desist because of those changing perceptions.[153]

Finally, research has shown that "going straight" or desisting from crime involves a change in self-concept. One researcher used life history narratives to explore desistance from offending and found that those who desisted adopted a "new outlook on life" that involved a greater sense of responsibility for their futures and a greater sense of control over their destinies.[154]

Advantages of the Developmental Perspective

In this section, we have highlighted five key themes of the developmental perspective: age of onset of problem behaviors, continuity and change in problem behaviors, progression of seriousness of offenses, generality of deviance or co-occurrence of problem behaviors, and desistance from offending. Considering these aspects of offending can advance understanding of delinquency and crime.

In the introduction to *Developmental Theories of Crime and Delinquency,* Terence Thornberry identifies four advantages of developmental theories.[155] First, developmental perspectives identify and attempt to explain several important and distinct dimensions of delinquency and crime, including prevalence, age of onset, duration of criminal career, frequency of offending, progression of seriousness of offenses, and desistance. One of the strengths of the developmental perspective is its recognition that these various dimensions may have quite different explanations and thus should be treated as distinct elements of offending.

Second, developmental perspectives identify different types of offenders based on developmental considerations.[156] For example, both Gerald Patterson and Terrie Moffitt (whose developmental models we discuss later in this chapter) propose that two types of offenders exist—those who begin offending early in the life course and those who begin later. If it is true that different types of offenders exist, then it is important to distinguish these types in empirical research, which might lead to inaccurate conclusions if researchers study all types of offenders together.

Third, developmental explanations focus on both the "precursors" and "consequences" of offending. That is, they consider both the paths—sometimes long—that lead to delinquency and the consequences of delinquency and crime for development later in life. For example, problem behaviors displayed early in life might spark the developmental progression that ultimately leads to delinquency in adolescence. Developmental perspectives consider these precursor behaviors and continuity over time to explain offending in adolescence or adulthood. Likewise, "involvement in delinquency and crime has consequences for other aspects of a person's development," such as relationships with family and friends and acquisition of adult roles in employment and marriage.[157] Developmental perspectives also consider these outcomes of offending to provide a "fuller and more accurate explanation for delinquent and criminal behavior."[158]

Finally, developmental perspectives "use systematically our understanding of the developmental changes that occur over the life-course to explain changing patterns of delinquent behavior."[159] Thornberry describes these developmental changes in terms of trajectories and transitions. *Trajectories* are defined as "long-term, age-graded patterns of development in major social institutions such as family, education, occupation, and crime."[160] *Transitions* are "events or short-term changes within these trajectories that can deflect the trajectory's arc or growth curve" (for example, getting married, starting a job, or dropping out of school).[161] As Thornberry writes, "an individual's development can be described in terms of the trajectories the person enters, the successful accomplishment of developmental tasks in those trajectories, and the timing of transitions along those trajectories."[162] These dynamic processes, in turn, relate to patterns of involvement in delinquency and crime, and, according to the developmental perspective, should be used in explanations of offending.

■ Developmental Models of Delinquent Behavior

The major components of the developmental perspective have been incorporated into various developmental models of delinquency, including (1) Loeber's pathways model, (2) Patterson's early- and late-starter routes to delinquency, and (3) Moffitt's work on "adolescence-limited" and "life-course-persistent" offenders. In this section, we describe these models and examine research that has tested them.

Loeber's Pathways Model

Loeber and his colleagues began their work on developmental pathways guided by a number of questions: "To what extent can different dimensions of problem behavior be distinguished?" "Is there 'coherence across different manifestations' of child problem behavior?" "Or are most of these behaviors only marginally related to each other?"[163] To address these questions, Paul Frick, along with Loeber and other colleagues, conducted a "meta-analysis" in which they analyzed the findings of 44 studies involving over 28,000 children and adolescents.[164] This study supported Loeber's earlier finding that child antisocial behavior could be represented in terms of a covert–overt dimension. At one end of the continuum for this dimension are covert or concealed antisocial behaviors such as lying, truancy, and stealing. At the other end are overt or confrontational behaviors such as temper tantrums, arguing, and fighting. Frick's study also revealed a second dimension to antisocial behavior—a destructive–nondestructive dimension. At one end of the continuum for this dimension are destructive behaviors such as cruelty to animals. At the other end are nondestructive behaviors such as swearing and rule breaking.[165]

Figure 6-2 illustrates the two-dimensional scaling of disruptive behavior that Loeber and his colleagues discovered. In this figure, "the distance between points on the matrix signifies the extent to which different behaviors correlate, or go together."[166] For example, running away and truancy are closely positioned in the lower left quadrant of the matrix, as these behaviors tend to "go together" relatively often. But truancy and assault, for example, occupy distant positions on the matrix, as they are less likely to occur together.[167] This cluster-

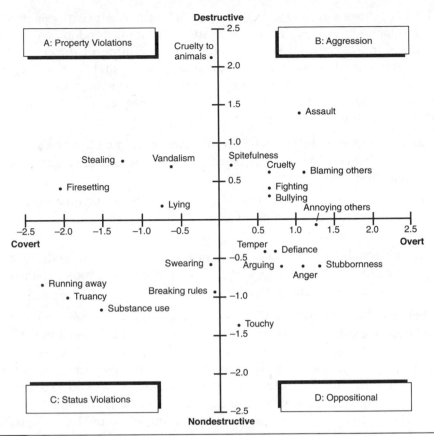

Figure 6-2 "Multidimensional Scale of Disruptive Behavior." Rolf Loeber and his colleagues developed this scale of disruptive behavior with two dimensions: destructive–nondestructive and covert–overt. Disobedient behaviors, such as defiance and rule breaking, are located near the center of the figure and are "shared" by the two dimensions.

Source: Kelley, Loeber, Keenan, and DeLamatre, "Developmental Pathways," 5. Originally published in Frick et al., "Oppositional Defiant Disorder." Copyright © 1993, Elsevier. All rights reserved. Reproduced with permission of the publisher.

ing of disruptive behavior along two dimensions results in four groups of behaviors that Frick and his colleagues label property violations (covert, destructive), aggression (overt, destructive), status violations (covert, nondestructive), and oppositional behavior (overt, nondestructive).[168]

The clusters of behavior in Figure 6-2 provided the foundation for Loeber's proposal of three distinct developmental pathways to offending.[169] Kelley, Loeber, and their colleagues describe the derivation of three pathways from the four behavior clusters in Figure 6-2:

> *Property violations, shown in the upper left quadrant, are considered part of the covert pathway. Aggression, shown in the upper right quadrant, is considered part of the overt pathway. These overt and covert behaviors are placed higher on the destructive axis, because they result in personal harm or property loss or damage. The authority conflict pathway encompasses status violations and oppositional behaviors under the horizontal axis, which represents disruptive behaviors that do not inflict the same degree of harm or distress on others as aggression and property violations.[170]*

pathway "A group of individuals who experience behavioral development that is distinct from the behavioral development of other group(s) of individuals" (Loeber, "Developmental Continuity," 14). The pathway concept incorporates both individuals' temporal sequences of problem behaviors and the increasing seriousness of problem behaviors over time.

Loeber defines a **pathway** as "a group of individuals who experience behavioral development that is distinct from the behavioral development of other group(s) of individuals."[171] He continues, "A key feature of the concept of a pathway is that it takes into account individuals' history and temporal sequence of problem behavior on a continuum of increasing seriousness of problem behavior over time."[172] Kelley, Loeber, and their colleagues describe these three features of a pathway approach:

1. Most individuals who advance to behaviors down a pathway will have displayed behaviors characteristic of the earlier stages in the temporal sequence.

2. Not all individuals progress to the most serious outcome(s); typically, increasingly smaller numbers of individuals reach more serious levels within a pathway.

3. Individuals who reach a more serious level in a pathway tend to continue to display behaviors typical of earlier levels, rather than replace them with the more serious acts.[173]

Loeber and several of his colleagues used data on boys from the Pittsburgh Youth Study to explore the extent to which the overt, covert, and authority conflict pathways characterized boys' disruptive and delinquent behaviors.[174] These three pathways are depicted in **Figure 6-3.** The *authority conflict pathway* begins at stage one with stubborn behavior. It can escalate at stage two to defiant and disobedient behavior and finally at stage three to "authority avoidance," manifested in behaviors such as staying out late, running away, and truancy. In terms of age of onset, this pathway is hypothesized to be the earliest of the three pathways, beginning before age twelve.[175] As the triangular shape in Figure 6-3 implies, the pathway approach suggests that a relatively large number of boys will display the stubborn behavior characteristic of the first stage of this pathway, but a much smaller number will display the types of authority avoidance behavior depicted at the end of this pathway. The *covert pathway* begins at stage one with relatively minor covert behaviors such as lying and shoplifting. It can escalate at stage two to property damage such as vandalism and fire setting and at stage three to serious property crimes such as burglary and serious theft. The *overt pathway* begins at stage one with minor aggression such as bullying. It can escalate at stage two to physical fighting and at stage three to serious violence such as rape and other forms of assault.

Loeber and his colleagues found that these three pathways represented the development of disruptive and delinquent behavior for the majority of boys they studied, and that they "fit" the data better than did a single pathway.[176] Boys tended to enter the pathways at stage one, so that "boys who had experienced the onset of more serious acts tended to have experienced the onset of less serious acts earlier in life."[177] As expected, the authority conflict pathway showed an earlier age of onset of problem behaviors than did the overt and covert pathways. In addition, the age range at which the stubborn behavior characterizing stage one of the authority conflict pathway emerged was quite wide, "indicating that authority conflict emerges over a wide period in childhood or adolescence."[178]

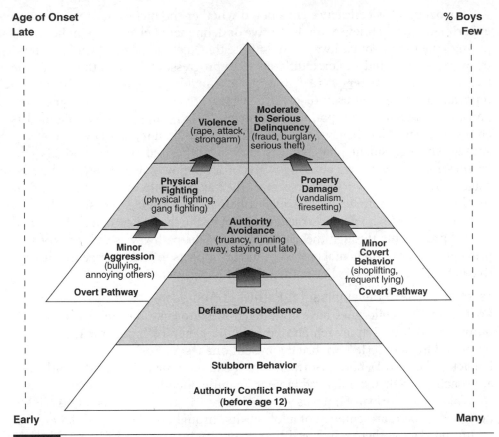

Figure 6-3 "Three Pathways to Boys' Disruptive Behavior and Delinquency." Rolf Loeber and his colleagues developed this pathways model of disruptive and delinquent behavior. The pathways represent "major dimensions of disruptive behavior." Disruptive youths may proceed along one or more pathways.

Source: Kelley, Loeber, Keenan, and DeLamatre, "Developmental Pathways," 9. Based on Loeber et al., "Developmental Pathways in Disruptive Child Behavior."

Loeber's developmental pathways model assumes that youths may travel down more than one pathway. Analysis of the extent to which escalation in one pathway was associated with escalation in another produced these results:

- Most of the boys who advanced to at least stage 2 in one pathway also had an onset of one or more behaviors in another pathway.

- Boys who reached more serious stages in the overt pathway were likely to advance in the covert pathway as well. . . .

- In contrast, many boys who engaged in covert behaviors did not engage in any stages of the overt pathway.[179]

Different combinations of pathways were associated with different rates of delinquent offending. Delinquency rates were lowest for youths progressing down only one pathway. Significantly higher rates of delinquency were found for those who occupied two pathways simultaneously—either the covert and overt pathways together or the covert and authority conflict pathways together. Youths who occupied all three pathways simultaneously showed the highest rates of delinquency.[180]

persisters Individuals involved in problem behaviors at a particular stage of a pathway at more than one point of assessment, according to Loeber's developmental model (Loeber et al., "Boys' Experimentation").

experimenters Individuals whose problem behaviors at a particular stage of a pathway did not persist to a later point of assessment, according to Loeber's developmental model (Loeber et al., "Boys' Experimentation").

Loeber and his colleagues considered whether the inclusion of youths who experimented only briefly with disruptive or delinquent behavior might be "hampering" their study of pathways. To do this, they distinguished "experimenters" from "persisters" and explored differences in progression down pathways for the two groups.[181] **Persisters** were defined as boys involved in problem behavior (as reported by the boy himself or his primary caregiver) within a given stage of a pathway at more than one point of assessment.[182] **Experimenters** were defined as "boys whose problem behavior within a given stage did not persist or recur at any subsequent assessment phase."[183] As they hypothesized, Loeber and his colleagues found that "boys entering a given developmental pathway at the first stage were more likely to be persisters, whereas boys entering at the second or third stage were more likely to be experimenters."[184] Among persisters, the vast majority of those who demonstrated behavior at the third stage of a pathway had moved predictably through the earlier stages of the pathway as well.[185] Not surprisingly, the developmental pathways model appears to characterize the disruptive behavior of persisters better than that of experimenters.

Tests of Loeber's Developmental Pathways Approach

Loeber and his colleagues have conducted numerous studies to test the developmental pathways approach and have found general support for their developmental model.[186] Here we briefly present one test of Loeber's theory in which Patrick Tolan and Deborah Gorman-Smith examine the ability of the pathways approach to predict serious and violent juvenile offending.[187]

Tolan and Gorman-Smith used two data sets in their analysis: (1) the NYS, designed to be representative of adolescents throughout the United States, and including data gathered at five different points in time; and (2) the Chicago Youth Development Study, a sample of high-risk boys from impoverished, crime-ridden Chicago neighborhoods, and including data gathered at four different points in time.[188] Tolan and Gorman-Smith addressed three issues. First, they tested the extent to which the developmental pathways model applied to both the nationally representative sample and the high-risk sample. Second, they separated out those in the two samples who could be classified as serious or violent offenders and examined the extent to which the developmental pathways model applied to more serious or violent offenders. Finally, they examined age of onset of delinquency and its relation to serious or violent offending.[189]

For both the nationally representative and high-risk samples, Tolan and Gorman-Smith found that delinquent behavior generally followed the hypothesized developmental pathways. Involvement in behavior at a later stage of a pathway tended to be preceded by involvement in behavior at an earlier stage. This was true for all three developmental pathways: overt, covert, and authority conflict.[190] When they distinguished serious and violent offenders from other delinquents, Tolan and Gorman-Smith found that the pathways model accounted well for the progression of offending for serious and violent offenders. In both samples, the development of delinquent behavior followed Loeber's pathways for more than 80% of serious or violent offenders.[191]

The findings regarding age of onset were more mixed. To explore this issue, Tolan and Gorman-Smith divided delinquents in each sample into three groups: non-serious, nonviolent delinquents; serious but nonviolent delinquents; and violent delinquents.[192] In the nationally representative sample, significant differences were found in age of onset across the three groups for authority avoidance, moderately serious delinquency, and serious delinquency. As expected, the violent delinquents tended to show earlier ages of onset of delinquent behavior than did the other two groups, particularly for authority avoidance and serious delinquency. In the high-risk sample, however, "no significant differences were found and the trends did not suggest a generally earlier age of onset for the violent or serious offenders."[193] This result is contrary to the findings of Loeber and his colleagues using data from the Pittsburgh Youth Study. It is clear, however, that studies testing Loeber's developmental pathways model have generally been supportive of the approach.

Patterson's Early- and Late-Starter Models

Gerald Patterson and his colleagues at the Oregon Social Learning Center (OSLC) began the Oregon Youth Study (OYS) in 1984.[194] Longitudinal data for all boys included in the OYS were gathered from multiple sources (parents, teachers, peers, and the boys themselves) using a variety of methods (interviews, questionnaires, home observations, videotapes of family problem solving, and peer nominations).[195]

Patterson and his colleagues used data from the extensive OYS to develop and test developmental models of delinquent behavior. They propose two routes to delinquency, each "defined by a different set of determinants and a different set of long-term outcomes."[196] The variable that distinguishes these two paths is age at which a child is first arrested. An **early starter** is defined as a child first arrested before age 14. A **late starter** is one who is first arrested at or after age 14.[197] For early starters, Patterson and his colleagues hypothesize that poor parenting practices lead *directly* to antisocial behavior in young children, which, in turn, puts them at risk for early arrest.[198] For late starters, the researchers hypothesize that a deviant peer group is the direct determinant of offending and that family processes are only *indirectly* involved in the process leading to arrest.[199] Moreover, Patterson and Yoerger hypothesize that early starters are at greater risk than late starters of becoming adult offenders.[200] According to Patterson's developmental model, "the path to chronic delinquency unfolds in a series of predictable steps."[201]

Developmental Model for Early Starters

Patterson and his colleagues develop a "coercion model" to explain the antisocial behavior of early starters. This model focuses on poor parenting skills as a primary cause of antisocial behavior in young children.[202] Home life for those who become antisocial at early ages is distressing—unskilled parents inadvertently but effectively reinforce children's problem behaviors. Three factors tended to characterize the parenting of antisocial boys: (1) no positive reinforcement for prosocial behaviors, (2) no effective punishment for coercive behaviors, and (3) "a *rich supply of reinforcement for coercive behaviors*."[203] This reinforcement occurs when an "aversive" behavior is directed at the child (e.g., parent says "no" to a request made by child), the child responds with coercive behavior (e.g., a temper tantrum), and the initial aversive behavior gets "turned off" (e.g., parent "gives in" to child's initial request in order to end the temper

early starter A child first arrested before the age of 14, according to the developmental model of Patterson and Yoerger. Patterson and his colleagues attribute risk for early arrest to poor parenting practices, which lead directly to antisocial behavior in young children (Patterson et al., "Developmental Perspective;" Patterson and Yoerger, "Developmental Models").

late starter A child first arrested at or after the age of 14, according to the developmental model of Patterson and Yoerger. Patterson and Yoerger propose that, for late starters, the deviant peer group is the direct determinant of offending and the process leading to arrest.

tantrum). Thus, the child's coercive behavior is reinforced by parents who give in to him because of it. Coercive behavior "works" for the child in the sense that it enables him to control the situation and stop others' aversive behaviors. In empirical research, Patterson found that parental disciplinary practices and monitoring explained a large portion (30%) of children's antisocial behavior.[204]

In presenting the key themes of the developmental perspective, we discussed continuity in problem behaviors and progression of seriousness of antisocial and delinquent acts. Patterson's model demonstrates both of these elements. The early-starter model assumes that, as antisocial acts in the home become more frequent, "trivial" coercive behaviors escalate to more severe delinquent acts. If individuals engage in serious delinquency, they are likely to have engaged in more minor forms as well.[205] Through this developmental process, "the repertoire of coercive and antisocial acts gradually expands to include new forms of antisocial acts such as fighting, stealing, fire setting, and delinquency."[206] Patterson's empirical research supports this hypothesis of progression of seriousness.[207]

Patterson and his colleagues postulate not only that reinforcement of coercion leads to antisocial behaviors, but also that parents' failure to reinforce prosocial behaviors hinders the development of children's prosocial skills. The presence of antisocial behaviors and absence of social skills produces problems for children that generalize to settings outside the home, leading to problems in school and with peers.[208]

According to Patterson and Yoerger, "from the perspective of the coercion model, the antisocial trait is merely the first step in a dynamic process"—setting in motion a "cascade" of actions and reactions.[209] "The child entering school initiates coercive actions, producing a predictable set of reactions from peers and teachers. The peers' and teachers' reactions produce predictable reactions from the problem child, and the sequence continues into adulthood."[210] Thus, according to this early-starter model, poor parenting skills lead to children's antisocial behavior, which, in turn, produces school failure and peer rejection. Similarly, rejection by peers contributes to "drift" toward deviant peers, who then become "partners in crime" for the antisocial child.[211]

Linking antisocial behavior to age at first arrest, Patterson and Yoerger reason that children who commit antisocial acts most frequently tend, at relatively young ages, to overwhelm parental attempts to "keep them off the street" and out of the company of deviant peers. In addition, those who commit delinquent acts at the highest rates are at greatest risk of being caught. Therefore, childhood antisocial behavior should significantly predict age at first arrest.[212] Patterson and Yoerger cite research findings to support this hypothesis.[213]

Finally, Patterson and Yoerger predict that early age of onset of delinquent behavior is related to chronic offending into adulthood.[214] Again, Patterson and Yoerger present empirical research findings to support their hypothesis.

Developmental Model for Late Starters

For late starters, the direct cause of delinquency is involvement with a deviant peer group. Family processes that lead to youths' unsupervised time with deviant peers are viewed as indirect causes of delinquency for late starters.[215] Patterson and Yoerger list three defining characteristics of late starters:

1. Unlike early starters, "they are not perceived as being antisocial when assessed at Grade 4."

2. During early adolescence, significant family conflicts result in disrupted parental supervision.

3. They begin involvement with a deviant peer group at about age thirteen or fourteen.[216]

To explain why some families permit adolescents more unsupervised time than do others, Patterson and his colleagues use the idea of "disrupters," which include "change of residence, pubescence, unemployment or financial loss, severe illness or death, and family transition (e.g., divorce, marital conflict)."[217] Patterson and Yoerger also cite studies showing increasing conflict between adolescents and their parents as the children age.[218] According to the late-starter model, a lack of parental supervision, caused by family disrupters, along with increases in family conflict, lead to a "flight to peers." Patterson and Yoerger found that family conflict and inadequate supervision together explained the majority (72%) of involvement with deviant peers.[219]

According to Patterson's developmental model, late starters are short-term offenders, less at risk for chronic offending than early starters.[220] Patterson's data also reveal a substantially higher frequency of serious Index offenses (according to self-reports) for early starters than for late starters. This difference in frequency of serious offending exists throughout adolescence, but becomes particularly great at ages fifteen and sixteen.[221]

Tests of Patterson's Early- and Late-Starter Models

Patterson and his associates have conducted numerous tests of their developmental models using OYS data. Other researchers have also tested the adequacy of Patterson's models for explaining delinquency. Ronald Simons and his colleagues used data from the Iowa Youth and Families Project to test the "two routes to delinquency" and found strong support for Patterson's models.[222] Consistent with Patterson's early-starter model, they found that, for youths defined as early starters, quality of parenting predicted coercive behavior, which in turn predicted association with deviant peers and involvement in the criminal justice system.[223] Consistent with Patterson's late-starter model, they found that, for youths defined as late starters, "quality of parenting predicted affiliation with deviant peers, which in turn predicted an increase in involvement with the criminal justice system."[224] Coercive behavior was unrelated to either association with deviant peers or delinquent behavior.

Moffitt's Adolescence-Limited and Life-Course-Persistent Offenders

Patterson's early- and late-starter models are similar in some ways to the developmental framework offered by Terrie Moffitt.[225] With this framework, Moffitt attempts to explain two contradictory findings in criminology: continuity in antisocial behavior across age, and a dramatic increase in the prevalence of offending during adolescence.[226] She proposes a pair of developmental theories—one to explain the antisocial behavior of "adolescence-limited" offenders, and one to explain the behavior of "life-course-persistent" offenders.

adolescence-limited offenders Offenders who participate in antisocial behavior for a relatively brief period of time during adolescence, according to Moffitt's developmental model.

Adolescence-limited offenders are those who participate in antisocial behavior for a relatively brief period of time during adolescence. For this relatively large group of offenders, involvement in antisocial behavior is temporary and situational.[227] **Life-course-persistent offenders** are those characterized by continuity of antisocial behavior from early childhood through adulthood. For this small group of offenders, antisocial behavior is stable across time and circumstance.[228] **Figure 6-4** illustrates the prevalence and timing of antisocial behavior for these two groups of offenders.

Life-Course-Persistent Offenders

life-course-persistent offenders Offenders characterized by continuity of antisocial behavior from early childhood through adulthood, according to Moffitt's developmental model.

Life-course-persistent offenders constitute a relatively small proportion of all offenders. According to Moffitt, "continuity is the hallmark of the small group of life-course-persistent antisocial persons."[229] Although the expression of antisocial behavior changes over the life course, the "underlying disposition" remains the same. As examples of the changing expressions of an antisocial disposition, Moffitt speaks of "biting and hitting at age 4, shoplifting and truancy at age 10, selling drugs and stealing cars at age 16, robbery and rape at age 22, and fraud and child abuse at age 30."[230] Although the behaviors change with age, the underlying antisocial disposition that gives rise to them is consistent over time and in diverse situations.

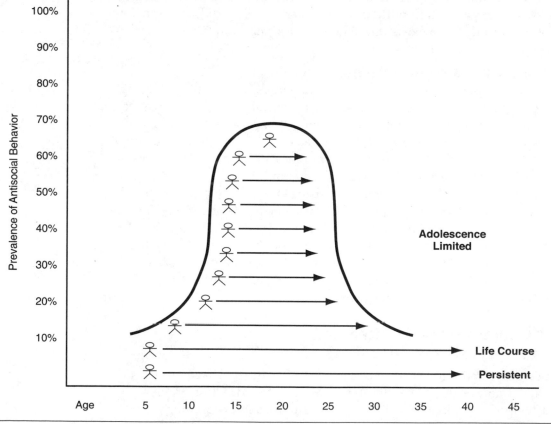

Figure 6-4 Moffitt's adolescence-limited and life-course-persistent offenders. Terrie Moffitt presents this illustration of "the changing prevalence of participation in antisocial behavior across the life course." In this figure, "the solid line represents the known curve of crime over age. The arrows represent the duration of participation in antisocial behavior by individuals."

Source: Moffitt, "Developmental Taxonomy," 677. Copyright © 1993, American Psychological Association. All rights reserved. Reproduced with permission of the publisher.

Moffitt's theory of life-course-persistent offending attributes the early onset of antisocial behavior to a combination of child's *neuropsychological deficits* and an adverse child-rearing environment. Neuropsychological refers to "anatomical structures and physiological processes within the nervous system [that] influence psychological characteristics such as temperament, behavioral development, cognitive abilities, or all three."[231] Neuropsychological deficits may begin before or shortly after birth, as a result of, for example, maternal drug use, poor prenatal nutrition, or deprivation of stimulation or affection following birth.[232] Moffitt reviews the literature demonstrating that "the link between neuropsychological impairment and antisocial outcomes is one of the most robust effects in the study of antisocial behavior."[233] Examples of manifestations of neuropsychological deficits include "difficult temperament," delays in language and motor development, difficulties with listening and problem solving, impaired memory, and learning disabilities.[234]

Children with neuropsychological deficits, already disadvantaged by the deficits themselves, also tend to be born into less than ideal family environments. Some of the same problems that contribute to the creation of neuropsychological deficits, such as maternal drug use, also impair the caregiver's ability to provide a supportive environment and deal with the particular challenges posed by a child with these deficits. The pairing of a difficult child with parents who are ill-equipped to deal with the challenges their child presents sets the stage for life-course-persistent antisocial behavior. Moffitt describes a process in which the demands of coping with a difficult child elicit a series of "failed" parent-child interactions.[235] She calls this *evocative interaction*—interaction in which children's difficult behaviors evoke particular responses from others.[236]

Two other types of interaction—reactive interaction and proactive interaction—help explain how the antisocial behavior of life-course-persistent offenders, once initiated, is sustained throughout the life course. *Reactive interaction* "occurs when different youngsters exposed to the same environment experience it, interpret it, and react to it in accordance with their particular style."[237] For example, an antisocial child might interpret an *ambiguous* situation as calling for an aggressive response, while another child might interpret it as calling for cooperation. *Proactive interaction* "occurs when people select or create environments that support their styles."[238] For example, antisocial children might choose as friends others who are also antisocial. These forms of interaction suggest that antisocial behavior is sustained as children consistently interpret, respond to, and create their own environment in antisocial ways. As they do so, they increasingly prevent opportunities for prosocial behaviors at each stage of development. Life-course-persistent antisocial behavior is maintained in part by "narrowing options for conventional behavior."[239]

Adolescence-Limited Offenders

Compared to life-course-persistent offenders, adolescence-limited offenders constitute a much larger group. According to Moffitt, "discontinuity is the hallmark of teenaged delinquents who have no notable history of antisocial behavior in childhood and little future for such behavior in adulthood."[240] Adolescence-limited offenders are inconsistent in their offending not only over time, but also across situations.[241] For example, they may engage in antisocial behavior in the company of their peers, but continue to obey the rules of parents and teachers. As Moffitt writes, "adolescence-limited delinquents are likely to engage in antisocial behavior in situations where such responses seem profitable to them, but they are also able to abandon antisocial behavior when prosocial styles are more rewarding."[242]

To answer the question of why adolescence-limited offenders begin involvement in antisocial behavior, Moffitt uses the concept of *social mimicry*. She points to the "maturity gap" between biological or sexual maturity in early adolescence and social maturity (characterized, for example, by freedom to work, drive, marry, vote, and buy alcohol) in late adolescence or early adulthood.[243] This maturity gap propels adolescents to search for means of achieving status other than those offered in the conventional world. Moffitt hypothesizes that adolescents view antisocial behavior as a way to achieve "mature status," power, and privilege. In attempts to acquire these resources, adolescence-limited delinquents mimic the behavior of life-course-persistent offenders, who appear relatively unaffected by the maturity gap.[244] This antisocial behavior is then reinforced by the responses it provokes. According to Moffitt, even responses that at first glance seem "negative," such as disruption of the parent–child relationship, are interpreted by adolescents as positive reinforcers for delinquency. This reinforcement of delinquent behavior explains its continued use.

So, why don't all youths become adolescence-limited delinquents? Moffitt proposes four possible answers to this question.[245] First, some adolescents experience delayed puberty, which presumably limits the maturity gap for them and thereby diminishes motivation for delinquency. Second, some adolescents may have few opportunities to mimic the antisocial behavior of life-course-persistent offenders. For example, the structure of the school setting, or residence in rural areas with smaller numbers of age peers than in urban areas, might limit contact with those who would be models of antisocial behavior. Third, personal characteristics might exclude some adolescents from antisocial peer networks. Finally, some adolescents have opportunities to play roles respected by adults.[246] These youths are able to gain "legitimate access to adult privileges" and thus do not need the type of adolescent "power" that accompanies delinquency.

The next obvious question is, "Why do adolescence-limited offenders desist from antisocial behavior in late adolescence or early adulthood?" The answer is that adolescence-limited offenders are able to adapt to *changing contingencies*. For adolescents caught in the maturity gap, delinquency is reinforced and rewarding. As individuals "age out" of that maturity gap and gain access to legitimate adult roles, however, delinquent behavior becomes costly, rather than rewarding. According to Moffitt, "adolescence-limited delinquents gradually experience a loss of motivation for delinquency as they exit the maturity gap. Moreover, when aging delinquents attain some of the privileges they coveted as teens, the consequences of illegal behavior shift from rewarding to punishing, *in their perception*."[247] Adolescence-limited offenders are able to recognize this shift, to understand that persisting in antisocial behavior beyond adolescence has social costs, and to respond with changes in behavior.[248] They have the "option for change" because, unlike life-course-persistent offenders, they have skills for prosocial behavior and academic success, and they are not restricted by underlying personality traits that entangle them in a life of antisocial behavior.

Moffitt writes, "At the crossroads of young adulthood, adolescence-limited and life-course-persistent delinquents go different ways. This happens because the developmental histories and personal traits of adolescence-limiteds allow them the option of exploring new life pathways. The histories and traits of life-course-persistents have foreclosed their options, entrenching them in the antisocial path."[249]

Tests of Moffitt's Developmental Theory

In a recent study, Bartusch, Moffitt, and their colleagues tested predictions derived from Moffitt's developmental theory against predictions derived from Gottfredson and Hirschi's general theory of crime (see Chapter 10).[250] In a nutshell, Gottfredson and Hirschi's theory proposes that consideration of age is unnecessary for understanding antisocial behavior because the factor underlying antisocial behavior—criminal propensity—is the same at all ages.[251] Moffitt's developmental theory, of course, assumes that age is essential to understanding antisocial behavior because the causes of antisocial behavior vary with age.[252]

To test these competing predictions, Bartusch and her colleagues used data from New Zealand, gathered by Moffitt as part of the Dunedin Multidisciplinary Health and Development Study. These data assessed antisocial behavior at ages 5, 7, 9, 11, 13, 15, and 18, through reports by parents, teachers, and the children themselves. Parent and teacher reports of antisocial behavior included measures of fighting, destroying property, bullying, having temper tantrums, being irritable, being disobedient, and lying. Self-reports of antisocial behavior, from ages 13, 15, and 18, included measures of burglary, carrying a weapon, damaging property, stealing, shoplifting, and using marijuana.[253]

Using parent reports of antisocial behavior, available for ages 5, 7, 9, 11, 13, and 15, Bartusch and her colleagues constructed two statistical models. In the first, a single underlying factor predicted antisocial behavior at all ages—consistent with general theory. In the second, separate factors predicted antisocial behavior in childhood and in adolescence—consistent with developmental theory. They tested these competing models and found that the model derived from developmental theory was more consistent with the data than the model derived from general theory.[254] Bartusch and her colleagues found additional support for the developmental model when they examined teacher reports and self-reports of antisocial behavior.

In a separate study, Nagin, Farrington, and Moffitt explored the *life-course trajectories,* or developmental progressions, of four distinct groups: adolescence-limited offenders, high-level chronic offenders, low-level chronic offenders, and those never convicted.[255] They examined self-reports of offending and official convictions at ages 10, 14, 18, and 32, as well as measures at multiple ages of impulsivity, family and peer relationships, and job instability.

Although none of the adolescence-limited offenders in this study had been convicted since age 22, their self-reports of theft, drug use, heavy drinking, and fighting at age 32 indicated that some continued to offend at high rates well beyond adolescence. Despite this continued involvement in some offenses, adolescence-limited offenders at age 32 were characterized by low levels of job instability and relatively strong relationships with their spouses.[256] Summarizing their findings regarding adolescence-limited offenders (categorized according to conviction rates), Nagin and his colleagues write, "The [adolescence-limiteds] appear to be engaged in what might be characterized as circumscribed deviance. At age 32 they seem to be careful to avoid committing crimes with a high risk of conviction, which might jeopardize their stable work careers, or to engage in behaviors, like spousal assault, that might harm their familial relationships. Instead, they seem to restrict their deviance to behaviors less likely to result in official sanction or disrupt intimate attachments."[257]

The findings of this study are consistent with some aspects of Moffitt's developmental theory. At the same time, though, this study suggests that desistance from crime for adolescence-limited offenders is not as clear-cut as Moffitt's theory might lead us to believe.

■ Summary and Conclusions

In this chapter, we have explored the nature of delinquent behavior in terms of developmental patterns of offending. Here we briefly summarize those patterns that theory should consider.

The research of Wolfgang and his colleagues produced what some consider to be one of criminology's most important findings. In *Delinquency in a Birth Cohort,* these researchers reported that a small portion (6.3%) of the cohort studied was responsible for over half (51.9%) of the offenses that resulted in police contact, and that the offenses of this small portion of the cohort tended to be more serious than those of other offenders. Wolfgang and his colleagues called this small portion of delinquents "chronic offenders." This study also revealed that delinquents who began offending at relatively young ages tended to commit a relatively large number of offenses and to accumulate lengthy criminal careers that extended well into adulthood.

"Chronic offenders" share similar characteristics with "career criminals," described by Blumstein and his colleagues. Career criminals are also characterized by a high frequency of offending and a long criminal career, as well as greater seriousness of offenses committed. Like chronic offenders, career criminals constitute a small portion of all offenders but commit a disproportionate share of offenses, especially serious ones. This pattern is one of those facts that theory should consider.

Blumstein and his colleagues also used the term "criminal career"—distinct from "career criminal"—to describe an individual's sequence of offenses during some period of time. Criminal careers are characterized by age of onset, age at termination of offending, duration of career, and frequency of offending. Research on criminal careers shows that most offenders commit only one or a limited number of offenses during late adolescence or early adulthood, and they desist in young adulthood without involvement in serious offending. Theory should consider this nonchronic nature of *most* delinquent behavior.

The developmental perspective of the 1990s incorporates some of the distinct elements of criminal careers, and signals other patterns for theory to consider. First, the developmental perspective suggests different patterns of offending based, in part, on age of onset of problem behaviors. Most offenders begin involvement in delinquency during adolescence and relatively quickly age out of crime in late adolescence or early adulthood. Other offenders, however, display problem behaviors early in the life course and are more likely to develop stable patterns of offending and continue problem behaviors into adulthood. Research has shown that age of onset is related to stability and frequency of offending and to diversity and seriousness of offenses committed. Compared to late-onset offenders, early-onset offenders show more stability in offending over time, a higher rate of offending, a wider variety of offenses, and more serious offenses.

Second, the developmental perspective considers different patterns of offending based on level of continuity or change in problem behaviors over time. For *most* offenders, problem behaviors are not stable over time. Involvement in delinquency is brief and represents a change in behavior patterns. But for a relatively small number of offenders with "extreme" behavior problems, antisocial behavior is stable over time, from childhood through adulthood.

A third pattern that developmental theories consider is progression of seriousness in offending. Research consistently reveals a progression from minor to serious offenses. Most individuals who engage in delinquency do not progress to the most serious offenses. Among those who do become serious delinquents, though, criminal careers tend to be characterized by an escalation in seriousness of offenses.

Fourth, the developmental perspective considers the generality of deviance or co-occurrence of problem behaviors. Studies suggest that delinquency is often a component of a larger group of problem behaviors—including drug and alcohol use, mental health problems, behavior problems and underachievement in school, and risky sexual behavior—that tend to co-occur.

Finally, the developmental perspective examines the process of desistance from offending. Research indicates that attachment to others who behave conventionally (especially a spouse), stable employment, the aging process, and changes in personal identity all significantly influence desistance from crime and delinquency.

We concluded this chapter by presenting three models that have incorporated these patterns of offending, which are central to the developmental perspective: Loeber's pathways model, Patterson's early- and late-starter routes to delinquency, and Moffitt's work on adolescence-limited and life-course-persistent offenders. Each of these models has received considerable empirical support.

THEORIES

Developmental theory
Loeber's developmental pathways model
Patterson's early- and late-starter models
Moffitt's adolescence-limited and life-course-persistent offenders

CRITICAL THINKING QUESTIONS

1. What makes a particular theory or perspective *developmental*? In other words, what are the defining elements of developmental perspectives?

2. Think again about the case of Mark and Shawn, presented at the beginning of this chapter. How would you explain their delinquency from a developmental perspective? Consider all five aspects of the developmental perspective: age of onset, continuity and change in problem behaviors, progression of seriousness, co-occurrence of problem behaviors, and desistance from offending. Are Mark's and Shawn's patterns of offending typical of most delinquents? Explain.

3. Explain the distinction between "criminal careers" and "career criminals."

4. What are the implications of empirical findings regarding chronic offenders for delinquency prevention and intervention strategies? What are the policy implications of the developmental perspective in general?

5. Think about the discussion in Chapter 3 of the relationship between theory and research method. Describe appropriate and adequate methods for conducting research based on a developmental perspective.

SUGGESTED READINGS

Loeber, Rolf. "Developmental Continuity, Change, and Pathways in Male Juvenile Problem Behaviors and Delinquency." In *Delinquency and Crime: Current Theories,* edited by J. David Hawkins, 1–27. Cambridge, England: Cambridge University Press, 1996.

Loeber, Rolf, and Marc LeBlanc. "Toward a Developmental Criminology." In *Crime and Justice: A Review of Research,* edited by Michael Tonry and Norval Morris, 375–473. Chicago: University of Chicago Press, 1990.

Moffitt, Terrie E. "Adolescence-Limited and Life-Course-Persistent Antisocial Behavior: A Developmental Taxonomy." *Psychological Review* 100 (1993):674–701.

Paternoster, Raymond, and Robert Brame. "Multiple Routes to Delinquency? A Test of Developmental and General Theories of Crime." *Criminology* 35 (1997):49–84.

Patterson, Gerald R., Barbara D. DeBaryshe, and Elizabeth Ramsey. "A Developmental Perspective on Antisocial Behavior." *American Psychologist* 44 (1989):329–335.

Sampson, Robert J., and John H. Laub. "Crime and Deviance in the Life Course." *Annual Review of Sociology* 18 (1992):63–84.

GLOSSARY

adolescence-limited offenders: Offenders who participate in antisocial behavior for a relatively brief period of time during adolescence, according to Moffitt's developmental model.

age of onset: The age at which an individual begins involvement in delinquent or criminal acts.

behavioral continuity: Patterns of behavior that are consistent and stable over time, resulting from stable individual attributes or styles of interaction.

career criminals: Characterized by "some combination of a high frequency of offending, a long duration of the criminal career, and high seriousness of offenses committed" (Blumstein et al., "Criminal Career Research," 22).

chronic offenders: The small proportion of offenders who engage in a disproportionate share of offenses, particularly serious and violent ones. In the research of Wolfgang and his colleagues, chronic offenders were defined as those with five or more police contacts (Wolfgang et al., *Delinquency*).

cohort: A group of "individuals (within some population definition) who experienced the same event within the same time interval" (Ryder, *Cohort,* 845). For example, all persons born in Chicago in 1985 form a cohort.

criminal career: "The longitudinal sequence of offenses committed by an offender who has a detectable rate of offending during some period" (Blumstein et al., "Criminal Career Research," 2).

desistance: "The causal process that supports the termination of offending" (Laub and Sampson, "Understanding Desistance," 11).

early starter: A child first arrested before the age of 14, according to the developmental model of Patterson and Yoerger. Patterson and his colleagues attribute risk for early arrest to poor parenting practices, which lead directly to antisocial behavior in young children (Patterson et al., "Developmental Perspective;" Patterson and Yoerger, "Developmental Models").

escalation: Progression of seriousness of offenses, or "the tendency for offenders to move to more serious offense types as offending continues" (Blumstein et al., *Criminal Careers,* 84).

experimenters: Individuals whose problem behaviors at a particular stage of a pathway did not persist to a later point of assessment, according to Loeber's developmental model (Loeber et al., "Boys' Experimentation").

frequency: "The average annual rate at which [the] subgroup of active offenders commits crimes" (Blumstein et al., "Criminal Career Research," 3). Sometimes called *incidence*.

generality of deviance: The extent to which problem behaviors—such as juvenile delinquency, drug and alcohol use, mental health problems, behavior problems and underachievement in school, and risky sexual behavior—tend to occur together or "co-occur."

late starter: A child first arrested at or after the age of 14, according to the developmental model of Patterson and Yoerger. Patterson and Yoerger propose that, for late starters, the deviant peer group is the direct determinant of offending and the process leading to arrest.

life-course-persistent offenders: Offenders characterized by continuity of antisocial behavior from early childhood through adulthood, according to Moffitt's developmental model.

participation: "The proportion of a population who are active offenders at any given time" (Blumstein et al., "Criminal Career Research," 3). Sometimes called *prevalence*.

pathway: "A group of individuals who experience behavioral development that is distinct from the behavioral development of other group(s) of individuals" (Loeber, "Developmental Continuity," 14). The pathway concept incorporates both individuals' temporal sequences of problem behaviors and the increasing seriousness of problem behaviors over time.

persisters: Individuals involved in problem behaviors at a particular stage of a pathway at more than one point of assessment, according to Loeber's developmental model (Loeber et al., "Boys' Experimentation").

recidivists: Individuals who commit multiple delinquent or criminal offenses resulting in police contact.

Chapter Resources

REFERENCES

Bartusch, Dawn R. Jeglum, Donald R. Lynam, Terrie E. Moffitt, and Phil Silva. "Is Age Important? Testing a General Versus a Developmental Theory of Antisocial Behavior." *Criminology* 35 (1997):13–48.

Blumstein, Alfred, and Jacqueline Cohen. "Characterizing Criminal Careers." *Science* 237 (1987): 985–991.

Blumstein, Alfred, Jacqueline Cohen, and David P. Farrington. "Criminal Career Research: Its Value for Criminology." *Criminology* 26 (1988):1–35.

———. "Longitudinal and Criminal Career Research: Further Clarifications." *Criminology* 26 (1988):57–74.

Blumstein, Alfred, Jacqueline Cohen, Jeffrey A. Roth, and Christy A. Visher. *Criminal Careers and "Career Criminals."* Volume I. Washington, DC: National Academy Press, 1986.

Bushway, Shawn D., Robert Brame, and Raymond Paternoster. "Assessing Stability and Change in Criminal Offending: A Comparison of Random Effects, Semiparametric, and Fixed Effects Modeling Strategies." *Journal of Quantitative Criminology* 15 (1999):23–61.

Bushway, Shawn D., Alex R. Piquero, Lisa M. Broidy, Elizabeth Cauffman, and Paul Mazerolle. "An Empirical Framework for Studying Desistance as a Process." *Criminology* 39 (2001):491–515.

Bushway, Shawn D., Terence P. Thornberry, and Marvin D. Krohn. "Desistance as a Developmental Process: A Comparison of Static and Dynamic Approaches." *Journal of Quantitative Criminology* 19 (2003):129–153.

Caspi, Avshalom. "Personality in the Life Course." *Journal of Personality and Social Psychology* 53 (1987):1203–1213.

Caspi, Avshalom, and Daryl J. Bem. "Personality Continuity and Change across the Life Course." In *Handbook of Personality: Theory and Research,* edited by Lawrence A. Pervin, 549–575. New York: Guilford, 1990.

Caspi, Avshalom, Daryl J. Bem, and Glen H. Elder, Jr. "Continuities and Consequences of Interactional Styles across the Life Course." *Journal of Personality* 57 (1989):375–406.

Caspi, Avshalom, Glen H. Elder, Jr., and Daryl J. Bem. "Moving Against the World: Life-Course Patterns of Explosive Children." *Developmental Psychology* 23 (1987):308–313.

Caspi, Avshalom, and Terrie E. Moffitt. "The Continuity of Maladaptive Behavior: From Description to Understanding in the Study of Antisocial Behavior." In *Developmental Psychopathology, Volume 2: Risk, Disorder, and Adaptation,* edited by Dante Cicchetti and Donald J. Cohen, 472–511. New York: Wiley, 1995.

Cohen, Jacqueline. "Research on Criminal Careers: Individual Frequency Rates and Offense Seriousness." In Blumstein, Cohen, Roth, and Visher, *Criminal Careers,* 292–418.

Dunford, Franklyn W., and Delbert S. Elliott. "Identifying Career Offenders Using Self-Report Data." *Journal of Research in Crime and Delinquency* 21 (1984):57–86.

Elliott, Delbert S., David Huizinga, and Suzanne A. Ageton. *Explaining Delinquency and Drug Use.* Beverly Hills, CA: Sage, 1985.

Elliott, Delbert S., David Huizinga, and Scott Menard. *Multiple Problem Youth: Delinquency, Substance Use, and Mental Health Problems.* New York: Springer-Verlag, 1989.

Farrington, David P. "Childhood Aggression and Adult Violence: Early Precursors and Later-Life Outcomes." In Pepler and Rubin, *Development and Treatment,* 5–29.

———. "Offending from 10 to 25 Years of Age." In *Prospective Studies of Crime and Delinquency,* edited by Katherine Teilmann Van Dusen, and Sarnoff A. Mednick. Boston: Kluwer-Nijhoff, 1983.

———. "Predictors, Causes, and Correlates of Male Youth Violence." In Tonry and Moore, *Youth Violence,* 421–475.

———. "Stepping Stones to Adult Criminal Careers." In *Development of Antisocial and Prosocial Behavior: Research, Theories, and Issues,* edited by Dan Olweus, Jack Block, and Marian Radke-Yarrow, 359–384. New York: Academic Press, 1986.

Farrington, David P., and Donald J. West. "Effects of Marriage, Separation, and Children on Offending by Adult Males." In *Current Perspectives on Aging and the Life Cycle,* Volume 4, edited by Zena Blau and John Hagan, 249–281. Greenwich, CT: JAI Press, 1995.

Fergusson, David M., L. John Horwood, and Daniel S. Nagin. "Offending Trajectories in a New Zealand Birth Cohort." *Criminology* 38 (2000):525–551.

Frick, Paul J., Benjamin B. Lahey, Rolf Loeber, Lynne Tannenbaum, Y. Van Horn, M. A. G. Christ, E. A. Hart, and K. Hanson. "Oppositional Defiant Disorder and Conduct Disorder: A Meta-Analytic

Review of Factor Analyses and Cross-Validation in a Clinical Sample." *Clinical Psychology Review* 13 (1993):319–340.

Gadd, David, and Stephen Farrall. "Criminal Careers, Desistance and Subjectivity: Interpreting Men's Narratives of Change." *Theoretical Criminology* 8 (2004):123–156.

Giordano, Peggy C., Stephen A. Cernkovich, and Donna D. Holland. "Changes in Friendship Relations Over the Life Course: Implications for Desistance from Crime." *Criminology* 41 (2002): 293–328.

Glueck, Sheldon, and Eleanor Glueck. *Delinquents and Nondelinquents in Perspective.* Cambridge, MA: Harvard University Press, 1968.

———. *Unraveling Juvenile Delinquency.* Cambridge, MA: Harvard University Press, 1950.

Gottfredson, Michael R., and Travis Hirschi. *A General Theory of Crime.* Stanford, CA: Stanford University Press, 1990.

———. "Science, Public Policy, and the Career Paradigm." *Criminology* 26 (1988):37–55.

———. "The True Value of Lambda Would Appear to Be Zero: An Essay on Career Criminals, Criminal Careers, Selective Incapacitation, Cohort Studies, and Related Topics." *Criminology* 24 (1986):213–233.

Hirschi, Travis, and Michael R. Gottfredson. "Age and the Explanation of Crime." *American Journal of Sociology* 89 (1983):552–584.

———. "The Generality of Deviance." In *The Generality of Deviance,* edited by Travis Hirschi and Michael R. Gottfredson, 1–22. New Brunswick, NJ: Transaction, 1994.

Horney, Julie, D. Wayne Osgood, and Ineke Haen Marshall. "Criminal Careers in the Short-Term: Intra-Individual Variability in Crime and Its Relation to Local Life Circumstances." *American Sociological Review* 60 (1995):655–673.

Huizinga, David, and Cynthia Jakob-Chien. "The Contemporaneous Co-Occurrence of Serious and Violent Juvenile Offending and Other Problem Behaviors." In Loeber and Farrington, *Serious and Violent Juvenile Offenders,* 47–67.

Huizinga, David, Rolf Loeber, Terence P. Thornberry, and Lynn Cothern. "Co-Occurrence of Delinquency and Other Problem Behaviors." In *Juvenile Justice Bulletin.* Washington, DC: US Department of Justice, Office of Juvenile Justice and Delinquency Prevention, 2000.

Junger, Marianne, and Maja Dekovic. "Crime as Risk-Taking: Co-Occurrence of Delinquent Behavior, Health-Endangering Behaviors, and Problem Behaviors." In *Control Theories of Crime and Delinquency, Advances in Criminological Theory,* Volume 12, edited by Chester L. Britt and Michael R. Gottfredson, 213–248. New Brunswick, NJ: Transaction, 2003.

Kelley, Barbara Tatem, Rolf Loeber, Kate Keenan, and Mary DeLamatre. "Developmental Pathways in Boys' Disruptive and Delinquent Behavior." In *Juvenile Justice Bulletin.* Washington, DC: US Department of Justice, Office of Juvenile Justice and Delinquency Prevention, 1997.

Laub, John H., Daniel S. Nagin, and Robert J. Sampson. "Trajectories of Change in Criminal Offending: Good Marriages and the Desistance Process." *American Sociological Review* 63 (1998): 225–238.

Laub, John H., and Robert J. Sampson. "The Sutherland–Glueck Debate: On the Sociology of Criminological Knowledge." *American Journal of Sociology* 96 (1991):1402–1440.

———. "Understanding Desistance from Crime." In *Crime and Justice: A Review of Research,* Volume 28, edited by Michael Tonry, 1–69. Chicago: University of Chicago Press, 2001.

LeBlanc, Marc, and Marcel Frechette. *Male Criminal Activity from Childhood through Youth: Multilevel and Developmental Perspectives.* New York: Springer-Verlag, 1989.

Loeber, Rolf. "Developmental Continuity, Change, and Pathways in Male Juvenile Problem Behaviors and Delinquency." In *Delinquency and Crime: Current Theories,* edited by J. David Hawkins, 1–27. Cambridge, England: Cambridge University Press, 1996.

Loeber, Rolf, and David P. Farrington, eds. *Serious and Violent Juvenile Offenders: Risk Factors and Successful Interventions.* Thousand Oaks, CA: Sage, 1998.

Loeber, Rolf, Kate Keenan, and Quanwu Zhang. "Boys' Experimentation and Persistence in Developmental Pathways Toward Serious Delinquency." *Journal of Child and Family Studies* 6 (1997):321–357.

Loeber, Rolf, and Marc LeBlanc. "Toward a Developmental Criminology." In *Crime and Justice: A Review of Research,* Volume 12, edited by Michael Tonry and Norval Morris, 375–473. Chicago: University of Chicago Press, 1990.

Loeber, Rolf, and Karen B. Schmaling. "Empirical Evidence for Overt and Covert Patterns of Antisocial Conduct Problems: A Meta-Analysis." *Journal of Abnormal Child Psychology* 13 (1985):337–353.

Loeber, Rolf, and Magda Stouthamer-Loeber. "Prediction." In *Handbook of Juvenile Delinquency,* edited by Herbert C. Quay, 325–382. New York: Wiley, 1987.

Loeber, Rolf, Magda Stouthamer-Loeber, and Stephanie M. Green. "Age at Onset of Problem Behavior in Boys, and Later Disruptive and Delinquent Behaviors." *Criminal Behavior and Mental Health* 1 (1991):229–246.

Loeber, Rolf, Phen Wung, Kate Keenan, Bruce Giroux, Magda Stouthamer-Loeber, Welmoet B. Van Kammen, and Barbara Maughan. "Developmental Pathways in Disruptive Child Behavior." *Development and Psychopathology* 5 (1993):103–133.

Maruna, Shadd. *Making Good: How Ex-Offenders Reform and Reclaim Their Lives.* Washington, DC: American Psychological Association Books, 2001.

Moffitt, Terrie E. "Adolescence-Limited and Life-Course-Persistent Antisocial Behavior: A Developmental Taxonomy." *Psychological Review* 100 (1993):674–701.

———. "Adolescence-Limited and Life-Course-Persistent Offending: A Complementary Pair of Developmental Theories." In Thornberry, *Developmental Theories,* 11–54.

———. "The Neuropsychology of Delinquency: A Critical Review of Theory and Research." In *Crime and Justice: A Review of Research,* Volume 12, edited by Michael Tonry and Norval Morris, 99–169. Chicago: University of Chicago Press, 1990.

Moore, Dennis R., and Judy L. Arthur. "Juvenile Delinquency." In *Handbook of Child Psychopathology,* 2nd ed., edited by Thomas H. Ollendick and Michel Hersen, 197–217. New York: Plenum Press, 1989.

Nagin, Daniel S., David P. Farrington, and Terrie E. Moffitt. "Life-Course Trajectories of Different Types of Offenders." *Criminology* 33 (1995):111–139.

Nagin, Daniel S., and Kenneth C. Land. "Age, Criminal Careers, and Population Heterogeneity: Specification and Estimation of a Nonparametric, Mixed Poisson Model." *Criminology* 31 (1993):327–362.

Nagin, Daniel S., and Raymond Paternoster. "On the Relationship of Past to Future Participation in Delinquency." *Criminology* 29 (1991):163–189.

———. "Population Heterogeneity and State Dependence: State of the Evidence and Directions for Future Research." *Journal of Quantitative Criminology* 16 (2000):117–144.

Olweus, Daniel. "Stability of Aggressive Reaction Patterns in Males: A Review." *Psychological Bulletin* 86 (1979):852–875.

Osgood, D. Wayne, Lloyd D. Johnston, Patrick M. O'Malley, and Jerald G. Bachman. "The Generality of Deviance in Late Adolescence and Early Adulthood." *American Sociological Review* 53 (1988):81–93.

Osgood, D. Wayne, and David C. Rowe. "Bridging Criminal Careers, Theory, and Policy through Latent Variable Models of Individual Offending." *Criminology* 32 (1994):517–554.

Paternoster, Raymond, and Robert Brame. "Multiple Routes to Delinquency? A Test of Developmental and General Theories of Crime." *Criminology* 35 (1997):49–84.

Patterson, Gerald R. "Maternal Rejection: Determinant or Product for Deviant Child Behavior?" In *Relationships and Development,* edited by W. W. Hartup and Z. Rubin, 73–94. Hillsdale, NJ: Lawrence Erlbaum, 1986.

Patterson, Gerald R., and Lou Bank. "Some Amplifying Mechanisms for Pathologic Process in Families." In *Systems and Development: The Minnesota Symposia on Child Psychology,* vol. 22, edited by M. R. Gunnar and E. Thelem, 167–210. Hillsdale, NJ: Lawrence Erlbaum, 1989.

Patterson, Gerald R., Deborah Capaldi, and Lou Bank. "An Early Starter Model for Predicting Delinquency." In Pepler and Rubin, *Development and Treatment,* 139–168.

Patterson, Gerald R., L. Crosby, and Samuel Vuchinich. "Predicting Risk for Early Police Arrest." *Journal of Quantitative Criminology* 8 (1992):335–355.

Patterson, Gerald R., Barbara D. DeBaryshe, and Elizabeth Ramsey. "A Developmental Perspective on Antisocial Behavior." *American Psychologist* 44 (1989):329–335.

Patterson, Gerald R., and Karen Yoerger. "Developmental Models for Delinquent Behavior." In *Mental Disorder and Crime,* edited by Sheilagh Hodgins, 140–172. Newbury Park, CA: Sage, 1993.

Pepler, Debra J., and Kenneth H. Rubin, eds. *The Development and Treatment of Childhood Aggression.* Hillsdale, NJ: Lawrence Erlbaum, 1991.

Pezzin, Liliana E. "Earning Prospects, Matching Effects, and the Decision to Terminate a Criminal Career." *Journal of Quantitative Criminology* 11 (1995):29–50.

Piquero, Alex R., Robert Brame, Paul Mazerolle, and Rudy Haapanen. "Crime in Emerging Adulthood." *Criminology* 40 (2002):137–169.

Piquero, Alex R., David P. Farrington, and Alfred Blumstein. "The Criminal Career Paradigm." *Crime and Justice* 30 (2003):359–506.

Piquero, Alex, Raymond Paternoster, Paul Mazerolle, Robert Brame, and Charles W. Dean. "Onset Age and Offense Specialization." *Journal of Research in Crime and Delinquency* 36 (1999):275–299.

Robins, Lee N. "Sturdy Childhood Predictors of Adult Antisocial Behaviour: Replications from Longitudinal Studies." *Psychological Medicine* 8 (1978):611–622.

Rojek, Dean G., and Maynard L. Erickson. "Delinquent Careers: A Test of the Career Escalation Model." *Criminology* 20 (1982):5–28.

Ryder, Norman B. "The Cohort as a Concept in the Study of Social Change." *American Sociological Review* 30 (1965):843–861.

Sampson, Robert J., and John H. Laub. "Crime and Deviance in the Life Course." *Annual Review of Sociology* 18 (1992):63–84.

———. "Crime and Deviance Over the Life Course: The Salience of Adult Social Bonds." *American Sociological Review* 55 (1990):609–627.

———. *Crime in the Making: Pathways and Turning Points Through Life.* Cambridge, MA: Harvard University Press, 1993.

———. "Life-Course Desisters? Trajectories of Crime among Delinquent Boys Followed to Age 70." *Criminology* 41 (2002):555–592.

———. "A Life-Course Theory of Cumulative Disadvantage and the Stability of Delinquency." In Thornberry, *Developmental Theories,* 133–161.

Shover, Neal. *Great Pretenders: Pursuits and Careers of Persistent Thieves.* Boulder, CO: Westview Press, 1996.

Shover, Neal, and Carol Y. Thompson. "Age, Differential Expectations, and Crime Desistance." *Criminology* 30 (1992):89–104.

Simons, Ronald L., Christine Johnson, Rand D. Conger, and Glen Elder, Jr. "A Test of Latent Trait Versus Life-Course Perspectives on the Stability of Adolescent Antisocial Behavior." *Criminology* 36 (1998):217–243.

Simons, Ronald L., Chyi-In Wu, Rand D. Conger, and Frederick O. Lorenz. "Two Routes to Delinquency: Differences Between Early and Late Starters in the Impact of Parenting and Deviant Peers." *Criminology* 32 (1994):247–276.

Smith, Douglas A., and Robert Brame. "On the Initiation and Continuation of Delinquency." *Criminology* 32 (1994):607–629.

Smith, Douglas A., Christy A. Visher, and G. Roger Jarjoura. "Dimensions of Delinquency: Exploring the Correlates of Participation, Frequency, and Persistence of Delinquent Behavior." *Journal of Research in Crime and Delinquency* 28 (1991):6–32.

Thornberry, Terence P., ed. *Developmental Theories of Crime and Delinquency.* New Brunswick, NJ: Transaction, 1997.

Thornberry, Terence P. "Introduction: Some Advantages of Developmental and Life-Course Perspectives for the Study of Crime and Delinquency." In Thornberry, *Developmental Theories,* 1–10.

Thornberry, Terence P., David Huizinga, and Rolf Loeber. "The Causes and Correlates Studies: Findings and Policy Implications." *Juvenile Justice* 9 (2004): 3–19.

———. "The Prevention of Serious Delinquency and Violence: Implications from the Program of Research on the Causes and Correlates of Delinquency." In *Serious, Violent, and Chronic Juvenile Offenders: A Sourcebook,* edited by James C. Howell, Barry Krisberg, J. David Hawkins, and John J. Wilson, 213–237. Thousand Oaks, CA: Sage, 1995.

Tittle, Charles R. "Two Empirical Regularities (Maybe) in Search of an Explanation: Commentary on the Age/Crime Debate." *Criminology* 26 (1988):75–85.

Tolan, Patrick H. "Implications of Age of Onset for Delinquency Risk." *Journal of Abnormal Child Psychology* 15 (1987):47–65.

Tolan, Patrick H., and Deborah Gorman-Smith. "Development of Serious and Violent Offending Careers." In Loeber and Farrington, *Serious and Violent Juvenile Offenders,* 68–85.

Tolan, Patrick H., and Peter Thomas. "The Implications of Age of Onset for Delinquency Risk II: Longitudinal Data." *Journal of Abnormal Child Psychology* 23 (1995):157–181.

Tonry, Michael, and Mark H. Moore, eds. *Youth Violence.* Chicago: University of Chicago Press, 1998.

Tracy, Paul E., Marvin E. Wolfgang, and Robert M. Figlio. *Delinquency Careers in Two Birth Cohorts.* New York: Plenum Press, 1990.

Uggen, Christopher. "Work as a Turning Point in the Life Course of Criminals: A Duration Model of Age, Employment, and Recidivism." *American Sociological Review* 65 (2000):529–546.

Uggen, Christopher, and Irving Piliavin. "Asymmetrical Causation and Criminal Desistance." *Journal of Criminal Law and Criminology* 88 (1998):1399–1422.

Warr, Mark. "Life-Course Transitions and Desistance from Crime." *Criminology* 36 (1998):183–216.

White, Helene Raskin. "Early Problem Behavior and Later Drug Problems." *Journal of Research in Crime and Delinquency* 29 (1992):412–429.

White, Helene Raskin, Robert J. Pandina, and Randy L. LaGrange. "Longitudinal Predictors of Serious Substance Use and Delinquency." *Criminology* 25 (1987):715–740.

White, Jennifer L., Terrie E. Moffitt, Felton Earls, Lee Robins, and Phil A. Silva. "How Early Can We Tell?: Predictors of Childhood Conduct Disorder and Adolescent Delinquency." *Criminology* 28 (1990):507–533.

Wiesner, Margit, Deborah M. Capaldi, and Gerald Patterson. "Development of Antisocial Behavior and Crime Across the Life-Span from a Social Interactional Perspective: The Coercion Model." In *Social Learning Theory and the Explanation of Crime: A Guide for the New Century,* edited by Ronald L. Akers and Gary F. Jensen, 317–337. New Brunswick, NJ: Transaction, 2003.

Wilson, James, and Richard Herrnstein. *Crime and Human Nature.* New York: Simon and Schuster, 1985.

Wolfgang, Marvin E., Robert M. Figlio, and Thorsten Sellin. *Delinquency in a Birth Cohort.* Chicago: University of Chicago Press, 1972.

Wolfgang, Marvin E., Terence P. Thornberry, and Robert M. Figlio. *From Boy to Man, From Delinquency to Crime.* Chicago: University of Chicago Press, 1987.

ENDNOTES

1. Loeber and LeBlanc, "Toward a Developmental Criminology."
2. Blumstein and Cohen, "Characterizing Criminal Careers;" Blumstein, Cohen, and Farrington, "Criminal Career Research," and "Further Clarifications;" and Blumstein, Cohen, Roth, and Visher, *Criminal Careers.*
3. Sampson and Laub, "Crime and Deviance in the Life Course," 64. See also Sampson and Laub, "Adult Social Bonds," and *Crime in the Making.*
4. Wolfgang, Figlio, and Sellin, *Delinquency;* Blumstein and Cohen, "Characterizing Criminal Careers;" Blumstein, Cohen, and Farrington, "Criminal Career Research," and "Further Clarifications;" and Blumstein, Cohen, Roth, and Visher, *Criminal Careers.*
5. Paternoster and Brame, "Multiple Routes."
6. Wolfgang, Figlio, and Sellin, *Delinquency.*
7. Ibid., 5.
8. Ryder, "Cohort," 845.
9. Wolfgang, Figlio, and Sellin, *Delinquency,* 244.
10. Ibid., 244–245.
11. Ibid., 245.
12. Ibid., 89.
13. Ibid.
14. Ibid.
15. Ibid.
16. Wolfgang, Figlio, and Sellin, *Delinquency,* derived by combining Table 5.3 on pages 68–69 and Table 6.16 on page 102.
17. Wolfgang, Thornberry, and Figlio, *From Boy to Man,* 3.
18. Ibid., 1.
19. Ibid., 201.
20. Tracy, Wolfgang, and Figlio, *Delinquency Careers,* 24–27.
21. See also Dunford and Elliott, who in "Identifying Career Offenders" explored chronic delinquency patterns using self-reports of delinquent behavior from the NYS. They identified a group of "serious career offenders" similar in size and offense characteristics (seriousness and frequency) to Wolfgang's chronic offenders. The startling finding in Dunford and Elliott's study was that only 24% of those who self-reported serious, persistent offending had ever been arrested.
22. Tracy, Wolfgang, and Figlio, *Delinquency Careers,* 38–39.
23. Ibid., 83.
24. Ibid., 90.
25. Blumstein, Cohen, Roth, and Visher, *Criminal Careers;* Blumstein and Cohen, "Characterizing Criminal Careers;" and Blumstein, Cohen, and Farrington, "Criminal Career Research."

26. Blumstein, Cohen, and Farrington, "Criminal Career Research," 2. See also Blumstein, Cohen, Roth, and Visher, *Criminal Careers;* and Blumstein and Cohen, "Characterizing Criminal Careers."

27. Blumstein and Cohen, "Characterizing Criminal Careers," 986; and Blumstein, Cohen, and Farrington, "Criminal Career Research," 2.

28. Blumstein, Cohen, Roth, and Visher, *Criminal Careers,* 2; Blumstein, Cohen, and Farrington, "Criminal Career Research," 4–6; and Piquero, Farrington, and Blumstein, "Criminal Career Paradigm."

29. Blumstein, Cohen, and Farrington, "Criminal Career Research," and "Further Clarifications."

30. Blumstein et al., "Criminal Career Research," 3.

31. Ibid.

32. Ibid.

33. Ibid.

34. Blumstein and Cohen, "Characterizing Criminal Careers;" Blumstein et al., "Criminal Career Research." See also Farrington, "Offending," cited in Blumstein, Cohen, and Farrington, "Criminal Career Research."

35. Hirschi and Gottfredson, "Age."

36. Smith, Visher, and Jarjoura, "Dimensions of Delinquency," 6.

37. Smith, Visher, and Jarjoura, "Dimensions of Delinquency." See also Smith and Brame, who in "Initiation" examined whether explanatory variables from several theories had different effects on two dimensions of criminal careers: initiation and continuation. They found mixed results: "while many variables predict initial and continued offending in a similar fashion, other variables predict only one of these decisions" (623).

38. Smith, Visher, and Jarjoura, "Dimensions of Delinquency," 21.

39. Blumstein, Cohen, and Farrington, "Criminal Career Research," 7.

40. Ibid., 22.

41. Blumstein, Cohen, Roth, and Visher, *Criminal Careers;* and Blumstein, Cohen, and Farrington, "Criminal Career Research."

42. Gottfredson and Hirschi, "True Value," and "Science;" and Blumstein, Cohen, and Farrington, "Criminal Career Research," and "Further Clarifications."

43. Gottfredson and Hirschi, "True Value," and "Science;" and Blumstein, Cohen, and Farrington, "Criminal Career Research," and "Further Clarifications."

44. Wolfgang, Figlio, and Sellin, *Delinquency;* Wolfgang, Thornberry, and Figlio, *From Boy to Man;* Blumstein, Cohen, Roth, and Visher, *Criminal Careers;* Blumstein and Cohen, "Characterizing Criminal Careers;" and Blumstein, Cohen, and Farrington, "Criminal Career Research," and "Further Clarifications."

45. Osgood and Rowe, "Bridging," 518.

46. Tittle, "Empirical Regularities," 78.

47. Blumstein, Cohen, and Farrington, "Criminal Career Research," 4.

48. Tittle, "Empirical Regularities."

49. Osgood and Rowe, "Bridging."

50. See Paternoster and Brame, "Multiple Routes" (51–52), and Osgood and Rowe, "Bridging" (519), for discussions of how the criminal career perspective has influenced and is compatible with developmental theories of the last decade.

51. Blumstein, Cohen, and Farrington, "Criminal Career Research," 4.

52. Paternoster and Brame, "Multiple Routes," 51–52.

53. Kelley, Loeber, Keenan, and DeLamatre, "Developmental Pathways," 1; Loeber and LeBlanc, "Toward a Developmental Criminology;" Nagin, Farrington, and Moffitt, "Trajectories;" and Thornberry, "Introduction."

54. Loeber and LeBlanc, "Toward a Developmental Criminology," 377.

55. Robins, "Sturdy Childhood Predictors," 611. Cited in Thornberry, "Introduction," 3.

56. Thornberry, "Introduction," 3.

57. Piquero, Brame, Mazerolle, and Haapanen, "Crime in Emerging Adulthood," 139.

58. Loeber and LeBlanc, "Toward a Developmental Criminology," 390–391.

59. Ibid., 390.

60. See, for example, Loeber and LeBlanc, "Toward a Developmental Criminology," 390–398; and Thornberry, "Introduction."

61. Loeber and LeBlanc, "Toward a Developmental Criminology," 395.

62. Loeber and LeBlanc, "Toward a Developmental Criminology," 395. See, for example, Tolan, "Implications."
63. Tolan, "Implications."
64. Rojek and Erickson, "Delinquent Careers;" and Cohen, "Research on Criminal Careers."
65. Piquero, Paternoster, Mazerolle, Brame, and Dean, "Onset Age."
66. Ibid., 294–295.
67. See, for example, Nagin, Farrington, and Moffitt, "Trajectories;" and Tolan and Thomas, "Longitudinal Data." But see also Cohen, "Research on Criminal Careers," who reports contradictory findings.
68. Tolan and Thomas, "Longitudinal Data."
69. See, for example, LeBlanc and Frechette, *Male Criminal Activity,* cited in Loeber and LeBlanc, "Toward a Developmental Criminology," 391.
70. White et al., "How Early."
71. Ibid., 522.
72. Ibid., 518–519.
73. Ibid., 521.
74. Loeber, Stouthamer-Loeber, and Green, "Age at Onset."
75. Moffitt, "Developmental Taxonomy."
76. Patterson, DeBaryshe, and Ramsey, "Developmental Perspective;" and Patterson and Yoerger, "Developmental Models."
77. Thornberry, "Introduction," 3.
78. Ibid.
79. Glueck and Glueck, *Unraveling Juvenile Delinquency,* and *Delinquents and Nondelinquents.* See Laub and Sampson, "Sutherland–Glueck Debate," 1431–1432, for a discussion of the Gluecks' early hypothesis of stability in offending.
80. Sampson and Laub, *Crime in the Making,* and "Life-Course Theory."
81. Caspi, "Personality;" Caspi, Bem, and Elder, "Continuities;" and Caspi and Bem, "Personality Continuity and Change."
82. See, for example, Elliott, Huizinga, and Ageton, *Explaining Delinquency;* Caspi, Elder, and Bem, "Continuities;" Loeber and Stouthamer-Loeber, "Prediction;" Farrington, "Childhood Aggression;" Gottfredson and Hirschi, *General Theory of Crime;* Sampson and Laub, *Crime in the Making;* and Bushway, Brame, and Paternoster, "Assessing Stability."
83. Olweus, "Stability," 854–855. Cited in Laub and Sampson, "Sutherland–Glueck Debate," 1431, and Sampson and Laub, *Crime in the Making,* 10.
84. Robins, "Sturdy Childhood Predictors."
85. Sampson and Laub, *Crime in the Making,* 13. See also Caspi and Moffitt, "Continuity of Maladaptive Behavior;" they reach a similar conclusion.
86. Farrington, "Predictors," 438; and Nagin and Paternoster, "Relationship."
87. Farrington, "Predictors," 438.
88. Bushway, Brame, and Paternoster, "Assessing Stability," 24.
89. Gottfredson and Hirschi, *General Theory of Crime.* This interpretation is also consistent with other criminological theories, such as Wilson and Herrnstein's theory of crime and human nature (*Crime and Human Nature*). See Nagin and Paternoster, "Relationship" (164–166) and Bushway, Brame, and Paternoster, "Assessing Stability" (24–25), for discussions of the compatibility of the population heterogeneity explanation with criminological theories.
90. Nagin and Paternoster, "Relationship."
91. Nagin and Paternoster, "Relationship," 166; and Bushway, Brame, and Paternoster, "Assessing Stability," 25.
92. See Nagin and Paternoster, "Relationship," 166, and Bushway, Brame, and Paternoster, "Assessing Stability," 26, for discussions of the compatibility of the state dependence explanation with criminological theories.
93. Kelley et al., "Developmental Pathways."
94. Loeber, "Developmental Continuity."
95. Ibid., 8.
96. Ibid., 8–10.
97. Sampson and Laub, *Crime in the Making.*
98. Ibid.

99. See, for example, Loeber and LeBlanc, "Toward a Developmental Criminology;" and Kelley et al., "Developmental Pathways."
100. Blumstein, Cohen, Roth, and Visher, *Criminal Careers,* 84.
101. Kelley et al., "Developmental Pathways," 1–2.
102. Farrington, "Stepping Stones."
103. Kelley et al., "Developmental Pathways," 3.
104. Kelley et al., "Developmental Pathways," 3; and Loeber, "Developmental Continuity," 3.
105. Kelley et al., "Developmental Pathways," 3.
106. Kelley et al., "Developmental Pathways," 6; Loeber et al., "Developmental Pathways;" Loeber, "Developmental Continuity;" and Loeber, Stouthamer-Loeber, and Green, "Age at Onset."
107. Elliott, Huizinga, and Menard, *Multiple Problem Youth,* 127–130.
108. Farrington, "Predictors," 428–431.
109. Ibid., 429.
110. Junger and Dekovic, in "Crime as Risk-Taking," offer a review of the literature on the co-occurrence of delinquency, health-endangering behaviors, and other problem behaviors.
111. Huizinga et al., "Co-Occurrence of Delinquency," 1. See Huizinga and Jakob-Chien, "Contemporaneous Co-Occurrence," for a list of these studies.
112. Elliott, Huizinga, and Menard, *Multiple Problem Youth.*
113. Ibid., 136.
114. See, for example, Farrington, "Predictors," 431–434; Huizinga and Jakob-Chien, "Contemporaneous Co-Occurrence;" and Thornberry, Huizinga, and Loeber, "Prevention of Serious Delinquency."
115. White, "Early Problem Behavior."
116. White, "Early Problem Behavior," 423–424. See also White, Pandina, and LaGrange, "Longitudinal Predictors."
117. White, "Early Problem Behavior," 424–426.
118. Huizinga et al., "Co-Occurrence of Delinquency;" and Thornberry, Huizinga, and Loeber, "Causes and Correlates Studies."
119. Huizinga et al., "Co-Occurrence of Delinquency," 5.
120. Ibid., 6.
121. Ibid.
122. Ibid.
123. White, "Early Problem Behavior," 414.
124. Ibid., 426.
125. Osgood et al., "Generality of Deviance."
126. Ibid., 81.
127. Gottfredson and Hirschi, *General Theory of Crime.* See also Hirschi and Gottfredson, "Age," and "Generality of Deviance;" and Gottfredson and Hirschi, "True Value."
128. Osgood et al., "Generality of Deviance," 81.
129. Ibid., 91.
130. Ibid., 89.
131. Ibid., 91.
132. White, Pandina, and LaGrange, "Longitudinal Predictors."
133. This section is based heavily on an excellent review of the literature on desistance from crime by John Laub and Robert Sampson, "Understanding Desistance."
134. Laub and Sampson, "Understanding Desistance," 1.
135. Ibid., 5–10.
136. Ibid., 11.
137. See also Bushway, Thornberry, and Krohn, "Desistance;" Bushway et al., "Empirical Framework;" and Maruna, *Making Good,* who share this view of desistance as a process.
138. Loeber and LeBlanc, "Toward a Developmental Criminology," 452.
139. Laub and Sampson, "Understanding Desistance," 5–6.
140. Uggen and Piliavin, "Asymmetrical Causation."
141. Laub and Sampson, "Understanding Desistance," 3.
142. Sampson and Laub, *Crime in the Making.*
143. Ibid., 21.

Chapter Resources

144. Laub and Sampson, "Understanding Desistance," 19–20, summarizing the findings of Sampson and Laub, *Crime in the Making.* See also Sampson and Laub, "Life-Course Desisters?"
145. See, for example, Horney, Osgood, and Marshall, "Criminal Careers;" and Farrington and West, "Effects of Marriage."
146. Laub, Nagin, and Sampson, "Trajectories of Change."
147. See also Giordano, Cernkovich, and Holland, who in "Changes in Friendship Relations" consider changes in the influence of friends over the life course, and how these changes contribute to the desistance process.
148. Warr, "Life-Course Transitions," 183.
149. Uggen, "Work."
150. Ibid.
151. Pezzin, "Earning Prospects."
152. Shover and Thompson, "Age."
153. See also Shover, *Great Pretenders.*
154. Maruna, *Making Good.* See also Gadd and Farrall, "Criminal Careers."
155. Thornberry, "Introduction," 2–5.
156. Ibid., 2.
157. Thornberry, "Introduction," 4.
158. Ibid.
159. Ibid.
160. Ibid.
161. Ibid.
162. Ibid., 5.
163. Loeber, "Developmental Continuity," 2.
164. Frick et al., "Oppositional Defiant Disorder." This study built on an earlier meta-analysis by Loeber and Schmaling, "Empirical Evidence," which revealed an overt–covert dimension to child antisocial behavior.
165. Frick et al., "Oppositional Defiant Disorder."
166. Kelley et al., "Developmental Pathways," 5.
167. Ibid., 5.
168. Frick et al., "Oppositional Defiant Disorder."
169. Kelley et al., "Developmental Pathways," 5.
170. Ibid.
171. Loeber, "Developmental Continuity," 14.
172. Ibid.
173. Kelley et al., "Developmental Pathways," 8.
174. Loeber et al., "Developmental Pathways."
175. Kelley et al., "Developmental Pathways," 8.
176. Loeber et al., "Developmental Pathways," 128–129; and Kelley et al., "Developmental Pathways," 9.
177. Loeber et al., "Developmental Pathways," 128.
178. Ibid.
179. Kelley et al., "Developmental Pathways," 13.
180. Loeber et al., "Developmental Pathways," 129; and Loeber, "Developmental Continuity," 15–16.
181. Loeber, Keenan, and Zhang, "Boys' Experimentation." See also Kelley et al., "Developmental Pathways," 9–10; and Loeber, "Developmental Continuity," 16.
182. Kelley et al., "Developmental Pathways," 9. See also Loeber, Keenan, and Zhang, "Boys' Experimentation."
183. Kelley et al., "Developmental Pathways," 10. See also Loeber, Keenan, and Zhang, "Boys' Experimentation."
184. Kelley et al., "Developmental Pathways," 10.
185. Ibid., 10–12.
186. See, for example, Loeber et al., "Developmental Pathways;" Loeber, "Developmental Continuity;" and Loeber, Keenan, and Zhang, "Boys' Experimentation."
187. Tolan and Gorman-Smith, "Development."
188. Ibid., 80.
189. Ibid.

190. Ibid., 81.
191. Ibid.
192. Ibid.
193. Ibid., 82.
194. Patterson and Yoerger, "Developmental Models," 143.
195. Ibid.
196. Ibid., 140.
197. Ibid.
198. Patterson, DeBaryshe, and Ramsey, "Developmental Perspective;" and Patterson and Yoerger, "Developmental Models."
199. Patterson and Yoerger, "Developmental Models," 140.
200. Ibid.
201. Patterson, DeBaryshe, and Ramsey, "Developmental Perspective," 329. See also Wiesner, Capaldi, and Patterson, "Development of Antisocial Behavior," who present an overview of Patterson's developmental approach.
202. Patterson and Yoerger, "Developmental Models."
203. Ibid., 141. Emphasis in original.
204. Patterson, "Maternal Rejection," cited in Patterson and Yoerger, "Developmental Models," 144.
205. Patterson and Yoerger, "Developmental Models," 142.
206. Ibid., 145.
207. Patterson and Bank, "Some Amplifying Mechanisms," cited in Patterson and Yoerger, "Developmental Models," 142.
208. Patterson, DeBaryshe, and Ramsey, "Developmental Perspective;" and Patterson and Yoerger, "Developmental Models."
209. Patterson and Yoerger, "Developmental Models," 145.
210. Ibid.
211. Ibid., 147.
212. Ibid.
213. Patterson, Crosby, and Vuchinich, "Predicting Risk," cited in Patterson and Yoerger, "Developmental Models," 150–151.
214. Patterson and Yoerger, "Developmental Models," 153.
215. Ibid., 163.
216. Patterson and Yoerger, "Developmental Models," 162–163. See also Patterson, DeBaryshe, and Ramsey, "Developmental Perspective," and Patterson, Capaldi, and Bank, "Early Starter Model."
217. Patterson and Yoerger, "Developmental Models," 164. See also Patterson, DeBaryshe, and Ramsey, "Developmental Perspective."
218. Patterson and Yoerger, "Developmental Models," 164.
219. Ibid., 166.
220. Patterson, DeBaryshe, and Ramsey, "Developmental Perspective," 331; and Patterson and Yoerger, "Developmental Models," 163.
221. Patterson and Yoerger, "Developmental Models," 159.
222. Simons, Wu, Conger, and Lorenz, "Two Routes to Delinquency."
223. Simons et al., "Two Routes to Delinquency," 269. Early starters were defined as those who reported committing two or more serious delinquent acts by the third wave of data collection, when the average respondent age was 14.
224. Simons et al., "Two Routes to Delinquency," 268. Late starters were defined as those who reported committing fewer than two serious delinquent acts during the first three waves of data collection.
225. Moffitt, "Developmental Taxonomy," and "Complementary Pair."
226. Moffitt, "Developmental Taxonomy," 674.
227. Ibid., 676.
228. Ibid.
229. Ibid., 679.
230. Ibid.
231. Ibid., 681.
232. Ibid., 680.
233. Moffitt, "Developmental Taxonomy," 680. See Moffitt, "Neuropsychology of Delinquency."
234. Moffitt, "Developmental Taxonomy," 680–681.

Chapter Resources

235. Ibid., 682.
236. Moffitt, "Developmental Taxonomy," 682. See also Caspi, Elder, and Bem, "Moving Against the World."
237. Moffitt, "Developmental Taxonomy," 683.
238. Ibid.
239. Ibid., 684.
240. Ibid., 685.
241. Ibid., 686.
242. Ibid.
243. Ibid., 686–687.
244. Ibid.
245. Ibid., 689–690.
246. Ibid., 689.
247. Ibid., 690. Emphasis in original.
248. Ibid.
249. Ibid., 691.
250. Bartusch et al., "Is Age Important?" Other studies have also tested developmental theories of crime and delinquency against Gottfredson and Hirschi's general theory. Simons et al. (1998) found support for developmental theories. But Paternoster and Brame, in "Multiple Routes," found more mixed results, and concluded, "We find that the evidence is not faithful to either a pure static/general model or a pure developmental model of crime" (49).
251. Gottfredson and Hirschi, *General Theory of Crime*.
252. Moffitt, "Developmental Taxonomy," and "Complementary Pair."
253. Bartusch et al., "Is Age Important?" 19–21.
254. Bartusch et al., "Is Age Important?" 27. See Fergusson, Horwood, and Nagin, "Offending Trajectories," who also found empirical support for Moffitt's developmental approach.
255. Nagin, Farrington, and Moffitt, "Trajectories." These four groups were first identified by Nagin and Land in "Age."
256. Nagin, Farrington, and Moffitt, "Trajectories."
257. Ibid., 132.

Explaining Delinquent Behavior

Classical and Positivist Criminology

7

Chapter Objectives

After completing this chapter, students should be able to:

- Identify and describe the key central ideas of classical and positivist criminology.
- More fully assess the degree to which punishment deters crime.
- Discuss the factors that enter into offending decisions.
- Provide an informed argument as to whether delinquent behavior is chosen or determined.
- Understand key theories and terms:

  Theories:

  classical criminology

  deterrence

  perceptual theory of deterrence

  rational choice

  criminal propensity

  positivist criminology

  neoclassical criminology

  drift

Terms:
the Enlightenment
will
free will
hedonism
social contract
utilitarianism
humanitarianism
legal rationality
statutory law
deterrent effect
rational choice
criminal propensity
positivism
determinism
differentiation
individual pathologies
social pathologies
individualized treatment
hard determinism
soft determinism
drift

CASE IN POINT

Stealing as Rational Choice

Clifford Shaw's classic case study of Stanley provides a first person account of a delinquent career. Growing up in poverty, Stanley's family resorted to theft to obtain food—theft was a rational response to their need. Though Stanley committed the theft, his stepmother told him to steal and his stepbrother, William, showed him how, planned, and coordinated the crime. It is apparent from Stanley's account that he gained far more than just food from the crime: he was proud of his skills and accomplishments, stealing allowed him to maintain his relationship with his stepmother, it was exciting, and it gave him status among his peers. Thus, stealing and other forms of delinquent behavior may involve elements of rational choice.

One day my stepmother told William to take me to the railroad yard to break into boxcars. William always led the way and made the plans. He would open the cars, and I would crawl in and hand out the merchandise. In the cars were foodstuffs, exactly the things my stepmother wanted. We filled our cart, which we had made for this purpose, and proceeded toward home. After we arrived home with our ill-gotten goods, my stepmother would meet us and pat me on the back and say that I was a good boy and that I would be rewarded. Rewarded, bah! Rewarded with kicks and cuffs.

After a year of breaking into box-cars and stealing from stores, my stepmother realized that she could send me to the market to steal vegetables for her. My stealing had proved to be very profitable to her, so why not make it even more profitable? I know it was for my own good to do what she wanted me to do. I was so afraid of her that I couldn't do anything but obey. Anyway, I didn't mind stealing, because William always went with me, and that made me feel proud of myself, and it gave me a chance to get away from home.

Every Saturday morning we would get up about three o'clock and prepare for the venture. William, Tony, and his two sisters and I would always go. We would board a street car, and the people on the car would always stare at us and wonder where such little kids were going so early in the morning. I liked to attract the attention of the people and have them look down upon me with curiosity. The idea of my riding a street car at that early hour appealed to my adventurous spirit and keyed me up to stealing. In the street car, William would give orders on what to steal and how to go about it. I listened to him with interest and always carried out his orders. He had me in the palm of his hand, so to speak. He got the satisfaction of ordering me, and I got the thrill of doing the stealing. He instructed me on how to evade peddlers and merchants if they gazed at me while I was stealing. After arriving at the market, William would lay out the plan of action and stand guard while I did the stealing. He knew what the stepmother wanted, and he always filled her orders to overflowing. All in all, I was a rather conceited little boy who thought himself superior to the other boys of his age; and I didn't miss impressing that little thing upon their minds. I was so little that the peddlers were not suspicious of me, and it didn't take long to fill our baskets and be ready for the journey home. All spring, summer, and fall did we go to the market, and never did I get caught and never did we go home with empty baskets.

Source: Shaw, *The Jack-Roller*, 52–54. Copyright 1930, University of Chicago Press. All rights reserved. Reproduced with permission of the publisher.

Some delinquent acts involve careful planning and calculated decision making. Potential benefits are weighed against possible costs—the chance of getting caught and punished, of damaging valued relationships, or of experiencing feelings of guilt. Do juveniles make a conscious, rational choice to get involved in delinquent acts, or is delinquent behavior a product of their backgrounds and experiences, over which they have little control? This fundamental question has long puzzled criminologists. The two sides of this question represent two primary schools of thought in criminology: *classical criminology* and *positivist criminology*. Each advances very different perspectives on the causes and control of juvenile delinquency. This chapter describes the central ideas that define each of these perspectives. We will also consider a contemporary version of classical thought, rational choice theory. In addition, we will examine the underlying debate that sharply divides classical and positivist thought: *Is delinquent behavior chosen or determined?*

■ Classical Criminology

Cesare Beccaria: *On Crimes and Punishments*

In 1764, at the age of twenty-six, Cesare Beccaria (1738–1794) published anonymously a short book entitled *On Crimes and Punishments*.[1] While its proposals about crime and justice were new and revolutionary, they clearly reflected ideas at the core of Enlightenment philosophy.[2] Beccaria's work was read widely and captured the attention of philosophers and "a large cross section of educated society," including religious, political, and governmental leaders.[3] In colonial America, for example, Thomas Jefferson and John Adams were well acquainted with Beccaria's work.[4] His ideas enthralled some people and appalled others. By 1800 the treatise had been published in twenty-three Italian editions, fourteen French editions, and eleven English editions—three of which were printed in the United States.[5]

On Crimes and Punishments proved to be tremendously influential both philosophically and politically, and it has been identified as the driving force behind a variety of reforms in criminal law and justice in the nineteenth-century Western world. One hundred years after the publication of this work, one of the founders of the sociology of law, Emile Durkheim, proclaimed that "it is incontestably the case that it was . . . *On Crimes and Punishments* which delivered the mortal blow to the old and hateful routines of the criminal law."[6]

Beccaria's work is an excursion into social thought of **the Enlightenment**, a period of active thought and action from the mid-seventeenth century to the last quarter of the eighteenth century. The Enlightenment was characterized by the prolific expression of ideas that were "enlightened" by reason, science, and a respect for humanity. Enlightenment thought dug deep into religion, science, politics, government, and economics—few areas of life were outside its purview.[7] Existing institutional arrangements were questioned, especially the concentration of wealth and power, and it was argued that improvement in the human condition was only possible through the application of new and diverse ideas, gained through experience and observation and guided by reason. Thus, the Enlightenment was reform-minded.

Beccaria was influenced greatly by the humanism of French Enlightenment thinkers and the scientific rationalism of English and Scottish Enlightenment writers.[8] Consequently, *On Crimes and Punishments* is most often seen as an essay that argued for the development of humane and rational legal systems at a time when these systems were incredibly brutal, repressive, and arbitrary. In fact, the tremendous influence of this work is often stated in terms of its practical application for legal reform.[9] The reforms that Beccaria advocated, however, did not

the Enlightenment A period of active thought and action from the mid-seventeenth century to the last quarter of the eighteenth century, "enlightened" by reason, science, and a respect for humanity.

deal with technical legal procedures—of which he apparently knew little—but with broader social and philosophic views toward crimes and punishments.[10]

Since *On Crimes and Punishments* is foundational to classical criminology, we will use passages from this work to identify four key components: will, utilitarianism, humanitarianism, and legal rationality. These elements cannot be derived with absolute certainty, since the essay is broad and sweeping, making it difficult to characterize and summarize. Also, Beccaria's work is not noted for its logical development, consistency, or clarity of expression.[11] In fact, the translator of the first French edition actually rearranged sentences, paragraphs, even entire sections, in order to make the argument develop more logically—an action of which Beccaria apparently did not disapprove (though he never adopted these revisions in subsequent editions for which he was responsible).[12]

Will

Drawing from Enlightenment thought, Beccaria emphasized that individuals naturally pursue their own interests. In fact, Beccaria's views on crimes and punishments build on the assumption that humans have a natural tendency for self-interest. Early in the essay he states: "No man ever freely gives up a part of his own liberty for the sake of the public good; such an illusion exists only in romances. If it were possible, each one of us would wish that the agreements binding on others were not binding on himself. Every man thinks of himself as the center of all the world's affairs."[13]

The classical notion of self-interest is usually referred to as **free will** and **hedonism**, in which individuals freely choose action based on a rational consideration of gains and losses, pleasure and pain, benefits and costs. However, Piers Beirne has argued that Beccaria's concept of **will** is not composed of freedom of choice and hedonism alone. Rather, will also involves forces that directly influence human thought, motivation, and action, including passion, sensations, individual temperament, ignorance, and characteristics of the situation.[14] Thus, the will to act includes both calculated choice ("free will") and nonrational, deterministic influences ("determined will").[15] Upon a close reading of Beccaria's work, Beirne contends that the interpretations of classical thought that are usually offered overemphasize the role of rational thought and fail to acknowledge these deterministic forces that are woven more subtly through *On Crimes and Punishments*. Nonetheless, the classical notion of will points to a volitional element in human action, in which action is taken based on the consideration of self-interest. Rarely is this judgment perfectly logical and rational; rather, the decision to act is influenced by a person's perceptions, feelings, and surroundings, including opportunity and peer pressure.

Utilitarianism

Beccaria's views on crime and justice are founded on a concept central to much Enlightenment thought: the **social contract**.[16] In fact, some philosophers of this era, including Hobbes, Locke, Montesquieu, Voltaire, and Rousseau, are sometimes referred to as "social contract thinkers."[17] "Social contract" refers to the mutual agreement among individuals in a community to relinquish a portion of their individual freedom and self-interest in order to promote interpersonal peace, order, and stability. It is a "peace treaty" of sorts in response to the natural inclination of individuals to pursue self-interest (will)—a condition that

free will The freedom to choose action.

hedonism Behavior in which a choice of action involves a rational consideration of gains and loses, pleasure and pain, benefits and costs.

will The volitional part of human conduct. Will involves rational choice but is also influenced by forces beyond the control of the individual, including passion, sensation, individual temperament, ignorance, and characteristics of the situation.

social contract An Enlightenment concept that refers to the mutual agreement among individuals in a political community to relinquish a portion of their individual freedom and self-interest in order to promote interpersonal peace, order, and stability.

Thomas Hobbes called the "war of all against all."[18] Near the beginning of his treatise, Beccaria provides a summary of the social contract, which is foundational to many ideas that follow.

> *The Origin of Punishments and The Right to Punish. Laws are the conditions by which independent and isolated men, tired of living in a constant state of war and enjoying a freedom made useless by the uncertainty of keeping it, unite in society. . . .*
>
> *It was necessity, then, that constrained men to give up part of their personal liberty; hence, it is certain that each man wanted to put only the least possible portion into the public deposit, only as much as necessary to induce others to defend it. The aggregate of these smallest possible portions of individual liberty constitutes the right to punish; everything beyond that is an abuse and not justice, a fact but scarcely a right.*[19]

utilitarianism The goal of the social contract is to promote *"the greatest happiness shared by the greatest number of people"* (Beccaria, *On Crimes and Punishments* [Paolucci], 8).

The goal of the social contract is to promote *"the greatest happiness shared by the greatest number"* of people, or what is referred to as **utilitarianism**.[20] With this utilitarian goal, Beccaria argued that crimes can be defined and classified in terms of the harm done to society.

> *The Measure of Crimes. We have seen what the true measure of crimes is—namely, the harm done to society. This is one of those palpable truths which, though requiring neither quadrants nor telescopes for their discovery, and lying well within the capacity of any ordinary intellect, are, nevertheless, because of a marvelous combination of circumstances, known with clarity and precision only by some few thinking men in every nation and in every age.*[21]

Because people have a natural inclination to pursue self-interest and are unwilling to participate in the collective good (utilitarianism), society must have a mechanism by which the social contract can be enforced.[22] Following the ideas of Enlightenment philosopher Thomas Hobbes, Beccaria believed that the proper role of government was to promote *"the greatest happiness,"* but to do so in a way that poses the least possible limits on individual freedom and self-interest. Given this utilitarian goal of the social contract, the state has the obligation, responsibility, and authority to define crime and respond to it in a humane and rational manner. Most of the legal reforms offered in *On Crimes and Punishments* follow from this logic and provide Beccaria's thoughts about how crimes should be classified in terms of seriousness and harm, and how systems of justice should respond to crime, especially the "utility" of punishment to deter crime.

Humanitarianism

humanitarianism A concern for the welfare of humanity. Beccaria's plea for legal reform was in tune with French humanists who sought better conditions for people.

On Crimes and Punishments deals mainly with what law and justice ought to be, instead of what law is—it is an "impassioned plea" for legal reform.[23] This emphasis on legal reform is often cited as the book's primary purpose.[24] Even critics who disagree with Beccaria's specific proposals for reform endorse his **humanitarian** mission.[25] Beccaria's arguments for legal reform are extremely wide-ranging, covering the need for codified and public laws, prohibition on the use of judicial torture to illicit confessions, the need for public trials, the use of witnesses and evidence, the role of jurors, and sentencing practices including imprisonment and the death penalty.[26]

In tune with Enlightenment thought, especially that of French humanists, Beccaria addressed the cruelty and brutality of European systems of justice. In this regard, his observations were directed primarily at the practices of judicial torture (involving coercive interrogation of the accused to induce confession) and the use of capital punishment.[27] The treatise also speaks generally of the need for humane punishment and the need to pursue the ultimate goal of preventing crime. This humanitarian theme runs throughout the book. As David Young summarizes: "Beccaria maintained that a mild legal system is both useful, in that it is likely to win widespread approval, and just, in that it is most in accord with basic human rights."[28] Beccaria states this most directly in a chapter entitled: "Mildness of Punishment."

> *Mildness of Punishment. . . . For a punishment to attain its end, the evil which it inflicts has only to exceed the advantage derived from the crime; in this excess of evil one should include the certainty of punishment and the loss of the good which the crime might have produced. . . .*
>
> *The severity of punishment of itself emboldens men to commit the very wrongs it is suppose to prevent; they are driven to commit additional crimes to avoid the punishments for a single one.*[29]

This humanitarian theme is also addressed near the very end of the book, where Beccaria argues that it is better to prevent crimes than to punish them.

> *How to Prevent Crimes. It is better to prevent crimes than to punish them. . . . Do you want to prevent crimes? See to it that the laws are clear and simple, that the entire strength of the nation is concentrated in their defense, and that no portion of that strength is employed in their destruction. See to it that the laws favor classes of men less than it favors men themselves. See to it that men fear the laws and only the laws. The fear of the laws is salutary, but the fear of one man for another is a fertile source of crimes.*[30]

Legal Rationality

Piers Beirne contends that there are two major themes in *On Crimes and Punishments:* the "right to punishment," that draws from utilitarianism and the social contract, and "how to punish."[31] Beccaria devoted much attention to "how to punish" not only in an attempt to encourage more humane legal procedures and outcomes, but also to promote greater **legal rationality**. By this he meant legal systems that are founded on **statutory law**, in which law is produced from a legislative process and is codified—written down in a systematic manner. Statutory law should define crime, specify an impartial and efficient judicial process, and stipulate punishment that is "measured" so that it deters future criminal acts. In this way, legal rationality fulfills the goal of utilitarianism: "the true foundation of the happiness I mentioned are security and freedom limited only by law."[32] This "rule of law" forces governments to operate in a rational and reasonable manner, especially when limiting individual freedoms in an effort to bring about the collective good.

legal rationality The view that statutory law should be used to define crime, require an impartial and efficient judicial process, and specify punishment that is proportionate to the crime and deters future criminal acts.

statutory law Law that is enacted by a legislative process and codified (written down in a systematic manner).

According to Beccaria, law must be codified, public, and predictable. In his words, the law must be a "fixed legal code" that is clear and understandable.

> *Interpretation of the Law.* . . . *When a fixed legal code that must be observed to the letter leaves the judge no other task than to examine a citizen's actions and to determine whether or not they conform to the written law, when the standard of justice and injustice that must guide the actions of the ignorant as well as the philosophic citizen is not a matter of controversy but of fact, then subjects are not exposed to the petty tyrannies of many men. . . . With fixed immutable laws, then, citizens acquire personal security. This is just because it is the goal of society, and it is useful because it enables them to calculate precisely the ill consequences of a misdeed.*[33]

> *Obscurity of Laws. If the interpretation of laws is an evil, their obscurity, which necessarily entails interpretation, is obviously another evil, one that will be all the greater if the laws are written in a language that is foreign to the common people. This places them at the mercy of a handful of men, for they cannot judge for themselves the prospect of their own liberty. . . . The greater the number of people who understand the sacred law code and who have it in their hands, the less frequent crimes will be, for there is no doubt that ignorance and uncertainty concerning punishments aid the eloquence of the passions.*[34]

The government's "right to punish," derived from the social contract and utilitarianism, requires that punishments be specified in the law. Beccaria points to this authority when discussing legal consequences.

> *Consequences. The first consequence of these principles is that only the law may decree punishments for crimes, and this authority can rest only with the legislator, who represents all of society united by a social contract.*[35]

The state's authority to punish, established in law, must be carried out thoughtfully, with reason and purpose. The primary purpose of punishment is to deter individuals and the general public from committing crime.

> *Purpose of Punishments. The purpose of punishment, then, is nothing other than to dissuade the criminal from doing fresh harm to his compatriots and to keep other people from doing the same. Therefore, punishments and the method of inflicting them should be chosen that, mindful of the proportion between crime and punishment, will make the most effective and lasting impression on men's minds and inflict the least torment on the body of the criminal.*[36]

Legal rationality also encompasses Beccaria's most famous ideas: in order to deter crime, punishment must be prompt, certain, and proportionate to the crime. In fact, much of the contemporary appeal of classical thought is derived from its emphasis on punishment. At the same time, Beccaria believed strongly that punishment must be humane. Just punishment requires only the minimum amount necessary to be proportionate to the crime and to deter future criminal acts.

> *Promptness of Punishment. The more prompt the punishment is and the sooner it follows the crime, the more just and useful it be. . . .*
>
> *I have said that the promptness of punishment is more useful, for the less time that passes between the misdeed and it chastisement, the stronger and more permanent is the human mind's association of the two ideas of crime and punishment, so that imperceptibly the one will come to be considered as the cause and the other as the necessary and inevitable result. It is well established that the association of ideas is the cement that shapes the whole structure of the human intellect; without it, pleasure and pain would be isolated feelings with no consequences. . . .*
>
> *The temporal proximity of crime and punishment, then, is of the utmost importance if one desires to arouse in crude and uneducated mined the idea of punishment in association with the seductive image of a certain advantageous crime.*[37]
>
> *The Certainty of Punishment. Mercy. One of the greatest curbs on crimes is not the cruelty of punishments, but their infallibility, and, consequently, the vigilance of magistrates, and the severity of an inexorable judge which, to be a useful virtue, must be accompanied by a mild legislation. The certainty of a punishment, even if it be moderate, will always make a stronger impression than the fear of another which is more terrible but combined with the hope of impunity. . . .*[38]
>
> *Proportion between Crimes and Punishments. Not merely is it in the common interest that crimes not be committed, but that they be more infrequent in proportion to the harm they cause society. Therefore, the obstacles that restrain men from committing crimes should be stronger according to the degree that such misdeeds are contrary to the public good and according to the motives which lead people to crimes. Thus, there must be a proportion between crimes and punishments.*[39]

Beccaria provides a short and to-the-point summary of the extended argument contained in *On Crimes and Punishments*. A single sentence, although lengthy, picks up the major themes of the book and casts them as goals to be pursued in order to bring about a just and rational legal system.

> *Conclusion. From all that has been seen hitherto, one can deduce a very useful theorem, but one that scarcely conforms to custom, the usual lawgiver of nations. It is this: In order that any punishment should not be an act of violence committed by one person or many against a private citizen; it is essential that it should be public, prompt, necessary, the minimum possible under the given circumstances, proportionate to the crimes, and established by law.*[40]

EXPANDING IDEAS — Understanding Classical Thought

See if you can describe classical criminology by summarizing, in your own words, the four key elements of Beccaria's work, *On Crimes and Punishments*. Classical criminology is often simplified to make it more understandable, but much of the appeal of this school of thought comes from its philosophical arguments—arguments that are not easily captured in a brief portrayal.

1. *Will:* including both "free will" and "determined will."
2. *Utilitarianism:* including the "social contract" and the "measure of crimes."
3. *Humanitarianism:* the need for "mildness of punishment" in order to "prevent crimes."
4. *Legal rationality:* a "fixed legal code" that is clear and understandable, with "consequences" specified in law. Legal punishment must be "prompt," "certain," and "proportionate to the crime."

Jeremy Bentham: The Utility of Punishment

Jeremy Bentham (1748–1832), a British contemporary of Beccaria, was apparently enthralled with Beccaria's single important work, *On Crimes and Punishments*. He wrote: "Oh, my master, first evangelist of Reason . . . you who have made so many useful excursions into the path of utility, what is there left for us to do?—Never to turn aside from that path."[41] Ironically, Bentham was a stark contrast to Beccaria. Criminologist Gilbert Geis described Bentham as an eccentric personality, an incredibly prolific writer, and a bold thinker who had the audacity to attempt to catalogue and label all varieties of human behavior and their motivations.[42] Captivated by Beccaria's efforts at legal reform, Bentham produced an huge collection of published and unpublished writings suggesting reform of English criminal law.[43]

Coleman Phillipson characterized the criminal codes of Bentham's time as "a mass of incongruities, absurdities, contradictions, and barbarities."[44] In response to these problems, Bentham emphasized and popularized Beccaria's basic tenets of utilitarianism and legal rationality. However, Bentham went further than Beccaria in trying to provide a moral and philosophical basis to legal reform, turning extensively to the work of Scottish philosopher David Hume (1711–1776).[45] According to Bentham, the *"utility of actions"* can be evaluated on the basis of whether they "produce benefit, advantage, pleasure, good, or happiness" or to the degree they "prevent the happening of mischief, pain, evil, or unhappiness."[46] Ultimately, the "goodness" or "badness" of an act must be judged in terms of the best interests of society as a whole—its utilitarian value.[47] This ethical principle of utilitarianism is the cornerstone of Bentham's writings.[48]

Bentham believed that the pleasure–pain ratio for different acts could be measured precisely through mathematical formulas that he called *"felicity cal-*

culus." However, felicity calculus was mathematical in name only. Gilbert Geis observes that Bentham's presentation of felicity calculus almost always deteriorated into "long-winded attempts" to explain verbally the pleasure–pain ratio for various offenses. Bentham never did calculate the felicity calculus for different offenses, nor did he consider the implications for such quantification.[49]

As Beccaria did, Bentham contended that people have a natural tendency to pursue self-interest, after deliberate and rational consideration of potential gains and losses.[50] "Men calculate," he stated, "some with less exactness, indeed, some with more; but all men calculate."[51] As a result, criminal law should set penalties that deter crime, but do so in the least punitive manner possible. Bentham, like Beccaria, saw the primary purpose of punishment as deterrence, not vengeance.[52]

The *utility of punishment,* according to Bentham, is to deter crime. It is the responsibility of the legislature to establish punishments that are specified in law. These legal sanctions must be considered carefully: sufficient to deter crime, but never excessive or unnecessary. Punishment should be just enough so that the pain outweighs the pleasure derived from committing crime, and the "evil of the punishment" should never exceed the evil of the offense.[53] In addition, punishment should never be used when deterrence can be achieved through less painful means, such as education.[54]

Application of Classical Criminology

Both Beccaria and Bentham provided wide-ranging and far-reaching arguments for rational and humane systems of law and justice.[55] As we have seen, their arguments for legal reform included the need for codified laws that are produced from a legislative process and then made public, the need for changes in judicial procedures, and the utility of "measured" punishment as a deterrent to crime.[56] The influence of classical thought was swift, extensive, and widespread, providing a philosophical basis for a variety of legal reforms in England, Western Europe, and the United States.[57] However, the arguments for legal reforms offered by classical thought were more philosophical than specific or practical.[58] Beccaria, for example, had little to say about the substance of law or the procedures to be followed in enforcing it.

Nonetheless, probably no area of legal reform was more influenced by classical criminology than was the development of statutory law in the form of legal codes. It is commonly noted that the writers of the US Constitution, as well as those of the French Code of 1791, paid close attention to the ideas that Beccaria expressed in *On Crimes and Punishments.*[59] A number of countries in Western Europe and each of the United States developed legal codes through legislative procedures. These statutory laws were codified—written down in a systematic fashion—and, as Beccaria advocated, they were made public. The substance of these statutory laws included definitions of crimes, legal procedures to be followed in determining whether the law had been violated, and sanctions that could be imposed for particular crimes. Thus, the widespread development of statutory law was derived from classical criminology's emphasis on legal rationality—the rationalization of law and the administration of justice.

Classical criminology also provides philosophical justification for the use of punishment to control crime.[60] Holding firmly to the basic tenets of classical thought, the deterrent effect of legal punishment became one of the corner-stones of modern systems of justice. The classical notion of utilitarianism provided justification for the state's "right to punish" under the pretense that the "greatest happiness" is the primary concern of government.[61] In addition, while classical humanitarianism contends that punishment should be "mild" (the "least possible"), legal rationality adds that in order for punishment to deter crime, it must be prompt, certain and proportionate to the crime.[62]

The deterrent effect of legal punishment dominates both the contemporary juvenile and adult systems of justice in the United States. Legal statutes and sanctions promote punishment as their primary, sometimes solitary, purpose. Classical criminology's stipulation that punishment be mild and proportionate to the crime seems lost in contemporary applications. As described in Chapter 2, the contemporary transformation of the juvenile justice system has made punishment a major goal of a system that was founded on the rehabilitative ideal.[63]

Punishment Deters Crime: Deterrence Theory

The resurrection of classical thought, with its emphasis on the deterrent effect of punishment, has been the subject of much debate in contemporary criminology. Classical criminology holds that in order to prevent crime, punishment must be prompt, certain, and proportionate to the crime.[64] The contemporary consideration of this classical notion is referred to as *deterrence theory*. Put simply, deterrence theory claims that punishment prevents crime when the cost of legal punishment outweighs the benefit of committing crime. This **deterrent effect** of punishment assumes that "human beings are both rational and self-interested beings."[65]

In the late 1960s and throughout the 1970s, deterrence research attempted to develop measures of the properties of legal penalties that have a deterrent effect on crime.[66] Certainty and severity of punishment were given extensive attention because of their central role in classical thought. Using aggregate data (characteristics of groups of people, not individuals), *certainty of punishment* was measured by state arrest and imprisonment rates for specific offenses. Similarly, *severity of punishment* was measured by the average length of incarceration served for the same offenses. These measures of certainty and severity were then compared (correlated) with officially recorded crime rates, obtained from the Uniform Crime Reporting System (UCR). Deterrence theory predicts an inverse relationship between these measures of legal punishment and crime rates: when levels of punishment certainty and severity are high, official crime rates are low. Research using this approach indicates that the deterrent effect of punishment generally holds true for certainty of punishment but that the relationship is modest at best. The findings regarding severity of punishment are far less supportive of deterrence theory, indicating consistently that level of severity was unrelated to crime rates, except in the case of homicide.[67]

Findings that punishment has little or no deterrent effect did not daunt researchers. Only a few years after deterrence research began, a new school of thought emerged, which claimed that the deterrent effect of punishment is not the result of actual levels of punishment, as most people are unaware of the real possibilities of punishment. Rather, punishment deters crime to the de-

deterrent effect The prevention of crime through the use of punishment, based on the assumption that "human beings are both rational and self-interested beings" (Paternoster and Bachman, "Introduction," 14). Beccaria held that in order to prevent crime, punishment must be prompt, certain, and proportionate to the crime.

gree that individuals *think* that it is certain and severe. The deterrence logic here is consistent with Beccaria's observations that crime is prevented to the degree that individuals fear punishment and that fear comes from their belief about the risk of being punished.[68] One of the originators of this point of view, Jack Gibbs, states: "The decisive factor in creating the deterrent effect is, of course, not the objective risk of detection but the risk as it is calculated by the potential criminal."[69] This approach to deterrence is referred to as the *perceptual theory of deterrence*.[70]

Since the first study of perceptual deterrence was published in 1972, a large number of studies have followed.[71] These studies attempted to test the deterrent effect of perceived certainty (likelihood) and severity (amount) of punishment.[72] Consistent with deterrence theory, research findings showed that perceived certainty of punishment was related inversely to various forms of self-reported misconduct: the greater the perceived risk of punishment, the less likely individuals are to report involvement in misconduct. However, the statistical association between perceived risk and involvement in crime found in these studies was only moderately strong, indicating that perceived certainty has only a modest deterrent effect on crime.

Perceived severity of punishment has received far less attention than certainty of punishment. The research that was conducted failed to provide consistent findings, and those few studies that found some deterrent effect of perceived severity found the relationship to be weak. Criminologist Raymond Paternoster, who thoroughly reviewed this research, concluded that "perceived severity plays virtually no role in explaining deviant/criminal conduct."[73]

Research on the deterrent effect of perceived certainty and severity of punishment is also plagued with significant methodological problems. The data and research methods employed in much of this research failed to allow temporal order to be established. When perceptions of risk (certainty and severity) and self-reported misconduct are measured at the same point in time, there is no way to establish which occurs first: perceptions or misconduct.[74] In addition, these studies did not consider other factors that might provide alternative explanations for involvement in crime.[75] When researchers studied perception of punishment with research methods that allowed for the consideration of additional causal variables, they found that perceptions of risk had only a weak deterrent effect. Relative to other variables, certainty of punishment did not have a consistent or strong deterrent effect on involvement in deviance.[76] These limitations make it difficult to establish the degree to which perceptions of risk influence involvement in crime and delinquency relative to other causal factors.

Deterrence theory and research has concentrated on the restraining effect of legal punishments either in terms of perceived risk or actual experience.[77] Some deterrence scholars have argued for a broader consideration of the deterrent effect of punishment to include not only legal costs, but also self-imposed shame and the social costs associated with embarrassment, disapproval from significant others, and social ostracism.[78] Additionally, some have argued that the classical notion of self-interest and hedonism involves consideration of not just costs, but also benefits, and that costs and benefits are both legal and social.[79] This expanded view of classical thought in contemporary criminology is taken up in *rational choice theory*.

Choosing Delinquency: Rational Choice Theory

Classical criminology holds that delinquent youth act deliberately and are motivated by self-interest. Delinquent acts result when the potential for personal gain is greater than the probable cost. However, few contemporary criminologists argue that teenagers carefully consider *all* the benefits and costs of delinquency—teenagers simply are not that rational and thoughtful, nor do they have full and accurate information.[80] Nonetheless, delinquent behavior involves at least some "measure of rationality."[81] In contemporary criminology, the rational element in delinquent behavior is referred to as **rational choice**.

> **rational choice** The conscious, calculated, purposive choice of action. The rational element in delinquent behavior.

Rational Choice Theory

Developed in the 1980s, rational choice theory draws on the economic principle of "expected utility:" people choose to engage in crime because *expected* benefits outweigh *expected* costs.[82] In advancing rational choice theory, Derek Cornish and Ronald Clarke argue that criminological theories have failed to acknowledge a rational component to crime—what they call a "measure of rationality."[83] They are careful, however, to point out that rationality is limited in that decisions to engage in crime are rarely based on full and accurate information, nor do decision makers have sufficient time, ability, and reason for judgment and choice to be completely logical.[84] Their theory, then, is an attempt to explain the influences and processes of criminal decisions.

Clarke and Cornish advance six basic propositions that summarize rational choice theory.

1. Crimes are purposive and deliberate acts, committed with the intention of benefitting the offender.

2. In seeking to benefit themselves, offenders do not always succeed in making the best decisions because of the risks and uncertainty involved.

3. Offender decision making varies considerably with the nature of crimes.

4. Decisions about becoming involved in particular kinds of crime (involvement decisions) are quite different from those relating to the commission of a specific criminal act (event decisions).

5. Involvement decisions can be divided into three stages—becoming involved for the first time (initiation), continued involvement (habituation), and ceasing to offend (desistance)—that must be separately studied because they are influenced by quite different sets of variables.

6. Event decisions include a sequence of choices made at each stage of the criminal act (e.g., preparation, target selection, commission of the act, escape, and aftermath).[85]

Following from these propositions, rational choice theory is composed of four models: initial involvement, the actual crime event, continuing involvement, and desistance. These models isolate different types of criminal decisions and the factors that influence each decision. Each of the four decision models is depicted in a separate "flow diagram" that summarizes the decision-making process.[86]

The *initial involvement model* depicts the individual's willingness or "readiness" to become involved in crime to satisfy individual needs. Needs launch the involvement decision and include desires such as money, sex, leadership, status, and excitement. The consideration of whether to engaging in crime to satisfy these needs is influenced by a variety of factors, especially the individual's background, including individual temperament, upbringing, and social and demographic characteristics (e.g., gender, social class, and neighborhood). **Figure 7-1** shows the decision sequence for initial involvement in residential burglary, beginning with background factors and moving to previous experience and current circumstances. Most factors said to influence initiation into crime were drawn from existing criminological theories, leading one criminologist to state that rational choice theory is really not all that different from these other theories.[87]

Decisions that are part of committing crime make up a second model: the *event model*. These decisions are specific to a particular crime, place, and time. Accordingly, the decision to engage in a particular crime is influenced heavily by the immediate situation—availability of goods, level of opportunity, ease of

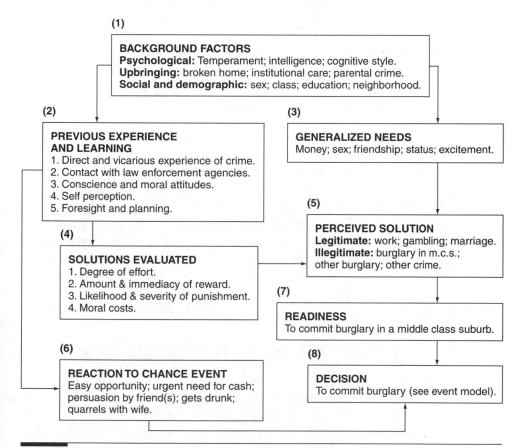

Figure 7-1 Initial involvement model of rational choice theory. Cornish and Clarke's first "flow diagram" depicts the initial involvement model for a suburban burglar. The potential offender must first be "ready and willing" for involvement in crime.

Source: Clarke and Cornish, "Modeling Offenders' Decisions," 168, Figure 1. Copyright © 1985, University of Chicago Press. All rights reserved. Reproduced with permission of the publisher.

committing the crime, amount of controls, and chance of detection. The event model suggests that the offender must be ready and willing (initial involvement model); then, if the situation is conducive to crime, the decision to commit crime follows. The implication of the event model is that crime can be prevented by limiting situations that are conducive to crime, such as controlling ease of access, increasing neighbor watch programs, and increasing police patrol.

Continuation and desistance from crime are considered in two additional models. The *continuing involvement model* includes the degree to which skills, knowledge, lifestyle, values, and peer groups either support or discourage criminal involvement. For example, becoming financially dependent on crime encourages continued involvement, as does the development of a criminal peer group. In contrast, the *desistance model* presents a series of "re-evaluations" that relate to life events such as getting married or getting a job, as well as factors more directly related to the crime event that might discourage continued involvement in crime, such as realizing that crime does not pay enough or that the income from crime is too irregular.[88]

Cornish and Clarke contend that an understanding of the rational elements of crime has far-reaching application for crime control efforts.[89] Three basic approaches to crime prevention using rational choice theory are identified in Theory into Practice, "Crime Prevention Using Rational Choice."

Rational choice theory has been most extensively applied to adult criminality. Delinquent behavior also has a rational component; however, it may be that rational choice operates differently within the adolescent population—a group not known for sophisticated decision making.[90] Drawing from the theoretical work of Cornish and Clarke, research into rational choice in delinquent behavior has focused most extensively on the considerations that enter into decisions to offend.

Choosing to Participate in Delinquent Acts

Criminologist Raymond Paternoster conducted an important study on the role of rational choice in delinquent behavior. He examined the degree to which different factors influence delinquent offending decisions.[91] Beginning with the premise that delinquent behavior is a product of "imperfectly informed choice," he argued that even if delinquency is only minimally rational, offending decisions involve far more than the simple consideration of potential costs of punishment (deterrence theory). Instead, offending decisions involve a variety of considerations including material gain, consistency with moral beliefs, and impact on social relationships.[92]

Paternoster's rational choice model includes six basic considerations ("controls") that enter into offending decisions: formal sanctions (certainty and severity of punishment), affective ties, material considerations, opportunities, informal sanctions, and moral beliefs.[93] Measures of these elements of rational choice are listed in Research in Action, "Measures of Rational Choice." The model also includes background factors of gender, household employment, and family structure. Paternoster analyzed whether these background factors influenced offending decisions directly or indirectly through the six basic considerations of rational choice.[94]

THEORY INTO **PRACTICE** **Crime Prevention Using Rational Choice**

Rational choice theory holds that "offenders benefit themselves by their criminal behavior" (Cornish and Clarke, "Introduction," 1). The decision to engage in crime involves a cost–benefit analysis. Efforts at crime prevention can incorporate this rational component of crime by reducing potential benefits and increasing costs. Ronald Clarke has applied this basic logic to crime prevention by offering a three-part framework that includes twelve different general strategies for preventing crime.

Increasing the Effort
1. **Target Hardening:** for example, door locks, window bars, steering column locks
2. **Access Control**
3. **Deflect Offenders:** provide alternatives to offending behavior such as skate parks and teen centers
4. **Control Facilitators:** situations or items that contribute to crime such as guns and alcohol

Increase the Risk
5. **Entry/Exit Screening**
6. **Formal Surveillance**
7. **Surveillance by Employees**
8. **Natural Surveillance:** conditions that aid in surveillance, such as street lights and open areas

Reducing the Rewards
9. **Target Removal:** reduce the potential reward for crime, such as limiting the amount of money a convenience store has on hand
10. **Identifying Property**
11. **Removing Inducements:** One example is to immediately clean up graffiti in order to eliminate public display.
12. **Rule Setting:** clear, public notice of rules and consequences

Source: Adapted from Clarke, "Introduction."

Four different types of delinquency were studied: marijuana use, drinking, petty theft, and vandalism. Drawing from rational choice theory, Paternoster speculated that the factors that influence offending decisions may vary depending on the type of offense being considered. Decisions regarding these types of offenses were also distinguished in terms of whether involvement was for the first time (initiation) or whether the decision was to continue or to desist involvement. Paternoster summarized his analytic approach as follows: "Looking at delinquency as involving a series of offending decisions provides an image of a much more active decision maker than presumed by most pure deterrence

Measures of Rational Choice

Raymond Paternoster argues that a variety of considerations are taken up in offending decisions. These include deterrence theory's focus on perceptions of certainty and severity of punishment, as well as material, moral, and affective considerations. His research included six factors with multiple measures. Most of the questions had response categories that ranged from "a little" to "a lot," or from "never" to "always." These response categories that lie along a continuum are called Likert items.

AFFECTIVE TIES

Attachment to parents:

If you think your father/mother would disapprove of something you wanted to do, how often would you go ahead and do it anyway?

Attachment to teachers:

In general, do you like your teachers?
Do you feel like your teachers understand you?

MATERIAL CONSIDERATIONS

Grades: self-reported grades that the respondent "typically received in school"

Conventional commitments:

How much would your chances of getting a good education be hurt if you were arrested for [each of four specific offenses] ?
How much would your chances of having good friends be hurt if you were arrested for [each of four specific offenses] ?

Importance of education:

How important is it that you get good grades in school?
How important is it that you finish high school?
How important is it that you get a college degree?

Expected education:

How likely is it that you will reach your desired educational goals?

Expected employment:

How likely is it that you will secure the job/career you are aspiring to?

OPPORTUNITIES

Parental supervision:

Do your parents know where you are when you are away from home?
Do your parents know who you are with when you are away from home?

Peer involvement and social activities: measures of proportion of friends that engaged in particular forms of delinquent behavior; amount of time spent with boyfriend or girlfriend and with friends.

INFORMAL SANCTIONS
Parental sanctions:
Questions were asked with regard to how their mother and father would react to each of four offenses.
Peer sanctions:
For each of the four offenses, questions were asked about whether respondents thought that their friends would feel it is morally wrong and would approve of participation in each of the offenses.

FORMAL SANCTIONS
Perceived certainty:
How likely is it that you would be caught by the police if you were to be involved in [each of four specific offenses]?
Perceived severity:
Would it create a problem if you were caught, taken to court, and punished for [each of four specific offenses]?

MORAL CONSIDERATIONS
Moral beliefs:
How wrong do you think it is to *[each of four specific offenses]?*

Source: Paternoster, "Decisions," 11, 15–19.

models. The informed decision maker of the rational choice perspective repeatedly evaluates information and makes behavior decisions on the basis of such information, re-evaluates that information, and makes new offending decisions that sometimes differ from ones previously made."[95]

The study's research findings showed that certainty of punishment is considered in relatively few offending decisions and that severity of punishment has virtually no effect on offending decisions.[96] For example, during the sophomore and junior years, involvement in marijuana use and vandalism was influenced only modestly by perceived certainty of punishment, whereas perceived severity had no effect on the decision to participate in these acts. When considering the decision to drink and to engage in petty theft, neither perceived certainty nor severity of punishment influenced the decision to offend.[97] In general, the various kinds of offending decisions and offenses that Paternoster studied showed that the deterrent effect of formal sanction was relatively unimportant to delinquent behavior and that other considerations were far more influential in offending decisions.[98]

Various social factors (unrelated to the law) were found to influence offending decisions in significant ways. Gender consistently played an important role in delinquent offending decisions; males and females make different offending decisions. Opportunities to engage in delinquency, measured by level of parental supervision and peer involvement, affected several offending decisions, especially decisions to get involved in delinquent activity.[99] Additionally, attachment to parents (affective ties) influenced some offending decisions when youths were not previously involved in delinquent behavior.[100] Across different types of offenses, moral beliefs had the strongest and most consistent influence on offending decisions. Taken together, these findings indicate that the decisions to offend, to continue offending, and to stop offending (desist) involve social considerations to a far greater degree than they do formal, legal considerations of certainty and severity of punishment.[101]

The study of rational choice in delinquent offending is complex and controversial.[102] Clearly, the decision to engage in delinquency is not a simple, "once and for all" choice. Rather, youths make a series of offending decisions over time that involve a variety of non-legal, social factors that are sometimes specific to particular types of crime. Furthermore, the deliberation that results in rational choice includes possible rewards as well as risks. These perceptions of risk and reward appear to change over time and vary according to the situation.[103] For example, an active social life, in which peers support and encourage delinquent involvement, provides opportunity to engage in delinquent acts and encouragement to do so. One group of researchers concluded that assessments of risk and reward are "to some extent situationally-induced, transitory, and unstable."[104] Past experience, too, influences perceptions of risk and reward. Surprisingly, past involvement in crime and experience with formal sanctions have been found to actually reduce perceptions of risk and thereby increase the possibility of future crime and delinquency.[105] This finding directly opposes the deterrence hypothesis, which states that punishment reduces future involvement in crime and delinquency. Finally, research shows that social costs in terms of impacted social relationships play a significant role in offending decisions.[106] These social costs include loss of respect and friendship, shame, and embarrassment.

Crime and Human Nature: Criminal Propensity

In a provocative and controversial book, boldly titled *Crime and Human Nature: The Definitive Study of the Causes of Crime,* James Q. Wilson and Richard Herrnstein focus on the choices people make to engage in crime. They advance a point of view that on the surface sounds very much like the concept of expected utility from the rational choice perspective: they contend that individuals choose to engage in crime when rewards outweigh costs. Wilson and Herrnstein's idea of choice, however, draws from behaviorism and social learning theory in psychology.[107]

Wilson and Herrnstein contend that the choice to engage in crime "cannot be understood without taking into account individual predispositions and their biological roots."[108] It is this biological and psychological emphasis that makes their work both innovative and controversial, especially among sociological criminologists. **Criminal propensity** refers to certain constitutional characteristics, which are present at birth, that are expressed in a variety of personality traits. "The average offender tends to be constitutionally distinctive though not extremely or abnormally so. The biological factors whose traces we see in faces, physiques, and correlations with the behavior of parents and siblings are predispositions toward crime that are expressed as psychological traits and activated by circumstances. It is likely that the psychological traits involve

criminal propensity
Individual predisposition to crime. Constitutional characteristics, especially low intelligence and certain personality traits, including impulsiveness, risk taking, present orientation, and low anxiety, produce a tendency to seek immediate benefit, sometimes through crime.

intelligence and *personality* and that the activating events include certain experiences within the family, in school, and in the community."[109]

Connecting choice with criminal propensity, Wilson and Herrnstein contend that "individuals differ in the degree to which they discount the future."[110] Because the rewards of crime tend to be immediate, whereas the rewards of "noncrime" are in the future, psychological traits that result in a desire for immediate gain and a tendency to "discount the future" create criminal propensity. Singled out in this regard are low intelligence and a number of personality traits, including impulsiveness, sensation seeking, the inability to learn from punishment, and low anxiety.[111] These psychological characteristics influence perceptions of rewards related to crime and noncrime and result in limited conscience, both of which predispose people to commit crime.[112]

Family life can restrain or magnify criminal propensity.[113] Socialization processes involving interaction between parent and child influence criminal propensity in three important ways: (1) by instilling, or failing to instill, a desire for the approval of others (attachment); (2) by cultivating internalized constraint—conscience—or the lack of it; and (3) by establishing time horizon—a present or future orientation. Each of these socialization outcomes bears on the individual's calculation of the rewards of crime relative to the rewards of noncrime.

Wilson and Herrnstein also discuss how other institutional contexts influence criminal propensity. Schools may affect criminal involvement by bringing youth together into groups that reinforce the value of crime and by generating a sense of inequality when school experiences fail to provide opportunity or when they are perceived as unfair. Communities also influence the expression of criminal propensity. The presence of criminal or delinquent subcultures in neighborhoods, especially in the form of street corner gangs, have strong influence on crime and noncrime values, thereby encouraging the expression of individual criminal propensity through criminal acts. The economy too has an indirect effect on involvement in crime. The condition of the economy affects both work aspirations and opportunity. The level of employment opportunity leads to adjustment in aspirations, and together these factors affect whether criminal propensity is displayed through involvement in crime. Thus, the institutional contexts of school, community, and economy provide either "activating" or inhibiting experiences that affect whether or not criminal propensity is exhibited through criminal acts.

Wilson and Herrnstein conclude that "there *is* a human nature that develops in intimate settings out of a complex interaction of constitutional and social factors, and that this nature affects how people choose between the consequences of crime and its alternatives."[114] Critics, however, have not been convinced: they view this "definitive study" as overly discursive—long-winded and meandering—and lacking in logical rigor. In addition, critics contend that the evidence cited provides only weak support, is selective, and is sometimes misinterpreted by the authors.[115] Nonetheless, Wilson and Herrnstein's point of view is consistent with contemporary considerations of rational choice that emphasize choice much more than rationality. According to Wilson and Herrnstein, rather than being entirely rational, the choice of crime involves a variety of influences, some biological, others psychological, and still others social.

Does the Perceived Risk of Punishment Deter Individuals Predisposed to Crime?

Wilson and Herrnstein's criminal propensity theory has a logical extension to the question of whether punishment deters crime. Their theory suggests that "impulsive, risk-taking, and present-oriented individuals"—those with criminal propensity—are less likely to be deterred by the prospect of punishment, because these individuals focus on immediate benefits rather than on long-term consequences.[116] In a 2004 study, Bradley Wright, Avshalom Caspi, Terrie Moffitt, and Ray Paternoster addressed the varying deterrent effect of perceived punishment on individuals, depending on their level of criminal propensity.[117] Their analysis of longitudinal data allowed for the measurement of criminal propensity in childhood, adolescence, and early adulthood; deterrence perceptions in late adolescence and early adulthood; and self-reported criminal behavior in early adulthood. The researchers found that (in contrast to the predictions of criminal propensity theory) individuals who were most prone to crime because of their impulsive, risk taking, and present-oriented natures were most deterred from crime by perceptions that crime was costly and risky. When criminal propensity was low, however, the deterrent effect of possible punishment was virtually nonexistent. The researchers interpreted this to mean that the deterrent effect of punishment was irrelevant when other inhibitions (such as moral beliefs) are strong. They concluded that criminal propensity theories "are incorrect in assuming that criminally prone individuals do not respond to the perceived risk of criminal sanctions; in fact, they should respond most strongly."[118]

■ Positivist Criminology

Contemporary versions of classical thought did not evolve from a continuous tradition within criminology. Rather, the dominance of classical criminology in the late 1700s and throughout much of the 1800s came to a relatively abrupt stop in the late 1800s. Only in the last quarter of the 1900s did classical thought reemerge in the form of deterrence and rational choice theory.

Positivist criminology is usually portrayed as a sharp departure from classical criminology.[119] Therefore, the shift from classical to positivist thought in the late nineteenth century is seen as a major change in theory and practice. In fact, criminologists customarily divide their field of study into two primary schools of thought: classical and positivist, each with a strikingly different approach to crime and justice.[120] In 1911, Gina Lombroso-Ferrero, the daughter of Cesare Lombroso, one of the founders of positivist criminology, contrasted classical and positivist thought in the following way.

> *The Classical School based its doctrines on the assumption that all criminals, except in a few extreme cases, are endowed with intelligence and feelings like normal individuals, and that they commit misdeeds consciously, being prompted thereunto by their unrestrained desire for evil. The offense alone was considered, and on it the whole existing penal system has been founded, the severity of the sentence meted out to the offender being regulated by the gravity of his misdeed.*

The Modern or Positive School of penal Jurisprudence, on the contrary, maintains that the antisocial tendencies of criminals are the result of their physical and psychic organization, which differs essentially from that of normal individuals; and it aims at studying the morphology and various functional phenomena of the criminal with the object of curing, instead of punishing him.[121]

Foundations of Positivist Criminology

Positivist criminology emerged through the ideas and writings of various European scholars in the nineteenth century.[122] There is disagreement over who founded positivist criminology.[123] Some credit the Italian physician Cesare Lombroso and his students, who argued that criminals can be distinguished from noncriminals by physical abnormalities that typify a more primitive person. Others contend that the physician Frans Joseph Gall was the first scientific criminologist. He systematically measured the skull, based on the premise that different parts of the brain control different mental functions and those parts that are most developed will be physically larger, influencing the shape of the skull. Still others argue that French lawyer and statistician André-Michel Guerry and Belgian astronomer and mathematician Adolphe Quetelet, working some forty years before the work of Lombroso, were first to provide systematic measurement of crime and the social characteristics of criminals. Regardless of who founded positivist criminology, the school of thought is characterized by at least four key components: positivism, determinism, differentiation and pathology, and treatment and rehabilitation.[124]

Positivism

The rapid rise of positivist criminology in the last half of the nineteenth century was in direct proportion to the level of criticism directed at classical thought. Most of this criticism centered on the belief that classical criminology was unscientific.[125] "**Positivism** represents the scientific approach to the study of crime, in which 'science' is characterized by methods, techniques, or rules of procedure rather than by substantive theory or perspective."[126]

As developed by August Comte (1798–1857), positivism is a method for observing and drawing conclusions about social behavior. Comte argued that the scientific methods and logic used in the natural sciences could be applied to the study of social relations and society. Scientific methods, including observation, measurement, description, and analysis, became the foundation for a new discipline that he called "sociology."[127]

The application of positivism to "scientific criminology" involved an "enduring commitment to measurement" that still characterizes criminology today.[128] The first efforts to collect national crime data were part of a broader movement in several European countries to develop and gather official records enumerating various aspects of living conditions and social well-being. Systematic recording of births and deaths, for example, was developed in the 1500s. Then, beginning in the 1600s, these records were compared to economic conditions, leading to a number of studies that numerically analyzed social life. In England, these studies were called "political arithmetic," while in Germany they were called "moral Statistik," and in France "statistique morale."[129] This

positivism The use of scientific methods to study phenomena. These methods include observation, measurement, description, and analysis.

empirical approach to "social matters" is generally referred to as "moral statistics."[130] National crime data were first collected in France beginning in 1825; these data were published two years later as the *Compte*.[131]

The collection of these criminological data was begun in order to justify and study a growing network of institutions of confinement that developed in France in the early 1800s. The inventory of institutions included "hospitals, workhouses, asylums, reformatories, houses of corrections, and prisons. Their official aim was moral reformation through the deprivation of liberty and prevention of crime through deterrence. Their 'delinquent' and 'pathological' inmates included syphilitics, alcoholics, idiots and eccentrics, vagabonds, immigrants, prostitutes, and petty and professional criminals."[132] The crime data that were gathered demonstrated that these institutions failed to "normalize" the behavior of these "dangerous classes" of people.[133]

The French Ministry of Justice, the agency responsible for gathering crime data in the *Compte*, hoped that the "hard facts" of crime could be used to understand the causes of crime and to alleviate them.[134] Among the social statisticians who studied these data was a young, but well-known Belgian astronomer and mathematician Adolphe Quetelet. In a visit to France in 1823, he was introduced to the growing use of statistics to study social conditions. Advocates of this scientific approach hoped that statistical findings could be used to bring about social reform through legislation and innovation in social welfare practice and commerce. For example, Quetelet's "first statistical work (1826) utilized Belgian birth and mortality tables as a basis for the construction of insurance rates." Later, in analyzing French crime data from 1826 to 1829, Quetelet observed great regularity in crime rates over this period of time, including the number and rate of people accused of crime and the number and rate of people convicted of crime.[135] While he acknowledged that this regularity did not allow for the prediction of individual criminal involvement, he argued that patterns of crime over time follow "the same law-like regularities as did physical phenomena."[136] He referred to these regularities as "social mechanics."[137]

Quetelet went on to study the social mechanics of crime more closely by systematically observing the number and rate of crimes across certain social categories, the most important of which, he said, were age and sex. Using data from the *Compte* between 1826 and 1829, he noted that the number and rate of crimes were highest among those who were young, male, poor, uneducated, and unemployed or employed in low-status jobs. Quetelet was cautious, however, in drawing conclusions about causation from these regularities. Instead, he went on to offer a theoretical, rather than empirical, explanation of the apparent relationship between various social characteristics and criminal involvement.[138]

Quetelet concluded that these social characteristics establish a propensity for crime but that few individuals actually translate propensity into criminal acts. According to Quetelet, the extent to which propensity is expressed in criminal acts is determined by level of moral temperance. For example, young males who are poor, uneducated, and unemployed are likely to commit crime when they lack moral character—they have little desire or option for "rational and temperate habits."[139] Though Quetelet pointed to the importance of these social characteristics in determining criminal propensity, his later work gave greater emphasis to the view that lack of moral temperance was a manifestation of biological defects.[140]

Quetelet's work provided foundation for positivist criminology by applying scientific methods to the study of crime. Not only did he provide empirical analysis of crime rates and trends, but he also systematically observed how those rates and trends vary across social categories such as age and sex. Beirne and Messerschmidt state that "Quetelet's placement of criminal behavior in a formal structure of causality was a remarkable advance over the unsystematic speculations of his contemporaries."[141]

The positivism of positivist criminology represents an attempt to incorporate scientific methods into the study of crime. Systematic observation, measurement, description, and analysis have allowed researchers to look for regularities and patterns of crime and delinquency. Beginning in the first half of the nineteenth century, the application of scientific methods to the study of crime gave positivist criminology a very different orientation from that of classical criminology, which was dominant at that time. Classical criminology soon fell out of favor because it was viewed as unscientific.

Determinism

The scientific approach advanced by positivist criminology adopts the central working assumption that crime and delinquency are *caused* or *determined* by identifiable factors that may be biological, psychological, or social.[142] This cause–effect relationship is called **determinism**.[143] Determinism holds that given the presence or occurrence of causal factors, crime and delinquency invariably follow. In contrast to the classical notion of will, determinism holds that the factors and forces that cause crime are usually beyond the control of the individual.

According to the scientific method, causal factors can be systematically observed and measured, and causal processes can be analyzed and described with the goal of predicting crime and delinquency. The causes of crime and delinquency, however, are multiple, not singular, and the task of scientific criminology is to identify these multiple causes through scientific methods. The scope of causal factors considered in positivist criminology is tremendously broad, ranging from genetic abnormalities to economic inequality. The chapters that follow examine this wide range of causal factors.

While the scientific exploration of causal factors is at the very heart of positivism, positivist criminology can also include factors that are part of classical thought. Michael Gottfredson and Travis Hirschi point this out when they say: "No deterministic explanation of crime can reasonably exclude the variables of the classical model on deterministic grounds. These variables may account for some of the variation in crime. If so, they have as much claim to inclusion in a 'positivistic' model as any other set of variables accounting for the same amount of variation."[144] Thus, scientific efforts to uncover the causes of crime and delinquency must consider classical elements, such as the deterrent effect of punishment and factors influencing offending decisions.

Differentiation and Pathology

The patterns and regularity of crime that Quetelet called the "social mechanics" of crime led him to conclude that certain categories of people were more likely

determinism A cause–effect relationship. Positivist criminology is based on the assumption that the causes of delinquency can be identified through the use of scientific methods.

to be involved in crime: young men, those with less education, and those who were unemployed or in low-status occupations. These empirical patterns of criminal propensity led Quetelet to compare and contrast the "average man," who did not commit crime, with people who did commit crime.[145]

Quetelet's speculation that criminals lacked moral character and were distinct from the average person was consistent with an emerging view in France in the early 1800s. Rising from growing public fear of crime, this view held that criminals constituted a large and expanding "dangerous class." Piers Beirne reports that classical criminology began to lose much of its influence during this time because the penal institutions and strategies were based on the classical idea that criminals were normal individuals who could be morally rehabilitated by depriving them of their freedom (the deterrent effect of incarceration).[146] Even before the development of crime data, it was reported that these penal strategies had failed: "In 1820 it was already understood that the prisons, far from transforming criminals into citizens, serve only to manufacture new criminals and drive existing criminals ever deeper into criminality."[147] Outside of penal institutions, the ever-present poor, unemployed thieves in urban areas generated much fear among law-abiding citizens, who concluded that crime was increasing and that it was committed by this "dangerous class."[148]

The view that criminals and delinquents are fundamentally different from the average person is a persistent theme within positivist criminology.[149] The study of these differences—**differentiation**—rapidly became the focus of attention in positivist criminology. For example, Cesare Lombroso's version of biological positivism, advanced in his 1876 book *The Criminal Man*, argued that criminals were physiologically less evolved than their noncriminal counterparts.

Early versions of positivist thought emphasized biological and psychological differences between criminals and noncriminals, claiming that criminals suffered from **individual pathologies** such as physiological defects, mental inferiority, insanity, and a tendency to give in to passion. These individual pathologies were usually thought to have biological roots. Theories of individual pathology will be discussed more fully in the next chapter.

The rise of sociology as a discipline in the late 1800s is associated with differentiation that emphasized **social pathologies** related to rapid urbanization and industrialization. Rapid social change was thought to break down social organization, resulting in lack of social control. The resulting social pathologies experienced by offenders include problematic social conditions such as divorce and family disruption, lack of parental supervision, poverty, cultural heterogeneity, and residential mobility. These social pathologies formed the basis for the various sociological theories of delinquency described in Chapter 12, all of which take a positivist approach to the study of crime and delinquency.

Rehabilitation and Individualized Treatment

The task of criminology, according to positivists, is to use scientific methods to uncover the causes of crime and delinquency and thereby understand what makes offenders different from non-offenders. In contrast to classical criminology, the focus in positivist criminology is on the criminal offender, rather than on the criminal law, and the implications concern offender rehabilitation rather than legal reform.

differentiation
Categorization of people based on common, identifiable characteristics. Positivist criminology holds that delinquents are fundamentally different from the average youth.

individual pathologies
Characteristics of the individual, such as physiological, mental, and psychological defects, that allegedly make it difficult for the person to function normally.

social pathologies
Characteristics of the social environment, such as divorce, poverty, cultural heterogeneity, and residential mobility, that allegedly make it difficult for a person to function normally.

EXPANDING IDEAS — Understanding Positivist Thought

Can you define the four key elements of positivist criminology and explain how they fit together to make a "school of thought?" These elements provide foundation to most contemporary theories of delinquency and encourage testing theory through research.

1. *Positivism*
2. *Determinism*
3. *Differentiation and pathology (individual and social)*
4. *Rehabilitation and individualized treatment*

Armed with an understanding of the causes of crime and a belief that offenders do not freely choose criminal behavior, positivist criminology contends that it is inappropriate to punish them for their crimes.[150] Rather, the logical extension of positivist criminology is **individualized treatment**.[151] Since the causes of crime are many and varied, involving biological, psychological, and social factors, rehabilitation must be founded on a scientific understanding of those causes, and it must be individualized. Enrico Ferri, an early positivist, argued that positivist criminology could learn from science and the practice of medicine: "As medicine teaches us that to discover the remedies for a disease we must first seek and discover the causes, so criminal science in the new form which it is beginning to assume [positivist criminology], seeks the natural causes of the phenomenon of social pathology which we call crime: it thus puts itself in the way to discover effective remedies."[152]

Thus, positivist criminology seeks a scientific understanding of the causes of crime and delinquency so that rehabilitative treatment can be effective and individualized. As described in Chapter 2, the early juvenile court was established on this *rehabilitative ideal* of positivist criminology.[153]

individualized treatment
Rehabilitative efforts with offenders that are specific to the individual causes of delinquency and that attempt to remedy those causes.

■ Is Delinquent Behavior Chosen or Determined?

The sharp and fundamental differences between classical and positivist criminology have long been framed in terms of the relative merits of *choice* and *determinism*.[154] Classical thought depicts a rational and self-serving offender who chooses crime, whereas positivist thought contends that factors beyond an individual's control cause criminal involvement.

Neoclassical Criminology

In his 1890 book, *Penal Philosophy*, French sociologist Gabriel Tarde (1843–1904), took issue with this debate between classical and positivist criminologists, arguing that neither perspective could be justified on theoretical or practical grounds.[155] The choice-or-free-will argument, he reasoned, is obviously inaccurate given the positivist findings that crime varied according to social factors such as age, gender, and social class. Tarde went on to describe how the ways that people think, feel, and act—including criminal behavior—are learned within groups through association and imitation.[156] He reasoned that the hard-nosed determinism of positivist criminology was also inaccurate because it grossly diminishes offender responsibility and accountability. Under determinism, criminal acts are beyond the control of offenders, thereby making individual treatment impossible, especially in the case of "born criminals."

Based on his critique, Tarde concluded that both classical and positivist criminology are inadequate as a basis for penal philosophy and practice. His solution to this unproductive debate was a compromise of sorts, taking the "partial truths" of classical and positivist thought, without accepting the extremes of either perspective.[157] The compromise is referred to as *neoclassical criminology,* a point of view that emerged in the late 1800s and early 1900s and sought "the best of both worlds," especially in terms of practical application. Four general points characterize the neoclassical perspective.[158] First, "the concept of *character* replaced the extremes of free will and determinism as the source of criminality. An offender's character was open to analysis by experts from the fields of law, medicine, psychiatry, probation, criminology, and social work. Because the links between character and crime can be influenced by an infinite variety of factors, crime should be understood through multi-causal (multifactorial) analysis."[159] Second, legal responses to crime and delinquency should be individualized and directed toward reducing or eliminating causal factors. Prior record, mental capacity, and social environment all must be considered in legal sanctioning. Third, certain groups of people (such as children, the insane, and the disabled) are not fully responsible for their actions and therefore should not be held legally accountable. Instead, these vulnerable groups must be given special treatment by legal systems. Fourth, deterrence is the goal of punishment, but punishment should be proportionate to the seriousness of the crime and should set an example not only for the offender but for others also.

Hard Determinism, Soft Determinism, and Drift

While neoclassical thought represents a blending of classical and positivist thought, this compromise has never been fully realized, nor has it been widely adopted in contemporary criminology. The distinction persists between classical and positivist criminology, and this distinction is framed largely in terms of the choice–determinism debate.

To resolve the conflict, some criminologists have proposed that exclusive attention to either choice or determinism fails to adequately capture delinquent behavior. Rather, a more thorough explanation must include elements of both choice *and* determinism. Delinquent behavior involves some choice, but that choice is influenced by both legal and social restraints, as well as by social and economic pressures and incentives. In addition, delinquent behavior is motivated and determined by individual characteristics, many of which are beyond the individual's control. Offenders choose crime, but choice is not the only factor to influence criminal involvement.

In an important book, *Delinquency and Drift,* David Matza challenged the overly deterministic view of delinquent behavior offered by positivist criminologist.[160] Such <u>hard determinism</u> portrays delinquent behavior as an inevitable outcome of biological, psychological, and sociological forces. In its quest for scientific status, positivist criminology rejected the classical emphasis on choice and replaced it with scientific determinism and differentiation.[161] Matza, however, contends that positivist criminology predicts "far more delinquency than actually occurs" and provides a "distorted and misleading picture" of delinquent youth and their patterns of behavior.[162] He reasons that if delinquent behavior is a product of deterministic forces, then delinquency should be "more permanent and less transient, more pervasive and less intermittent than is apparently the case."[163] He also argues that the heavy emphasis on determinism and differentiation in positivist criminology has failed to incorporate new perspectives in the social sciences that concede that human behavior involves at least some element of freedom of choice and reason—what he refers to as a "voluntaristic conception" of human behavior.[164]

Matza offers a view of delinquent behavior that is far less deterministic—a view he calls <u>soft determinism</u>. Soft determinism acknowledges that delinquent behavior involves an element of choice, but contends that youth are not entirely free in choosing their behavior nor are they compelled to commit delinquency by their biological and psychological makeup or their social background and experiences.[165]

The logical extension of soft determinism is an image of delinquent youth and delinquent acts that Matza calls <u>drift</u>: "The image of the delinquent I wish to convey is one of drift, an actor neither compelled nor committed to deeds nor freely choosing them; neither different in any simple fundamental sense from the law abiding, nor the same; conforming to certain traditions in American life while partially unreceptive to other more conventional traditions. . . ."[166] In other words, delinquent youth are not all that different from other youth, and their choices and actions are neither totally free nor totally determined—they live in a state of drift. As a result, delinquent behavior is far less predictable than the deterministic models of positivist criminology would contend.

■ Summary and Conclusions

Is delinquency chosen or determined? The answer to this fundamental question is not easy or straightforward. Delinquent behavior certainly involves elements of self-interest, choice, and rationality. The classical ideas of will and hedonism contend that people choose their actions based on a rational consideration of gains and losses, of pleasure and pain. Few criminologists, however, would argue that delinquent behavior is purely rational.

A contemporary version of classical thought, *deterrence theory,* is primarily concerned with the restraining effect of legal punishment. Focusing on the likelihood and amount of punishment, deterrence theory holds that certainty and severity of punishment deters crime. Research support for this notion is modest at best, with certainty of punishment having some deterrent effect, but severity having virtually no deterrent effect. Much research reports that this deterrent effect is only a small part of rational choice.[167]

hard determinism A view holding that delinquent behavior is an inevitable outcome of biological, psychological, and sociological forces.

soft determinism The view that delinquent behavior is difficult to predict because people have choice and because their thoughts, beliefs, and actions are not completely the product of their background, experiences, and present life conditions.

drift An image of delinquents advanced by David Matza "in which youth are neither compelled nor committed to deeds nor freely choosing them" (*Delinquency and Drift,* 28).

The study of rational choice in delinquent offending considers a broad range of factors relating to material gain, legal costs, the normative controls of relationships, moral beliefs, and guilt and shame. *Rational choice theory* contends that a series of offending decisions are made over time, involving a variety of individual, social, and legal factors that are sometimes specific to particular types of crime. Furthermore, these perceptions of risk and reward appear to change over time and vary according to the situation.[168] Thus, the study of rational choice is complex and controversial.[169]

Positivist criminology is usually depicted as a sharp departure from classical criminology. In fact, criminologists customarily divide their field of study into two schools of thought: classical and positivist, each with a strikingly different approach to crime and justice. Positivist criminology seeks to apply the scientific methods of observation, measurement, description, and analysis to the study of delinquency (positivism) in an attempt to understand cause and effect (determinism).

Early crime data provided important insight into how crime and delinquency were distributed socially. After analyzing these data, positivist criminologists contended that criminals and delinquents were fundamentally different from the average person (differentiation and pathology). A scientific understanding of the causes of delinquency was also used to individualize correctional treatment and became the basis of the rehabilitative ideal in positivist criminology.

The sharp differences between classical and positivist criminology have long been expressed in terms of the choice versus determinism debate. However, the difference between choice and determinism may not be as great as normally depicted, especially when these concepts are considered in the context of their respective schools of thought: classical and positivist. Piers Beirne has argued that Beccaria's classical notion of will is not based solely on freedom of choice and hedonism.[170] Rather, criminal involvement is influenced by nonrational, deterministic forces such as passion, temperament, ignorance, and characteristics of the situation. Similarly, framing their work in terms of positivist criminology, Michael Gottfredson and Travis Hirschi have pointed out that the causes of crime necessarily include both social and legal restraints.[171] As a result, deterministic explanations may include classical elements of deterrence. Perhaps the transformation from classical to positivist criminology was not as drastic as scholars typically portray.

THEORIES

- classical criminology
- deterrence
- perceptual theory of deterrence
- rational choice
- criminal propensity
- positivist criminology
- neoclassical criminology
- drift

CRITICAL THINKING QUESTIONS

1. Using information from the opening case study, explain how Stanley's involvement in theft, while a rational response to poverty, was based on a variety of individual, social, and economic "reasons."

2. Compare and contrast classical criminology with positivist criminology by identifying the key elements of each school of thought.

3. According to Bentham, what is the *utility of punishment?* How does the utility of punishment reflect Beccaria's ideas of *legal rationality?*

4. In what ways does the contemporary juvenile justice system reflect classical thought?

5. What is the *deterrent effect* of punishment?

6. According to Paternoster's research, what factors influence delinquent offending decisions?

7. Provide several examples of how delinquent behavior is a product of *soft determinism.*

SUGGESTED READINGS

Beccaria, Cesare. *On Crimes and Punishments.* 1764. Translated by David Young. Indianapolis, IN: Hackett, 1986.

Beirne, Piers. *Inventing Criminology: Essays on the Rise of Homo Criminalis.* Albany, NY: State University of New York, 1993.

Gottfredson, Michael, and Travis Hirschi. "The Positive Tradition." In *Positive Criminology,* edited by Michael Gottfredson and Travis Hirschi, 9–22. Newbury Park, CA: Sage, 1987.

Matza, David. *Delinquency and Drift.* New York: Wiley, 1964.

Paternoster, Raymond. "The Deterrent Effect of Perceived Certainty and Severity of Punishment: A Review of Evidence and Issues." *Justice Quarterly* 42 (1987):173–217.

GLOSSARY

criminal propensity: Individual predisposition to crime. Constitutional characteristics, especially low intelligence and certain personality traits, including impulsiveness, risk taking, present orientation, and low anxiety, produce a tendency to seek immediate benefit, sometimes through crime.

determinism: A cause–effect relationship. Positivist criminology is based on the assumption that the causes of delinquency can be identified through the use of scientific methods.

deterrent effect: The prevention of crime through the use of punishment, based on the assumption that "human beings are both rational and self-interested beings" (Paternoster and Bachman, "Introduction," 14). Beccaria held that in order to prevent crime, punishment must be prompt, certain, and proportionate to the crime.

differentiation: Categorization of people based on common, identifiable characteristics. Positivist criminology holds that delinquents are fundamentally different from the average youth.

drift: An image of delinquents advanced by David Matza "in which youth are neither compelled nor committed to deeds nor freely choosing them" (*Delinquency and Drift,* 28).

the Enlightenment: A period of active thought and action from the mid-seventeenth century to the last quarter of the eighteenth century, "enlightened" by reason, science, and a respect for humanity.

free will: The freedom to choose action.

hard determinism: A view holding that delinquent behavior is an inevitable outcome of biological, psychological, and sociological forces.

hedonism: Behavior in which a choice of action involves a rational consideration of gains and loses, pleasure and pain, benefits and costs.

humanitarianism: A concern for the welfare of humanity. Beccaria's plea for legal reform was in tune with French humanists who sought better conditions for people.

individual pathologies: Characteristics of the individual, such as physiological, mental, and psychological defects, that allegedly make it difficult for the person to function normally.

individualized treatment: Rehabilitative efforts with offenders that are specific to the individual causes of delinquency and that attempt to remedy those causes.

legal rationality: The view that statutory law should be used to define crime, require an impartial and efficient judicial process, and specify punishment that is proportionate to the crime and deters future criminal acts.

positivism: The use of scientific methods to study phenomena. These methods include observation, measurement, description, and analysis.

rational choice: The conscious, calculated, purposive choice of action. The rational element in delinquent behavior.

social contract: An Enlightenment concept that refers to the mutual agreement among individuals in a political community to relinquish a portion of their individual freedom and self-interest in order to promote interpersonal peace, order, and stability.

social pathologies: Characteristics of the social environment, such as divorce, poverty, cultural heterogeneity, and residential mobility, that allegedly make it difficult for a person to function normally.

soft determinism: The view that delinquent behavior is difficult to predict because people have choice and because their thoughts, beliefs, and actions are not completely the product of their background, experiences, and present life conditions.

statutory law: Law that is enacted by a legislative process and codified (written down in a systematic manner).

utilitarianism: The goal of the social contract is to promote *"the greatest happiness shared by the greatest number of people"* (Beccaria, *On Crimes and Punishments* [Paolucci], 8, italics in translation).

will: The volitional part of human conduct. Will involves rational choice but is also influenced by forces beyond the control of the individual, including passion, sensation, individual temperament, ignorance, and characteristics of the situation.

REFERENCES

Agnew, Robert. "Determinism, Indeterminism, and Crime: An Exploration." *Criminology* 33 (1995):83–109.

———. *Juvenile Delinquency.* Los Angeles: Roxbury, 2001.

Akers, Ronald L., and Christine S. Sellers. *Criminological Theories: Introduction, Evaluation, and Application.* 4th ed. Los Angeles: Roxbury, 2004.

———. "Rational Choice, Deterrence, and Social Learning Theory in Criminology: The Path not Taken." *Journal of Criminal Law and Criminology* 81 (1990):653–676.

Beccaria, Cesare. *On Crimes and Punishments.* 1764. Translated by Henry Paolucci. Indianapolis, IN: Bobbs-Merrill, 1963.

———. *On Crimes and Punishments.* 1764. Translated by David Young. Indianapolis, IN: Hackett, 1986.

Beirne, Piers. "Adolphe Quetelet and the Origins of Positivist Criminology." *American Journal of Sociology* 92 (1987):1140–1169.

———. *Inventing Criminology: Essays on the Rise of Homo Criminalis.* Albany, NY: State University of New York Press, 1993.

———. "Inventing Criminology: The 'Science of Man' in Cesare Beccaria's *Dei Delitti e Delle Pene* (1764)." *Criminology* 29 (1991):777–820.

Beirne, Piers, and James Messerschmidt. *Criminology.* 3rd ed. Boulder, CO: Westview Press, 2000.

Becker, Gary. "Crime and Punishment: An Economic Approach." *Journal of Political Economy* 76 (1968):169–217.

Bentham, Jeremy. *An Introduction to the Principles of Morals and Legislation.* 1780. Edited by Laurence J. Lafleur. New York: Hafner, 1948.

Binder, Arnold, Gilbert Geis, and Dickson D. Bruce, Jr. *Juvenile Delinquency: Historical, Cultural, and Legal Perspectives.* 2nd ed. Cincinnati, OH: Anderson, 1997.

Bohm, Robert M. *A Primer on Crime and Delinquency Theory.* 2nd ed. Belmont, CA: Wadsworth, 2000.

Chiricos, Theodore G., and Gordon P. Waldo. "Punishment and Crime: An Examination of Some Empirical Evidence." *Social Problems* 18 (1972):200–217.

Clarke, Ronald V. "Introduction." In *Situational Crime Prevention,* edited by Ronald V. Clarke, 3–36. New York: Harrow and Heston, 1992.

Clarke, Ronald V., and Derek B. Cornish. "Modeling Offenders' Decisions: A Framework for Research and Policy." In *Crime and Justice: An Annual Review of Research,* Vol. 6, edited by Michael Tonry and Norval Morris, 147–185. Chicago: University of Chicago Press, 1985.

———. "Rational Choice." In *Explaining Criminals and Crime: Essays in Contemporary Criminological Theory,* edited by Raymond Paternoster and Ronet Bachman, 23–42. Los Angeles: Roxbury, 2001.

Cornish, Derek, and Ronald Clarke. "Introduction." In *The Reasoning Criminal: Rational Choice Perspectives on Offending,* edited by Derek B. Cornish and Ronald V. Clarke, 1–16. New York: Springer-Verlag, 1986.

———. "Understanding Crime Displacement: An Application of Rational Choice Theory." *Criminology* 25 (1987):933–947.

Cullen, Francis T., and Robert Agnew. *Criminological Theory: Past to Present—Essential Readings.* Los Angeles: Roxbury, 1999.

Cullen, Francis T., and Karen E. Gilbert. *Reaffirming Rehabilitation.* Cincinnati, OH: Anderson, 1982.

Curran, Daniel J., and Claire M. Renzetti. *Theories of Crime.* 2nd ed. Boston: Allyn and Bacon, 2001.

Durkheim, Emile. "Two Laws of Penal Evolution." 1901. In *Durkheim and the Law,* edited by Steven Lukes and Andrew Scull. New York: St. Martin's Press, 1983.

Feld, Barry C. *Bad Kids: Race and the Transformation of the Juvenile Court.* New York: Oxford, 1999.

Ferri, Enrico. *Criminal Sociology.* New York, NY: Appleton, 1917.

Foucault, Michel. *Discipline & Punish: The Birth of the Prison.* New York: Vintage, 1979.

———. *Power/Knowledge: Selected Interviews and Other Writings, 1972–1977.* Edited by Colin Gordon, translated by Gordon et al. New York: Pantheon, 1980.

Geis, Glibert. "Jeremy Bentham." In Mannheim, *Pioneers in Criminology,* 51–67.

Gibbs, Jack P. "Crime, Punishment, and Deterrence." *Southwestern Social Science Quarterly* 9 (1968):9–14.

———. *Crime, Punishment, and Deterrence.* New York: Elsevier, 1975.

Gibbons, Don C. *Society, Crime, and Criminal Behavior.* 6th ed. Englewood Cliffs, NJ: Prentice Hall, 1992.

Gottfredson, Michael, and Travis Hirschi. *A General Theory of Crime.* Stanford, CA: Stanford University Press, 1990.

———. "The Positive Tradition." In *Positive Criminology,* edited by Michael Gottfredson and Travis Hirschi, 9–22. Newbury Park, CA: Sage, 1987.

Grasmick, Harold G., and Robert J. Bursik, Jr. "Conscience, Significant Others, and Rational Choice: Extending the Deterrence Model." *Law and Society Review* 24 (1990):837–861.

Hirschi, Travis. "On the Compatibility of Rational Choice and Social Control Theories of Crime. In *The Reasoning Criminal: Rational Choice Perspectives on Offending,* edited by Derek B. Cornish and Ronald V. Clarke, 105–118. New York, NY: Springer-Verlag, 1986.

Hagan, John. "The Assumption of Natural Science Methods: Criminological Positivism." In *Theoretical Methods in Criminology,* edited by Robert F. Meier, 75–92. Beverly Hills, CA: Sage, 1985.

Jenkins, Philip. "Varieties of Enlightenment Criminology." *British Journal of Criminology* 24 (1984):112–130.

Lombroso-Ferrero, Gina. *Criminal Man, According to the Classification of Cesare Lombroso.* New York: G. P. Putnam, 1979.

Mannheim, Herman, ed. *Pioneers in Criminology.* Chicago: Quadrangle Books, 1960.

Matza, David. *Delinquency and Drift.* New York: Wiley, 1964.

McCarthy, Bill. "Not Just 'For the Thrill of It': An Instrumentalist Elaboration of Katz's Explanation of Sneaky Thrill Property Crimes." *Criminology* 33 (1995):519–538.

Meier, Robert F., Steven R. Burkett, and Carol A. Hickman. "Sanctions, Peers, and Deviance: Preliminary Models of a Social Control Process." *Sociological Quarterly* 25 (1984):67–82.

Michalowski, Raymond L. "Perspectives and Paradigms: Structuring Criminological Thought." In *Theory in Criminology: Contemporary Views,* edited by Robert F. Meier, 17–39. Beverly Hills, CA: Sage, 1977.

Miethe, Terance D., and Robert F. Meier. "Opportunity, Choice, and Criminal Victimization: A Test of a Theoretical Model." *Journal of Research in Crime and Delinquency* 27 (1990):243–266.

Monachesi, Elio. "Cesare Beccaria." In Mannheim, *Pioneers in Criminology,* 36–50.

Nagin, Daniel S., and Raymond Paternoster. "Enduring Individual Differences and Rational Choice Theories of Crime." *Law and Society Review* 27 (1993):467–496.

Newman, Graeme, and Pietro Marongiu. "Penological Reforms and the Myth of Beccaria." *Criminology* 28 (1990):325–346.

Paternoster, Raymond. "Decisions to Participate and Desist from Four Types of Common Delinquency: Deterrence and the Rational Choice Perspective." *Law and Society Review* 23 (1989): 7–40.

———. "The Deterrent Effect of Perceived Certainty and Severity of Punishment: A Review of Evidence and Issues." *Justice Quarterly* 42 (1987):173–217.

Paternoster, Raymond, and Ronet Bachman. "Introduction" to Chapter 2, "Classical and Neuve Classical Schools of Criminology: Deterrence, Rational Choice, and Situational Theories of Crime." In *Explaining Criminals and Crime,* edited by Raymond Paternoster and Ronet Bachman, 11–22. Los Angeles: Roxbury, 2001.

Paolucci, Henry. "Translator's Introduction." In Beccaria, *On Crimes and Punishments,* translated by Henry Paolucci, ix–xxiii.

Phillipson, Coleman. *Three Criminal Law Reformers: Beccaria, Bentham, and Romilly.* Montclair, NJ: Patterson Smith, 1970.

Piquero, Alex, and Greg Pogarsky. "Beyond Stafford and Warr's Reconceptualization of Deterrence: Personal and Vicarious Experiences, Impulsivity, and Offending Behavior." *Journal of Research in Crime and Delinquency* 39 (2002):153–186.

Piquero, Alex, and Stephen Tibbetts. "Specifying the Direct and Indirect Effects of Low Self-Control and Situational Factors in Offenders' Decision Making: Toward a More Complete Model of Rational Offending." *Justice Quarterly* 13 (1996):481–510.

Piliavin, Irving, Rosemary Gartner, Craig Thornton, and Ross L. Matsueda. "Crime Deterrence and Choice." *American Sociological Review* 51 (1986):101–119.

Pogarsky, Greg. "Identifying Deterrable Offenders: Implications for Deterrence Research." *Justice Quarterly* 16 (2002):451–471.

Quetelet, Adolphe. *Research on the Propensity of Crime at Different Ages.* 1831. Translated by Sawyer Sylvester. Cincinnati: Anderson, 1984.

Roshier, Bob. *Controlling Crime: The Classical Perspective in Criminology.* Chicago: Lyceum Books, 1989.

Tarde, Gabriel. *The Laws of Imitation.* 1903. Translated by F. Parsons. New York: Henry Holt, 1912.

———. *Penal Philosophy.* Translated by R. Howell. Boston: Little, Brown, 1912.

Tittle, Charles R. "Crime Rates and Legal Sanctions." *Social Problems* 16 (1969):409–422.

———. *Sanctions and Social Deviance.* New York: Praeger, 1980.

Tuck, Mary, and David Riley. "The Theory of Reasoned Action: A Decision Theory of Crime." In *The Reasoning Criminal: Rational Choice Perspectives on Offending,* edited by Derek B. Cornish and Ronald V. Clarke, 156–169. New York: Springer-Verlag, 1986.

Vold, George B., and Thomas J. Bernard. *Theoretical Criminology.* 2nd ed. New York: Oxford University Press, 1986.

Vold, George B., Thomas J. Bernard, and Jeffrey B. Snipes. *Theoretical Criminology.* 5th ed. New York: Oxford University Press, 1998.

Whitehead, John T., and Steven P. Lab. *Juvenile Justice: An Introduction.* 4th ed. Cincinnati, OH: Anderson, 2004.

Williams, Kirk R., and Richard Hawkins. "Perceptual Research on General Deterrence: A Critical Review." *Law and Society Review* 20 (1986):545–572.

Wilson, James Q., and Richard J. Herrnstein. *Crime and Human Nature: The Definitive Study of the Causes of Crime.* New York: Simon and Schuster, 1985.

Winfree, L. Thomas, Jr., and Howard Abadinski. *Understanding Crime: Theory and Practice.* Chicago: Nelson-Hall, 1996.

Wright, Bradley R.E., Avshalom Caspi, Terrie E. Moffitt, and Phil A. Silva. "The Effects of Social Ties on Crime Vary by Criminal Propensity: A Life-Course Model of Interdependence." *Criminology* 39 (2001):321–351.

Wright, Bradley R. E., Avshalom Caspi, Terrie E. Moffitt, and Ray Paternoster. "Does the Perceived Risk of Punishment Deter Criminally Prone Individuals? Rational Choice, Self-Control, and Crime." *Journal of Research in Crime and Delinquency* 41 (2004):180–213.

Young, David. "Introduction." In Beccaria's *On Crimes and Punishments,* translated by David Young, xi–xvi.

ENDNOTES

1. Beccaria, *On Crimes and Punishments* (Paolucci and Young translations).
2. Beirne, "Science of Man;" Newman and Marongiu, "Penological Reforms;" and Young, "Introduction," xv.
3. Beirne, "Science of Man," 780; see also 781–782.
4. Ibid., 780–781.
5. Ibid., 780.
6. Durkheim, "Two Laws," 113, quoted in Beirne, "Science of Man," 779. See also Newman and Marongiu, "Penological Reforms," 329.
7. Beirne, "Science of Man," 784.
8. Ibid.
9. A number of sociologists and historians of penology have contended that Beccaria's work is neither revolutionary in content or in application (Foucault, *Discipline & Punish*; Jenkins, "Varieties of Enlightenment Criminology;" Newman and Marongiu, "Penological Reforms;" and Roshier, *Controlling Crime*). Newman and Marongiu, for example, argue that Beccaria's ideas for legal reform reflect fairly specific convictions already advanced by Enlightenment philosophers (333).
10. Young, "Introduction," xi.
11. Beirne, "Science of Man;" Newman and Marongiu, "Penological Reforms;" and Young, "Introduction," xv.
12. Beirne, "Science of Man," 780, footnote 5; Paolucci, "Translator's Introduction," xvii; and Young, "Introduction," xvii–xviii.
13. Beccaria, *On Crimes and Punishments* (Young), 8.
14. Beirne, "Science of Man," 801, 802, 806, 807.
15. Ibid., 807, 812.

Chapter Resources

16. Monachesi, "Cesare Beccaria," 40.
17. Vold, Bernard, and Snipes, *Theoretical Criminology*, 14.
18. Beccaria, *On Crimes and Punishments* (Paolucci), 6, footnote 7; Beirne and Messerschmidt, *Criminology*, 63; and Vold, Bernard, and Snipes, *Theoretical Criminology*, 16.
19. Beccaria, *On Crimes and Punishments* (Young), 7, 8–9.
20. Beccaria, *On Crimes and Punishments* (Paolucci), 8 (italics in translation).
21. Ibid., 64–65 (italics in translation).
22. Vold, Bernard, and Snipes, *Theoretical Criminology*, 16.
23. Paolucci, "Translator's Introduction," xviii; see also xx.
24. Beccaria, *On Crimes and Punishments* (Paolucci), 9. Beirne, in "Science of Man" (782, 812) acknowledges this humanitarian reform aspect of the book; however, he argues that a number of other themes go largely unexamined.
25. Beirne, "Science of Man," 781.
26. Coleman Phillipson's *Three Criminal Law Reformers* review of *On Crimes and Punishments* (56) uses six topical themes to summarize its content: (1) measures of crimes and punishments; (2) certainty of punishment, and the right of pardon; (3) nature and division of crimes, and relative punishments; (4) consideration of certain punishments; (5) procedures including secret accusations and torture; and (6) prevention of crimes.
27. Beirne, "Science of Man," 788.
28. Young, "Introduction," xv.
29. Beccaria, *On Crimes and Punishments* (Paolucci), 43.
30. Beccaria, *On Crimes and Punishments* (Young), 74–75.
31. Beirne, "Science of Man," 812. See also Monachesi, "Cesare Beccaria," 43.
32. Beccaria, *On Crimes and Punishments* (Young), 62; and Beirne, "Science of Man," 812.
33. Beccaria, *On Crimes and Punishments* (Young), 12.
34. Ibid., 12–13.
35. Ibid., 9.
36. Ibid., 23.
37. Ibid., 36–37, italics in translation.
38. Beccaria, *On Crimes and Punishments* (Paolucci), 58.
39. Beccaria, *On Crimes and Punishments* (Young), 14.
40. Ibid., 81, emphasis in original.
41. Paolucci, "Translator's Introduction," x–xi.
42. Geis, "Jeremy Bentham," 51.
43. Binder, Geis, and Bruce, *Juvenile Delinquency*, 81.
44. Phillipson, *Three Criminal Law Reformers*, 166.
45. Beirne and Messerschmidt, *Criminology*, 67–68.
46. Bentham, *Introduction*, 2.
47. Geis, "Jeremy Bentham," 55, 58–59.
48. Beirne and Messerschmidt, *Criminology*, 68.
49. Geis, "Jeremy Bentham," 55.
51. Cited in Binder, Geis, and Bruce, *Juvenile Delinquency*, 82.
52. Geis, "Jeremy Bentham," 59, 61–62.
53. Geis, "Jeremy Bentham," 62; and Curran and Renzetti, *Theories of Crime*, 9.
54. Bentham, *Introduction*, 170–178; and Geis, "Jeremy Bentham," 62.
55. Beirne and Messerschmidt, *Criminology*, 67.
56. Newman and Marongiu, in "Penological Reforms" (326–329) note that Beccaria failed to campaign for the reforms he advocated; therefore, his image as a "great reformer" is overplayed.
57. Newman and Marongiu, "Penological Reforms," 329.
58. Young, "Introduction," xi.
59. Beirne, "Science of Man."
60. Vold, Bernard, and Snipes, *Theoretical Criminology*, 18–20.
61. Beccaria, *On Crimes and Punishments* (Paolucci), 11–13.
62. Beccaria, *On Crimes and Punishments* (Paolucci), 43, 55–58, 62. Graeme Newman and Pietro Marongiu, in "Penological Reforms" (331), argued that the notion that punishment should be reasonable and proportionate was expressed by various Enlightenment thinkers before the published work of Beccaria and Bentham.

63. Feld, *Bad Kids.*

64. Beccaria, *On Crimes and Punishments* (Paolucci), 55–62.

65. Paternoster and Bachman, "Introduction," 14.

66. Ibid., 16.

67. Chiricos and Waldo, "Punishment and Crime;" Gibbs, "Crime, Punishment, and Deterrence," and *Crime, Punishment, and Deterrence;* and Tittle, "Crime Rates," and *Sanctions and Social Deviance.* See also Grasmick and Bursik, "Conscience;" Paternoster, "Deterrent Effect," and "Decisions;" and Piliavin et al., "Crime Deterrence and Choice."

68. Beccaria, *On Crimes and Punishments* (Paolucci), 58–59, 94–95.

69. Gibbs, *Crime, Punishment, and Deterrence,* 115.

70. Paternoster, "Decisions," 8.

71. Chiricos and Waldo, "Punishment and Crime." See Paternoster, "Deterrent Effect," 175–179.

72. Grasmick and Bursik, "Conscience," 839.

73. Paternoster, "Deterrent Effect," 191.

74. Paternoster, "Deterrent Effect," 179–181; and Piliavin et al., "Crime Deterrence and Choice," 103.

75. See Paternoster, "Deterrent Effect," for a thorough review of this research.

76. Paternoster, "Deterrent Effect," 184–186. Grasmick and Bursik, "Conscience," 838–839.

77. Gibbs, *Crime, Punishment, and Deterrence.*

78. Grasmick and Bursik, "Conscience;" Meier, Burkett, and Hickman, "Sanctions;" Piliavin et al., "Crime Deterrence and Choice;" and Williams and Hawkins, "Perceptual Research."

79. Paternoster and Bachman, "Introduction," 18.

80. Agnew, *Juvenile Delinquency,* 204. See also Akers and Sellers, *Criminological Theories,* 26–29; and Paternoster, "Decisions," 7, 10.

81. Cornish and Clarke, "Introduction," 1.

82. The idea of "expected utility" was first advanced by Gary Becker in a 1968 article titled "Crime and Punishment: An Economic Approach." Mary Tuck and David Riley provide a more psychologically oriented approach by referring to "subjective expected utility." This approach emphasizes individual perception of expected benefits and costs of crime. We will focus on the perceptual literature rather than that which presents economic models. See also Clarke and Cornish, "Modeling," and Paternoster, "Decisions," 8.

83. Cornish and Clarke, "Introduction," 1.

84. Ibid.

85. Clarke and Cornish, "Rational Choice," 24.

86. Cornish and Clarke, "Introduction," 2.

87. Akers, "Rational Choice."

88. Cornish and Clarke, "Introduction," 7.

89. Ibid., 1.

90. Paternoster, "Decisions," 10.

91. Ibid.

92. Ibid., 7, 37.

93. By including social costs in his rational choice model, Paternoster draws on social control theory. Social control theory will be discussed more fully in Chapter 10, but in terms of rational choice, it is argued that one of the main reasons that people conform is because of commitments they develop in social relationships (Hirschi, "Compatibility"). The factors that are assumed to influence the decision to participate in an offense are often referred to as the "choice-structuring properties of offenses" (Cornish and Clarke, "Understanding Crime Displacement," 935; Miethe and Meier, "Opportunity," 245; and Paternoster, "Decisions," 12).

94. Paternoster, "Decisions;" see also Cornish and Clarke, "Introduction," 167.

95. Paternoster, "Decisions," 37.

96. Ibid., 7, 37.

97. Ibid., 23.

98. Williams and Hawkins, in "Perceptual Research" (568), suggest incorporating social and material considerations into deterrence theory, thereby making the approach more useful for crime prevention. However, Paternoster ("Deterrent Effect," 211–212) argues that rational choice theory already includes both legal and non-legal factors. Thus, rational choice theory incorporates the deterrent effect of legal or formal sanctions together with non-legal factors, including informal sanctions, affective ties, and material costs.

99. Paternoster, "Decisions," 22–23, 26–28.

100. Paternoster, "Decisions," 33–34.
101. Paternoster, "Decisions," 38.
102. Akers, "Rational Choice;" Akers and Sellers, *Criminological Theory*, 24–27; Paternoster, "Deterrent Effect;" and Williams and Hawkins, "Perceptual Research."
103. Piliavin et al., "Crime Deterrence and Choice," 115. Change in perception of risk has been studied by Paternoster ("Deterrent Effect," 194–205, and "Decisions," 31–37), Piliavin et al. ("Crime Deterrence and Choice," 115), and Williams and Hawkins ("Perceptual Research," 552–554).
104. Piliavin et al., "Crime Deterrence and Choice," 116.
105. Paternoster, "Deterrent Effect," 179–181; and Piliavin et al., "Crime Deterrence and Choice," 103.
106. Grasmick and Bursik, "Conscience;" McCarthy, "Instrumentalist Elaboration;" and Piquero and Tibbetts, "Specifying."
107. Wilson and Herrnstein, in *Crime and Human Nature,* state that their assumptions about choice are "commonplace in philosophy and social sciences. Philosophers speak of hedonism or utilitarianism, economists of value or utility, and psychologists of reinforcement or reward" (34).
108. Wilson and Herrnstein, *Crime and Human Nature,* 103.
109. Ibid., 102–103, emphasis added.
110. Ibid., 54, 61.
111. Ibid., 199–207.
112. Ibid., 207.
113. Ibid., 217.
114. Ibid., 508.
115. Gibbons, *Society,* 147–148.
116. Wright, Caspi, Moffitt, and Paternoster, "Perceived Risk," 182.
117. Ibid.
118. Wright, Caspi, Moffitt, and Paternoster, "Perceived Risk," 208. See also Nagin and Paternoster, "Enduring Individual Differences;" Piquero and Pogarsky, "Beyond;" and Pogarsky, "Identifying Deterrable Offenders."
119. As discussed later, Beirne, in "Science of Man" and *Homo Criminalis,* has argued that there are elements of positivist criminology in Beccaria's writings.
120. Akers and Sellers, *Criminological Theory,* 45–46; Beirne and Messerschmidt, *Criminology,* 73; Cullen and Agnew, *Criminological Theory,* 7; Curran and Renzetti, *Theories of Crime,* 6–15; and Gottfredson and Hirschi, "Positive Tradition," 9–14.
121. Lombroso-Ferrero, *Criminal Man,* 75.
122. Beirne and Messerschmidt, *Criminology,* 72.
123. Curran and Renzetti, *Theories of Crime,* 15–16.
124. Roshier, *Controlling Crime,* 21–22.
125. Gottfredson and Hirschi, "Positive Tradition," 14.
126. Ibid., 10, emphasis added.
127. Winfree and Abadinski, *Understanding Crime,* 28; and Bohm, *Primer,* 22.
128. Hagan, "Assumption," 78.
129. Beirne, "Adolphe Quetelet," 1150. See also Vold, Bernard, and Snipes, *Theoretical Criminology,* 22.
130. Beirne, "Adolphe Quetelet," 1150.
131. Ibid., 1148.
132. Beirne and Messerschmidt, *Criminology,* 73; see also Beirne, "Adolphe Quetelet," 1143–1144. Foucault (*Power/Knowledge,* 47–49) is referenced in Beirne, "Adolphe Quetelet."
133. Beirne, "Adolphe Quetelet," 1144–1147; and Beirne and Messerschmidt, *Criminology,* 74–75.
134. Beirne and Messerschmidt, *Criminology,* 75.
135. Beirne, "Adolphe Quetelet," 1153–1154; and Beirne and Messerschmidt, *Criminology,* 75–76.
136. Beirne and Messerschmidt, *Criminology,* 75; see also Beirne, "Adolphe Quetelet," 1150.
137. Quetelet, *Research,* 69, quoted in Beirne, "Adolphe Quetelet," 1153.
138. Beirne, "Adolphe Quetelet," 1155.
139. Ibid., 1159.
140. Ibid., 1160.
141. Beirne and Messerschmidt, *Criminology,* 76.
142. Gottfredson and Hirschi, "Positive Tradition," 12.
143. Michalowski, "Perspectives and Paradigms," 28.

144. Gottfredson and Hirschi, "Positive Tradition," 12.

145. Beirne, "Adolphe Quetelet," 1159–1160.

146. Beirne, "Adolphe Quetelet," 1144–1145.

147. Foucault, *Power/Knowledge,* 40.

148. Beirne, "Adolphe Quetelet," 1145–1146.

149. Cullen and Gilbert, *Reaffirming Rehabilitation,* 33; Matza, *Delinquency and Drift,* 11; and Roshier, *Controlling Crime,* 21–22.

150. Cullen and Gilbert, *Reaffirming Rehabilitation,* 33.

151. Whitehead and Lab, *Juvenile Justice,* 51.

152. Ferri, *Criminal Sociology,* 18–19.

153. Feld, *Bad Kids,* Chapter 2.

154. Beirne and Messerschmidt, *Criminology,* 93; and Gottfredson and Hirschi, "Positive Tradition," 11.

155. The discussion of Gabriel Tarde's work draws heavily from Beirne and Messerschmidt, *Criminology,* 83–84.

156. Tarde, *Penal Philosophy,* 322, 340; and Tarde, *Laws of Imitation.*

157. Beirne and Messerschmidt, *"Criminology,"* 84.

158. Beirne and Messerschmidt, *Criminology,* 84; and Curran and Renzetti, *Theories of Crime,* 9–10.

159. Beirne and Messerschmidt, *Criminology,* 84.

160. Matza, *Delinquency and Drift.*

161. Ibid., 5, 11.

162. Ibid., 2, 21.

163. Ibid., 22.

164. Ibid., 11, see 7–11.

165. See also Agnew, "Determinism."

166. Matza, *Delinquency and Drift,* 28.

167. Paternoster, "Deterrent Effect," 175–194, and "Decisions," 28; and Piliavin et al., "Crime Deterrence and Choice," 104.

168. Paternoster, "Deterrent Effect," 194–205, and "Decisions," 31–37; Piliavin et al., "Crime Deterrence and Choice," 115; and Williams and Hawkins, "Perceptual Research," 552–554.

169. Akers, "Rational Choice;" Akers and Sellers, *Criminological Theories,* 26–29; Paternoster, "Deterrent Effect;" and Williams and Hawkins, "Perceptual Research."

170. Beirne, "Science of Man," 801–802, 806–807, and *Homo Criminalis,* Epilogue.

171. Gottfredson and Hirschi, "Positive Tradition," 12–13.

Biological and Psychological Approaches

8

Chapter Objectives

After completing this chapter, students should be able to:

- Provide an overview of the historical development of thought regarding biological factors and criminal behavior.
- Identify a number of physical characteristics associated with criminality that were advanced by early biological approaches.
- Describe nature–nurture interaction as it applies to contemporary biosocial criminology.
- Identify three contemporary approaches in biosocial criminology that are used to study antisocial behavior.
- Identify and describe key dimensions of personality and their roles in delinquency causation.
- Assess the role of intelligence in delinquent behavior.
- Understand key terms:
  atavism
  stigmata
  somatotypes

reductionistic

biological determinism

environmental determinism

nature–nurture debate

biosocial criminology

nature–nurture interaction

neurophysiology

neuropsychology

central nervous system (CNS)

peripheral nervous system

somatic nervous system

autonomic nervous system (ANS)

executive cognitive functions (ECF)

conditionability

low arousal theory

neurotransmitters

testosterone

behavioral genetics

heritability

personality

agreeableness

conscientiousness

mental age

intelligence quotient (IQ)

CASE IN POINT

The Neurological Underpinning of Psychopathy in Two Case Illustrations

Juvenile delinquency is sometimes attributed to abnormal brain structure and function. The following medical cases, offered by David Rowe in his book *Biology and Crime,* illustrate how the physical structure of the brain affects psychological functioning. This important link between the body's neurological structure—including the central and peripheral nervous systems—and psychological functioning is referred to as *neuropsychology*. This chapter addresses the role of biological and psychological factors in delinquent behavior.

Two medical cases have been studied by the neurologist Antonio Damasio and his colleagues that illustrate a possible neurological underpinning for psychopathy. Their subjects, a man and a woman, had both suffered injuries to the prefrontal cortex during infancy. The prefrontal cortex, located just behind the eye sockets and above the bridge of the nose, is involved in planning a sequence of actions and in anticipating the future. The female subject was run over by a car when she was 15 months old. The male subject had a brain tumor removed from his prefrontal area when he was 3 months old. Both subjects grew up in stable, middle-class families with college-educated parents and had normal biological siblings, but neither made a satisfactory social adjustment; neither had friends and both were dependent on support from their parents. Neither subject had any plans for the future. The woman was a compulsive liar; she stole from her parents and shoplifted; her early and risky sexual behavior led to a pregnancy by age 18. By age 9, the male subject had committed minor theft and aggressive delinquent acts; he had no empathy for others.

The researchers tested the two subjects on a computerized gambling test used to detect how people respond to the uncertainty of rewards and punishments. The task is designed so that payoffs to the 'bad' card deck are high and immediate while payoffs to the 'good' card deck are low immediately but better in the long term. Most people quickly learn to draw cards from the 'good' deck that offers better long-term payoff. Neither subject was able to learn to use the long-term payoff deck.

Most surprisingly, these brain-injured victims failed to understand the difference between right and wrong; they lacked a sense of social norms and of how to act in social situations. Their moral blindness contrasts with the thought processes of adults who have brain damage in the same region and who display symptoms of psychopathy but understand without any difficulty the moral difference between right and wrong.

Source: Rowe, *Biology and Crime,* 69–70.

The individual has long been the center of attention when trying to explain delinquent behavior. As we noted in the last chapter, early versions of positivist criminology adopted the view that delinquents and criminals are fundamentally different from the average person. Using scientific methods, researchers sought to uncover and study these biological and psychological differences. Scholars commonly claimed that criminals were marked by individual pathologies such as physiological abnormalities, mental inferiority, insanity, and a lack of "rational and temperate habits."[1]

The connection between biological and psychological characteristics of individuals and their behavior has been explored extensively since the latter part of the eighteenth century. The fundamental theme of this work has been that individual traits and characteristics predispose some people to delinquency and crime.[2] While past work focused on individual traits and characteristics, much contemporary work in this area is integrated and interdisciplinary, attempting to consider the interrelationships of biological, psychological, and sociological factors and processes.[3]

This chapter explores scientific efforts to discover biological and psychological forces at work in the individual as they interface with the environment and influence behavior. We begin with biological approaches because early positivist criminology tried to use scientific methods to discover biological predispositions to crime, as they were reflected in physical appearance. Psychological approaches are discussed in the second half of this chapter. The interplay between biological factors and processes and mental and emotional functioning of the individual is a persistent theme in the theory and research that make up this area of study. As we mentioned, contemporary work also considers how biological and psychological predispositions to antisocial behavior interface with environmental conditions.

■ Early Biological Approaches: Focusing on Physical Characteristics

Physical Appearance and Biological Differences

Physical appearance has long been thought to reveal the nature and character of the individual. Unusual or atypical physical characteristics were once thought to indicate biological defects, abnormalities, and overall inferiority. According to this line of reasoning, criminals have a physical appearance that sets them apart from law-abiding citizens, and this difference is indicative of biological difference between criminals and noncriminals. Criminal behavior, then, is a manifestation of these biological differences.

Since ancient times, the physical characteristics of the face were held to be of particular importance and were given special attention. The arguably most complete coverage of the connection between facial features and behavior was offered by Johan Caspar Lavater (1741–1801), a Swiss scholar and theologian who in 1775 published a four-volume work on *physiognomy* (the interpretation of facial features and expressions). This work organized and classified many popular observations of that day, connecting facial features to human conduct. Interestingly, this publication received "nearly as much favorable attention as Beccaria's work had only eleven years earlier."[4]

While *physiognomy* studied facial features, *phrenology* studied the external shape of the skull.[5] The most famous phrenologist was Italian physician and anatomist Franz Joseph Gall (1758–1828), who systematically measured the skull, basing his research on the premises that mental functioning is localized in different parts of the brain and that those parts that are most developed will be physically larger, affecting the shape of the skull. Gall identified three major regions of the brain: intellectual faculties, moral faculties, and base or animal faculties. This last group of faculties controlled traits such as secretiveness, destructiveness, and aggression. According to Gall, it was this group of traits that was overdeveloped in criminals, resulting in a variety of crimes, including violent behavior.[6]

Phrenology gained considerable popularity in Europe and the United States in the first half of the 1800s, and as it grew in popularity, the scope of its explanation also expanded. Phrenologists claimed to identify an increasing variety of behaviors by the shape of the skull.[7] Beginning in the 1830s, however, the number of empirical phrenological studies began to decrease, and "scientists were displaced by charlatans who practiced a popularized version of phrenology in which they 'read' heads much the same way other fortune tellers read palms."[8] Not long after this movement, the work of Cesare Lombroso began to capture the attention of the academic community.

Cesare Lombroso's Search for "Criminal Man"

Cesare Lombroso (1835–1909), an Italian physician and professor of forensic medicine, is often called "the father of modern criminology."[9] He too argued that there is a relationship between the physical characteristics of individuals and their behavior, but his exploration went much further than that of physiognomy and phrenology to include the anatomy and physiology of the human body, especially the brain.[10] He also tied objective biological observations to the theory of evolution advanced by Charles Darwin just a few years prior to Lombroso's first major publications. In fact, Darwin was the first to use the term "atavistic man," a concept that Lombroso developed into a theory of crime.[11]

Lombroso's thesis, as initially advanced in *L'Uomo delinquente* (*Criminal Man,* 1876), was that criminals are physically distinct from noncriminals—criminals are a particular physical type. According to Lombroso, this difference arises because criminals are not as evolved as other individuals, and their physical differences represent a reversion to a more primitive stage of evolution. Using Darwin's term, Lombroso referred to this biological reversion as **atavism**.

The application of atavism to criminals came to Lombroso during a postmortem examination of a convicted thief. He excitedly observed:

> *At the sight of that skull, I seemed to see all of a sudden, lighted up as a vast plain under a flaming sky, the problem of the nature of the criminal—an atavistic being who reproduces in his person the ferocious instincts of primitive humanity and the inferior animals. Thus were explained anatomically the enormous jaws, high cheek-bones, prominent superciliary arches, solitary lines in the palms, extreme size of orbits, handle-shaped or sessile ears found in criminals, savages, and apes, insensibility to pain, extremely acute sight, tatooing, excessive idleness, love of orgies, and the irresistible craving for evil for its own sake. . . .*[12]

These observations were given full expression in *L'Uomo delinquente,* which was based on more extensive study involving autopsies of 66 deceased male criminals and examination of 832 living criminals, both male and females, selected from among the "most notorious and depraved" Italian criminals.[13] This latter group of criminals was then compared to noncriminal Italian soldiers and a group of "lunatics."[14] Lombroso observed that criminals display distinctive physical characteristics, called **stigmata**, that resemble "savages" and lower animals. Criminal stigmata, according to Lombroso, include asymmetry of the face, large jaws and cheekbones, unusually large or small ears that stand out from the head, fleshy lips, abnormal teeth, flattened nose, angular form of the skull, scanty beard but general hairiness of the body, and excessively long

atavism Reappearance in an individual of physical characteristics associated with a more primitive stage of evolution that has been absent in intervening generations. Lombroso claimed that such individuals are "born criminals" because of these characteristics.

stigmata The distinctive physical characteristics (observed in criminals) that identify atavism.

arms.[15] Lombroso contended that while these stigmata do not cause criminality, they are useful for identifying individuals with atavism—the inherited predisposition to crime.[16]

Lombroso expanded his views throughout subsequent editions of *L'Uomo delinquente*. While the first edition, published in 1876, was 252 pages, the fifth edition required three volumes and covered almost 2,000 pages.[17] Even though each successive edition gave greater attention to environmental explanations, Lombroso never abandoned the idea of a born criminal type.[18] Gradually, he developed four major categories of criminals:[19] (1) *born criminals*—people with atavistic characteristics, who account for about one third of all offenders; (2) *criminals by passion,* who commit crime for anger, love, or honor, being propelled to crime by "irresistible forces"; (3) *insane criminals,* who commit crime as a "consequence of an alteration of the brain . . . [which] makes them unable to discriminate between right and wrong";[20] and (4) *occasional criminals,* "who do not seek the occasion for the crime, but are almost drawn into it, or fall into the meshes of the code for very insignificant reasons."[21] This last criminal category was the broadest and included three subtypes: *pseudocriminals*—those who commit crime involuntarily, such as in self-defense; *criminaloids*—those whose predisposition to crime is stimulated by environmental circumstances or opportunities; and *habitual criminals*—those who regularly violate the law as a part of their day-to-day life, with little guilt or remorse.[22]

Two of Lombroso's students extended his work into new realms. Raffaele Garofalo (1851–1934) made extensive legal application of Lombroso's theory, and Enrico Ferri (1856–1929) developed a sociological perspective on it. Both were famous as criminologists in their own right.[23]

Raffaele Garofalo

Throughout his career, Garofalo was a lawyer, prosecutor, magistrate, professor of criminal law, and prominent member of the Italian government.[24] Because of his background in law, Garofalo was most interested in the legal implications of Lombroso's ideas, particularly as they related to legal definitions of crime and appropriate sanctions for criminals. Following Lombroso, Garofalo concluded that criminals have "regressive characteristics" indicating a "lower degree of advancement."[25] Garofalo, however, was more concerned with psychological deficiencies than with physical anomalies of criminals. More specifically, Garofalo argued that criminals lacked basic moral sentiments that result from fear of punishment ("pity") and self-control ("probity").[26] Such moral deficiencies, he said, are a result of heredity and tradition. Because of these fundamental psychological faults, Garofalo believed that criminals were unfit for the society in which they lived, and the only solution to this evolutionary problem was the incapacitation or elimination of the unfit through legislated criminal sanctions.[27]

Enrico Ferri

Another of Lombroso's students, Enrico Ferri, was a politically active scholar who advocated social and political change to deal with crime and criminality. While Lombroso emphasized biological characteristics of the individual, Ferri's *Criminal Sociology* gave attention to the interrelatedness of "three different sets of factors in crime: namely those in the physical or geographic environment, those in the constitution of the individual, and those in the social environment."[28] The constitution of the individual referred to both inherited and acquired factors such as age, sex, and psychological makeup. The physical or geographic environment included factors such as land forms, natural resources, climate, and vegetation. The social environment was composed of factors such as economic status, racial composition, population density, and culture. Because of the interrelatedness of these causal factors, Ferri argued that crime could be controlled by social changes. "He advocated subsidized housing, birth control, freedom of marriage, divorce, public recreation facilities, each reflective of his socialistic belief that the state is responsible for creating better living and working conditions. It is not surprising that Ferri was also a political activist."[29]

The work of Lombroso, Garofalo, and Ferri—three Italian positivists—is noteworthy because it pushed the study of crime away from its classical roots "to a scientific study of the criminal and the conditions under which he commits crime."[30] While Lombroso initially emphasized biological characteristics that distinguished criminals from noncriminals, all three criminologists moved toward "a multi-factor explanation of crime that included not only heredity but social, cultural, and economic variables."[31] Ironically, even though Lombroso was one of the first to advance a positivistic perspective, the most damaging criticism of his approach concerned methodological problems.

Charles Goring: Lombroso Assessed

Charles Goring (1870–1919), a medical officer in the English prison service, "often is regarded as the prime debunker of Lombroso's ideas. In fact, although he attacked Lombroso harshly for the sloppiness of his methods and thinking, Goring ended up taking a position that was not very different from Lombroso's."[32]

In *The English Convict* (1913), Goring reports on a statistical study that compared 2,348 English, male convicts to a large group of noncriminals.[33] The study involved objective measures of physical and mental characteristics. Goring concluded that *"no evidence has emerged confirming the existence of a physical criminal type, such as Lombroso and his disciples have described—our inevitable conclusion must be that there is no such thing as a physical criminal type."*[34]

Despite the apparent decisiveness of this statement, Goring also observed that criminals were "physically inferior" in height, weight, and mental capacity as compared to noncriminals. He further argued that criminality was inherited and that inheritance was one of the most significant causal factors of crime.[35] Piers Beirne and James Messerschmidt offer the following appraisal: "In sum, *The English Convict* failed to refute Lombroso's concept of the born criminal. . . . Goring's dubious achievement was simply to replace Lombroso's atavistic criminal with one born with inferior weight, stature, and mental capacity."[36] A few decades later, heredity and physical characteristics were given renewed attention by an American academic, Earnest Hooton, who revived some of Lombroso's ideas and Goring's findings.

Earnest Hooton: Lombroso Revisited

Earnest Hooton (1887–1954) was a Harvard University physical anthropologist who late in his career ventured into criminology. In 1939 he published two books, *Crime and the Man* and *The American Criminal*, which were based on a

12-year study of the relationship between physical characteristics and criminal behavior.[37] He examined 14,873 male convicts in 10 different state prisons and 3,203 "civilians" using an assortment of bodily (anthropometric) measures.[38] Hooton claimed that there were significant physical differences between criminals and noncriminals. Criminals have "straighter hair, more mixed patterns of eye color, more of the various kinds of skin folds in the upper eyelid, lower and more sloping foreheads, more pointed chins, more extreme variations in the projection of cheek bones, ears with less roll of the rim, and a greater frequency of Darwin's tubercle—a cartilaginous module on the fee margin of the ear."[39]

Similarly to Lombroso and Goring, Hooton contended that these differences indicated that criminals are biologically inferior to noncriminals. Hooton took this argument to its logical conclusion in terms of public policy.[40] The last paragraph of *The American Criminal* states: "Criminals are organically inferior. Crime is the resultant of the impact of environment upon low grade human organisms. It follows that the elimination of crime can be effected only by extirpation [extermination] of the physically, mentally, and morally unfit, or by their complete segregation in a socially aseptic environment."[41] Hooton offers a chilling prescription in light of the Nazis' imminent efforts to exterminate those deemed to be "unfit."[42]

Criticism of Hooton's work flared. The most systematic critique was offered by sociologist Robert Merton and anthropologist M. F. Ashley Montague in an article in *American Anthropologist*.[43] Merton and Montague criticized Hooton's sampling that compared incarcerated offenders with a control group composed mainly of Tennessee fire fighters and Massachusetts militia members.[44] They also drew attention to Hooton's assertion that minority groups display biological inferiority, for which he provided no empirical data.[45] Another criticism addressed Hooton's circular reasoning. He claimed that criminals are marked by physical inferiority and that such physical inferiority results in criminality. Despite the sharp criticism of Hooton's work, the search for biological explanations of crime and delinquency continued into the mid-1900s.

Physique, Temperament, and Delinquency

William Sheldon (1898–1977) was perhaps the last great believer in biological determinism, maintaining that the primary determinants of behavior are constitutional and inherited.[46] Sheldon believed that body physique was an accurate and reliable indicator of personality and consequently a predictor of behavior.[47] He spent much of his career as a psychologist developing a classification system of body types and the personality patterns, or temperaments, that are associated with them.[48]

Derived from detailed measurements from photographs of volunteer male college students, Sheldon advanced three primary structures of human physique, or **somatotypes**: endomorphy, mesomorphy, and ectomorphy.[49] *Endomorphy* is indicated by a soft roundness throughout the regions of the body. According to Sheldon, the digestive system is "massive and highly developed," resulting in a tendency for obesity.[50] Bones tend to be small and limbs short and tapering.[51] In contrast, bone and muscle dominate *mesomorphy*, with a physique characterized by "uprightness and sturdiness of structure."[52] *Ectomorphy* is char-

somatotypes Structures of human physique, or body types. William Sheldon identified three somatotypes: endomorphy, mesomorphy, and ectomorphy. All people possess each of these body types to varying degrees.

acterized by fragility, linearity, and delicacy. The extremities are "poorly mus-cled" with "delicate bones."[53]

Sheldon extended his somatotype classification scheme by connecting physique to personality temperament. Beginning with a large inventory of per-sonality traits, Sheldon identified three major clusters of temperament: viscero-tonia, somatotonia, and cerebrotonia.[54] A *viscerotonic* temperament is relaxed and outgoing and includes a desire for comfort. A *somatotonic* temperament is active, assertive, motivated, and achievement-oriented. A *cerebrotonic* tempera-ment is introverted, inhibited, and restrained. Sheldon went on to argue that there is a close correspondence between these three temperament clusters and the three major somatotypes that he advanced earlier (see Research in Action, "Body Physique and Temperament").

Sheldon made direct application of this two dimensional classification sys-tem to delinquency behavior in *Varieties of Delinquent Behavior* (1949).[55] Here he

RESEARCH IN ACTION — Body Physique and Temperament

Through a series of studies, William Sheldon isolated three primary structures of human physique that he said were associated with three personality tem-perament types.

SOMATOTYPES	TEMPERAMENT
Endomorphic: A soft roundness of the body. The digestive system is large and highly developed, while other features of the body are weak and underdeveloped. Small bones; soft and smooth skin.	**Viscerotonic** is "characterized by general relaxation, love of comfort, sociability, conviviality, gluttony for food, for people and for affection."
Mesomorphic: Bone and muscle predominate. The physique is hard, firm, upright, strong, and sturdy. Large blood vessels. Skin is "thick, with large pores."	**Somatotonic** is "a predominance of muscular activity and of vigorous bodily assertiveness. The motivational organization seems dominated by the soma [physical]. These people have vigor and push. . . . Action and power define life's primary purpose."
Ectomorphic: Fragile, thin, and delicate. "Poorly muscled extremities" with delicate bones. Muscles are poorly developed. Relative to mass, individuals characterized by ectomorphy have "the greatest surface area and hence the greatest sensory exposure to the outside world."	**Cerebrotonic** is "a predominance of the element of restraint, inhibition, and of the desire for concealment. Cerebrotonic people shrink away from sociality as from too strong a light. They . . . avoid attracting attention to themselves."

Sources: Sheldon, *Human Physique,* 8; Sheldon, *Temperament,* 10–11.

reported research findings from a study over a 10-year period of 200 delinquent boys placed in the Hayden Goodwill Inn, a small private residential facility in Boston. Sheldon compared these boys to a group of male college students whom he had examined previously. The difference between the two groups was statistically significant, with the delinquent boys being more mesomorphic and less ectomorphic than the male college students.[56] Sheldon went on to compare the delinquency of the boys in residential care with that of their parents. Observing similarity, he concluded that the tendency to become delinquent is inherited.[57]

Sheldon and Eleanor Glueck

In 1950, Sheldon and Eleanor Glueck published an extensive study called *Unraveling Juvenile Delinquency*, which is now considered classic but controversial.[58] "The Gluecks sought to answer a basic and enduring question: What factors differentiate boys reared in poor neighborhoods who become serious and persistent delinquents from boys raised in the same neighborhoods who do *not* become delinquent or antisocial? To address this question, the Gluecks studied in meticulous detail the lives of 500 delinquents and 500 nondelinquents who were raised in the same low-income environments of central Boston during the Great Depression era (circa 1928–1940)."[59] Among the factors found to distinguish these two groups was physique. Using two independent methods of physique classification, the Gluecks found that delinquent boys were disproportionately mesomorphic.[60]

In light of these findings, the Gluecks devoted an entire book to the topic of *Physique and Delinquency* (1956). Here they reported that 60% of the delinquents, but only 31% of the nondelinquents, were mesomorphic.[61] Their study also considered a large number of personality traits and sociocultural factors in connection with physique and delinquency.[62] The Gluecks observed that mesomorphs were characterized by traits particularly suitable to the commission of aggressive acts, such as physical strength, energy, insensitivity, and the tendency to express tension and frustration in action. Mesomorphs also experience few inhibitions that would restrain aggressive behavior.[63] Thus, the interaction of mesomorphic physique with an antisocial temperament and a poor social environment makes for a "delinquency potential."

reductionism The view that a complex phenomenon can be explained by reducing it to its most basic element.

■ Contemporary Biological Approaches: Biosocial Criminology

While early biological approaches attempted to connect physical characteristics with behavior, they often acknowledged the importance of environmental conditions. Nonetheless, their heavy biological emphasis made it seem as if all behavior—including criminal behavior—could be reduced to its biological roots. As such, early biological perspectives are usually characterized as **reductionism**. As we have seen, the pioneering versions of positivist criminology sought to identify, through scientific methods, the physical characteristics of criminals that differentiate them from noncriminals and allegedly produce their criminal behavior. This form of reductionism is referred to as **biological determinism**.

biological determinism Identifying, through scientific methods, the biological characteristics of individuals that are the primary causes of behavior. People are the products of their biological makeup.

When psychology and sociology came on the positivist scene in the late 1800s and early 1900s, scientific attention turned to environmental conditions. In the years following World War II, emphasis on the biological underpinnings of criminology weakened. The *nature* emphasis in biological determinism was, to a great degree, replaced with a focus on *nurture,* in which behavior was seen as a product of social and environmental factors and influences.[64] Over the course of this transformation, biological determinism was displaced by **environmental determinism**, which concentrated on the social conditions that cause behavior. The **nature–nurture debate** that ensued over this period was emotionally charged and discipline-driven. Each side adamantly held its ground, alleging that the causes of criminal behavior were either biological *or* environmental, but certainly not both.[65]

It is now recognized that "the 'nature *or* nurture' debate on the origins of criminal behavior is sadly outdated."[66] As a result, the reductionism of the nature–nurture controversy has faded and contemporary biological perspectives commonly incorporate environmental factors.[67] Sociological theories and research, however, have rarely followed suit by adopting biological elements.[68]

The new, integrated approach of **biosocial criminology** contends that delinquency and other forms of "antisocial behavior result from a combination of social, psychological and biological causes."[69] The approach is deliberately *interdisciplinary* and *integrated,* and it includes factors from a variety of disciplines.[70] A biosocial approach points out that biological factors do not operate alone—they are dynamically related to one another and to environmental conditions, all of which interact to influence behavior.[71]

Biosocial research reveals that biological and social variables are often related in complex ways. It is not enough simply to acknowledge that both biology and environment may independently contribute to delinquent behavior; rather, various biological and environmental factors influence each other, as together they affect the likelihood of delinquent behavior.[72] Diana Fishbein refers to this biosocial process as **nature–nurture interaction** (nature *plus* nurture), while Kristen Jacobson and David Rowe use the term *joint influence of nature and nurture.*[73]

While biological and environmental characteristics and conditions are powerful predictors of antisocial behavior, they are not causal in a deterministic sense. Rather, they are *probabilistic,* increasing the likelihood or "risk" of antisocial behavior.[74] Antisocial behavior is not an inevitable outcome, but people who are vulnerable by virtue of their biological makeup and are exposed to an adverse environment are at greater risk.[75] A variety of behavioral outcomes are possible, not just delinquent behavior. Researchers in this field usually use the general term *antisocial behavior.*[76]

The discussion that follows will not try to inventory all the findings on biological influences in antisocial behavior; instead, we will offer just a sampling, organized in order to point to the major areas of theory and research.[77] In many cases these biological factors reveal nature–nurture interaction. Key biological influences on antisocial behavior are categorized into three groups: (1) the nervous system: **neurophysiology** and **neuropsychology**; (2) biochemical factors; and (3) heredity: behavioral genetics. The first two categories relate to the human body's neurological structure, chemistry, and function—often referred to as

environmental determinism Identifying, through scientific methods, the social conditions that cause behavior. People are the products of their environment.

nature–nurture debate The argument that grew out of biological determinism and environmental determinism. The causes of behavior are either biological or environmental, but not both.

biosocial criminology The approach to the study of criminality that holds that crime and other forms of "antisocial behavior result from a combination of social, psychological, and biological causes" (Brennan and Raine, "Biosocial Bases," 590).

nature–nurture interaction The perspective that it is not enough to say that biological and environmental factors contribute to behavior; rather, each influences the other, as together they produce behavior.

neurophysiology The physical structure of the nervous system, including the central and peripheral nervous systems.

neuropsychology The link between the body's neurological structure, including the central and peripheral nervous systems, and psychological functioning.

central nervous system (CNS) The part of the human nervous system that includes the brain and spinal cord.

peripheral nervous system The part of the human nervous system that emanates from and lies outside of the brain and spinal cord.

"anatomy" and "physiology." The last category considers research findings on heritable contributions to delinquent behavior.

The Nervous System: Neurophysiology and Neuropsychology

The human nervous system has two major divisions: the **central nervous system (CNS)** and the **peripheral nervous system**.[78] Each is composed of cells called *neurons* or *neural cells*. The central nervous system includes the brain and spinal cord, whereas the peripheral nervous system includes those neural cells emanating from the brain and spinal cord, which carry information to and from the CNS. The peripheral nervous system can be broken down further into two subsystems: somatic and autonomic. The **somatic nervous system** primarily consists of sensory and motor nerves, providing awareness or consciousness of sensations such as light, touch, sound, and smell, and the resulting activation of muscle groups. The **autonomic nervous system (ANS)** controls bodily functions that are usually beyond people's conscious control, including blood pressure, heart activity, breathing, and hormone levels.[79] **Figure 8-1** shows the human nervous system. The neuronal structures (*anatomy*) and functions (*physiology*) of the nervous system have powerful influences on behavior, but, as we will see, these influences must be understood in the context of environmental conditions.

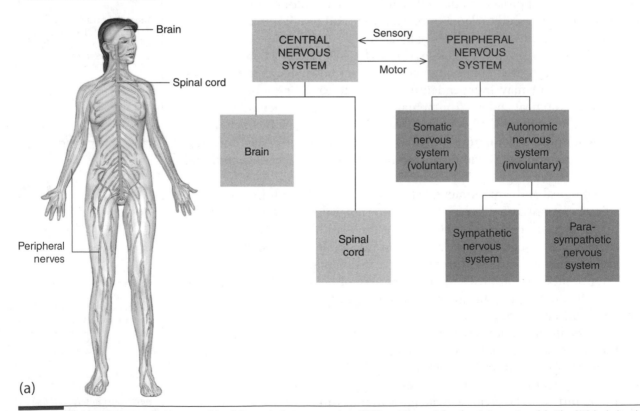

Figure 8-1 The human nervous system consists of the central nervous system (CNS) and the peripheral nervous system (a). The CNS includes the brain and spinal cord, and the peripheral nervous system consists of the somatic and autonomic nervous systems. The nervous system includes such important components as the brain stem (b) and the reticular activating system (c).

Sources: (a) Alters, *Biology,* 286; (b) Clark, *Anatomy and Physiology,* 196; (c) Chiras, *Human Biology,* 187.

Figure 8-1 Continued

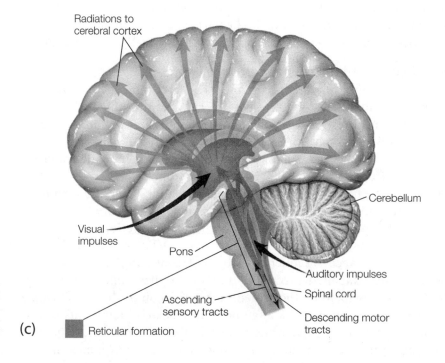

(b)

(c)

The human brain has three primary components: the brain stem, the limbic system, and the cerebral cortex. The *brain stem* is positioned at the top of the spinal cord and is continuous with it. A number of structures make up the brain stem, including the medulla, the pons (see Figure 8-1b), and the reticular formation or reticular activating system (RAS) (see Figure 8-1c). The RAS functions as an information filter, arousing and alerting the individual to environmental stimuli. Collectively, the structures of the brain stem are responsible for physical survival, controlling reflexive and instinctual behaviors.

Surrounding the brain stem is a set of linked brain regions referred to as the *limbic system*. The limbic system includes the hypothalamus, hippocampus, olfactory bulb, and numerous other structures that evoke and regulate motivation and emotion in response to environmental stimuli. In general, damage or disturbance

somatic nervous system
The part of the human nervous system that consists of sensory and motor nerves, providing awareness or consciousness of sensations such as light, touch, sound, and smell, and the resulting activation of muscle groups.

autonomic nervous system (ANS) The part of the peripheral nervous system that controls unconscious bodily functions such as blood pressure, heart activity, breathing, and hormone levels.

of the limbic system results in problems related to motivation and inhibition, including "affective disorders (depression), violence, sleep problems, memory lapses, sexual disorders, and other so-called mental disturbances."[80]

The *cerebral cortex* (the outer layer of the cerebrum; Figure 8-1b) is the largest part of the brain. "This structure is responsible for higher intellectual functioning, such as problem solving, logic, forethought, insight, information processing, and decision making."[81] It is referred to as the "thinking" or "higher" brain because it analyzes and organizes information from other parts of the brain and relays back responses.[82] The cerebral cortex consists of two hemispheres, divided into four lobes: frontal, temporal, parietal, and occipital. Each hemisphere has its own specialized functions. The right hemisphere is the center of perception, creativity, and emotion (particularly negative emotion). The left hemisphere is responsible for analytic, verbal, and sequential thinking, in addition to positive emotion. The frontal lobes instigate goal-directed behavior and are responsible for behavioral sequencing. The temporal lobes provide auditory perception, temporal sequencing, memory, and emotions. The parietal and occipital lobes are involved in sensory perception, language, and abstract information processing.[83]

The Prefrontal Lobes of the Cerebral Cortex

Biosocial scientists who are interested in antisocial behavior have focused attention on the prefrontal lobes of the cerebral cortex because this is the region of the brain that is most responsible for impulses, emotions, and goal-directed actions: "Many individuals with conduct disorders, antisocial behavior, hyperactivity, and other traits that place an individual at risk for delinquency or criminal behavior and drug abuse are believed to suffer from defects in the cortex, particularly within the frontal lobes."[84] The prefrontal lobe is about one third of the cerebral cortex and it serves an integrative and supervisory role in the brain.[85] In particular, problems with **executive cognitive functions** (ECF) have been associated with prefrontal lobe abnormalities: "Deficits in the prefrontal cortex may reduce the executive function—that is, the ability to plan and to reflect on one's actions. Impaired executive function implies impulsiveness and disorganized behavior, a focus on the present rather than on the future."[86] Diana Fishbein observes that "reviews of a large body of research unanimously conclude that impairment in cognitive or neuropsychological functions, which involve information processing, memory, and assessment of environmental cues [ECF], are implicated in poor self-regulation of behavior."[87]

executive cognitive functions (ECF) A person's ability to plan and reflect on his or her actions. This involves information processing, memory, assessment, and self-regulation.

Brain Imaging

Advances in brain-imaging techniques allow for direct assessment of brain structure and function. In particular, new imaging techniques such as *positron emission tomography* (PET) and *functional magnetic resonance imaging* (fMRI) have been used to detect deficits in brain structure and function.[88] After conducting a comprehensive review of brain-imaging studies, Adrian Raine concludes: "An integration of findings from these studies gives rise to the hypothesis that frontal [lobe] dysfunction may characterize violent offenders while temporal lobe dysfunction may characterize sexual offending; offenders with conjoint violent and sexual behavior are hypothesized to be characterized by both frontal and temporal lobe dysfunction."[89]

Electroencephalograph (EEG)

Another method of measuring brain abnormalities is the electroencephalograph (EEG), which has a long history of use. Researchers have used the EEG to identify differences in brain activity when comparing people with behavioral disorders and those without.[90] The EEG measures the brain's electrical activity as indicated by brain waves. When an individual is presented with a stimulus, the brain's response can be tracked as it travels through the various regions of the brain. The signals that are received and recorded by the EEG monitor are called "evoked potentials."[91] "Individuals with a history of drug abuse and those with impulsive aggression tend to show relatively slow wave activity in their EEGs and delay in their evoked potentials (EPs). . . . Individuals who have a greater amount of the slow waves may not process information as efficiently or effectively; thus, such slowing may be related to cognitive deficits."[92]

Autonomic Nervous System

Low levels of central nervous system activity, indicated by brain-imaging and EEG studies, suggest a neurophysiological basis to antisocial behavior. Still another indication of this is offered by research on the autonomic nervous system (ANS). The ANS is a part of the peripheral nervous system, and it controls involuntary bodily functions such as blood pressure, heart rate, intestinal activity, and hormone levels. When confronted with stress, the limbic system activates the ANS to produce a number of physiological responses, designed to motivate and mobilize the body for an efficient and effective behavioral response.[93] This "fight-or-flight" response involves increasing heart rate, blood pressure, respiration, and skin electricity conductance resulting from stimulated sweat glands.

The polygraph (lie detector) is a method of detecting nervous arousal that is based on this fight-or-flight response. The measure of nervous arousal is used to determine whether a subject is telling the truth. "The theory is that, as children, most people have been conditioned to anticipate punishment when they tell a lie. The anticipation of punishment produces the involuntary 'fight or flight' response, which results in a number of measurable changes in heart, pulse, and breathing rate, and, because sweat itself conducts electricity, in the electric conductivity of the skin."[94]

The theory behind the polygraph provides the basic reason that researchers believe the ANS plays an important role in antisocial behavior. Hans Eysenck, a well-known English psychiatrist, argues that children with habitually low cortical arousal "condition poorly" because they do not anticipate punishment. As a result, they fail to develop a conscience that prevents them from committing crime. According to Eysenck, the conscience is a conditioned response that is at the core of socialization.[95] Thus, "because the anxiety reaction in anticipation of punishment is essentially an autonomic nervous system function related to the fight or flight response, the level of socialization in children may depend at least in part the functioning of that system."[96] **Conditionability**, then, is based on anxiety reaction—the desire and ability to avoid or to quickly reduce punishment. When a child's ANS is characterized by low levels of arousal, there is little conditionability. David Rowe describes this **low arousal theory** in terms of "fearlessness"—the lack of fear could "predispose toward crime because fearless

conditionability The variable tendency to anticipate punishment. Anxiety reaction (fear), in anticipation of punishment, is the primary mechanism of socialization. People with low ANS activity condition poorly because they do not anticipate punishment and subsequently fail to develop a conscience.

low arousal theory Low levels of cortical arousal are associated with lack of fear. Since anxiety reaction in anticipation of punishment is the basis for conditioning, individuals with low arousal levels have little desire or ability to avoid punishment. Therefore, they are often poorly socialized and engage in antisocial behavior without restraint.

children would be more difficult to socialize than fearful ones—punishment would arouse a less intense emotion, and the lesson [is] inadequately learned."[97]

Research has found that, in a resting state, antisocial individuals have lower levels of skin conductance (SC) and lower heart rates as compared with normal individuals.[98] These findings suggest that antisocial individuals have a lower level of ANS activity, motivating them to seek arousal through a "variety of means, such as risk-taking, sensation-seeking, impulsive action, socializing with many other people, drug abuse, multiplicity of sexual partners, etc. These activities are likely to lead such a person toward criminal activity, but not inevitably; risky sports activities may take the place of criminality in middle- and upper-class persons."[99]

The relationship between skin conductance and aggressiveness and antisocial behavior has been studied most extensively with regard to a subgroup of criminal offenders classified as "psychopathic"—individuals with personality and behavioral traits that indicate a lack of emotion and empathy for others (see Expanding Ideas, "Psychopathy").[100] As anticipated by low arousal theory, "deficits in measures of SC arousal are believed to be associated with low autonomical arousal levels which are, in turn, related to low emotionality, poor conditionability, lack of empathy and remorse, and ability to lie easily."[101] Expanding Ideas, "Psychopathy," expands on the concept of psychopathy.

Resting heart rate provides another measure of ANS functioning.[102] "Overall, there is considerable evidence for a link between a low resting heart rate and higher rates of antisocial behavior, especially violence."[103] Low heart rate is said to be indicative of low ANS arousability.[104] Consistent with low arousal theory, these findings have been interpreted to show that antisocial behavior is, to some degree, sensation-seeking activity.

Biochemical Factors

The biochemistry of the human body influences behavior by affecting the central and peripheral nervous systems.[105] We will focus on two key aspects of biochemistry that have been connected extensively with antisocial behavior: neurotransmitters and the hormone testosterone.[106]

Neurotransmitters

The intricate communication system of the brain is composed of billions of nerve cells called neurons.[107] All thoughts, feelings, emotions, and behaviors are the result of this system.[108] Each neuron consists of three basic structures. "The *soma* is the cell body, including the nucleus of the cell. *Dendrites* extend from the soma in the form of several branches to receive and respond to electrical activity of other neurons. The *axon* also extends from the soma to transmit electrical activity from the soma to other neurons, muscles, or glands. When an electrical impulse is conducted from the soma down the axon, it will reach the *synapse* which is the gap between cell bodies."[109] The axon then releases specific chemicals, called **neurotransmitters**, into the synapse. These neurotransmitters carry the signal to a receptor site on the dendrite of the neighboring neuron. After transmitting the signal, the neurotransmitter is released back into the synapse, where it is metabolized or reabsorbed back into the sending neuron.[110] This neuronal anatomy and physiology is shown in **Figure 8-2.**

neurotransmitters
Chemical compounds, found in the synapse between nerve cells, that carry signals from one neuron to another. Neurotransmitters provide a crucial link in the neural communications system.

Psychopathy

Some juvenile delinquents are loosely referred to as *psychopaths* because they lack anxiety and guilt for the wrongs they do. In the past, this term clinically designated a distinctive pattern of behaviors and personality traits. Hervey Cleckley characterized psychopaths as "chronically antisocial individuals who are always in trouble, profiting neither from experience nor punishment, and maintaining no real loyalties to any person, group, or code. They are frequently callous and hedonistic, showing marked emotional immaturity with lack of responsibility, lack of judgement and an ability to rationalize their behavior so that it appears warranted, reasonable, and justified."

Today, this clinical term has been subsumed under the diagnostic category of *antisocial personality disorder* in the *Diagnostic and Statistical Manual (DSM-IV)* of the American Psychiatric Association. The DSM-IV says that "the essential feature of antisocial personality disorder is a pervasive pattern of disregard for, and violation of, the rights of others that begins in childhood or early adolescence and continues into adulthood." Some of the characteristics of antisocial personality disorder include:

- repeatedly performing acts that are grounds for arrest
- deceitfulness
- impulsivity
- irritability and aggressiveness
- reckless disregard for safety of self or others
- consistent irresponsibility
- lack of remorse

A person must be 18 years of age or older, however, to be diagnosed with antisocial personality disorder. Juveniles who display similar characteristics are diagnosed with *conduct disorder.* This diagnostic category refers to "a repetitive and persistent pattern of behavior in which the basic rights of others or major age-appropriate societal norms or rules are violated. . . ." The following are some of the criteria:

- often bullies, threatens, or intimidates others
- often initiates physical fights
- has been physically cruel to animals
- has deliberately engaged in fire setting
- has broken into someone's house, building, or car
- often stays out night despite parental prohibitions, beginning before age 13
- is often truant from school, beginning before age 13

Sources: American Psychiatric Association, *DSM-IV,* 90–91, 645; Cleckley, "Psychopathic States," 567–569; Widiger and Lynam, "Psychopathy."

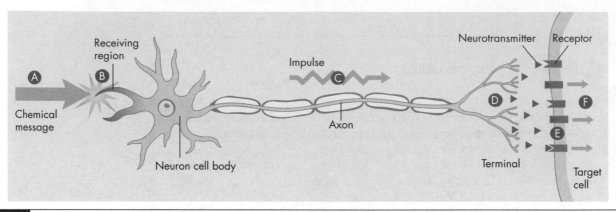

Figure 8-2 The process of sending messages by neurons. The receiving region (B) of the neuron is activated by an incoming message (A) near the neuronal cell body. The neuron sends an electricity-like impulse down the axon to its terminal (C). The impulse causes the release of neurotransmitters from the terminal to transmit the message to the target (D). This is done when the neurotransmitter molecule activates the receptors on the membranes of the target cell (E). The activated receptors then cause a change in intracellular function to occur (F).

Source: Hanson et al., *Drugs and Society,* 115.

Three specific neurotransmitters—dopamine, serotonin, and norepinephrine—have been linked to the likelihood of antisocial behavior. In addition, the enzyme monoamine oxidase is also implicated in antisocial behavior, especially aggression and violence, because of its critical role in regulating neurotransmitter concentrations and activity.

Dopamine affects an individual's ability to associate cues (events or objects) in the environment with rewards and punishments. As a result, conditioned responses are influenced by levels of dopamine. Increased dopamine levels and activity appear to evoke emotions, as part of the fight-or-flight response, which in turn motivate behavior. As a result, the overproduction of dopamine has been associated with psychotic behavior and with aggression and violence. Antipsychotic drugs that are administered to decrease dopamine levels tend to reduce aggressive behaviors.[111] However, research findings have been inconsistent and researchers have not uncovered a direct effect of dopamine levels on aggression.[112]

Abnormally low levels of *serotonin* have been connected with impulsive and aggressive behavior in both juveniles and adults. While research specifically relates low serotonin activity with impulsivity, the expression of that impulsivity depends on a variety of predisposing and environmental factors: "A deficit in serotonin activity jeopardizes the ability to inhibit urges, increasing the likelihood that underlying hostility or negative mood will lead to aggression or another inappropriate behavior."[113]

Norepinephrine is a neurotransmitter that is produced from dopamine. It has been found to play a central role in the fight-or-flight response by causing the release of stress hormones from the adrenal glands and by stimulating the central and peripheral nervous systems. Generally, norepinephrine is related to arousal, agitated mood states, and behavioral activation. While several studies seem to establish a link between violence and norepinephrine, others do not. The "actual behavioral outcome depends upon the circumstance, setting, and individual predisposition."[114]

Monoamine oxidase (MAO) is an enzyme responsible for the breakdown of neurotransmitters, including dopamine, serotonin, and norepinephrine. As a result, MAO plays a vital role in regulating neurotransmitter concentrations and activity levels. Low MAO activity results in excessive levels of dopamine and norepinephrine, both of which seem to contribute to the fight-or-flight response, aggression, low impulse control, and loss of self-control.[115] In a review of the literature, Lee Ellis reports that extensive research has linked irregular MAO levels with antisocial behaviors, especially psychopathy, aggression, and violence.[116]

Testosterone

Hormones are chemical compounds secreted into the bloodstream by various glands that make up the endocrine system. Hormones are carried throughout the body, where they regulate or control certain cells and organs. One particular "sex" hormone, secreted by the testes, ovaries, and adrenal glands, has received the vast majority of attention with regard to antisocial behavior: **testosterone**. Sometimes identified as a male sex hormone, testosterone is also produced by females, but in much smaller amounts and with different effects.[117] This sex difference in testosterone level is presumed to be partly responsible for higher levels of aggressive and criminal behavior among males.[118]

testosterone Chemical compounds secreted into the bloodstream by various glands. This hormone plays an important role in aggression and violence.

Animal studies have clearly and consistently demonstrated that testosterone plays an important role in the expression of aggressive and violent behavior.[119] Human research is far less conclusive. In particular, a cause and effect relationship between testosterone level and subsequent behavior has not been established and issues abound related to the measurement of such a cause and effect relationship.[120]

A number of research findings qualify the relationship between testosterone and aggression. The relationship appears strongest for older adolescent and adult males, but it is weak and inconsistent for male children and younger adolescents.[121] For example, while one study found that higher serum testosterone levels differentiate violent offenders from nonviolent offenders in a sample of 15–17 year-old males, another study found that serum testosterone levels failed to differentiate a group of highly aggressive 4–10 year-old boys from a carefully matched control group.[122] It is also the case that social variables may intervene in the relationship between testosterone and behavior. In a widely cited study, Alan Booth and Wayne Osgood examined the relationship between testosterone, social integration, prior involvement in juvenile delinquency, and adult deviance.[123] While testosterone level was strongly associated with adult deviance, the relationship was reduced significantly for individuals who were socially integrated and who were involved in crime as juveniles. These findings indicate that the effect of testosterone on deviance is conditioned by social factors. Similarly, a study of military veterans found that high levels of testosterone were associated with multiple measures of aggression, including retrospective reports of juvenile delinquency, adult crime, and substance abuse. However, this association was true only for veterans from a lower social class, which indicates again the importance of social context to antisocial behavior.[124] Finally, research shows that aggressive behavior may increase testosterone production, calling into question the causal order of these variables.[125] Recent research shows that testosterone levels rise and fall in the course of involvement in competitive sport or in response to a competitive challenge. In one study, levels of testosterone rose prior to a competitive tennis match, declined while the match was being played, and increased dramatically for players who won but dropped for players who lost.[126]

Heritability: Behavioral Genetics

Until recently, the role of genetic influences in criminal behavior was heavily discounted and, for the most part, ignored by social scientists.[127] Genetic contributions are still misunderstood and sometimes distorted. "Identification of genetic contributions does not reduce behavior to a gene level," nor does identifying the role of genetics in criminal behavior imply that there is a "crime gene."[128] Moreover, a genetic predisposition toward criminal behavior does not mean that an individual is destined to become criminal, regardless of environmental conditions.[129] Rather, **behavioral genetics** seeks to "estimate the relative contributions of heredity and environment" on behavioral traits, such as impulsivity, aggressiveness, and negativity.[130] Whether these traits are given full expression in the form of delinquent and criminal behavior depends on environmental conditions. Thus, as a biosocial approach, behavioral genetics incorporates the basic perspective of nature–nurture interaction.[131]

Behavioral genetics research attempts to sort out influences on aggression and criminal behavior in terms of three sources: shared environmental factors, nonshared environmental factors, and genetic factors.[132] *Shared environmental factors* are conditions and experiences shared by family members. Family structure and socioeconomic status are examples of shared environmental factors when they have the same type of influence on all family members. *Nonshared environmental influences* are conditions and experiences that family members do not have in common or dissimilar responses to shared experiences. Siblings may respond to divorce, for example, in very different ways. Also, siblings often do not share the same friends. *Genetic factors* refer to traits and characteristics that are handed down from parent to child, based on the influence of genes. The measure of this genetic influence is called **heritability**.[133]

Researchers have found it impossible to disentangle environmental and genetic influences by looking just at traditional families. Two contemporary research designs have been developed in an effort to control for genetic influences and at the same time consider shared and nonshared environments: adoption studies and twin studies.

Adoption Studies

Adoption studies obtain data from adopted individuals and their adoptive and biological parents. Similarity between adoptive parents and their adopted children is likely due to shared environmental influences, because adoptive parents and their children for the most part do not share genes.[134] If adopted individuals are similar to their biological parents, then genetic factors are implied for those traits.

The most famous adoption study of antisocial behavior used data from an extensive Danish Adoption Register. These data on 14,427 Danish adoptees and their biological and adoptive parents showed that rates of court convictions were higher for males when their biological parents were convicted.[135] If neither the

behavioral genetics The science that tries to estimate the relative contributions of heredity and environment to personality and behavioral traits such as impulsivity, constraint, aggressiveness, and negative emotionality.

heritability The measure of genetic influence.

biological nor the adoptive parents were convicted, 13.5% of the sons were convicted. When the adoptive parents were convicted and the biological parents were not, 14.7% of the sons were convicted. In cases where the biological parents were convicted and the adoptive parents were not, 20% of the adopted sons were convicted. The highest rates occurred when both biological and adoptive parents had criminal convictions: 24.5% of these sons were convicted.[136] The number of convictions for biological parents was also related to the likelihood of criminal convictions for adopted sons such that a greater number of convictions for biological parents was associated with a greater number of convictions for biological sons. This relationship was much stronger for property offenses than for violent offenses.

Twin Studies

Twin studies compare identical twins to fraternal twins. Identical or *monozygotic twins* (MZ) are genetically identical, sharing 100% of their genes. Fraternal or *dizygotic twins* (DZ) share, on average, 50% of their genes. Genetic influences are implicated to the degree that the similarity observed in MZ twins is greater than that observed in DZ twins.[137] In twin studies such similarity is referred to as *concordance rates*.

Jasmine Tehrani and Sarnoff Mednick report that ten twin studies have examined a genetic effect on criminal behavior.[138] Tehrani and Mednick conclude that a greater concordance rate for criminal behavior is observed for MZ twins than for DZ twins. However, they also point out that "some researchers believe that the twin methodology may be flawed in that MZ twins, in addition to sharing more genetic information than DZ twins, are also more likely to be treated more similarly than DZ twins."[139] One methodological solution to this problem is to study twins reared apart. Grove and his colleagues studied concordance rates of antisocial behavior in a sample of thirty-two sets of MZ twins that were reared apart, having been adopted by non-relatives shortly after birth.[140] These researchers found that MZ twins reared apart displayed concordance in both childhood conduct disorders and adult antisocial behaviors. While the sample size was small and concordance rates were less than in previous twin studies, these findings substantiate that antisocial behavior has a genetic component.

Molecular Genetic Studies

Adoption and twin studies suggest that there is a heritable component to traits that underlie delinquent and criminal behavior—traits such as impulsivity, aggressiveness, and negativity. These studies do not, however, identify the genetic mechanism that may contribute to these traits.[141] Diana Fishbein provides a brief and useful overview of the contemporary study of molecular genetics related to delinquent and criminal behavior: "Characteristics of genes that have been linked to psychological and behavior traits are called *markers*; they 'mark' a location of genes that may be actively involved in contributing to a trait. Genetic *variants* are structural differences in genes, also called *polymorphism*. Eventually, when variants and markers have been identified for relevant traits, we will understand better how genes are expressed, or become active, in response to environmental input, and how their activity (or lack thereof) contributes to a behavioral trait."[142]

Gene-Based Evolutionary Theories

Research evidence supporting a genetic basis to crime and delinquency is often advanced without theory. The chief aim is simply to identify genetic factors and processes. Modern gene-based evolutionary theories attempt to fill this void. Lee Ellis and Anthony Walsh, two leading advocates of this perspective, present five gene-based evolutionary theories, all of which "rest on the assumption that genetic factors influence criminal behavior."[143] As the name states, this group of theories combines evolutionary theory with the field of genetics. The different versions of gene-based evolutionary theory converge on "a simple but powerful idea: To the degree a particular characteristic is prevalent in a population, it is likely to have contributed to the reproduction success of the ancestors of the individual currently living."[144]

In attempting to explain crime and delinquency, gene-based evolutionary theories acknowledge that, in all likelihood, there is not a "crime gene." Rather, genes contribute to the development of individual traits that are conducive to exploitive behavior. Ellis and Walsh contend that two traits are especially relevant to the study of antisocial behavior: *deception* and *cheating*.[145] "When human deception and victimizing behavior reach high levels of harm, the term *criminal* is often applied to the behavior, and when offenders are sufficiently chronic in these activities, they are said to be *antisocial* or *psychopathic* (or *sociopathic*)."[146] The evolutionary part of these theories claims that people with these traits are able to reproduce at relatively high rates, thereby passing on their genes to others and perpetuating a genetic predisposition to antisocial behavior.

Some gene-based evolutionary theories are directed at particular types of crime, such as rape and child abuse, while others attempt to explain crime in general. One theory focuses on the high rates of reproduction among antisocial individuals and their neglect of parenting. Ellis and Walsh offer a series of hypotheses for each of the theories, but many of them defy testing because their components are unmeasurable or unverifiable. For example, one hypothesis offered in their gene-based evolutionary theory of rape states: "Rape should be strongly resisted by female victims because it denies them the opportunity to choose sex partners who are most likely to help care for offsprings."[147]

Biosocial criminologist do *not* argue that biological factors lead directly to crime and delinquency. Rather, they argue that biological factors, in combination with environmental conditions, produce personality traits that are conducive to involvement in delinquency and crime.[148] As we have seen, the biosocial approach contends that biological characteristics and processes influence *conditionability,* thereby affecting individuals' ability to respond to the environment in a socially acceptable manner. We now turn to the psychological study of personality, here considered as an extension of the biosocial developmental perspective of antisocial behavior.[149]

■ Personality and Biosocial Development

While the discipline of psychology has not come to a consensus on a definition of personality, the concept has acquired greater precision and shared meaning in the last thirty years. Advances in personality theory and research have resulted in significant strides in understanding antisocial behavior in general and juvenile delinquency in particular.[150]

In their 1950 book, *Unraveling Juvenile Delinquency*, Sheldon and Eleanor Glueck itemized the personality traits that distinguish delinquent youth from nondelinquent youth, as indicated by their extensive study of 500 delinquents and 500 nondelinquents. "On a whole, delinquents are more extroverted, vivacious, impulsive, and less self-controlled than the non-delinquents. They are more hostile, resentful, defiant, suspicious and destructive. They are less fearful of failure or defeat than the non-delinquents. They are less concerned about meeting conventional expectations, and more ambivalent toward or far less submissive to authority. They are, as a group, more socially assertive. To a greater extent than the control group, they express feelings of not being recognized or appreciated."[151]

While out of step with dominant sociological approaches of the time, the Gluecks gave extended attention to individual "temperamental traits" and applied these traits to a variety of behaviors, especially delinquency.[152] In addition, by associating temperament with physique, they suggested that personality traits may have biological roots. Their approach, then, was biosocial and developmental: "The factors involved [in juvenile delinquency] are neither essentially biological nor essentially sociologic, but biosocial. We are concerned, for example, with the results of such a dynamic process as the introjection of certain childhood experiences and the effect of such activity on the development of personality and character."[153] Contemporary theory and research have, to a great degree, revived the Gluecks' point of view by defining personality in terms of traits and by examining both biological and environmental aspects of personality.

<u>Personality</u> refers to reasonably stable patterns of perceiving, thinking, feeling, and responding to the environment.[154] "Traits are the basic building blocks of personality," providing foundation to thought, emotion, character, and behavior.[155] The Gluecks inventoried sixty-seven traits in their study. Contemporary theory and research contend that there are "a finite number of basic traits, and that these traits provide comprehensive coverage of human personality."[156] Contemporary researchers have found that personality can be characterized along a number of key dimensions, sometimes called "*superfactors*," that organize the array of personality traits into a limited number of categories according to their interrelatedness.[157] Four theoretical models of personality are most widely used. Because these theories group personality traits into a limited number of superfactors, they are referred to as "structural models of personality."[158]

personality The reasonably stable patterns of perceiving, thinking, feeling, and responding to the environment. Traits are the basic building blocks of personality, providing foundation to thought, emotion, character, and behavior.

Trait-Based Personality Models[159]

Eysenck's PEN Model

Hans Eysenck associated three personality dimensions with crime and delinquency: *extraversion (E), neuroticism (N),* and *psychoticism (P)*.[160] Each of these superfactors represents a collection of temperament traits that are often expressed together and that typify an individual's responses to environmental stimuli. Eysenck also argued that there is an underlying biological basis to these superfactors.

Eysenck's pioneering work originally advanced a two-factor model consisting of *extraversion* and *neuroticism*.[161] This basic conceptualization of personality has been used repeatedly. Extraversion is contrasted with introversion, and neuroticism is contrasted with emotional stability, resulting in a personality model that can be visualized by two intersecting dimensions laying at right angles. **Figure 8-3** portrays the various temperament traits that compose each of the four quadrants. Extraverts high in neuroticism, for example, tend to be touchy, restless, aggressive, and excitable, whereas introverts who are emotionally stable are likely to be passive, careful, thoughtful, and controlled.

Additional analysis led Eysenck to the identification of a third superfactor, *psychoticism,* because certain types of crime and offenders were not adequately captured in the original two-factor model.[162] Psychoticism is characterized by social insensitivity, self-centeredness, unemotionality, impersonality, impulsiveness, and aggressiveness (see Expanding Ideas, "Trait-Based Personality Models," for a summary of these superfactors).

Eysenck contended that, in general, "crime and antisocial conduct are positively and causally related to high psychoticism, high extraversion, and high

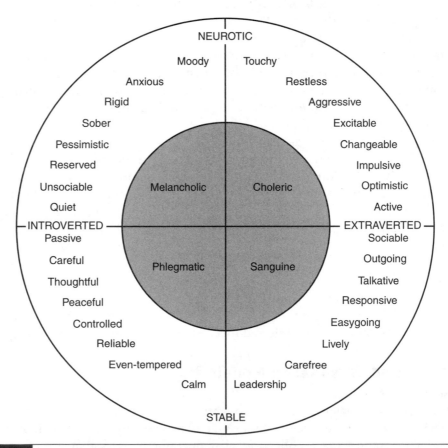

Figure 8-3 Eysenck's superfactors of extraversion and neuroticism. Extraversion is contrasted with introversion, while neuroticism is contrasted with emotional stability. The relationship between these two personality dimensions results in four temperaments identified by ancient Greek physicians.

Source: Ewen, *Personality,* 139. Copyright © 1998, Lawrence Erlbaum Associates, Inc. All rights reserved. Reproduced with permission of the publisher.

Trait-Based Personality Models

Eysenck's PEN Model

Extraversion	Sociability, impulsivity, optimism, and need for excitement and new experiences
Neuroticism	Emotionality, intense and persistent reaction, nervousness, anxiety, and insecurity
Psychoticism	Social insensitivity, self-centeredness, unemotionality, impersonality, impulsiveness, and aggressiveness

Tellegen's Structures of Mood and Personality

Positive emotionality	Optimism, social well-being, social potency, social closeness, and achievement orientation
Negative emotionality	Experience of negative emotions including fear, anxiety, and anger; results in aggression, alienation, and stress reaction
Constraint	Ability to control impulses, plan and act deliberately; harm and danger avoidance; traditionalism

Five-Factor Model of Personality

Extroversion	Talkative, assertive, and energetic; seeks excitement; in contrast to introversion
Neuroticism	Nervousness, anxiety, and insecurity; in contrast with emotional stability
Openness to experience	Interest and willingness to try new activities, ideas, and beliefs; intellectual curiosity
Agreeableness	Manner of social relationships, ranging from compliant to antagonistic
Conscientiousness	Ability to organize, plan, and complete tasks; control impulses and delay gratification

Cloninger's Biosocial Theory

Novelty seeking	Tendency to pursue excitement in new situations; impulsiveness and excitability
Harm avoidance	Propensity to avoid punishment and inhibit behavior; anxiousness, cautiousness, and apprehensiveness
Reward dependence	Penchant for behavior that has been rewarded or provided relief from punishment
Persistence	Perseverance in the face of frustration and fatigue
Self-directedness	Desire for self-determination and displays of will power
Cooperativeness	Agreeableness; in contrast to antagonism and hostility
Self-transcendence	Spirituality

Source: Drawn from Miller and Lynam, "Structural Models," 769 (Table 1).

neuroticism."[163] Further, he theorized that, while psychoticism is related to antisocial behavior for all ages, extraversion applies more readily to children and adolescents, and neuroticism is more relevant for older offenders.[164]

Advancing his personality model as a biosocial theory, Eysenck provides an extended argument that extraversion and neuroticism have biological roots.[165] Extraversion is connected with basic functions of the central nervous system. In particular, he holds that the reticular activating system (RAS) of an extravert excessively filters stimuli, resulting in a habitually underaroused cerebral cortex. (The cerebral cortex is responsible for thinking, memory, and decision making.) In the case of introversion, the RAS fails to filter stimuli adequately, resulting in a perpetually overaroused cerebral cortex. Thus, the extravert, who is cortically underaroused, seeks stimulation to achieve an optimally aroused cortex, and the introvert, who is cortically overaroused, tries to avoid stimulation.[166] Earlier we described the two primary results of low cortical arousal: (1) poor conditionability and the inability to develop a conscience, and (2) sensation-seeking behavior such as crime and delinquency.[167]

Neuroticism, according to Eysenck, reflects the functioning of the autonomic nervous system. Neuroticism is characterized by an unusually sensitive limbic system in which emotionality is achieved quickly and lasts long.[168] Such heightened and prolonged emotionality is likely to lead to criminal activity, mainly as a result of individual agitation, excitability, and impulsiveness. Finally, psychoticism, which relates to impersonality, lack of empathy, aggressiveness, and impulsiveness, is connected to high testosterone levels.[169]

Tellegen's Structures of Mood and Personality[170]

Auke Tellegen's structural model of personality was developed from the study of self-reported mood ratings. The model was then refined through work on a personality questionnaire called the Multidimensional Personality Questionnaire (MPQ).[171] Three superfactors are offered, each consisting of related traits (see Expanding Ideas, "Trait-Based Personality Models"). *Positive emotionality* is the tendency to strive for achievement and to exhibit optimism. It includes aspects of social well-being, potency, closeness, and achievement orientation. *Negative emotionality* involves the tendency to experience negative emotions—most notably fear, anxiety, and anger—resulting in aggression, alienation, and stress reaction. *Constraint* is the tendency to control impulses, act decisively, and avoid harm and danger. This last dimension also involves the endorsement of traditional values and beliefs ("traditionalism").[172]

Five-Factor Model

The five-factor model (FFM) was developed originally from studies of the English language, under the assumption that the most important human traits are encoded in language as single words: the *lexical hypothesis*.[173] Using a statistical technique that identifies common dimensions in data (factor analysis), researchers derived five basic personality factors—sometimes identified as the "big five trait taxonomy": extroversion, agreeableness, conscientiousness, neuroticism, and openness to experience.[174] *Extroversion* has to do with being talkative, assertive, energetic, and seeking excitement. *Agreeableness* involves an individual's approach to social relationships, ranging from compliant to antagonistic. *Conscientiousness* relates to one's ability to organize, plan, and complete tasks, and to control impulses and delay gratification. *Neuroticism* is in contrast to emotional stability and even-temperedness and is characterized by nervousness, anxiety, and insecurity. *Openness to experience* pertains to intellectual curiosity, imagination, conventionality, and the willingness to try new activities (see Expanding Ideas, "Trait-Based Personality Models").[175] Research in Action, "The Big Five Inventory," offers the Big Five Inventory (BFI), developed by Oliver John.

Cloninger's Biosocial Theory

Like Eysenck and Tellegen, C. Robert Cloninger originally advanced three personality dimensions. While each superfactor is alleged to be genetically independent, they represent inherited tendencies that are interrelated in terms of how they are expressed in behavior.[176] *Novelty seeking* is a "tendency toward intense exhilaration or excitement in response to novel stimuli or cues."[177] Novelty seeking has been characterized as impulsivity.[178] *Harm avoidance* is a relates to "aversive conditioning"—the process of learning to inhibit behavior or avoid punishment, novelty, and the frustration of not receiving a reward.[179] Anxiety is the motivation for harm avoidance that leads to conditionability.[180] *Reward dependence* seeks to maintain behavior that has been associated with reward (especially social approval, support, and sentiment) or relief from punishment.[181]

Four additional personality dimensions were added to the original three. *Persistence* applies to the tendency to persevere despite frustration and fatigue.[182] *Self-directedness* refers to an individual's "self determination and willpower."[183] *Cooperativeness* has to do with agreeability in contrast to antagonism and hostility, and *self-transcendence* relates to spirituality (see Expanding Ideas, "Trait-Based Personality Models").

While biological in origin, these personality dimensions are assessed clinically through the use of measurement scales. Cloninger offers separate scales that classify each of the personality dimensions into one of seven categories from "severely high" to "severely low," with an average midpoint.[184] Various combinations of these seven personality dimensions are hypothesized to yield predictable temperament and character traits. In general, Cloninger hypothesizes that antisocial individuals will be high in novelty seeking and low in harm avoidance and reward dependence.[185]

Personality Traits and Antisocial Behavior

After thorough review of the research findings related to these four major models of personality, Joshua Miller and Donald Lynam concluded that "all four models have received sufficient empirical attention to warrant confidence in their reliability and validity."[186] They then ask, how can so many different models be so well validated? "The answer lies in recognizing that there is actually substantial agreement across the models in terms of the traits that are represented."[187] In fact, by examining the eighteen different superfactors (see Expanding Ideas, "Trait-Based Personality Models"), Miller and Lynam conclude that the structural models advancing them are "actually congruent," and that "a strong pattern emerges when the models are integrated."[188] Using labels from the FFM, **agreeableness** exhibited the strongest relationship with antisocial behavior. Agreeableness corresponds to psychoticism in Eysenck's PEN model, negative

agreeableness A dimension of personality that is related strongly to antisocial behavior. Antisocial individuals tend to be hostile, self-centered, spiteful, jealous, and indifferent to others.

RESEARCH IN  The Big Five Inventory (BFI)

One of the personality questionnaires designed to measure the five-factor model of personality (FFM) is the Big Five Inventory (BFI). The BFI is a 44-item inventory, making it short and efficient.

Here are a number of characteristics that may or may not apply to you. For example, do you agree that you are someone who *likes to spend time with others*? Please write a number next to each statement to indicate the extent to which you agree or disagree with that statement.

1. Disagree strongly
2. Disagree a little
3. Neither agree nor disagree
4. Agree a little
5. Agree strongly

I See Myself as Someone Who . . .

____1. Is talkative
____2. Tends to find fault with others
____3. Does a thorough job
____4. Is depressed, blue
____5. Is original, comes up with new ideas
____6. Is reserved
____7. Is helpful and unselfish with others
____8. Can be somewhat careless
____9. Is relaxed, handles stress well
____10. Is curious about many different things
____11. Is full of energy
____12. Starts quarrels with others
____13. Is a reliable worker
____14. Can be tense
____15. Is ingenious, a deep thinker
____16. Generates a lot of enthusiasm
____17. Has a forgiving nature
____18. Tends to be disorganized
____19. Worries a lot
____20. Has an active imagination

____21. Tends to be quiet
____22. Is generally trusting
____23. Tends to be lazy
____24. Is emotionally stable, not easily upset
____25. Is inventive
____26. Has an assertive personality
____27. Can be cold and aloof
____28. Perseveres until the task is finished
____29. Can be moody
____30. Values artistic, aesthetic experiences
____31. Is sometimes shy, inhibited
____32. Is considerate and kind to almost everyone
____33. Does things efficiently
____34. Remains calm in tense situations
____35. Prefers work that is routine
____36. Is outgoing, sociable
____37. Is sometimes rude to others
____38. Makes plans and follows through with them
____39. Gets nervous easily
____40. Likes to reflect, play with ideas
____41. Has few artistic interests
____42. Likes to cooperate with others
____43. Is easily distracted
____44. Is sophisticated in art, music, or literature

Please check: Did you write a number in front of each statement?

BFI scale scoring ("R" denotes reverse-scored items):
Extroversion: 1, 6R, 11, 16, 21R, 26, 31R, 36; Agreeableness: 2R, 7, 12R, 17, 22, 27R, 32, 37R, 42; Conscientiousness: 3, 8R, 13, 18R, 23R, 28, 33, 38, 43R; Neuroticism: 4, 9R, 14, 19, 24R, 34R, 39; Openness: 5, 10, 15, 20, 25, 30, 35R, 40, 41R, 44.

Source: John and Srivastava, "Big Five Taxonomy," 132, see also 114–115, 120.

EXPANDING IDEAS — Two Key Dimensions of Personality

After thorough review of the research, Joshua Miller and Donald Lynam conclude that two dimensions of personality are most strongly related to antisocial behavior: agreeableness and conscientiousness. Antisocial youth tend to be low in each. These two dimensions are identified consistently across different theoretical models of personality and are supported consistently in research.

Five-Factor Model Dimension	Corresponding Dimensions
Agreeableness: Antisocial individuals tend to be hostile, self-centered, spiteful, jealous, and indifferent to others.	*Psychoticism* (Eysenck) *Negative emotionality* (Tellegen) *Cooperativeness* (Cloninger)
Conscientiousness: Antisocial individuals tend to lack ambition, motivation, and perseverance; have difficulty controlling their impulses; seek out activity; and hold nontraditional values and beliefs.	*Psychoticism* (Eysenck) *Constraint* (Tellegen) *Novelty seeking* (Cloninger) *Self directedness* (Cloninger)

Source: Miller and Lynam, "Structural Models," 780.

emotionality in Tellegen's model, and cooperativeness in Cloninger's model (see Expanding Ideas, "Two Key Dimensions of Personality"). Personality dimensions having to do with **conscientiousness** were related to antisocial behavior at lower, but still significant levels. Conscientiousness corresponds to Eysenck's psychoticism, Tellegen's constraint, and Cloninger's novelty seeking and self-directedness (see Expanding Ideas, "Two Key Dimensions of Personality"). Miller and Lynam contend that these two dimensions appear to capture and summarize the most important aspects of personality in relation to antisocial behavior. Antisocial individuals are low in agreeableness and conscientiousness—they tend to be "hostile, self-centered, spiteful, jealous, and indifferent to others," and "they tend to lack ambition, motivation, and perseverance, have difficulty controlling their impulses, and hold nontraditional and unconventional values and beliefs."[189]

While the research evidence linking these basic dimensions of personality and antisocial behavior is substantial and quite consistent, the process by which personality leads to antisocial behavior is not well understood. So far the study of personality traits associated with antisocial behavior is more descriptive than explanatory.[190] Drawing from the work of Avshalom Caspi and Daryl Bem, Miller and Lynam suggest that there are three person–environment transactions that conceivably explain the connection between personality and antisocial behavior: *reactive transactions, evocative transactions,* and *proactive transactions.*

conscientiousness A dimension of personality that is related to antisocial behavior. Antisocial individuals tend to lack ambition, motivation, and perseverance; they have difficulty controlling their impulses; and they hold nontraditional values and beliefs (Miller and Lynam, "Structural Models," 780).

In terms of reactive transactions, individuals low in Agreeableness may be expected, similar to aggressive individuals, to make hostile attributions in ambiguous situations, generate more aggressive responses, and be more likely to believe that aggressive responses will work; these responses will increase the likelihood of violence in any given situation. In terms of evocative transactions, children who are difficult to manage (i.e., low in Agreeableness and Conscientiousness) evoke typical reactions from parents and peers that include harsh and erratic parental discipline, reduction of parental efforts at socialization, increases in permissiveness for later aggression, and peer rejection. In terms of proactive transactions, individuals who are low in Conscientiousness will likely have poorer educational and occupational histories than will those who are high in Conscientiousness; these kinds of decisions will limit the prosocial opportunities for advancement of these individuals.[191]

■ Intelligence and Delinquency

The idea that crime and delinquency are expressions of low intelligence has a long history, but it is a history of confusion, controversy, and in some ways contempt. The systematic study of intelligence began in earnest in the early 1900s, and it quickly connected with the growing field of criminology. From the very beginning, intelligence was viewed as an inherited attribute, but there was much debate about whether it was fixed at birth or could be changed.[192] Criminological interest in intelligence faded by the 1930s and was largely dormant until the late 1970s.

Early Intelligence Tests, Crime, and Delinquency

While attempts to measure intelligence quantitatively began in the latter part of the 1800s, Alfred Binet (1857–1911), a French psychologist, is credited with the development of the intelligence test in the early 1900s. Binet was commissioned to develop a method to identify public school students with learning problems so that they could be placed in special education classes to receive help. Ironically, Binet began his career believing that intelligence was an inborn ability, rather than a learned skill.[193] His approach was both practical and innovative. He assembled a series of small "tasks involving basic reasoning skills with each task assigned an age level indicating its difficulty."[194] In his book *The Measurement of Man*, Stephen Jay Gould summarizes Binet's strategy: "A child began the Binet test with tasks for the youngest age and proceeded in sequence until he [sic] could no longer complete the tasks. The age associated with the last tasks he could perform became his '**mental age**,' and his general intellectual level was calculated by subtracting this mental age from his true chronological age. Children whose mental ages were sufficiently behind their chronological ages could then be identified for special education programs. . . . ."[195]

Binet collaborated with Theodore Simon, the medical officer of the Paris public schools, to develop the *Binet-Simon Scale of Intelligence*, first published in 1905. Several years later, in 1912, the German psychologist W. Stern originated

mental age In the context of an intelligence test, this refers to the age associated with the last task an individual can successfully perform.

the **intelligence quotient (IQ)**—the score derived by dividing mental age by chronological age and multiplying by 100. Thus, a typical 11-year-old who had a mental age of 11 would have an IQ of 100 (mental age of 11 divided by chronological age of 11 equals 1, multiplied by 100 equals an IQ of 100).[196] Binet, however, cautioned that the intelligence test was developed only to identify children who needed help in school, not as a measure of intelligence. He stated directly that the test "does not permit the measure of intelligence," because intelligence "is not a single, scorable thing like height."[197]

As a result of this work, Binet developed a strong commitment to the view that students who did poorly in school could be identified and assisted to improve their academic performance. He helped set up special classes in Paris and reported enthusiastically that "we have increased the intelligence of a pupil: the capacity to learn and to assimilate instruction."[198] This view stands in sharp contrast to his earlier belief that intelligence was a fixed, inborn quality.

While Binet's testing strategies crossed the Atlantic and became popular, his interpretation and application of intelligence testing did not. "Translators and popularizers of the test took it as a measure of innate intelligence and used the scores to lobby for discriminatory and repressive social policies. For students of criminology, one of the most important proponents of these ideas was Henry Goddard."[199]

intelligence quotient (IQ) The score derived by dividing mental age by chronological age and then multiplying by 100.

"Feeblemindedness" and Delinquency

Henry Goddard was director of research at the New Jersey Training School for Feebleminded Boys and Girls in Vineland in the early 1900s. Goddard adapted Binet's intelligence test and administered it to residents at the school, finding that no student scored above the mental age of 12. He concluded that the mental age of 13 marked the lower limit of normal intelligence, with 16 the mental age of full mental capacity, and 12 and below the mental age of "feeblemindedness," "morons," or "high-grade defectives."[200]

Goddard went on to conduct a well-known study of a large family group of "defectives" named the "Kallikaks," who lived in "the pine barrens of New Jersey."[201] He traced their lineage back to a man who had an illegitimate child by a "feebleminded" barmaid. According to Goddard, the 480 descendants of this union included a long line of sexual perverts, alcoholics, prostitutes, and criminals who were disproportionately feebleminded. The man later married a "righteous" Quaker woman and this union ultimately produced 496 "normal" and socially upstanding descendants. From this "research," Goddard concluded that feeblemindedness is inherited.

Goddard also reviewed the results of research conducted by psychologists who administered intelligence tests to inmates in prisons, jails, hospitals, and other public institutions. The average proportion of criminals who were diagnosed as feebleminded in these studies was 70%, leading Goddard to conclude that in general criminals were feebleminded.[202]

Since Goddard believed that crime and feeblemindedness go hand-in-hand and are inherited, he argued that the only credible crime prevention approach was to prohibit the feebleminded from reproducing. This basic strategy led to the implementation of routine intelligence testing for various groups, including draftees during World War I, convicted criminals, and immigrants to the United States. Intelligence tests of these groups identified a disturbingly high percentage of feebleminded draftees, criminals, and immigrants. It was estimated at that time that "about half of the U.S. population was made up of morons."[203] The absurdity of this observation led even advocates of testing programs to question the validity and reliability of intelligence tests, and, as a result, to refute the hypothesized relationship between feeblemindedness and crime. Shortly after World War I, Goddard observed, "The most extreme limit that anyone has dared to suggest is that one percent of the population is feebleminded."[204] He also conceded that "feeblemindedness might be remedied by education," and it is "not necessary to segregate the feebleminded in institutions and to prevent them from reproducing."[205]

In 1926, psychologist Carl Murchison published *Criminal Intelligence*, a book that disputed early findings from intelligence testing.[206] Murchison presented data showing that intelligence test scores of enlisted men during World War I were actually lower than those earned by prisoners in the federal penitentiary at Fort Leavenworth. Additional group comparisons from around the country convinced him that prisoners were no less intelligent than the general population. He also noted that prisoners probably represent a less intelligent sector of the criminal population—more intelligent criminals are more likely to avoid detection and to manipulate the system to their advantage. The "intellectual deficits of criminals no longer seemed so large or so certain."[207]

By the 1930s, the idea that most offenders were feebleminded had lost its stronghold in criminology, in a field that was increasingly dominated by sociologists and a concern with social and societal factors.[208] In a critique of intelligence testing, sociologist Edwin Sutherland observed that "as the methods of testing improved, the apparent intellectual deficits of offenders shrank."[209] He then jumped to the conclusion that any remaining difference was a lingering result of flaws in research—offenders and non-offenders do not differ in intelligence. The discrediting of intelligence testing methods and findings led criminologists to ignore any possible link between intelligence and delinquency.

Rebirth of Interest in the IQ–Delinquency Relationship

Interest in the link between intelligence and delinquency was rekindled in 1977 with an article by Travis Hirschi and Michael Hindelang entitled "Intelligence and Delinquency: A Revisionist Review."[210] They first exposed criminology's tendency to discredit and ignore any link whatsoever between low intelligence and delinquency. Based on a review of research findings, they claimed that IQ scores are significantly related to both self-reported and officially recorded delinquency, although the relationship is considerably stronger using self-report measures. Hirschi and Hindelang concluded that low intelligence is an important causal factor in delinquent behavior, independent of social class and race, although the relationship is primarily indirect. They contended that IQ affects delinquent behavior through school performance: youths with low IQs do poorly in school, and poor school performance leads to frustration and anger and subsequently to delinquent behavior.

While Hirschi and Hindelang argued convincingly that low intelligence influences delinquency indirectly through school failure, the causal connections may be more complex than they suggested. More recent research, for example, points out that the relationship between intelligence and delinquency may differ by gender, race, and class; for various types of offenders; and according to temperament.[211] Also, research has recently established that low IQ is related to a host of experiences, both in the preschool and school years, that place youths at greater risk for delinquency. Psychologist Herbert Quay provides an overview of this developmental perspective on intelligence and delinquency.

> *Lower intelligence is one of many factors which may put a child at a disadvantage with respect to success in a variety of situations which children face in the process of development. In the early years lower IQ may make a child more vulnerable to poor parenting and, in fact, even act with a predisposed parent to make poor parenting more likely. Such an interaction would be more likely if the IQ deficit was accompanied by a fussy or difficult temperament, motor overactivity, and poor inhibitory control. The result of all these forces can be the early onset of troublesome behavior. The affected child is now at a double disadvantage when he [sic] does enter school; he has both less intellectual ability, particularly in the verbal sphere, to cope with academic tasks, and he has oppositional and aggressive behavior problems that are alienating to teachers and peers. As development proceeds, both are, in combination, likely to lead to school failure, the results of which, in turn, reinforce more conduct-disordered behavior. At the same time, those higher cognitive functions (e.g., verbal self-regulation, social problem solving, moral judgment) fail to develop adequately. This is likely due both to the limited intellectual capacity and to the mutually reinforcing social interactions that now characterize the child's relations with others. All of these factors, and others as well (e.g., deviant parental and peer models) interact to produce behavior which is legally proscribed.*[212]

Lingering Issues with IQ

While data suggest that individuals with low intelligence are more likely to engage in delinquent behavior,[213] at least three controversial issues remain: (1) What does IQ measure and what does it mean? (2) Does the effect of IQ on delinquency reflect a direct, indirect, or spurious relationship? (3) To what degree is intelligence inherited?[214]

What Does IQ Measure and What Does it Mean?

As we have seen, Binet's original intelligence test was developed to identify students with learning problems—those who did poorly in school. He did not consider the test to be a precise measure of innate intelligence. Most educators today agree that IQ tests are good measures of children's ability to perform in school. According to Ronald Simons, the IQ is best viewed as "a broad set of verbal and problem-solving skills which are better labeled academic aptitude or scholastic readiness."[215]

IQ scores also indicate the degree to which a child's socialization experiences are conducive to learning and academic performance.[216] The availability of books in the home, amount of time spent reading, and parents' encouragement of verbal expression, for example, appear to affect IQ scores.[217] In particular, delinquent youth have been found to perform poorly on "verbal" elements of intelligence tests, while scoring "average" on "performance" elements. The so-called "verbal IQ" measures language comprehension, whereas "performance IQ" measures nonverbal, concrete operation skills. As such, family and school experiences that fail to encourage development of verbal skills are likely to account for some of the discrepancy in these scores for delinquent youth. Herbert Quay speculates that low verbal abilities may inhibit development of higher-order cognitive processes such as moral reasoning, empathy, and problem solving.[218]

IQ scores are also influenced by cultural context.[219] A common complaint about intelligence tests is that they are culturally biased, providing IQ scores that are heavily determined by cultural background. If IQ scores actually measure academic preparedness and aptitude, rather than innate intelligence, then the degree to which a youth's social and cultural contexts are relevant to academic achievement will influence test performance. Similarly, cultures vary in terms of what they hold to be important, what they value as knowledge, the cognitive abilities they encourage, and how they conceptualize time and space. As a result, the degree to which intelligence test questions are culturally based will affect IQ scores. Supporters of IQ testing claim that, since the 1970s, test makers have sought to eliminate cultural bias in questions.[220] Nonetheless, Hirschi and Hindelang point out that IQ differences between delinquent and nondelinquent youth have not disappeared entirely and seem to have stabilized at about an eight-IQ-point difference.[221] More recent studies have also found that serious offenders have lower scores than minor offenders and that low IQ scores of young children are related to later offending as adolescents and young adults.[222]

Does the Effect of IQ on Delinquency Reflect a Direct, Indirect, or Spurious Relationship?

The way in which low IQ scores affect delinquency is also not completely understood. Hirschi and Hindelang claimed that low IQ scores are indirectly related to delinquency, influencing delinquency through poor school performance. This is also the interpretation offered by Donald Lynam and his colleagues in a study of the relationship between IQ and delinquency that considers a variety of other factors including class, race, and school failure.[223] This study contends that the relationship between IQ and delinquency is complex and must be considered in the context of other potential influences. Deborah Denno, for example, found that intelligence scores did not have a direct effect on delinquency for either males or females, but for males, poor verbal ability indirectly affected delinquent behavior through poor school achievement.[224] Thus, there is reason to believe that the effect of IQ on delinquency is influenced by other factors—its relationship to delinquency is largely indirect.

To What Degree is Intelligence Inherited?

There is much debate about the degree to which intelligence is inherited. Earlier in this chapter, we described how genetic influences on delinquent behavior are studied primarily through adoption and twin studies. The same methods are used to try to determine the genetic basis of intelligence.[225] Some research shows that IQ levels of children more closely resemble those of their biological mothers than those of their adoptive parents, but other research finds that IQ scores of adopted children are influenced significantly by the conditions in which they are reared.[226] Twin studies indicate that, while there is likely a heritable component to IQ, the influence of heritability is far less than once believed. Some researchers have claimed that IQ is 70–80% inherited. Recent research on identical twins, however, shows that only about half of the similarity in IQ scores is a result of genetic influence.[227]

■ Summary and Conclusions

Since its beginning, the study of juvenile delinquency has concentrated on individual offenders and tried to distinguish them from non-offenders. Using scientific methods, early researchers searched for biological and psychological differences that predispose some people to delinquency and crime.

The roots of positivist criminology are grounded in the study of physical differences between offenders and non-offenders. Early researchers noted that criminals "look different" in terms of facial features, shape of the skull, and a host of other physical characteristics. The nineteenth-century Italian physician Cesare Lombroso, for example, based his theory of atavism on autopsies of deceased criminals and on physical examination of some of the most "notorious and depraved" criminals alive in Italy during his day. The observation of physical distinctions between criminals and noncriminals led him to conclude that criminals were a biological throwback to an earlier stage of evolution. He called such biological reversion *atavism*.

According to early biological approaches, physical differences were inherited, but not a direct cause of criminality. Rather, physical characteristics were commonly connected with the "psychology of the criminal," as Lombroso called it: impulsive aggression, revenge, lack of foresight, idleness, and "a passion for play and activity."[228] Similarly, Raffaele Garofalo, a student of Lombroso, argued that criminals lacked basic "moral sentiment"—they did not fear punishment, and they lacked self-control. He speculated that these fundamental psychological flaws were a result of heredity and tradition, passed on from one generation to the next. Later attempts by William Sheldon to classify physical characteristics (physique) connected body type to personality temperament, indicating that the temperament associated with a particular body type was the mechanism that made some people predisposed to delinquent behavior.

Early biological approaches also recognized that physical characteristics of the individual are couched in a physical and social environment. Another of Lombroso's students, Enrico Ferri, examined the interrelatedness of "three different sets of factors in crime: namely those in the physical or geographic environment, those in the constitution of the individual, and those in the social environment."[229] Nonetheless, the heavy biological emphasis of early positivist criminology made it seem as if criminal behavior could be reduced to its biological roots. This reflects the reductionistic perspective of *biological determinism*.

Contemporary *biosocial criminology* contends that delinquency and other forms of antisocial behavior result from a combination of biological, psychological, and social causes. This approach is interdisciplinary and integrated, and it is based on the fundamental concept of *nature–nurture interaction*. Nature–nurture interaction refers to the process by which biological, psychological, and environmental factors influence each other, as together they affect the likelihood of delinquent behavior.

We described three biological aspects in this chapter. First, the anatomy and physiology of the nervous system have been studied extensively in relation to antisocial behavior. Of special importance are nervous system activities related to arousal, impulsiveness, and self-regulation. Second, several key neurotransmitters and the hormone testosterone have been associated with antisocial behavior. Neurotransmitters are the chemical compounds that carry signals between neurons in the intricate communication system of the brain. Their concentration and metabolism play a critical role in how various parts of the brain evoke and regulate behavior. Levels of testosterone are linked to displays of aggression and violence. Third, behavioral genetics seeks to establish the contributions of heredity and environment to personality and behavioral traits, such as impulsivity, constraint, aggressiveness, and negative emotionality.

The biosocial developmental model does not contend that biological factors have direct effects on delinquency. Rather, biological factors, in combination with environmental conditions, produce personality traits that are conducive to delinquent behavior. Research has established clearly that antisocial individuals are low in *agreeableness* and *conscientiousness*—they tend to be "hostile, self-centered, spiteful, jealous, and indifferent to others," and "they tend to lack ambition, motivation, and perseverance, have difficulty controlling their impulses, and hold nontraditional and unconventional values and beliefs."[230]

The connection between intelligence and delinquency has been investigated for nearly 100 years, but few clear conclusions are available. On average, delinquents score about eight IQ points lower than nondelinquents, yet IQ scores are probably most accurately viewed as a measure of academic preparedness and aptitude, rather than of innate intelligence. As such, low IQ places youths at risk for poor school performance and frustration, leading indirectly to delinquent behavior. Research documents a heritable component to intelligence, but intelligence is also influenced in significant ways by environment.

CRITICAL THINKING QUESTIONS

1. According to Lombroso, the physical differences that distinguish criminals from noncriminals are inherited but do not directly cause criminality. What are some of the distinctive physical characteristics of criminals? How are these physical characteristics related to crime?

2. Several early criminologists like Lombroso, Garofalo, and Ferri argued that other factors in addition to physical characteristics are necessary to understand crime. What are some of these other factors?

3. Explain nature–nurture interaction and provide an example.

4. How does low arousal theory explain antisocial behavior?

5. Describe the neuropsychological impact of the prefrontal cortex injuries reported in the two case illustrations that opened the chapter.

6. Develop your own working definition of personality. What key dimensions of personality are related to delinquency? How and why?

7. According to Hirschi and Hindelang, IQ scores are significantly related to delinquency, but the relationship is indirect. How does Quay explain this indirect relationship across the life course?

SUGGESTED READINGS

Caspi, Avshalom, Terrie E. Moffitt, Phil A. Silva, Magda Stouthamer-Loeber, Robert F. Krueger, and Pamela S. Schmutte. "Are Some People Crime-Prone? Replications of the Personality-Crime Relationship across Countries, Genders, Races, and Methods." *Criminology* 32 (1994):163–195.

Hirschi, Travis, and Michael J. Hindelang. "Intelligence and Delinquency: A Revisionist Review." *American Sociological Review* 42 (1977):571–587.

Miller, Joshua D., and Donald Lynam. "Structural Models of Personality and Their Relation to Antisocial Behavior: A Meta-Analytic Review." *Criminology* 39 (2001):765–798.

Rowe, David C. *Biology and Crime.* Los Angeles, CA: Roxbury, 2002.

Walsh, Anthony. *Biosocial Criminology: Introduction and Integration.* Cincinnati, OH: Anderson, 2002.

GLOSSARY

agreeableness: A dimension of personality that is related strongly to antisocial behavior. Antisocial individuals tend to be hostile, self-centered, spiteful, jealous, and indifferent to others.

atavism: Reappearance in an individual of physical characteristics associated with a more primitive stage of evolution that has been absent in intervening generations. Lombroso claimed that such individuals are "born criminals" because of these characteristics.

autonomic nervous system (ANS): The part of the peripheral nervous system that controls unconscious bodily functions such as blood pressure, heart activity, breathing, and hormone levels.

behavioral genetics: The science that tries to estimate the relative contributions of heredity and environment to personality and behavioral traits such as impulsivity, constraint, aggressiveness, and negative emotionality.

biological determinism: Identifying, through scientific methods, the biological characteristics of individuals that are the primary causes of behavior. People are the products of their biological makeup.

biosocial criminology: The approach to the study of criminality that holds that crime and other forms of "antisocial behavior result from a combination of social, psychological, and biological causes" (Brennan and Raine, "Biosocial Bases," 590).

central nervous system (CNS): The part of the human nervous system that includes the brain and spinal cord.

conditionability: The variable tendency to anticipate punishment. Anxiety reaction (fear), in anticipation of punishment, is the primary mechanism of socialization. People with low ANS activity condition poorly because they do not anticipate punishment and subsequently fail to develop a conscience.

conscientiousness: A dimension of personality that is related to antisocial behavior. Antisocial individuals tend to lack ambition, motivation, and perseverance; they have difficulty controlling their impulses; and they hold nontraditional values and beliefs (Miller and Lynam, "Structural Models," 780).

environmental determinism: Identifying, through scientific methods, the social conditions that cause behavior. People are the products of their environment.

executive cognitive functions (EFC): A person's ability to plan and reflect on his or her actions. This involves information processing, memory, assessment, and self-regulation.

heritability: The measure of genetic influence.

intelligence quotient (IQ): The score derived by dividing mental age by chronological age and then multiplying by 100.

low arousal theory: Low levels of cortical arousal are associated with lack of fear. Since anxiety reaction in anticipation of punishment is the basis for conditioning, individuals with low arousal levels have little desire or ability to avoid punishment. Therefore, they are often poorly socialized and engage in antisocial behavior without restraint.

mental age: In the context of an intelligence test, this refers to the age associated with the last task an individual can successfully perform.

nature–nurture debate: The argument that grew out of biological determinism and environmental determinism. The causes of behavior are either biological or environmental, but not both.

nature–nurture interaction: The perspective that it is not enough to say that biological and environmental factors contribute to behavior; rather, each influences the other, as together they produce behavior.

neurophysiology: The physical structure of the nervous system, including the central and peripheral nervous systems.

neuropsychology: The link between the body's neurological structure, including the central and peripheral nervous systems, and psychological functioning.

neurotransmitters: Chemical compounds, found in the synapse between nerve cells, that carry signals from one neuron to another. Neurotransmitters provide a crucial link in the neural communications system.

peripheral nervous system: The part of the human nervous system that emanates from and lies outside of the brain and spinal cord.

personality: The reasonably stable patterns of perceiving, thinking, feeling, and responding to the environment. Traits are the basic building blocks of personality, providing foundation to thought, emotion, character, and behavior.

reductionism: The view that a complex phenomenon can be explained by reducing it to its most basic element.

somatic nervous system: The part of the human nervous system that consists of sensory and motor nerves, providing awareness or consciousness of sensations such as light, touch, sound, and smell, and the resulting activation of muscle groups.

somatotypes: Structures of human physique, or body types. William Sheldon identified three somatotypes: endomorphy, mesomorphy, and ectomorphy. All people possess each of these body types to varying degrees.

stigmata: The distinctive physical characteristics (observed in criminals) that identify atavism.

testosterone: Chemical compounds secreted into the bloodstream by various glands. This hormone plays an important role in aggression and violence.

REFERENCES

Albert, D. J., R. H. Jonik, N. V. Watson, B. B. Gorzalka, and M. L. Walsh. "Hormone-Dependent Aggression in Male Rates is Proportional to Serum Testosterone Concentration but Sexual Behavior is Not." *Physiology and Behavior* 48 (1990):409–416.

Alters, Sandra. *Biology: Understanding Life.* 3rd ed. Sudbury, MA: Jones and Bartlett, 2000.

American Psychiatric Association. *Diagnostic and Statistical Manual of Mental Disorders.* 4th ed. (DSM-IV). Washington, DC: American Psychiatric Association, 1994.

Archer, J. "The Influence of Testosterone on Human Aggression." *British Journal of Psychology* 82 (1991):1–28.

Arseneault, L., B. Boulerice, R. C. Tremblay, and J. F. Saucier. "A Biosocial Exploration of the Personality Dimensions That Predispose to Criminality." In *Biosocial Bases of Violence,* edited by Raine et al., 313–316, New York: Plenum, 1997.

Bartusch, Dawn R. Jeglum, Donald R. Lynam, Terrie E. Moffitt, and Phil Silva. "Is Age Important? Testing a General Versus a Developmental Theory of Antisocial Behavior." *Criminology* 35 (1997):13–48.

Barrett, G. V., and R. I. Depinet. "A Reconsideration of Testing for Competence Rather than Intelligence." *American Psychologist* 46 (1999):1012–1024.

Bartol, Curt R. *Criminal Behavior: A Psychosocial Approach.* 4th ed. Englewood Cliffs, NJ: Prentice Hall, 1995.

Beirne, Piers. "Adolphe Quetelet and the Origins of Positivist Criminology." *American Journal of Sociology* 92 (1987):1140–1169.

———. "Heredity Versus Environment: A Reconsideration of Charles Goring's *The English Convict* (1913)." *British Journal of Criminology* 28 (1988):315–339.

Beirne, Piers, and James Messerschmidt. *Criminology.* 3rd ed. Boulder, CO: Westview, 2000.

Binder, Arnold, Gilbert Geis, and Dickson D. Bruce, Jr. *Juvenile Delinquency: Historical, Cultural, and Legal Perspectives.* 2nd ed. Cincinnati, OH: Anderson, 1997.

Blumstein, Alfred, David P. Farrington, and Soumyo Moitra. "Delinquent Careers." In *Crime and Justice: A Review of Research,* vol. 7, edited by Michael H. Tonry and Norval Morris, 187–219. Chicago: University of Chicago Press, 1985.

Block, Jack. "On the Relation Between IQ, Impulsivity, and Delinquency: Remarks on The Lynam, Moffitt, and Stouthhamer-Loeber (1993) Interpretation." *Journal of Abnormal Psychology* 104 (1995):395–398.

Booth, Alan, and D. Wayne Osgood. "The Influence of Testosterone on Deviance in Adulthood: Assessing and Explaining the Relationship." *Criminology* 31 (1993):93–117.

Booth, Alan, G. Shelley, A. Mazure, G. Tharp, and R. Kittok. "Testosterone and Winning and Losing in Human Competition." *Hormones and Behavior* 23 (1989):556–571.

Bouchard, J. T., Jr., D. T. Lykken, M. McGue, N. L. Degal, and A. Tellegen. "Sources of Human Psychological Differences: The Minnesota Study of Twins Reared Apart." *Science* 250 (1990): 223–228.

Brain, Paul. "Hormonal Aspects of Aggression and Violence." In *Understanding and Preventing Violence,* vol. 2, edited by Albert J. Reiss, Jr., Klaus A. Miczek, and Jeffrey A. Roth, 173–244. Washington, DC: National Academy, 1993–1994.

Brennan, Patricia A., and Adrian Raine. "Biosocial Bases of Antisocial Behavior: Psychophysiological, Neurological, and Cognitive Factors." *Clinical Psychology Review* 17 (1997):589–604.

Brooks, J. H., and J. R. Reddon. "Serum Testosterone in Violent and Nonviolent Young Offenders." *Journal of Clinical Psychology* 52 (1996):475–483.

Caspi, Avshalom, Terrie E. Moffitt, Phil A. Silva, Magda Stouthamer-Loeber, Robert F. Krueger, and Pamela S. Schmutte. "Are Some People Crime-Prone? Replications of the Personality–Crime Relationship across Countries, Genders, Races, and Methods." *Criminology* 32 (1994):163–195.

Chiras, Daniel D. *Human Biology.* 5th ed. Sudbury, MA: Jones and Bartlett, 2005.

Chretien, R. D., and M. A. Persinger. " 'Prefrontal Deficits' Discriminate Young Offenders from Age-matched Cohorts: Juvenile Delinquency as an Expected Feature of the Normal Distribution of Prefrontal Cerebral Development." *Psychological Reports* 87 (2000):1196–1202.

Clark, Lee Anna, and David Watson. "Temperament: A New Paradigm for Trait Psychology." In *Handbook of Personality: Theory and Research,* 2nd ed., edited by Lawrence A. Pervin and Oliver P. John, 399–423. New York: Guilford, 1999.

Clark, Robert K. *Anatomy and Physiology: Understanding the Human Body.* Sudbury, MA: Jones and Bartlett, 2005.

Cleckley, Hervey. "Psychopathic States." In *American Handbook of Psychiatry,* edited by S. Aneti, 567–569. New York: Basic Books, 1959.

Cloninger, C. Robert. "A Systematic Method for Clinical Description and Classification of Personality Variants." *Achieves of General Psychiatry* 44 (1987):573–588.

———. "A Unified Biosocial Theory of Personality and Its Role in the Development of Anxiety States." *Psychiatric Development* 3 (1986):167–226.

Cloninger, C. Robert, Dragon M. Svrakic, and Thomas R. Prybeck. "A Psychobiological Model of Temperament and Character." *Achieves of General Psychiatry* 50 (1993):975–990.

Constantino, J. N., D. Grosz, P. Saenger, D. W. Chandler, R. Nandi, and F. J. Earls. "Testosterone and Aggression in Children." *Journal of the American Academy of Child and Adolescent Psychiatry* 32 (1993):1217–1222.

Cortés, Juan B., and F. M. Gatti. *Delinquency and Crime: A Biopsychological Approach.* New York: Seminar Press, 1972.

Costra, Paul T., and Robert R. McCrae. "Primary Traits of Eysenck's P-E-N Model: Three- and Five-Factor Solutions." *Journal of Personality and Social Psychology* 69 (1995):308–317.

Cullen, Francis T., and Robert Agnew. *Criminological Theory: Past to Present: Essential Readings.* Los Angeles: Roxbury, 2003.

Curran, Daniel J., and Claire M. Renzetti. *Theories of Crime.* 2nd ed. Boston: Allyn and Bacon, 2001.

Dabbs, J. M., and R. Morris. "Testosterone, Social Class, and Antisocial Behavior in a Sample of 4,462 Men." *Psychological Science* 1 (1990):209–211.

Denno, Deborah. *Biology and Violence: From Birth to Adulthood.* New York: Cambridge University Press, 1990.

———. "Sociological and Human Developmental Explanations of Crime: Conflict or Consensus." *Criminology* 23 (1985):141–174.

Ellis, Havelock. *The Criminal.* 2nd ed. New York: Scribner, 1900.

Ellis, Lee. "Monoamine Oxydase and Criminality: Identifying an Apparent Biological Marker for Antisocial Behavior." *Journal of Research in Crime and Delinquency* 28 (1991):277–251.

Ellis, Lee, and Anthony Walsh. "Gene-Based Evolutionary Theories in Criminology." *Criminology* 35 (1997):229–276.

Ewen, Robert B. *Personality: A Topical Approach: Theories, Research, Major Controversies, and Emerging Findings.* Mahwah, NJ: Erlbaum, 1998.

Eysenck, H. J. *Crime and Personality.* Boston: Houghton Mifflin, 1964.

———. "Dimensions of Personality: 16, 5, or 3—Criteria for a Taxonomic Paradigm." *Personality and Individual Differences* 8 (1991):773–790.

———. *A New Look at Intelligence.* London: Transaction, 1998.

———. "Personality and the Biosocial Model of Anti-social and Criminal Behavior." In *Biosocial Bases of Violence,* edited by Raine et al., 21–37, New York: Plenum, 1997.

Eysenck, H. J., and Gisli H. Gudjonsson. *The Causes and Cures of Criminality.* New York: Plenum, 1989.

Farrington, David P. "Individual Differences and Offending." In *Handbook of Crime and Punishment,* edited by Michael Tonry, 241–268. New York: Oxford University Press, 1988.

———. "The Relationship Between Low Resting Heart Rate and Violence." In *Biosocial Bases of Violence,* edited by Raine et al., 89–105, New York: Plenum, 1997.

Fishbein, Diana. *Biobehavioral Perspectives in Criminology.* Belmont, CA: Wadsworth, 2001.

———. "Biological Perspectives in Criminology." *Criminology* 28 (1990):27–72.

Fishbein, Diana, and Susan E. Pease. *The Dynamics of Drug Abuse.* Boston: Allyn and Bacon, 1996.

Ferri, Enrico. *Criminal Sociology.* Boston: Little, Brown, 1917.

Garofalo, Raffaele. *Criminology.* Translated by R. Millar. Boston: Little, Brown, 1914.

Glueck, Sheldon, and Eleanor Glueck. *Physique and Delinquency.* New York: Harper, 1956.

———. *Unraveling Juvenile Delinquency.* New York: Commonwealth Fund, 1950.

Goddard, Harry H. *Feeblemindedness: Its Causes and Consequences.* New York: Macmillan, 1914.

————. "Feeblemindedness: A Question of Definition." *Journal of Psycho-Asthenic* 33 (1928):225.

————. *The Kallikak Family: A Study in the Heredity of Feeble-Mindedness.* New York: Macmillan, 1912.

Goring, Charles. *The English Convict: A Statistical Study.* Montclair, NJ: Patterson Smith, 1972.

Gottfredson, Michael, and Travis Hirschi. "The Positive Tradition." In *Positive Criminology,* edited by Michael Gottfredson and Travis Hirschi, 9–22. Newbury Park, CA: Sage, 1987.

Gould, Stephen Jay. *The Measurement of Man.* New York: Norton, 1981.

Grove, W. M., E. D. Eckert, L. Heston, T. J. Bouchard, N. Segal, and D. Y. Lykken. "Heritability of Substance Abuse and Antisocial Behavior: A Study of Monozygotic Twins Reared Apart." *Biological Psychiatry* 27 (1990):1293–1304.

Hagan, John. *Modern Criminology: Crime, Criminal Behavior, and Its Control.* New York: McGraw-Hill, 1985.

Hanson, Glen R., Peter J. Venturelli, and Annette E. Fleckenstein. *Drugs and Society.* 8th ed. Sudbury, MA: Jones and Bartlett Publishers, 2004.

Harris, Julie Aitken. "Review and Methodological Considerations in Research on Testosterone and Aggression." *Aggression and Violent Behavior* 4 (1985):273–291.

Hirschi, Travis, and Michael J. Hindelang. "Intelligence and Delinquency: A Revisionist Review." *American Sociological Review* 42 (1977):571–587.

Hooton, Earnest A. *The American Criminal: An Anthropological Study.* Cambridge, MA: Harvard University Press, 1939.

————. *Crime and the Man.* Cambridge, MA: Harvard University Press, 1939.

Jacobson, Kristen C., and David C. Rowe. "Nature, Nurture, and the Development of Criminality." In *Criminology,* 3rd ed., edited by Joseph F. Sheley, 323–347. Belmont, CA: Wadsworth, 2000.

John, Oliver P., and Sanjay Srivastava. "The Big Five Taxonomy: History, Measurement, and Theoretical Perspectives." In *Handbook of Personality: Theory and Research,* 2nd ed., edited by Lawrence A. Pervin and Oliver P. John, 102–138. New York: Guilford, 1999.

Kleinmuntz, Benjamin. *Personality and Psychological Assessment.* New York: St. Martin's, 1982.

Laub, John H., and Robert J. Sampson. "The Sutherland–Glueck Debate: On the Sociology of Criminological Knowledge." *American Journal of Sociology* 96 (1991):1402–1440.

Lilly, J. Robert, Francis T. Cullen, and Richard A. Ball. *Criminological Theory: Context and Consequences.* Newbury Park, CA: Sage, 1989.

Loeber, Rolf, David P. Farrington, Magda Stouthamer-Loeber, and Welmoet B. VanKammen. *Antisocial Behavior and Mental Health Problems.* Mahwah, NJ: Erlbaum, 1998.

Lombroso, Cesare. *Crime: Its Causes and Remedies.* Translated by H. P. Horton. Boston: Little, Brown, 1918.

————. *Criminal Man (L'Uomo delinquents).* 1911. Translated by Gina Lombroso-Ferrero. Montclair, NJ: Patterson Smith, 1972.

Lynam, Donald, and Terrie Moffitt. "Delinquency and Impulsivity and IQ: A Reply to Block (1995)" *Journal of Abnormal Psychology* 104 (1995):399–401.

Lynam, Donald, Terrie Moffitt, and Magda Stouthamer-Loeber. "Explaining the Relationship Between IQ and Delinquency: Class, Race, Test Motivation, School Failure, or Self-Control?" *Journal of Abnormal Psychology* 102 (1993):187–196.

Martin, Randy, Robert J. Mutchnick, and W. Timothy Austin. *Criminological Thought: Pioneers Past and Present.* New York: Macmillan, 1990.

Matza, David. *Delinquency and Drift.* New York: Wiley, 1964.

Mazur, Allan, and Alan Booth. "Testosterone and Dominance in Men." *Behavior and Brain Science* 21 (1998):353–363.

McCrae, Robert R., and Paul T. Costra, Jr. "A Five-Factor Theory of Personality." In *Handbook of Personality: Theory and Research,* 2nd ed., edited by Lawrence A. Pervin and Oliver P. John, 139–153. New York: Guilford, 1999.

McGloin, Jean Marie, Travis Pratt, and Jeff Maahs. "Rethinking the IQ–Delinquency Relationship: A Longitudinal Analysis of Multiple Theoretical Models." *Justice Quarterly* 21 (2004):603–635.

Mednick, Sarnoff. "A Biosocial Theory of the Learning of Law-Abiding Behavior." In *Biosocial Bases of Criminal Behavior,* edited by Sarnoff Mednick and Karl O. Christiansen, 1–8. New York: Gardner, 1977.

Mednick, Sarnoff A., William F. Gabrielli, and Bernard Hutchings. "Genetic Influences in Criminal Conviction: Evidence from an Adoption Cohort." *Science* 224 (1984):891–894.

Menard, Scott, and Barbara J. Morse. "A Structural Critique of the IQ–Delinquency Hypothesis: Theory and Evidence." *American Journal of Sociology* 89 (1984):1347–1378.

Merton, Robert, and M. F. Ashley Montague. "Crime and the Anthropologist." *American Anthropologist* 42 (1940):384–408.

Miller, Joshua D., and Donald Lynam. "Structural Models of Personality and Their Relation to Antisocial Behavior: A Meta-Analytic Review." *Criminology* 39 (2001):765–798.

Moffitt, Terrie. "The Neuropsychology of Juvenile Delinquency: A Critical Review." In *Crime and Justice: A Review of Research,* vol. 12, edited by Michael H. Tonry and Norval Morris, 99–169. Chicago: University of Chicago Press, 1990.

Moffitt, Terrie E., Gary L. Brammer, Avshalom Caspi, Paul Fawcett, Michael Raleigh, Arthur Yuwiler, and Phil Silva. "Whole Blood Serotonin Relates to Violence in an Epidemiological Study." *Biological Psychiatry* 43 (1998):446–457.

Moffitt, Terrie E., Avshalom Caspi, Paul Fawcett, Gary L. Brammer, Michael Raleigh, Arthur Yuwiler, and Phil Silva. "Whole Blood Serotonin and Family Background Relate to Male Violence." In *Biosocial Bases of Criminal Behavior,* edited by Sarnoff Mednick and Karl O. Christiansen, 231–249. New York: Gardner, 1997.

Moffitt, Terrie E., Donald R. Lynam, and Phil A. Silva. "Neuropsychological Testing Predicting Persistent Male Delinquency." *Criminology* 32 (1994):277–300.

Murchison, Carl. *Criminal Intelligence.* Worcester, MA: Clark University Press, 1926.

Petrill, S. A., R. Plomin, S. Berg, B. Johanson, N. L. Pedersen, F. Ahern, and G. E. McClean. "The Genetic and Environmental Relationship Between General and Specific Cognitive Abilities in Twins Age 80 and Older." *Psychological Science* 9 (1998):183–189.

Pitts, Traci Bice. "Reduced Heart Rate Levels in Aggressive Children." In *Biosocial Bases of Violence,* edited by Raine et al., 317–320. New York: Plenum, 1997.

Plomin, R., D. W. Fulker, R. Corley, and J. C. DeFries. "Nature, Nurture, and Cognitive Development from 1 to 16 years: A Parent–Offspring Adoption Study." *Psychological Science* 8 (1997):442–447.

Quay, Herbert C. "Intelligence." In *Handbook of Juvenile Delinquency,* edited by Herbert C. Quay, 106–117. New York: Wiley, 1987.

Rafter, Nicole Hahn. "Criminal Anthropology in the United States." *Criminology* 30 (1992):525–545.

Raine, Adrian. *The Psychopathology of Crime: Criminal Behavior as a Clinical Disorder.* New York: Academic Press, 1993.

Raine, Adrian, Patricia Brennan, and David P. Farrington. "Biosocial Bases of Violence: Conceptual and Theoretical Issues." In *Biosocial Bases of Violence,* edited by Raine et al., 1–20. New York: Plenum, 1997.

Raine, Adrian, Patricia A. Brennan, David P. Farrington, and Sarnoff A. Mednick, eds. *Biosocial Bases of Violence.* New York: Plenum, 1997.

Rowe, David C. *Biology and Crime.* Los Angeles: Roxbury, 2003.

Rutter, Michael, Henri Giller, and Ann Hagell. *Antisocial Behavior by Young People.* Cambridge, England: Cambridge University Press, 1998.

Sampson, Robert J., and John H. Laub. "Unraveling the Social Context of Physique and Delinquency: A New, Long-Term Look at the Glueck's Classic Study." In *Biosocial Bases of Violence,* edited by Raine et al., 175-188. New York: Plenum, 1997.

Scarr, S., and R. A. Weinberg. "Education and Occupation Achievements of Brothers and Sisters in Adopted and Biologically Related Families." *Behavior-Genetics* 24 (1994):301–325.

———. "The Minnesota Adoption Studies: Genetic Differences and Malleability." *Child Development* 54 (1983):260–267.

Shah, Saleem, and Loren H. Roth. "Biological and Psychophysiological Factors in Criminality." In *Handbook of Criminology,* edited by Daniel Glaser, 101–173. Chicago: Rand McNally, 1974.

Sheldon, William H. *Varieties of Delinquent Youth.* New York: Harper, 1949.

———. *The Varieties of Human Physique: An Introduction to Constitutional Psychology.* New York: Harper, 1940.

———. *The Varieties of Temperament: A Psychology of Constitutional Differences.* New York: Harper, 1942.

Simons, Ronald. "The Meaning of the IQ–Delinquency Relationship." *American Sociological Review* 43 (1978):268–270.

Steinberg, Laurence. "Adolescent Development." *Annual Review of Psychology* 52 (2001):83–110.

Tehrani, Jasmine, and Sarnoff A. Mednick. "Genetic Factors and Criminal Behavior." *Federal Probation* 64, issue 2 (2000):24–27.

Tellegen, Auke. "Structures of Mood and Personality and Their Relevance to Assessing Anxiety, with an Emphasis on Self-Report." In *Anxiety and the Anxiety Disorders,* edited by A. Hussain Tuma and Jack D. Maser, 681–706. Hillsdale, NJ: Erlbaum, 1985.

Tremblay, Richard E., Robert O. Pihl, Frank Vitaro, and Patricia Dobkin. "Predicting Early Onset of Male Antisocial Behavior from Preschool Behavior." *Archives of General Psychiatry* 51 (1994):732–739.

Udry, Richard J. "Biological Predispositions and Social Control in Adolescent Sexual Behavior." *American Sociological Review* 53 (1988):709–722.

———. "Biosocial Models of Adolescent Problem Behaviors." *Social Biology* 37 (1990):1–10.

———. "Sociology and Biology: What Biology Do Sociologists Need to Know?" *Social Forces* 73 (1995):1267–1278.

Vold, George B., Thomas J. Bernard, and Jeffrey B. Snipes. *Theoretical Criminology.* 5th ed. New York: Oxford University Press, 2002.

Walsh, Anthony. *Biosocial Criminology: Introduction and Integration.* Cincinnati, OH: Anderson, 2002.

Weinberg, R. A. "Intelligence and IQ: Landmark Issues and Great Debates." *American Psychologist* 44 (1989):98–104.

Widiger, Thomas A., and Donald R. Lynam. "Psychopathy as a Variant of Common Personality Traits: Implications for Diagnosis, Etiology, and Pathology." In *Psychopathy: Antisocial, Criminal, and Violent Behavior,* edited by T. Million, 171–187. New York: Guilford, 1998.

Wilson, James Q., and Richard J. Herrnstein. *Crime and Human Nature: The Definitive Study of the Causes of Crime.* New York: Simon and Schuster, 1985.

Wolfgang, Marwin K. "Cesare Lombroso." In *Pioneers in Criminology,* 2nd ed., edited by Hermann Mannheim, 232–291. Montclair, NJ: Patterson Smith, 1973.

Yaralian, Pauline S., and Adrian Raine. "Biological Approaches to Crime." In *Explaining Criminals and Crime,* edited by Raymond Paternoster and Ronet Bachman, 57–72. Los Angeles: Roxbury, 2001.

ENDNOTES

1. Beirne, "Adolphe Quetelet," 1159.
2. Jacobson and Rowe, "Nature," 323.
3. Jacobson and Rowe, "Nature;" Fishbein, *Biobehavioral Perspectives in Criminology,* 2–7; and Walsh, *Biosocial Criminology.*
4. Vold, Bernard, and Snipes, *Theoretical Criminology,* 32.
5. Ibid.
6. Curran and Renzetti, *Theories of Crime,* 28.
7. Lilly, Cullan, and Ball, *Criminological Theory.*
8. Curran and Renzetti, *Theories of Crime,* 29.
9. Wolfgang, "Cesare Lombroso," 232.
10. Ibid., 234.
11. Martin, Mutchnick, and Austin, *Criminological Thought,* 27.
12. Lombroso, *Criminal Man,* xxiv–xxv.
13. Beirne and Messerschmidt, *Criminology,* 77; and Vold, Bernard, and Snipes, *Theoretical Criminology,* 33.
14. Ibid.
15. Vold, Bernard, and Snipes, *Theoretical Criminology,* 33; and Lombroso, *Criminal Man.*
16. Curran and Renzetti, *Theories of Crime,* 30.
17. Martin, Mutchnick, and Austin, *Criminological Thought,* 28.
18. Lilly, Cullen, and Ball, *Criminological Theory,* 28.
19. Lilly, Cullen, and Ball, *Criminological Theory,* 28; and Martin, Mutchnick, and Austin, *Criminological Thought,* 30–32.
20. Lombroso, *Criminal Man,* 74, quoted in Martin, Mutchnick, and Austin, *Criminological Thought,* 30.
21. Lombroso, *Crime,* 376, quoted in Martin, Mutchnick, and Austin, *Criminological Thought,* 31.
22. Curran and Renzetti, *Theories of Crime,* 31; and Martin, Mutchnick, and Austin, *Criminological Thought,* 31.
23. Martin, Mutchnick, and Austin, *Criminological Thought,* 32.
24. Hagan, *Modern Criminology,* 21; and Martin, Mutchnick, and Austin, *Criminological Thought,* 32.
25. Hagan, *Modern Criminology,* 21.

26. Garofalo, *Criminology*, 23, 31, referenced in Martin, Mutchnick, and Austin, *Criminological Thought*, 33.
27. Garofalo, *Criminology*, 219–220, quoted in Hagan, *Modern Criminology*, 22.
28. Ferri, *Criminal Sociology*, xxxi, quoted in Martin, Mutchnick, and Austin, *Criminological Thought*, 38.
29. Lilly, Cullen, and Ball, *Criminological Theory*, 30.
30. Wolfgang, "Cesare Lombroso," 286.
31. Lilly, Cullen, and Ball, *Criminological Theory*, 29. See also Matza, *Delinquency and Drift*, 3–12.
32. Binder, Geis, and Bruce, *Juvenile Delinquency*, 91. This discussion draws heavily from Beirne and Messerschmidt, *Criminology*, 80–82. See also Martin, Mutchnick, and Austin, *Criminological Thought*, 125–126.
33. Binder, Geis, and Bruce, *Juvenile Delinquency*, 92; and Beirne and Messerschmidt, *Criminology*, 81.
34. Goring, *English Convict*, 173, emphasis in original, as quoted in Beirne and Messerschmidt, *Criminology*, 81.
35. Goring, *English Convict*, 368, as quoted in Beirne and Messerschmidt, *Criminology*, 81.
36. Beirne and Messerschmidt, *Criminology*, 82.
37. Hooton, *American Criminal*, and *Crime and the Man*.
38. Binder, Geis, and Bruce, *Juvenile Delinquency*, 92–94; and Martin, Mutchnick, and Austin, *Criminological Thought*, 127.
39. Hooton, *American Criminal*, 238.
40. Curran and Renzetti, *Theories of Crime*, 34.
41. Hooton, *Crime and the Man*, 309.
42. Binder, Geis, and Bruce, *Juvenile Delinquency*, 93.
43. Merton and Montague, "Crime and the Anthropologist."
44. Curran and Renzetti, *Theories of Crime*, 34.
45. Ibid.
46. Martin, Mutchnick, and Austin, *Criminological Thought*, 129, 134.
47. Ibid., 129.
48. Ibid., 121.
49. Ibid., 130–131.
50. Sheldon, *Human Physique*, 8.
51. Vold, Bernard, and Snipes, *Theoretical Criminology*, 36.
52. Sheldon, *Human Physique*, 8.
53. Ibid.
54. Martin, Mutchnick, and Austin, *Criminological Thought*, 132.
55. Sheldon, *Delinquent Youth*.
56. Vold, Bernard, and Snipes, *Theoretical Criminology*, 37.
57. Sheldon, *Delinquent Youth*, 836; and Curran and Renzetti, *Theories of Crime*, 37–38.
58. Laub and Sampson, "Sutherland–Glueck Debate."
59. Sampson and Laub, "Unraveling," 175.
60. Glueck and Glueck, *Unraveling Juvenile Delinquency*, 187, 33; and Sampson and Laub, "Unraveling," 177, 180.
61. Glueck and Glueck, *Physique and Delinquency*, 9.
62. Ibid., 27–31.
63. Ibid., 226.
64. Jacobson and Rowe, "Nature," 323–324.
65. Fishbein, "Biological Perspectives in Criminology," and *Biobehavioral Perspectives in Criminology*; Jacobson and Rowe, "Nature;" Rowe, *Biology and Crime*; Shah and Roth, "Biological;" and Walsh, *Biosocial Criminology*.
66. Jacobson and Rowe, "Nature," 324. See also Fishbein, "Biological Perspectives in Criminology," and *Biobehavioral Perspectives in Criminology*; Rowe, *Biology and Crime*; Shah and Roth, "Biological;" and Walsh, *Biosocial Criminology*.
67. Fishbein, "Biological Perspectives in Criminology," *Biobehavioral Perspectives in Criminology*; Jacobson and Rowe, "Nature;" Rowe, *Biology and Crime*; Shah and Roth, "Biological;" and Walsh, *Biosocial Criminology*.
68. Brennan and Raine, "Biosocial Bases," 590; Gottfredson and Hirschi, 70; Udry, "Sociology and Biology;" and Walsh, *Biosocial Criminology*. Perhaps the most significant exception to this is the work of Richard Udry in "Biological Predispositions" and "Biosocial Models."

69. Brennan and Raine, "Biosocial Bases," 590, italic omitted. See also Raine, Brennan, and Farrington, "Biosocial Bases," 3; Fishbein, *Biobehavioral Perspectives in Criminology,* 9; and Walsh, *Biosocial Criminology.*

70. Fishbein, *Biobehavioral Perspectives in Criminology,* 2–7, 12–13; and Walsh, *Biosocial Criminology,* 11.

71. Fishbein, *Biobehavioral Perspectives in Criminology,* 2, 62.

72. Brennan and Raine, "Biosocial Bases;" Fishbein, "Biological Perspectives in Criminology," and *Biobehavioral Perspectives in Criminology*; Jacobson and Rowe, "Nature;" and Raine, Brennan, and Farrington, "Biosocial Bases."

73. Jacobson and Rowe, "Nature," 338.

74. Fishbein, *Biobehavioral Perspectives in Criminology,* 63.

75. Ibid.

76. Ibid., 9–11.

77. The discussion draws heavily from Fishbein, *Biobehavioral Perspectives in Criminology.* Two other excellent overviews of biosocial criminology are Rowe, *Biology and Crime,* and Walsh, *Biosocial Criminology.*

78. Fishbein and Pease, *Dynamics of Drug Abuse,* 27–36.

79. Fishbein and Pease, *Dynamics of Drug Abuse,* 27–28; and Fishbein, *Biobehavioral Perspectives in Criminology.*

80. Fishbein and Pease, *Dynamics of Drug Abuse,* 30–31.

81. Ibid., 34.

82. Walsh, *Biosocial Criminology,* 77.

83. Fishbein and Pease, *Dynamics of Drug Abuse,* 34–35.

84. Fishbein and Pease, *Dynamics of Drug Abuse,* 35. See also Raine, *Psychopathology of Crime,* 103–127; and Walsh, *Biosocial Criminology,* 77.

85. Walsh, *Biosocial Criminology,* 77.

86. Rowe, *Biology and Crime,* 70. See also Fishbein, *Biobehavioral Perspectives in Criminology,* 56, 55–60.

87. Fishbein, *Biobehavioral Perspectives in Criminology,* 55; see, for example, Chretien and Persinger, "Prefrontal Deficits."

88. Rowe, 80–84.

89. Raine, *Psychopathology of Crime,* 155. See also Rowe, 80–84.

90. Fishbein, *Biobehavioral Perspectives in Criminology,* 49.

91. Ibid.

92. Ibid.

93. Ibid.

94. Vold, Bernard, and Snipes, *Theoretical Criminology,* 47.

95. Eysenck, "Biosocial Model," 24. See also Eysenck and Gudjonsson, *Causes and Cures.*

96. Vold, Bernard, and Snipes, *Theoretical Criminology,* 48.

97. Rowe, *Biology and Crime,* 80. See also Mednick, "Biosocial Theory."

98. Vold, Bernard, and Snipes, *Theoretical Criminology,* 48; Jacobson and Rowe, "Nature;" Raine, *Psychopathology of Crime*; Rowe, *Biology and Crime*; and Yaralian and Raine, "Biological Approaches to Crime."

99. Eysenck, "Biosocial Model," 24; Eysenck and Gudjonsson, *Causes and Cures*; and Rowe, *Biology and Crime.*

100. Fishbein, *Biobehavioral Perspectives in Criminology,* 51; and Raine, *Psychopathology of Crime.*

101. Fishbein, *Biobehavioral Perspectives in Criminology,* 51.

102. Rowe, *Biology and Crime,* 79.

103. Jacobson and Rowe, "Nature," 328; see Farrington, "Relationship;" Pitts, "Reduced Heart Rate;" and Raine, *Psychopathology of Crime.*

104. Fishbein, *Biobehavioral Perspectives in Criminology,* 52.

105. Ibid., 35.

106. This discussion draws heavily from Diana Fishbein's excellent book, *Biobehavioral Perspective in Criminology* (36–41). For further information, refer to her book.

107. Fishbein and Pease, *Dynamics of Drug Abuse,* 40; and Walsh, *Biosocial Criminology,* 77.

108. Walsh, *Biosocial Criminology,* 77; and Fishbein, *Biobehavioral Perspectives in Criminology,* 36.

109. Fishbein and Pease, *Dynamics of Drug Abuse,* 40, emphasis added; and Walsh, *Biosocial Criminology,* 77–78.

Chapter Resources

110. Fishbein and Pease, *Dynamics of Drug Abuse,* 40.
111. Fishbein, *Biobehavioral Perspectives in Criminology,* 41.
112. Fishbein, *Biobehavioral Perspectives in Criminology,* 37; Raine, *Psychopathology of Crime.*
113. Fishbein, *Biobehavioral Perspectives in Criminology,* 38; Moffitt et al., "Male Violence;" and Moffitt et al., "Epidemiological Study."
114. Fishbein, *Biobehavioral Perspectives in Criminology,* 39.
115. Ibid., 40.
116. Lee Ellis, "Monoamine Oxydase and Criminality."
117. Fishbein, *Biobehavioral Perspectives in Criminology,* 42.
118. Jacobson and Rowe, "Nature," 329.
119. Albert et al., "Hormone-Dependent;" Brain, "Hormonal Aspects;" Jacobson and Rowe, "Nature."
120. Archer, "Influence of Testosterone;" Brain, "Hormonal Aspects;" Harris, "Review;" Jacobson and Rowe, "Nature;" and Raine, *Psychopathology of Crime.*
121. Jacobson and Rowe, "Nature."
122. Brooks and Reddon, "Serum Testosterone;" Constantino et al., "Testosterone and Aggression."
123. Booth and Osgood, "Influence."
124. Dabbs and Morris, "Testosterone."
125. Brain, "Hormonal Aspects," 221.
126. Booth et al., "Winning and Losing;" and Mazur and Booth, "Testosterone and Dominance."
127. Tehrani and Mednick, "Genetic Factors."
128. Fishbein, *Biobehavioral Perspectives in Criminology,* 26; and Tehrani and Mednick, "Genetic Factors."
129. Walsh, *Biosocial Criminology,* 26–28.
130. Fishbein, *Biobehavioral Perspectives in Criminology,* 26; and Walsh, *Biosocial Criminology,* 23.
131. Walsh, *Biosocial Criminology,* 38–40.
132. Jacobson and Rowe, "Nature," 331–332.
133. Ibid., 333.
134. Ibid., 332.
135. Mednick, Gabrielli, and Hutchings, "Genetic Influences;" Jacobson and Rowe, "Nature," 333.
136. Tehrani and Mednick, "Genetic Factors;" and Jacobson and Rowe, "Nature."
137. Tehrani and Mednick, "Genetic Factors."
138. Ibid.
139. Ibid., 25.
140. Grove et al., "Heritability."
141. Fishbein, *Biobehavioral Perspectives in Criminology,* 29.
142. Fishbein, *Biobehavioral Perspectives in Criminology,* 29, emphasis in original. See Rowe, *Biology and Crime,* for an excellent overview of molecular genetics and criminal behavior.
143. Ellis and Walsh, "Gene-Based," 230.
144. Ibid., 232.
145. Ibid., 233.
146. Ibid., 234, emphasis in original.
147. Ibid., 236.
148. Cullen and Agnew, *Criminological Theory,* 73.
149. We should note that several areas of study are conspicuously absent from our discussion of psychological approaches to delinquency, including psychoanalytic theory, cognitive and moral development, and "criterion-keyed" personality tests like the Minnesota Multiphasic Personality Inventory (MMPI) and the California Psychological Inventory (CPI). While traditionally discussed in juvenile delinquency texts, the study of individual, psychological factors has moved in a decidedly different direction: one that we have called biosocial criminology (Steinberg, "Adolescent Development;" and Miller and Lynam, "Structural Models").
150. Miller and Lynam, "Structural Models."
151. Glueck and Glueck, *Unraveling Juvenile Delinquency,* 275, italics in original.
152. Ibid., 215, Chapters 18 and 19.
153. Ibid., 215.
154. Kleinmuntz, *Personality and Psychological Assessment,* 7.
155. Miller and Lynam, "Structural Models," 767.

156. Miller and Lynam, "Structural Models," 767; Glueck and Glueck, *Physique and Delinquency,* 27–30; and Glueck and Glueck, *Unraveling Juvenile Delinquency.*

157. Miller and Lynam, "Structural Models," 767; and Eysenck, "Dimensions."

158. Miller and Lynam, "Structural Models," 767.

159. Caspi et al., "Replications," 165.

160. Eysenck spelled "extraversion" with an "a," rather than an "o."

161. Clark and Watson, 403; and Eysenck, *Crime and Personality.*

162. Clark and Watson, 403.

163. Eysenck and Gudjonsson, *Causes and Cures,* 55.

164. Ibid., 56.

165. Ibid.

166. Bartol, *Criminal Behavior,* 42–43; and Ewen, *Personality,* 138.

167. Eysenck, "Biosocial Model," 24.

168. Bartol, *Criminal Behavior,* 45.

169. Miller and Lynam, "Structural Models," 770.

170. Tellegen, "Structures."

171. Miller and Lynam, "Structural Models," 770; and Tellegen, "Structures."

172. Miller and Lynam, "Structural Models," 770. See also Caspi et al., "Replications."

173. John and Srivastava, "Big Five Taxonomy," 103; and Miller and Lynam, "Structural Models," 767–768.

174. John and Srivastava, "Big Five Taxonomy."

175. Ewen, *Personality,* 140; John and Srivastava, "Big Five Taxonomy," 105, 121; and Miller and Lynam, "Structural Models," 768.

176. Cloninger, "Systematic Method," 574, 575; and Cloninger, "Unified Biosocial Theory."

177. Cloninger, "Systematic Method," 574–575.

178. Tremblay et al., "Predicting Early Onset," 735.

179. Cloninger, "Systematic Method," 575.

180. Tremblay et al., "Predicting Early Onset," 735.

181. Cloninger, "Systematic Method," 575.

182. Cloninger, Svrakic, and Prybeck, "Psychobiological Model," 978.

183. Ibid., 979.

184. Cloninger, "Systematic Method," 576–578.

185. Arseneault et al., "Biosocial Exploration," 313; Caspi et al., "Replications," 164–165; and Miller and Lynam, "Structural Models," 772.

186. Miller and Lynam, "Structural Models," 768, 771.

187. Miller and Lynam, "Structural Models," 771. See also Clark and Watson, 403–406; Eysenck and Gudjonsson, *Causes and Cures*; and John and Srivastava, "Big Five Taxonomy," 122.

188. Miller and Lynam, "Structural Models," 772, 778.

189. Miller and Lynam, "Structural Models," 780. See also Caspi et al., "Replications."

190. Miller and Lynam, "Structural Models," 781.

191. Ibid., 782, references omitted.

192. Vold, Bernard, and Snipes, *Theoretical Criminology.*

193. Ibid., 58.

194. Curran and Renzetti, *Theories of Crime,* 68.

195. Gould, *Measurement of Man,* 149–150.

196. Vold, Bernard, and Snipes, *Theoretical Criminology,* 58.

197. Binet, quoted in Gould, *Measurement of Man,* 151.

198. Ibid., 154.

199. Curran and Renzetti, *Theories of Crime,* 68.

200. Gould, *Measurement of Man,* 158–164.

201. Goddard, *Kallikak Family.*

202. Goddard, *Causes and Consequences*; and Vold, Bernard, and Snipes, *Theoretical Criminology.*

203. Curran and Renzetti, *Theories of Crime,* 69.

204. Goddard, *Kallikak Family,* 173.

205. Vold, Bernard, and Snipes, *Theoretical Criminology,* 61; and Goddard, "Question of Definition," 225.

Chapter Resources

206. Murchison, *Criminal Intelligence*.
207. Wilson and Herrnstein, *Crime and Human Nature*, 152.
208. Hirschi and Hindelang, "Intelligence and Delinquency."
209. Wilson and Herrnstein, *Crime and Human Nature*, 153.
210. Hirschi and Hindelang, "Intelligence and Delinquency."
211. Block, "Remarks;" Denno, "Sociological;" Lynam and Moffitt, "Reply to Block;" and Lynam, Moffitt, and Stouthamer-Loeber, "Explaining."
212. Quay, "Intelligence," 114–115.
213. Bartusch et al., "Is Age Important?;" Farrington, "Individual Differences and Offending;" Hirschi and Hindelang, "Intelligence and Delinquency;" Lynam and Moffitt, "Reply to Block;" Moffitt, "Neuropsychology;" Moffitt, Lynam, and Silva, "Neuropsychological Testing;" and Rutter, Giller, and Hagell, *Antisocial Behavior*.
214. Curran and Renzetti, *Theories of Crime*, 71–74; Hirschi and Hindelang, "Intelligence and Delinquency," 581–584; Menard and Morse, "Structural Critique;" and Weinberg, "Intelligence and IQ."
215. Simons, "Meaning," 269.
216. Curran and Renzetti, *Theories of Crime*, 71.
217. Barrett and Depinet, "Reconsideration."
218. Quay, "Intelligence."
219. Simons, "Meaning;" and Weinberg, "Intelligence and IQ."
220. Curran and Renzetti, *Theories of Crime*, 72–73.
221. Hirschi and Hindelang, "Intelligence and Delinquency," 581.
222. Blumstein, Farrington, and Moitra, "Delinquent Careers."
223. Lynam, Moffitt, and Stouthamer-Loeber, "Explaining." See also McGloin, Pratt, and Maahs, "Rethinking."
224. Denno, *Biology and Violence*.
225. This discussion is drawn from Curran and Renzetti, *Theories of Crime*, 71.
226. Eysenck, *New Look*; Plomin et al., "Parent–Offspring Adoption Study;" Scarr and Weinberg, "Minnesota Adoption Studies," and "Education."
227. Bouchard et al., "Sources;" and Petrill et al., "Genetic and Environmental Relationship."
228. Lombroso, *Criminal Man*, xiv, quoted in Binder, Geis, and Bruce, *Juvenile Delinquency*, 89.
229. Ferri, *Criminal Sociology*, xxxi, quoted in Martin, Mutchnick, and Austin, *Criminological Thought*, 38.
230. Miller and Lynam, "Structural Models," 780. See also Caspi et al., "Replications."

Situational and Routine Dimensions of Delinquency

9

Chapter Objectives

After completing this chapter, students should be able to:

- Identify characteristics of situations that motivate and provide opportunity for delinquent behavior.
- Describe how the routine activities of adolescents provide opportunity for delinquency.
- Explain how the adversity of homelessness is related to delinquent behavior.
- Distinguish the *subculture of delinquency* from *delinquent subculture,* and be able to explain how each is relevant to an understanding of delinquent behavior.
- Describe how an individual's perception and interpretation of the immediate situation are related to delinquency.
- Understand key terms:
  Theories:
  situational inducements and commitment to conformity
  routine activities theory
  drift theory

techniques of neutralization
moral self transcendence

suitable targets
capable guardians
routine activities

Terms:

situational explanations
developmental explanations
objective content of situations
subjective content of situations
situational inducements
commitment to conformity
criminogenic situations
motivated offenders

hard determinism
soft determinism
drift
subculture of delinquency
subterranean values
techniques of neutralization
will

CASE IN POINT

The Sneaky Thrill of Pizza Theft: A Spontaneous, Situational Act

In his provocative book, *Seductions of Crime,* Jack Katz provides a series of personal accounts of crime that were offered by students in his criminology class. These accounts offer vivid description of the experience of crime—"what it means, feels, sounds, tastes, or looks like to commit a particular crime." The immediate situation of delinquency not only provides sensation, but also elicits interpretation, meaning, and response. Certain situations also motivate and provide opportunity for delinquent behavior. This chapter considers such situational and routine dimensions of delinquent behavior.

I grew up in a neighborhood where at 13 everyone went to Israel, at 16 everyone got a car and after high school graduation we were all sent off to Europe for the summer. . . . I was 14 and my neighbor was 16. He had just gotten a red Firebird for his birthday and we went driving around. We just happened to drive past the local pizza place and we saw the delivery boy getting into his car. . . . We could see the pizza boxes in his back seat. When the pizza boy pulled into a high rise apartment complex, we were right behind him. All of a sudden, my neighbor said, "You know, it would be so easy to take a pizza!" . . . I looked at him, he looked at me, and without saying a word I was out of the door . . . got a pizza and ran back. . . .

The feeling I got from taking the pizza, the thrill of getting something for nothing, knowing I got away with something I never thought I could, was wonderful. . . . I'm 21 now and my neighbor is 23. Every time we see each other, I remember and relive a little bit of that thrill.

Source: Katz, *Seductions of Crime,* 3, 52, 64.

Almost 60 years ago, renown criminologist Edwin Sutherland made the observation that "scientific explanations of criminal behavior may be stated either in terms of the processes operating at the moment of the occurrence of crime or the processes operating in the earlier history of the criminal."[1] He referred to the first type of explanation as **situational**, and the second as **developmental**.[2] He went on to acknowledge that, of the two, situational explanations are undoubtedly "superior" as an explanation of criminal behavior.[3]

Despite this concession, Sutherland is most well known for advancing a developmental theory of crime and delinquency—differential association theory. We'll look at his theory in Chapter 11. With few notable exceptions, criminological theory and research has followed Sutherland's lead in emphasizing developmental explanations of delinquency to the almost total exclusion of situational explanations.[4] This trend is ironic in that adolescent behavior in general, and delinquent behavior in particular, is often characterized as spontaneous and as being driven by the "spur of the moment."

Basing his assertions on his well-known study of delinquent behavior, Martin Gold contends that most delinquent acts are unplanned and rarely develop into a repetitive pattern of delinquent behavior.[5] In addition, those who participate in delinquent acts have varying levels of commitment and involvement. These observations led Gold to compare delinquent behavior to a pickup game of basketball in which participation is desired, but casual, unplanned, and short-term. Teams are rarely set beforehand, and "boys who want to play know where the game is usually played . . . and that is where other likely players hang out."[6] The game's competitiveness depends on the commitment of those who play: some players are invested in the game, while others are casual participants. Gold's observations support an image of delinquency that is far more spontaneous and situational than most developmental explanations of delinquency.

The reluctance of criminologists to consider situational explanations of delinquency is due largely to the fact that the perceptions, activities, and opportunities of the current situation are much more difficult to incorporate into theory and research than are developmental characteristics. Probably the most significant difficulty is that situational aspects include both objective and subjective content in which the characteristics of the current situation must be considered together with the perceived understanding of the actor.[7]

Characteristics of the immediate setting are referred to as the **objective content of situations**.[8] These are characteristics of the situation that motivate and provide opportunities for delinquent acts. Encountering interpersonal conflict, for example, may stimulate an aggression response. Similarly, poverty and hunger are often thought to lead to property offending. In addition, the routine daily activities of many adolescents provide ample opportunity for involvement in delinquency.

The **subjective content of situations** encompasses the "individual's perception and interpretation of the immediate setting."[9] The various rationalizations that run through a youth's mind before committing a delinquent act are one example of subjective content. Subjective content of situations also includes how delinquent acts are experienced by youths and what these experiences mean to

situational explanations
Explanations of criminal behavior that focus on processes operating at the moment of the occurrence of crime, including characteristics of the current setting that motivate and provide opportunity for crime.

developmental explanations
Explanations of criminal behavior that focus on processes operating in the earlier history of the criminal. A wide range of social and psychological characteristics of offenders and their environments are considered.

objective content of situations Characteristics of a given situation that provide motivation and opportunity for crime. Poverty and hunger, for example, may lead to property offending.

subjective content of situations The "individual's perception and interpretation of the immediate setting" (Birkbeck and LaFree, "Situational Analysis," 129).

them. Shoplifting, for example, sometimes takes on a game-like quality, pitting the shoplifter against the store clerk. In these cases, shoplifting provides a sense of thrill and accomplishment, rather than the need for the particular item being stolen.[10]

This chapter will consider a number of approaches that offer situational explanations of delinquent behavior. In each of these, the situational context of delinquent acts is the unit of analysis and the immediate circumstances are the center of attention.[11] Several of these approaches emphasize the objective content of situations—characteristics of the setting that are thought to motivate and to provide opportunity for delinquent acts. Other approaches emphasize subjective elements of situations by focusing on an individual's perception and interpretation of the current situation. A word of warning: while these various approaches all provide situational explanations of delinquency, there is not a great deal of continuity between them. We will simply describe each approach in turn, with little carryover in discussion.

■ Characteristics of the Current Situation

Current settings of a situation "can be described in terms of who is there, what is going on, and where it takes place."[12] Viewed in this way, situations take on an objective quality, being composed of characteristics that motivate delinquent acts and provide opportunity.[13] Situations can motivate delinquent behavior by "imposing negative experiences such as frustration, threats, humiliation, and boredom; by offering positive attractions such as money, property, image-building, thrills, and sexual satisfaction; or by providing models to be imitated."[14] Situations can also provide opportunity by affecting the extent to which criminal motivations can be fulfilled.[15] The possibility of success, absence of detection, availability of goods, and access to victims all relate to the degree to which a situation provides opportunity for delinquency. For criminologists, the primary task is to identify the link between these situational characteristics and subsequent delinquent behavior.

Situational Correlates of Aggression

One large body of research, conducted primarily by psychologists, has examined the situational factors that are related to aggressive behavior. Experimental studies reveal five situational correlates of aggression.[16]

Blocked Goals and Frustration

Research has examined the degree to which aggression is a consequence of frustration, commonly identified as the frustration–aggression hypothesis. This research reveals that individuals sometimes respond aggressively to situations that they find frustrating, especially when these frustrations are thought to be intentionally produced by others.[17]

Physical or Verbal Provocation

Research has also found that verbal and physical threats and attacks provoke aggressive responses. Daniel Lockwood's study of 250 violent incidents reported by 110 middle and high school students found that the most common provocation for violence was "offensive touching," including grabbing, pushing, and hitting. Provocations were also nonphysical; these included teasing, insulting, and saying something negative about a person to a third party. Typically, these provocations led first to an argument, then escalated into an exchange of insults, and finally resulted in a violent response by one of the youths.[18]

It is important to point out, however, that aggressive responses are not simply a product of innocent victims being provoked by others. Research shows consistently that aggressive responses sometimes are instigated by the offender's mistreatment of others, which elicits negative reactions.[19] In turn, these reactions are taken as provocations for aggression. The classic example of this is a school bully who picks on students and, when these students respond in kind, the bully claims that he was provoked into an aggressive response.[20]

Aggressive Models

Exposure to violent models, particularly in the media, appears to produce aggressive behavior in some individuals. An extensive review of the literature by Wendy Wood and her associates concludes: "Our results demonstrate that media violence enhances children's and adolescents' aggression in interaction with strangers, classmates, and friends. Our findings cannot be dismissed as representing artificial constructions because studies included in our review evaluated media exposure on aggression as it naturally emerged in unconstrained social interaction."[21] However, they point out that the studies they reviewed focused on short-term media effects, since they measured aggressive behavior shortly after the media portrayal of violence. It is also the case that researchers, using experimental methods, have not been able to establish conclusively a causal relationship between media violence and subsequent violent behavior.[22] It may be that violent children watch more media violence, identify with violent characters, or believe that violence is a normal part of everyday life and thus an appropriate response to interpersonal conflict.[23]

Cues for Aggression

In some instances, people or objects may prompt aggressive behavior. "Crimes of obedience" sometimes result from the physical presence or instruction of authority figures. Milgram's famous experiment found that subjects would administer what they thought to be a near-fatal electric shock to another subject when told to do so by an experimenter.[24] Similarly, delinquent youths often claim that they were just "following the leader" in committing delinquent offenses. Research also indicates that the presence of weapons and certain clothing (for example, displaying gang colors) may function as a cue for aggression.

Low Levels of Restraint on Aggression

Situations also vary in the level of restraint on aggression. In some situations, the presence of rules and sanctions, together with the threat of punishment (either physical or nonphysical), operates as a control for aggressive action. However, the influence of these situational characteristics on aggressive behavior depends on the perception of the individual. When restraints are perceived as unjust or as lacking authority, restraint is reduced.

Situational Inducements

The reluctance of sociological criminologists to consider situational factors in delinquent behavior was first exposed in 1965 by Scott Briar and Irving Piliavin in an article entitled "Delinquency, Situational Inducements, and Commitment to Conformity." They argued that most criminological theory is based on the assumption that some youths develop long-lasting predispositions to engage in illegal acts because of "certain individual, interpersonal, or social conditions . . . which propel them into illegal behavior."[25] According to Briar and Piliavin, these "motivational theories"—what Sutherland would call developmental explanations—share a number of serious problems.[26] Most significantly, the conditions that supposedly lead to the development of delinquent dispositions do not operate "uniformly." Many youths who experience these conditions never develop delinquent dispositions. Even among youths who exhibit these dispositions, many never engage in delinquent acts. In the end, most delinquent youths, regardless of whether they develop delinquent predispositions, become law-abiding in late adolescence and early adulthood. Thus, Briar and Piliavin question whether the causal conditions identified by motivational theories adequately explain why some youths become involved in delinquent offenses and others do not. Furthermore, Briar and Piliavin claim that motivational theories are unable to account for the short-term duration of most delinquent involvement.

In contrast to motivational theories, Briar and Piliavin turn to characteristics of the current situation that provide "conflicts, opportunities, pressures, and temptations which may influence the actors' actions and views."[27] They refer to these as "situationally induced stimuli of relative short duration," or, more simply, **situational inducements**.[28] The acts that follow from these situational inducements serve a purpose—they allow youth "to obtain valued goods, to portray courage in the presence of, or be loyal to peers, to strike out at someone who is disliked, or simply to 'get kicks.' "[29]

Briar and Piliavin point out that all youths experience these situationally induced stimuli—there is nothing unique about experiencing these "inducements." Such situational experiences influence what youths think and do, and delinquent acts are sometimes an expression of these situational inducements: "There is considerable basis for assuming that the immediate situation in which a youth finds himself can play an important role in his decision to engage in delinquency behavior. Obviously, however, this is not to say that the situation offering inducement or pressure to a youth to deviate will necessarily lead him to take such action."[30]

This last comment by Briar and Piliavin leads to a second element of their theory. While drawing attention to the importance of situational inducements, they also acknowledge that not all youths succumb to such pressures and that situational stimuli may not be sufficient in themselves to produce delinquent behavior. All youths experience situational motivation and opportunity to engage in delinquent acts; however, the probability that these situational inducements will be acted upon depends on an individual's **commitment to conformity**.[31] Those youths who have little commitment to conformity are heavily influenced by situational pressures, whereas those who are highly committed are not. In ad-

situational inducements Characteristics and processes of the immediate situation that make crime and delinquency appealing.

commitment to conformity Logical reasons to conform, such as the desire to maintain or even enhance self image, status, valued relationships, and future activities.

dition, commitment to conformity influences the youth's response to adult authority figures and to the choice of friends.

Commitment to conformity is due not only to fear of punishment, but also to a youth's desire to "maintain a consistent self image, to sustain valued relationships, and to preserve current and future statuses and activities."[32] In other words, commitment is derived not only from direct controls but also from controls that arise from interpersonal relationships.

The development of commitment is most heavily influenced by the youth's relationship with his or her parents.[33] Affection, discipline, expectations, and willingness to conform to parental authority are key aspects of the relationship between parents and youths. Youths with strong commitments to parents are less likely to act upon situational inducements to deviate than are youths with minimal commitments. Thus, the parent–child relationship, as a primary determinant of commitment, influences the effect that situational pressures have on behavior and the likelihood of involvement in delinquency.

Mean Streets: Adverse Situations and Delinquency

The motivational theories of delinquent behavior that Briar and Piliavin criticize have also been taken to task more recently by Bill McCarthy and John Hagan. Similar to Briar and Piliavin, they point out that "these theories assume that crime and delinquency are predetermined by background experiences and developments that cumulate over time."[34] Further, they contend that criminological theory in the last fifty years has become increasingly focused on background consideration to the almost total exclusion of the "foreground" of criminal experience.[35] It is their intent "to push sociological criminology in precisely the opposite direction: toward foreground causes of delinquency and crime through the study of criminogenic situations."[36]

<u>Criminogenic situations</u> refer to characteristics of a given situation that motivate and provide opportunity for delinquent acts. According to Don Gibbons: "Lawbreaking behavior may arise out of some combination of situational pressures and circumstances, along with opportunities for criminality which are totally outside the actor."[37] Instead of emphasizing characteristics of the current situation that provide opportunities, pressures, and temptations to engage in delinquent acts (as do Briar and Piliavin), McCarthy and Hagan draw attention to the adverse situations in which some youths find themselves. In particular, they consider the problems of everyday life for homeless youths. McCarthy and Hagan contend that living on the streets provides not only temptations and opportunities for involvement in delinquent behavior, but the harsh living conditions also produce much strain: "The homeless youth who live on the streets of our cities confront desperate situations on a daily basis. Often without money, lacking shelter, hungry, and jobless, they frequently are involved in crime as onlookers, victims, and perpetrators."[38] Crime is all around them—they see it, experience it, and resort to it as a means of coping with the difficulties of being homeless. It is also the case that street life involves few controls to restrain crime and delinquency. For homeless youths, the excitement and freedom of street life goes hand in hand with the harsh living conditions of "mean streets." Desperate

criminogenic situations
Situational pressures and circumstances that motivate and provide opportunity for crime and delinquency.

situations require desperate means, and street crime is an ever-present part of life for homeless youths.

Not only has contemporary criminology neglected the role that adverse situations play in delinquent behavior, but it has also failed to study the street youths who disproportionately experience such hardships. The dominance of self-report research methods over the past fifty years has led to a focus on delinquency committed by adolescents living at home and attending school—adolescents living in "institutionally protected situations."[39] As a result, self-report methods may minimize the causal importance of situational factors, especially the strain that results from difficult situations. McCarthy and Hagan's solution is to take criminology back to the streets, where delinquency research first began. In what are now classic studies, "sociologists of the street studied whether the rapidly building forces of urbanization, immigration, and industrialization were exposing the youth of growing cities to criminogenic conditions. The street seemed like a natural place to study these youth, especially as they gathered on street corners and in gangs."[40]

McCarthy and Hagan conducted a survey of 390 homeless youths in the city of Toronto. They deliberately choose not to refer to these street youths as "runaways" because "it implies that leaving home is inappropriate and that a return is both possible and desirable. Yet previous research reveals that many youths who leave home are forced out by parents (i.e., they are 'throwaways') or are escaping from abusive environments; not surprisingly, many are unwilling to return home."[41] The survey was administered primarily in two contexts: at social service agencies that provided assistance to homeless youths (e.g., hostels, shelters, drop-in centers) and at street locations where homeless youths panhandled or spent the night (e.g., inner-city parks, street corners, and bus stations).

McCarthy and Hagan supplement their street data with data drawn from a sample of youths living at home and attending school. These data were gathered at three different high schools in the Toronto area from randomly selected 9th- through 12th-grade classes. Using a questionnaire designed to parallel the instrument used with the street youth, 562 students provided information about their families, friends, and school. These data were then combined with those gathered from street youths, allowing the researchers to investigate the factors that compel youths to leave home and "take to the streets."[42]

Taking to the Streets

McCarthy and Hagan found that the likelihood of leaving home increased with a number of distinguishing factors: age, coercive controls by parents, sexual abuse, conflict with teachers, and having delinquent friends. Youths who leave home tend to be older, to have little desire to achieve in school, and to have frequent conflict with teachers. Their parents are often divorced and characterized as abusive both physically and sexually. They are also more likely to have delinquent friends. These empirical findings are supported by the narrative accounts of family and school life offered by homeless youths. Jeremy, a young boy who left home at the age of 12, describes his situation this way:

> My dad was an alcoholic, and he always abused me—physically. He'd punch me and stuff like that—throw me up against the walls.

And like one night we were going at it, and I turned around, like he punched me a couple of times. I turned around and got a baseball bat out of the bedroom, and I hit him in the head, and then he got back up, and he started pounding on me big time. Well, the cops came and they took him, and they said "You can go live with your mother, right?" My mother had already said, "We don't want you," so I said, "Okay, I'm going to go to my mother's" and [instead] I went out in the streets.[43]

The Adversity of Homelessness: A Criminogenic Situation?

Youths who leave home are exposed almost immediately to harsh living conditions. One of the first problems they confront is finding a safe place to sleep. The narrative accounts of initial experiences on the street frequently tell of youths walking for hours with no place to go.[44] Jeremy expressed the problem of shelter like this: "I mostly didn't sleep the first couple days. Then I got to a bridge, and I just slept underneath the bridge. Then I kept on walking, sleeping at night in people's sheds and stuff like that."[45] The lack of shelter is an almost immediate strain for most homeless youth: "The worst thing was not knowing where you were going to sleep. Sleeping in a storefront or on the street. Who knows what's going to happen to you?"[46]

It also is not long before homeless youths face the problem of hunger. Hagan and McCarthy report that over three-fourths of the youths surveyed revealed that they had gone an entire day without eating, and more than half said they frequently went hungry.[47] Without resources of money or employment, homeless youths confront the unrelenting problems of food and shelter. As a result, homelessness is a criminogenic situation, and street crimes—principally theft, robbery, selling drugs, prostitution, and violence—are a means of coping with adversity. "Many street youth, particularly those with lengthy street careers, move in and out of these activities to sustain and support their existence on the street."[48] Some criminologists, however, argue just the opposite: involvement in delinquency leads to leaving home, and these youths simply continue their involvement in crime once they are on the streets. This alternative point of view claims: "Homelessness does not lead to crime; crime leads to homelessness."[49]

McCarthy and Hagan's study of street youth found support for the view that homelessness leads to crime, especially street crimes associated with problems of shelter, hunger, and the lack of legitimate resources (most importantly, money and employment). A significant proportion of homeless youth were involved in criminal activities only after leaving home. This was true for crimes typically associated with street life: "A significantly and substantially greater number of adolescents chose to use hallucinogens and cocaine, steal, and work as prostitutes once they became homeless."[50] The effect of street life on criminal involvement held true for both males and females and was true regardless of age. McCarthy and Hagan also found that youths who were homeless for a year or more were more likely to turn to crime. They concluded that motivation for crime exists in the criminogenic conditions of homelessness, rather than in a person's background.[51]

In a study that followed, McCarthy and Hagan attempted to explore more fully how homelessness was connected to criminal involvement, particularly to the types of crime most commonly associated with street life: theft of food, serious theft, and prostitution.[52] Using the same data provided by homeless youth, they investigated how three specific conditions of homelessness—hunger, lack of shelter, and unemployment—were related to these forms of street crime. A large number of background factors were incorporated into their study, including social class, strain within the family, family structure, parental supervision and controls, school experiences, involvement in delinquency at home, delinquent associates at home and on the streets, and time on the streets. The simultaneous consideration of these background factors, together with conditions of homelessness, allowed the researchers to see whether street crime was a response to the adversity of homelessness or a result of characteristics that homeless youth bring with them to the streets. McCarthy and Hagan used a statistical method (ordinary least squares regression) that allowed them to determine whether situational adversity, as measured by hunger, lack of shelter, and unemployment, had a measurable direct effect on street crime, independent of youth's background when living at home. They found that each measure of situational adversity was "at least as (if not more) prominently associated with street crime as background and developmental variables."[53] In separate analysis of these three types of street crime, they found that hunger played the strongest role in street theft; problems of hunger and shelter were significantly related to serious theft; and problems of shelter and unemployment predict involvement in prostitution. They concluded that the adverse situations of homelessness have strong direct effects on street crime.[54]

The Situation of Company

One of the most enduring areas of study in criminology is the role that peers play in learning and reinforcing patterns of delinquent behavior. Chapter 11 will address the major questions, issues, and theories related to the nature and extent of peer influences. The consideration of peer influence, however, is based on a developmental explanation in which delinquent attitudes and behaviors are learned from delinquent peers over time. The peer relationships that produce such strong influence on adolescents are thought to be strong, close, and long-lasting.

A contrasting point of view on peer influence has been offered by LaMar Empey and Steven Lubeck. They contend that situations involving peers—the "situation of company"—generate certain expectations for behavior, and these "normative expectations in group contexts" are the primary mechanisms of peer influence.[55]

Empey and Lubeck's research involved a survey of youths in two different locations: one urban, one rural. The survey instrument included sixteen different "situated choices," asking youths what they were "prepared to do."[56] For example:

- "Suppose you were home alone watching television one night and your friends called and asked you to go mess around at a cafe or bowling alley, would you go with them?"
- "If a friend of yours were in some kind of trouble and the police asked you about him, would you tell them what you know?"

- "Suppose when you and your friends were messing around one night, they decided to break into a place and steal some stuff, do you think you would go with them?"[57]

Those who completed the survey chose one of five response categories for each question: (1) every time, (2) most of the time, (3) about half the time, (4) some of the time, and (5) never. The pattern of responses indicated the degree to which youths were subjected to normative expectations of particular situations.

Empey and Lubeck found that all youths—both urban and rural, delinquent and nondelinquent—reported a willingness to conform to situational expectations. However, delinquent youth were more inclined to associate with peers than were nondelinquent youth.[58] As a result, delinquent youth were more likely to expose themselves to the "situation of company" and the associated expectations for behavior. In contrast, urban, nondelinquent youth were least inclined to involve themselves with peers in situations that involve deviance.[59]

Routine Activities of Adolescents: Opportunities for Delinquency

Situations are important largely to the degree that they provide opportunity for delinquency.[60] Obviously, some situations are more favorable for delinquency than others, but what characteristics of situations provide greater opportunity?[61] Opportunity includes not only the actual possibility of delinquent acts that some situations provide, but also the youth's consideration of whether a delinquent act is likely to bring about an expected reward and the probability of getting caught.[62] Viewed in this way, opportunity involves choice, but it is a choice prescribed by the situation. Marcus Felson observes: "People make choices, but they cannot choose the choices available to them."[63] Together with Lawrence Cohen, Felson developed *routine activities theory* in an attempt to address the characteristics of the situation that influence the range of choices available to individuals.[64]

According to the theory, three basic elements of the situation are necessary for crime or delinquency to occur: **motivated offenders** must come in contact with **suitable targets** in the absence of **capable guardians**.[65] The argument here is that situations that provide opportunity for crime are not necessarily unique or unusual in any way; rather, what must be considered are the **routine activities** of everyday life.[66] Crime and delinquency are more common when the routines of daily living—at home, on the job, and activities away from home—provide motivated offenders, suitable targets, and absence of capable guardians. Cohen and Felson go on to point out that significant changes in lifestyle since the end of World War II have resulted in "the dispersion of activities away from households."[67] Changes include increases in female labor force participation, in the proportion of households unattended during the day, in out-of-town travel, and in sales of consumer goods. These changes in routine activities are, in turn, related to the rise in predatory crimes such as robbery, burglary, larceny, and murder. The thesis of routine activities theory is that "the dramatic increase in the reported crime rates in the U.S. since 1960 is linked to changes in the routine activity structure of American society and to a corresponding increase in target suitability and decrease in guardian presence."[68]

motivated offenders The inclination to engage in crime for self-benefit.

suitable targets How advantageous a crime is to commit, depending on value, visibility, access, and ease of committing.

capable guardians The level of control and protection provided to people and property.

routine activities The repetitive patterns of daily living—at home, school, work, and leisure.

Motivated Offenders

The great paradox of this increase in crime is that it has occurred despite improvements in the structural conditions that are usually thought to motivate individuals to commit crime.[69] Prosperity, including an improved standard of living, should reduce individuals' motivation for crime, but the post-war boom was accompanied by an increase in crime, not a decrease. This increase was most pronounced between 1960 and 1975, when a variety of social and economic indicators showed substantial improvement, even in most metropolitan areas.

Cohen and Felson, however, assume that people are always motivated to commit crime. While motivation is one of the three "minimal elements" that must "converge in space and time" for predatory crime to occur, it is considered "a given" and therefore requires no explanation.[70] What must be explained, then, is change in the other two elements: suitable targets and absence of capable guardians.

Suitable Targets

Cohen and Felson observe that "it is ironic that the very factors which increase the opportunity to enjoy the benefits of life also may increase the opportunity for predatory violations."[71] Electronic goods, for example, have greatly enhanced quality of life, but they are also relatively easy to steal and have high value. The suitability of targets for crime depends on a number of characteristics, including value, physical visibility, access, and inertia (weight, size, and protective features).[72] Technological advances have tended to make consumer goods smaller, lighter, and more valuable, thereby increasing opportunity for crime. Additionally, greater availability and value have increased consumption of consumer products, making them more available for theft and robbery.

Absence of Capable Guardians

Crime prevention is often associated with police activity such as preventive patrol, but routine activities theory is more concerned with the daily activities of ordinary people that provide or fail to provide guardianship of one another and of property.[73] The dispersion of activities away from the home is a major recent change that reduces level of guardianship of homes, making them vulnerable to predatory crimes. The proportion of households unattended during the day, for example, has increased dramatically since 1960. Additionally, routine activities theory specifies that individuals engaged in activities away from home, without family members, are at greater risk of criminal victimization than those who are at home with family. A variety of propositions related to risk of criminal victimization follow from a consideration of guardianship. For example, homes left unattended for long periods of time are susceptible to theft,[74] and single adults employed outside the home and adolescents who engage in peer group activities have high rates of criminal victimization.[75] The element of guardianship affects opportunities for crime that are related to the routine activities of daily life that diminish social control.

Routine Activities of Adolescents

Routine activities theory holds that the likelihood of experiencing crime depends on the degree to which the routine activities of people's everyday lives provide opportunity for crime. However, the original theory made no attempt to explain individual offending. It simply assumed that given the opportunity for crime, there will always be people motivated to commit it. Wayne Osgood and his colleagues have recently taken up the consideration of how routine activities of adolescents influence involvement in a broad range of deviant behavior, including delinquency, heavy alcohol use, marijuana use, use of other illicit drugs, and dangerous driving.[76] They were especially interested to see if the routine activities perspective might provide insight into the relationship between age and criminal involvement. Chapter 5 described how criminal involvement is especially pronounced in the adolescent and young adult years and rises and falls rapidly before and after these years. Do "age-related changes in the activities of everyday life" account for this age–crime connection?[77] Popular opinion contends that extensive involvement with peers during the adolescent years provides frequent opportunity for delinquent behavior. The application of routine activities theory to this contention, however, is an attempt to better understand the types of adolescent activities that provide opportunity for deviance.

Osgood and his associates frame their study within a revised routine activities theory. Instead of "motivated offenders," the researchers contend that the motivation for deviant behavior resides in the situation rather than in the person. Some situations to which youths are exposed provide both opportunity and reward for deviant behavior. This is referred to as *situational motivation*. Variation in the daily activities of adolescents and the corresponding variation in levels of exposure to situations that provide opportunity and reward for deviance are seen as the equivalent of Cohen and Felson's second element of "suitable targets."

Of special importance during the adolescent years is the role that peer influence plays in deviant behavior. Osgood and his colleagues contend that "being with peers can increase the situational potential for deviance by making deviance easier and rewarding."[78] Consequently, they include *time with peers* as a second key dimension of their routine activities theory.

In applying the "absence of guardian" concept to the adolescent years, Osgood and his associates use a situational factor that they refer to as the *absence of authority figures*. Settings of work, school, and family provide authority figures that limit the possibility of deviance. In contrast, situations conducive to deviance are most common in leisure activities with peers, away from authority figures.[79]

The final element of their revised routine activities theory is *unstructured activities*. While structured activities such as participation in athletics, clubs, and work often involve authority figures who exercise social control and leave little time for deviant behavior, these activities do not necessarily eliminate opportunity for deviant activity. John Hundleby, for example, found that participation in athletic activities is positively associated with substance use, sexual behavior, and delinquency.[80] Routine activities theory, however, draws attention to activities that provide opportunity for deviant behavior. In particular, opportunity for deviance increases to the degree that activities involve unstructured, unsupervised socializing with peers.[81]

Taken together, these four key aspects of adolescents' routine activities demonstrate that "situations conducive to deviance are especially prevalent in unstructured socializing activities with peers that occur in the absence of authority figures. The lack of structure leaves time available for deviance; the presence of peers makes it easier to participate in deviant acts and makes them more rewarding; and the absence of authority figures reduces the potential for social control responses to deviance."[82]

Testing this thesis involved distinguishing which routine activities are most related to deviant behavior, while controlling for other possible explanations. It may be that the types of routine activities in which youths are involved are largely a function of structural characteristics such as age, sex, and social class. If so, the relationship between routine activities and deviant behavior will no longer exist once these other variables are introduced into the analysis. Using longitudinal data from the Monitoring the Future study, the researchers measured thirteen different routine activities and five types of deviant behavior: delinquent behavior, heavy alcohol use, marijuana use, use of illicit drugs, and dangerous driving. Research in Action, "Measuring the Routine Activities of Adolescents," lists the activities that were included in the analysis.

Four of these thirteen routine activities involve unstructured socializing with peers without authority figures present, as specified by the revised theory: riding around in a car for fun, getting together with friends informally, going to parties, and spending evenings out for fun and recreation.[83] When all thirteen of these routine activities are considered together, they explained a significant amount of the variation in involvement in deviant behavior. However, unstructured socializing activities accounted for the largest share of the explained variance.[84] The researchers conclude that "it is not merely spending time outside the home or socializing that leads to deviant behavior."[85] Rather, it is the unstructured and unsupervised nature of these activities that is associated with involvement in deviant behavior.

Further analysis revealed that unstructured socializing activities accounted for a substantial portion of the association that age, sex, and socioeconomic status have with these different types of deviant behaviors.[86] This means that much of the reason age, sex, and socioeconomic status are related to deviance is because of their relation to routine activities. These structural characteristics are related to youth's routine activities, and these routine activities are, in turn, linked to level of involvement in deviant acts.

hard determinism
Delinquent behavior is caused by factors that can be identified by science. Given these factors, delinquency is a predictable outcome.

drift A view of youths that "stands midway between freedom and control," in which youths hold both conventional and deviant values, resulting in both conforming and delinquent behavior (Sykes and Matza, "Techniques of Neutralization," 28).

■ Drifting into Delinquency

In his classic book, *Delinquency and Drift,* David Matza challenged the view that delinquent behavior is determined by factors that can be identified through science—a view called **hard determinism**.[87] The deterministic model is associated with positivist criminology, a school of thought described in Chapter 7. Hard determinism characterizes delinquent youth as being driven into delinquent acts because of biological or psychological conditions or because of social structural influences, such as "the effects of social class, ethnic affiliation, family, and neighborhood."[88] In contrast, Matza advances the concept of **drift** to convey a

Measuring the Routine Activities of Adolescents

The next questions ask about the kinds of things you might do. How often do you do each of the following?

(1) Never
(2) A few times a year
(3) Once or twice a week
(4) At least once a week
(5) Almost everyday

_____ Watch TV

_____ Go to movies

_____ Ride around in a car (or motorcycle) just for fun

_____ Participate in community affairs or volunteer work

_____ Actively participate in sports, athletics, or exercising

_____ Work around the house, yard, garden, car, etc.

_____ Get together with friends, informally

_____ Go shopping or window-shopping

_____ Spend at least an hour of leisure time alone

_____ Read books, magazines, or newspapers

_____ Go to parties or other social affairs

During a typical week, on how many evenings do you go out for fun and recreation?

(1) Less than one   **(4)** Three
(2) One   **(5)** Four or five
(3) Two   **(6)** Six or seven

On the average, how often do you go out with a date (or your spouse, if you are married)?

(1) Never   **(4)** Once a week
(2) Once a month or less   **(5)** 2 or 3 times a week
(3) 2 or 3 times a month   **(6)** Over 3 times a week

Source: Osgood et al., "Routine Activities," 653.

soft determinism
Delinquent behavior is not reliably predictable because humans have choice and their thoughts, beliefs, and actions are not completely produced by their background, experiences, and present life conditions.

far less deterministic view of delinquent youth and their involvement in delinquent behavior—a view he calls <u>soft determinism</u>. Soft determinism incorporates the element of choice from classical criminology but acknowledges that youths are not entirely free in choosing their behavior, nor are they compelled to commit delinquency by their background and experiences. Matza describes the idea of drift or soft determinism: "Drift stands midway between freedom and control. . . . The delinquent *transiently* exists in a limbo between convention and crime responding in turn to the demands of each, flirting now with one, now the other, but postponing commitment, evading decision. Thus, he drifts between criminal and conventional action."[89]

Many of Matza's observations about drift and delinquency were developed with Gresham Sykes. *Drift theory* incorporates three fundamental elements: the subculture of delinquency and subterranean values, techniques of neutralization, and the development of a "will" for delinquency.

Subculture of Delinquency and Subterranean Values

A variety of positivist theories of criminology emphasize that some youths engage in delinquent acts simply because they are conforming to the norms and values of a delinquent subculture—delinquent behavior is a direct result of membership in a *delinquent subculture*.[90] Sykes and Matza reject such hard determinism, reasoning that if delinquent subcultures define delinquency as the expected and "right" thing to do, then delinquent youth should hold values and norms that are in complete opposition to those of conventional society. Instead, they offer four fairly simple observations that question whether delinquent subcultures exert such powerful influence over youths.[91] First, rather than experiencing pride and fulfillment from following the values and norms of the delinquent subculture, most delinquents experience a sense of guilt and shame over illegal acts. This indicates that delinquents are not completely separated from the standards and expectations of conventional society. Second, delinquent youths often respect and admire honest, law-abiding individuals, indicating that delinquents do not completely reject the legitimacy and "rightness" of the traditional normative system. Third, delinquents often draw a "sharp line between who can be victimized and those who cannot."[92] The notions that you "don't steal from friends" and "don't commit vandalism against the church of your own faith" reflect such distinction.[93] However, even if the delinquent subculture values offending, the mere act of making distinctions between who can be victimized and who cannot implies that the wrongfulness of delinquent acts is understood by delinquent youth. Fourth, delinquents are not immune to the demands of conformity made by the dominant social order. Parents, for example, usually agree with "respectable society" and view delinquency as wrong. Delinquent youth are subject to related expectations and pressures.

subculture of delinquency In contrast to a delinquent subculture, in which a group of adolescents has values and norms that are in sharp contrast with those of the larger society, a subculture of delinquency refers to values and norms that are less than conventional, balanced between convention and delinquency, that allow and sometimes encourage delinquent acts. The subcultural standards are usually not publicly acknowledged.

Matza goes on to make what appears to be a very fine distinction: "there is a subculture of delinquency, but it is not a delinquent subculture."[94] In other words, the <u>subculture of delinquency</u> does not represent an entirely separate set of values and norms that distinguish delinquents from the rest of society; rather, delinquent traditions consist of less conventional and less publicized standards of behavior that are still a part of the dominant culture.

Contemporary culture, then, is not as simple and uniform as often depicted. It is composed of conventional values and norms, but also of contrasting beliefs and expectations, including those that allow and encourage delinquent acts. Matza and Sykes refer to these subtle, underlying, and alternative traditions as <u>subterranean values</u>.[95] The subculture of delinquency and its subterranean values are embraced by most adolescents, making delinquent activity possible but not required. Matza and Sykes observe that "there is a subterranean tradition of deviance to which adolescents are attuned, depending on its relevance for various groups or situational contexts."[96] Links, "Subterranean Values and the Subculture of Delinquency," allows you to explore whether young people adopt subterranean values as a part of the subculture of delinquency.

The subculture of delinquency emphasizes and encourages a search for thrills and excitement ("kicks"), a reluctance for work, conspicuous consumption, an image of toughness, and a taste for aggression; however, these are no less a part of conventional society.[97] It seems that all adolescents are attuned to the subculture of delinquency, but the degree to which delinquent subterranean values influence youths depends on the relevance of those values to particular

subterranean values
Group values that lie beneath and in contrast to those of the larger society. The youth culture is sometimes said to have values that are not openly in opposition to the larger society but that are not entirely consistent with them. For example, teenagers' desires for fun, thrill, and irresponsibility are often condoned, but not encouraged, by adults, and these desires are valued by the youth culture.

Subterranean Values and the Subculture of Delinquency

Do young people profess subterranean values? David Matza and Gresham Sykes claim that the subculture of delinquency (and its associated subterranean values) is widespread among adolescents, making delinquency possible but not required. A number of surveys attempt to measure attitudes and values of young people. Several of these surveys are reported in the *Sourcebook of Criminal Justice Statistics*. Section 2 of the *Sourcebook* is titled "Public Attitudes Toward Crimes and Criminal Justice-Related Topics." Included here are opinion data from high school seniors and college freshmen.

High School Seniors
- Perceptions of harmfulness of drug and alcohol use
- Disapproval of drug use, alcohol use, and cigarette smoking
- Legalization of marijuana

College Freshmen
- Legalization of marijuana
- Drug testing
- Gun control
- Pornography

Consider the attitudes and values expressed by young people regarding the above issues. Do they reflect subterranean values? The *Sourcebook* is available online and can be accessed via a link provided at:

http://criminaljustice.jbpub.com

situations. For example, while violence is generally not socially approved, it can be argued that "the dominant society exhibits a widespread taste for violence in books, magazines, movies, and television are everywhere at hand."[98] Additionally, violence may be justified in certain situations in which subterranean values condone or even expect it, such as when a male adolescent has his masculinity questioned. The processes that allow subterranean values to come to the forefront in certain situations are described in a second aspect of drift: techniques of neutralization.

Techniques of Neutralization

techniques of neutralization The mental justifications used to deactivate conventional norms and values, thereby releasing a youth from social controls. Neutralizations are temporary and episodic.

If the subculture of delinquency does not entirely abandon the values and norms of conventional society, and in fact generally disapproves of delinquent acts, how is it that subterranean values come to have influence in certain situations? According to Sykes and Matza, subterranean values include a number of justifications that can be used prior to delinquent acts in order to neutralize conventional values and norms and thereby render them nonbinding. These **techniques of neutralization** are "episodic" and temporary, and they allow a youth to drift out of society's moral constraints and into delinquency.[99] The use and usefulness of these techniques depend on how appropriate they are to a given situation. Sykes and Matza describe five techniques of neutralization.[100]

Denial of Responsibility

While most people distinguish between harm that results from an accident and that which is intended, denial of responsibility moves beyond the claim that a delinquent act was just an accident. Personal responsibility can also be denied because of unloving parents, bad companions, or a ghetto neighborhood—any assertion that makes the delinquent act seem as if it were beyond the offender's control.[101]

Denial of Injury

"For the delinquent . . . wrongfulness may turn on the question of whether or not anyone has clearly been hurt by his deviance, and this matter is open to a variety of interpretations. Vandalism, for example, may be defined by the delinquent simply as 'mischief'—after all, it may be claimed, the persons whose property has been destroyed can well afford it. Similarly, auto theft may be viewed as 'borrowing,' and gang fighting may be seen as a private quarrel, an agreed upon duel between two willing parties, and thus of no concern to the community at large."[102]

Denial of the Victim

Even if the delinquent youth admits that his or her actions involve harm or injury, moral responsibility may be neutralized by the youth's insistence that the actions are justified in light of the circumstances and that therefore the harm or injury is also justified. Rightful retaliation moves the delinquent "into the position of an avenger and the victim is transformed into a wrong-doer. Assaults on homosexuals or suspected homosexuals, attacks on members of minority groups who are said to have gotten 'out of place,' vandalism as revenge on an unfair teacher or school official, thefts from a 'crooked' store owner—all may be hurts inflicted on a transgressor, in the eyes of the delinquent."[103]

Condemnation of the Condemner

Similarly, the delinquent may shift the focus of attention from his own delinquent behavior to the motives and action of those who disapprove of him or her. The delinquent may claim that these condemners are hypocrites or are motivated by self-righteousness or self-interest. For example, "police . . . are corrupt, stupid, and brutal. Teachers always show favoritism and parents always 'take it out' on their children."[104]

Appeal to Higher Loyalties

The delinquent may believe that a group of friends or a gang demands allegiance, even if it means violating the law. The delinquent does not necessarily denounce the law, despite his or her failure to follow it; rather, loyalty to the group takes priority.[105]

Techniques of neutralization theory is based on the assumption that most adolescents generally accept conventional beliefs and that they are subject to the social controls that result from the acceptance of these beliefs. If this is true, delinquent behavior is only possible when the binding power of social control is neutralized. In contrast, if delinquents are actually uncommitted and uninfluenced by conventional values and norms, then neutralization is unnecessary. This fundamental assumption has resulted in much research.

After reviewing this research, Robert Agnew concluded that "at minimum, it would appear that there is good reason to believe that the techniques of neutralization may not be relevant to a sizable portion of the individuals engaging in violence."[106] However, "mixed results" and methodological problems in prior neutralization research caused him to not abandon the theory completely. As a result, he conducted his own research using longitudinal data from the National Youth Survey and focusing on juvenile violence—specifically on attitudes toward violence and neutralization of conventional beliefs against the use of violence. In contrast to prior research, Agnew found that a vast majority of the youths surveyed disapprove of violence but that a large percentage of these youth accept one or more neutralizations that justify the use of violence in particular situations. As shown in Research in Action, "Measuring Techniques of Neutralization that Justify the Use of Violence," four measures of neutralization were used. In terms of Sykes and Matza's techniques of neutralization, Agnew classified all four measures as "denial of the victim," with the third question also indicating "denial of responsibility." The fairly wide acceptance of these neutralizations implied the existence of subterranean values in which the prevailing beliefs against violence are qualified by beliefs held simultaneously that justify the use of violence in certain situations.

Agnew found that, as Sykes and Matza suggest, neutralization preceded violence among youths who disapprove of violence and contributed to the likelihood of violent behavior. However, youths who had weak conventional beliefs about the appropriateness of violence were more likely to justify their delinquency after they committed the act, rather than before. Thus, neutralization occurs both before and after violent acts. For some youths, the binding power of conventional beliefs must be neutralized before delinquent behavior is even possible. For others, neutralization occurs only after they commit delinquent acts, serving more as a justification for the act than as a

Measuring Techniques of Neutralization That Justify the Use of Violence

The National Youth Survey includes four questions that Robert Agnew used to measure techniques of neutralization. These neutralizations represent beliefs that justify the use of violence in particular situations. While Agnew found that almost all youth disapprove of violence, a large percentage also accepts one or more neutralizations for violence.

1. It's alright to beat up people if they started the fight.
2. It's alright to physically beat up people who call you names.
3. If people do something to make you really mad, they deserve to be beaten up.
4. If you don't physically fight back, people will walk all over you.

Source: Agnew, "Techniques of Neutralization," 565.

---

contributing factor. Agnew also found that neutralization was most likely to lead to violent behavior among youths who associated with delinquent peers, suggesting that peers play a role in learning techniques of neutralization.[107]

A "Will" for Delinquency

The image of most juvenile delinquents offered by Matza and Sykes is that they are not immune to the demands for conformity made by society, nor do they wholeheartedly reject conventional beliefs. Modern societies are complex, being composed of conventional beliefs and expectations and of subterranean values that sometimes allow and encourage delinquent behavior. Neutralization allows for the "episodic release" from "moral constraints," resulting in drift, in which the individual is "neither committed nor compelled" to delinquent action.[108] The "moral vacuum" of drift, however, is not enough to explain the occurrence of delinquency: "Drift makes delinquency possible or permissible by temporarily removing the restraints that ordinarily control members of society," but it does not require delinquent action.[109] Matza observes that there is a missing element here: an element of "thrust or impetus" that results in delinquency. He calls this missing element <u>will</u>.[110]

Since Matza rejects the hard determinism of positivist criminology, he turns to classical criminology for a less deterministic approach to account for the driving force that leads to delinquent behavior. He finds a close fit to his idea of drift in classical criminology's central concept of will. The potential for delinquent behavior brought on by drift can be realized only when this potential is acted upon. In classical criminology, this decision or desire to engage in crime is referred to as *will*. The choice to engage in a delinquent act may or may not be exercised; it is an option. The concept of will is therefore consistent with Matza's

will The desire or decision to engage in crime. This desire or decision is put into action through past experiences that make crime seem possible ("preparation") and by feelings of lacking self-determination ("desperation").

desire to provide for soft determinism in a model of delinquency. However, the classical idea of will fails to account for "how, why, or when the will to crime becomes activated."[111]

Matza offers two conditions that activate the will: *preparation* and *desperation*.[112] Preparation refers to learning through experience that delinquent acts *can* be done, and in the absence of constraint, *may* be done. In a state of drift, illegal behavior becomes possible in a sense that youths realize that delinquent acts are something that they are capable of doing and for which there are no moral barriers.[113] Desperation arises primarily from a "mood of fatalism"—a key component of the subculture of delinquency. This mood of fatalism involves a sense, common among adolescents, that they have little or no control over their surroundings or their destiny—a feeling that they are being "pushed around."[114] While the mood of fatalism does not always lead to desperation, it is sufficiently common among adolescents so as to draw them together in peer groups in which desperation is experienced collectively. Together they seek to regain a sense of meaning and control over their lives. Within peer groups, the desperation that they experience serves to neutralize the legal constraints by making delinquent behavior permissible and desirable. Furthermore, delinquent acts allow youths to gain a sense of accomplishment—to allow desperate youth to "make something happen" and thereby to regain a sense of meaning and control over their lives.[115]

The three elements of Matza's drift theory are not often considered together. However, as we have seen, the subculture of delinquency, with its subterranean values, together with techniques of neutralization and the motivational element of will, provides a view of delinquency that emphasizes the immediate situation.

■ The Experience of Delinquency

The perspectives discussed so far have considered characteristics of current situations as if they are entirely objective. Being hungry, homeless, and in the "situation of company," for example, are thought to motivate and provide opportunity for delinquent acts. Characteristics of the situation, however, also include subjective content, relating to the individual's perception and interpretation of the current circumstances—the meaning that situations have for those involved.[116]

In a provocative book, *Seductions of Crime: Moral and Sensual Attractions in Doing Evil,* sociologist Jack Katz focuses on the subjective content of crime and delinquency. His central thesis is that delinquent and criminal behavior cannot be understood or explained without grasping how it is experienced or what it means to the offender.[117] He argues that "the social science literature contains only scattered evidence of what it means, feels, sounds, tastes, or looks like to commit a particular crime. Readers of research on homicide and assault do not hear the slaps and curses, see the pushes and shoves, or feel the humiliation and rage that may build toward the attack, sometimes persisting after the victim's death. How adolescents manage to make the shoplifting or vandalism of cheap and commonplace things a thrilling experience has not been intriguing to many students of delinquency."[118]

The understanding of the immediate situation, drawn from experience, is the focal point of a branch of sociology called phenomenology, and Katz presents one of the few phenomenological studies of crime and delinquency.[119] He refers to his approach as a theory of *moral self transcendence* in which crime and delinquency are responses to morally or emotionally compelling situations.[120] The crime situation is experienced as "magical" or "transcendent" in that the offender comes to view the criminal or delinquent act as "sensible, even sensually compelling."[121] Central to the "experience of criminality" is one or more "moral emotions: humiliation, righteousness, arrogance, ridicule, cynicism, defilement, and vengeance."[122] The act of crime or delinquency is done in retaliation or in defense of these moral emotions—to defend the "good" or to reestablish justice. As an "extreme example," Katz provides the story of the killing of a five-week old infant by his father. "The victim . . . started crying early in the morning. The offender, the boy's father, ordered the victim to stop crying. The victim's crying, however, only heightened in intensity. The . . . persistent crying may have been oriented *not* toward challenging his father's authority, but toward acquiring food or a change of diapers. Whatever the motive for crying, *the child's father defined it as purposive and offensive.*"[123]

For those involved, the current situation provides both reason and motive for crime and delinquency. It is on this subjective but dynamic process that Katz concentrates: "something causally essential happens in the very moments in which a crime is committed."[124] The "something" that Katz wants to draw attention to is the "moral and sensual attraction of doing evil" (to quote the book's subtitle). Katz proposes that the situational dynamics of crime involve three stages. Each stage is necessary for crime to occur, and together the three stages are sufficient to explain crime; however, the content of each stage varies for different types of crime.

1. A *path of action*—distinctive practical requirements for successfully committing the crime;

2. a *line of interpretation*—unique ways of understanding how one is and will be seen by others; and

3. an *emotional process*—seduction and compulsions that have special dynamics.[125]

These moral and sensual dynamics of crime situations are difficult to analyze through traditional data; therefore, Katz turns to various accounts of illegal activity, including biographies, autobiographies, ethnographies, observational studies, and journalism.[126] These stories relate real-life experiences that make his theory of moral self transcendence remarkably compelling. Several of these stories will be offered here in the context of Katz's explanation of three types of juvenile offenses: the "sneaky thrill" of property crime; the feeling of being a "badass"—being tough, alien, and mean; and involvement in street gangs and gang delinquency.

Sneaky Thrills: Juvenile Property Crime

"Various property crimes share an appeal to young people, independent of material gain or esteem from peers. Vandalism defaces property without satisfying a desire for acquisition. During burglaries, young people sometimes break in and exit successfully but do not try to take anything. Youthful shoplifting, especially by older youth, often is a solitary activity retained as a private memory. 'Joyriding' captures a form of auto theft in which getting away with something in celebratory style is more important than keeping anything or getting anywhere in particular."[127] Katz contends that these property offenses are essentially emotional events that provide sneaky thrills. Rather than reflecting a youth's upbringing or material living conditions, property crimes are determined almost exclusively by situational experiences. The sneaky thrill of property crime is created in the three-stage process just described, although the content of these stages is unique to juvenile property crimes.

Constructing an Object as Seductive

The "path of action" for sneaky thrills often begins with a situation that is simply exciting.[128] Even though juvenile property offenders may set out to commit a crime, there is usually little forethought or planning; rather, spontaneity and excitement prevail. The possibility of property crime emerges when the future offender realizes that an act or an object has appeal, that the act can be done, and that the item can be taken with relative ease.

> There we were, in the most lucrative department Mervyn's had to offer two curious (but very mature) adolescent girls: the cosmetic and jewelry department. . . . We didn't enter the store planning to steal anything. In fact, I believed we had "given it up" a few weeks earlier; but once my eyes caught sight of the beautiful white and blue necklaces alongside the counter, a spark inside me was once again ignited. . . . Those exquisite puka necklaces were calling out to me, "Take me! Wear me! I can be yours!" All I needed to do was take them to make it a reality.[129]

Remaining in Rational Control

The second stage of juvenile property crime—what Katz refers to as "the emergence of practiced reason"—involves an attempt to appear normal and legitimate in an effort to successfully carry out the crime.

Efforts to appear normal and legitimate are especially relevant to shoplifting. Katz provides a recollection of one of his students who shoplifted with her sister while using their mother as "cover."

> I can clearly remember when we coaxed my mom into taking us shopping with the excuse that our summer trip was coming and we just wanted to see what the stores had so we could plan on getting it later. We walked over to the section that we were interested in, making sure that we made ourselves seem "legitimate" by keeping my mom close and by showing her items that appealed to us. We thought "they won't suspect us, two girls in school uniforms with their mom, no way." As we carried on like this, playing this little game "Oh, look how pretty, gee, I'll have to tell dad about all these pretty things."[130]

Avoiding suspicion is a challenge that constantly confronts would-be property offenders. Shoplifters, for example, become extraordinarily conscious of their actions and how they appear to others. Avoiding suspicion involves at least two "layers of work": analytic attention to every detail of behavior and the attempt to remain in rational control.[131] Katz provides the following account of a juvenile shoplifter.

> She [the store clerk] stopped me about 5 ft. from the door, my heart was beating so hard, not fast just hard like it was going to jump out of my chest. The lady asked me "Didn't you find anything you liked?" I knew she was trying to see if I was nervous and to let me know she had seen me earlier. I said no, that I hadn't discovered anything that I couldn't live without. I remember trying to phrase the sentence as grownup as possible so she wouldn't think I was a dumb little kid. Then she said "What about that green necklace I saw you holding." . . . I said "I simply don't own anything to go with it so I hung it back on the rack." She said "Oh" and started toward the rack, so I continued out the door.[132]

Being Thrilled

The third stage to the moral and sensual attraction of juvenile property offending involves the emotional sensation of being thrilled by a delinquent act. Thrill is derived from the secretive and cunning nature of these offenses and the sense of accomplishment after being able to "pull it off." Katz summarizes a lengthy account of nonacquisitive burglary provided by one of his students.

> When she was 13, she would enter neighbors' homes and roam around. Somehow being in a neighbor's house without express permission made the otherwise mundane environment charged. She had been invited into all these homes before but by entering without notice through an unlocked door or an open window, she found that a familiar kitchen or living room was magically transformed into a provocative environment. The excitement was distinctly sensual. . . . But she rarely took anything. Instead she might simply rearrange the furniture. It seems she was not so much "playing house" or decorating to fit her tastes as she was trying to leave evidence that someone had been there.[133]

An Assessment of Sneaky Thrill Property Crime

Bill McCarthy offers an extension to Katz's phenomenological explanation of sneaky thrill property crime. He acknowledges that criminology has neglected the importance of situational factors, especially those moral and sensual dynamics of the immediate situation that make theft compelling to juveniles. McCarthy contends, however, that Katz's emphasis on the emotional elements of situations is overstated: "Katz portrays sneaky thrill offenses as emotional events determined almost exclusively by situational contingencies."[134] McCarthy argues that not all people respond to the seductive quality of situations by resorting to shoplifting: "They simply purchase the articles; delay gratification of ownership to a future, more affordable time; or accept that possession will remain a fantasy."[135]

McCarthy proposes that the seductive quality of theft is influenced by an individual's background, including age, gender, and social class. Shoplifting, for example, may be tied to a low socioeconomic status in which the lack of economic means to purchase an item leads to theft.

McCarthy also claims that processes of rational thought, or what he calls *instrumental cognitive processes,* play an important a role in the degree to which theft is morally and sensually compelling to the individual. Juvenile property crime is not simply an emotional response to an enticing situation, devoid of reason. To support his claims, McCarthy offers two of the accounts originally presented by Katz. One relates the theft of clothing: "I was . . . trying on leotards and thinking to myself what a 'rip-off!' $29.00 just for a basic leotard for aerobics. The more I thought about it the more infuriated I became—the price was absurd and I wasn't going to pay it. . . ."[136] Another respondent felt justified in stealing flu tablets: "I thought the manufacturers deserved to be punished for charging a ridiculously . . . high price. If those lousy, profiteering manufacturers had charged a fair price of $2, I would have been glad to pay for it. This way they could be punished . . . the bastards."[137]

Using survey data, McCarthy tested a model of sneaky thrill property crime that included background and instrumental cognitive influences. Questionnaire items were used to measure background factors, including gender, age, race, socioeconomic status, and strain resulting from lack of opportunity. Still other questionnaire items were used to measure instrumental cognitive processes that indicated rational thought, such as concern for being sanctioned, fear of loss of respect, and moral commitment.

Developing empirical measures for the subjective and experiential processes identified by Katz proved to be a difficult task. Seduction to crime was measured through questions that asked, "How often would you like to take something that does not belong to you?" and "If you were in a situation tomorrow where you had an extremely strong desire or need to, what are the chances that you would take something that does not belong to you?"[138]

Analysis of the survey data revealed that the reported desire to steal both in the past and in the future was not simply an emotional response to the current situation. Rather, the desire to steal was influenced in important ways by instrumental cognitive processes and by background characteristics of the individual. In terms of rational thought, people who thought that theft was not morally wrong and who felt that their detection would not lead to a loss of respect were more likely to report that they had experienced a desire to commit theft in the past and that they would likely commit theft in the future, given a desire to do so. In addition, having committed theft in the past and having been exposed to sanctions for crime did not inhibit consideration of future theft. In fact, these experiences actually seemed to encourage consideration of future theft.

Gender was also an important influence; males were more likely to report that they experienced a desire to steal in the past and that they would steal given the desire sometime in the future. Thus, the desire or seduction of theft is not simply an emotional response. Rather, it is influenced by instrumental cognitive thought and by an individual's gender.[139]

Ways of the Badass

Another of Katz's chapters deals with mannerisms rather than with action that is directly delinquent. In many adolescent circles, being "bad" is often regarded as good.[140] Being bad, however, is not necessarily being delinquent. The image of a badass is sought by many delinquent youth and is, in fact, one of the first images that come to mind when adults think of delinquent youths. Furthermore, being bad, tough, or macho is one of the subterranean values that constitute the subculture of delinquency.[141] As such, it is a characteristic of which most adolescents are aware. But "how does one go about being a badass?"[142] According to Katz's three-fold scheme, the situational dynamics that allow badness to be socially constructed include being tough as a path of action, being alien as a line of interpretation, and being mean as an emotional process.

Being Tough

"Someone who is 'real bad' must be tough, not easily influenced, highly impressionable, or anxious about the opinions that others hold of him."[143] A tough appearance is accomplished both through clothing and face-to-face interactions in public. Dark clothes, preferably leather, and lots of metal adornments provide a tough appearance. Given that the eyes are windows to the soul, the use of dark glasses effectively "pulls the shades" and makes the individual impenetrable. Certain words and styles of interaction also perpetuate a tough-guy image.

Being Alien

Being tough is not enough because respectable people can also be tough. The task for the badass is to construct a way of living and interacting that is incomprehensible to conventional people—not just foreign, but disturbingly different. To be alien is to be "unnerving."[144] Being alien involves a wide range of deviant "esthetics": language, body posture, clothing, fashions, car styles, and graffiti.[145] These esthetics make a statement of not only toughness, but of defiance and pride.

Being Mean

Being mean is the culmination and extension of being tough and being alien—it "completes the project of becoming a badass."[146] Being mean involves the use of physical violence to back up the image of being tough and being alien—to leave no question that "I mean it!" Violence is used in an unpredictable manner, devoid of rational application, in order to promote chaos. For obvious reasons, weapons of all kinds play an important role in conveying meanness. Physical contact through the "accidental" bump is also used as a prompt for violence. Manny Torres recalls the use of the bump in his adolescent year in Spanish Harlem: "walking around with your chest out, bumping into people and hoping they'll give you a bad time so you can pounce on them and beat 'em into the goddamn concrete."[147]

Being mean is also communicated through the use of various profanities, some of which become very popular and are used with astounding frequency. Often the same profanity is used in the same sentence numerous times, taking on different meaning with each usage. Profanities not only project a tough-guy image, but also allow the badass to convey an attitude superiority and dominance—in other words, meanness.[148]

Street Elites

The third form of juvenile crime that Katz turns to is gang violence. First of all, he points out that gang members usually do not refer to their own group as a "gang" because the term is considered derogatory and belittling. " 'Gang' is a troublesome term not because it conveys an impression of evil, but because the evil implied by an admission of gang membership is not sufficiently grand."[149] Gangs are for kids and the reference is reserved for rival gangs. Their own "collective commitment is to a 'club,' an 'organization,' a 'clique,' a 'barrio,' a 'mob,' a 'brotherhood,' a 'family,' an ethnic 'nation,' a 'team,' or a 'crew.'"[150]

While gang violence is unique and requires its own explanation, the moral and sensual dynamics of gang violence can be explored by means of the same three-step process that Katz uses to analyze other forms of crime.

The Posture of the Street Elite

The first stage of gang violence—what Katz calls a "path of action"—has to do with how groups of adolescent males in lower- and working-class neighborhoods construct the appearance of being "street elites" through physical intimidation. Street gangs demarcate a local geographic area as the place of importance. Ties to home (the homeboy) and the neighborhood are valued, and gang members set that area and its people apart from other areas. Gang members seek to protect and control the area and to be seen as rulers, or what Katz calls "aristocrats." As one gang member indicates: "You was out there. You was holding that street twenty-four hours a day. And you just had to constantly fight life and death out there. Lots of time the police even threatened us. . . . We was a little city. That's why they called it Vice Lord City."[151]

The naming of street gangs suggests their desire to be viewed as nobility: "Lords, Nobles, Knights, Pharaohs, Kings, Emperors, Viceroys, Crusaders, and Dukes."[152] Elevated status, protection, and control of a particular geographic area are achieved largely through threats and use of violence.

Violence and the Generation of Dread

The second stage in the moral and sensual dynamics of gang violence relates to how violence is understood by the gang members themselves and by others. Violent acts are used not only to exert control, but also to perpetuate fear and even dread. Street "gang members liken themselves to feudal knights, ancient warlords, barbarian hordes who extract tribute and respect from their vassals as a consequence of their reckless bravery, their awesome and beastly power of destruction, their symbols of intimidation. Their very survival—indeed their triumph—in a rude and savage environment attests to their savage grace."[153]

Evoking the Spirit of Street Elites

The final stage of gang violence relates to an emotional process whereby allegiance to the gang is promoted. Events and experiences that edge on violence serve to test the loyalty of gang members and their willingness to respond with violence. Katz provides an example of two members of the Vice Lords gang in enemy territory: "There weren't but two of us—me and Ringo. So Ringo, he yells, 'We're Vice Lords, mighty Vice Lords!' He yelling to a man coming up the street. This stud said, 'Well you all Vice Lords, huh!' We said, 'Yeah!' And by the time he was reaching for his back pocket . . . I guess he didn't get out what he wanted to get out before I got what I got out. I shot him in the leg or foot or something, and he ran down the street . . . hollering like a dog! I don't know if he a Cobra, but it was in the Cobra's hood."[154]

The spirit of street elites is aroused and enhanced through threats of violence and actual use of violence. Violence may be directed toward the gang or used by the gang; in either case, the spirit of street elites is boosted. Thus, gang violence serves to promote an aristocratic image, to secure power and control "on the streets" (the only place that really matters), and to provide a spirit of being street elites.

Katz's depiction of these three types of delinquent activity draws attention to how the immediate situation provides both reason and motive for delinquent acts. An individual's perception and interpretation gives definition to the situation in what has been called the subjective content of situations.

■ Summary and Conclusions

The immediate setting in which behavior occurs is referred to simply as the *situation.*[155] Situational explanations of delinquent behavior include characteristics of the current setting that motivate and provide opportunity for delinquent acts, and the perceptions and interpretations of those involved in the situation. While criminologists have long neglected *situational explanations* of delinquent behavior in favor of *developmental explanations,* a broad range of situational factors have been advanced. Rare, however, is the attempt to come to grips with the role that situational factors play in delinquent behavior. Gary LaFree and Christopher Birkbeck refer to this as the "neglected situation."[156]

Briar and Piliavin were among the first criminologists to suggest that the current setting plays an important role in the attitudes and actions of adolescents. They contended that *situational inducements* include desires to obtain valued goods, attempts to portray courage and loyalty, efforts to strike out at someone who is disliked, and the pursuit of "kicks." All youths experience situational inducements that provide opportunity for delinquency and make it appealing. However, the extent to which situational inducements produce delinquent behavior depends on the youth's *commitment to conformity.* Those youths who have little commitment to conformity are heavily influenced by situational inducements, whereas those who are highly committed are not.

McCarthy and Hagan considered a particular situational inducement: living on the streets. They found that the harsh and desperate living conditions of homeless youth are associated with crime—experiencing crime as a victim and committing crime as an offender. While life on the streets is exciting and provides freedom, it is also dangerous and *criminogenic.*

One of the most important forces present in many of the situations in which youths are involved is peer influence. The "situation of company" carries with it certain expectations for action, or what sociologists call norms. Empey and Lubeck found that these norms are experienced by virtually all youths and that they vary from situation to situation. Delinquent youth, however, were found to be more inclined to participate in situations involving peers than were nondelinquent youth, suggesting that delinquent youth are more likely to be subjected to the peer influence associated with situational norms.

Just as chance encounters provide situational influences, so do the *routine activities* of people's daily lives. The likelihood and extent of crime depend on the degree to which these routine activities provide opportunity for crime. In routine activities theory, opportunity for crime involves three key elements: *motivated offenders, suitable targets,* and the *absence of capable guardians.* The routine activities of adolescents and young adults have been found to be especially conducive to crime, thereby providing explanation for the age–crime connection. Involvement in crime is more likely to the extent that the routine activities of adolescents' daily lives lack structure and controls and involve peers.

The concept of adolescents *drifting* into delinquency also provides a situational explanation of delinquent behavior. Matza and Sykes contend that all adolescents are exposed to a *subculture of delinquency* that exists alongside conventional society. This subculture of delinquency is based on *subterranean values* that emphasize and encourage a search for thrills and excitement, a reluctance for work, conspicuous consumption, an image of toughness, and a taste for aggression. Subterranean values make delinquency possible but not required. The degree to which subterranean values come into play in a given situation depends on the use of *techniques of neutralization* that effectively neutralize the controls of conventional values and norms. For example, a youth may deny responsibility for a possible delinquent act and thereby become open to values and norms that define the situation as one in which delinquency can and should occur.

Situational explanations of delinquent behavior also include the perceptions and interpretations of the situation by those involved. According to Jack Katz, crime and delinquency are responses to morally or emotionally compelling situations—what he calls "moral and sensual attractions of doing evil." Situations provide both reason and motive for criminal and delinquent acts, and the dynamics of a given situation include a path of action by which delinquency comes to be viewed as an act that is possible, a line of interpretation in which the offender considers how he or she will be seen by others, and an emotional process by which the delinquent act becomes emotionally desirable. Katz applies this situational, interpretive process to three types of delinquent activity: sneaky thrill property crime, being a "badass" (tough, alien, and mean), and gang violence.

THEORIES

- situational inducements and commitment to conformity
- routine activities theory
- drift theory
- techniques of neutralization
- moral self transcendence

CRITICAL THINKING QUESTIONS

1. According to the opening case example, how does the immediate situation provide sensation and meaning, and how does it require interpretation and response?

2. What situational characteristics provide inducements for delinquency?

3. How are adverse situations related to delinquency?

4. In what ways might peer expectations in a given situation encourage delinquency?

5. How does *routine activities theory* explain criminal opportunity?

6. Provide examples of routine activities of adolescents that provide opportunity for delinquency.

7. Distinguish *subculture of delinquency* from *delinquent subculture*.

8. How does Matza's idea of *will* extend our understanding of delinquent behavior?

9. Provide several examples of how individual perception and interpretation of the immediate situation are related to delinquent behavior.

SUGGESTED READINGS

Briar, Scott, and Irving Pillivin. "Delinquency, Situational Inducements, and Commitment to Conformity." *Social Problems* 13 (1965):35–45.

Hagan, John, and Bill McCarthy. *Mean Streets: Youth Crime and Homelessness.* Cambridge: Cambridge University Press, 1997.

Katz, Jack. *Seductions of Crime: Moral and Sensual Attractions in Doing Evil.* New York: Basic Books, 1988.

Osgood, D. Wayne, Janet K. Wilson, Patrick M. O'Malley, Jerald G. Bachman, and Lloyd D. Johnston. *American Sociological Review* 61 (1996):635–655.

Sykes, Gresham M. and David Matza. "Techniques of Neutralization: A Theory of Delinquency." *American Sociological Review* 22 (1957):664–70.

GLOSSARY

capable guardians: The level of control and protection provided to people and property.

commitment to conformity: Logical reasons to conform, such as the desire to maintain or even enhance self image, status, valued relationships, and future activities.

criminogenic situations: Situational pressures and circumstances that motivate and provide opportunity for crime and delinquency.

developmental explanations: Explanations of criminal behavior that focus on processes operating in the earlier history of the criminal. A wide range of social and psychological characteristics of offenders and their environments are considered.

drift: A view of youths that "stands midway between freedom and control," in which youths hold both conventional and deviant values, resulting in both conforming and delinquent behavior (Sykes and Matza, "Techniques of Neutralization," 28).

hard determinism: Delinquent behavior is caused by factors that can be identified by science. Given these factors, delinquency is a predictable outcome.

motivated offenders: The inclination to engage in crime for self-benefit.

objective content of situations: Characteristics of a given situation that provide motivation and opportunity for crime. Poverty and hunger, for example, may lead to property offending.

routine activities: The repetitive patterns of daily living—at home, school, work, and leisure.

situational explanations: Explanations of criminal behavior that focus on processes operating at the moment of the occurrence of crime, including characteristics of the current setting that motivate and provide opportunity for crime.

situational inducements: Characteristics and processes of the immediate situation that make crime and delinquency appealing.

soft determinism: Delinquent behavior is not reliably predictable because humans have choice and their thoughts, beliefs, and actions are not completely produced by their background, experiences, and present life conditions.

subculture of delinquency: In contrast to a delinquent subculture, in which a group of adolescents has values and norms that are in sharp contrast with those of the larger society, a subculture of delinquency refers to values and norms that are less than conventional, balanced between convention and delinquency, that allow and sometimes encourage delinquent acts. The subcultural standards are usually not publicly acknowledged.

subjective content of situations: The "individual's perception and interpretation of the immediate setting" (Birkbeck and LaFree, "Situational Analysis," 129).

subterranean values: Group values that lie beneath and in contrast to those of the larger society. The youth culture is sometimes said to have values that are not openly in opposition to the larger society but that are not entirely consistent with them. For example, teenagers' desires for fun, thrill, and irresponsibility are often condoned, but not encouraged, by adults, and these desires are valued by the youth culture.

suitable targets: How advantageous a crime is to commit, depending on value, visibility, access, and ease of committing.

techniques of neutralization: The mental justifications used to deactivate conventional norms and values, thereby releasing a youth from social controls. Neutralizations are temporary and episodic.

will: The desire or decision to engage in crime. This desire or decision is put into action through past experiences that make crime seem possible ("preparation") and by feelings of lacking self-determination ("desperation").

REFERENCES

Agnew, Robert. "Determinism, Indeterminism, and Crime: An Exploration." *Criminology* 33 (1995):83–109.

———. *Juvenile Delinquency: Causes and Control.* Los Angeles: Roxbury, 2001.

———. "The Techniques of Neutralization and Violence." *Criminology* 32 (1994):555–580.

Agnew, Robert, and David M. Petersen. "Leisure and Delinquency." *Social Problems* 36 (1989): 332–350.

Chapter Resources

Akers, Ronald and Christine S. Sellers. *Criminological Theories: Introduction and Evaluation.* 4th ed. Los Angeles: Roxbury, 2004.

Argyle, Michael, Adrian Furnham, and Jean Ann Graham. *Social Situations.* Cambridge: Cambridge University Press, 1981.

Bartol, Curt R. *Criminal Behavior: A Psychosocial Approach.* 6th ed. Englewood Cliffs, NJ: Prentice Hall, 2002.

Birbeck, Christopher, and Gary LaFree. "The Situational Analysis of Crime and Deviance." *Annual Review of Sociology* 19 (1993):113–137.

Briar, Scott, and Irving Piliavin. "Delinquency, Situational Inducements, and Commitment to Conformity." *Social Problems* 13 (1965):35–45.

Cohen, Lawrence E., and Marcus Felson. "Social Change and Crime Rate Trends: A Routine Activity Approach." *American Sociological Review* 44 (1979):588–608.

Cullen, Francis T., and Robert Agnew, eds. *Criminological Theory: Past to Present.* Los Angeles: Roxbury, 1999.

Douglas, Jack, ed. *Deviance and Respectability: The Social Construction of Moral Meanings.* New York: Basic Books, 1970.

Empey, LaMar T., and Steven G. Lubeck. "Conformity and Deviance in the 'Situation of Company.' " *American Sociological Review* 26 (1968):760–774.

Felson, Marcus. "Linking Criminal Choices, Routine Activities, Informal Control, and Criminal Outcomes." In *The Reasoning Criminal,* edited by Derek B. Cornish and Ronald V. Clark, 119–128. New York: Springer-Verlag, 1986.

Freedman, Jonathan. "Effect of Television Violence on Aggression." *Psychological Bulletin* 96 (1988):227–246.

Gibbons, Don C. "Observations on the Study of Crime Causation." *American Journal of Sociology* 77 (1971):262–278.

Gold, Martin. *Delinquent Behavior in an American City.* Belmont, CA: Brooks-Cole, 1970.

Goode, Eric. "Phenomenology and Structure in the Study of Crime and Deviance: Crime Can be Fun; The Deviant Experience." Review of *Seductions of Crime* by Jack Katz. *Contemporary Sociology* 19 (1990):5–12.

Hagan, John, Gerd Hefler, Cabriele Classen, Klaus Boehnke, and Hans Merkens. "Subterranean Sources of Subcultural Delinquency Beyond the American Dream." *Criminology* 36 (1998): 309–341.

Hagan, John, and Bill McCarthy. *Mean Streets: Youth Crime and Homelessness.* Cambridge: Cambridge University Press, 1997

———. "Streetlife and Delinquency." *British Journal of Sociology* 43 (1992):533–561.

Hundleby, John D. "Adolescent Drug Use in a Behavioral Matrix: A Confirmation and Comparison of the Sexes." *Addictive Behaviors* 12 (1987):103–112.

Jensen, Gary F., and David Brownfield. "Gender Lifestyle and Victimization: Beyond Routine Activities Theory." *Violence and Victims* 1 (1986):85–99.

Katz, Jack. *Seductions of Crime: Moral and Sensual Attractions in Doing Evil.* New York: Basic Books, 1988.

Keiser, R. Lincoln. *The Vice Lords.* New York: Holt, Rinehart, and Winston, 1979.

LaFree, Gary, and Christopher Birkbeck. "The Neglected Situation: A Cross-National Study of the Situational Characteristics of Crime." *Criminology* 29 (1991):73–98.

Lockwood, Daniel. "Violence Among Middle School and High School Students: Analysis and Implications for Prevention." Washington, DC: National Institute of Justice, 1997.

Matsueda, Ross L. "The Dynamics of Moral Beliefs and Minor Delinquency." *Social Forces* 68 (1989):428–457.

Matza, David. *Becoming Deviant.* Englewood Cliffs, NJ: Prentice Hall, 1969.

———. *Delinquency and Drift.* New York: Wiley, 1964.

Matza, David, and Gresham M. Sykes. "Delinquency and Subterranean Values." *American Sociological Review* 26 (1961):712–719.

McCarthy, Bill. "Not Just 'For the Thrill of It': An Instrumentalist Elaboration of Katz's Explanation of Sneaky Thrill Property Crimes." *Criminology* 33 (1995):519–538.

McCarthy, Bill, and John Hagan. "Getting Into Street Crime: The Structure and Process of Criminal Embeddedness." *Social Science Research* 24 (1995):63–95.

———. "Homelessness: A Criminogenic Situation?" *British Journal of Criminology* 31 (1991):393–410.

———. "Mean Streets: The Theoretical Significance of Situational Delinquency Among Homeless Youths." *American Journal of Sociology* 98 (1992):597–627.

Messner, Steven. "Television Violence and Violent Crime: An Aggregate Analysis." *Social Problems* 33 (1986):218–235.

Miethe, Terance D., and Robert F. Meier. "Opportunity, Choice, and Criminal Victimization: A Test of a Theoretical Model." *Journal of Research in Crime and Delinquency* 27 (1990):243–266.

Milgram, S. *Obedience to Authority.* New York: Harper and Row, 1974.

Osgood, D. Wayne, and Amy L. Anderson. "Unstructured Socializing and Rates of Delinquency." *Criminology* 42 (2004):519–549.

Osgood, D. Wayne, Janet K. Wilson, Patrick M. O'Malley, Jerald G. Bachman, and Lloyd D. Johnston. "Routine Activities and Individual Deviant Behavior." *American Sociological Review* 61 (1996):635–655.

Pervin, Lawrence A. "Definitions, Measurements, and Classifications of Stimuli, Situations, and Environments." *Human Ecology* 6 (1978):71–105.

Rettig, Richard P., Manual J. Torres, and Gerald R. Garret. *Manny: A Criminal Addict's Story.* Boston: Houghton Mifflin, 1977.

Riley, David. "Time and Crime: The Link Between Teenager Lifestyle and Delinquency." *Journal of Qualitative Criminology* 3 (1987):339–354.

Sampson, Robert, and John D. Wooldredge. "Linking the Micro- and Macro-Level Dimensions of Lifestyle—Routine Activity and Opportunity Models of Predatory Victimization." *Journal of Quantitative Criminology* 3 (1987):371–393.

Scott, Robert A., and Jack D. Douglas, eds. *Theoretical Perspectives on Deviance.* New York: Basic Books, 1972.

Snyder, Scott. "Movies and Juvenile Delinquency: An Overview." *Adolescence* 26 (1991):121–132.

Sutherland, Edwin H. *Principles of Criminology.* 4th ed. Chicago: Lippincott, 1947.

Sutherland, Edwin H., and Donald R. Cressey. *Principles of Criminology.* 9th ed. Philadelphia: Lippincott, 1974.

Sutherland, Edwin H., Donald R, Cressey, and David F. Luckenbill. *Principles of Criminology.* 11th ed. Dix Hills, NY: General Hall, 1992.

Sykes, Gresham M., and David Matza. "Techniques of Neutralization: A Theory of Delinquency." *American Sociological Review* 22 (1957):664–670.

Wood, Wendy, Frank Wong, and J. Gregory Chachere. "Effects of Media Violence on Viewers' Aggression in Unconstrained Social Interaction." *Psychological Bulletin* 109 (1991):371–383.

ENDNOTES

1. Sutherland, Cressey, and Luckenbill, *Principles of Criminology,* 88. This is the most current statement of Sutherland's original with only slight editing (*Principles of Criminology,* 5).
2. Sutherland (*Principles of Criminology,* 5; Sutherland and Cressey, *Principles of Criminology,* 74) used several other terms to refer to these two types of explanations. He also called situational explanations *mechanistic* and *dynamic,* and developmental explanations *genetic* and *historical.*
3. Sutherland, *Principles of Criminology,* 5; and Birkbeck and LaFree, "Situational Analysis," 113.
4. Birkbeck and LaFree, "Situational Analysis," 113; and Gibbons, "Observations," 271.
5. Gold, *Delinquent Behavior,* 92–99.
6. Ibid., 92.
7. Birkbeck and LaFree, "Situational Analysis," 115.
8. Sutherland, Cressey, and Luckenbill, *Principles of Criminology,* 88; and Birkbeck and LaFree, "Situational Analysis," 129.
9. Birkbeck and LaFree, "Situational Analysis," 129.
10. Katz, *Seductions of Crime.*
11. Gibbons, "Observations," 270; and Matsueda, "Dynamics," 447.
12. LaFree and Birkbeck, "Neglected Situation," 75; and Pervin, "Definitions," 79–80.
13. Birkbeck and LaFree, "Situational Analysis," 129.
14. Ibid., 130.
15. Ibid.
16. This section is based extensively on Birkbeck and LaFree's ("Situational Analysis," 117–119) summary of Argyle, Furnham, and Graham's (1981) literature review of the experimental research on aggression.

Chapter Resources

17. Birkbeck and LaFree, "Situational Analysis," 117.
18. Lockwood, "Violence."
19. Jensen and Brownfield, "Gender Lifestyle and Victimization."
20. Agnew, *Juvenile Delinquency*, 203.
21. Wood, Wong, and Chachere, "Effects of Media Violence," 380.
22. Freedman, "Effect of Television Violence;" Messner, "Television Violence;" and Snyder, "Movies and Juvenile Delinquency."
23. Bartol, *Criminal Behavior*, 200–203.
24. Milgram, *Obedience to Authority*.
25. Briar and Piliavin, "Delinquency," 35.
26. Ibid.
27. Ibid., 36–37.
28. Ibid., 37, 35.
29. Ibid., 36.
30. Ibid., 38.
31. Ibid., 39, 45.
32. Ibid., 39.
33. Ibid., 41.
34. McCarthy and Hagan, "Mean Streets," 598.
35. Ibid.
36. Ibid., 623.
37. Gibbons, "Observations," 268.
38. Hagan and McCarthy, *Mean Streets*, 1.
39. McCarthy and Hagan, "Mean Streets," 598. See also Hagan and McCarthy, "Streetlife and Delinquency."
40. Hagan and McCarthy, *Mean Streets*, 3.
41. McCarthy and Hagan, "Mean Streets," 602. See also Hagan and McCarthy, *Mean Streets*, 7–8.
42. McCarthy and Hagan, "Mean Streets," 604, 619–620.
43. Hagan and McCarthy, *Mean Streets*, 26.
44. Ibid., 36–46.
45. Ibid., 38.
46. Ibid., 41.
47. Ibid., 49.
48. Ibid., 115.
49. McCarthy and Hagan, "Homelessness," 395.
50. Ibid., 407–408; see also 400–401.
51. Ibid., 406–407.
52. McCarthy and Hagan, "Mean Streets."
53. Ibid., 614.
54. Ibid., 624.
55. Empey and Lubeck, "Conformity and Deviance."
56. Ibid., 761.
57. Ibid.
58. Ibid., 767.
59. Ibid., 769.
60. Sutherland, Cressey, and Luckenbill, *Principles of Criminology*, 88.
61. Birkbeck and LaFree, "Situational Analysis," 123.
62. Ibid., 125.
63. Felson, "Linking," 119. Miethe and Meier, in "Opportunity," refer to this as "structural-choice" (245).
64. Cullen and Agnew, *Criminological Theory*, 249.
65. Cohen and Felson, "Social Change," 590; and Cullen and Agnew, *Criminological Theory*, 249.
66. Osgood et al., "Routine Activities," 635; and Cohen and Felson, "Social Change," 593. Sampson and Wooldredge argue in "Linking" that this emphasis on routines of daily living—the lifestyles or routine activities—fails to adequately consider the structural context of such activities. Marital status, family disruption (single-parent households), and unemployment influence routine activities and are, in fact, directly related to victimization.

67. Cohen and Felson, "Social Change," 598.
68. Ibid.
69. Ibid., 588.
70. Cohen and Felson, "Social Change," 589, 605. Akers and Sellers point out in *Criminological Theories* (35, 39) that most research on routine activities theory has failed to measure variation in motivation for crime or variation in the presence of motivated offenders. While Cohen and Felson take motivation as a given, they do not rule out that their approach might be "applied to the analysis of offenders and their inclinations" (605).
71. Cohen and Felson, "Social Change," 605.
72. Cohen and Felson, "Social Change," 591; and Miethe and Meier, "Opportunity."
73. Cohen and Felson, "Social Change," 590.
74. Ibid., 598.
75. Ibid., 596.
76. Osgood et al., "Routine Activities." See also Osgood and Anderson, "Unstructured Socializing."
77. Osgood et al., "Routine Activities," 641.
78. Ibid., 639.
79. Ibid., 640.
80. Hundleby, "Adolescent Drug Use."
81. Osgood et al., "Routine Activities," 641.
82. Ibid., 651.
83. Ibid., 642.
84. Osgood et al., "Routine Activities," 645, 651. A study by Riley, "Time and Crime," revealed similar results. His interpretation emphasized that peer group activities, away from home, provide opportunity for crime. However, his data were not longitudinal, making it difficult to determine the causal ordering of peer activities and involvement in delinquency. Agnew and Petersen, in "Leisure and Delinquency," also studied the role played by leisure time in delinquent behavior. They found that unsupervised leisure activities with peers are related to delinquency.
85. Osgood et al., "Routine Activities," 645.
86. Osgood et al., "Routine Activities," 652. But Sampson and Wooldredge, in "Linking," found that demographic and structural factors such as age, marital status, family disruption (single-parent households), and unemployment had larger direct effects on victimization than did factors related to lifestyle.
87. Agnew, "Determinism," 83; and Matza, *Delinquency and Drift*, Chapter 1.
88. Matza, *Delinquency and Drift*, 4.
89. Ibid., emphasis in original.
90. Sykes and Matza, "Techniques of Neutralization," 664.
91. Sykes and Matza, "Techniques of Neutralization," 664–666. See also Matza, *Delinquency and Drift*, 40–48.
92. Sykes and Matza, "Techniques of Neutralization," 665.
93. Sykes and Matza, "Techniques of Neutralization," 665. See also Hagan et al., "Subterranean Sources."
94. Matza, *Delinquency and Drift*, 33.
95. Matza and Sykes, "Delinquency and Subterranean Values."
96. Empey and Lubeck, "Conformity and Deviance," 766, drawn from Matza, *Delinquency and Drift*, 35–59.
97. Matza and Sykes, "Delinquency and Subterranean Values," 713.
98. Ibid., 717.
99. Matza, *Delinquency and Drift*, 69.
100. Sykes and Matza, "Techniques of Neutralization." In a later work (*Delinquency and Drift*), Matza describes the subterranean values of the subculture of delinquency in terms of three encompassing forms of neutralization: the negation of offense; the sense of injustice; and custom, tort, and injustice. These more general forms include the fives techniques neutralization advanced first by Sykes and Matza.
101. Sykes and Matza, "Techniques of Neutralization," 667.
102. Ibid.
103. Ibid., 668.
104. Ibid.
105. Ibid., 669.

Chapter Resources

106. Agnew, "Techniques," 558.
107. Ibid., 555.
108. Matza, *Delinquency and Drift,* 69, 28, 181.
109. Ibid., 181.
110. Ibid.
111. Ibid.
112. Ibid., 183.
113. Ibid., 184.
114. Matza, *Delinquency and Drift,* 188. Agnew, "Techniques," 561.
115. Matza, *Delinquency and Drift,* 188–191.
116. Birkbeck and LaFree, "Situational Analysis," 113, 120; and Sutherland, Cressey, and Luckenbill, *Principles of Criminology,* 88.
117. Goode, "Phenomenology," 7.
118. Katz, *Seductions of Crime,* 3.
119. Douglas, *Deviance and Respectability*; Matza, *Becoming Deviant*; and Scott and Douglas, *Theoretical Perspectives on Deviance.*
120. Katz, *Seductions of Crime,* 10.
121. Ibid., 4, 3.
122. Ibid., 9.
123. Ibid., 12, emphasis added.
124. Ibid., 4.
125. Ibid., 9, emphasis added.
126. Goode, "Phenomenology," 7.
127. Katz, *Seductions of Crime,* 52.
128. Ibid., 54.
129. Ibid.
130. Ibid., 59.
131. Ibid., 63.
132. Ibid., 64.
133. Ibid., 69–70.
134. McCarthy, "Instrumentalist Elaboration," 520.
135. Ibid., 522.
136. Ibid., 524.
137. Ibid.
138. Ibid., 527, 537.
139. Ibid., 533.
140. Katz, *Seductions of Crime,* 80.
141. Matza and Sykes, "Delinquency and Subterranean Values."
142. Katz, *Seductions of Crime,* 80.
143. Ibid.
144. Ibid., 81.
145. Ibid., 90.
146. Ibid., 99.
147. Rettig, Torres, and Garret, *Manny,* 18, in Katz, *Seductions of Crime,* 107.
148. Katz, *Seductions of Crime,* 107.
149. Ibid., 115.
150. Ibid.
151. Keiser, *Vice Lords,* 68, in Katz, *Seductions of Crime,* 133.
152. Katz, *Seductions of Crime,* 120.
153. Goode, "Phenomenology," 9.
154. Keiser, *Vice Lords,* 61, in Katz, *Seductions of Crime,* 142.
155. Birkbeck and LaFree, "Situational Analysis," 115.
156. LaFree and Birkbeck, "Neglected Situation." This article and a subsequent one (Birkbeck and LaFree, "Situational Analysis") provide two of the very few considerations of situational dimensions of delinquent behavior.

Social Control Theories: Family Relations

10

Chapter Outline

- Informal social control
- Studying family relations and delinquent behavior
- Social control theories
- Characteristics of family life and informal social control

Chapter Objectives

After completing this chapter, students should be able to:

- Describe how families operate as institutions of both socialization and social control.
- Distinguish various forms of social control: formal, informal, direct, indirect, and internalized.
- Identify and describe the key concepts of the different versions of control theory: social bond, life course, and self-control.
- Describe how family structure and process are related to delinquent behavior.
- Explain the connections between parents, peers, and delinquent behavior.
- Understand key terms:
  Theories:
  social bond theory
  life-course theory
  self-control theory or a general theory of crime

Terms:

informal social control

formal social control

broken home

direct controls

indirect controls

internalized controls

attachment

commitment

involvement

belief

life course

trajectory or pathway

transition

turning point

social capital

low self-control

family disruption

CASE IN POINT

Brothers in Crime: Family Life and Delinquent Behavior

Clifford Shaw's third and final "life-history," *Brothers in Crime,* is a study of five brothers in the Martin family. Shaw describes their social backgrounds and each of their delinquent careers. Included is a chapter on family life entitled "Family Disorganization and Culture Conflict," in which he describes the social environment of the family. Born to rural European immigrants, the Martin brothers and their parents lived in different social worlds. The "old world" roots of their parents were in stark contrast to the daily life experiences of the brothers, who grew up in the rapidly changing urban environment of Chicago. Shaw describes the family as "disorganized" because it was plagued by poverty, frequent unemployment, lack of supervision, and culture conflict. As a result, "the family lacked the unity which is usually regarded as being essential to normal family life."

As an agency for providing training, guidance, emotional satisfaction and security for the brothers, its [the family's] ineffectiveness is amply demonstrated by the repeated delinquencies of the brothers. These delinquencies occurred despite the good intentions and continuous effort of the parents to instill in their children conventional ideals. Probably these efforts were particularly ineffective because they were not supported by a corresponding sentiment, public opinion, and practice in the community. . . .

Presumably, the problems which the parents faced in rearing their children in the American community were quite different from what they would have been in the Old World community from which they emigrated. There they would have had the support of the large family of relatives and the common sentiments and public opinion of the community in their effort to train, guide, and control their children. On coming to Chicago they settled in a community in which primary group controls had largely disintegrated and the social groupings were characterized by widely divergent traditions, moral norms, and practices, and where traditions of delinquency were already established. As the brothers became incorporated into the life of the neighborhood groups, whose norms and mode of life were opposed to those of the parents, parental control was weakened and the parents were unable to transmit the value of their tradition. The social heritage of the parents was not carried over to the children; there was a sharp break in family tradition.

Source: Shaw, *Brothers in Crime,* 139–140. Copyright © 1938, University of Chicago Press. All rights reserved. Reproduced with permission of the publisher.

Two of the most commonly identified sources of delinquent behavior are family and friends. Intuition tells us that bad families and bad companions produce bad kids. Social relationships have a powerful influence on behavior. An old and widely used introductory sociology textbook states this clearly: "Social relations are at the foundation of both motivation and control. The goals and aspirations that set people into motion are greatly influenced by their social relations. Social relations are also instruments of control, for they limit action and restrain impulses that might threaten the orderly arrangement of independent lives."[1]

The distinction between motivation and control is the dividing point for two groups of theories, both of which consider social relations to be central to delinquent behavior. One group of theories, called *social learning theories,* adopts the common theme that social relations with delinquent peers motivate and perpetuate delinquent behavior by providing the primary group context in which delinquent attitudes and behaviors are learned and reinforced. This will be the topic of the Chapter 11. A second group of theories emphasizes the importance of social relations in controlling behavior. Referred to as *social control theories,* this group of theories pays close attention to the family as an institution of social control. In fact, considerations about the role of the family in delinquent behavior are left largely to the control perspective.[2] Strong, stable, and nurturing family relations are fundamental not only to normal child development and socialization, but also to social control. Thus, the control perspective is distinctive not only because of the important role given to the family, but also because of its attempts to explain conformity rather than what motivates delinquent behavior.

Informal Social Control

informal social controls
Characteristics of social relations that bring about conformity, including parental supervision; sensitivity to others (their feelings, wishes, and expectations); identification with others; emotional attachment; and informal sanctions, including withdrawal of affection, disregard, and ridicule.

The social controls that originate from family relations are more accurately referred to as **informal social controls**. Informal social controls refer to characteristics of social relations that bring about conformity, including parental supervision; sensitivity to others (their feelings, wishes, and expectations); identification with others; emotional attachment; and informal sanctions, including withdrawal of affection, disregard, and ridicule. These relational controls can be contrasted with **formal social controls** that are based in institutions such as schools, churches, and justice systems. Institutional controls are deliberately established to promote conformity and to sanction deviance. For example, schools established rules and procedures to suspend students, as do churches to excommunicate members and justice systems to imprison serious offenders. When most people think about social control, they usually have in mind only the bigger picture of formal social control, especially criminal or juvenile justice systems.[3] Sociologists have long claimed, however, that informal social controls are likely more important in explaining conformity than are formal social controls.[4]

formal social controls
Mechanisms used within the context of institutions such as schools, churches, and justice systems that promote and encourage conformity and sanction deviance.

This chapter addresses both the theoretical explanations and empirical findings on informal social control. Three different versions of social control theory are described: social bond theory, life-course theory, and self-control theory. The remainder of the chapter considers how characteristics of family life are related to informal social control, including the quality of parent–child relations, parental monitoring, supervision, discipline, and parental criminality. Family structure is also discussed in order to point to the various living arrangement of contemporary families. First, however, let's turn to a brief historical account of the study of family relations and delinquent behavior.

Studying Family Relations and Delinquent Behavior

The role of the family in delinquent behavior seems to be a greater mystery to social scientists than to parents, the public, or juvenile justice officials. Because stable family relations are generally held to be important to child and adolescent socialization, it is commonly believed that the family plays a vital role in delinquent behavior. Social scientists, however, do not see the link between the family and delinquent behavior as so simple, clear, or compelling. This mystery for scholars is certainly not due to lack of attention. The literature on the family and delinquent behavior spans nearly a century. Despite such extensive attention, the scholarly literature is inconclusive and reveals little cumulative understanding. Furthermore, there is a great deal of controversy over the significance of the family's causal role. Perspectives range "from the view that the family is the single most important determinant of delinquent behavior to the view that while some association may exist, there is no real causal link between the two."[5]

The family has fallen into and out of theoretical favor as a causal factor in delinquent behavior. In the late 1800s and early 1900s, most people, including social scientists, were convinced that one of the most important causes of juvenile misconduct was the breakdown of traditional family life, especially in terms of the __broken home__—families left with only one parent as the result of separation, divorce, or death. Family breakdown was most often thought to have been caused by the disruptive effects of urbanization and industrialization. In reviewing studies conducted during this period, Thomas Monahan observed that "early writers saw the broken home to be an important if not the greatest single proximate factor in understanding delinquency."[6] It was argued that when the family relationships are disrupted or broken, children do not develop adequately, and parents' control over children is reduced, making delinquent behavior a likely outcome.

broken home Families with only one parent as a result of separation, divorce, or death. The term "broken home" was used in early research on family structure and delinquency. The contemporary term is "family disruption," as we discuss later in this chapter.

Early studies usually reported a strong connection between broken homes and delinquent behavior. However, the research methods of these studies were unsophisticated, most often comparing the proportion of broken homes among delinquents to that of a control group. In addition, Karen Wilkinson claims that the biases of early researchers toward the stable family and their desire to promote social reform overrode scientific concerns. As a result, the findings of these studies were questioned: "the subjectivity and the methodology of these earlier studies were rejected; therefore the [broken home] explanation itself was also rejected. Instead of improving the objectivity and methodology, the assumption was made that the explanation was of no value, and sociologists began examining other variables."[7]

The work of Clifford Shaw and Henry McKay was especially influential to the change in perspective regarding the family and delinquent behavior that began in the 1930s. They offered the first empirical challenge to the accepted importance of broken homes to delinquent behavior. In a comparative study of juvenile court cases and Chicago schoolboys, Shaw and McKay found that the percentage of the delinquent group from broken homes was only slightly greater than that of the control group (42% and 36%, respectively). However, they did *not* conclude that the family was irrelevant to delinquent behavior. Instead, they contended that the family's influence "must be sought in more subtle aspects of family relationships rather than in the formal break in family organization."[8] As a result, Shaw and McKay encouraged a broader consideration of family factors, not just the broken home.

The move to consider a broader range of causal factors also led to the development of theoretical explanations of delinquent behavior.[9] The theories that were advanced were mainly sociological, focusing on the larger social environments of peer groups, neighborhoods, subcultures, and society. As a result, the family was considered to be just one of many causal factors and was given far less attention.[10]

At the same time, sociologists began to view the family as an institution of declining importance. Many of the traditional functions of the family—economic, protective, religious, recreational, educational, and status-related—were gradually being taken over by other institutions, especially schools, in a process referred to as the "transfer of function." The function of the family was thought to be reduced to that of affection.[11] Wilkinson describes the significance of this view to explanations of delinquent behavior: "With other institutions gaining control over the development of children, the family was considered less capable of influencing the behavior of its children and was therefore less likely to be considered responsible for juvenile delinquency."[12]

The growing disinterest in the family that began in the 1930s was not without exception. In the course of more than forty years of research, Sheldon and Eleanor Glueck vigorously argued that family relations were key to understanding delinquent behavior. However, their perspective was often ignored or discredited by other scholars in the field.[13] Their emphasis on the family was dismissed because their studies were judged to be inadequate in terms of theory and statistical analysis. Nonetheless, the Gluecks' data and perspective were unique and innovative in many ways and were unencumbered by disciplinary bias. Their interest in explaining the development of delinquent behavior from infancy to young adulthood required data that provided measures across the full span of life, rather than at a single point during adolescence. As a result, the Gluecks were among the first criminologists to design and gather longitudinal data. The Gluecks' work anticipated many of the key issues in criminology today, including the relationship between age and crime, patterns of criminal careers, the stability of crime and deviance, interrelationships between individual characteristics and interpersonal relations, and the importance of family relationships and informal social controls.

As described in Chapter 3, the Gluecks are most well known for their book *Unraveling Juvenile Delinquency,* a work based on a comparative, longitudinal study of 500 officially defined delinquents and 500 nondelinquent boys.[14] The data set includes a wide range of information on biological, psychological, and sociological characteristics, as well as significant life events. The Gluecks came to the conclusion that the most important factor that distinguished delinquents from nondelinquents in early life was family life experiences.[15] Drawing from their research, the Gluecks developed a prediction instrument that included five social factors that most effectively distinguished delinquents from nondelinquents. All of these five social factors dealt with various aspects of family life: paternal discipline, maternal supervision, paternal affection, maternal affection, and family cohesiveness.[16] Robert Sampson and John Laub summarize the Gluecks' emphasis on the family: "Those families with lax discipline, combined with erratic and threatening punishment, poor supervision, and weak emotional ties between parent and child generated the highest probability of delinquency. Although a focus on the family was to become extremely unpopular in the 1950s and 1960s, it was one of the Gluecks' major interests."[17]

Another exception to the general lack of interest in the family's role in delinquent behavior was the work of Ivan Nye. Nye conducted an extensive study of family characteristics and delinquent behavior that resulted in the book, *Family Relationships and Delinquent Behavior,* published in 1958.[18] In each chapter, Nye considered a particular aspect of family life and how it was related to delinquent behavior. His consideration was extensive and included parental acceptance and rejection, discipline and punishment, freedom and responsibility, family recreation, parental disposition and character, value agreement, and communication. Nye's study was one of the first to use self-reported measures of delinquent behavior and family characteristics. Like the Gluecks, Nye came to the conclusion that the family plays an important role in preventing and controlling delinquent behavior.

While Nye did not advance a formal theory, he did differentiate three major forms of control and demonstrate how family relations play a vital role in the development and application of these controls. <u>Direct controls</u> are those restrictions and punishments imposed by others that restrict behavior and those rewards that encourage and reinforce behavior. <u>Indirect controls</u> are based upon affectional identification with others, especially parents, such that the youth conforms in order to maintain the relationship bond and to avoid disappointing others. <u>Internalized controls</u> are exercised from within the individual through conscience or sense of guilt. Nye also argued that when youths' relational needs are not met within the family, then they are more likely to turn outside the family. The absence of family relational controls leads to greater likelihood of delinquent behavior. Even though Nye recognized that some social controls are formal, operating through institutions like the school and the juvenile justice system, he emphasized the informal social controls of family relations. Nye's conceptualization of social control is central to the development of social control theories.

direct controls
Restrictions and punishments imposed by others that limit behavior, and rewards that encourage and reinforce behavior.

indirect controls
Pressures and reasons to conform that are based on affectional identification with others, especially parents, such that people conform in order to maintain relationship bonds and to avoid disappointing others.

■ Social Control Theories

internalized controls A person's adoption of standards of behavior that are exercised from within the individual through conscience or sense of guilt.

Beginning in the 1950s, the development of social control theories rekindled interest in the family's role in delinquent behavior. Joseph Rankin and Roger Kern summarize the common themes of control theories: "Various social control theories focus on the family as the primary source of attachments, commitments, and disciplinary controls in preventing delinquency. According to these models, the family acts as a buffer against deviant influences by providing a source of basic ties and commitments to the conventional order. Parents not only furnish a source of ongoing motivations to conform and normative definitions of appropriate behavior, they also provide an important coercive function in the supervision and punishment of children's compliance through parental motivations to conform as well as restraints against deviance."[19]

Among the various social control theories, Travis Hirschi's social bond theory is arguably most responsible for the renewed interest in the family and delinquent behavior.[20] His theory, as we will see, led to significant developments in theory and research.

Hirschi's Social Bond Theory

In developing his version of social control theory, Hirschi makes frequent reference to a classic question posed long ago by Thomas Hobbes: "Why do men obey the rules of society?"[21] Hirschi claims that this question has never been adequately answered and that the primary task of delinquency theory is to explain conformity, not delinquent behavior. In fact, this is a basic theme of all social control theories. Hirschi answered the Hobbesian question by advancing four elements of an individual's social bond to society: *attachment, commitment, involvement,* and *belief.* These elements provide a "stake in conformity," or the reason to conform.[22]

The thesis of social bond theory is that "delinquent acts result when an individual's bond to society is weak or broken."[23] As is consistent with the broader range of control theories, Hirschi does *not* address what motivates young people to get involved in delinquent behavior; rather, he holds that when the bond to society is weak or broken, they are free to engage in delinquent acts. Delinquent behavior is not an automatic outcome of weakened or broken social bonds, but it is an appealing option.

The full statement of Hirschi's social bond theory is presented in the book *Causes of Delinquency,* published in 1969. The theory incorporated elements of previous control theories, but these early theories were for the most part replaced by Hirschi's version of social control theory.[24] Hirschi's presentation of social bond theory is unique in a number of ways. First, he clearly lays out the theory's assumptions, concepts, and propositions. The presentation is complex but intriguing and very readable. Second, his theory presents a strong argument about the "causes of delinquency," as the book's title suggests. The theoretical argument develops logically and is consistent and concise. In addition, the elements of the theory fit together closely. Third, Hirschi developed empirical measures of his major concepts and then systematically tested the theory, using data from the Richmond Youth Project, a self-report survey of more than 4,000 youths.[25] Hirschi's empirical analysis deliberately attempted to assess how different measures of elements of the social bond are related to delinquent behavior and to compare social bond theory with other competing theories. In these four ways, Hirschi's work is unique. Criminologists Ronald Akers and Christine Sellers observe: "His combination of theory construction, conceptualization, operationalization, and empirical testing was virtually unique in criminology at the time and stands as a model today."[26]

Hirschi repeatedly argues that these elements of the bond to society are predominantly social, not internal. As such, they arise in the context of social relationships, and, as a result, the mechanisms of control are predominantly *informal* in nature.

Attachment

attachment The affectional identification one has with others—sensitivity to others' opinions, intimate communication, mutual respect, identification, and valuing of relationship ties.

<u>**Attachment**</u> is generally regarded as the primary element of the social bond.[27] The essence of attachment is the affectional identification that the youth has with others, through which he or she is sensitive to their opinions, communicates intimately with them, mutually respects and identifies with them, and values his or her relationship with them.[28] Hirschi has referred to attachment as the "bond of affection."[29] Joseph Rankin and Roger Kern point out that "attachment is essentially a social–psychological concept involving the motivational value of *social approval.*"[30] Attachment results in conformity because of the vested interest that the youth has in a relationship. As a result, social control is based in the relationship bond itself, rather than in some process of internalization through which the youth develops self-control or a conscience.[31]

While Hirschi considered three forms of attachment—attachments to parents, to school, and to peers—he presents the relational controls of the parent–child attachment as the most important in inhibiting delinquent behavior.[32] The emphasis on relationship bonds means that direct parental controls, such as monitoring and supervision, are relatively unimportant in bringing about conformity.[33] Rather, relationship attachments bring about a sensitivity to the wishes and feelings of parents, and it is this sensitivity that results in social control.[34]

Using self-reported data, Hirschi's test of social bond theory revealed that youths who were strongly attached to their parents were less likely to be delinquent. The analysis of parental attachment was based most extensively on two measures: *intimacy of communication* and *affectional identity*. Intimacy of communication included questions about communication from child to parent and from parent to child, such as: "Do you share your thoughts and feelings with your mother [father]?"[35] Affectional identity was measured using questionnaire items such as: "Would you like to be the kind of person your mother [father] is?"[36] Using these measures, Hirschi found that youths with strong attachments to their parents were significantly less likely to engage in delinquent acts.

Hirschi also investigated whether this association between attachment and delinquent behavior persists across different social classes and racial categories. He found that regardless of the class or racial status of the parent, the closer the youth was to his parent, the less likely he was to commit delinquent acts.[37]

In contrast to social bond theory, which emphasizes parental attachment, other theories contend that delinquent peer groups are the major determinant of involvement in delinquent acts. Delinquent behavior is learned from delinquent peers. While we will examine this view more completely in the next chapter, Hirschi's analysis revealed that parental attachment was directly related to delinquent behavior such that youths attached to their parents were less likely to be delinquent regardless of the number of delinquent friends.[38]

Social bond theory advances two other dimensions of attachment: attachments to school and attachment to peers. In fact, Hirschi devoted separate chapters to "Attachment to the School" (Chapter VII) and "Attachment to Peers" (Chapter VIII). Attachment to school is more of an attitudinal social bond than a relational social bond. Hirschi portrayed and measured attachment to school through questions that gauged whether the youth liked school and cared what teachers thought of him.[39] Another question addressed the youth's views about the legitimacy of the school's authority.[40] Hirschi found that youths who do poorly in school are usually unattached to school and frequently involved in delinquent acts. Put in terms of social bond theory, when the social bond of school attachment is weakened, youths are free to commit delinquent acts. Research in Action, "Schools and Delinquency," summarizes research on the relationships between delinquency and various aspects of school experiences.

As it does with attachment to parents and school, social bond theory stresses that peer attachment is "conducive to conformity," *not* to delinquent behavior: "the more one respects or admires one's friends, the less likely one is to commit delinquent acts."[41] While Hirschi recognized that delinquent acts are usually committed with companions and that delinquent youths are likely to have delinquent friends, he argued that delinquent behavior is related to weak attachment to peers.[42] His research showed that youths who were strongly attached to friends were least likely to be delinquent. This finding suggests that peer attachment operates as a control of delinquent involvement, not as motivation to it. Hirschi claims that it is the presence or absence of attachment that is related to delinquency, not whether peers are involved in delinquent acts. His

RESEARCH IN ACTION — Schools and Delinquency

Research has shown that various aspects of school experiences are related to delinquency. These include poor academic performance, school social bonds (including commitment and attachment to school), curriculum-tracking experiences, and truancy and dropping out of school.

Poor academic performance. Maguin and Loeber conducted a meta-analysis of more than 100 studies of the relationship between academic performance and delinquency. They found that poor academic performance consistently predicted delinquent behavior. More specifically, children who performed poorly academically committed more frequent, serious, and violent offenses and persisted longer in their offending than children who showed higher academic achievement. The relationship between poor academic performance and delinquency was stronger for males than for females, and stronger for whites than for blacks.[43]

School social bonds. More than three decades ago, Travis Hirschi theorized that attachment to school, commitment to education, and involvement in school-related activities would decrease involvement in delinquency.[44]

- **Attachment to school.** Many have argued, from the perspective of social bond theory, that students who are strongly attached to school—who, for example, like school, "feel a part of their school," and care what teachers think of them—are less likely than others to engage in delinquency.[45] Numerous studies have supported this hypothesis.[46] Some researchers found, though, that the relationship between attachment to school and delinquency was reciprocal, meaning that attachment influenced delinquency, but delinquency also affected attachment.[47]

- **Commitment to education.** Criminologists have also argued, from a social bond perspective, that weak commitment to education (i.e., weak valuing of educational goals) and low educational aspirations are related to delinquency. Research has strongly supported this hypothesis, showing that weak school commitment is strongly related to school misconduct and various forms of delinquency, including violence.[48] Commitment to education is likely related to academic performance. Students who perform well academically are likely to be more committed to education and have higher aspirations and expectations for success than those who perform poorly.[49]

- **Involvement in school-related activities.** Research generally has not supported the hypothesis that involvement in school-related activities will inhibit delinquency.[50]

Curriculum-tracking experiences. Curriculum tracking is "the practice of assigning students to certain curricula (honors, vocational, etc.) according to perceived student ability, teacher recommendations, parental approval, and other criteria."[51] Research has shown links between lower track placement (e.g., remedial or vocational tracks) and delinquency, drug use, and school misconduct.[52] Compared to higher track students, those in lower tracks tend to experience more academic failure. Thus, the relationship between tracking and delinquency is likely due in part to the strong relationship between academic performance and delinquent behavior.

Truancy and dropping out of school. Research has shown that truancy is a precursor to more serious delinquency (both violent and nonviolent offenses), and is also related to substance abuse and marital and job problems later in life.[53] Truancy signifies lack of commitment to school, which puts children at risk for many negative behaviors, including delinquency. Dropping out of school also indicates weak bonds to school, and some research has shown that dropping out is related to delinquent behavior, including violence.[54]

research revealed that even for youths who were attached to delinquent peers, the stronger the attachment to those friends, the less likely they were to be delinquent.[55]

Taking the argument one step further, Hirschi argued that youths with weak social bonds tend to develop friendships with delinquent youths, like themselves.[56] Furthermore, he portrayed the relationship bond of delinquent youths as weak and therefore uninfluential.[57]

Commitment

Most people in the course of everyday life acquire material possessions, reputations, and positions that they do not want to risk losing through involvement in illegal acts. "These accumulations are society's insurance that they will abide by the rules."[58] **Commitment** builds up or accumulates—the more that is acquired, the more invested the individual is in conforming. Thus, commitment is sometimes referred to as a "stake in conformity."[59]

Commitment is the "rational component" of the social bond. A bond to society is generated when a person invests time and energy in conventional activities such as school and work. Whatever possible gains come from deviant behavior must then be weighed against the risk of losing the investment that has been made in conventional behavior.[60]

While commitment implies an attitudinal bond to society, the concept more accurately refers to attitudes that are put into action. In fact, the full name given to this concept is "commitment to conventional lines of action."[61] During adolescence, conventional lines of action most commonly involve actions and aspirations related to school and work.

How is commitment demonstrated in the lives of adolescents? Hirschi first explored the relationship between commitment to "unconventional activity," such as smoking, drinking, frequent dating, and the importance of having a car—and delinquent behavior. His analysis revealed that such commitments are positively related to delinquent behavior and that their influence was cumulative—the greater the amount of unconventional activity in which the youth was

commitment The investments one has that provide reason to conform—a stake in conformity.

involved, the greater the number of delinquent acts.[62] Hirschi went on to explore the link between various measures of commitment to conventional activity and delinquent behavior. Included here were youths' aspirations, expectations for education and high-status occupation, and "achievement orientation"—a measure of how much a youth values good grades and works hard in school and other activities. Social bond theory holds that youths with high educational and occupational aspirations and expectations and higher achievement orientations are unlikely to pursue delinquent activities. Hirschi's research findings support the importance of these various forms of commitment—the more committed the youth is to conventional pursuits, the less likely he is to be delinquent. Not surprisingly, Hirschi's research also showed that the educational and occupational expectations of delinquent youth tended to be lower than those of nondelinquent youth.

Hirschi's findings also provide some interesting clarification on the role of commitment in delinquent behavior. His data showed that few youths had educational or occupational aspirations that exceeded their expectations. Therefore frustration resulting from high aspirations and low expectations for success cannot be viewed as a motivating force behind delinquent behavior. Hirschi also found that the educational and occupation expectations of delinquent youths tend to be lower than those of nondelinquent youths. In addition, Hirschi's findings revealed that lack of commitment to conventional activity was more strongly related to delinquent behavior when such commitment was indicated by current attitudes and activities, rather than by "hopes, plans, and prospects for the future."[63] Apparently commitment is fairly immediate and originates through social relations, rather than through long-term, deeply held attitudes.

Involvement

The old adage, "idle hands are the devil's workshop," reflects the belief that too much leisure time provides opportunity for delinquent acts.[64] This relates to the popular notion that **involvement** in conventional activities controls delinquency simply by consuming a youth's time and energy.[65] As was the case with the social bond of commitment, involvement does not necessarily require a deep, personal investment in conventional activity. Rather, involvement operates as a social bond by occupying the youth's time and thereby reducing the opportunity for delinquency.[66]

Hirschi does not argue, however, that any activity that consumes a youth's time and energy will prevent delinquent behavior. In fact, Hirschi's data showed that the more time a boy spent working, dating, watching television, reading, and playing games, the more—not less—likely he was to be delinquent (although the relationship was very weak).[67] Hirschi goes on to observe that most conventional activities are actually neutral with respect to delinquency—they neither inhibit nor promote it. The reason for this is that most delinquent acts require very little time to commit. Accordingly, we must consider *what* the youth is doing to see if involvement inhibits delinquent behavior.

Hirschi's research points to two specific types of activities that are related to delinquent behavior: school–educational activities and "working-class adult activities." Most of his analysis of conventional activities is devoted to time spent on homework. The relevance to social control is straightforward and is sup-

involvement
Conventional activities that not only consume time and energy, but also inhibit involvement in delinquency.

ported by Hirschi's findings: the more time spent on homework, the less involvement in delinquency. Involvement in education-related activities is the primary way that adolescent time is structured, thereby reducing the leisure time and opportunity for involvement in delinquent acts. School-related involvement also reflects the social bond of commitment, through which educational aspirations are translated into action.

Hirschi refers to a second group of activities as "working-class adult activities" because they provide adult-like activities for youth with limited educational and occupational aspirations and opportunities. His research measured working-class adult activities in terms of "boys who smoked, drank, dated frequently, rode around in cars, and had feelings of boredom." These boys "were more likely to commit delinquent acts than boys who did not have these attitudes and did not engage in these activities."[68] Thus, youths who have little interest in education-related activities fall outside of the structure provided by schools and become involved in activities in which opportunities for delinquent acts are readily available.

Belief

One of the distinguishing features of control theories is the idea that "delinquency is not caused by beliefs that require delinquency, but rather made possible by the absence of beliefs that forbid delinquency."[69] A variety of other theories attempt to explain the development of beliefs that allow and motivate delinquent behavior. Delinquent peer groups and subcultures, for example, provide beliefs that are opposed to the values and norms of conventional society. Delinquent behavior is a natural outcome when such beliefs are adopted. In contrast, control theories consider the strength of **belief** in the law and the legal system: "the less a person believes he should obey the rules, the more likely he is to violate them."[70] Thus, control theories do not focus on the beliefs that encourage delinquent behavior, but rather on the beliefs that encourage conformity.

belief The attitudes, values, and norms one has that forbid delinquency.

Some control theories emphasize the internalization of belief, in which deeply held values inhibit delinquent behavior. Such internalized beliefs are referred to as personal or internal controls.[71] More commonly, self-control and conscience are thought to restrain delinquent behavior. Hirschi's social bond theory, however, stresses the degree to which beliefs are socially and collectively reinforced. This places the concept of belief at the social, interpersonal level rather than at the internal, individual level.

The aspect of belief that is given the greatest emphasis in social bond theory is belief in the moral validity of the law. Hirschi argues that the most relevant measure of belief is the individual's attitude about the law, and he provides much analysis based on the questionnaire item: "It is alright to get around the law if you can get away with it." His findings revealed that there is a great deal of variation in the extent to which youths believe they should obey the law, and that the more they believe that they should, the less likely they are to get involved in delinquent acts.[72] However, Hirschi also found that almost two thirds of the youths who engage in delinquent behavior do *not* agree that it is "alright to get around the law if you can get away with it." Apparently belief in the legitimacy of the law does not always inhibit delinquent behavior; instead, the lack of belief may free the youth to consider delinquent acts as a reasonable option.

So what determines the degree to which the youth desires to obey the laws of society? Hirschi argues that the level of belief is closely connected to parental attachment and commitment: the youth who is unattached to parents and teachers and who has few aspirations for success is unlikely to feel that the law is something to obey.[73]

Connections Between Elements of the Social Bond

Hirschi's study demonstrated that the four elements of the social bond are additive—the more bonds individuals have, the more likely they are to conform.[74] He also held that people tied to conventional society through one of the social bonds are likely to display other elements of the social bond. This suggests that the elements of the social bond are interrelated and that the social bond cannot be adequately described or understood without considering such interconnections. Hirschi identifies three combinations of social bond elements that are particularly important.[75] First, *attachment* and *commitment* tend to vary together. Those attached to others also tend to be committed to conventional "lines of action." Second, *commitment* and *involvement* are closely related. Those who are committed to educational and occupational success are also likely to be involved in related conventional activities, such as doing homework. Third, *attachment* and *belief* are interrelated. When discussing belief, we referred to Hirschi's finding that youths who are attached to parents and who care what their parents think and feel are likely to express belief in the moral validity of the law.

Research Findings on the Social Bond

As mentioned earlier, one of the unique aspects of Hirschi's social bond theory is the fact that he provides an extensive test of the theory using data from the Richmond Youth Project. Although the relationship between the measures of the social bond elements has been questioned, Hirschi's analysis provided rich and consistent support for the theory. After its original presentation and testing, social bond theory has undergone testing by several researchers. The resulting evaluation has generally supported the core concepts of social bond theory and provided a number of important qualifications. Ronald Akers, a leading theorist in the social learning tradition, states that social bond theory "has come to occupy a central place in criminological theory. Indeed, it is the most frequently discussed and tested of all theories in criminology."[76]

A number of important qualifications to social bond theory have been suggested by research.

- Using similar research methods, Michael Hindelang replicated many of Hirschi's findings. He (and others in more recent research) found that, in contrast to the predictions of social bond theory, attachment to peers leads to conformity only when those peers are not delinquent, while attachment to delinquent peers is related to delinquent behavior.[77]

- Allen Liska and Mark Reed used longitudinal data analysis to gain a more complete understanding of the role of parental and school attachments in delinquent behavior. Their findings indicate that parental attachment is associated with lower levels of delinquent behavior and that involvement in delinquent acts appears to precede, not follow, weak school attachment.

Those youths who are involved in delinquent behavior tend to show less attachment to school following their involvement in delinquent acts. Weak attachment to school, in turn, influenced parental attachment.[78]

- Hirschi specified that the bond of commitment involves attitudes that are put into action—"commitment to conventional lines of action."[79] However, research that follows this specification has usually found only modest support for commitment as an element of the social bond.[80]

- Hirschi observed that the key to the bond of involvement is *what* a youth is involved in. Research by Robert Agnew and David Peterson found that organized leisure activities (e.g., band, school newspaper, scouts, church activities), passive entertainment (e.g., reading, radio, TV, movies), and noncompetitive sports (e.g., bike riding, horseback riding, roller skating) were negatively related to delinquent behavior (higher involvement was related to lower levels of delinquent behavior). In contrast, hanging out with friends, unsupervised activities with friends, and least favorite activities with parents were positively related with delinquent behavior (higher involvement, higher delinquency).[81] Many forms of leisure activities, however, were found to be unrelated to delinquent behavior. Agnew and Peterson concluded that the unsupervised nature of some leisure activities seems to be one of the most important determinants of freeing a youth for involvement in delinquent behavior.

- While research generally supports the contention that belief effectively inhibits delinquent behavior, Ross Matsueda sought to study this relationship further.[82] Using longitudinal data on high school boys from a national sample, Matsueda examined the link between moral belief and five measures of non-serious deviance, including being suspended or expelled from school, skipping school, running away from home, staying out past curfew, and arguing with parents ("fought with your parents"). In contrast to social bond theory and to previous research, Matsueda's research led him to conclude that the effect of belief on deviance was relatively small, especially when compared to the effect of deviance on belief. He concluded that belief is largely a product of participation in deviance, rather than a cause of deviance.

- Researchers have found that the different elements of the social bond may have greater importance at different ages.[83] Findings indicate that parental attachment is most important in early adolescence, commitment to school is most important in early to middle adolescence, and belief is most important in middle to late adolescence.

- Much research indicates that the elements of the social bond are more strongly related to minor delinquency than to serious delinquency.[84]

- Research findings show that social bond variables are probably better able to explain female delinquency than male delinquency and that the elements of the social bond may operate differently for girls than for boys.[85]

- Research using longitudinal data reveals that social bond variables are related to delinquent behavior in the expected direction but that these relationships are relatively weak and fail to significantly predict delinquent behavior.[86] The effects of social bond variables are weaker than those of variables specified by competing theories, and their effect is only indirect.[87] Basing their work on these findings, a number of scholars have developed causal models suggesting that social bond variables are indirectly related to delinquent behavior because they increase the probability that some adolescents will associate with delinquent peers from whom they learn delinquent attitudes and behaviors.[88]

Sampson and Laub's Life-Course Theory

Two prominent criminologists, Robert Sampson and John Laub, have recently introduced life-course theory, bringing a sense of Indiana Jones-like intrigue to their work by talking about "puzzles" and "dusty cartons of data" in the "sub-basement" of the Harvard Law School Library.[89] Rich data are to sociologists what artifacts are to archaeologists. Sampson and Laub convey this excitement: "As we began to sort through the case files, we soon discovered that these were not conventional data. And, as we went on, we found out the Gluecks [the original researchers who gathered the data] were not conventional researchers."[90] The data that Sampson and Laub had discovered were those of Sheldon and Eleanor Glueck's famous longitudinal and comparative study of 500 delinquents and 500 nondelinquents (described in Chapter 3). The Gluecks published their findings in the 1950 book *Unraveling Juvenile Delinquency*. This far-reaching study provided Sampson and Laub with both the data and perspective for their contemporary *life-course theory*.

Life-course theory has its roots in familiar ground: "Our general organizing principle is that the probability of deviance increases when an individual's bond to society is weak or broken."[91] Life-course theory is most extensively advanced in Sampson and Laub's book, *Crime in the Making: Pathways and Turning Points Through Life*. Like Hirschi, Sampson and Laub focus on the importance of social bonds as the primary means of informal social control. Rather than elaborating on the elements of the social bond, life-course theory examines social bonds over the course of life, considering how changes influence informal social control and, ultimately, behavior. In brief, Sampson and Laub refer to their theory as an "age-graded theory of informal social control."[92]

The Life-Course Perspective

Sampson and Laub argue that sociological criminology has traditionally "concentrated on the teenage years" and thereby "neglected the theoretical significance of childhood characteristics and the link between early childhood behaviors and later adult outcomes."[93] Life-course theory is an attempt to bring childhood and adulthood into the criminological picture. Like the Gluecks, Sampson and Laub seek to study criminal behavior from its beginning to end—what we referred to earlier as the *criminal career* (Chapter 6).[94] Thus, the frame of reference is the entire life span, and primary attention is given to the different factors that explain the onset, persistence, and desistance of criminal involvement over the life course. To study these factors, Sampson and Laub adopt the conceptual tools of the life-course perspective and the causal principles of social bond theory.[95] **Life course** refers to the age-graded sequence of culturally defined roles and social transitions.[96] Sociologically, roles are behavioral expecta-

life course The age-graded sequence of culturally defined roles and social transitions.

tions of a given position. These roles are age-graded in that the expectations for behavior vary across the course of life—expectations for a child are different from those for an adolescent, and expectations of an adolescent are different from those of an adult. The life course also involves a general <u>trajectory</u> or <u>pathway</u>, which is a long-term pattern of behavior relating to education, work, parenting, conformity or deviance, and the like. The life-course trajectory is marked by significant life events, or <u>transitions</u>, that may alter the established patterns of behavior or trajectory. When a transition results in significant change in the life-course trajectory, it is referred to as a <u>turning point</u>. Sampson and Laub's life-course theory relates these basic concepts to the development and alteration of relationship bonds, or <u>social capital</u>, and informal social control. In doing so, the theory attempts to account for not only the onset of delinquent behavior, but also the continuation and desistance of delinquent behavior. These life-course concepts become the "building blocks" of life-course theory.

trajectory or pathway The pattern or direction of life over a number of years, relating to areas of life such as education, work, parenting, and conformity or deviance.

transitions Significant life events that occur within the life trajectory or pathway, such as marriage, enlistment in the military, first job, or change in employment.

Pathways: Behavioral Continuity over the Life Course

Basing their conclusions on four decades of research, Sheldon and Eleanor Glueck argued that child temperament and family socialization were the factors that most distinguished delinquents from nondelinquents in early life.[97] Sampson and Laub's life-course theory adopts this same emphasis and expands upon it. Following the Gluecks, Sampson and Laub propose four family socialization factors that significantly increase the likelihood of delinquent behavior: (1) erratic, threatening, and harsh or punitive discipline by both mothers and fathers; (2) low parental supervision; (3) parental rejection of the child; and (4) weak emotional attachment of the child to his parents.[98] These key dimensions of family relations influence the likelihood of delinquent behavior mainly because they determine the effectiveness of informal social control. Unlike Hirschi's presentation of social bond theory, which emphasizes the *indirect controls* of the parent–child relational bond, life-course theory also includes *direct controls* in the form of discipline and supervision.

turning point A transition that results in significant change in the life-course trajectory.

social capital Expectations, obligations, and restraints that are part of adult relationship bonds and that result in informal social control.

Life-course theory also recognizes that parenting styles, especially discipline and emotional attachment, are responsive to the personality and behavior of the child. While the Gluecks referred to child temperament, Sampson and Laub expanded the notion to include a difficult personality ("restless and irritable"), a predisposition toward aggression and fighting (where violence was a "predominant mode of response"), and the early beginning or "onset" of misbehavior. Together, these characteristics are referred to as "early childhood predisposition toward disruptive behavior."[99]

Sampson and Laub argue that family socialization factors must be considered in conjunction with a child's predisposition toward disruptive behavior. At least in part, parenting styles are a response to troublesome behaviors of the child.[100] Ronald Simons and his colleagues have referred to this as the "corrosive effect" that childhood disruptive behavior has on the "social relationships that serve as informal social controls."[101]

Sampson and Laub call their theory a "sociogenic developmental theory"— a reference that distinguishes their point of view from that of developmental psychology. While both views use the idea of developmental stages in the life course, a sociogenic approach also emphasizes the structural context of such developmental stages, rather than the individual's adaptation to these stages. In particular, Sampson and Laub point to the importance of structural factors such as poverty, residential mobility, family disruption, family size, household crowding, and mother or father deviance. They propose that "poverty and structural disadvantage influence delinquency in large part by reducing the capacity of families to achieve effective informal social control."[102]

The notion that pathways are established early in life implies continuity in behavior over time. Such *behavioral continuity* finds strong and consistent support in research findings from a variety of longitudinal studies that show that antisocial behavior tends to be persistent across the life course.[103] Ronald Simons and his colleagues observe: "Children who are aggressive and noncompliant during elementary school are at risk for serious delinquency during adolescence, and serious delinquency during adolescence significantly increases the probability that a person will engage in criminal acts as an adult."[104]

The pathway to criminal involvement that stretches from early childhood to the adult years implies not only behavioral continuity, but also progression and escalation of deviance. Disruptive behavior in childhood leads to greater likelihood of delinquent behavior in adolescence and to more serious crimes in later adolescence and early adulthood. In addition, the continuity of problem behavior is not restricted simply to deviance of increasing degrees, but it also reaches into other areas of life, "including economic dependence, educational failure, employment instability, and marital discord."[105]

Life-Course Change

The continuity of behavior throughout the life course is only half the story. While early childhood misconduct increases the probability of future delinquency and criminality, the causal sequence is far from being perfectly deterministic. Research reveals that at least half of all antisocial children do not become delinquent during adolescence.[106] Furthermore, most antisocial children do not become antisocial adults, and a majority of adult criminals have no history as juvenile delinquents. Behavioral stability is true in a "relatively small number of males whose behavioral problems are quite extreme."[107] Because of this potential for change in the pathway to crime, life-course theory attempts to identify the processes whereby childhood disruptive behavior escalates to delinquency and crime, and to discover those processes that enable some antisocial children to assume a more conventional lifestyle during adolescence.[108] Thus, Laub and Sampson attempt to "examine crime and deviance in childhood, adolescence, and adulthood in a way that recognizes the significance of both *continuity* and *change* over the life-course."[109]

Social Capital and Informal Social Control

Sampson and Laub's search for an explanation for both continuity and change returned them to the importance of social bonds as the primary vehicle of informal social control. Weak social bonds within the family lessen the informal social control of parents, allowing for childhood misconduct to escalate into adolescent delinquency. The social bonds of adulthood are also central to Sampson and Laub's explanation of adult criminal behavior. Sampson and Laub introduce the concept of social capital to refer to the "obligations and restraints" that are a part of relationship bonds.[110] These obligations and restraints are connected to the institutional roles that are a part of adult life—the behavioral expectations associated with positions in the family, school, place of employment, and community. The result of these role obligations and restraints is informal social control.

Turning Points: Explaining Change in the Life Course

Life-course theory holds that behavioral continuity occurs throughout life. "Individuals show persistent involvement in antisocial behavior to the extent that their deviant actions have an attenuating effect on the social and institutional bonds linking them to society. This attenuating process is labeled *cumulative continuity*."[111] Thus, changes in social bonds will influence informal social control and the level of involvement in deviance.

What accounts for changes in social bonds? Significant life events, called transitions, constitute important milestones in the life-course trajectory or pathway. Transitions are usually connected to institutions such as the family, education, and work. When these transitions significantly alter social capital and informal social control, they are referred to as turning points. Sampson and Laub depict their full life-course model in **Figure 10-1.**

Research Findings on Life-Course Theory

The dusty cartons of old data not only provided Sampson and Laub with the intrigue that fueled their development of life-course theory, but they also provided the data to test that theory. Those cartons contained the original case files on 500 delinquents and 500 nondelinquents that became the basis for the Gluecks' well-known longitudinal study, *Unraveling Juvenile Delinquency*.[112] Sampson and Laub resurrected these data by computerizing, supplementing, and validating them. In an extensive reanalysis of the data, Sampson and Laub claimed support for their life-course theory.[113]

Using measures of an early childhood predisposition to disruptive behavior (difficult disposition, early onset, and tantrums), family process (supervision, attachment, and discipline), and structural background (residential mobility, family disruption, social class, and parental deviance), Sampson and Laub found that family process variables had a strong and direct influence on delinquent behavior, as did early childhood difficulties. As the theory holds, early childhood difficulty also influenced parental supervision and discipline. Surprisingly, early childhood difficulties did not affect the emotional bond between parent and child. The data showed that early childhood difficulty was not sufficient in itself to explain delinquent behavior. It was also necessary to consider the effect of childhood difficulty in conjunction with family processes. The effect of structural background variables was found to occur indirectly, largely through their influence on family processes. In fact, almost 75% of the total effect of structural background variables on delinquent behavior was through their effect on family processes.[114]

Sampson and Laub also considered turning points in adult life that changed social capital and informal social control. They found that social capital in the form of social bonds in marriage (marital attachment) and work (job stability) significantly reduced deviant behavior during adulthood, even among those with a history of delinquent behavior in childhood and adolescence. Both continuity and change in behavior were strongly predicted by prior social bonds and the development of adult social bonds.

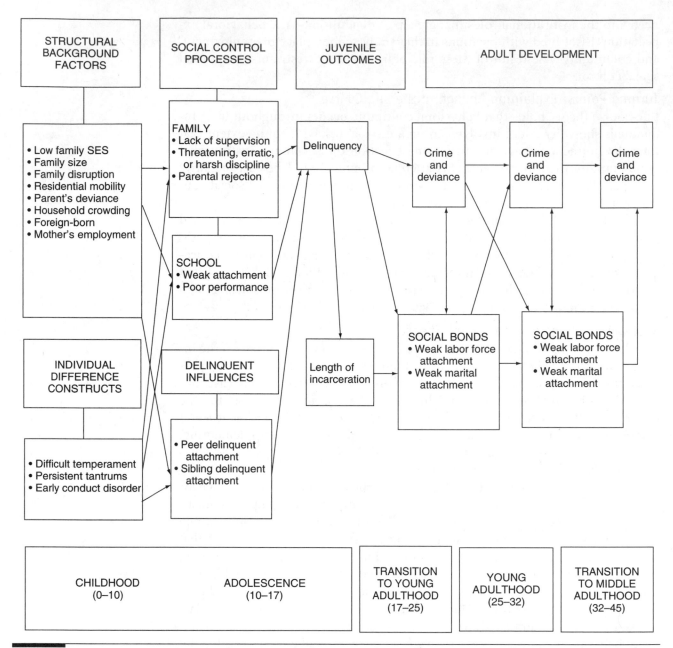

Figure 10-1 The full life-course model of delinquency, deviance, and adult criminality.

Source: Sampson and Laub, *Crime in the Making,* 244–245. Copyright © 1993, Harvard University Press. All rights reserved. Reproduced with permission of the publisher.

In a separate study of life-course theory, Ronald Simons and his colleagues investigated whether childhood antisocial behavior had a disruptive influence on key dimensions of the social bond (parenting, school commitment, affiliation with conventional peers) and, in turn, with deviant behavior.[115] Using longitudinal data from 179 boys and their parents, Simons and his colleagues found that oppositional and defiant behavior during late childhood reduced the quality of parenting and commitment to school and increased affiliation with deviant

Stability of Antisocial Behavior Over the Life Course

Ronald Simons and his associates found that the continuation of problem behavior into the adolescent years is a function of the loss of informal social control, especially in terms of diminished quality of parenting, lack of school commitment, and involvement with deviant peers.

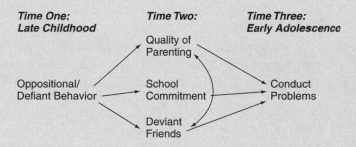

Source: Simons et al., "Test," 223. Copyright © 1998, American Society of Criminology. All rights reserved. Reproduced with permission of the publisher.

friends. Poor parenting, low school commitment, and affiliation with deviant peers, in turn, predicted behavioral problems during adolescence. In addition, reductions in the quality of parenting increased the chances of low school commitment and affiliation with deviant peers. The effect of oppositional and defiant behavior in late childhood on problem behavior in adolescence was indirect, influencing the quality of parenting, school commitment, and affiliation with deviant peers, which then affected problem behavior in adolescence. The researchers concluded that the continuation of antisocial behavior into the adolescent years ("cumulative continuity") must be understood in terms of the reduction in informal social control that results from early antisocial behavior. Research in Action, "Stability of Antisocial Behavior Over the Life Course," shows this life-course model.

Gottfredson and Hirschi's General Theory of Crime— Self-Control Theory

Hirschi's newest version of control theory is a theory of self-control, developed with Michael Gottfredson and presented in their book *A General Theory of Crime.*[116] Gauged by its reception, the theory is undoubtedly an important contribution to criminological theory; however, it has generated a tremendous amount of controversy because of its unique point of view: "Its emphasis on enduring individual differences and its disdain for social causation are out of step with the ways of thinking about crime now dominant in the discipline."[117]

Consistent with other control theories, self-control theory attempts to explain conformity, not why people engage in crime. However, this contemporary control theory argues that the critical dimension of control is self-control, an internal control that is instilled largely through effective parenting.

Gottfredson and Hirschi's presentation of their theory is based upon a logically developed argument, sustained throughout the book, which covers a number of critical issues in criminology. Gottfredson and Hirschi begin with the contention that criminologists have oddly ignored studying the *nature of crime,* which (Gottfredson and Hirschi believe) can provide important insight into the *nature of criminals:* "We started with a conception of crime, and from it attempted to *derive* a conception of the offender."[118]

The Nature of Crime

After study of the crimes that occur most frequently, Gottfredson and Hirschi claim that a "vast majority of criminal acts are trivial and mundane affairs" that require "little in the way of effort, planning, preparation, or skill."[119] Because most crimes are similar, they argue that it is theoretically unproductive for criminology to make distinctions among crimes.[120] Instead, it is possible, and in fact advantageous, to offer a general theory of crime—a theory that accounts for all types of crime.

By the same reasoning, Gottfredson and Hirschi observe that virtually all crimes are "short lived, immediately gratifying, easy, simple, and exciting."[121] These characteristics make crime appeal to everyone, and consistent with most other control theories, self-control theory makes no effort to explain the motivation that is necessary for involvement in crime—crime is appealing by its very nature. Instead, what needs to be explained is what keeps people from getting involved in crime.

The Nature of Criminality: Low Self-Control

low self-control
Individuals with low self-control "tend to be impulsive, insensitive, physical (as opposed to mental), risk-taking, short-sighted, and nonverbal" (Gottfredson and Hirschi, *General Theory of Crime,* 90). Together these characteristics constitute a single latent trait.

Gottfredson and Hirschi go on to argue that "the properties of criminal acts are closely connected to the characteristics of people likely to engage in them." Their theory holds that crimes are committed by persons with **low self-control**. Such people are "impulsive, insensitive, physical (as opposed to mental), risk taking, short-sighted, and nonverbal." People who lack self-control are "vulnerable to the temptations of the moment" because they are unable to calculate or consider the negative consequences of their acts.[122]

Gottfredson and Hirschi stress that the lack of self-control does not predispose or motivate a person to crime, and, as a result, crime is not an automatic consequence of low self-control.[123] Rather, the probability that low self-control will lead to crime depends on other conditions, especially opportunity.[124] Similarly, they argue that self-control should *not* be considered "a personality concept or 'an enduring criminal predisposition'" that compels a person to crime.[125] The line of reasoning here is consistent with the fundamental assumption of control theories that people are not predisposed to crime. In fact, Hirschi and Gottfredson argue that particular personality traits such as temper and lack of cautiousness actually reflect the more underlying individual characteristic of self-control. In this regard, self-control is the most fundamental cause of crime.

As a "general" theory of crime, self-control is used to explain not only crime, but also a variety of "analogous acts" such as accidents, smoking, and alcohol use.[126] Since lack of self-control frees the individual to pursue self-interest and personal gain, crime is just one of many possible behaviors. Individuals with low self-control continually "succumb to life's temporary temptations."[127]

Corresponding to involvement in crime and analogous acts are a wide variety of "social consequences" of low self-control. Gottfredson and Hirschi argue that low self-control fosters failure in interpersonal relations, social activities, and social institutions. Individuals with low self-control have difficulty making and keeping friends, and when they do develop friendships, they tend to end up in the company of others who similarly lack self-control and are involved in deviance. In addition, they are less able to succeed in school and the workplace, and they tend to enter into marriages that are destined to fail.[128] Finally, Gottfredson and Hirschi claim that "people who lack self-control tend to dislike settings that require discipline, supervision, or other constraints on their behavior; such settings include school, work, and for that matter, home. These people therefore tend to gravitate to 'the street' or, at least in adolescence, to the same-sex peer group" in which opportunities for crime abound.[129] This tendency to seek out similar individuals and to place oneself in social and institutional situations that are ripe for problems in general and crime in particular is referred to as *selection*.[130]

The logical conclusion to this line of thought provides self-control theory with a unique viewpoint that has generated considerable controversy: many of the traditional causes of crime are in fact consequences of low self-control.[131] Traditional causes of crime include such factors as deviant peers, dysfunctional family background, school failure, and weak social bonds. According to Gottfredson and Hirschi, the apparent link between these traditional causes and criminal involvement is actually evidence of the true causal connection between low self-control and these apparent causes. Low self-control is the true cause of crime. The argument here is depicted in Research in Action, "The Priority of Low Self-Control," in which low self-control is shown to have a variety of consequences that are traditionally viewed as causes of crime. However, low self-control is shown to be the true cause of crime and analogous acts.

Origins of Self-Control

According to Gottfredson and Hirschi, self-control is instilled early in life primarily through effective child-rearing. Drawing extensively from Gerald Patterson's work on *family management skills*, life-course theory advances three key components of effective parenting: (1) monitoring of the child's behavior, (2) recognition of deviant behavior when it occurs, and (3) consistent and proportionate punishment of deviance when it is recognized.[132] The influence of these direct controls on self-control is apparent when Gottfredson and Hirschi state that parents "who care for the child will watch his behavior, see him doing things he should not do, and correct him. The result may be a child more capable of delaying gratification, more sensitive to the interests and desires of others, more independent, more willing to accept restraints on his activity, and more unlikely to use force or violence to attain his ends."[133] In contrast, "the characteristics associated with low self-control tend to show themselves in the absence of nurturance, discipline, or training."[134]

RESEARCH IN ACTION — The Priority of Low Self-Control

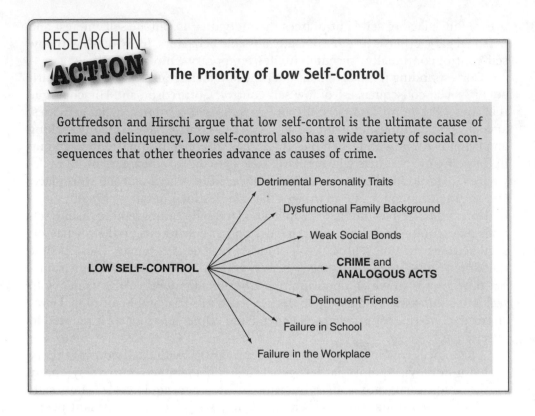

Gottfredson and Hirschi argue that low self-control is the ultimate cause of crime and delinquency. Low self-control also has a wide variety of social consequences that other theories advance as causes of crime.

LOW SELF-CONTROL →
- Detrimental Personality Traits
- Dysfunctional Family Background
- Weak Social Bonds
- CRIME and ANALOGOUS ACTS
- Delinquent Friends
- Failure in School
- Failure in the Workplace

This emphasis on the importance of direct controls in self-control theory is similar to that of life-course theory, which also considers parental supervision and discipline early in life to be central to the development of controls.[135] However, self-control theory, as its name states, emphasizes self-control, whereas life-course theory centers on the informal social controls of relationship bonds. In addition, self-control theory emphasizes that self-control is established early in life and remains stable throughout life. This is called the "stability postulate."[136] Life-course theory, in contrast, considers informal social controls throughout life, stressing (for the adult years) informal social control resulting from social relationship bonds or social capital.

Gottfredson and Hirschi also recognize that schools may have some importance as an institution of socialization and may, in fact, have several advantages over the family in this regard.[137] Gottfredson and Hirschi identify teachers' ability to closely monitor students, their ability to recognize deviant and disruptive behavior, and their authority and interest in maintaining order and discipline. However, Gottfredson and Hirschi claim that schools have a difficult time teaching self-control beyond that which has already been establishing through child-rearing. Since self-control is established early in life and remains stable, the experiences of school occur too late in life to have much influence on self-control.

Research Findings on the Self-Control Theory

The first problem confronted by researchers attempting to assess self-control theory is how to measure self-control. While Gottfredson and Hirschi note that it can be operationalized with attitudinal measures (such as impulsivity, sensitivity, and short time horizon), they advocate for behavioral measures. These are measures of actual behavior, especially in terms of what Gottfredson and Hirschi refer to as "social consequences"—measures such as school failure, relationship problems, and delinquent friends.[138] Ronald Akers argues that such measures are not independent from self-control and, as a result, self-control theory is tautological in that the cause and effect are the same.[139] Hirschi and Gottfredson, however, claim that this is really not a problem because self-control is a *characteristic* of the individual whereas the social consequences of self-control are *behaviors*.[140]

Harold Grasmick and his colleagues have developed a widely used attitudinal measure of low self-control, made up of six key components: impulsivity, preference for simple rather than complex tasks, risk seeking, preference for physical rather than cerebral activities, self-centered orientation, and a volatile temper. Each of these components was drawn from Hirschi and Gottfredson's definition of low self-control and was measured by several survey questions, as indicated in Research in Action, "Measures of Low Self-Control."[141]

Grasmick and his colleagues contend that the different components that make up low self-control can best be understood collectively, as a single characteristic of the individual.[142] Consistent with self-control theory, they also argued that these different components of self-control should be considered in conjunction with opportunity for crime. In support of this, their research found that when considered together, self-control and opportunity for crime significantly predicted self-reported crime.

Despite the controversy over how self-control should be measured, researchers using a variety of measures of self-control have found that low self-control is consistently related to self-reported delinquency, adult crime, and analogous acts such cutting classes, drinking, smoking, and gambling.[143] Also consistent with the theory, research by T. David Evans and his colleagues found that low self-control was associated with a variety of social consequences. Evans and his colleagues summarize: "We find that low self-control is related to diminished quality of interpersonal relationships with family and friends, reduced involvement in church, low levels of educational and occupational attainment, and possible poor marriage prospects. Individuals with low self-control also are more likely to reside in disorderly neighborhoods. Further, persons with low self-control are more likely to have criminal associates (a 'flocking together' effect) and to internalized criminal values."[144]

The most controversial aspect of self-control theory is its assertion that "many of the traditional causes of crime are in fact consequences of low self-control."[145] In this view, many of the individual and social factors identified by traditional causal theories are consequences of low self-control, not causes of delinquency and crime. While low self-control is directly related to crime, it is also the primary influence on a number of factors that have traditionally been viewed as causes of crime. It is further argued that these traditional causes of crime are rendered inconsequential when low self-control is considered. Gottfredson and Hirschi argue that the investigation of such spurious relationships (see Chapter 3) "provides a crucial test" of self-control theory in contrast to traditional criminological theories.[146]

Research directed at this "crucial test" finds that low-self control is not *the* single, fundamental cause of delinquency, crime, or analogous behavior. Rather,

RESEARCH IN ACTION — Measures of Low Self-Control

In research conducted shortly after self-control theory was introduced, Harold Grasmick and his colleagues advanced a measure of low self-control that was composed of six components: impulsivity, preference for simple rather than complex tasks, risk seeking, preference for physical rather than cerebral activities, self-centered orientation, and a volatile temper. Each of these components was measured by several questions.

Impulsivity

I often act on the spur of the moment without stopping to think.
I don't devote much thought and effort to preparing for the future.
I often do whatever brings me pleasure here and now, even at the cost of some distant goal.

Simple Tasks

I frequently try to avoid projects that I know will be difficult.
When things get complicated, I tend to quit or withdraw.
The things in life that are easiest to do bring me the most pleasure.

Risk Seeking

I like to test myself every now and then by doing something a little risky.
I sometimes find it exciting to do things for which I might get in trouble.
Excitement and adventure are more important to me than security.

Physical Activities

If I had a choice, I would almost always rather do something physical than something mental.
I almost always feel better when I am on the move than when I am sitting and thinking.
I like to get out and do things more than I like to read or contemplate.

Self-Centered

I try to look out for myself first, even if it means making things difficult for other people.
I'm not very sympathetic to other people when they are having problems.
If things I do upset people, it's their problem not mine.

Temper

I lose my temper pretty easily.
Often, when I'm angry at people I feel more like hurting them than talking to them about why I am angry.
When I'm really angry, other people better stay away from me.

Source: Grasmick et al., "Testing," 14–15.

many of the traditional causes of crime (e.g., parenting, school experiences, peers, and neighborhood conditions) are found to be significantly related to delinquent behavior, even after controls for self-control are introduced.[147] Moreover, in at least one study, when the effects of these traditional causes of crime are considered in causal models, the relationship between measures of low self-control and delinquent behavior is found to be no longer significant.[148] It is probably safe to conclude that low self-control is one among many causal factors related to delinquency behavior, though its influence is substantial.[149]

Still another area of growing research deals with the origins of low self-control. As already mentioned, Gottfredson and Hirschi attribute self-control to effective family management skills—careful monitoring, recognition of problem behaviors, and consistent and proportionate punishment of such behaviors. While not entirely consistent in results, this research generally indicates that parenting is an important, but not exclusive, determinant of self-control.[150] Recent research, for example, found that self-control was predicted by neighborhood conditions having to do with informal social control—the level of caring about what goes on in the neighborhood.[151] The effect of adverse neighborhood conditions on self-control was just as strong as the effect of poor family management skills.

■ Characteristics of Family Life and Informal Social Control

All three versions of social control theory point to the central importance of family relations in generating informal social control. In doing so, they attempt to explain conformity. Despite this common theme, each version of social control theory characterizes the informal social controls of families in a different way, especially in terms of the relative importance of direct and indirect controls. Velmer Burton and his colleagues provide a clear distinction of these two types of informal social controls: "While indirect controls are the constraint on the youth that flow from the quality of their affective attachment to parents, direct controls are the actions that parents take to limit misconduct."[152]

Social bond theory stresses indirect controls that result from family relationship bonds, or what are referred to as "attachments." In fact, Hirschi argued that direct controls are relatively unimportant.[153] Life-course theory too emphasizes indirect controls stemming from family relationship bonds and bonds to other institutions such as school and work. However, the theory also includes direct controls in the form of parental supervision and discipline. In addition, life-course theory incorporates the structural context of informal social controls, especially social class background, residential mobility, family disruption, and parental deviance, contending that these structural characteristics influence both direct and indirect social controls.[154] The third social control theory—self-control theory—holds that the direct controls of parents in the form of monitoring behavior and discipline are foundational to the development of self-control. While self-control originates socially, it is an internal, individual characteristic.

While social control theories are useful in describing how informal social controls within families generate conformity, they fail to fully consider how characteristics of family life influence these informal social controls. Recent research has studied how a number of family life characteristics influence child well-being and socialization, including the quality of the parent–child relationship, family management, family structure and processes, family social class, and parental criminality. Since we are considering social control theory, the discussion here will focus on how these family life characteristics influence direct and indirect controls.

Quality of the Parent–Child Relationship

Hirschi and Sampson and Laub all found that the strength of the relationship bond between parent and child was closely related to delinquent behavior. Social bond theory and life-course theory explain this purported link by referring to the informal social controls that result from the affectional identification that a youth has with parents such that the youth is sensitive to their opinions, desires to please them, values their relationship, identifies with them, and communicates actively and intimately with them. In other words, conformity is generated by the desire for *social approval*.[155]

Researchers have tried to measure parental attachments and the resulting indirect control in a variety of ways, including questions about "affection and love, interest and concern, support and help, trust, encouragement, lack of rejection, desire for physical closeness, amount of interaction or positive communication, and 'identification.'"[156] Psychological research has tended to focus on attachment in terms of *family affect*, involving the warmth, affection, and acceptance among family members.[157] In their review and analysis of this literature, Rolf Loeber and Magda Stouthamer-Loeber report that delinquent behavior is related to low levels of parental acceptance,[158] a child's sense of belonging to the family,[159] identification with parents,[160] parental caring and trust,[161] positive communication,[162] and parental support.[163]

While the indirect controls derived from parent–child attachment have been found to inhibit delinquent behavior at very similar levels for both boys and girls, the various forms of attachment appear to operate somewhat differently.[164] Research by Steven Cernkovich and Peggy Giordano found that for boys, the most important indirect controls are intimate communication and instrumental communication (sharing problems and future plans with parents); whereas for girls, identity support, conflict, instrumental communication, and parental disapproval of peers held the greatest influence.[165] Based on these findings, Cernkovich and Giordano concluded: "This seems to suggest that while family attachment is important in inhibiting delinquency among all adolescents, the various dimensions of this bond operate somewhat differently among males and females."[166]

In contrast to attachment, parent–child relationships characterized by hostility and conflict have been found to result in limited parental control. Rejection of the child by the parent and rejection of the parent by the child are both related to delinquent behavior.[167] Similarly, self-reported delinquency has been found to be associated with high levels of parent–adolescent conflict.[168] In a study of street kids, drugs, and crime in Miami, James Inciardi and his colleagues report that "greater drug involvement is correlated with more arguments with families about drug use and more crime is correlated with more arguments with their families about crime."[169] They go on to observe: "These associations suggest that families are aware of and struggle against the serious delinquency of teenagers." These findings indicate that not only does family conflict reduce the indirect controls of parents, but also that delinquency and drug use may produce still more relationship conflict.

The relationship between parent–child conflict, child-rearing, and delinquent behavior has been studied extensively by Gerald Patterson and his colleagues at the Oregon Social Learning Center. According to Patterson, some families are characterized by a high degree of irritable and hostile exchanges between family members due to stress experienced by the family unit.[170] Patterson refers to such family interaction as "coercive family processes."[171] Stress may come from a variety of sources, including economic hardship, parental conflict, health difficulties, and child behavior problems (e.g., difficult temperament, early childhood disruptive and antisocial behavior, and delinquency). Patterson argues that frequent negative exchanges, especially those involving emotional manipulation, hostile interaction, neglect, and abuse, are used in an attempt to control other family members. Such manipulative interaction tends to escalate, leading to a cycle of negative interaction among family members. Because such coercive family processes result in limited informal controls, Patterson advocates for family effectiveness training that centers on direct controls—an approach derived from theory and research on family management skills.

Family Management: Monitoring, Supervision, and Discipline

Until recently, the direct controls of parents have been largely dismissed as uninteresting and unimportant.[172] As already noted, Hirschi's social bond theory argues that the indirect controls of relationship bonds are the primary mechanisms of parental social control. Direct controls are viewed as relatively unimportant.[173] Supporting this theoretical point of view, research has found that the degree to which youths are involved in the family and therefore under the direct control of their parents is only weakly related to delinquent behavior.[174]

More recently, however, the notion of direct controls has been reconceptualized to go beyond the single consideration of family involvement to the more comprehensive consideration of "family management," which relates to effective parenting practices.[175] Gerald Patterson and his colleagues at the Oregon Social Learning Center state that effective family management involves at least seven key skills, as identified in Theory into Practice, "Effective Family Management" (See also Links, "Family-Based Prevention Programs").

Research shows that these direct parental controls are related to low levels of delinquent behavior, even when other causal factors are statistically controlled.[176] Moreover, this research reveals that direct controls by parents have at least as great an effect on delinquent behavior as do indirect relational controls. Moreover, greater levels of family involvement have been found to reduce exposure to delinquent peers and opportunities for involvement in delinquent acts.[177]

The relationship between direct control and delinquent behavior, however, is not always simple and direct. This is most evident in the case of discipline.

THEORY INTO PRACTICE — Effective Family Management

Gerald Patterson and his colleagues at the Oregon Social Learning Center take a comprehensive approach to delinquent behavior that includes theory, research, and practice. Their developmental theory of delinquency, called coercive family processes, was described in Chapter 6. Patterson and his colleagues have conducted research to test their theory, and they have applied the theory to a delinquency intervention program that provides parent training in family management. According to this approach, effective parenting includes seven key "family management skills."

- Notice what the child is doing.
- Monitor the child's behavior over long periods.
- Model social skill behavior.
- Clearly state house rules.
- Consistently provide for sane punishments for transgressions.
- Provide reinforcement for conformity.
- Negotiate disagreements so that conflicts and crises do not escalate.

Source: Patterson, "Children Who Steal," 81.

Links — Family-Based Prevention Programs

Gerald Patterson's theory of coercive family processes is the basis for two family-based prevention programs that have been the subject of extensive evaluation research. Both programs have been deemed "model programs" because of their documented success.

The Links for The Incredible Years Training Series and The Strengthening Families Program are provided at:

http://criminaljustice.jbpub.com/burfeind

Research shows that strict and punitive discipline increases the likelihood of delinquent behavior, as does discipline that is lax and erratic; consistent and certain discipline, by contrast, is related to lower levels of delinquent behavior.[178] It is also the case that the direct controls of parents are influenced by characteristics of the child. Individual traits during childhood, such as difficult temperament, tantrums, and early childhood disruptive and antisocial behaviors, have been found to influence parents' direct control efforts, especially efforts to monitor and discipline children.[179] Sampson and Laub, for example, found that parents provided lower levels of supervision of children who were "restless and irritable," showed a predisposition to angry and aggressive responses, and who engage in misbehavior at an early age.[180] Such childhood difficulties also predicted the use of inconsistent and harsh discipline by parents.

The application of direct controls by parents has also been found to be different for boys and girls. Put simply, parents often deal differently with boys than with girls. Research shows that females are subjected to greater control and supervision than are males.[181] Researchers have found, however, that direct controls are more strongly related to male delinquency than to female delinquency.[182]

The research is very clear on one problematic aspect of direct and indirect control: child maltreatment (physical abuse, sexual abuse, and neglect) is a risk factor for delinquency.[183] Research shows consistently that maltreated children and adolescents are more likely to be involved in delinquent behavior, especially serious and violent delinquency, as measured by arrest and self-reports.[184] Furthermore, "there is a tendency for adults to repeat the abuse they experienced as a child. This phenomenon is often labeled 'the cycle of violence.' Although most victims of childhood abuse do not go on to abuse their offspring, they are 10 to 15 times more likely to be abusive parents than persons who were not exposed to abusive parenting."[185]

Family Structure and Family Processes

As we noted near the beginning of this chapter, early social science researchers saw the "broken home" as the single most important factor in understanding juvenile delinquency.[186] In 1932, however, Clifford Shaw and Henry McKay asserted that the family's role in delinquent behavior must be studied by looking at "more subtle aspects of family relationships rather than in the formal break in the family organization."[187] More recent research follows their suggestion and has found consistently that while "children from broken homes are more delinquent than those from intact families . . . the effects of family structure are largely mediated by family processes, such as parental monitoring, supervision, and closeness."[188] Thus, the influence of family structure on delinquency must be understood in terms of its impact on family processes.

The terminology of *family structure* or *family disruption* has been introduced to more accurately capture the diverse living arrangements of today's children. We must not only distinguish between two-parent and single-parent families, but also consider the structure of two-parent families (two-biological-parent, mother-stepfather, father-stepmother), as well as the structure of single-parent families (single-father or single-mother).[189] Studying the influence of family structure on delinquency becomes more complex, but also more accurate and more insightful. In addition, contemporary research adds *family processes* into the family structure–delinquency equation, assessing how family structure influences family interaction, especially in terms of direct and indirect social controls. The proposition advanced in this regard is that family structure influences delinquency largely because of its impact on direct and indirect control within the family. Single-parent families, for example, may provide less supervision, monitoring, and discipline and may involve weaker relationship ties between parent and child.

Along this line of thought, Sampson and Laub's life-course theory argues that family structure influences family processes, which in turn influence the likelihood of delinquent behavior. Their measure of family structure was labeled **family disruption** and referred to families in which one or both parents were absent due to divorce, separation, desertion, or death. Family processes, including discipline, attachments, and supervision, were found to intervene between family disruption and delinquent behavior. The causal sequence runs from family disruption to family processes to delinquent behavior. More specifically, family disruption was related to relationship attachment within the family—what we have been referring to as indirect controls—but not to direct controls of discipline and supervision. Moreover, it was found that family disruption did not have a direct effect on delinquent behavior, but instead, its effect on delinquency was determined by how such disruption influenced family processes.[190]

In a recent study, Stephen Demuth and Susan Brown were able to compare single-mother, single-father, stepfamilies, and two-biological-parent families in terms of delinquency involvement.[191] Mean levels of delinquency were highest among adolescents in single-father families and lowest in two-biological-parent families, with single-mother and stepfamilies in the middle. Parental absence alone was not a significant predictor of delinquency. Rather, family structure was related to relationship ties and informal social control. The researchers concluded that "parental absence undermines direct and indirect control, which in turn accounts for the higher levels of delinquency among adolescents residing in single-mother and single-father families versus two-parent-married families. Parental absence is negatively associated with involvement, supervision, monitoring, and closeness."[192]

Family Size and Birth Order

It has long been noted that delinquents are more likely to come from large families than from small families.[193] In addition, middle children have been found to be more likely to engage in delinquent acts than are first or last born children.[194] These two family structural characteristics are thought to be related to delinquent behavior largely because of how they influence the capacity of parents for effective child-rearing, especially supervision and discipline. The larger the family, the more limited the parent's time and energy for supervision and discipline. Ivan Nye also observed that parents of larger families have less opportunity for interaction and involvement with their children.[195] As a result, the attachment between parent and child is attenuated. The same can be said for middle children: middle children experience less attachment, supervision, and discipline than do older and younger children.[196] Thus, the family structural characteristics of size and birth order affect not only the parent's capacity for supervision and discipline, but also the bonds of attachment. As a result, both direct and indirect social controls are influenced by the family structures of size and ordinal position.

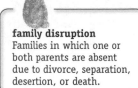

family disruption
Families in which one or both parents are absent due to divorce, separation, desertion, or death.

Family Social Class

Much research and debate have surrounded the relationship between social class and delinquent behavior. This controversy was discussed in Chapter 5. Our concern here, however, is with the link between social class background and different aspects of informal control by parents. Do the informal controls of relationship bonds or the direct controls of monitoring, supervision, and discipline vary by social class? Michael Rutter and Henri Giller propose that "serious economic disadvantage has an adverse effect on the parents, such that parental disorders and difficulties are more likely to develop and good parenting is impeded."[197] Robert Larzelere and Gerald Patterson argue that many lower-class parents have marginal parenting skills because they experience greater stress and have fewer resources than do middle-class parents.[198] Similarly, life-course theory proposes that the effect of social class on delinquent behavior is indirect, in that social class is related to diminished parental discipline and monitoring, and that the lack of these direct controls promotes delinquent behavior. Research supports this causal sequencing.[199]

Parental Criminality

Control theories share the theme that individuals are inherently deviant and must be socialized into conformity.[200] If such socialization does not occur, youths are free to deviate. Thus, control theories are not designed to explain how delinquent attitudes and techniques are learned. Applying this theme to parental criminality, control theories are not concerned with how parents who engage in criminal acts may teach their children attitudes and techniques that encourage and allow for criminal involvement. Here's how Sampson and Laub state the argument: "We argue that parents who commit crimes and drink excessively are likely to use harsh discipline in an inconsistent manner or to be lax in disciplining their children. A central characteristic of deviant and criminal life styles is the rejection of restrictions and duties—especially those that involve planning, patience, and investment in the future. Parenting is perhaps the most demanding of conventional roles, and we expect that deviance in the adult world will manifest itself in disrupted styles of child socialization. Namely, supervision and discipline will be haphazard or nonexistent, and the parent–child/child–parent attachments will be tenuous."[201]

Research has supported this theoretical link between parental criminality and diminished direct and indirect controls of parents. Citing numerous researchers, Scott Henggeler concludes that the association between parental criminality and delinquent behavior is one of the most consistent findings in the literature. He offers the following interpretation: "Although the modeling of aggressive and antisocial parental behavior is probably a component of this process [of socialization], the literature seems clear that adult criminals rarely involve their children directly in criminal activity. It also seems highly likely that antisocial parents possess cognitive and interpersonal deficits that interfere with their capacity for positive parenting. Here, the delinquent behavior of their offspring might be linked more directly with poor parent–adolescent affective relations and the use of ineffective parental control strategies than with the modeling of parental deviance."[202] Sampson and Laub's research also supports this point of view. They offer a concise summary: "Parental deviance of both mother and father strongly disrupts family processes of social control, which in turn increases delinquency."[203]

Parents, Peers, and Delinquent Behavior

The different version of social control theories also address how informal social controls within the family influence association with peers. Social bond theory holds that weak attachments—low indirect control—free a youth to associate with delinquent peers, which in turn leads to greater probability of delinquent behavior. Hirschi reported that boys with low "stake in conformity" (i.e., low indirect control) were more susceptible to delinquent peer influence, while boys with a high stake in conformity were relatively immune to peer influences.[204] Thus, the influence of delinquent peers depends on family relationship bonds and the indirect controls that come from these relationships. Life-course theory similarly assumes that relational controls influence the acquisition of delinquent friends. However, the theory also points to the importance of the direct controls of monitoring and discipline in inhibiting the formation of friendships with delinquent youth. Self-control theory also emphasizes the importance of parental monitoring and discipline; however, instead of considering how these direct controls limit exposure to delinquent peers during late childhood and adolescence, this theory stresses that these direct controls during the early childhood years are the key source of self-control. Without effective parenting, youths fail to develop self-control, and without self-control they are inclined to develop friendships with youths who also lack self-control.

Most research investigating the relationship of parent and peers to delinquent behavior finds that parents and peers each exert direct and independent influence on delinquent behavior.[205] Gary Jensen, for example, found that the likelihood of engaging in delinquent acts decreased as parental support and supervision increased, regardless of whether the youth had been exposed to the delinquent behavior of peers. The same independent effect was found for delinquent friends: youth who were exposed to delinquent behavior of friends were more likely to be delinquent, regardless of the level of parental support and supervision. However, Jensen's analysis also revealed that the effect of parental supervision depended on whether the youth had delinquent attitudes and values.[206] It is reasonable to conclude that the effect of parental informal controls is conditioned by whether the youth has delinquent friends.[207] The counterpart is also true: the effect of delinquent friends on delinquent behavior depends on the informal controls of parents.

Gerald Patterson and his colleagues found that parents who fail to provide direct controls in the form of monitoring and discipline increase the likelihood that their children will associate with deviant peers. As described by life-course theory, early childhood antisocial behavior (e.g., whining, hitting, difficult temperament) reduced the amount and effectiveness of parental monitoring and discipline. Lack of direct controls were linked to poor social skills of the youth and subsequently to increased likelihood of association with deviant peers.[208]

Mark Warr also considered the interrelationships of parents, peers, and delinquent behavior. He proposes that "parents who spend time with their children may reduce the likelihood of delinquent behavior, either by reducing opportunities for delinquent acts or by maximizing their effect as positive role models."[209] Warr's notion of time is very similar to the direct control of involvement. He proposes that adolescents with strong parental attachment may be less prone to develop delinquent friends. Parental attachment, as we have seen repeatedly, is the primary dimension of indirect control. Warr labels these controls "access barriers" because they reduce exposure to delinquent peers. He also argues that investment in the relationship bond with parents makes the youth more likely to internalize the parents' moral inhibitions to delinquent behavior—an indirect control. He states, "Among adolescents with strong bonds to their parents, the potential loss of parental approval or affection may be sufficient to deter delinquency even when pressure from peers is intense."[210]

Warr's research reveals that while attachment to parents is not directly related to delinquent behavior, it has an indirect effect, inhibiting the formation of delinquent peers. These findings suggest that access barriers reduce exposure to and affiliation with delinquent peers, thereby reducing delinquent behavior. Once a youth has delinquent friends, however, the indirect controls of parental attachment do little to reduce delinquent behavior.

■ Summary and Conclusions

The essence of informal social control is the extent to which the child is linked to the family and ultimately to society through bonds of attachment and through socially integrative forms of direct control, such as clearly established rules, monitoring, and punishment.[211] Control theories are largely concerned with these mechanisms of informal social control, especially the central role that the family plays in generating and implementing such controls.

Early studies of the family and delinquent behavior focused almost exclusively on the breakdown of social controls in families that were "broken" through death, desertion, or divorce. As Thomas Monahan observed: "Early writers saw the broken home to be an important if not the greatest single proximate factor in understanding delinquency."[212] Beginning in the 1930s, this focus on the structure of the family was replaced with an emphasis on the informal social controls of the family. In recent years, this emphasis has dominated most criminological theory and research with regard to the family's role in delinquent behavior.

This chapter described three different versions of control theory. In his *social bond theory,* Travis Hirschi identified four elements of an individual's social bond to society. Relationship *attachments* within the family, school, and peer group provide indirect controls, based on affectional identification and the desire for social approval. *Commitment,* the rational component of the social bond, has to do with an individual's investment in conventional activities or "lines of action." Individuals are controlled to the degree that they desire to participate in conventional activities. *Involvement* too relates to activities, but focuses on what youths actually do with their time. Some activities limit opportunities for involvement in delinquent acts, whereas other activities provide few restraints for delinquent behavior. The final element of the social bond, *belief,* promotes conformity when the youth adopts beliefs that forbid delinquent behavior and when those beliefs are collectively held and reinforced.

Life-course theory also emphasizes the importance of social bonds as the primary vehicle of informal social control. Rather than elaborating on the elements of the social bond as Hirschi did, Sampson and Laub's life-course theory examines social bonds over the course of life, considering their origins and how changes in these bonds influence informal social control and subsequently behavior. Family processes of supervision, attachment, and discipline are key to the development of social bonds in childhood; these family processes are influenced by the family's structural background and by early childhood difficulties (whining, difficult temperament, disruptive and antisocial behavior). Sampson and Laub also considered social bonds in adult life, which they refer to as social capital—a term that more explicitly links relationship bonds to institutional roles within the family, school, work, and community. The informal social controls of social capital involve not only affective attachment, but also social obligations and restraints that are attached to particular roles. Significant changes in life, or turning points (such as marriage, military enlistment, or work), affect social capital and thereby influence informal social control.

Self-control theory veers from this emphasis on the informal controls of relationship bonds to focus on controls that are within the individual. As a result of effective child-rearing practices of monitoring and discipline, some children develop skills to respond to situations that require delayed gratification, planning, sensitivity to others, independence, cognitive and verbal skills, and a willingness to accept restraints on their activities. In contrast, individuals with low self-control are "impulsive, insensitive, physical (as opposed to mental), risk taking, short-sighted, and nonverbal."[213] They are unable to resist temptations not only of crime, but also of analogous acts such as reckless driving, smoking, and alcohol and drug use. As a result, low self-control fosters problematic interpersonal relations, social activities, and involvement in social institutions. Individuals with low self-control have difficulty making and keeping friends, and when they do develop friendships, they tend to end up in the company of others who similarly lack self-control and are involved in deviance. In addition, they are less able to succeed in school and the workplace, and they tend to enter into marriages that are destined to fail.

These control theories detail family processes of indirect and direct social control. At the same time, researchers have assessed the conceptual accuracy and causal arguments of control theories. Social bond theory and life-course theory stress the importance of indirect relational controls, and research supports the importance of attachment to delinquent behavior. Life-course theory, together with self-control theory, introduces direct controls such as rule setting, monitoring, and discipline into the causal equation. Rather than revealing indirect or direct controls as more important, research indicates that both are significant restraints on behavior. Additionally, it is apparent that controls operate at both the social and individual level and that family structural characteristics influence informal social controls of the family. Thus theory and research have taken us back to Ivan Nye's depiction of social control as involving *direct controls, indirect controls,* and *internalized controls.*[214] His study demonstrated that the structure of the family, together with the interactions and relationships among family members, plays a vital role in the development and implementation of these controls.

THEORIES

- social bond theory
- life-course theory
- self-control theory or a general theory of crime

CRITICAL THINKING QUESTIONS

1. In the opening case example, Clifford Shaw described the Martin family as "disorganized." In what ways was their family life disorganized, and how did this influence the parents' efforts to "train, guide, and control their children?"

2. Distinguish the various forms of social control—formal, informal, direct, indirect, and internalized—and provide an example of each.

3. Describe how family relations are related to delinquent behavior according to Sheldon and Eleanor Glueck.

4. In your own words, describe each element of the social bond: attachment, commitment, involvement, and belief. How are these elements of the social bond interconnected?

5. From the life-course perspective, why is it important not to concentrate simply on the adolescent years, as sociological criminology has traditionally done?

6. What do Gottfredson and Hirschi mean when they say that "many of the traditional causes of crime are in fact consequences of low self-control"? (*General Theory of Crime*, 119)

7. To what does *family structure* refer? Offer several examples of how family structure is related to delinquent behavior.

8. What does Mark Warr mean when he says that both family and friends exert considerable influence on adolescent behavior, but that their influence "does not operate independently of one another"? ("Parents, Peers, and Delinquency," 258)

SUGGESTED READINGS

Gottfredson, Michael, and Travis Hirschi. *A General Theory of Crime.* Stanford, CA: Stanford University Press, 1990.
Hirschi, Travis. *Causes of Delinquency.* Berkeley, CA: University of California Press, 1969.
Sampson, Robert J., and John H. Laub. *Crime in the Making: Pathways and Turning Points Through Life.* Cambridge, MA: Harvard University Press, 1993.

GLOSSARY

attachment: The affectional identification one has with others—sensitivity to others' opinions, intimate communication, mutual respect, identification, and valuing of relationship ties.

belief: The attitudes, values, and norms one has that forbid delinquency.

broken home: Families with only one parent as a result of separation, divorce, or death. The term "broken home" was used in early research on family structure and delinquency. The contemporary term is "family disruption."

commitment: The investments one has that provide reason to conform—a stake in conformity.

direct controls: Restrictions and punishments imposed by others that limit behavior, and rewards that encourage and reinforce behavior.

family disruption: Families in which one or both parents are absent due to divorce, separation, desertion, or death.

formal social controls: Mechanisms used within the context of institutions such as schools, churches, and justice systems that promote and encourage conformity and sanction deviance.

indirect controls: Pressures and reasons to conform that are based on affectional identification with others, especially parents, such that people conform in order to maintain relationship bonds and to avoid disappointing others.

informal social controls: Characteristics of social relations that bring about conformity, including parental supervision; sensitivity to others (their feelings, wishes, and expectations); identification with others; emotional attachment; and informal sanctions, including withdrawal of affection, disregard, and ridicule.

internalized controls: A person's adoption of standards of behavior that are exercised from within the individual through conscience or sense of guilt.

involvement: Conventional activities that not only consume time and energy, but also inhibit involvement in delinquency.

life course: The age-graded sequence of culturally defined roles and social transitions.

low self-control: Individuals with low self-control "tend to be impulsive, insensitive, physical (as opposed to mental), risk-taking, short-sighted, and nonverbal" (Gottfredson and Hirschi, *General Theory of Crime,* 90). Together these characteristics constitute a single latent trait.

social capital: Expectations, obligations, and restraints that are part of adult relationship bonds and that result in informal social control.

trajectory or pathway: The pattern or direction of life over a number of years, relating to areas of life such as education, work, parenting, and conformity or deviance.

transitions: Significant life events that occur within the life trajectory or pathway, such as marriage, enlistment in the military, first job, or change in employment.

turning point: A transition that results in significant change in the life-course trajectory.

REFERENCES

Agnew, Robert. "A Longitudinal Test of Social Control Theory and Delinquency." *Journal of Research in Crime and Delinquency* 28 (1991):126–156.

———. "Social Control Theory and Delinquency: A Longitudinal Test." *Criminology* 23 (1985):47–62.

Agnew, Robert, and D. M. Peterson. "Leisure and Delinquency." *Social Forces* 36 (1989):332–350.

Akers, Ronald L., and John K. Cochran. "Adolescent Marijuana Use: A Test of Three Theories of Deviant Behavior." *Deviant Behavior* 6 (1985):323–346.

Akers, Ronald L., and Christine S. Sellers. *Criminological Theories: Introduction and Evaluation.* 3rd ed. Los Angeles: Roxbury, 2004.

Arneklev, Bruce J., Harold G. Grasmick, Charles R. Tittle, and Robert J. Bursik, Jr. "Low Self-Control and Imprudent Behavior." *Journal of Quantitative Criminology* 9 (1993):225–247.

Aseltine, Robert H., Jr. "A Reconsideration of Parental and Peer Influences on Adolescent Deviance." *Journal of Health and Social Behavior* 36 (1995):103–121.

Baker, Myriam L., Jane Nady Sigmon, and M. Elaine Nugent. *Truancy Reduction: Keeping Students in School.* Washington, DC: Office of Juvenile Justice and Delinquency Prevention, 2001.

Broom, Leonard, and Philip Selznick. *Introduction to Sociology.* New York: Harper and Row, 1968.

Brownfield, David, and Ann Marie Sorenson. "Latent Structure Analysis of Delinquency." *Journal of Quantitative Criminology* 3 (1987):103–124.

Burgess, Ernest W., Harvey J. Locke, and Mary Margaret Thomas. *The Family: From Institution to Companionship.* 3rd ed. New York: American Book, 1963.

Burton, Velmer S., Jr., Francis T. Cullen, T. David Evans, Leanne Fiftal Alarid, and R. Gregory Dunaway. "Gender, Self-Control, and Crime." *Journal of Research in Crime and Delinquency* 35 (1998):123–147.

Burton, Velmer, Francis Cullen, T. David Evans, R. Gregory Dunaway, Sesha Kethineni, and Gary Payne. "The Impact of Parental Controls on Delinquency." *Journal of Criminal Justice* 23 (1995): 111–126.

Canter, Rachelle J. "Family Correlates of Male and Female Delinquency." *Criminology* 20 (1995): 149–167.

Caspi, Avshalom, Glen H. Elder, Jr., and Ellen S. Herbener. "Childhood Personality and the Prediction of Life-Course Patterns." In *Straight and Devious Pathways from Childhood to Adulthood,* edited by Lee Robins and Michael Rutter, 13–35. New York: Cambridge University Press, 1990.

Cernkovich, Steven A., and Peggy G. Giordano. "Family Relationships and Delinquency." *Criminology* 25 (1987):295–321.

Cochran, John K., Peter B. Wood, Christine S. Sellers, Wendy Wilkerson, and Mitchell B. Chamlin. "Academic Dishonesty and Low Self-Control: An Empirical Test of a General Theory of Crime." *Deviant Behavior* 19 (1998):227–255.

Coleman, James S. *Foundations of Social Theory.* Cambridge, MA: Harvard University Press, 1990.

———. "Social Capital in the Creation of Human Capital." *American Journal of Sociology* 94 (1988):95–120.

Conger, Rand. "Social Control and Social Learning Models of Delinquency: A Synthesis." *Criminology* 14 (1976):17–40.

Conger, Rand D., Xiaojia Ge, Glen H. Elder, Jr., Frederick O. Lorenz, and Ronald L. Simons. "Economic Stress, Coercive Family Process, and Developmental Problems of Adolescents." *Child Development* 65 (1994):541–561.

Crosnoe, Robert. "High School Curriculum Track and Adolescent Association With Delinquent Friends." *Journal of Adolescent Research* 17 (2002):143–167.

Cullen, Francis T., and Robert Agnew, eds. *Criminological Theory: Past to Present.* Los Angeles: Roxbury, 2003.

Demuth, Stephen, and Susan L. Brown. "Family Structure, Family Processes, and Adolescent Delinquency: The Significance of Parental Absence Versus Parental Gender." *Journal of Research in Crime and Delinquency* 41 (2004):52–81.

Elliott, Delbert S., David Huizinga, and Suzanne S. Ageton. *Explaining Delinquency and Drug Use.* Newbury Park, CA: Sage, 1985.

Elliott, Delbert S., David Huizinga, and Scott Menard. *Multiple Problem Youth: Delinquency, Substance Use, and Mental Health Problems.* New York: Cambridge University Press, 1989.

Evans, T. David, Francis T. Cohen, Velmer S. Burton, Jr., R. Gregory Dunaway, and Michael Benson. "The Social Consequences of Self-Control: Testing the General Theory of Crime." *Criminology* 35 (1997):475–504.

Farnworth, Margaret, Lawrence J. Schweinhart, and John R. Berrueta-Clement. "Preschool Intervention, School Success and Delinquency in a High-Risk Sample of Youth." *American Educational Research Journal* 22 (1985):445–464.

Farrington, David P. "Early Predictors of Adolescent Aggression and Adult Violence." *Violence and Victims* 4 (1989):79–100.

Feldman, S. Shirley, and Daniel A. Weinberger. "Self-Restraint as a Mediator of Family Influences on Boys' Delinquent Behavior: A Longitudinal Study." *Child Development* 65 (1994):195–211.

Gamoran, Adam, and M. Berends. "The Effects of Stratification in Secondary School: Synthesis of Survey and Ethnographic Research." *Review of Research in Education* 57 (1987):415–435.

Gibbons, Don C. *The Criminological Enterprise: Theories and Perspectives.* Englewood Cliffs, NJ: Prentice Hall, 1979.

Gibbons, Don C., and Marvin D. Krohn. *Delinquent Behavior.* 5th ed. Englewood Cliffs, NJ: Prentice Hall, 1979.

Gibbs, John J., and Dennis Giever. "Self-Control and Its Manifestations Among University Students: An Empirical Test of Gottfredson and Hirschi's General Theory." *Justice Quarterly* 12 (1995): 231–235.

Gibbs, John J., Dennis Giever, and Jamie S. Martin. "Parental Management and Self-Control: An Empirical Test of Gottfredson and Hirschi's General Theory." *Journal of Research in Crime and Delinquency* 35 (1998):40–70.

Glueck, Sheldon, and Eleanor Glueck. *Delinquents and Nondelinquents in Perspective.* Cambridge, MA: Harvard University Press, 1968.

———. *Unraveling Delinquency.* Cambridge, MA: Harvard University Press, 1950.

Gottfredson, Michael, and Travis Hirschi. *A General Theory of Crime.* Stanford, CA: Stanford University Press, 1990.

Gove, Walter R. "The Effect of Age and Gender on Deviant Behavior: A Biopsychosocial Perspective." In *Gender and the Life Course,* edited by Alice S. Rossi, 115–144. New York: Aldine, 1985.

Gove, Walter R., and Robert D. Crutchfield. "The Family and Juvenile Delinquency." *The Sociological Quarterly* 23 (1987):301–319.

Grasmick, Harold G., Charles R. Tittle, Robert J. Bursik, Jr., and Bruce J. Arneklev. "Testing the Core Empirical Implications of Gottfredson and Hirschi's General Theory of Crime." *Journal of Research in Crime and Delinquency* 30 (1993):5–29.

Hagan, John. *Structural Criminology.* New Brunswick, NJ: Rutgers University Press, 1989.

Hawkins, J. David, Todd I. Herrenkohl, David P. Farrington, Devon Brewer, Richard F. Catalano, and Tracy W. Harachi. "A Review of Predictors of Youth Violence." In *Serious and Violent Juvenile Offenders: Risk Factors and Successful Interventions,* edited by Rolf Loeber and David P. Farrington, 106–146. Thousand Oaks, CA: Sage, 1998.

Hawkins, J. David, Todd I. Herrenkohl, David P. Farrington, Devon Brewer, Richard F. Catalano, Tracy W. Harachi, and Lynn Cothern. "Predictors of Youth Violence." Washington, DC: Office of Juvenile Justice and Delinquency Prevention, 2000.

Hay, Carter. "Parenting, Self-Control, and Delinquency: A Test of Self-Control Theory." *Criminology* 39 (2001):707–736.

Henggeler, Scott W. *Delinquency in Adolescence.* Newbury Park, CA: Sage, 1989.

Hindelang, Michael. "Causes of Delinquency: A Partial Replication and Extension." *Social Problems* 21 (1973):471–487.

Hirschi, Travis. *Causes of Delinquency.* Berkeley, CA: University of California Press, 1969.

———. "Crime and the Family." In *Crime and Public Policy,* edited by James Q. Wilson, 53–68. San Francisco, CA: Institute for Contemporary Studies, 1983.

Hirschi, Travis, and Michael Gottfredson. "Commentary: Testing the General Theory of Crime." *Journal of Research in Crime and Delinquency* 30 (1993):47–54.

Hoffman, John P., and Jiangmin Xu. "School Activities, Community Service, and Delinquency." *Crime and Delinquency* 48 (2002):568–591.

Inciardi, James A., Ruth Horowitz, and Anne E. Pottieger. *Street Kids, Street Drugs, and Street Crime: An Examination of Drug Use and Serious Delinquency in Miami.* Belmont, CA: Wadsworth, 1993.

Ireland, T., Carolyn A. Smith, and Terence P. Thornberry. "Developmental Issues in the Impact of Child Maltreatment on Later Delinquency and Drug Use." *Criminology* 40 (2002):359–399.

Jang, Sung Joon. "Age-Varying Effects of Family, School, and Peers on Delinquency: A Multilevel Modeling Test of Interactional Theory." *Criminology* 37 (1999):643–685.

Janowitz, Morris. "Sociological Theory and Social Control." *American Journal of Sociology* 81 (1975):82–108.

Jarjoura, G. Roger. "Does Dropping Out of School Enhance Delinquent Involvement? Results from a Large-Scale National Probability Sample." *Criminology* 31 (1993):149–172.

Jenkins, Patricia H. "School Delinquency and School Commitment." *Sociology of Education* 68 (1995):221–239.

———. "School Delinquency and the School Social Bond." *Journal of Research in Crime and Delinquency* 34 (1997):337–368.

Jensen, Gary F. "Parents, Peers, and Delinquent Action: A Test of the Differential Association Perspective." *American Sociological Review* 78 (1972):562–575.

Johnson, Monica Kirkpatrick, Robert Crosnoe, and Glen H. Elder, Jr. "Students' Attachment and Academic Engagement: The Role of Race and Ethnicity." *Sociology of Education* 74 (2001):318–340.

Johnson, Richard E. "Family Structure and Delinquency: General Patterns and Gender Differences." *Criminology* 24 (1986):65–84.

———. *Juvenile Delinquency and Its Origins: An Integrated Theoretical Approach.* Cambridge, MA: Cambridge University Press, 1979.

Johnstone, W. C. John. "Juvenile Delinquency and the Family: A Contextual Interpretation." *Youth and Society* 9 (1978):299–313.

Keane, Carl, Paul S. Maxim, and James J. Teevan. "Drinking and Driving, Self-Control, and Gender: Testing A General Theory of Crime." *Journal of Research in Crime and Delinquency* 30 (1993): 30–46.

Kendal, Denise B. "The Parental and Peer Contexts of Adolescent Deviance: An Algebra of Interpersonal Influences." *Journal of Drug Issues* 26 (1996):289–315.

Krohn, Marvin D. "The Web of Conformity: A Network Approach to the Explanation of Delinquent Behavior." *Social Problems* 33 (1986):581–593.

Krohn, Marvin D., and James L. Massey. "Social Control and Delinquent Behavior: An Examination of the Elements of the Social Bond." *Sociological Quarterly* 21 (1980):529–543.

Krohn, Marvin D., Susan Stern, Terence Thornberry, and Sung Joon Jang. "The Measurement of Family Process Variables: The Effect of Adolescent and Parent Perceptions of Family Life on Delinquent Behavior." *Journal of Quantitative Criminology* 3 (1992):287–315.

LaGrange, Randy, and Helen Raskin White. "Age Differences in Delinquency: A Test of Theory." *Criminology* 23 (1985):19–45.

LaGrange, Teresa C., and Robert A. Silverman. "Low Self-Control and Opportunity: Testing the General Theory of Crime as an Explanation for Gender Differences in Delinquency." *Criminology* 37 (1999):41–72.

Lamborn, Susie D., Nina S. Mounts, Laurence Steinberg, and Sanford M. Dornbusch. "Patterns of Competence and Adjustment Among Adolescents from Authoritative, Authoritarian, Indulgent, and Neglected Homes." *Child Development* 62 (1991):1049–1065.

Larzelere, Robert E., and Gerald R. Patterson. "Parental Management: Mediator of the Effect of Socioeconomic Status on Early Delinquency." *Criminology* 28 (1990):301–323.

Laub, John H., and Robert J. Sampson. "The Sutherland–Glueck Debate: On the Sociology of Criminological Knowledge." *American Journal of Sociology* 96 (1991):1402–1440.

———. "Turning Points in the Life Course: Why Change Matters to the Study of Crime." *Criminology* 31 (1993):301–325.

———. "Unraveling Families and Delinquency: A Reanalysis of the Gluecks' Data." *Criminology* 26 (1988):355–380.

Lazarsfeld, P., and N. W. Henry. *Latent Structural Analysis.* New York: Houghton-Mifflin, 1965.

Leonard, Kimberly Kempf, and Scott Decker. "The Theory of Social Control: Does It Apply to the Very Young?" *Journal of Criminal Justice* 22 (1994):89–105.

Linden, Eric, and James Hackler. "Affective Ties and Delinquency." *Pacific Sociological Review* 16 (1973):27–46.

Liska, Allen E., and Mark D. Reed. "Ties to Conventional Institutions and Delinquency: Estimating Reciprocal Effects." *American Sociological Review* 50 (1985):547–560.

Loeber, Rolf. "The Stability of Antisocial Child Behavior: A Review." *Child Development* 53 (1982):1431–1446.

Loeber, Rolf, and Thomas J. Dishion. "Early Predictors of Male Delinquency: A Review." *Psychological Bulletin* 94 (1983):68–99.

Loeber, Rolf, and Marc LeBlanc. "Toward a Developmental Criminology." In *Crime and Justice: An Annual Review of Research,* vol. 12, edited by Michael Tonry and Norval Morris, 375–437. Chicago: University of Chicago Press, 1990.

Loeber, Rolf, and Magda Stouthamer-Loeber. "Family Factors as Correlates and Predictors of Juvenile Conduct Problems and Delinquency." In *Crime and Justice: An Annual Review of Research,* vol. 7, edited by Michael Tonry and Norval Morris, 29–149. Chicago: University of Chicago Press, 1986.

Lytton, Hugh. "Child and Parent Effects in Boys' Conduct Disorder: A Reinterpretation." *Developmental Psychology* 26 (1990):683–697.

Maguin, Eugene, and Rolf Loeber. "Academic Performance and Delinquency." In *Crime and Justice: A Review of Research,* vol. 20, edited by Michael Tonry, 145–264. Chicago: University of Chicago Press, 1996.

Massey, James L., and Marvin D. Krohn. "A Longitudinal Examination of An Integrated Social Process Model of Deviant Behavior." *Social Forces* 65 (1986):106–134.

Matsueda, Ross L. "The Dynamics of Moral Beliefs and Minor Delinquency." *Social Forces* 68 (1989):428–457.

———. "Testing Control Theory and Differential Association: A Causal Modeling Approach." *American Sociological Review* 47 (1982):489–504.

Chapter Resources

McCord, Joan. "Patterns of Deviance." In *Human Functioning in Longitudinal Perspective*, edited by S. B. Sells, Rick Crandall, Merrill Roff, John S. Strauss, and William Pollin, 157–165. Baltimore: Williams and Wilkins, 1980.

McCord, William, and Joan McCord. *Origins of Crime: A New Evaluation of the Cambridge–Sommerville Youth Study*. New York: Columbia University Press, 1959.

Miller, Brent J., Kelly McCoy, Terrence Olson, and Christopher Wallace. "Parental Discipline and Control in Relation to Adolescent Sexual Attitudes and Behavior." *Journal of Marriage and the Family* 48 (1986):503–512.

Monahan, Thomas P. "Family Status and the Delinquent Child: A Reappraisal and Some New Findings." *Social Forces* 35 (1957):250–258.

Nagin, Daniel S., and Raymond Paternoster. "Enduring Individual Differences and Rational Choice Theories of Crime." *Law and Society Review* 3 (1993):467–496.

Nye, Ivan F. *Family Relationships and Delinquency Behavior*. New York: Wiley, 1958.

Ogburn, William F. "The Changing Family." *Family* 19 (1938):139–143.

———. "The Family and Its Functions." In *Recent Social Trends in the United States*, Report of the President's Research Committee on Social Trends, 661–708. New York: McGraw-Hill, 1933.

Parsons, Talcott, and Robert F. Bales. *Family Socialization and Interaction Process*. New York: Free Press, 1955.

Patterson, Gerald R. "Children Who Steal." In *Understanding Crime*, edited by Travis Hirschi and Michael Gottfredson, 73–90. Beverly Hills, CA: Sage, 1980.

———. *Coercive Family Process*. Eugene, OR: Castalia, 1982.

Patterson, Gerald R., Barbara D. DeBaryshe, and Elizabeth Ramsey. "A Developmental Perspective on Antisocial Behavior." *American Psychologist* 44 (1989):329–335.

Patterson, Gerald R., and Thomas J. Dishion. "Contributions of Families and Peers to Delinquency." *Criminology* 23 (1985):63–79.

Patterson, Gerald R., J. G. Reid, and Thomas J. Dishion. *Antisocial Boys*. Eugene, OR: Castalia, 1992.

Patterson, Gerald R., and Magda Stouthamer-Loeber. "The Correlation of Family Management Practices and Delinquency." *Child Development* 55 (1984):1299–1307.

Piquero, Alex, and Stephen Tibbetts. "Specifying the Direct and Indirect Effect of Low Self-Control and Situational Factors in Decision-Making: Toward a More Complete Model of Rational Offending." *Justice Quarterly* 13 (1996):481–510.

Polakowski, Michael. "Linking Self- and Social Control with Deviance: Illuminating the Structure Underlying a General Theory of Crime and Its Relation to Deviant Activity." *Journal of Quantitative Criminology* 10 (1994):41–78.

Polk, Ken, and Walter E. Schafer. *Schools and Delinquency*. Englewood Cliffs, NJ: Prentice Hall, 1972.

Poole, Eric D., and Robert M. Regoli. "Parental Support, Delinquent Friends, and Delinquency: A Test of Interaction Effects." *Journal of Criminal Law and Criminology* 70 (1979):188–193.

Pratt, Travis C., and Francis T. Cullen. "The Empirical Status of Gottfredson and Hirschi's General Theory of Crime: A Meta-Analysis." *Criminology* 39 (2000):931–964.

Pratt, Travis C., Michael G. Turner, and Alex R. Piquero. "Parental Socialization and Community Context: A Longitudinal Analysis of the Structural Sources of Low Self-Control." *Journal of Research in Crime and Delinquency* 41 (2004):219–243.

Rankin, Joseph H., and Roger Kern. "Parental Attachments and Delinquency." *Criminology* 32 (1994):495–515.

Rankin, Joseph H., and L. Edward Wells. "The Effect of Parental Attachments and Direct Controls on Delinquency." *Journal of Research in Crime and Delinquency* 27 (1990):140–165.

Reckless, Walter. *The Crime Problem*. New York: Appleton-Century-Crofts, 1967.

———. "A New Theory of Delinquency and Crime." *Federal Probation* 25 (1961):42–46.

Reiss, Albert, Jr. "Delinquency as the Failure of Personal and Social Controls." *American Sociological Review* 16 (1951):196–207.

Rollins, Boyd, and Darwin Thomas. "Parental Support, Power, and Control Techniques in the Socialization of Children." In *Contemporary Theories about the Family*, edited by W. Burr, R. Hill, F. I. Nye, and I. Reiss, 317–364. New York: Free Press, 1979.

Rosen, Lawrence, Leonard Savitz, Michael Lalli, and Stanley Turner. "Early Delinquency, High School Graduation, and Adult Criminality." *Sociological Viewpoints* 7 (1991):37–60.

Rosenbaum, Jill L. "Social Implications of Educational Grouping." *Review of Research in Education* 8 (1980):361–401.

Rosenbaum, Jill L., and James R. Lasley. "School, Community Context, and Delinquency: Rethinking the Gender Gap." *Justice Quarterly* 7 (1990):493–513.

Rutter, Michael. *Changing Youth in a Changing Society: Patterns of Adolescent Development and Disorder.* Cambridge, MA: Harvard University Press, 1980.

Rutter, Michael, and Henri Giller. *Juvenile Delinquency: Trends and Perspectives.* New York: Guilford, 1983.

Sampson, Robert J., and John H. Laub. *Crime in the Making: Pathways and Turning Points Through Life.* Cambridge, MA: Harvard University Press, 1993.

———. "Urban Poverty and the Family Context of Delinquency: A New Look at Structure and Process in a Classic Study." *Child Development* 65 (1994):523–540.

Shaw, Clifford R., and Henry D. McKay. "Are Broken Homes a Causal Factor in Juvenile Delinquency?" *Social Forces* 10 (1932):514–524.

Simons, Ronald L. Christine Johnson, Rand D. Conger, and Glen Elder, Jr. "A Test of Latent Trait versus Life-Course Perspectives on the Stability of Adolescent Antisocial Behavior." *Criminology* 36 (1998):217–243.

Simons, Ronald L., Leslie Gordon Simons, and Lora Ebert Wallace. *Families, Delinquency, and Crime: Linking Society's Most Basic Institution to Antisocial Behavior.* Los Angeles: Roxbury, 2004.

Simons, Ronald L., Chyi-In Wu, Rand D. Conger, and Frederick O. Lorenz. "Two Routes to Delinquency: Differences Between Early and Late Starters in the Impact of Parenting and Deviant Peers." *Criminology* 32 (1994):247–276.

Smith, Carolyn A., and Thornberry, T. P. "The Relationship Between Childhood Maltreatment and Adolescent Involvement in Delinquency." *Criminology* 33 (1995):451–481.

Snyder, J., and Gerald R. Patterson. "Family Interaction and Delinquent Behavior." In *Handbook of Juvenile Delinquency,* edited by H. C. Quay, 216–243. New York: Wiley, 1987.

Stanfield, Robert Everett. "The Interaction of Family Variables and Gang Variables in the Aetiology of Delinquency." *Social Problems* 13 (1966):411–117.

Stewart, Eric A. "School Social Bonds, School Climate, and School Misbehavior: A Multilevel Analysis." *Justice Quarterly* 20 (2003):575–604.

Stitt, B. Grant, and David J. Giacopassi. "Trends in the Connectivity of Theory and Research in Criminology." *Criminologist* 17 (1992):1, 3–6.

Thornberry, Terence P. "Toward an Interactional Theory of Delinquency." *Criminology* 25 (1987): 863–891.

Thornberry, Terence, David Huizinga, and Rolf Loeber. "The Causes and Correlates Studies: Findings and Policy Implication." *Juvenile Justice Journal* 9 (2004):3–19.

Thornberry, T. P., T. O. Ireland, and Carolyn A. Smith. "The Importance of Timing: The Varying Impact of Childhood and Adolescent Maltreatment on Multiple Problem Outcomes." *Development and Psychopathology* 13 (2001):957–979.

Thornberry, Terence P., Carolyn A. Smith, Craig Rivera, David Huizinga, and Magda Stouthamer-Loeber. "Family Disruption and Delinquency." *Juvenile Justice Bulletin.* Washington, DC: Office of Juvenile Justice and Delinquency Prevention, 1999.

Tittle, Charles R. *Sanctions and Social Deviance.* New York: Praeger, 1980.

Tittle, Charles R., and Robert F. Meier. "Specifying the SES/Delinquency Relationship." *Criminology* 28 (1990):271–300.

Toby, Jackson. "Social Disorganization and Stake in Conformity: Complimentary Factors in the Predatory Behavior of Hoodlums." *Journal of Criminal Law, Criminology, and Police Science* 48 (1957):12–17.

Tremblay, Richard E., Bernard Boulerice, Louise Arseneault, and Marianne Junger Niscale. "Does Low Self-Control During Childhood Explain the Association Between Delinquency and Accidents in Early Adolescence?" *Criminal Behavior and Mental Health* 5 (1995):439–451.

Van Voorhis, Patricia, Francis T. Cullen, Richard A. Mathers, and Connie Chenoweth Garner. "The Impact of Family Structure and Quality on Delinquency: A Comparative Assessment of Structural and Functional Factors." *Criminology* 26 (1998):235–261.

Vazsonyi, Alexander T., Lloyd E. Pickering, Marianne Junger, and Dick Hessing. "An Empirical Test of a General Theory of Crime: A Four-Nation Comparative Study of Self-Control and the Prediction of Deviance." *Journal of Research in Crime and Delinquency* 38 (2001):91–131.

Warr, Mark. "Life-Course Transition and Desistance from Crime." *Criminology* 36 (1998):183–215.

———. "Parents, Peers, and Delinquency." *Social Forces* 72 (1993):247–264.

Wasserman, Gail A., Kate Keenan, Richard E. Tremblay, John D. Cole, Todd I., Herrenkohl, Rolf Loeber, and David Petechuk. "Risk and Protective Factors of Child Delinquency." Washington, DC: Office of Juvenile Justice and Delinquency Prevention, 2003.

Wasserman, Gail A., L. Miller, E. Pinner, and B. S. Jaramillo. "Parenting Predictors of Early Conduct Problems in Urban, High-Risk Boys." *Journal of the American Academy of Child and Adolescent Psychiatry* 35 (1996): 1227–1236.

Wells, Edward L., and Joseph H. Rankin. "Direct Parental Controls and Delinquency." *Criminology* 26 (1988):263–285.

————. "Families and Delinquency: A Meta-Analysis of the Impact of Broken Homes." *Social Problems* 38 (1991):71–93.

White, Jennifer L., Terrie E. Moffitt, Felton Earls, Lee N. Robbins, and Phil A. Silva. "How Early Can We Tell? Predictors of Childhood Conduct Disorder and Adolescent Delinquency." *Criminology* 28 (1990):507–533.

Wiatrowski, Michael, and Kristine L. Anderson. "The Dimensionality of the Social Bond." *Journal of Quantitative Criminology* 3 (1987):65–81.

Wiatrowski, Michael D., David B. Griswold, and Mary K. Roberts. "Social Control Theory and Delinquency." *American Sociological Review* 46 (1981):525–541.

Widom, Cathy S. "The Cycle of Violence." *Science* 244 (1989):160–166.

Wilkinson, Karen. "The Broken Home and Juvenile Delinquency: Scientific Explanations or Ideology?" *Social Problems* 21 (1974):726–739.

Wilson, James Q., and Richard J. Herrnstein. *Crime and Human Nature: The Definitive Study of the Causes of Crime.* New York: Simon and Schuster, 1985.

Winfree, L. Thomas, and Frances P. Bernat. "Social Learning, Self-Control, and Substance Abuse by Eighth Grade Students: A Tale of Two Cities." *Journal of Drug Issues* 28 (1998):539–558.

Wright, John Paul, and Francis T. Cullen. "Parental Efficacy and Delinquency Behavior: Do Control and Support Matter?" *Criminology* 39 (2001):677–705.

Wood, Peter B., Betty Pfefferbaum, and Bruce J. Arneklev. "Risk-Taking and Self-Control: Social Psychological Correlates of Delinquency." *Journal of Crime and Justice* 16 (1993):111–130.

Zingraff, M. T., Leiter, J., Myers, K. A., and Johnsen, M. C. "Child Maltreatment and Youthful Problem Behavior." *Criminology* 31 (1993):173–202.

ENDNOTES

1. Broom and Selznick, *Introduction to Sociology,* 15.
2. Johnson, *Juvenile Delinquency,* 5.
3. Janowitz, "Sociological Theory."
4. See, for example, Tittle, *Sanctions and Social Deviance.*
5. Johnstone, "Juvenile Delinquency," 299.
6. Monahan, "Family Status," 250.
7. Wilkinson, "Broken Home," 734.
8. Shaw and McKay, "Broken Homes," 524.
9. Gibbons, *Criminological Enterprise.*
10. Wilkinson, "Broken Home." See also Cernkovich and Giordano, "Family Relationships and Delinquency."
11. Burgess, Locke, and Thomas, *The Family;* Ogburn, "Family and Its Functions;" and Parsons and Bales, *Family Socialization.*
12. Wilkinson, "Broken Home," 732.
13. Laub and Sampson, "Sutherland–Glueck Debate."
14. Glueck and Glueck, *Unraveling Delinquency.*
15. Sampson and Laub, *Crime in the Making,* 36.
16. Glueck and Glueck, *Unraveling Delinquency,* 260–262.
17. Sampson and Laub, *Crime in the Making,* 36.
18. Nye, *Family Relationships.*
19. Rankin and Kern, "Parental Attachments and Delinquency," 495.
20. Hirschi, *Causes of Delinquency;* and Rankin and Kern, "Parental Attachments and Delinquency."
21. Hirschi, *Causes of Delinquency.*
22. Hirschi, *Causes of Delinquency,* 19; and Toby, "Social Disorganization."
23. Hirschi, *Causes of Delinquency,* 16, see also 10–11.

24. Nye, *Family Relationships;* Reckless, "New Theory," and *Crime Problem;* Reiss, "Delinquency;" and Akers and Sellers, *Criminological Theories.*

25. Since most of Hirschi's (*Causes of Delinquency,* 79) analysis was conducted on a subsample of 1,588 white males, his reporting of findings makes frequent use of the term "he" and "boys" to refer more specifically to the subsample that was used most.

26. Akers and Sellers, *Criminological Theories,* 117.

27. Simons, Wu, Conger, and Lorenz, "Two Routes to Delinquency."

28. Hirschi, *Causes of Delinquency,* 18, 83, 92–94. See also Rankin and Kern, "Parental Attachments and Delinquency," 494.

29. Hirschi, *Causes of Delinquency,* 83.

30. Rankin and Kern, "Parental Attachments and Delinquency," 496, emphasis added.

31. Hirschi, *Causes of Delinquency,* 94, see also 18 and 88.

32. Hirschi, *Causes of Delinquency,* 86; Rankin and Kern, "Parental Attachments and Delinquency;" and Sampson and Laub, *Crime in the Making,* 67.

33. Hirschi, *Causes of Delinquency,* 89–90. See also Cernkovich and Giordano, "Family Relationships and Delinquency;" Larzelere and Patterson, "Parental Management;" Loeber and Stouthamer-Loeber, "Family Factors;" and Nye, *Family Relationships.*

34. Hirschi, *Causes of Delinquency,* 88–90.

35. Ibid., 90.

36. Ibid., 92.

37. Ibid., 97.

38. Ibid., 99.

39. Ibid., 121, 123.

40. Ibid., 129.

41. Ibid., 152.

42. Ibid., 135, 145–158.

43. Maguin and Loeber, "Academic Performance and Delinquency," 145.

44. Hirschi, *Causes of Delinquency.*

45. See studies cited in Johnson et al., "Students' Attachment."

46. For example: Stewart, "School Social Bonds;" Jang, "Age-Varying Effects;" and Farnworth, Schweinhart, and Berrueta-Clement, "Preschool Intervention."

47. Liska and Reed, "Ties."

48. For example: Jenkins, "School Commitment;" Stewart, "School Social Bonds;" Hawkins et al., "Review;" and Simons et al., "Test."

49. Wasserman et al., "Risk and Protective Factors."

50. See, for example, Stewart, "School Social Bonds;" Jenkins, "School Social Bond;" and Hoffman and Xu, "School Activities."

51. Crosnoe, "High School," 144.

52. For example: Crosnoe, "High School;" Gamoran and Berends, "Effects of Stratification;" Rosenbaum, "Social Implications;" and Polk and Schafer, *Schools and Delinquency.*

53. See studies cited in Baker, Sigmon, and Nugent, *Truancy Reduction.*

54. Rosen et al., "Early Delinquency;" Farrington, "Early Predictors;" but see also Jarjoura, "Dropping Out."

55. Hirschi, *Causes of Delinquency,* 151–152.

56. Ibid., 159.

57. Ibid., 145.

58. Ibid., 21.

59. Hirschi, *Causes of Delinquency,* 19; and Toby, "Social Disorganization."

60. Hirschi, *Causes of Delinquency,* 20.

61. Ibid., 162.

62. Ibid., 168–169.

63. Ibid., 186.

64. Ibid., 22, 187.

65. Ibid., 22.

66. Ibid., 111.

67. Ibid., 190.

68. Ibid., 196.

69. Ibid., 198.

70. Ibid., 26.

71. Reiss, "Delinquency;" and Nye, *Family Relationships.*

72. Hirschi, *Causes of Delinquency,* 203.

73. Ibid., 203.

74. Krohn, "Web of Conformity."

75. Hirschi, *Causes of Delinquency,* 27–30.

76. Akers and Sellers, *Criminological Theories,* 116. Akers and Sellers cite Stitt and Giacopassi, "Trends."

77. Hindelang, "Causes of Delinquency;" Conger, "Social Control;" Elliott, Huizinga, and Ageton, *Explaining Delinquency;* and Linden and Hackler, "Affective Ties and Delinquency."

78. Liska and Reed, "Ties," 558. See also: Kendal, "Parental and Peer Contexts;" Krohn et al., "Family Process Variables;" Wiatrowski and Anderson, "Dimensionality;" and Wiatrowski, Griswold, and Roberts, "Social Control Theory."

79. Hirschi, *Causes of Delinquency,* 178.

80. Agnew, "Longitudinal Test;" Krohn and Massey, "Social Control;" and Wiatrowski, Griswold, and Roberts, "Social Control Theory."

81. Agnew and Peterson, "Leisure and Delinquency." See also Burton et al., "Impact of Parental Controls."

82. Matsueda, "Dynamics." See also Kohn and Massey, "Social Control;" Wiatrowski and Anderson, "Dimensionality;" and Wiatrowski, Griswold, and Roberts, "Social Control Theory."

83. Agnew, "Social Control Theory;" LaGrange and White, "Age Differences;" Leonard and Decker, "Theory of Social Control;" Liska and Reed, "Ties;" Thornberry, "Toward an Interactional Theory;" and Wiatrowski and Anderson, "Dimensionality."

84. Agnew, "Social Control Theory;" Krohn and Massey, "Social Control;" and Matsueda, "Dynamics."

85. Burton et al., "Impact of Parental Controls;" Cernkovich and Giordano, "Family Relationships and Delinquency;" Johnson, *Juvenile Delinquency;* Krohn and Massey, "Social Control;" Rosenbaum and Lasley, "Rethinking the Gender Gap."

86. Agnew, "Longitudinal Test;" and Elliott, Huizinga, and Ageton, *Explaining Delinquency.*

87. Akers and Cochran, "Adolescent Marijuana Use;" and Matsueda, "Testing Control Theory."

88. Elliott, Huizinga, and Ageton, *Explaining Delinquency;* Massey and Krohn, "Longitudinal Examination;" and Warr, "Parents, Peers, and Delinquency."

89. Sampson and Laub, *Crime in the Making,* 1.

90. Ibid., 1.

91. Sampson and Laub, "Urban Poverty," 524.

92. Sampson and Laub, *Crime in the Making;* and Laub and Sampson, "Turning Points."

93. Sampson and Laub, *Crime in the Making,* 6.

94. Ibid., 35.

95. Sampson and Laub, *Crime in the Making,* 8–9, 17–19; and Warr, "Life-Course Transition," 183.

96. Caspi, Elder, and Herbener, "Childhood Personality," 15.

97. Glueck and Glueck, *Delinquents and Nondelinquents,* 167; and Sampson and Laub, *Crime in the Making,* 36.

98. Sampson and Laub, *Crime in the Making,* 65.

99. Ibid., 85–88.

100. Sampson and Laub, *Crime in the Making.* See also Sampson and Laub, "Urban Poverty."

101. Simons et al., "Test," 218, see also p. 222.

102. Sampson and Laub, "Urban Poverty," 523–524. See also Sampson and Laub, *Crime in the Making,* 65.

103. Loeber, "Stability;" Patterson, Reid, and Dishion, *Antisocial Boys;* Sampson and Laub, *Crime in the Making;* and White et al., "How Early."

104. Simons et al., "Test," 218.

105. Sampson and Laub, *Crime in the Making,* 20.

106. Ibid., 12.

107. Sampson and Laub, *Crime in the Making,* 13. Sampson and Laub provide the following reference support to their critique of behavior continuity: Gove, "Biopsychosocial Perspective;" Loeber and LeBlanc, "Toward a Developmental Criminology;" and McCord, "Patterns of Deviance."

108. Simons et al., "Test," 221.
109. Laub and Sampson, "Turning Points," 302, emphasis added.
110. Laub and Sampson, "Turning Points," 311. The concept of *social capital* is drawn from Coleman, "Social Capital," and *Foundations of Social Theory.*
111. Simons et al., "Test," 222, emphasis added.
112. Glueck and Glueck, *Unraveling Delinquency.*
113. Laub and Sampson, "Unraveling Families and Delinquency;" Laub and Sampson, "Turning Points;" Sampson and Laub, *Crime in the Making;* and Sampson and Laub, "Urban Poverty."
114. Sampson and Laub, *Crime in the Making,* 85.
115. Simons et al., "Test," 222.
116. Gottfredson and Hirschi, *General Theory of Crime.*
117. Evans et al., "Social Consequences," 495.
118. Hirschi and Gottfredson, "Commentary," 52, emphasis in original; and Evans et al., "Social Consequences," 476.
119. Gottfredson and Hirschi, *General Theory of Crime,* 16–17.
120. Ibid., 42–44.
121. Ibid., 14.
122. Ibid., 14, 90, 87, 95, respectively.
123. Ibid., 91.
124. Hirschi and Gottfredson, "Commentary," 50. See also Gottfredson and Hirschi, *General Theory of Crime,* 137.
125. Hirschi and Gottfredson, "Commentary," 49. See also Gottfredson and Hirschi, *General Theory of Crime,* 88, 94, 96, 111.
126. Gottfredson and Hirschi, *General Theory of Crime,* 91.
127. Cullen and Agnew, *Criminological Theory,* 240.
128. Gottfredson and Hirschi, *General Theory of Crime,* Chapter 7.
129. Ibid., 157.
130. Evans et al., "Social Consequences;" and Laub and Sampson, "Turning Points;" and Sampson and Laub, *Crime in the Making.*
131. Gottfredson and Hirschi, *General Theory of Crime,* 119. See also Brownfield and Sorenson, "Latent Structure Analysis;" Lazarsfeld and Henry, *Latent Structural Analysis;* and Simons et al., "Test."
132. Patterson, "Children Who Steal;" Patterson, *Coercive Family Process;* and Gottfredson and Hirschi, *General Theory of Crime,* 97.
133. Gottfredson and Hirschi, *General Theory of Crime,* 97.
134. Ibid., 95.
135. Sampson and Laub, *Crime in the Making.*
136. Gottfredson and Hirschi, *General Theory of Crime,* 118.
137. Ibid., 105.
138. Hirschi and Gottfredson, "Commentary," 49.
139. Akers and Sellers, *Criminological Theories,* 6–7.
140. Hirschi and Gottfredson, "Commentary," 52.
141. Grasmick et al., "Testing," 7–8; and Gottfredson and Hirschi, *General Theory of Crime,* 89.
142. Grasmick et al., "Testing."
143. Arneklev et al., "Low Self-Control;" Burton et al., "Impact of Parental Controls;" Gibbs and Giever, "Self-Control;" Grasmick et al., "Testing;" Keane, Maxim, and Teevan, "Drinking and Driving;" LaGrange and Silverman, "Low Self-Control and Opportunity;" Nagin and Paternoster, "Enduring Individual Differences;" Piquero, and Tibbetts, "Specifying;" Pratt and Cullen, "Empirical Status;" Tremblay et al., "Low Self-Control;" and Wood, Pfefferbaum, and Arneklev, "Risk-Taking and Self-Control."
144. Evans et al., "Social Consequences," 493.
145. Gottfredson and Hirschi, *General Theory of Crime,* 119.
146. Evans et al., "Social Consequences," 481; and Gottfredson and Hirschi, *General Theory of Crime,* 119.
147. Evans et al., "Social Consequences;" Simons et al., "Test;" and Winfree and Bernat, "Social Learning."
148. Simons et al., "Test," 217.
149. Pratt and Cullen, "Empirical Status."

Chapter Resources

150. Cochran et al., "Academic Dishonesty;" Feldman and Weinberger, "Self-Restraint;" Gibbs, Giever, and Martin, "Parental Management;" Hay, "Parenting;" and Polakowski, "Linking."
151. Pratt, Turner, and Piquero, "Parental Socialization."
152. Burton et al., "Impact of Parental Controls." See also Nye, *Family Relationships.*
153. Hirschi, *Causes of Delinquency,* 88.
154. Sampson and Laub, *Crime in the Making.*
155. Rankin and Kern, "Parental Attachments and Delinquency," 496; and Rankin and Wells, "Effect of Parental Attachments," 142.
156. Rankin and Wells, "Effect of Parental Attachments," 142.
157. Henggeler, *Delinquency in Adolescence,* 36; and Wilson and Herrnstein, *Crime and Human Nature.* See also Simons, Simons, and Wallace, *Families, Delinquency, and Crime.*
158. Loeber and Stouhamer-Loeber, "Family Factors;" Cernkovich and Giordano, "Family Relationships and Delinquency;" and Loeber and Dishion, "Early Predictors."
159. Canter, "Family Correlates."
160. Rankin and Wells, "Effect of Parental Attachments."
161. Cernkovich and Giordano, "Family Relationships and Delinquency."
162. Rankin and Wells, "Effect of Parental Attachments."
163. Patterson and Stouthamer-Loeber, "Correlation."
164. Cernkovich and Giordano, "Family Relationships and Delinquency;" and Rankin and Kern, "Parental Attachments and Delinquency."
165. Cernkovich and Giordano, "Family Relationships and Delinquency."
166. Ibid., 315.
167. Henggeler, *Delinquency in Adolescence;* Loeber and Stouthamer-Loeber, "Family Factors;" Patterson, *Coercive Family Process;* and Sampson and Laub, *Crime in the Making.*
168. Cernkovich and Giordano, "Family Relationships and Delinquency."
169. Inciardi, Horowitz, and Pottieger, *Street Kids,* 137.
170. Conger et al., "Economic Stress."
171. Patterson, *Coercive Family Process;* and Patterson, Reid, and Dishion, *Antisocial Boys.*
172. Wells and Rankin, "Direct Parental Controls," 263.
173. Hirschi, *Causes of Delinquency,* 88. See also Nye, *Family Relationships,* 7.
174. Hirschi, *Causes of Delinquency;* Cerkovich and Giordano, "Family Relationships and Delinquency;" and Wells and Rankin, "Direct Parental Controls."
175. Patterson, "Children Who Steal;" and Patterson, *Coercive Family Process.*
176. Burton et al., "Impact of Parental Controls;" Cernkovich and Giordano, "Family Relationships and Delinquency;" Hawkins et al., "Predictors of Youth Violence;" Loeber and Dishion, "Early Predictors;" Loeber and Stouthamer-Loeber, "Family Factors;" Wright and Cullen, "Parental Efficacy;" Wasserman et al., "Parenting Predictors;" Wasserman et al., "Risk and Protective Factors;" and Wells and Rankin, "Direct Parental Controls."
177. Canter, "Family Correlates;" Loeber and Stouthamer-Loeber, "Family Factors;" and Warr, "Parents, Peers, and Delinquency."
178. Burton et al., "Impact of Parental Controls;" Henggeler, *Delinquency in Adolescence;* Loeber and Stouthamer-Loeber, "Family Factors;" and Sampson and Laub, *Crime in the Making.*
179. Laub and Sampson, "Unraveling Families and Delinquency;" Lytton, "Child and Parent Effects;" Patterson, DeBaryshe, and Ramsey, "Developmental Perspective;" Sampson and Laub, *Crime in the Making;* Simons et al., "Test;" and Snyder and Patterson, "Family Interaction."
180. Sampson and Laub, *Crime in the Making,* 89–91.
181. Cernkovich and Giordano, "Family Relationships and Delinquency;" Hagan, *Structural Criminology;* and McCord and McCord, *Origins of Crime.*
182. Cernkovich and Giordano, "Family Relationships and Delinquency;" and Burton et al., "Impact of Parental Controls."
183. Thornberry, Huizinga, and Loeber, "Causes and Correlates Studies;" Wasserman et al., "Risk and Protective Factors;" Widom, "Cycle of Violence;" and Zingraff et al., "Child Maltreatment."
184. Ireland, Smith, and Thornberry, "Developmental Issues;" Smith and Thornberry, "Relationship;" Thornberry, Ireland, and Smith, "Importance of Timing;" Thornberry, Huizinga, and Loeber, "Causes and Correlates Studies;" Widom, "Cycle of Violence;" and Zingraff et al., "Child Maltreatment."
185. Simons, Simons, and Wallace, *Families, Delinquency, and Crime,* 168–169.
186. Monahan, "Family Status," 250; and Wells and Rankin, "Families and Delinquency," 73.

187. Shaw and McKay, "Broken Homes," 524. See also Cernkovich and Giordano, "Family Relationships and Delinquency."

188. Demuth and Brown, "Family Structure," 61; references omitted in quote. Cernkovich and Giordano, "Family Relationships and Delinquency;" Gove and Crutchfield, "Family and Juvenile Delinquency;" Lamborn et al., "Patterns of Competence;" Miller et al., "Parental Discipline;" Rankin and Kern, "Parental Attachments and Delinquency;" Rollins and Thomas, "Parental Support;" Thornberry et al., "Family Disruption and Delinquency;" Van Voorhis et al., "Impact of Family Structure;" Wasserman et al., "Risk and Protective Factors;" and Wells and Rankin, "Direct Parental Controls."

189. Demuth and Brown, "Family Structure," 62.

190. Sampson and Laub, *Crime in the Making,* 79–80. For an opposing view, see Cernkovich and Giordano, "Family Relationships and Delinquency," 316. See also Van Voorhis et al., "Impact of Family Structure," and Simons, Simons, and Wallace, *Families, Delinquency, and Crime.*

191. Demuth and Brown, "Family Structure."

192. Ibid., 77–78.

193. Glueck and Glueck, *Delinquents and Nondelinquents,* Hirschi, *Causes of Delinquency;* Loeber and Stouthamer-Loeber, "Family Factors;" and Rutter, *Changing Youth.*

194. Hirschi, *Causes of Delinquency;* McCord and McCord, *Origins of Crime;* and Nye, *Family Relationships.*

195. Nye, *Family Relationships.*

196. Hirschi, *Causes of Delinquency;* McCord and McCord, *Origins of Crime;* and Nye, *Family Relationships.*

197. Rutter and Giller, *Juvenile Delinquency,* 185.

198. Larzelere and Patterson, "Parental Management," 307.

199. Laub and Sampson, "Unraveling Families and Delinquency;" Larzelere and Patterson, "Parental Management," 307; Sampson and Laub, *Crime in the Making,* 70; Sampson and Laub, "Urban Poverty;" and Tittle and Meier, "Specifying."

200. Wiatrowski and Anderson, "Dimensionality."

201. Sampson and Laub, *Crime in the Making,* 69. The quotation provides two references: Gottfredson and Hirschi, *General Theory of Crime,* 101; and Hirschi, *Causes of Delinquency,* 53–60.

202. Henggeler, *Delinquency in Adolescence,* 46.

203. Sampson and Laub, *Crime in the Making,* 96.

204. Hirschi, *Causes of Delinquency,* 157–158.

205. Aseltine, "Reconsideration;" Jensen, "Parents;" and Poole and Regoli, "Parental Support."

206. Jensen, "Parents."

207. Hirschi, *Causes of Delinquency;* Jensen, "Parents;" Poole and Regoli, "Parental Support;" Simons et al., "Test;" and Stanfield, "Interaction."

208. Patterson and Dishion, "Contributions." See also Elliott, Huizinga, and Menard, *Multiple Problem Youth;* Patterson, DeBaryshe, and Ramsey, "Developmental Perspective;" Patterson, Reid, and Dishion, *Antisocial Boys;* Simons et al., "Test;" and Snyder and Patterson, "Family Interaction."

209. Warr, "Parents, Peers, and Delinquency," 248.

210. Ibid., 249.

211. Sampson and Laub, "Urban Poverty," 525.

212. Monahan, "Family Status," 250.

213. Gottfredson and Hirschi, *General Theory of Crime,* 90.

214. Nye, *Family Relationships.*

Social Learning Theories: Peer-Group Influences

11

Chapter Objectives

After completing this chapter, students should be able to:

- Describe the connection between peers and delinquent behavior according to social learning theories.
- Describe and illustrate how peer-group association varies in terms of *frequency, priority, duration,* and *intensity.*
- Describe how delinquent attitudes and behaviors are acquired.
- Explain the role that modeling and reinforcement play in peer-group influences.
- Distinguish the *socialization* and *selection* perspectives on peer-group involvement.
- Discuss gender differences in peer-group influences.
- Explain the role that peers play in drug use.
- Identify key elements of a definition of street gangs.
- Describe the group processes of street gangs.
- Understand theories and key terms:

Theories:
differential association
social learning
social bond
self-control

Terms:
group offending or co-offending
social learning
differential association
definitions of the law
operant conditioning
differential reinforcement
imitation
socialization perspective
selection perspective
gang
group processes
social facilitation model
enhancement model

CASE IN POINT

Sidney's "Companions in Crime"

The biographical account of Sidney Blotzman's delinquent career, offered by Clifford Shaw in *The Natural History of a Delinquency Career,* provides a good deal of information on his "companions in crime." Criminologist Mark Warr offers a summary of the group nature of Sidney's delinquency.

Sidney grew up in a highly "deteriorated" neighborhood west of the Loop in a family marked by frequent desertion of the father. He moved with his family to a nearby neighborhood when he was 10 and moved yet again when he was 15. Sidney was first arrested in 1916 (at about age 8) for petty theft and last arrested in 1925 for armed robbery and rape. All but 2 of the 13 offenses for which he was arrested were committed with accomplices. Over the course of his delinquent career, Sidney was affiliated with three delinquent groups, each of which inhabited the neighborhood in which he resided at the time. Although these groups were fairly large, and although Sidney was arrested with a total of 11 different co-offenders during his career, he never committed an offense with more than 3 companions on any one occasion. In his first group, which contained six members, Sidney committed offenses with four distinct subsets (triads) of the group. The two groups to which he later belonged were considerably larger (more than a dozen), but in both cases Sidney committed offenses with only a small portion of the group (three members), usually with no more than two of them at any one time.

Shaw and McKay emphasize the fact that the larger groups with which Sidney was affiliated existed before he joined them, that each had its own unique repertoire of offenses, and that Sidney's own history of offending closely paralleled the activities of the groups to which he belonged: "The successive types of delinquent activity in which Sidney engaged, beginning with pilfering in the neighborhood and progressing to larceny of automobiles and robbery with a gun, show a close correspondence with the delinquent patterns prevailing in the successive groups with which he had contact."

Sources: Warr, "Organization and Instigation," 15; Shaw, *Natural History*.

■ Companions in Crime: The Group Character of Delinquency

One of the distinctive features of the adolescent years is the degree to which peers take on added importance and influence. It therefore is no surprise that most delinquent acts are committed with friends. Yet the powerful influence of peers during adolescence is often taken for granted, with little consideration of its extent and nature. We often assume that adolescents spend an inordinate amount of time with friends and that their thoughts and actions are almost totally dictated by peer pressure. In terms of delinquency, we assume that delinquent friends are the major force behind the initiation and persistence of delinquent offending. While these assumptions may be true, they tell us little about how and to what degree peers exert their influence.

Criminologists have long known that delinquency occurs most often in the company of peers.[1] This is known as **group offending** or **co-offending**.[2] As indicated in the life history at the beginning of the chapter, empirical study of group offending began in the 1930s with the research of Shaw and McKay, who found that 90% of the offenses reported in juvenile court records involved two or more participants.[3] More recently, in a review of research on the group character of delinquency, Albert J. Reiss concluded that "co-offending is most characteristic of what we think of as juvenile delinquency and characterizes juvenile careers."[4] Moreover, research indicates that juvenile offenders who commit crime with others commit more crime, and are involved in more serious offenses, than juveniles who commit crime alone.[5]

group offending or co-offending Commitment of delinquent acts in the company of peers.

A related finding is that adolescents who have delinquent peers are more likely to be delinquent themselves. Travis Hirschi's research on social bond theory found that 83% of youths reporting two or more delinquent acts had at least one close friend who had been picked up by the police, whereas only 25% of youths with no delinquent friends had committed a delinquent act.[6] As Mark Warr observes: "For decades, criminologists have recognized that the number of delinquent friends an adolescent has is the strongest known predictor of delinquent behavior."[7]

While these findings are strong and consistent, Robert Agnew points out that "the research on delinquent peers . . . has been rather simplistic. In most studies, researchers simply measure the number of the adolescent's friends who are delinquent and/or the frequency with which friends commit delinquent acts. No effort is made to examine other dimensions of peer interaction."[8] Thus, the research findings linking delinquent behavior to involvement with delinquent peers tell us little about the nature of adolescent peer groups or the processes through which peers exert their influence.

A leading gang researcher, Malcolm Klein, points out that just because a delinquent youth has delinquent friends and commits delinquent acts with friends does not mean that these actions are an outcome of group processes.[9] Instead, delinquent acts may be committed by "contagious individuals" who happen to be in the same place engaging in the same behavior, with little or no peer influence involved. Along these lines, Martin Gold likens group offending to a pickup game of basketball, in which participation is spontaneous, desired but unplanned, and short-term. Those who participate hang out where the game is played, and teams are made up of whoever happens to be on the playground at the time. When the game is over, players go their own ways.[10]

Peggy Giordano and her colleagues contend that delinquency theorists and researchers have not given enough consideration to the nature of peer-group relations and the role of peer influences in delinquent behavior.[11] Moreover, Giordano claims that theorists have ignored findings on peer relations reported in the literature of developmental and social psychology. Giordano and her colleagues identify three dimensions of peer relations that need to be considered more fully: (1) the rewards or benefits gained from a relationship, such as sharing of confidences, self-disclosure, caring, trust, and identity support; (2) patterns of peer interaction and influence, such as amount of time spent with friends, longevity of relationships, and peer pressure; and (3) friendship dynamics, including levels of conflict, interaction reciprocity, and loyalty.

This chapter will first explore several theories that offer dramatically different views of the connection between peers and delinquency. Differential association and social learning theories point to peer groups as the context in which delinquent behavior is learned and reinforced. In sharp contrast to these views, control theories argue that peers are largely irrelevant to why youth become delinquent.[12] This theoretical debate gives rise to some important questions regarding peer-group influences on delinquent behavior that we will consider later in the chapter. After exploring these questions, we will turn to an aspect of peer influence that has received a great deal of attention in recent years: delinquent gangs.

Theoretical Views of Peers and Delinquent Behavior

Social Learning Theories

Sutherland's Differential Association Theory

Peer-group relations take center stage in differential association theory. Edwin Sutherland developed the theory in an attempt to explain how group relations influence people's attitudes and behavior. The first formal statement of this theory appeared in 1939 in the third edition of Sutherland's textbook, *Principles of Criminology*. A revised version appeared in the fourth edition, published in 1947. This version remains the best-known formal statement of the learning processes that occur in delinquent peer groups.[13]

When applied to delinquent behavior, the theory contends that a youth becomes delinquent by learning from others in the course of their social interaction. This learning involves not only "techniques" for committing delinquent acts but also "definitions favorable to the violation of the law." The latter concept refers to the attitudes that encourage delinquent behavior. Sutherland's theory is stated in the form of nine propositions, each with a brief explanation (see Expanding Ideas, "Sutherland's Theory of Differential Association"). The propositions are phrased in terms of criminal behavior, but Sutherland intended his theory to explain a broad range of crimes, including traditional street crime, white-collar crime, and delinquent behavior. In fact, the theory has been applied most extensively to delinquent behavior.[14]

Differential association theory is a theory of **social learning**: it holds that criminal behavior is learned through social interaction in groups. According to Sutherland, the vehicle for learning is verbal communication, and the learning is within *intimate personal groups* (Proposition #3). The theory goes on to state that association with different groups—**differential association**—varies in "frequency, duration, priority, and intensity" (Proposition #7). In other words, the influence of relationships within groups is greatest when interaction occurs frequently (*frequency*), for long periods (*duration*), and early in life (*priority*), and when those relationships are highly valued (*intensity*).[15] Through such interaction, individuals learn techniques of committing crime and motives, drives, rationalizations, and attitudes that go along with involvement in crime (Proposition #4).

At the heart of the theory are **definitions of the law** (Proposition #6). A person defines the law as favorable or unfavorable—either as rules to obey or as rules to violate—and such definitions are learned through verbal communication in intimate personal groups. The "direction" of these definitions involves motives, drives, rationalizations, and attitudes that either support obedience to the law or encourage violation of it (Proposition #5).

Sutherland assigned special importance to delinquent peer groups as a context in which adolescents learn "definitions favorable to the violation of the law." In fact, he contended that adolescent peer groups are much more important than the family in teaching delinquent behavior. The family is important, in Sutherland's view,

social learning
Delinquent and criminal behavior is learned through social interaction in groups.

differential association
Association with different groups. According to differential association theory, association with groups varies in frequency, duration, priority, and intensity.

definitions of the law A person defines the law as favorable or unfavorable—either as rules to obey or to violate. A person's attitudes and beliefs toward the law, its legitimacy, and authority.

Sutherland's Theory of Differential Association

Edwin Sutherland's well-known theory, differential association, is stated in the form of nine propositions. While only the propositional statements are provided here, Sutherland included a brief explanation for each statement. Differential association theory is one of the most succinct theoretical statements about crime and delinquency.

1. Criminal behavior is learned.
2. Criminal behavior is learned in interaction with other persons in a process of communication.
3. The principal part of the learning of criminal behavior occurs within intimate personal groups.
4. When criminal behavior is learned, the learning includes (a) techniques of committing the crime, which are sometimes very complicated, sometimes very simple; (b) the specific direction of motives, drives, rationalizations, and attitudes.
5. The specific direction of motives and drives is learned from definitions of the legal codes as favorable or unfavorable.
6. A person becomes delinquent because of an excess of definitions favorable to violation of law over definitions unfavorable to violation of law.
7. Differential association may vary in frequency, duration, priority, and intensity.
8. The process of learning criminal behavior by association with criminal and anticriminal patterns involves all the mechanisms that are involved in any other learning.
9. While criminal behavior is an expression of general needs and values, it is not explained by those general needs and values, since noncriminal behavior is an expression of the same needs and values.

Source: Sutherland, Cressey, and Luckenbill, *Principles of Criminology,* 88–90.

mainly because where the family lives determines the degree to which the youth is exposed to patterns of delinquent behavior outside of the home. In addition, unpleasant family experiences may drive the youth out of the home and encourage association with delinquent peers. These delinquent peers, in turn, provide the primary group context in which delinquency is learned.[16]

Akers' Social Learning Theory

As we have just seen, differential association theory argues that delinquent behavior is learned through interaction with others. The youth learns both techniques for committing crime and definitions that favor violation of the law. However, differential association theory has little to say about how such learning occurs other than noting that the "process of learning criminal behavior . . .

incorporates all the mechanisms that are involved in any other learning" (Proposition #8). Social learning theory was developed in an attempt to overcome this shortcoming.

Robert Burgess and Ronald Akers extended differential association theory and restated it in the form of seven propositions.[17] The revised theory, called *differential association–reinforcement theory*, incorporates principles from behavioral learning theory, especially **operant conditioning**, in which behavior is shaped by rewards and punishments. To emphasize the group context of learning, Burgess and Akers refer to the social processes of rewards and punishments as **differential reinforcement**. Other concepts from behavioral learning theory were also incorporated in the revised propositions, but Sutherland's emphasis on group interaction was maintained.

Ronald Akers continued to develop the theory, which is now called *social learning theory*, by elaborating on the processes of learning that he and Burgess first identified. Akers emphasizes four particular aspects of learning: differential association, definitions, differential reinforcement, and imitation. Akers contends that delinquent behavior is a "function of the balance of these influences on behavior."[18]

The idea of *differential association* builds on Sutherland's use of the term to include not only the group context in which delinquent attitudes and behaviors are learned, but also the group's ability to model and reinforce these attitudes and behaviors. Also drawing on Sutherland's original theory, social learning theory points to the importance of delinquent *definitions*. Attitudes and beliefs that encourage delinquent acts are acquired in groups through imitation and reinforcement. Akers asserts that delinquent definitions are "basically positive or neutralizing. Positive definitions are beliefs or attitudes that make the behavior morally desirable or wholly permissible. Neutralizing definitions favor the commission of crime by justifying or excusing it."[19]

The third aspect of learning is *differential reinforcement*. This term refers to learning processes that involve rewards and punishments. Rewards and punishments may be actual or anticipated, social or nonsocial.[20] However, "the theory proposes that most of the learning in criminal and deviant behavior is the result of social exchange in which the words, responses, presence, and behavior of other persons directly reinforce behavior . . . or serve as the conduit through which other social rewards and punishments are delivered or made available."[21] Clifford Shaw's case study provides many personal accounts of Sidney's involvement in delinquent acts that were motivated by thrill and excitement (nonsocial reinforcement) and by the social reinforcement of his companions. Regarding his stealing, Sidney said: "The fact is that I never stole when I was by myself. The kick came when there was someone with me and the fun could be mutual. It was a merry exciting pastime that interested me to the exclusion of all others."[22]

The fourth aspect of learning that social learning theory incorporates is **imitation**, in which the behavior modeled by others is copied. Sometimes people behave in particular ways after observing other people's behavior. Imitation may result in new kinds of behavior or serve to maintain current behaviors. In some cases, it can result in the discontinuation of behavior. Akers, for example, argues that imitation is key to learning how to use drugs. Seeing how other people actually take drugs and then observing the effect that they experience

operant conditioning The process by which behavior is shaped by rewards and punishments.

differential reinforcement The social processes and dynamics of rewards and punishments that are the basis of learning.

imitation The copying of behavior displayed or modeled by others.

provides an important first step into drug use.[23] Similarly, after observing someone take drugs, a person may perceive the effects to be undesirable and refrain or discontinue further drug use.

Akers emphasizes that groups are the main context in which these learning processes occur. Like differential association theory, social learning theory holds that "those associations which occur first (priority), last longer (duration), occur more frequently (frequency), and involve others with whom one has the more important or closer relationships (intensity) will have the greater effect."[24]

The processes of learning that social learning theory spells out are nicely illustrated in Sidney's self-description of how he learned to shoplift, found in Case in Point, "Sidney Learns to Shoplift." Sidney's story indicates not only that he learned how to steal by imitating the techniques of a close friend, but that he also learned what to do with the stolen goods and how to respond if he were caught by the store detective. After acquiring these skills, Sidney began to view shoplifting as a means of making money. He also became confident in his abilities and viewed his accomplishments with much pride—shoplifting was a real skill that was rewarding to him.

CASE IN POINT   Sidney Learns to Shoplift

After many trips to the big stores in the Loop I became an expert shoplifter. It took much practice but under Joseph's teaching I made good progress. He would be walking in a store, me following, and take a ring or two from a counter, or a bottle of perfume, or a large carton of gum, or cigarettes and cigars, or toys, stuff them under his belt and leave the store. Then we sold the things to a fence. We could always find fences to buy our stolen goods; and then go to a show, buy something to eat, and there you are. I finally got so I could shoplift almost as well as Joseph did. I got so I could not only spot a house detective a mile away, but I could almost smell him. I learned all of the little tricks of the game. If we got caught, and we did several times, a few tears and a promise never to steal again would be enough to make the house detective turn us loose. Being young and little it was easy to win the sympathy of the detective. So on I went, becoming interested in stealing and not knowing anything else. Crime became a business, I began to steal as a means of making money. I wanted to learn all there was to be known about the game, how to steal, how to evade the police, and how to sell the stolen goods. I became cocky and self-confident and had a real pride in my ability to steal.

Source: Shaw, *Natural History*, 228.

Robert Agnew points out that "several dimensions of peer interaction . . . may condition the impact of delinquent peers on delinquency. At the most basic level, this theory suggests that the impact of peers will be conditioned by the extent to which peers differentially reinforce delinquency."[25] He goes on to identify three other aspects of peer interaction that affect the degree of peer influence on delinquency: whether peers display (or model) delinquent behavior, whether peers have delinquent definitions, and how strong the attachment to peers is. Thus, social learning theory attempts to specify the learning processes that are a part of peer interaction.

Socialization versus Selection

Differential association and social learning theories contend that delinquency friends are the main cause of delinquent behavior: "A youngster becomes delinquent by learning attitudes favorable to law violation and then being socially rewarded for delinquent behaviors. Both the learning and the rewards come from the people the youth spends most time with; for most adolescents, this means friends."[26] Thus, peer groups provide the context in which delinquent attitudes and behavior are learned and rewarded. In order to produce such marked changes in attitudes and behavior, peer-group relationships are seen as strong, cohesive, and long-lasting.[27] It is also apparent that peer-group interaction occurs before delinquency. This process of learning delinquent behavior through peer-group interaction is the primary theme of social learning theories; it is sometimes referred to as the **socialization perspective**.[28]

In contrast, social bond theory contends that delinquent behavior is a product of weak social bonds, rather than association with delinquent peers. Youths with weak social bonds, including weak attachment to peers, experience few social controls that would prevent them from getting involved in delinquency. The theory goes on to argue that youths who are involved in delinquency tend to develop friendships with other delinquents—a process that Sheldon and Eleanor Glueck described as "birds of a feather flock together."[29] James Q. Wilson and Richard Hernnstein summarize: "Instead of being led into a life of crime by the influence of peers, they [delinquent youth] merely seek out those peers who share their interest in delinquency."[30] This tendency to seek out individuals like oneself is referred to as the **selection perspective**.[31]

The selection perspective holds that individual attitudes, values, beliefs, and behaviors are established early in life, largely through family attachments, and that these characteristics become the criteria for choosing friends, rather than the product of friendship. As Travis Hirschi notes, "There is a very strong tendency for boys to have friends whose activities are congruent with their own attitudes."[32] Social bond theory goes on to take issue with "the idea that delinquents have comparatively warm, intimate social relations with each other." Hirschi contends that this image of delinquent youth having close peer relationships is a "romantic myth"; instead, delinquent youth have relationships that are "cold and brittle."[33] In his later work with Gottfredson, Hirschi describes peer group relations among delinquent peers as "short-lived, unstable, unorganized collectivities, whose members have little regard for one another."[34]

socialization perspective The view that peer-group relations are strong, cohesive, and long-lasting and therefore able to exert strong influence on those involved. Peer-group relations and interactions are the context in which attitudes and behavior are learned.

selection perspective The tendency to seek out individuals like oneself, with similar attitudes, values, beliefs, and behaviors. These characteristics become the criteria for choosing friends, rather than the product of friendship.

As you probably anticipate, these very different views on peers and delinquent behavior have stimulated much controversy. We will explore six questions stemming from this theoretical debate.

1. Is delinquency learned from delinquent peers?
2. Is delinquency a group activity?
3. What is the nature of delinquent groups?
4. How does social interaction within peer groups influence youth?
5. Are there gender differences in peer-group influence?
6. What role do peers play in drug use?

■ Peer-Group Influences

Is Delinquency Learned from Delinquent Peers?

The lively debate over the role of peers in delinquent behavior has given rise to a great deal of research on the question of *causal* or *temporal ordering*.[35] Travis Hirschi, a central figure in this debate, portrays this question as follows: "A major point of contention between control and learning theories is the causal ordering of delinquency and involvement with delinquent friends. Control theory says delinquency comes first."[36] In contrast, learning theories contend that relations with delinquent peers come first.[37]

Though people who associate with each other tend to display similar attitudes and behaviors, this does not clarify whether the similarity of attitudes and behaviors existed before the relationship or if they are products of the relationship.[38] Research on adolescent friendships in general—not just friendships among delinquent youth—reveals that "birds of a feather" do in fact "flock together."[39] Youths with similar values, attitudes, and behaviors are drawn together; in terms of causal ordering, there is evidence that similarity in values precedes friendship. Such findings provide support for control theories. This sequencing is illustrated in Clifford Shaw's case study, in which Sidney discloses that in each neighborhood to which his family moved, he quickly and easily developed friendships with delinquent youths in the area[40] (see Case in Point, "Sidney's Companions in Crime," pages 424–425).

Studies of Causal Ordering

One of the first studies to address the question of causal ordering was conducted by Sheldon and Eleanor Glueck.[41] Their data, drawn from a matched sample of 500 delinquent and 500 nondelinquent boys, revealed that the onset of delinquent behavior precedes involvement in delinquent peer groups. Almost 20 years later, Hirschi took the argument a step further. He contended that the question of causal ordering is not relevant because the connection between delinquent peers and delinquent behavior is spurious: both delinquent friends and delinquent behavior are products of the lack of social bonds.[42] The same argument is made by Gottfredson and Hirschi in their general theory of crime, this time using the concept of self-control instead of social bonds.[43] In their view, association with delinquent friends is not a cause of delinquency; instead, the connection is spurious once low self-control is considered. Here too, association with delinquent youth and delinquent behavior are both products of the same factor—in this case, low self-control.

The question of causal ordering has been examined by Delbert Elliott and Scott Menard through an extensive analysis of the National Youth Survey, a longitudinal study of 1,725 youth aged 11 through 17.[44] Elliott and Menard found that youths typically make delinquent friends before getting involved in delinquent behavior. This general temporal sequence was more true for serious forms of delinquency than it was for minor delinquent acts, and it was more true for younger adolescents than for older adolescents.[45]

These findings support the temporal ordering offered by learning theories. However, Elliott and Menard also found that after a youth is exposed to minor delinquent acts by peers, more serious delinquent behavior precedes association with a more delinquent peer group—a finding that is consistent with control theories. Moreover, Elliott and Menard's analysis revealed that *both* peer association and delinquency increase from early to middle adolescence and decrease in late adolescence and adulthood. This finding indicates a close connection between level of peer group involvement and level of delinquent behavior across the adolescent years.[46] In Chapter 5 we identified this as the *age–crime curve*.

So which view is correct? Do delinquents merely seek out friends like themselves, as most control theories contend, or do they become delinquent because they associate with delinquent friends, as learning theories contend?[47] The most likely answer is that both are partially correct: youths who lack social bonds or self-control are more likely to engage in delinquent acts and then seek out delinquent friends, whereas others engage in delinquent behavior only after they have been exposed to such behavior through their peer-group relations.[48] This complementary process was studied by Denise Kandel, who found that adolescents tend to choose similar friends, but that friends also influence each other in developing attitudes and behaviors.[49] In the same vein, Terence Thornberry contends that delinquent behavior and association with delinquent peers influence each other.[50]

Actions Speak Louder than Words

The debate over the temporal ordering of delinquent friends and delinquent behavior leaves another issue unsettled: how do delinquent peers exert their influence? As Mark Warr and Mark Stafford put it, "Although the association between delinquent friends and delinquent behavior is well established, the mechanisms by which delinquency is socially transmitted remains unclear."[51]

As we have seen, differential association theory contends that the mechanism for transmitting delinquency is communication within peer groups. It is through such communication that delinquent definitions are acquired. To this basis, social learning theory adds imitation and reinforcement of behavior by peers. The central place of peer attitudes and behavior leads to another key question: "Is it what they think or do?"[52] Put another way, which has greater influence: friends' attitudes or their actions?

Warr and Stafford point out that differential association theory stresses the role of attitudes. "Definitions favorable to the violation of law" are the key factor in delinquent behavior. Delinquent attitudes are derived from the attitudes and behaviors of friends, and delinquent acts occur only after a youth develops attitudes that are consistent with such behavior.[53] As a result, these attitudes determine the effect of all other factors on delinquent behavior. Even if a youth has delinquent friends who model delinquent behavior and attitudes, he or she will not engage in delinquent acts until delinquent attitudes are adopted. In contrast, social learning theory proposes that youths imitate both the attitudes and the behavior of peers. As such, the attitudes and behaviors of peers influence a youth's behavior, regardless of the youth's attitude.

Using longitudinal data from the National Youth Survey, Warr and Stafford analyzed the effects of these factors on three types of delinquency: cheating, larceny, and marijuana use. For each of type of delinquency, the attitude of friends was found to influence the youth's attitude. However, friends' attitudes had only a weak effect on delinquent behavior. The researchers also found that friends' behavior influenced both the youth's attitude and the youth's behavior. In fact, the effect of friends' behavior on the youth's behavior was much stronger than the effect of friends' attitude. These findings led Warr and Stafford to conclude that "the effect of peers' attitudes is small in comparison to that of peers' behavior, and the effect of peers' behavior remains strong even when peers' attitudes and the adolescent's own attitude are controlled. . . . These findings suggest that delinquency is not primarily a consequence of attitudes acquired from peers. Rather, it more likely stems from other social learning mechanisms such as imitation or vicarious reinforcement, or from group pressures to conform."[54] Thus "the actions of peers . . . speak louder than their attitudes."[55]

Is Delinquency a Group Activity?

Unlike the question of how peer groups influence delinquent behavior, the idea that delinquent youth have delinquent friends provokes general support. This gives the impression that the vast majority of all delinquencies are committed in a group context. As Albert Reiss notes, "Group offending is most characteristic of what we think of as juvenile delinquency."[56] However, there is considerable evidence that not all types of delinquency are typically group offenses. While some offenses (such as drug and alcohol use, burglary, and vandalism) are committed mainly in groups, others (such as assaults, robberies, and most status offenses) are committed as often or even more often by solitary offenders as by groups.[57] Moreover, the majority of delinquent careers are characterized by a mix of offenses committed alone and offenses committed with accomplices.[58]

What Is the Nature of Delinquent Groups?

The composition of delinquent groups tends to change in the course of the adolescent years. Reiss reports that group offending is more common in early adolescence and then decreases in later adolescence.[59] These findings on the group nature of delinquency have led Elliott and Menard to conclude, "With regard to the argument that delinquency is group behavior, we believe the case is overstated. Studies of the group nature of delinquency provide mixed evidence at best."[60]

The variable nature of group offending points to the need for a better understanding of the groups in which delinquent offenses are committed. Are co-offenders drawn from the youth's peer group? What is the nature of these groups? Mark Warr set out to discover the "essential features of delinquent groups," using a data set that was gathered by Martin Gold, known as the National Survey of Youth.[61] The data were drawn from interviews that asked about particular offenses with close attention to characteristics of co-offenders. Warr found that offenders usually committed offenses with only two or three co-offenders. The composition of these groups changed frequently, and group members were drawn from a larger network of delinquent companions.[62] Warr also found that the size of delinquent groups decreased with age. By middle and late adolescence, offending groups were made up of only two or three youths.[63] Together, these findings indicate that "delinquents ordinarily belong to multiple groups over the course of their delinquent careers and . . . they are exposed to a substantially larger number of delinquent companions than one would surmise from the small size of offending groups."[64]

Warr also found that it was unusual for an offender to commit more than three or four offenses with the same group. These findings led him to describe delinquent groups as "transitory."[65] Similarly, Albert Reiss found that most delinquent groups are "short lived" and "unstable," largely because youths move from the area ("transience") and because group members mature.[66]

In the same study, Warr also examined the structure of delinquent groups. He points out that the group nature of delinquency is often taken for granted: "By concentrating on the delinquency of individuals, conventional theories of delinquency seem to portray delinquent groups as mere aggregates of like-minded or similarly motivated individuals. Yet it is difficult to believe that all members of offending groups are equally motivated or inclined to break the law on any given occasion. It is equally difficult to believe that members of offending groups are prepared to follow any and all other members of the group into illicit activities."[67]

Warr's research revealed that delinquent groups tend to have two distinct types of members, based on their roles within the group: instigators and joiners. "Instigators" are identifiable leaders who recruit others to engage in delinquent acts and who coordinate group activities. Other group members can be classified as "joiners." Instigators tend to be older, more experienced, and are more emotionally close to others in the delinquent group than are joiners. Because of the transitory nature of delinquent groups, the role that a youth assumes in one group may not be the same as his or her role in another group. The youth's role is determined by individual traits in comparison to those of other group members. As a result, there is role stability within groups but not necessarily between groups—a youth may be an instigator in one group and a joiner in another.[68]

How Does Social Interaction within Peer Groups Influence Youth?

In addition to studying the roles and structure of delinquent groups, researchers have investigated several aspects of group interaction. As mentioned earlier, the number of delinquent friends a youth has and the extent of their delinquency are closely related to the level of a youth's own delinquency.[69] This has led criminologists to conclude that peer influence plays a key role in the initiation and persistence of delinquent behavior.[70] Peers expose youths to delinquent attitudes and behaviors in a group context. As already described, this is the basic tenet of differential association theory: delinquent behavior is learned through group interaction (Proposition #2). According to differential association theory, group interaction varies in *frequency, priority, duration,* and *intensity* (Proposition #7); these variations are referred to as *differential association.*

Frequency and Duration of Association

It is usually assumed that more frequent and prolonged exposure to delinquent attitudes and behaviors results in a greater likelihood of delinquent behavior. Researchers measure these aspects of group relations by asking questions about the amount of time a youth spends with friends. Another important dimension of these group relations is the extent to which peers present delinquent patterns. The concept of delinquent patterns is derived from differential association theory, but it also has implications for social learning theory. The idea here is that peers provide, or model, both delinquent attitudes and delinquent behaviors that can be imitated and that are reinforced by the peer group. Research consistently shows that the more involved a youth is with delinquent friends, the more likely he or she is to engage in delinquent behavior.[71] Going one step further, Robert Agnew, in an analysis of data from the National Youth Survey, found that the influence of delinquent peers on delinquent behavior depends not only on the amount of time spent with friends but also on the extent to which delinquent patterns are presented in group interaction.[72] The more peers are involved in delinquent behavior, the more likely a youth will be to engage in similar forms of delinquent behavior. Delinquency is modeled by peers and imitated by the youth.

The frequency and duration of peer association naturally depend on the longevity of peer relationships. Mark Warr found that youths who acquire delinquent friends are likely to retain them. He used the term "sticky friends" to describe this tendency.[73] This does not mean, however, that delinquent youths necessarily keep the same friends; rather, Warr's finding was that their friends are consistently delinquent. More to the point is a study conducted by Giordano, Cernkovich, and Pugh.[74] In a comparison of delinquent and nondelinquent youth, these researchers found that delinquent youths report very similar levels of contact (frequency) and stability (duration) in their friendships.

The concepts of frequency and duration are often presented as static traits of peer association, with little consideration of how they may change during the adolescent years or throughout life. Warr recently applied life-course theory (discussed in Chapter 10) to this question.[75] Life-course theory contends that major changes in life ("transitions") have a strong influence on informal social control. Changes in the level of informal control, in turn, affect the youth's degree of involvement in delinquency. The theory does not, however, consider how such transitions alter relationships with delinquent peers. Warr's study attempted to do just that by examining how marriage, as a transition, influences time spent with friends, as well as exposure to delinquent patterns. While not surprising, his findings reveal that such a major life event dramatically reduces the amount of time spent with friends and the exposure to delinquent patterns.

Priority: Connecting Age and Peer-Group Influence

Sutherland's concept of priority refers to association with delinquent peers early in life. Although exposure is distinct from duration, Warr found that these concepts are highly correlated: youths who associated with delinquent peers early in life did so for longer periods of time. However, he also found that delinquency was more strongly correlated with the delinquency of a youth's current friends than with the delinquency of friends at an early age. He concluded that "it is the *recency,* not the priority, of delinquent friends that affects delinquent behavior at a particular age."[76]

The concept of priority assumes not only that delinquent peer associations precede delinquent behavior but also that associations that occur earlier in life have a stronger influence on delinquent behavior than do those that occur later in life. The second assumption is not often directly analyzed. Instead, the simple question of whether a youth's "first friends" were delinquent is used to measure priority.[77] Measured in this way, priority has been found to be less closely linked to delinquency than are frequency, duration, and intensity. This measure, however, is incomplete.

In a study investigating the relationships between age, peers, and delinquency, Warr found that over the course of adolescence, youth are increasingly exposed to delinquent behaviors by peers.[78] Data from the National Youth Survey revealed a very strong relationship between age and exposure to delinquent peers. Across the adolescent years, youths reported fewer "close friends" who were *not* involved in delinquency. For most types of delinquency, this trend continued until middle or late adolescence and then was reversed.[79] As Warr states, "During their early life, individuals frequently undergo rapid and enormous changes in exposure to delinquent peers, from a period of relative innocence in the immediate preteen years to a period of heavy exposure in the middle-to-late teens. This intense exposure to delinquent peers begins to decline, however, for many, but not all offenses as individuals leave their teens and enter young adulthood."[80]

This pattern of exposure to delinquent peers is strikingly similar to the age curve of most delinquent offenders that we examined in Chapters 5 and 6. Noting this similarity, Warr concluded that the strong connection between age and criminal involvement is explained at least in part by changing levels of peer association.[81]

Intensity: The Strength of Delinquent Peer Relationships

The strength of the relationship bonds between delinquent peers is critical in determining the extent to which peers influence one another. Sutherland used the concept of intensity to refer to the strength of peer relationships. Peer-group relationships are able to have influence to the degree that they are strong and cohesive—in other words, their intensity.[82] Still, there is heated debate over this issue, and this debate follows theoretical lines.

Learning theories contend that delinquent attitudes and behaviors are transmitted through interaction within "intimate personal groups" (Sutherland's Proposition #3). The communication, modeling, imitation, and reinforcement of delinquent attitudes and behaviors are possible only in peer groups in which the member's relationships are "intimate." Intimate personal groups are characterized by relationships that are close, warm, strong, cohesive, loyal, solidary (united in purpose), and long-lasting. Such intimacy leads to conformity. It is well known that highly cohesive groups are far more able to obtain conformity from members than are less cohesive groups. Clifford Shaw's classic case study of a juvenile robber or "jack-roller" included the following description of the youth's peer group: "We were like brothers and would stick by each other through thick and thin. We cheered each other in our troubles and loaned each other dough. A mutual understanding developed, and nothing could break our confidence in each other."[83]

In sharp contrast with learning theories is the point of view offered by control theorists, including Travis Hirschi. In his social bond theory, Hirschi states that "since delinquents are less strongly attached to conventional adults than nondelinquents, they are less likely to be attached to each other. . . . It seems reasonable to conclude that persons whose social relations are cold and brittle, whose social skills are severely limited, are incapable of influencing each other in the manner suggested by those who see the peer group as the decisive factor in delinquency."[84]

Hirschi's statement demonstrates the considerable disagreement between theorists over the true nature of relationships within delinquent peer groups. Robert Agnew, however, points out the dichotomy between delinquent and nondelinquent peer groups in terms of intimacy is misleading.[85] There is much variation in intimacy within each type of group. As a result, it is difficult to determine a level of intimacy that is typical of delinquent or nondelinquent peer groups. Moreover, in the study discussed earlier, Warr found that the value a youth places on peer relations and loyalty to friends—both of which are aspects of intimacy—peaks in the middle to late teens.[86] The influence of peer relationships decreases sharply in later adolescence.

Giordano, Cernkovich, and Pugh carried out an extensive study of relationships among adolescent friends by identifying and measuring different dimensions of friendship.[87] They found that delinquent youth, at least as much as other adolescents, find their friendships meaningful and beneficial. Delinquent and nondelinquent youth report similar levels of contact, stability, caring and trust, and identity support (self-confirmation) in their friendships. Moreover, as compared to nondelinquents, delinquent youth report higher levels of self-disclosure (how often they talk with their friends about personal issues or problems). They are also more likely to believe that they receive rewards from their friendship and that they are influenced by friends and in turn influence their friends. It was also found that the friendships of delinquent youth are characterized more by conflict and imbalance and less by loyalty than are the friendships of nondelinquent youth. Thus, while friendship styles may differ, the friendships of delinquent youth reveal a strong level of attachment and intimacy, and these friendships are perceived as rewarding and influential in their lives. Giordano and her colleagues conclude: "The data present a picture more complex than that provided by control theorists, who have depicted the friendships of delinquents as 'exploitive rather than warm and supportive' or, alternatively, by earlier subcultural theorists who may have idealized the gang as a noble fraternity characterized only by camaraderie and we-feeling. Overall, we find that youth who are very different in their levels of involvement in delinquency are nevertheless quite similar in the ways in which they view their friendship relations."[88]

Similarly, a study of seriously delinquent youth in Miami conducted by Inciardi, Horowitz, and Pottieger found that the friendships of these youth were not simply "short-term, conflict-ridden alliances of convenience." Rather, they were "close and meaningful."[89] In contrast, Hirschi found that the more delinquent a boy was, the less he identified with and respected the opinions of his friends.

In sum, most research indicates that the friendships of delinquent youth, even seriously delinquent youth, are not very different from those of nondelinquent youth. This is especially true in terms of the strength of relationship bonds between friends, for which delinquent and nondelinquent youth report very similar levels.

Are There Gender Differences in Peer-Group Influence?

Research has revealed that peer-group influences operate differently for males and females. Peggy Giordano and her colleagues, for instance, found that the friendships of adolescent males allow them to gain prestige, status, and self-identity more than do the friendships of adolescent females.[90] Males also experience more conflict in their friendship groups and more peer pressure to conform than do females. On the other hand, females were found to interact with friends in ways that promote disclosure, caring and trust, and loyalty. In other words, the friendships of females are characterized by greater levels of attachment and intimacy and less by conflict and peer pressure as compared to males. These findings have important theoretical significance. The friendships of adolescent males may promote delinquency (which would be consistent with differential association theory), whereas the friendships of females may inhibit delinquency (as proposed by social bond theory).

Though there are notable differences in the friendships of males and females, the critical issue with regard to peer influence and delinquent behavior is association with delinquent peers. Are there differences in the degree to which males and females are exposed to delinquent peers and affected by peer influence? This twofold question was taken up by Daniel Mears, Matthew Ploeger, and Mark Warr.[91] They found that males are substantially more likely than females to be exposed to delinquent friends. Males spend more time with friends and are roughly twice as likely to have friends who have broken the law. This difference in exposure to delinquent friends partially, but not completely, explains gender differences in involvement in delinquent acts. Mears and his colleagues also found that males are more strongly affected by delinquent friends than are females—having delinquent peers was related to delinquent behavior more strongly for males than it was for females.

Why then are males and females affected differently by exposure to delinquent peers? Mears and his associates explore one possible answer: differences in moral evaluation of illegal conduct. Both differential association theory and social bond theory hold that an individual's attitudes toward the law influence the likelihood of delinquent behavior. Carol Gilligan has suggested that females are socialized in such a way that they are more restrained by moral evaluations of behavior than are males.[92] Using a series of questions that asked how wrong it would be to engage in various delinquent acts ("moral evaluations"), Mears and his colleagues found that females were consistently more likely than males to rate the offenses as "very wrong." For both males and females, having delinquent friends was less influential for those with high levels of moral disapproval, but for females this constraint was far greater than it was for males. Females with strong moral disapproval of delinquent acts were immune to the influence of delinquent peers, whereas for males with the same level of belief, the influence of delinquent peers was reduced but not eliminated.

At first glance, these findings seem consistent with differential association theory, because moral evaluations are parallel to the theory's notion of "definitions of the law." More precisely, the theory holds that "definitions favorable to the violation of the law" acquired through peer group association are the motivating force behind delinquent behavior. Mears and his colleagues, however, found that moral constraints *inhibit* delinquency, even when a youth is exposed to delinquent peers. In this way, beliefs operate as a control for involvement in delinquency behavior. This line of thought is more consistent with social bond theory.

What Role Do Peers Play in Drug Use?

Nowhere is the controversy over peer influence more evident than with adolescent drug use. Even though association with drug-using friends is the single best predictor of adolescent drug use, the *extent* of peer influence is better understood than the *processes* through which peers exert their influence.[93] As a result, the role of peer influence in drug use is the subject of much debate. One point of view—the *socialization perspective*—draws from learning theories and holds that drug use most commonly occurs in groups and is best characterized as a group experience. Peers not only supply drugs, but they also define the experience as desirable and pleasant, thereby providing motivation and reinforcement for drug use.[94] The competing *selection perspective* is based on social control theories and contends that both drug use and association with peers who use drugs are a result of weak social bonds or lack of self-control. Instead of being led into drug use by friends, drug-using youth merely seek out those who share their same interest in drug use. As a result, the correlation between drug-using friends and drug use is spurious.

Research reveals that both views are partially correct. Youths who use drugs tend to choose friends who also use drugs, and once friendships are established, peer-group interaction promotes attitudes and behaviors that encourage continued drug use.[95] In this regard, Marvin Krohn and his colleagues found that a youth's drug use increases the likelihood of that youth associating with peers who also use drugs.[96] Association with drug-using peers increases the likelihood of developing beliefs supportive of drug use and, in turn, such beliefs encourage greater drug use and more association with drug-using friends. Krohn and his colleagues describe this as a "spiraling process." In addition, relationships among drug users have been found to be strong and intimate and therefore capable of reinforcing attitudes and behaviors that favor drug use.[97]

Researchers have also found that drug use is associated with delinquent behavior—the frequency and seriousness of drug use is strongly related to the frequency and seriousness of delinquent acts.[98] Indeed, it is commonly assumed that drug use and delinquent behavior go hand-in-hand. Despite the apparent connection, the causal ordering and causal processes are difficult to establish: does drug use lead to delinquent behavior or does involvement in delinquent acts lead to drug use? A third possibility is that drug use and delinquent behavior are products of some other factor that is responsible for both behaviors. In an effort to untangle the causal connections, researchers have begun to explore the developmental patterns of drug use and delinquency. A number of general observations can be made based on the findings of this research.

First, both drug use and delinquent acts appear to be components of a larger group of problem behaviors that often occur together. Delbert Elliott, David Huizinga, and Scott Menard refer to youth who engage in such behaviors as "multiple problem youth."[99] These problem behaviors involve a range of deviant, rebellious, and antisocial activities, including drug use, delinquency, academic failure and dropping out of school, and premature ("precocious") sexual activity. A study by Helene Raskin White found that five problem behaviors cluster together for both males and females: delinquency, substance abuse, school misconduct and underachievement, precocious sexual behavior, and suicide.[100] These problem behaviors are displayed early in life and continue on into the adult years. Problem behaviors occur together because they are thought to share a common set of causes, referred to as risk factors. Risk factors operate at the individual, family, school, and community level and include characteristics such as impulsiveness, family conflict, attendance at a school with high delinquency rates, and poverty. In this view, since drug use and delinquency share a "common cause," the relationship between these two problem behaviors is spurious. Critics argue that the common cause hypothesis is far too general and is, in fact, inaccurate. White found that while problem behaviors occur together, they are not as closely related as this perspective holds, and "various problem behaviors follow different developmental paths, for example, delinquency peaks between ages 15 and 17 and then declines, whereas polydrug use increases through adolescence into youth adulthood."[101] Nonetheless, it is reasonable to conclude that both drug use and delinquency are a part of a more general pattern of problem behavior and, as a result, that drug use should not be viewed as a cause of delinquency.[102]

Second, recent research exploring the developmental progression of drug use and delinquency has found that minor forms of delinquency precede drug use.[103] This may seem to be a simple finding, but it provides straightforward insight into the temporal ordering of these two factors. However, this does not mean that all youths involved in minor delinquency proceed to drug use; rather, the research findings indicate that the transition from delinquency to drugs is much more common than the drugs-to-delinquency transition. Eric Wish, for example, found that youths who were arrested most commonly began their criminal career by committing petty crimes and drinking alcohol and then proceeded to both harder drugs and more serious crimes.[104]

Third, research on developmental patterns indicates that most youths do *not* progress to more serious forms of drug use or delinquency and that the patterns of escalation (when it does occur) are somewhat different for each of these two types of deviance.[105] The development of serious drug use tends to follow a pattern of escalation from beer or wine to hard liquor, to marijuana, to other illicit drugs (most commonly prescription drugs). This pattern is cumulative when illicit drug users still use alcohol and marijuana. Furthermore, different social factors appear to be related to each of these progressive stages of drug use: minor forms of delinquency precede the use of hard liquor; peer influences and belief predict marijuana use; and parental influences (especially the quality of the parent–child relationship) are related to illicit drug use, although peer influences also remain strong. Finally, it may be surprising that most youths who become serious drug users have no history of serious involvement in delinquent acts. This calls into question the importance of delinquent behavior to serious drug use, other than as an entry point to drug use.[106]

Those youths who eventually engage in serious delinquencies show a similar pattern of progression in terms of delinquent behavior. Minor exploratory delinquencies precede serious delinquent acts.[107] Contrary to the findings on serious drug use, almost all serious delinquents are also drug users and those youths who are chronic and multiple drug users report more frequent and more serious delinquent acts.[108] After an extensive review of the drug–crime connection, Inciardi, Horowitz, and Pottieger conclude that "drug use may be a critical factor in the move from trivial to serious delinquency, increasing the chances that those few youth completing the transition will stay involved in serious crime."[109] Therefore, drug use appears more important to the development of serious delinquency than delinquent behavior is to the development of serious drug use.

Fourth, both drug use and delinquency appear to be defining characteristics of a preferred lifestyle for those few youths who develop chronic patterns of offending and substance abuse.[110] The peer groups of these youths model and reinforce attitudes and behaviors that encourage drug use and involvement in delinquent acts. When these peer groups are involved in drug sales, the link between drug use and crime becomes even more explicit: drug sales are engaged in to support drug use.[111] The peer-group context of serious delinquency and drug use has been found to be true for a wide range of youth: inner-city youth,[112] gangs involved in drug trafficking,[113] street kids,[114] and suburban youth.[115]

The study of peer-group influences on adolescents provides valuable insight into delinquent behavior. We have considered the degree to which delinquent behavior is learned from delinquent peers, the degree to which delinquent acts are group offenses, and the degree to which peer-group association varies when comparing delinquent and nondelinquent peer groups. We also examined gender differences in peer-group influences and the role of peers in drug use. We now turn to a particular group context in which social learning occurs: street gangs.

■ Street Gangs, Group Processes, and Delinquency

To most people, delinquent groups are synonymous with gangs. Gangs, however, account for only a small fraction of delinquent groups, and not all delinquent behavior occurs in gangs.[116] Still, there is much evidence that gang membership increases the likelihood, frequency, and seriousness of involvement in crime.[117] Moreover, gang members commit a disproportionate percentage of both property and violent crimes.[118]

The Persistent Gang Problem

"The violent gang is not a new phenomenon. Yet its contemporary form reflects a brand and intensity of violence that differentiate it from earlier gang patterns. The 'kill for kicks' homicide is today a source of concern not only in the large city but also in the suburb and the small towns."[119] These sobering words appear at the beginning of Lewis Yablonsky's book *The Violent Gang,* published in 1962. Gang violence was not new then, nor is it new today. Yablonsky's concern over a new "brand and intensity of violence" can easily be said to have reached a new level with today's heavily armed gangs. Yablonsky was one of many criminologists who studied gangs during the 1960s. These researchers sought to provide insight for intervention efforts into the egregious gang problem of that day. With few apparent solutions to the gang problem, the wave of theory and research quietly receded in the 1970s and 1980s, only to rise again in the late 1980s.[120] Once again, gang theory and research, with a focus on intervention, has reemerged as one of the hot topics in criminology. The resulting gang literature is enormous and varied, far beyond the scope of what can be discussed here. Given this chapter's focus on peer influence, we will restrict our attention to a consideration of group processes in gang delinquency. Two other chapters in this book deal with gangs. Chapter 9 took up situational factors in delinquency and described the social dynamics of gang involvement as perceived by the youth. Chapter 12 examines the larger societal context in which gangs emerge and endure.

Defining Gangs

Despite social scientists' strong interest in gangs, there is no widely agreed-upon definition of a gang. Nor is there agreement on the point at which a delinquent group becomes a gang. In fact, this question has been debated for decades, with no clear conclusion.[121] As result, the term *gang* is employed with a great deal of variation and imprecision, and it is often used interchangeably with *delinquent group.*[122]

While the distinction between delinquent groups and gangs is not entirely clear, Muzafer Sherif and Carolyn Sherif have argued that the study of gangs should not be separated from that of adolescent groups in general.[123] They contend that delinquent gangs can only be understood in the context of what is known about adolescent groups. In fact, the first systematic study of delinquent gangs, conducted by Frederic Thrasher in the 1920s, took this point of view. Thrasher defined the adolescent **gang** as "an interstitial group originally formed spontaneously and then integrated through conflict. It is characterized by the following types of behavior: meeting face to face, milling, movement through space as a unit, conflict, and planning. The result of this collective behavior is the development of tradition, unreflective internal structure, esprit de corps, solidarity, morale, group awareness, and attachment to a local territory."[124]

Several aspects of Thrasher's definition should be noted. First, he viewed gangs as a specific type of group. Second, the term *interstitial* as used here has two meanings. Literally, interstitial means "between." In one sense, the term refers to the period between childhood and adulthood, which is marked by adjustment and transition, during which gang involvement is most common. Thrasher also used the term to refer to the location of gangs in neighborhoods between the central business district and better residential areas. He described these lower-class areas as socially disorganized and therefore lacking in social control. Third, Thrasher said that gangs formed spontaneously. Gangs are made up of youths of about the same age who live in a common neighborhood where

gang "An interstitial group originally formed spontaneously and then integrated through conflict. It is characterized by the following types of behavior: meeting face to face, milling, movement through space as a unit, conflict, and planning. The result of this collective behavior is the development of tradition, unreflective internal structure, esprit de corps, solidarity, morale, group awareness, and attachment to a local territory" (Thrasher, *The Gang,* 57).

they are likely to come into contact with each other and form playgroups, which sometimes evolve into gangs. Fourth, conflict is the force behind gang formation and cohesion. Fifth, gangs are involved in a variety of activities that are engaged in collectively. It should be noted, however, that Thrasher did not include delinquency as a defining activity of the gang. Finally, the gang's collective behavior results in a group structure and culture involving "esprit de corps, solidarity, morale, group awareness, and attachment to a local territory." Thus, Thrasher's definition focused on the social processes that give rise to gangs and those which allow gangs to flourish.

Thrasher based these observations on a study of 1,313 adolescent groups in Chicago—groups that he identified as gangs. He recognized that these gangs varied greatly in terms of "membership, type of leaders, mode of organization, interests and activities, and finally as to its status in the community."[125]

Thrasher's definition of gangs was dominant until the mid-1960s. At that time numerous studies sought to describe gangs and their activities in terms of group processes and their wider community context. Robert Bursik and Harold Grasmick refer to this approach as a "process-based definition" of gangs.[126] Rather than being stable or permanent, gangs were portrayed as constantly changing in terms of membership, activities, structure, and cohesion.

Bursik and Grasmick also note that most current research no longer takes this approach.[127] Instead, gangs are defined in terms of involvement in illegal behavior. In fact, Walter Miller has argued that the term *gang* should be reserved for a group that is formally organized and engages in serious crime, as opposed to a street group. Bursik and Grasmick refer to this as a "delinquency-based definition."[128]

In a widely cited study, Miller attempted to determine whether there was any consensus among officials who work with youth as to what constitutes a gang.[129] He surveyed a national sample of police officers, prosecutors, defense attorneys, educators, city council members, state legislators, and even past and present gang members, asking them "What is your conception of a gang?" and "Exactly how would you define gang?" Six characteristics were identified by at least 85% of the 309 respondents. Miller used these to compose the following definition of gangs: "a self-formed association of peers, bounded together by mutual interests, with identifiable leadership, well-developed lines of authority, and other organizational features, who act in concert to achieve a specific purpose or purposes which generally include the conduct of illegal activity and control over a particular territory, facility or type of enterprise."[130]

Some criminologists have argued that such a definition provides an overly organized picture of gangs. Lewis Yablonsky, for example, describes gangs as "near groups."[131] By this he means that the structure and organization of most delinquent gangs are somewhere between a disorganized mob and a highly organized social group. In sharp contrast to Miller, Yablonsky argues that violent gangs are characterized by diffuse leadership, limited cohesion, impermanence, little agreement on norms, shifting membership, unclear membership expectations, and leaders who are sociopathic.

The definitional distinction between the group processes of gangs and their delinquent activity continues to be a source of controversy in the study of gangs.[132] While no resolution appears forthcoming, efforts to understand and respond to gangs must consider both aspects of gangs, especially in terms of violent crime. Expanding Ideas, "Defining 'Gangs,'" provides several contemporary definitions of gangs. We now turn to a more extensive consideration of the group processes of gangs.

EXPANDING IDEAS   Defining "Gangs"

Even though gangs have been a topic of enduring interest, there still is not a widely agreed-upon definition of a gang. "The lack of a consistent definition of gangs creates problems, not the least of which is the ability to compare information about gangs across cities and across different periods of times. For example, many of the groups regarded as gangs in the 1890s would not be so identified at the current time. Since not all of the illegal groups activity of young people has a similar motivation or character, it is useful to have a less rigid definition of gangs. In this way, the term can capture variation across time, cities, ethnic, and age groups. However, the lack of consistent definition of gangs creates problems for public officials who must formulate a response to what is perceived as the 'gang problem.' Without a clear concept of what is a gang and who is a gang member, public officials find themselves responding to an amorphous, ill-defined problem. This often leads, on one hand, to denial that gangs exist or, on the other hand, to the overidentification of gangs."[133] Below are several definitions of gangs offered by contemporary experts in the field.

Malcolm Klein: "Any denotable adolescent group of youngsters who (a) are generally perceived as a distinct aggregation by others in the neighborhood; (b) recognize themselves as a denotable group (almost invariably with a group name); and (c) have been involved in a sufficient number of delinquent incidents to call forth a consistent negative response from neighborhood residents and/or law enforcement agencies."[134]

Irving Spergel: "A group or collectivity of persons with a common identity whose members interact on a fairly regular basis in a clique or sometimes as a whole group. The activities of the gang may be regarded as legitimate, illegitimate, or criminal in varying combinations."[135]

Scott Decker and Barrik Van Winkle: "An age-graded peer group that exhibits some permanence, engages in criminal activity, and has some symbolic representation of membership."[136]

James Short: "Gangs are groups of youth people whose members meet together with some regularity, over time, on the basis of group-defined criteria of membership and group-defined organizational characteristics. In the simplest terms, gangs are unsupervised (by adults), self-determining groups that demonstrate continuity over time."[137]

Group Processes and Gang Delinquency

As already mentioned, there is considerable research evidence that gang members have higher rates of involvement in crime, especially serious and violent crime, than do non-gang members.[138] The link between gang involvement and delinquency suggests that gang membership and social interaction within gangs have a strong influence on the behavior and attitudes of members. Because of this powerful effect, we now turn to a fuller discussion of **group processes** and gang delinquency.

group processes The ways in which relationships and interaction within groups influence the attitudes and behaviors of individuals.

Terence Thornberry and his colleagues found that gang members did not have higher rates of delinquent behavior or drug use than other youth *before* joining the gang. However, once they became gang members, their rates of delinquency increased markedly.[139] Moreover, when gang members left the gang their rates of delinquency and drug use usually declined. The researchers concluded that the link between gang membership and delinquency can best be explained in terms of gang values and expectations that encourage involvement in crime by gang members. This is referred to as the **social facilitation model**.

social facilitation model The view that the normative support and groups processes of gangs encourage, support, and reinforce commission of crime. The social facilitation model incorporates the *socialization perspective* of peer-group relations.

However, both association with delinquent peers and gang membership are related to delinquent behavior.[140] Do the norms and group processes of gang membership have an effect on delinquent behavior beyond that of providing delinquent friends? Sara Battin and her colleagues recently investigated this question in a study of gangs in the Seattle area.[141] Battin and her fellow researchers found that gang membership has an effect on delinquent behavior that is separate from having delinquent friends. They also found that youth who were previously involved in delinquent acts were more likely to have delinquent friends and to join gangs. Unlike Thornberry and his colleagues, who supported a social facilitation model, Battin's research team interpreted their findings as supporting an **enhancement model** of gang formation. In the enhancement model, youth who are already involved in delinquent acts are attracted to and recruited by gangs because the norms and values of the gang are consistent with their own.[142] Once in the gang, violence and a variety of criminal acts are expected, encouraged, and reinforced.[143]

enhancement model The view that youth who are already involved in delinquent acts are attracted to and recruited by gangs because the norms and values of the gang are consistent with their own. The enhancement model incorporates the *selection perspective* of peer-group relations.

While *group cohesion* is often viewed as an essential attribute of gangs and one that distinguishes gangs from groups, Malcolm Klein and Lois Crawford depict the cohesion of gangs as fragile and as generated more by external forces, such as threats from rival gangs, than by internal processes of the group.[144] In their view, what distinguishes gangs from social groups is not the level of cohesion but how cohesion is generated. Social groups derive cohesiveness from internal sources such as interpersonal attraction, common goals, shared norms and values, and stability of membership. Klein and Crawford argue that, in contrast, there are few internal sources of cohesion in gangs. Gangs have few group goals, their membership is unstable, and they have sparse norms and limited role differentiation. Even the names of gangs change often, reflecting minimal group identity.

Especially important in Klein and Crawford's view are the social interactions that promote gang cohesion. As an example, Klein and Crawford point to the communication involved in delinquent acts: "Some of the jargon, the 'tough' talk, and the recounting of delinquent exploits engaged in by gang members probably serve the function of reinforcing the weak affiliative bonds within the group."[145] Thus, group offending and the social interaction that is a part of it are key to the development of gang cohesion. One of the reasons that gang members are involved in delinquency and violence is that such involvement enhances gang cohesion. At the same time, gang cohesion is a motivating factor for gang delinquency, including violence.

Similarly, James Short and Fred Strodtbeck observed that gang leaders acquired and defended their leadership status largely through aggressive responses to verbal or physical challenges to their leadership—what they call "status threats."[146] Status threats sometimes come from within the gang when gang members challenge the leader's authority, but threats may also come from external sources such as rival gangs, or even from larger institutions like schools and from adults in the neighborhood. Internally, the status of leaders is continually called into question because of changing group membership.[147] On the basis of an analysis of systematic notes drawn from observations by youth workers assigned to delinquent gangs, Short and Strodtbeck conclude that verbal and physical aggression directed outside the gang ("out-group aggression") is one of the few responses available to gang leaders facing status threats from within the gang. Aggressive responses to status threats are an important group process that reaffirms leadership status and generates cohesion in gangs whose structure and boundaries are typically limited and changing.

Group processes related to violence are key to understanding the formation, growth, and power of gangs. In an important field study of ninety-nine gang members in St. Louis, Scott Decker and Barrik Van Winkle use the concept of *threat* to refer to the broad and sweeping role that violence plays in the group processes of gangs. Through interviews and observation, the researchers identified three group processes involving threats of violence that are key to understanding gangs.[148]

First, in keeping with the perspective of Klein and Crawford, the researchers found that threats of violence, whether real or perceived, increase the cohesiveness of gangs.[149] Gang members perceived that threats and acts of violence originate primarily outside of the gang, often from rival gangs. External threats serve to strengthen relationship ties among gang members, increase commitment and loyalty to other gang members and to the gang itself, break down constraints against violence, and compel local youth to join neighborhood gangs.[150]

Second, threats of violence have a "contagion" effect that allows gangs to increase in size and territory.[151] Some new members join for protection because of threats of violence—threats that are either perceived or experienced. Others join gangs in search of the excitement that violence provides. In either case, violence is a contagion for gang membership growth. In addition, field research has repeatedly shown that gangs seek to maintain or extend their territory ("turf"), especially under the threat of violence by rival gangs.[152] Here too, gangs expand as a result of actual or perceived violence by outsiders.

THEORY INTO PRACTICE

Gang Cohesion: The Fundamental Group Process

Malcolm Klein has long argued that gang "cohesiveness is the quintessential group process." However, the role of gang cohesiveness in gang intervention efforts has surprised a number of gang experts, including Klein.

Beginning in the late 1940s, the Los Angeles Probation Department established a special gang unit called the *Group Guidance Section*. Originally established to work with Hispanic gangs, the unit also began to work with four large African American gangs in 1961. This new effort, funded by the Ford Foundation, had an evaluation component of which Klein was the head.

Five "street workers" engaged in three primary tasks:

1. Carry out group activities with their assigned gangs, such as weekly club meeting, sports activities, money-raising events (car washes, dances, etc.), and skills development, most notably a tutoring project.
2. Carry out individual counseling on family, school, and interpersonal (and sexual) relations and on delinquent behavior.
3. Open up the institutional doors in schools, court, job opportunities, welfare agencies, and the like in order to 'deisolate,' as they put it, the gang members from their own community institutions.

The workers were apparently successful in these tasks: "Here were five street workers in direct contact for several years with almost eight hundred of the most agency-resistant, delinquently oriented youth in the city. The workers were accepted by the gang members; they were respected by them; they were listened to; and the gang members seemed to respond."

The efforts of these street workers, when successful, increased gang cohesiveness as reflected in increasing gang size and group activities. Associated with this raise in gang cohesiveness was an increase in gang delinquency: "Using officially recorded arrest data, over the four years of the project, we found a significant increase in the offenses committed by gang members. The increase included both prevalence (more gang members, therefore more offenses overall) and individual offense frequency (more offenses per member, on average)." Evaluation research also distinguished between "low-companion" and "high-companion" offenses. Low-companion offenses remained stable for the evaluation period. "However, there was a very significant increase during the project among high-companion offenses. It was this latter effect that produced the overall finding of increased crime during the project." Gang cohesiveness not only bred crime, but more specifically, it bred co-offending crime. More to the point: "increased group programming leads to increased cohesiveness (in both gang growth and gang 'tightness'), and increased cohesiveness leads to increased gang crime."

Source: Klein, *American Street Gang*, 43–49.

The third group process identified by Decker and Van Winkle turns the table and deals with threats of violence by gang members to conventional groups and individuals, especially in neighborhoods where gangs operate. In this instance, gangs also instigate threats of violence. Such violence isolates gang members from local institutions such as families, schools, and businesses, thereby cutting them off from legitimate activities, roles, and opportunities. Decker and Van Winkle found that as gang members became increasingly immersed in gang activities, such gang involvement was to the "virtual exclusion of all else." This social isolation, then, results in greater gang cohesiveness.[153]

Decker and Van Winkle contend that violence is probably the single most important aspect of gang culture and a key ingredient in group processes of gangs.[154] "The centrality of violence to gang culture is evident in gang membership (initiation) through drive-by shooting, shootouts with rival gangs, and (for many gang members) the decision to leave the gang. Life in the gang is a life under threat of violence coupled with the willingness to use violence."[155]

While violence is part of everyday life for gang members, it does not always result in violent acts. Decker and Van Winkle use the term threat to refer to the potential for violence, either in the form of bravado, intimidation, or action. Additionally, threat of violence can be either real or perceived—actually expressed through words and action, or merely perceived by the gang member. Malcolm Klein argues that the cultural orientation of gangs is not solely directed toward threat of violence.[156] He observes that some street gangs are heavily involved in property crime, others in drug sales. In fact, he points out, as do Decker and Van Winkle, that "the most customary activities are sleeping, eating, and hanging around."[157] Nonetheless, threat of violence—sometimes acted upon, sometimes not—is a defining orientation of gang culture.

Gang Culture, Organizational Structure, and Group Processes

Gang violence, then, is *collective action* resulting from group processes.[158] The degree to which gang culture influences the behavior of members depends on the organizational structure of gangs. Field research in various cities (and notably different cultural contexts) reveals that urban gangs differ significantly in level of organization, leadership, roles, and rules.[159] Classic studies by Malcolm Klein in Los Angeles and James Short and Fred Strodtbeck in Chicago, for example, revealed that the gangs were only minimally organized. Most activities were not purposive or well organized. Leadership was not firmly established, but challenged frequently, resulting in limited power and authority. Membership was shifting, and allegiances and loyalty faltered over time. Cohesiveness was a largely a product of external, rather than internal, forces.[160]

In contrast, Joan Moore's study of gang members in the Mexican American neighborhoods of East Los Angeles found strong ethnic culture influences on the structure and activities of gangs in this area.[161] Moore reported three distinctive characteristics of Chicano gangs: they were (1) territorially based; (2) strongly age-graded, resulting in smaller age cohort groups called "klikas"; and (3) predominated by fighting. The ethnic culture of Chicano gangs resulted in an emphasis on protection of gang territory or "turf," physical toughness or "machismo," and status, often referred to as "honor." Grounded in this common culture, Moore found that Chicano gangs were highly structured, with distinct leadership and roles and extensive behavioral expectations (norms). Thus, gang culture and organizational structure appear to go hand-in-hand: cultural cohesiveness provides foundation to the organizational structure of gangs, while gang structure institutionalizes gang culture.

While Moore's research focused on the cultural context of Chicano gangs, Martin Sanchez Jankowski's ambitious field study included thirty-seven different gangs of several different ethnic groups (Chicano, Dominican, Puerto Rican, Central American, African American, and Irish). Based on observation of these urban gangs in low-income areas of New York, Boston, and Los Angeles, Jankowski claimed that gangs are more organized than they are traditionally portrayed to be and that gang culture was the organizing force behind gang structure. He found that low-income neighborhoods revolve around an "intense competition for, and conflict over, the scarce resources that exist in these areas."[162] With such intense competition, gang members had to be "defiant individualists" who were aggressive and self-reliant, with strong survival skills. Most of the gangs that Jankowski studied displayed a hierarchical structure with leadership and roles, and a system of norms and values that he called "formal codes" and "collective ideology."[163] These "mechanisms of internal organization" regulated the behavior of gang members and provided them with a common "picture of the world."[164] Unified with an organizational structure, together with common values and norms, Jankowski described gangs as "formal-rational" organizations. In addition, gang structure usually involved organized means for acquiring both legal and illegal income. These urban gangs were found to be in conflict (culturally and economically) not only with rival gangs, but also with legitimate neighborhood organizations such as schools and churches. Conflict not only solidified internal organization, but also generated gang cohesiveness in ways similar to those found by Klein and Crawford.[165]

Gang culture, however, has not always been found to be the primary force behind gang organizational structure. Ruth Horowitz studied a single Chicano gang in Chicago, called the "Lions."[166] As Moore found in the gangs of East Los Angeles, Horowitz found that the Lions placed great emphasis on "honor," achieved primarily through fighting. Given the limited educational and occupational opportunities normally available to lower-class youth, gang membership provided the primary arena in which honor could be achieved through demonstrations of toughness or machismo. While the Lions displayed collective goals, distinct roles, and stable membership, there were few membership rules and an unclear structure, especially in terms of leadership. Horowitz characterized the gang as having considerable flexibility, and, at the same time, as being a strong presence in the neighborhood that appeared to reinforce Chicano culture.

Similar to the works of Moore and of Horowitz, James Diego Vigil's study of Hispanic gang members in Los Angeles found Chicano culture to have a strong influence on barrio (neighborhood) gangs.[167] According to Vigil, barrio youth are "marginalized" from American society because of their low economic status and distinctive Chicano culture, a process called "choloization." As a result, barrio youth are not socialized into mainstream society and they do not aspire to the "American Dream" of economic success, nor do they have opportunities to achieve success. Instead, barrio youth turn to gangs for identity. Fighting and threats of violence were a constant feature of barrio gang life, primarily because violence provided a means to achieve identity in Chicano culture. Violence was expressed through threats and acts toward both fellow gang members and people outside the gang, sometimes rival gang members, sometimes others in the barrio. As a result, the strong cultural emphasis on violence appeared to limit development of gang structure, in that the barrio gangs studied by Vigil displayed only limited structure in terms of leadership, roles, and rules.

The organizational structure of contemporary gangs is also strongly influenced by the degree to which gangs are involved in drug sales.[168] In this regard, Felix Padilla conducted an important study of a Chicago Puerto Rican gang, called the "Diamonds," that he described as an "ethnic enterprise," engaged in street drug sales.[169] While gang members participated in a variety of activities, such as "hanging out" and playing basketball, street drug sales proved to be the primary organizing force of the gang. Drug sales were a collective activity requiring a hierarchical organizational structure with authoritative leadership, distinct roles, and extensive but specific rules. Padilla reported that gang members progressed through different roles in drug sales, graduating from "mules," who transported drugs, to "runners," who sold drugs on the street. Leadership was assumed when gang members became involved in the supply of drugs. Career development of gang members was governed by a clear set of rules for drug sales—the enterprise of the gang.

Padilla interprets the elaborate organizational structure of the Diamonds as an "ethnic enterprise" because drug trade serves as a "rational" response to the low status and limited opportunities of Puerto Rican youth living in poverty. Like Moore and Vigil, Padilla explains that gang membership provides a viable solution to ethnic youth who are marginalized by Anglo society and by the negative evaluations of schools, neighborhood groups, even their own families. Gang members saw the decision to join a gang as a constructive decision, not a matter of coercion.

Among Diamond gang members, street drug sales were referred to as "work," and the organizational structure of the gang allowed the work to be operated like a business. The gang's drug sales did not appear to result in tremendous profit; instead, proceeds funded typical adolescent pursuits, such as food, clothing, and parties.[170]

All these studies share in common an emphasis on the social context in which street gangs emerge and operate. Their existence and persistence are based on a community environment conducive to gangs. As a result, gang prevention and intervention strategies are directed at gaining a better understanding of the social context of street gangs. Theory into Practice, "Gang Intervention Strategies," lists the key social factors associated with gang membership. Theory into Practice, "Gang Intervention Strategies: The *Comprehensive Gang Model*," summarizes a gang intervention strategy sponsored by the Office of Juvenile Justice and Delinquency Prevention.

THEORY INTO PRACTICE

Gang Intervention Strategies

Efforts to deal with street gangs have long been informed by theory and research. Thrasher's research on Chicago gangs in the 1920s lead him to propose a community "restoration" strategy focusing on integrated and widespread neighborhood social services (*The Gang,* 364). The work of Short and Strodtbeck and Klein was based on the observations and experiences of detached street workers in Chicago and Los Angeles (Short and Strodtbeck, *Group Processes;* Klein, *Street Gangs*).

Over the years, theory and research have provided a number of key factors associated with gang membership. James Howell, of the National Youth Gang Center, recently compiled the following list of factors related to gang membership. The factors are divided into five categories.

Community
- Social disorganization, including poverty and residential mobility
- Organized lower class communities
- Underclass communities
- Presence of gangs in neighborhoods
- Availability of drugs in neighborhoods
- Availability of firearms
- Barriers to and lack of social and economic opportunities
- Lack of social capital
- Cultural norms supporting gang behavior
- Feeling unsafe in neighborhood; high crime
- Conflict with social control institutions

Family
- Family disorganization, including broken homes and parental drug/alcohol abuse
- Troubled families, including incest, family violence, and drug addiction
- Family members in a gang
- Lack of adult male role models
- Lack of parental role models
- Low socioeconomic status
- Extreme economic deprivation, family management problems, parents with violent attitudes, sibling antisocial behavior

School
- Academic failure
- Low educational aspirations, especially among females
- Negative labeling by teachers
- Trouble at school

- Few teacher role models
- Educational frustration
- Low commitment to school, low school attachment, high levels of antisocial behavior in school, low achievement test scores, and identification as being learning disabled

Peer Group
- High commitment to delinquent peers
- Low commitment to positive peers
- Street socialization
- Gang members in class
- Friends who use drugs or who are gang members
- Friends who are drug distributors
- Interaction with delinquent peers

Individual
- Prior delinquency
- Deviant attitudes
- Street smartness; toughness
- Defiant and individualistic character
- Fatalistic view of the world
- Aggression
- Proclivity for excitement and trouble
- *Locura* (acting in a daring, courageous, and especially crazy fashion in the face of adversity)
- Higher levels of normlessness in the context of family, peer group, and school
- Social disabilities
- Illegal gun ownership
- Early or precocious sexual activity, especially among females
- Alcohol and drug use
- Drug trafficking
- Desire for group rewards such as status, identity, self-esteem, companionship, and protection
- Problem behaviors, hyperactivity, externalizing behaviors, drinking, lack of refusal skills, and early sexual activity
- Victimization

Source: Howell, "Youth Gangs."

THEORY INTO PRACTICE

Gang Intervention Strategies: The *Comprehensive Gang Model*

In 1987 the Office of Juvenile Justice and Delinquency Prevention established a program to study and respond to the gang problem throughout the United States. The program, called the National Youth Gang Suppression and Intervention, was headed by Irving A. Spergel, from the School of Social Service Administration at the University of Chicago. The first task accomplished was a review of theory and research related to gang intervention. Drawing from this review, the program identified five strategies that provided a comprehensive approach to gangs and that were evaluated as "promising." These strategies make up the *Comprehensive Gang Model,* sometimes referred to as the "Spergel Model."

1. Mobilizing community leaders and residents to plan, strengthen, or create new opportunities or linkages to existing organizations for gang-involved and at-risk youth;
2. Using outreach workers to engage gang-involved youth;
3. Providing and facilitating access to academic, economic, and social opportunities;
4. Conducting gang suppression activities and holding gang-involved youth accountable; and
5. Facilitating organizational change and development to help community agencies better address gang problems through a team 'Problem-solving' approach that is consistent with the philosophy of community oriented policing.

Beginning in 1995, the Office of Juvenile Justice and Delinquency Prevention awarded grants to five communities to implement and test model programs based upon this model. These model programs and a large number of other resources can be accessed by visiting the Office of Juvenile Justice and Delinquency Prevention homepage and by following the links to publications and then to a topical listing that includes "juvenile gangs."

Sources: Burch and Kane, "Implementing," 1; Spergel and Chance, "National Youth Gang Suppression," 21–24. See also Spergel, *Youth Gang Problem,* 171–188; and Curry and Decker, *Confronting Gangs,* 141–151.

■ Summary and Conclusions

Much delinquent behavior is engaged in with peers, and youth who commit crimes with peers commit more crime and are involved in more serious crimes than youth acting alone. Delinquent youth also tend to have delinquent friends. In fact, the number of delinquent friends a youth has and the extent of their delinquency is one of the best predictors of that youth's delinquency. Despite these clear and consistent findings, the extent and the nature of peer influences are subjects of much debate.

Learning theories such as differential association theory and social learning theory contend that peer groups provide an intimate social context in which delinquent behavior is learned. Such learning occurs through communication, modeling, imitation, and reinforcement among peers. Peer-group association has the greatest influence when it occurs often, for longer periods, early in life, and when the youth values the relationship. In contrast, control theories minimize the importance of peer-group influences, arguing that delinquency results from weak social bonds or from lack of self-control. Delinquent youth seek out peers with similar attitudes and behaviors, and therefore delinquent peers are a result of delinquent involvement, not a cause.

Analysis of peer-group influence partially resolves these divergent views. Adolescents appear to develop friendships with delinquent peers before becoming involved in delinquent acts. Exposure to delinquent attitudes and behaviors thus is an important component of peer influence. More frequent and extensive association with delinquent peers leads to greater involvement in delinquency. Delinquent youth engage in the types of acts in which their delinquent group is most involved. Peer-group processes are also relevant to involvement in delinquent behavior, especially in gangs. Group processes related to cohesion, status threats, and group norms and values have a powerful influence on group delinquency, especially on aggression against other groups.

While peers appear to play an important role in the initiation and continuation of delinquent behavior, not all delinquent acts are group offenses. Youth who develop patterns of delinquency commit offenses both alone and with accomplices. Moreover, because delinquent groups tend to be small and transitory, delinquent youths commit offenses with only two or three other youths, and these companions change frequently.

Gangs constitute an important aspect of peer influence. The influence of gangs appears to go beyond that of delinquent peers. Gang members commit more frequent and more serious crime as compared with delinquent youth who are not gang members. Group processes related to group cohesion, status, and threat of violence are central to an understanding of how gangs influence the behavior of members. In addition, group characteristics such as gang culture and organizational structure provide the social context for group processes.

THEORIES

- differential association
- social learning
- social bond
- self-control

CRITICAL THINKING QUESTIONS

1. As described in the opening case example, what does Sidney's delinquent career tell us about the nature of involvement in delinquent groups?

2. According to differential association theory, how does group association vary? How is this variation important to peer group influences?

3. How does social learning theory extend differential association theory?

4. Distinguish the *socialization* and *selection perspectives* on peer group involvement.

5. What role do peers play in drug use?

6. What group processes are part of gang involvement?

SUGGESTED READINGS

Decker, Scott H., and Barrik Van Winkle. *Life in the Gang: Family, Friends, and Violence.* New York: Cambridge University Press, 1996.

Giordano, Peggy C., Stephen A. Cernkovich, and M. D. Pugh. "Friendship and Delinquency." *American Journal of Sociology* 5 (1986):1170–1202.

Howell, James C. "Youth Gangs: An Overview." Washington, DC: Office of Juvenile Justice and Delinquency Prevention, 1998.

Warr, Mark. *Companions in Crime: The Social Aspects of Criminal Conduct.* New York: Cambridge University Press, 2002.

GLOSSARY

definitions of the law: A person defines the law as favorable or unfavorable—either as rules to obey or to violate. A person's attitudes and beliefs toward the law, its legitimacy, and authority.

differential association: Association with different groups. According to differential association theory, association with groups varies in frequency, duration, priority, and intensity.

differential reinforcement: The social processes and dynamics of rewards and punishments that are the basis of learning.

enhancement model: The view that youth who are already involved in delinquent acts are attracted to and recruited by gangs because the norms and values of the gang are consistent with their own. The enhancement model incorporates the *selection perspective* of peer-group relations.

gang: "An interstitial group originally formed spontaneously and then integrated through conflict. It is characterized by the following types of behavior: meeting face to face, milling, movement through space as a unit, conflict, and planning. The result of this collective behavior is the development of tradition, unreflective internal structure, esprit de corps, solidarity, morale, group awareness, and attachment to a local territory" (Thrasher, *The Gang*, 57).

group offending or co-offending: Commitment of delinquent acts in the company of peers.

group processes: The ways in which relationships and interaction within groups influence the attitudes and behaviors of individuals.

imitation: The copying of behavior displayed or modeled by others.

operant conditioning: The process by which behavior is shaped by rewards and punishments.

selection perspective: The tendency to seek out individuals like oneself, with similar attitudes, values, beliefs, and behaviors. These characteristics become the criteria for choosing friends, rather than the product of friendship.

social facilitation model: The view that the normative support and groups processes of gangs encourage, support, and reinforce commission of crime. The social facilitation model incorporates the *socialization perspective* of peer-group relations.

social learning: Delinquent and criminal behavior is learned through social interaction in groups.

socialization perspective: The view that peer-group relations are strong, cohesive, and long-lasting and therefore able to exert strong influence on those involved. Peer-group relations and interactions are the context in which attitudes and behavior are learned.

REFERENCES

Agnew, Robert. "The Interactive Effects of Peer Variables on Delinquency." *Criminology* 29 (1991):47–72.

Akers, Ronald. *Deviant Behavior: A Social Learning Approach.* 3rd ed. Belmont, CA: Wadsworth, 1985.
———. "Self-Control as a General Theory of Crime." *Journal of Qualitative Criminology* 7 (1991):201–211.

Akers, Ronald, and Christine S. Sellers. *Criminological Theories: Introduction and Evaluation.* 4th ed. Los Angeles: Roxbury, 2004.

Altschuler, David M., and Paul J. Brounstein. "Patterns of Drug Use, Drug Trafficking, and Other Delinquency Among Inner-City Adolescent Males in Washington, DC." *Criminology* 29 (1991): 581–621.

Ball, Richard A., and G. David Curry. "The Logic of Definition in Criminology: Purposes and Methods for Defining 'Gangs.'" *Criminology* 33 (1995):225–245.

Battin, Sara R., Karl G. Hill, Robert D. Abbott, Richard F. Catalano, and J. David Hawkins. "The Contribution of Gang Membership to Delinquency Beyond Delinquent Friends." *Criminology* 36 (1998):93–115.

Bjerregaard, B., and C. Smith. "Gender Differences in Gang Participation, Delinquency, and Substance Use." *Journal of Quantitative Criminology* 9 (1993):329–355.

Burch, Jim, and Candice Kane. "Implementing the OJJDP Comprehensive Gang Model." *OJJDP Fact Sheet #12.* Washington DC: Office of Juvenile Justice and Delinquency Prevention, 1999.

Burgess, Robert L., and Ronald L. Akers. "A Differential Association–Reinforcement Theory of Criminal Behavior." *Social Problems* 14 (1966):128–147.

Bursik, Robert J., Jr., and Harold G. Grasmick. *Neighborhoods and Crime: The Dimensions of Effective Community Control.* New York: Lexington, 1995.

Coughlin, Brenda C., and Sudhir Alladi Venkatesh. "The Urban Street Gang After 1970." *Annual Review of Sociology* 29 (2003):41–64.

Curry, G. David, and Scott H. Decker. *Confronting Gangs: Crime and Community.* Los Angeles: Roxbury, 1998.

Decker, Scott H., and Barrik Van Winkle. *Life in the Gang: Family, Friends, and Violence.* New York: Cambridge University Press, 1996.

Elliott, Delbert S., David Huizinga, and Suzanne S. Ageton. *Explaining Delinquency and Drug Use.* Newbury Park, CA: Sage, 1985.

Elliott, Delbert, David Huizinga, and Scott Menard. *Multiple Problem Youth: Delinquency, Substance Abuse and Mental Health.* New York: Springer-Verlag, 1989.

Elliott, Delbert S., and Scott Menard. "Delinquent Friends and Delinquent Behavior: Temporal and Developmental Patterns." In *Delinquency and Crime,* edited by J. David Hawkins, 28–67. New York: Cambridge University Press, 1996.

Erickson, Maynard L. "The Group Context of Delinquent Behavior." *Social Problems* 19 (1971): 114–129.

Erickson, Maynard L., and LaMar T. Empey. "Class Position, Peers, and Delinquency." *Sociology and Social Research* 49 (1965):268–282.

Erickson, Maynard L., and Gary F. Jensen. "Delinquency is Still Group Behavior!: Toward Revitalizing the Group Premise in the Sociology of Deviance." *Journal of Criminal Law and Criminology* 68 (1977):262–273.

Esbensen, Finn-Aage, and David Huizinga. "Gangs, Drugs, and Delinquency in a Survey of Urban Youth." *Criminology* 31 (1993):565–589.

Fagan, Jeffrey. "The Social Organization of Drug Use and Drug Dealing Among Urban Gangs." *Criminology* 27 (1989):633–669.

Fagan, Jeffrey, Joseph G. Weis, and Yu-The Cheng. "Delinquency and Substance Use Among Inner-City Students." *Journal of Drug Issues* 20 (1990):351–402.

Gilligan, Carol. *In a Different Voice: Psychological Theory and Women's Development.* Cambridge, MA: Harvard University Press, 1982.

Giordano, Peggy C., Stephen A. Cernkovich, and M. D. Pugh. "Friendship and Delinquency." *American Journal of Sociology* 5 (1986):1170–1202.

Glueck, Sheldon, and Eleanor Glueck. *Unraveling Juvenile Delinquency.* New York: Commonwealth Fund, 1950.

Gold, Martin. *Delinquent Behavior in an American City.* Belmont, CA: Brooks-Cole, 1970.

Gordon, Rachel A., Benjamin B. Lahey, Eriko Kawai, Rolf Loeber, Magda Stouthamer-Loeber, and David P. Farrington. "Antisocial Behavior and Youth Gang Membership: Selection and Socialization." *Criminology* 42 (2004):55–89.

Gottfredson, Michael, and Travis Hirschi. *A General Theory of Crime.* Stanford, CA: Stanford University Press, 1990.

Hagedorn, John M. *People and Folks: Gangs, Crime, and the Underclass in a Rustbelt City.* Chicago: Lakeview Press, 1988.

Hepburn, John R. "Testing Alternative Models of Delinquency Causation." *Journal of Criminal Law and Criminology* 67 (1976):450–460.

Hindelang, Michael J. "The Social Versus Solitary Nature of Delinquent Involvements." *British Journal of Criminology* 11 (1971):167–175.

Hirschi, Travis. *Causes of Delinquency.* Berkeley, CA: University of California, 1969.

———. Review of *Explaining Delinquency and Drug Use,* by Delbert S. Elliott, David Huizinga, and Suzanne S. Ageton. *Criminology* 25 (1987):193–201.

Hochstetler, Andy, Heith Copes, and Matt DeLisi. "Differential Association in Group and Solo Offending." *Journal of Criminal Justice* 30 (2002):559–566.

Horowitz, Ruth. *Honor and the American Dream.* New Brunswick, NJ: Rutgers University Press, 1983.

Howell, James C. "Youth Gangs: An Overview." Washington, DC: Office of Juvenile Justice and Delinquency Prevention, 1998.

Huff, C. Ronald. "Comparing the Criminal Behavior of Youth Gangs and At-Risk Youths." Washington, DC: Office of Justice Programs, 1998.

Huizinga, David, Rolf Loeber, and Terence Thornberry. "Urban Delinquency and Substance Abuse." Washington DC: Office of Juvenile Justice and Delinquency Prevention, 1991.

Huizinga, David, Scott Menard, and Delbert Elliott. "Delinquency and Drug Use: Temporal and Developmental Patterns." *Justice Quarterly* 6 (1989):419–455.

Inciardi, James A., Ruth Horowitz, and Anne E. Pottieger. *Street Kids, Street Drugs, and Street Crime: An Examination of Drug Use and Serious Delinquency in Miami.* Belmont, CA: Wadsworth, 1993.

Jankowski, Martin Sanchez. *Islands in the Street: Gangs in American Urban Society.* Berkeley, CA: University of California Press, 1991.

Jensen, Gary F. "Parents, Peers, and Delinquent Action." *American Journal of Sociology* 78 (1972):562–575.

Johnson, Bruce, Eric Wish, J. Schmeidler, and David Huizinga. "Concentration of Delinquent Offending: Serious Drug Involvement and High Delinquency Rates." *Journal of Drug Issues* 21 (1991):205–229.

Chapter Resources

Johnson, Lloyd D., Patrick M. O'Malley, and Leslie K. Eveland. "Drugs and Delinquency: A Search for Causal Connections." In *Longitudinal Research on Drug Use,* edited by Denise B. Kandel, 137–156. Washington, DC: Hemisphere, 1978.

Johnson, Richard E. *Juvenile Delinquency and Its Origins: An Integrated Theoretical Approach.* New York: Cambridge University Press, 1978.

Johnson, Richard E., Anastasios C. Marcos, and Stephen J. Bahr. "The Role of Peers in the Complex Etiology of Adolescent Drug Use." *Criminology* 25 (1987):323–339.

Kandel, Denise B. "Homophily, Selection, and Socialization in Adolescent Friendships." *American Journal of Sociology* 84 (1978):427–436.

———. "Issues of Sequencing of Adolescent Drug Use and Other Problem Behaviors." *Drugs and Society* 3 (1988):55–76.

Kandel, Denise, and Mark Davies. "Friendship Networks, Intimacy, and Illicit Drug Use in Young Adulthood: A Comparison of Two Competing Theories." *Criminology* 29 (1991):441–467.

Kandel, Denise B., R. C. Kessler, and R. Z. Margulies. "Antecedents of Adolescent Initiation into Stages of Drug Use: A Developmental Analysis." *Journal of Youth and Adolescence* 7 (1978):13–40.

Kandel, Denise B., Ora Simcha-Fagan, and Mark Davies. "Risk Factors for Delinquency and Illicit Drug Use from Adolescence to Young Adulthood." *Journal of Drug Issues* 16 (1986):67–90.

Klein, Malcolm W. *The American Street Gang: Its Nature, Prevalence and Control.* New York: Oxford University Press, 1995.

———. "On the Group Context of Delinquency." *Sociology and Social Research* 54 (1969):63–71.

———. *Street Gangs and Street Workers.* Englewood Cliffs, NJ: Prentice Hall, 1971.

Klein, Malcolm W., and Lois Y. Crawford. "Groups, Gangs, and Cohesiveness." *Journal of Research in Crime and Delinquency* 4 (1967):63–75.

Krohn, Marvin D., Alan J. Lizotte, Terence P. Thornberry, Carolyn Smith, and David McDowall. "Reciprocal Causal Relationships Among Drug Use, Peers, and Beliefs: A Five-Wave Panel Model." *Journal of Drug Issues* 26 (1996):405–428.

Lazarsfeld, Paul F., and Robert K. Merton. "Friendship as a Social Process." In *Freedom and Control in Modern Society,* edited by M. Berger, T. Abel, and C. H. Page, 18–66. Princeton, NJ: Van Nostrand, 1954.

Loftin, Colin. "Assaultive Violence as a Contagious Process." *Bulletin of the New York Academy of Medicine* 62 (1984):550–555.

Longshore, Douglas, Eunice Chang, and Shih-chao Hsieh. "Self-Control and Social Bonds: A Combined Control Perspective on Deviance." *Crime and Delinquency* 50 (2004):542–564.

Marcos, Anastasios C., Stephen J. Bahr, and Richard E. Johnson. "Test of a Bonding/Association Theory of Adolescent Drug Use." *Social Forces* 65 (1986):135–161.

Matsueda, Ross L. "The Current State of Differential Association Theory." *Crime and Delinquency* 34 (1988):277–306.

———. "The Dynamics of Moral Beliefs and Minor Deviance." *Social Forces* 68 (1989):428–457.

———. "Testing Control Theory and Differential Association: A Causal Modeling Approach." *American Sociological Review* 47 (1982):489–504.

Matsueda, Ross L., and Kathleen Anderson. "The Dynamics of Delinquent Peers and Delinquency Behavior." *Criminology* 36 (1998):269–308.

Matsueda, Ross L., and Karen Heimer. "Race, Family Structure and Delinquency: A Test of Differential Association and Control Theories." *American Sociological Review* 52 (1987):826–840.

Mears, Daniel, Matthew Ploeger, and Mark Warr. "Explaining the Gender Gap in Delinquency: Peer Influence and Moral Evaluations of Behavior." *Criminology* 35 (1998):251–266.

Menard, Scott, and Delbert S. Elliott. "Delinquent Bonding, Moral Belief, and Illegal Behavior: A Three-Wave Panel Model." *Justice Quarterly* 11 (1994):173–188.

———. "Longitudinal and Cross-Sectional Data Collection and Analysis in the Study of Crime & Delinquency." *Justice Quarterly* 7 (1990):11–55.

Miller, Walter B. "Gangs, Groups, and Serious Youth Crime." In *Critical Issues in Juvenile Delinquency,* edited by David Schichor and Delos H. Kelly, 115–138. Lexington, MA: D.C. Heath, 1980.

Monti, Daniel J. *Wannabe: Gangs in Suburbs and Schools.* United Kingdom: Blackwell, 1994.

Moore, Joan W. *Going Down to the Barrio: Homeboys and Homegirls in Change.* Philadelphia, PA: Temple University Press, 1991.

———. *Homeboys: Gangs, Drugs, and Prisons in the Barrios of Los Angeles.* Philadelphia, PA: Temple University Press, 1978.

Morash, Merry. "Gangs, Groups, and Delinquency." *British Journal of Criminology* 23 (1983):309–331.

Chapter Resources

Padilla, Felix M. *The Gang as an American Enterprise.* New Brunswick, NJ: Rutgers University Press, 1992.

Reiss, Albert J., Jr. "Co-Offender Influences on Criminal Careers." In *Criminal Careers and "Career Criminals,"* vol. II, edited by Alfred Blumstein, Jacqueline Cohen, Jeffrey A. Roth, and Christy Visher, 121–160. Washington DC: National Academy Press, 1986.

———. "Co-Offending and Criminal Careers." In *Crime and Justice: A Review of Research,* vol. 10, edited by Michael Tonry and Norval Morris, 117–170. Chicago: University of Chicago Press, 1988.

Sampson, Robert J., and John H. Laub. *Crime in the Making: Pathways and Turning Points Through Life.* Cambridge, MA: Harvard University Press, 1993.

Shaw, Clifford T. *The Jack-Roller: A Delinquent Boy's Own Story.* Chicago: University of Chicago Press, 1930.

———. *The Natural History of a Delinquent Career.* Chicago: University of Chicago Press, 1931.

Shaw, Clifford R., Frederick M. Zorbaugh, Henry D. McKay, and Leonard S. Cottrell. *Delinquency Areas: A Study of the Geographic Distribution of School Truants, Juvenile Delinquents, and Adult Offenders in Chicago.* Chicago: University of Chicago Press, 1929.

Sherif, M., and C. W. Sherif. *Reference Groups.* New York: Harper and Row, 1964.

Short, James F., Jr. "Differential Association as a Hypothesis: Problems of Empirical Testing." *Social Problems* 8 (1960):14–25.

———. "Differential Association with Delinquent Friends and Delinquent Behavior." *Pacific Sociological Review* 1 (1958):20–25.

———. "The Level of Explanation Problem Revisited." *Criminology* 36 (1998):3–36.

Short, James F., and Fred L. Strodtbeck. *Group Process and Gang Delinquency.* Chicago: University of Chicago Press, 1965.

———. "The Response of Gang Leaders to Status Threats: An Observation on Group Process and Delinquent Behavior." *American Journal of Sociology* 68 (1963):571–579.

Spergel, Irving A. "Youth Gangs: Continuity and Change." In *Crime and Justice: A Review of Research,* vol. 12, edited by Michael Tonry and Norval Morris, 171–275. Chicago: University of Chicago Press, 1990.

———. *The Youth Gang Problem: A Community Approach.* New York: Oxford University Press, 1995.

Spergel, Irving A., and Ronald Chance. "National Youth Gang Suppression and Intervention Program." *National Institute of Justice Reports* (June 1991):21–24.

Sutherland, Edwin, Donald R. Cressey, and David F. Luckenbill. *Principles of Criminology.* 11th ed. Dix Hills, NY: General Hall, 1992.

Suttles, Gerald. *Social Construction of Communities.* Chicago: University of Chicago Press, 1972.

Taylor, Carl. *Dangerous Society.* East Lansing, MI: Michigan State University Press, 1990.

Thornberry, Terence P. "Toward an Interactional Theory of Delinquency." *Criminology* 25 (1987): 863–891.

Thornberry, Terence P., and J. H. Burch. "Gang Members and Delinquent Behavior." Washington, DC: Office of Juvenile Justice and Delinquency Prevention, 1997.

Thornberry, Terence P., Marvin D. Krohn, Alan J. Lizotte, and Deborah Chard-Wierschem. "The Role of Juvenile Gangs in Facilitating Delinquent Behavior." *Journal of Research in Crime and Delinquency* 30 (1993):55–87.

Thornberry, Terence P., Marvin D. Krohn, Alan J. Lizotte, Carolyn A. Smith, and Kimberly Tobin. *Gangs and Delinquency in Developmental Perspective.* New York: Cambridge University Press, 2003.

Thornberry, Terence P., Alan J. Lizotte, Marvin D. Krohn, Margaret Farnworth, and Sung Joon Jang. "Delinquent Peers, Beliefs, and Delinquent Behavior: A Longitudinal Test of Interaction Theory." *Criminology* 32 (1994):47–84.

Thrasher, Frederic M. *The Gang: A Study of 1,313 Gangs in Chicago.* Chicago: University of Chicago Press, 1927.

Venkatesh, Sudhir Alladir. *American Project: The Rise and Fall of a Modern Ghetto.* Cambridge, MA: Harvard University Press, 2000.

Vigil, James Diego. *Barrio Gangs.* Austin, TX: University of Texas Press, 1988.

Warr, Mark. "Age, Peers, and Delinquency." *Criminology* 31 (1993):17–40.

———. *Companions in Crime: The Social Aspects of Criminal Conduct.* New York: Cambridge University Press, 2002.

———. "Life-Course Transitions and Desistance From Crime." *Criminology* 36 (1998):183–215.

———. "Organization and Instigation in Delinquent Groups." *Criminology* 34 (1996):11–37.

Warr, Mark, and Mark Stafford. "The Influence of Delinquent Peers: What They Think or What They Do?" *Criminology* 29 (1991):851–866.

White, Helene Raskin. "Early Problem Behavior and Later Drug Problems." *Journal of Research in Crime and Delinquency* 29 (1992):412–429.

White, Helene Raskin, Robert J. Pandina, and Randy L. LaGrange. "Longitudinal Predictors of Serious Substance Use and Delinquency." *Criminology* 25 (1987):715–740.

Wilson, James Q., and Richard J. Herrnstein. *Crime and Human Nature*. New York: Simon and Schuster, 1985.

Wish, Eric. "U.S. Drug Policy in the 1990's: Insights from New Data from Arrestees." *International Journal of the Addictions* 25 (1990):1–15.

Wood, Peter B., Walter R. Gove, James A. Wilson, and John K. Cochran. "Nonsocial Reinforcement and Habitual Criminal Conduct: An Extension of Learning Theory." *Criminology* 35 (1997): 335–366.

Yablonsky, Lewis. "The Delinquent Gang as a Near Group." *Social Problems* 7 (1959):108–117.

———. *The Violent Gang*. New York: Macmillan, 1962.

ENDNOTES

1. Erickson, "Group Context;" Erickson and Jensen, "Delinquency;" Hindelang, "Social Versus Solitary;" Reiss, "Co-offending and Criminal Careers;" Shaw et al., *Delinquency Areas*; and Warr, "Organization and Instigation," and *Companions in Crime*.
2. Reiss, "Co-offending and Criminal Careers."
3. Shaw et al., *Delinquency Areas*.
4. Reiss, "Co-offending and Criminal Careers."
5. Erickson, "Group Context;" Erickson and Jensen, "Delinquency;" Hindelang, "Social Versus Solitary;" Inciardi, Horowitz, and Pottieger, *Street Kids*; and Reiss, "Co-offending and Criminal Careers."
6. Hirschi, *Causes of Delinquency*.
7. Warr, "Life-Course Transitions," 184. See, for example, Elliott, Huizinga, and Ageton, *Explaining Delinquency*; Erickson and Empey, "Class Position;" Erickson and Jensen, "Delinquency;" Hepburn, "Testing Alternative Models;" Hindelang, "Social Versus Solitary;" Johnson, *Juvenile Delinquency*; and Johnson, Marcos, and Bahr, "Role of Peers."
8. Agnew, "Interactive Effects," 47.
9. Klein, "Group Context."
10. Gold, *Delinquent Behavior*, 83–94.
11. Giordano, Cernkovich, and Pugh, "Friendship and Delinquency."
12. Warr, *Companions in Crime*.
13. Matsueda, "Current State;" and Sutherland, Cressey, and Luckenbill, *Principles of Criminology*.
14. Agnew, "Interactive Effects."
15. Akers and Sellers, *Criminological Theories*, 86.
16. Sutherland, Cressey, and Luckenbill, *Principles of Criminology*, 211–214.
17. Burgess and Akers, "Differential Association–Reinforcement Theory," 128–147.
18. Akers and Sellers, *Criminological Theories*, 85.
19. Ibid., 86.
20. Akers and Sellers, *Criminological Theories*; and Wood et al., "Nonsocial Reinforcement."
21. Akers and Sellers, *Criminological Theories*, 88.
22. Shaw, *Natural History*, 59–60.
23. Akers, *Deviant Behavior*.
24. Akers and Sellers, *Criminological Theories*, 86.
25. Agnew, "Interactive Effects," 50.
26. Inciardi, Horowitz, and Pottieger, *Street Kids*, 150.
27. Hirschi, *Causes of Delinquency*, 139.
28. Thornberry et al., "Role of Juvenile Gangs;" and Thornberry et al., "Delinquent Peers."
29. Glueck and Glueck, *Unraveling Juvenile Delinquency*, 164.
30. Wilson and Herrnstein, *Crime and Human Nature*.
31. Kandel, "Homophily;" and Sampson and Laub, *Crime in the Making*.

32. Hirschi, *Causes of Delinquency,* 159. Gottfredson and Hirschi's *General Theory of Crime* takes a very similar position.
33. Hirschi, *Causes of Delinquency,* 159, 141.
34. Gottfredson and Hirschi, *General Theory of Crime,* 159.
35. Hirschi, "Review;" and Warr, "Life-Course Transitions."
36. Hirschi, "Review," 198.
37. Hepburn, "Testing Alternative Models."
38. Lazarsfeld and Merton, "Friendship."
39. Giordano, Cernkovich, and Pugh, "Friendship and Delinquency;" and Kandel, "Homophily."
40. Shaw, *Natural History,* 53–76.
41. Glueck and Glueck, *Unraveling Juvenile Delinquency.*
42. Hirschi, *Causes of Delinquency,* 138, emphasis in original.
43. Gottfredson and Hirschi, *General Theory of Crime.*
44. Elliott and Menard, "Delinquent Friends." See also Menard and Elliott, "Longitudinal."
45. Elliott and Menard, "Delinquent Friends," 62–63.
46. Warr, "Age, Peers, and Delinquency."
47. Matsueda and Anderson, "Dynamics;" and Warr, "Life-Course Transitions."
48. Inciardi, Horowitz, and Pottieger, *Street Kids,* 168.
49. Kandel, "Homophily." See also Akers, "Self-Control;" Krohn et al., "Reciprocal Causal Relationships;" and Matsueda and Anderson, "Dynamics."
50. Thornberry, "Toward an Interactional Theory." See also Krohn et al., "Reciprocal Causal Relationships;" Matsueda and Anderson, "Dynamics;" and Thornberry et al., "Delinquent Peers."
51. Warr and Stafford, "Influence of Delinquent Peers," 851. See also Hochstetler, Copes, and DeLisi, "Differential Association."
52. Warr and Stafford, "Influence of Delinquent Peers," 851.
53. Matsueda, "Dynamics;" and Matsueda, "Testing Control Theory."
54. Warr and Stafford, "Influence of Delinquent Peers," 851.
55. Ibid., 862.
56. Reiss, "Co-Offender Influences," 152.
57. Elliott and Menard, "Delinquent Friends;" Erickson and Jensen, "Delinquency;" and Reiss, "Co-Offending and Criminal Careers," 121–122, 136, 137.
58. Reiss, "Co-Offending and Criminal Careers," 123, 151; and Warr, "Organization and Instigation."
59. Reiss, "Co-Offending and Criminal Careers," 145.
60. Elliott and Menard, "Delinquent Friends," 31.
61. Warr, "Organization and Instigation," 13.
62. See also Reiss, "Co-Offending and Criminal Careers," 125–126.
63. Warr, "Organization and Instigation," 16.
64. Ibid., 22.
65. Warr, "Organization and Instigation," 33. This finding that delinquent groups are transitory is reminiscent of Lewis Yablonsky's characterization of delinquent gangs as "near groups" ("Delinquent Gang").
66. Reiss, "Co-Offending and Criminal Careers," 117, 120, 129. See also Gottfredson and Hirschi, *General Theory of Crime,* 157–159.
67. Warr, "Organization and Instigation," 17.
68. Ibid., 33.
69. Agnew, "Interactive Effects;" Elliott and Menard, "Delinquent Friends," 29; Matsueda and Heimer, "Race," 831; Morash, "Gangs, Groups, and Delinquency," 319, 321; Reiss, "Co-Offending and Criminal Careers," 128; and Warr, "Age, Peers, and Delinquency," 19.
70. Elliott and Menard, "Delinquent Friends;" Kandel, Kessler, and Margulies, "Antecedents;" and Reiss, "Co-offending and Criminal Careers."
71. Agnew, "Interactive Effects;" Elliott, Huizinga, and Ageton, *Explaining Delinquency*; Erickson and Jensen, "Delinquency;" Reiss, "Co-Offending and Criminal Careers;" Short, "Problems of Empirical Testing;" and Warr, "Age, Peers, and Delinquency."
72. Agnew, "Interactive Effects." See also Menard and Elliott, "Delinquent Bonding."
73. Warr, "Age, Peers, and Delinquency," 31.
74. Giordano, Cernkovich, and Pugh, "Friendship and Delinquency."
75. Warr, "Life-Course Transitions."

76. Warr, "Age, Peers, and Delinquency," 34, emphasis in original.
77. Short, "Delinquent Friends."
78. Warr, "Age, Peers, and Delinquency."
79. See also Elliott and Menard, "Delinquent Friends."
80. Warr, "Age, Peers, and Delinquency," 24.
81. Ibid., 35.
82. Giordano, Cernkovich, and Pugh, "Friendship and Delinquency," 1172.
83. Shaw, *Jack-Roller,* 96.
84. Hirschi, *Causes of Delinquency,* 140–141.
85. Agnew, "Interactive Effects."
86. Warr, "Age, Peers, and Delinquency."
87. Giordano, Cernkovich, and Pugh, "Friendship and Delinquency."
88. Ibid., 1191.
89. Inciardi, Horowitz, and Pottieger, *Street Kids,* 163; and Kandel and Davies, "Friendship Networks."
90. Giordano, Cernkovich, and Pugh, "Friendship and Delinquency."
91. Mears, Ploeger, and Warr, "Explaining."
92. Gilligan, *In a Different Voice*; and Mears, Ploeger, and Warr, "Explaining," 254.
93. Kandel and Davies, "Friendship Networks;" and Marcos, Bahr, and Johnson, "Test."
94. Longshore, Chang, and Hsieh, "Self-Control and Social Bonds;" and Matsueda and Anderson, "Dynamics."
95. Kandel, Kessler, and Margulies, "Antecedents."
96. Krohn et al., "Reciprocal Causal Relationships."
97. Kandel and Davies, "Friendship Networks." See also Inciardi, Horowitz, and Pottieger, *Street Kids.*
98. Huizinga, Menard, and Elliott, "Delinquency and Drug Use."
99. Elliott, Huizinga, and Menard, *Multiple Problem Youth.*
100. White, "Early Problem Behavior." See also Huizinga, Loeber, and Thornberry, "Urban Delinquency."
101. White, "Early Problem Behavior," 414.
102. Elliott, Huizinga, and Ageton, *Explaining Delinquency*; and Elliott, Huizinga, and Menard, *Multiple Problem Youth.*
103. Elliott, Huizinga, and Menard, *Multiple Problem Youth*; Johnson, O'Malley, and Eveland, "Drugs and Delinquency;" Kandel, Simcha-Fagan, and Davies, "Risk Factors;" Kandel, Kessler, and Margulies, "Antecedents;" and Wish, "U.S. Drug Policy."
104. Wish, "U.S. Drug Policy."
105. Elliott, Huizinga, and Ageton, *Explaining Delinquency*; Kandel, "Issues of Sequencing;" and White, Pandina, and LaGrange, "Longitudinal Predictors."
106. Kandel, "Issues of Sequencing."
107. Elliott, Huizinga, and Menard, *Multiple Problem Youth.*
108. Elliott, Huizinga, and Menard, *Multiple Problem Youth*; Inciardi, Horowitz, and Pottieger, *Street Kids*; Johnson et al., "Concentration of Delinquent Offending;" and White, Pandina, and La-Grange, "Longitudinal Predictors." It should be pointed out that the number of youth who are both multiple-drug users and serious delinquents is very small. In a study of more than 1,700 youth, Elliott, Huizinga, and Menard (*Multiple Problem Youth*) found just 23 youth (1.4% of the sample) who were multiple drug users and serious delinquents.
109. Inciardi, Horowitz, and Pottieger, *Street Kids.* See also Elliott, Huizinga, and Menard, *Multiple Problem Youth,* 190.
110. Inciardi, Horowitz, and Pottieger, *Street Kids.*
111. Fagan, "Social Organization;" and Inciardi, Horowitz, and Pottieger, *Street Kids,* 46.
112. Altschuler and Brounstein, "Patterns of Drug Use;" Fagan, Weis, and Cheng, "Delinquency and Substance Use;" and Wish, "U.S. Drug Policy."
113. Fagan, "Social Organization;" and Padilla, *Gang.*
114. Inciardi, Horowitz, and Pottieger, *Street Kids.*
115. Monti, *Wannabe.*
116. Warr, "Organization and Instigation," 14. See also Morash, "Gangs, Groups, and Delinquency;" Reiss, "Co-Offending and Criminal Careers."

117. Battin et al., "Contribution;" Bjerregaard and Smith, "Gender Differences;" Esbensen and Huizinga, "Gangs, Drugs, and Delinquency;" Fagan, "Social Organization;" Gordon et al., "Antisocial Behavior;" Huff, "Comparing;" Spergel, "Youth Gangs;" Thornberry and Burch, "Gang Members;" Thornberry et al., "Role of Juvenile Gangs," and "Gangs and Delinquency;" and Venkatesh, *American Project*.

118. Battin et al., "Contribution;" Bjerregaard and Smith, "Gender Differences;" Esbensen and Huizinga, "Gangs, Drugs, and Delinquency;" Gordon et al., "Antisocial Behavior;" Thornberry and Burch, "Gang Members;" and Thornberry et al., *Gangs and Delinquency*.

119. Yablonsky, *Violent Gang,* 3.

120. Howell, "Youth Gangs."

121. Warr, "Organization and Instigation," 14.

122. Bursik and Grasmick, *Neighborhoods and Crime*; and Decker and Van Winkle, *Life in the Gang*.

123. Sherif and Sherif, *Reference Groups*.

124. Thrasher, *The Gang,* 57.

125. Ibid., 45.

126. Bursik and Grasmick, *Neighborhoods and Crime,* 121.

127. Bursik and Grasmick, *Neighborhoods and Crime*. This observation is based on Hagedorn, *People and Folks*. See also Spergel, "Youth Gangs," 179.

128. Bursik and Grasmick, *Neighborhoods and Crime,* 121.

129. Miller, "Gangs."

130. Ibid., 121.

131. Yablonsky, "Delinquent Gang;" and Yablonsky, *Violent Gang*.

132. Ball and Curry, "Logic of Definition;" Bursik and Grasmick, *Neighborhoods and Crime*; Curry and Decker, *Confronting Gangs*; Klein, *American Street Gang*; and Spergel, *Youth Gang Problem*.

133. Decker and Van Winkle, *Life in the Gang,* 2–3.

134. Klein, *Street Gangs,* 111.

135. Spergel, *Youth Gang Problem,* 309.

136. Decker and Winkle, *Life in the Gang,* 31.

137. Short, "Level of Explanation," 16.

138. Battin et al., "Contribution;" Bjerregaard and Smith, "Gender Differences;" Esbensen and Huizinga, "Gangs, Drugs, and Delinquency;" Fagan, "Social Organization;" Gordon et al., "Antisocial Behavior;" Huff, *Comparing*; Spergel, "Youth Gangs;" Thornberry and Burch, "Gang Members;" Thornberry et al., "Role of Juvenile Gangs," and *Gangs and Delinquency*; and Venkatesh, *American Project*.

139. Thornberry et al., "Role of Juvenile Gangs." See also Gordon et al., "Antisocial Behavior."

140. Thornberry et al., *Gangs and Delinquency,* 161–162.

141. Battin et al., "Contribution."

142. Decker and Van Winkle, *Life in the Gang,* 184.

143. Battin et al., "Contribution," 108. See also Esbensen and Huizinga, "Gangs, Drugs, and Delinquency."

144. Klein and Crawford, "Groups, Gangs, and Cohesiveness;" Klein, *Street Gangs*; Klein, *American Street Gang*; and Spergel, *Youth Gang Problem*.

145. Klein and Crawford, "Groups, Gangs, and Cohesiveness," 70.

146. Short and Strodtbeck, *Group Process*. Also Short and Strodtbeck, "Response of Gang Leaders."

147. Short and Strodtbeck, *Group Process,* 187, 196.

148. Decker and Van Winkle, *Life in the Gang,* 20–26.

149. Decker and Van Winkle, *Life in the Gang,* 22. This observation is consistent not only with Klein and Crawford ("Groups, Gangs, and Cohesiveness"), but also with a wide range of research on gangs, including Hagedorn (*People and Folks*) and Padilla (*Gang*).

150. Decker and Van Winkle, *Life in the Gang,* 22.

151. Loftin, "Assaultive Violence."

152. Decker and Van Winkle, *Life in the Gang*; Hagedorn, *People and Folks*; Short and Strodtbeck, *Group Process*; and Suttles, *Social Construction of Communities*.

153. Decker and Van Winkle, *Life in the Gang,* 188, see also 23–24, 224, 228.

154. Decker and Van Winkle, *Life in the Gang,* 273. Spergel (*Youth Gang Problem,* 103–104) and Howell ("Youth Gangs," 8–11) similarly contend that violence is a distinctive feature of gang life.

155. Decker and Van Winkle, *Life in the Gang,* 273, see also 173, 178–179. Short and Strodtbeck (*Group Process,* Chapter 9) also describe the importance of values and norms in gang culture. See also Short, "Level of Explanation."

156. Klein, *American Street Gang,* 23–30.

157. Klein, *American Street Gang,* 29. See also Decker and Van Winkle, *Life in the Gang,* 117–143; and Short and Strodtbeck, *Group Process.*

158. Decker and Van Winkle, *Life in the Gang,* 24.

159. Coughlin and Vernkatesh, "Urban Street Gang;" Decker and Van Winkle, *Life in the Gang,* 100; Hagedorn, *People and Folks*; Klein, *Street Gangs*; Short and Strodtbeck, "Response of Gang Leaders;" Vigil, *Barrio Gangs*; Thrasher, *The Gang*; and Yablonsky, "Delinquent Gang," and *Violent Gang.*

160. Klein, *Street Gangs*; and Short and Strodtbeck, "Response of Gang Leaders." See Thornberry et al., *Gangs and Delinquency,* for a different point of view.

161. Moore, *Homeboys* and *Going Down.*

162. Jankowski, *Islands in the Street,* 22.

163. Ibid., 78, 84.

164. Ibid., 78, 84.

165. Klein and Crawford, "Groups, Gangs, and Cohesiveness."

166. Horowitz, *Honor.*

167. Vigil, *Barrio Gangs.*

168. Howell, "Youth Gangs;" Padilla, *Gang*; and Taylor, *Dangerous Society.*

169. Padilla, *Gang,* 3. Similar observations are made by Taylor in a study of Detroit gangs (*Dangerous Society*).

170. Decker and Van Winkle, *Life in the Gang,* 16–18.

Social Structure Theories: Community, Strain, and Subcultures

12

Chapter Outline

- The foundation of social structure theories: Durkheim's *social solidarity*
- Social disorganization theory
- Anomie and strain theories

Chapter Objectives

After completing this chapter, students should be able to:

- Offer several examples of *social structures*—organizational features of the social environment.
- Explain how *societal* characteristics exert influence on *individuals*.
- Describe how rapid social change influences social integration and social regulation.
- Explain how and why social disorganization leads to a breakdown in social control.
- Summarize the social ecology of delinquency in Chicago as portrayed in the research of Clifford Shaw and Henry McKay.
- Distinguish anomie and strain theories.
- Describe the "strains" that lead to gang delinquency among urban, lower-class males.

- Understand theories and key terms:

Theories:
social solidarity
social disorganization
collective efficacy
anomie
institutional anomie
strain
general strain
reaction formation
differential opportunity

Terms:
social structures
solidarity
mechanical solidarity
organic solidarity
anomie
collective efficacy
cultural goals
institutional means
modes of adaptation
status frustration
reaction formation

CASE IN POINT   Living in "Delinquent Areas"

Clifford Shaw's life history of Sidney Blotzman describes the social forces that led to Sidney's delinquent career. One of Shaw's chief concerns was with the deteriorated and disorganized neighborhood environments in which delinquent peer groups flourished and influenced young people. He referred to these neighborhoods as "delinquency areas." Shaw provided the following description of the neighborhoods in which Sidney lived and how those areas failed to control the behavior of youth. Think about the various ways in which the characteristics of these local neighborhoods might have influenced Sidney's delinquent career.

[Sidney] lived in one of the most deteriorated and disorganized sections of the city. . . . The most obvious characteristic of the neighborhood is the marked physical deterioration of its buildings and the presence of large accumulations of debris in the streets and alleys. Everywhere are unpainted, dilapidated wooden tenements, interspersed with warehouses, junk yards, and factories. With the rapid growth of the city, the central business and industrial district has gradually encroached upon this neighborhood, resulting in physical deterioration and a change in use of the land. . . .

Everywhere throughout the neighborhood is evidence of the inferior economic status of the inhabitants. Rents are universally low. . . . The deteriorated condition of the neighborhood renders it extremely undesirable for residential purposes. The low rents attract population groups representing the lowest economic class, which is composed largely of the most recent immigrants. Families and individuals escape from the neighborhood as rapidly as they prosper sufficiently to do so. . . .

For the purpose of understanding Sidney's delinquent behavior it is important to draw attention to the social confusion and disorganization which characterize this area. The successive changes in the composition of population, the disintegration of the alien cultures, the diffusion of divergent cultural standards, and the gradual industrialization of the area have resulted in a dissolution of the neighborhood culture and organization. The continuity of conventional neighborhood traditions and institutions is broken. Thus, the effectiveness of the neighborhood as a unit of control and as a medium for the transmission of the moral standards of society is greatly diminished. . . .

In this area the conventional traditions, neighborhood institutions, and public opinion, through which neighborhoods usually effect a control over the behavior [of the] child, were largely disintegrated. Consequently, Sidney had little access to the cultural heritages of conventional society and he was not subject to the constructive and restraining influences which surround the child in the more highly integrated and conventional residential neighborhoods of the city. In the area in which he lived neighborhood control was limited largely to the control that was exerted through such formal agencies as the school, the courts, and the police.

This community situation was not only disorganized and thus ineffective as a unit of control, but it was characterized by a high rate of juvenile delinquency and adult crime, not to mention the widespread political corruption which had long existed in the area. Various forms of stealing and many organized delinquent and criminal gangs were prevalent in the area. These groups exercised a powerful influence and tended to create a community spirit which not only tolerated but actually fostered delinquent and criminal practices.

Source: Shaw, *Natural History,* 229, 13–15, 229. Copyright © 1931, University of Chicago Press. All rights reserved. Reproduced with permission of the publisher.

The neighborhoods in which Sidney Blotzman grew up were places of physical deterioration and social disorder. Through Sidney's life history, well-known criminologist Clifford Shaw tried to show that such neighborhoods lacked social control and that, as a result, their youth were exposed to delinquent and criminal behavior patterns. These neighborhoods had high rates of delinquency, which is why Shaw referred to them as *delinquency areas*.[1]

While Shaw carried out case studies to document "the natural history of a delinquent career," he also conducted empirical studies of delinquency areas with another criminologist, Henry McKay. Rather than trying to account for an individual's involvement in delinquent acts, these studies examined the social characteristics of communities that were linked to high rates of delinquent offenses. Shaw and McKay identified three key structural characteristics that disrupt community social organization and, in turn, are related to high rates of crime and delinquency: *low economic status, ethnic heterogeneity,* and *residential mobility*.[2] Sociologists refer to such organizational features of the social environment as **social structures**. Beyond the three structural characteristics that Shaw and McKay emphasized, the concept of social structure also refers to characteristics such as age and gender distributions; cultural traditions; the vitality of institutions like the family, school, and church; and statuses and roles.[3] When applied to delinquent behavior, theoretical approaches that emphasize elements of the social structure search for causes that are external to the individual.

social structures Organizational features of the social environment, such as neighborhood socioeconomic status, ethnic heterogeneity, residential mobility, cultural traditions, and age and gender distributions.

■ The Foundation of Social Structure Theories: Durkheim's Social Solidarity

Theory and research that focus on structural characteristics of the social environment have their roots in the work of French sociologist Emile Durkheim. Writing in the late 1800s, Durkheim concentrated on the social factors that integrate and regulate individuals and groups.[4] The time period in which Durkheim wrote was a period of rapid urbanization and industrialization; Durkheim was especially intrigued with how such rapid social change influenced social integration and regulation. Within the concept of **social solidarity** (the social structures of society that provide integration and regulation), Durkheim distinguished two subtypes of solidarity: mechanical and organic.[5]

Mechanical solidarity refers to social integration and regulation that develop in small, cohesive societies in which people do the same type of work and have the same values and norms. Integration is based upon social similarity, or what Durkheim called *collective consciousness*—"the totality of beliefs and sentiments common to average citizens."[6] Collective consciousness results in a strong (Durkheim used the word "passionate") and unified community response when rules or customs of society are broken.[7] Durkheim referred to this type of social control as "penal" or "repressive law."[8] Sanctions such as banishment and corporal and capital punishments are common under penal law.

Organic solidarity is based upon the complementary nature of social life in modern, complex societies. Large, urban populations with cultural diversity and a "diversion of labor" make interdependence necessary. As a result, organic sol-

social solidarity Social structures of society that integrate and regulate individuals and groups.

mechanical solidarity Social integration and regulation based on social similarity; develops in small, cohesive societies in which people do the same type of work and have the same values and norms.

organic solidarity Social integration and regulation based on the complementary nature of social life; develops in modern, complex societies in which interdependence is necessary.

idarity is more fragile than mechanical solidarity. Durkheim characterized the law associated with organic solidarity as "cooperative" or "restitutive" law because of its complementary character. Fines and other civil law compensations are examples of comparative law; they do not usually imply a crime against society, nor does the sanction typically involve social condemnation or censure.[9]

Regardless of the form of solidarity, societies that experience rapid social change are often unable to effectively regulate individuals. Durkheim referred to the breakdown in societal regulation as **anomie**. Anomie is sometimes simply defined as normlessness. The concept of rapid social change resulting in anomie is an important legacy of Durkheim that appears in social structure theories of delinquency.

Based on Durkheim's work, a number of theoretical perspectives have developed that emphasize structural features of the social environment in relation to delinquency.[10] This chapter will describe two prominent theoretical traditions within the structural perspective: social disorganization theories, and anomie and strain theories. *Social disorganization theories* focus on the breakdown of traditional patterns of social life at the local, neighborhood level as a result of rapid change in urban environments. *Anomie* and *strain theories* deal with availability of legitimate opportunities for material success at both the individual and societal levels.

anomie According to Durkheim, a societal condition of normlessness in which the mechanisms of social control break down as a result of rapid social change.

■ Social Disorganization Theory

Rapid social change and physical deterioration have long been associated with the breakdown in community social control and consequently with high rates of delinquent offenses. This perspective, referred to as *social disorganization theory,* was most extensively advanced by faculty members and people affiliated with the Department of Sociology at the University of Chicago. Established in 1892, this department was the first academic unit in the United States devoted to sociology. The development of sociology as a discipline was connected strongly with the research, writings, and theories of individuals associated with this department.[11] As a result, the proponents of social disorganization theory are often referred to as the *Chicago School.*

The Chicago School

In these formative years for the discipline, sociological attention centered on the social problems that plagued urban communities, and sociologists sought to study and explain these social problems so that practical ways of eliminating them could be developed.[12] The problems of urban life were associated with rapid population growth, residential mobility, poverty, physical deterioration, and increased cultural diversity. Special attention was given to the poor immigrants and migrants who flocked to the industrial centers of the United States in the late 1800s and early 1900s, hoping to find work and prosperity in the major cities. Chicago sociologists believed that this infusion of people, together with the associated increase in cultural diversity and physical deterioration, led to a decline in *moral order,* or what Durkheim had referred to as solidarity—the interconnectedness and cohesiveness of people.[13]

Two pioneering Chicago sociologists, Robert Park and Ernest Burgess, observed that a variety of social problems were distributed *ecologically*—in a geographic pattern associated with the growth of cities.[14] In particular, Ernest Burgess developed the *concentric zone model* of urban development, in which he argued that urban areas grow through progressive expansion from the central city outwards, in a series of concentric zones.[15] **Figure 12-1** depicts Burgess' model of urban growth.

As cities grow larger, business and industry of the central city spill over into adjacent residential areas, producing an area of *mixed use*. This area just outside the central city is referred to as the *zone in transition*, where newcomers settle, attracted by factory jobs and inexpensive housing. The *zone of workingmen's homes* lies beyond these central areas and is populated by second-generation immigrants who have been able to escape the zone of transition. Further out are the *residential zone* and *commuters' zone*, where white, middle- and upper-class homeowners reside.

Associated with the concentric zone model of urban development were a number of identifiable patterns of the distribution of people and their activities. These patterns are referred to as *human ecology*. Using Chicago as a "laboratory," early sociologists observed that immigrant groups tended to move into the city in waves—first one ethnic group, then another. Often poor, these immigrant groups tended to locate in areas with the least expensive housing and in neighborhoods with individuals of similar ethnicity. In the early part of the twentieth

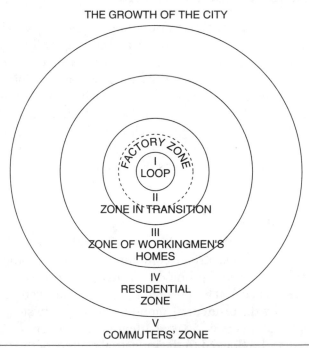

Figure 12-1 Burgess' concentric zone model of urban development. Ernest Burgess hypothesized that urban places grow in a systematic, progressive way, both in terms of population distribution and land use. His colleagues at the University of Chicago also argued that urban growth has social consequences and that they too are distributed ecologically. This is referred to as *social ecology*.

Source: Shaw, Zorbaugh, McKay, and Cottrell, *Delinquency Areas,* 19. Copyright © 1929, University of Chicago Press. All rights reserved. Reproduced with permission of the publisher.

century, these areas constituted the zone of transition. This ecological process was referred to as *ethnic invasion*, which resulted in *segregation*. As individuals achieved greater economic success, they moved into outlying areas, and new immigrant groups moved into the transitional area—a process called *residential succession*. These ecological processes were viewed as common to all cities that experienced rapid urban and industrial growth at the turn of the century.[16]

After gathering data, Chicago sociologists observed that a wide variety of social problems—alcoholism, mental disorder, crime and delinquency, even infant mortality—were highest in zones closest to the center of the city. The central business district and the zone in transition accounted for the highest rates of social problems, with rates declining progressively in zones farther from the city center. Advocates of the social disorganization perspective argued that characteristics of local neighborhoods, not characteristics of individuals in those areas, were the primary influence on rates of social problems.

More recently, Rodney Stark advanced five structural characteristics that he believes summarize the idea of social disorganization: (1) density (many people in a small area); (2) poverty; (3) mixed use (a combination of residence, industry, and retail); (4) transience (people moving in and out of the area); and (5) dilapidation (physical deterioration of buildings).[17] Like earlier Chicago sociologists, Stark argues that these key dimensions of social disorganization break down traditional patterns of social life. Using the concept introduced by Robert Park and Ernest Burgess, Stark describes this breakdown as decline in the moral order of the community. More specifically, he proposes that the declining moral order involves moral cynicism (a skepticism about the moral character of others), increased opportunities for crime and deviance, increased motivation to deviate, and diminished social control.[18] Stark describes these areas of declining moral order as "deviant places."

Shaw and McKay's *Delinquency Areas*

The application of social disorganization theory to delinquency was carried out most extensively by Clifford Shaw and Henry McKay. Though Shaw and McKay received graduate training in sociology at the University of Chicago, both left the graduate program before earning doctorate degrees. Nonetheless, they were influenced heavily by the sociological point of view of the Chicago School, especially the writings of W. I. Thomas, Ellsworth Faris, Robert Park, and Ernest Burgess. Shaw spent his entire career at the Institute for Juvenile Research, where he directed the Sociology Department, and McKay was this department's lead researcher for much of his career. The Institute was heavily psychological in orientation, but under Shaw's leadership, the Sociology Department conducted sociological research on delinquency and engaged in delinquency prevention based on the findings of that research. As a result, much of Shaw and McKay's research resulted in both theory and practice.

Drawing on the themes of the Chicago School, Shaw and McKay studied how the urban social environment influenced delinquency rates in the city of Chicago in the first half of the 1900s. They observed that delinquency rates varied extensively across the city, a finding they referred to as the "geographic distribution" of delinquency.

While their empirical studies documented the geographic distribution of delinquency and isolated key community characteristics associated with the ecology of delinquency, Shaw and McKay used case studies of delinquent youth to illustrate the social, psychological and cultural processes that invited involvement in delinquent behavior in those areas with high rates of delinquent offending.

The Ecology of Delinquency

Shaw and McKay's findings on the geographic distribution of delinquency were first presented in their book *Delinquency Areas* (1929).[19] The primary purpose of the study was to investigate the *ecological distribution* of delinquent behavior and to discover the community characteristics that were associated with high rates of delinquency. The researchers used Chicago police and juvenile court records to develop measures of delinquency, including truancy, juvenile court appearances, felony adjudications, and recidivism. These measures were depicted geographically on a series of Chicago maps. Four types of maps were used to depict the different measures of delinquency: (1) *spot maps,* which depicted offenders' places of residence; (2) *rate maps,* which were graphic representations of delinquency rates in square mile areas, (3) *radial maps,* which provided delinquency rates along radials emanating from the center of the city, and (4) *zone maps,* which depicted delinquency rates in concentric zones. **Figure 12-2** shows the zone map for the rate of male juvenile court appearances from 1900 to 1906. The resulting collection of maps—40 maps in all—is an unparalleled study of the social ecology of delinquency in a particular city.

Shaw and McKay's quantitative data were supplemented with a number of autobiographical accounts from delinquent youth. A delinquent's "own story" was used to illustrate the link between community background and involvement in delinquency.[20] These case studies were drawn from the large number of life histories that Shaw compiled through interviews and autobiographies of delinquent youth. The case that opened this chapter was part of one such life history.

Shaw and McKay drew a number of conclusions from their study. We highlight four here because they speak directly to the link between community characteristics and rates of delinquency.

1. There are marked variations in the rate of school truants, juvenile delinquents, and adult criminals between areas in Chicago. Some areas are characterized by very high rates, while others show very low rates . . .

2. Rates of truancy, delinquency, [recidivism], and adult crime tend to vary inversely in proportion to the distance from the center of the city. In general, the nearer to the center of the city a given locality is, the higher will be its rates of delinquency and crime . . .

3. Differences in rates of truancy, delinquency, and crime reflect differences in community backgrounds. High rates occur in areas which are characterized by physical deterioration and declining population. . . .

4. The main high rate areas of the city . . . have been characterized by high rates over a long period. . . .[21]

The ecological distribution of delinquency was explored more fully in Shaw and McKay's subsequent work, *Juvenile Delinquency and Urban Areas,* first published in 1942 and revised in 1969.[22] This work also provided extensive study

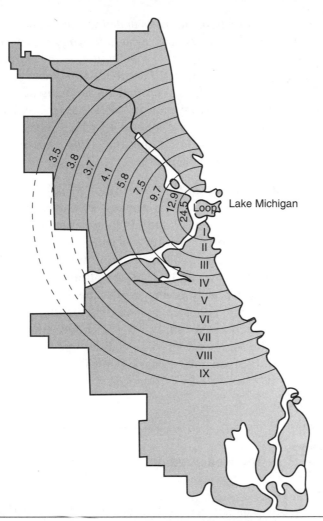

Figure 12-2 Zones of delinquency in Chicago, 1900–1906. This map identifies the percent of boys (males under the age of 18) who were brought before the juvenile court from 1900 to 1906. This period covers the first seven years of the juvenile court's existence. The Chicago Juvenile Court was the first juvenile court in the United States, established in 1899.

Source: Shaw, Zorbaugh, McKay, and Cottrell, *Delinquency Areas,* 99. Copyright © 1929, University of Chicago Press. All rights reserved. Reproduced with permission of the publisher.

of the association between characteristics of the local community and rates of delinquency. In fact, the subtitle of the book is *A Study of Rates of Delinquency in Relation to Differential Characteristics of Local Communities in American Cities.* Shaw and McKay examined the ecological distribution of delinquency in several US cities in this work.

Shaw and McKay's findings in *Juvenile Delinquency and Urban Areas* confirmed their earlier work in *Delinquency Areas,* showing that rates of delinquency vary widely across different areas of Chicago and that there was stability to such variation—*ecological stability.*[23] As before, they argued that the ecological distribution of delinquency is most closely associated with the characteristics of the local community, not with characteristics of individuals:

> It is clear from the data included in this volume that there is a direct relationship between conditions existing in local communities of

American cities and differential rates of delinquents and criminals. Communities with high rates have social and economic characteristics which differentiate them from communities with low rates . . . Moreover, the fact that in Chicago the rates of delinquents for many years have remained relatively constant in the areas adjacent to centers of commerce and heavy industry, despite successive changes in the nativity and nationality composition of the population, supports emphatically the conclusion that the delinquency-producing factors are inherent in the community.[24]

Social Disorganization and the Breakdown of Social Control

Shaw and McKay's basic argument is that processes of urban growth influence community characteristics. Three structural characteristics are especially disruptive of community social organization: low economic status, ethnic heterogeneity, and residential mobility.[25] These community characteristics do not directly cause delinquency, however. Instead, they limit informal social control and thereby allow high rates of delinquency and crime to occur.[26] Juvenile delinquency, then, is most directly a consequence of the breakdown of social control among primary groups such as the family, neighborhood, church, and school.[27] Research in Action, "The Measurement of Social Disorganization," provides an example of how social disorganization has been measured in recent research.

RESEARCH IN ACTION

The Measurement of Social Disorganization

As part of the contemporary resurrection of social disorganization theory, Robert Sampson and W. Byron Groves in 1989 published an influential article in which they empirically tested Clifford Shaw and Henry McKay's social disorganization theory. Sampson and Groves hypothesized that "low economic status, ethnic heterogeneity, residential mobility, and family disruption lead to community social disorganization, which, in turn, increases crime and delinquency rates" (page 774). To test their hypotheses, they used data gathered in two separate national surveys of residents of Great Britain. Each survey included approximately 11,000 respondents. Sampson and Groves found support for their hypothesis and thus for social disorganization theory.

Sampson and Groves clearly distinguish social disorganization from the factors that produce it, which include poverty, ethnic heterogeneity, and residential mobility. Sampson and Groves view social disorganization and social organization as two ends of a continuum. Thus, to assess social disorganization, they used three measures:

Sparse local friendship networks
Respondents were asked how many of their friends resided in the local community, which was defined as the area within a 15-minute walk of the respondent's home. (Responses were on a five-point scale ranging from none to all.)

Low organizational participation

Respondents were asked about their attendance at meetings of committees or clubs in the week before the interview.

Unsupervised youth peer groups

Respondents were asked how common it was for groups of teenagers to hang out in public in the neighborhood and make nuisances of themselves. (Responses were on a four-point scale.)

Source: Sampson and Groves, "Community Structure and Crime," 774, 777–778, 783–784.

Even though Shaw and McKay did not clearly specify how social disorganization inhibits social control, they identified several social processes that begin with social disorganization and end in diminished social control.

1. Youths fail to acquire internal controls when they are not socialized into the normative expectations of the local community. Instead, youths may be exposed to delinquent and criminal behavior patterns.

2. Youths lack primary relationships with conforming friends and adults, thereby limiting informal controls of relationship bonds.

3. Lack of commitment and involvement in the local community preclude direct controls, especially the supervision that goes along with active community membership.

4. Limited interaction and communication inhibit the development of common community values and norms and hinder community controls and efforts to solve common community problems, including delinquency.[28]

Research in Action, "Testing the 'Broken Windows' Thesis," discusses a contemporary interpretation of the relationships among disorder, informal social control, and crime. This explanation is called the "broken windows" thesis.

Cultural Aspects of Social Disorganization

Over the course of their work, Shaw and McKay emphasized different aspects of social disorganization. In its initial form, social disorganization theory emphasized the loss of a common culture in the community—"the customs, codes, taboos, and traditions" of the local neighborhood.[29] In their first work, *Delinquency Areas*, Shaw and McKay explained ecological variation in delinquency rates by considering the cultural consequences of social disorganization: "In the areas close to the central business district, and to a less extent in the areas close to industrial developments, the neighborhood organization tends to disintegrate. For in these areas the mobility of population is so great that there is little opportunity for the development of common attitudes and interests [a common culture]."[30] Robert Bursik comments that social disorganization in its purest form "refers to the inability of local communities to realize the common values of their residents or solve commonly experienced problems."[31] With regard to delinquent behavior, "the absence of common community ideals and standards prevents cooperative social action ... to ... suppress delinquency."[32] Shaw's

RESEARCH IN ACTION — Testing the "Broken Windows" Thesis

Social disorganization and social and physical disorder are separate but related. *Social disorder* refers to "behavior usually involving strangers and considered threatening, such as verbal harassment on the street, open solicitation for prostitution, public intoxication, and rowdy groups of young males in public." *Physical disorder* refers to "the deterioration of urban landscapes, for example, graffiti on buildings, abandoned cars, broken windows, and garbage in the streets" (Sampson and Raudenbush, "Systematic Social Observation," 603–604).

Disorder signals to observers both inside and outside the neighborhood that residents are unwilling or unable to exercise informal social control. In other words, disorder can be seen as a symptom or outcome of social disorganization.

The "broken windows" thesis, first advanced by James Wilson and George Kelling, proposes that if social and physical disorder in urban neighborhoods goes unchecked, it can lead to serious crime. "The reasoning is that even such minor public incivilities as drinking in the street, spray-painting graffiti, and breaking windows can escalate into predatory crime because prospective offenders assume from these manifestations of disorder that area residents are indifferent to what happens in their neighborhood" (Sampson and Raudenbush, "Disorder in Urban Neighborhoods," 1).

The broken windows thesis has generated changes in policing strategies in many cities, most notably in New York City, where police have become more vigilant in responding to public signs of disorder, including even minor "incivilities." The goal of these "zero tolerance" policies is to curb serious crime by addressing the disorder that might otherwise promote serious offending.

Sampson and Raudenbush recently tested the broken windows thesis using observational data from the Project on Human Development in Chicago Neighborhoods. The innovative method of gathering these data consisted of having trained observers videotape (using video recorders mounted to a sport utility vehicle) the "face blocks" of more than 23,000 streets in 196 Chicago neighborhoods. A "face block" is defined as a one-block segment on only one side of a street. Trained observers also kept written logs of their observations on each face block. Sampson and Raudenbush refer to this type of measurement as *systematic social observation,* in which "the means of observation are independent of what is observed." Using this method, places and actions are observed as they "naturally" occur (Sampson and Raudenbush, "Systematic Social Observation," 616; "Disorder in Urban Neighborhoods," 3).

As an alternative to the broken windows thesis that disorder produces serious crime, Sampson and Raudenbush propose that "disorder and crime are manifestations of the same phenomenon," and that both stem from structural characteristics of neighborhoods, such as concentrated disadvantage and residential instability, and from low levels of "collective efficacy" among neighbors. Collective efficacy is defined as "cohesion among neighborhood residents combined with shared expectations for informal social control of public space" (Sampson and Raudenbush, "Disorder in Urban Neighborhoods,"

1–2). (See Research in Action, "The Measurement of Collective Efficacy," on pages 485–486, for a description of collective efficacy.)

In support of their hypothesis and in contrast to the broken windows thesis, Sampson and Raudenbush found that disorder was directly linked only to robbery, but not to other types of crime. Instead, structural characteristics of neighborhoods (especially concentrated disadvantage) and collective efficacy predicted both disorder and crime. Once collective efficacy and structural characteristics of neighborhoods were taken into account, the relationship between disorder and crime disappeared in most cases.

Sampson and Raudenbush note the implication of their findings for policing policies: "The findings strongly suggest that policies intended to reduce crime by eradicating disorder solely through tough law enforcement tactics are misdirected" (Sampson and Raudenbush, "Disorder in Urban Neighborhoods," 5).

Sources: Sampson and Raudenbush, "Systematic Social Observation;" Sampson and Raudenbush, "Disorder in Urban Neighborhoods;" Sampson and Raudenbush, "Seeing Disorder;" Wilson and Kelling, "Police and Neighborhood Safety."

description of Sidney's neighborhood, which introduced this chapter, emphasized the cultural consequences of social disorganization.

Social Disorganization and Delinquent Subcultures

The absence of effective community social control also allows for the development of delinquency subcultures whereby delinquent values are transmitted through association with delinquent youths and adult criminals in the neighborhood. Delinquent subcultures that spring up in socially disorganized areas are the vehicles for passing on delinquent traditions, values, and expectations. As Shaw and McKay observe:

> The presence of a large number of older offenders in a neighborhood is a fact of great significance for the understanding of the problem of juvenile delinquency. It indicates, in the first place, that the possibility of contact between the children and hardened offenders is very great. The older offenders, who are well known and have prestige in the neighborhood, tend to set standards and patterns of behavior for the younger boys, who idolize and emulate them. In many cases the "big shot" represents for the young delinquent an ideal around which his own hopes and ambitions are crystallized. His attainment of this coveted ideal means recognition in his group and the esteem of his fellows.[33]

Shaw and McKay's emphasis on group association clearly foreshadowed the development of *differential association theory*, discussed in Chapter 11, and *subcultural theories* within the strain tradition, which we address later in this chapter. The transmission of a delinquent culture from older youths and adults in the neighborhood to younger, aspiring delinquents implies a level of organization that might seem contrary to the basic idea of social disorganization.[34] However, the absence of a common culture in socially disorganized neighborhoods and the subsequent breakdown in effective community control provides fertile ground for delinquent subcultures. Here too, Shaw and McKay's emphasis on group association is based in the socially disorganized community context in which delinquent peers and adult offenders are present and flourishing.

Strain in Social Disorganization Theory

So far we have seen that Shaw and McKay's work focused on the consequences of urban growth in terms of culture, social control, group associations, and the development and transmission of delinquent subcultures. The rapid social change associated with urban growth came to an abrupt halt with the nationwide economic collapse (1929–1933) that precipitated the Great Depression. Widespread unemployment resulted in declining immigration and migration to cities and greatly reduced residential mobility. As a result, the ecological processes of ethnic invasion, segregation, and residential succession that characterized urban growth in the late nineteenth century and early twentieth century slowed dramatically. With this change, Shaw and McKay gave greater emphasis in their research to issues of poverty, unemployment, and social class.[35] The social disorganization of inner-city neighborhoods was conceived not so much in terms of culture and social control as in terms of lack of economic opportunity, relative deprivation, and the resulting personal frustration and strain that might motivate involvement in criminal and delinquent acts. This is expressed clearly in a closing passage of *Juvenile Delinquency and Urban Areas:*

> *Despite these marked differences in relative position [income and status] in different communities, children and young people in all areas, both rich and poor, are exposed to the luxury values and success patterns of our culture. In school and elsewhere they are also exposed to ideas of equality, freedom, and individual enterprise. Among children and young people residing in low-income areas, interests in acquiring material goods and enhancing personal status are developed which are often difficult to realize by legitimate means because of limited access to the necessary facilities and opportunities.[36]*

Shaw and McKay's point of view here is strikingly similar to Robert Merton's *anomie theory,* another structural theory discussed later in this chapter.[37]

Delinquency Rates in Other Urban Areas: Applications of Social Disorganization Theory

Shaw and McKay's second major ecological study, *Juvenile Delinquency and Urban Areas,* included analysis of Philadelphia, Boston, Cincinnati, Cleveland, and Richmond, Virginia. From studies conducted in these cities, Shaw and McKay observed that delinquent offenses were distributed in an ecological pattern similar to the one in Chicago, and that this persistent pattern reflected common conditions of urban communities.[38] Following the work of Shaw and McKay, a number of other researchers conducted area studies in Baltimore, Detroit, and Indianapolis.[39] These studies found that, while some neighborhood characteristics, such as substandard housing and residential overcrowding, were unrelated to delinquency rates, other socioeconomic factors of neighborhoods were consistently related to delinquency rates.

While researchers have connected the community characteristics associated with social disorganization to high rates of delinquent behavior, the theory of social disorganization has experienced only limited direct testing.[40] The theory of social disorganization argues that these structural community characteristics affect social organization, especially primary group controls, thereby allowing higher rates of delinquency. According to Robert Sampson and W. Byron Groves, it is this intervening variable of social organization that must be considered to directly test the full theory.[41] Many studies that have tried to test social disorganization theory have measured only whether structural characteristics of communities are related to their rates of crime. While valuable, this research does not assess whether these community characteristics are related to the exercise of social control. In contrast, Sampson and Groves developed a causal model of social disorganization in which *community structure* (economic status, ethnic heterogeneity, residential mobility, family disruption, and urbanization) influences *social organization* (friendship networks, supervision of teenage peer groups, and participation in local organizations). Social organization, in turn, determines level of crime and delinquency. The model is depicted in **Figure 12-3.**

Using two surveys conducted in Great Britain, Sampson and Groves found that "communities characterized by sparse friendship networks, unsupervised teenage peer groups, and low organizational participation had disproportionately high rates of crime and delinquency. Moreover, variations in these dimensions of community social disorganization were shown to mediate, in large part,

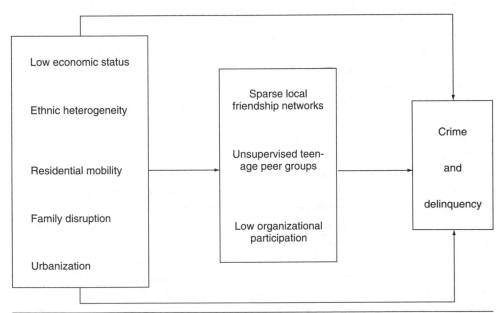

Figure 12-3 Sampson and Grove's model of social disorganization. Robert Sampson and W. Byron Groves hypothesized that measures of community structure (listed in the box on the left) influence social disorganization (middle box), which in turn affects the rate of crime and delinquency.

Source: Sampson and Groves, "Community Structure and Crime," 783. Copyright © 1989, University of Chicago Press. All rights reserved. Reproduced with permission of the publisher.

the effects of community structural characteristics (low socioeconomic status, residential mobility, ethnic heterogeneity, and family disruption) in the manner predicted by our theoretical model."[42]

Delinquency Prevention in Delinquency Areas

The life histories and ecological studies associated with social disorganization theory provide insight into the problem of juvenile delinquency in large American cities and how more effective "control techniques" can be developed.[43] In fact, Clifford Shaw is responsible for a widely recognized delinquency prevention program called the Chicago Area Project. Theory into Practice, "A Program Based on Social Disorganization Theory: Shaw's Chicago Area Project," provides a description of this program. Theory into Practice, "Community-Based Crime Prevention Efforts," gives a snapshot of contemporary efforts at community-based crime prevention.

A Contemporary Version of Social Disorganization Theory: Collective Efficacy

Social disorganization theory was vitally important in directing criminological attention toward community and societal factors. By the 1970s, however, the theory had lost much of its appeal and influence.[44] The theory was flatly dismissed by some, while others contended that social disorganization "should be seen as a descriptive convenience rather than a model of criminogenic behavior."[45] Additionally, new theories appeared and captured criminological interest. Beginning in the 1980s, though, renewed interest in social disorganization theory stirred.

Robert Sampson has played a leading role in revitalizing social disorganization theory by showing its relevance for understanding and responding to high rates of crime in inner-city neighborhoods.[46] He points to the importance of both structural and organizational characteristics of neighborhoods, as well as relationship bonds, social interaction, and social control efforts among neighbors.

Sampson adopts the basic thesis of social disorganization theory that structural features of a community affect its social organization, especially in terms of the ability of primary groups to exercise social control. Poverty, family disruption, and residential instability are structural characteristics that impede relations between neighbors and reduce their involvement in community organizations. Because of this low social capital, these communities are unable to exert effective collective control in public areas such as streets and parks.[47] In neighborhoods characterized by social disorganization, teenagers are free to roam and get involved in delinquent and criminal groups. High rates of crime and delinquency are a natural consequence.

Working with William Julius Wilson, Sampson also resurrected the cultural emphasis of social disorganization by arguing that structural conditions affect the culture of communities. The harsh conditions of inner-city life for many African Americans include social isolation—"the lack of contact or of sustained interaction with individuals and institutions that represent mainstream society."[48] In this context, cultural norms and values develop that do not necessarily approve of violence and crime, but instead define such actions as an unavoidable expression of ghetto life.[49] The social isolation, lack of opportunity,

THEORY INTO PRACTICE

A Program Based on Social Disorganization Theory: Shaw's Chicago Area Project

Convinced that delinquency prevention could be accomplished only through community organization and "community action," Clifford Shaw established the Chicago Area Project (CAP) in 1932 under the auspices of the Department of Sociology at the Institute for Juvenile Research. In philosophy and in purpose, the CAP drew from the ecological findings and case histories that are the basis of social disorganization theory. Under Shaw's energetic and charismatic leadership, the program grew to include twenty-two neighborhood centers that promoted community organization, integration, and stability by developing and using community resources and leadership. The CAP sought to generate a sense of community through "self-help from within."

The CAP attempted to organize local neighborhoods by having community members participate in various activities and projects. Such involvement was expected to create relationship bonds and investment in the local community. It was assumed that such community social controls were more effective than the formal controls of the juvenile justice system.

A range of programs were implemented, including local counseling services, recreation facilities and programs, educational tutoring, improved sanitation, and having community members advocate for youths in trouble with police and courts. Perhaps the most innovative but controversial element of the CAP was the use of gang "outreach workers" whose mission was to mobilize and organize neighborhood youth in generating community organization and change.

The Chicago Area Project operated in its original form for twenty-five years, until Shaw's death in 1957. At that time, the project personnel were transferred to the Illinois Youth Commission. James Short notes that this administrative change effectively separated the action and research aspects of the CAP. The program still exists today, relying on a three-pronged approach including direct service, community organizing, and advocacy. The CAP presently provides services throughout the Chicago area and the state of Illinois through dozens of affiliated programs.

While no other delinquency prevention program has combined theory and practice for such a long time, the effectiveness of the CAP has not been evaluated directly. Critics have argued that the original program neglected the political and economic nature of urban problems and that it was not confrontational enough with local government, business, and industry to generate change. Jon Snodgrass has argued that the original CAP was "cosmetic rather than surgical," failing to "deal directly with the forces destroying and disorganizing the community" ("Chicago Sociologists," 15–16). In this vein, Snodgrass was critical of Shaw's efforts to maintain positive relations with local government and to secure funding from business leaders in the Chicago area. Such close allegiances, according to Snodgrass, made significant change impossible in delinquency areas.

Sources: Alinsky, *Reveille for Radicals*; Empey, Stafford, and Hay, *American Delinquency*; Kobrin, "Chicago Area Project;" Lundman, *Prevention and Control;* Short, "Introduction;" Snodgrass, "Chicago Sociologists."

THEORY INTO
PRACTICE   Community-Based Crime Prevention Efforts

Several US Department of Justice publications discuss community-based crime prevention efforts and evaluate "what works, what doesn't, what's promising" (Sherman et al., "Preventing Crime," 1). Many of these community-based efforts are grounded in social disorganization theory.

Community-based crime prevention programs that "work" or show promise for reducing crime include the following:

- *Community-based afterschool recreation programs that provide supervised, structured recreational opportunities.*
- *Community policing strategies that bring together police departments, other government agencies, and community residents to solve crime issues.* Studies have shown that community policing reduces physical and social disorder and increases residents' satisfaction with the police in areas where it is used, but researchers have observed less consistently reduced crime rates in areas served by community policing programs.
- *Situational prevention efforts, such as the use of access control mechanisms or physical barriers to crime.* Examples include electronic personal identification numbers to control access to buildings and anti-theft locking devices for automobile steering wheels (Sherman et al., "Preventing Crime;" Catalano et al., "School and Community Interventions").

Based on preliminary research, Jeffrey Roth views the following strategies as promising and calls for "further development" in these areas ("Understanding and Preventing Violence," 9):

- *Housing policies to reverse the geographic concentration of low-income families.*
- *Economic revitalization in urban neighborhoods to restore opportunities for economic self-advancement through prosocial, nonviolent activities.*

Based on statistical evaluation of specific crime-prevention programs, researchers have found the following prevention efforts to be largely ineffective in reducing neighborhood crime and delinquency:

- *Citizen mobilization programs, such as neighborhood block watch programs and citizen patrols, in high-crime, inner-city areas of concentrated poverty.*
- *Community-based mentoring programs in which adult mentors spend time with and serve as role models for youth. (One study, however, showed that the Big Brothers and Big Sisters of America program reduced drug abuse among youth.)* (Sherman et al., "Preventing Crime;" Catalano et al., "School and Community Interventions.")

Sources: Catalano, Loeber, and McKinney, "School and Community Interventions;" Roth, "Understanding and Preventing Violence;" Sherman et al., "Preventing Crime."

and fatalistic cultural values of inner-city ghetto areas are experienced disproportionately by African Americans. Sampson and Wilson refer to such structured inequality as *concentrated disadvantage,* which results in the breakdown of community controls needed to restrain criminal behavior. The effects of concentrated disadvantage, however, depend on **collective efficacy**—the willingness of community residents to be involved with each other and to exercise informal control. Communities, then, vary not only in level of concentrated disadvantage, but also in cohesiveness of relationships and the collective capacity for informal social control. The degree to which concentrated disadvantage produces crime and delinquency in a community is dependent on that community's collective efficacy.

Sampson, together with Stephen Raudenbush and Felton Earls, tested this theory of collective efficacy by examining rates of violence across 343 Chicago neighborhoods.[50] Three measures of neighborhood social structure were analyzed in conjunction with collective efficacy: concentrated disadvantage (measured by combining a community's poverty, race and age composition, and level of family disruption), immigrant concentration, and residential stability. Raudenbush and Earls found the social composition of neighborhoods in terms of these three structural characteristics to be related strongly to neighborhood rates of violence. This relationship was heavily dependent, however, on the collective efficacy of neighborhoods—the degree to which neighborhood residents were interdependent, cohesive, and willing to exercise informal social control. Those neighborhoods that were characterized by collective disadvantage and residential instability, yet which still maintained cohesive social relationships and engaged readily in informal social control, had low rates of crime and delinquency.[51] Sampson and his associates concluded that the effect of concentrated disadvantage and residential instability is mediated by collective efficacy. These findings support Shaw and McKay's observation that "strong communities can act to quell disorder while communities weakened by structural problems will be fertile soil for the growth of crime."[52] Research in Action, "The Measurement of Collective Efficacy," lists the measures that Sampson and his colleagues used to assess collective efficacy.

collective efficacy The willingness of community residents to be involved with each other and to exercise informal social control; collective efficacy influences the degree to which concentrated disadvantage produces crime and delinquency in a community.

RESEARCH IN ACTION

The Measurement of Collective Efficacy

Robert Sampson, Stephen Raudenbush, and Felton Earls developed and tested the theory of collective efficacy using data from the Project on Human Development in Chicago Neighborhoods (PHDCN). The PHDCN is an enormous research undertaking that includes several components: a community survey of residents from all Chicago neighborhoods; a series of coordinated longitudinal studies of children, adolescents, and young adults; and a videotaped "systematic social observation" of neighborhoods. In the community survey, begun in 1994, approximately 8,800 people residing in 343 Chicago neighborhoods were interviewed about a variety of neighborhood factors, including neighborhood cohesion, social capital, informal social control, social disorder, and availability of programs and services, as well as activism, organizational involvement, and criminal victimization of residents.

Sampson, Raudenbush and Earls used measures from the community survey portion of the PHDCN to construct "informal social control" and "social cohesion and trust" scales that they used to measure collective efficacy, defined as mutual trust among neighbors combined with a willingness to intervene on behalf of the common good using informal methods of social control. The following measures were used to construct the informal social control and social cohesion "scales."

Informal social control

Residents were asked to respond to these questions using a five-point scale, ranging from very likely to very unlikely.

If a group of neighborhood children were skipping school and hanging out on a street corner, how likely is it that your neighbors would do something about it?

If some children were spray-painting graffiti on a local building, how likely is it that your neighbors would do something about it?

If a child was showing disrespect to an adult, how likely is it that people in your neighborhood would scold that child?

If there was a fight in front of your house and someone was being beaten or threatened, how likely is it that your neighbors would break it up?

Suppose that because of budget cuts the fire station closest to your home was going to be closed down by the city. How likely is it that neighborhood residents would organize to try to do something to keep the fire station open?

Social cohesion and trust

Residents were asked to respond to these questions using a five-point scale, ranging from strongly agree to strongly disagree.

People around here are willing to help their neighbors.

This is a close-knit neighborhood.

People in this neighborhood can be trusted.

People in this neighborhood generally don't get along with each other. (This item was reverse coded.)

People in this neighborhood do not share the same values. (This item was reverse coded.)

Source: Sampson, Raudenbush, and Earls, "Neighborhoods and Violent Crime."

■ Anomie and Strain Theories

While Shaw and McKay emphasized how social disorganization in inner-city neighborhoods disrupted traditional forms of social control and allowed delinquent peer groups to flourish, they also pointed out that legitimate opportunities for success are significantly limited in these areas and that this too is conducive to crime and delinquency.

> *The groups in the areas of lowest economic status find themselves at a disadvantage in the struggle to achieve the goals idealized in our civilization. . . . Those persons who occupy a disadvantageous position are involved in conflict between the goals assumed to be attainable in a free society and those actually attainable for a large portion of the population. It is understandable, then, that the economic position of persons living in the areas of least opportunity should be translated at times into unconventional conduct, in an effort to reconcile the idealized status and their prospects of attaining this status.*[53]

This emphasis on cultural goals of success in conjunction with the availability of opportunities for success is the central theme for a second group of social structure theories, referred to as *anomie* and *strain theories*. These alternate names distinguish two parts to this theoretical tradition: the first part explains why societies such as the United States have such high rates of crime, and the second part explains why some people and groups within society are more likely to engage in crime.[54]

The origins of anomie and strain theories can be traced to the work of Robert K. Merton, a famous sociological theorist, who in 1938 published a widely read and cited article entitled "Social Structure and Anomie."[55] Merton has revised and extended his anomie theory at least eight times since then.[56]

Though Merton based his theory on Durkheim's concept of *anomie,* he significantly reshaped the concept. Durkheim argued that much of what people need and desire can only be achieved socially, in the context of relationships. The desire for status, wealth, and power, for example, can only be obtained through human interaction. As a result, social needs and desires must be regulated externally through social control. When societies change rapidly, however, social control tends to break down, and behavior goes unregulated—a societal condition Durkheim called *anomie*. Accordingly, anomie refers to the breakdown or absence of social regulation.

In comparison, Merton argued that social needs and desires are not only regulated by society, but also defined and established by society. Social aspirations are not an innate human drive, as Durkheim suggested, but are instilled through socialization. Merton referred to these culturally defined "goals, purposes, and interests" as **cultural goals**—those which are commonly held to be desirable, worthwhile, and meaningful.[57] Cultural goals are learned socially in the context of families, schools, churches, and through all forms of media (advertisement, television, newspapers, magazines, etc.). People desire these goals because they are rewarding socially and economically and because they provide influence, power, and control. Merton, however, emphasized that a single cultural goal is most prominent in the United States: the goal of monetary success. While using a variety of terms for this cultural goal, he frequently referred to it as "pecuniary success"—the "accumulation of wealth as a symbol of success."[58]

A society also "defines, regulates, and controls the acceptable means of reaching out for these goals."[59] These accepted avenues for achieving cultural goals are referred to as **institutional means**. Like cultural goals, the norms or

cultural goals Goals that are socially learned and commonly held to be desirable, worthwhile, and meaningful.

institutional means Accepted avenues for achieving cultural goals.

RESEARCH IN ACTION   The Measurement of Anomie

In research using data from the community survey portion of the Project on Human Development in Chicago Neighborhoods, Robert Sampson and Dawn Jeglum Bartusch provide a contemporary example of the measurement of "anomie."

Sampson and Bartusch define anomie "in the classic Durkheimian sense" as "a state of normlessness in which the rules of the dominant society (and hence the legal system) are no longer binding in a community or for a population subgroup" ("Legal Cynicism," 782). According to Sampson and Bartusch, "Normlessness and powerlessness tend . . . to go hand in hand, breeding cynicism about the rules of the society and their application" ("Legal Cynicism," 782). Thus, they view "legal cynicism" as a component of anomie or normlessness. Sampson and Bartusch measure legal cynicism using five variables designed to assess general beliefs about the legitimacy of law and social norms.

Legal Cynicism

Residents were asked to respond to these questions using a five-point scale, ranging from strongly agree to strongly disagree.

Laws were made to be broken.

It's okay to do anything you want as long as you don't hurt anyone.

To make money, there are no right and wrong ways anymore, only easy ways and hard ways.

Fighting between friends or within families is nobody else's business.

Nowadays a person has to live pretty much for today and let tomorrow take care of itself.

Sampson and Bartusch created their legal cynicism scale by drawing on and modifying an anomie scale created much earlier by Leo Srole. To measure "anomia" or "interpersonal alienation," Srole used five items, with responses on an agree–disagree scale.

Srole's Measures of "Anomia"

There's little use writing to public officials because often they aren't really interested in the problems of the average man.

Nowadays a person has to live pretty much for today and let tomorrow take care of itself.

In spite of what some people say, the lot of the average man is getting worse, not better.

It's hardly fair to bring children into the world with the way things look for the future.

These days a person doesn't really know whom he can count on.

Sources: Sampson and Bartusch, "Legal Cynicism;" Srole, "Social Integration," 712–713.

means for obtaining goals are instilled and enforced within the context of institutions such as the family, schools, and work. Thus, Merton emphasized and examined the relationship between two aspects of society: cultural goals and institutional norms to attain those goals.

Merton went on to observe that some societies emphasize certain cultural goals without a corresponding emphasis on the institutional means to obtain those goals. This is a societal condition that he called **anomie**. In this sense, anomie refers to the normlessness that results when societal goals are stressed to a much greater degree than are the institutionalized means for achieving those goals. Within this framework, Merton advanced two explanations for deviance. While Merton did not use these terms, others have called the explanations *anomie theory* and *strain theory*.[60] Research in Action, "The Measurement of Anomie," provides examples of how anomie has been measured in contemporary and earlier research.

anomie According to Merton, a societal emphasis on certain cultural goals without a corresponding emphasis on the institutional means to obtain those goals.

Anomie Theory

Anomie theory attempts to explain why some societies have a high rate of deviance. Merton stated the main idea of anomie theory in one long sentence: "It is only when a system of cultural values extols, virtually above all else, certain *common* symbols of success [cultural goals] *for the population at large* while its social structure rigorously restricts or completely eliminates access to approved modes of acquiring these symbols *for a considerable part of the same population,* that antisocial behavior ensues on a considerable scale."[61]

Merton's anomie theory focuses on the relative emphasis a society places on the cultural goal of success in correspondence to opportunities for obtaining symbols of success.[62] He argued that, in the United States, economic success is stressed far more than any other goal and that there is not an equivalent emphasis on the institutional norms or means to achieve material success. This is often referred to as a "disjunction" or "imbalance" between goals and means.[63] Some societies, such as the United States, are characterized by a state of anomie in which "the goal-seeking behavior of individuals is subject to little regulation."[64] As a result, people are likely to pursue economic success by whatever means necessary, including crime.

Merton also points out that anomie associated with an overemphasis on economic success and an underemphasis on the acceptable means to be successful is structured into American society through the class system. The goal of material success is adopted by virtually everyone in the United States, whereas the means to achieve success are not equally available to all classes of people.[65] This structured inequality is what produces a higher rate of crime and delinquency among the lower class. Merton clarifies that poverty in itself does not produce a high rate of criminal behavior, nor does relative deprivation necessarily lead to a high rate of crime. Rather, it is only when poverty or relative deprivation is associated with a desire for material success and a lack of opportunity that criminal behavior is a likely outcome.[66]

Merton's anomie theory argues that "the very nature of American society generates considerable crime and deviance."[67] By focusing on the structural characteristics of American society, he deliberately rejected causal factors at the individual level.

Crime and the American Dream: Institutional Anomie Theory

A contemporary version of Merton's anomie theory has been offered by Steven Messner and Richard Rosenfeld in their book *Crime and the American Dream*.[68] Like Merton, Messner and Rosenfeld argue that the United States has a high crime rate because of the disproportionate emphasis on the goal of material success, without a corresponding emphasis on the normative means for achieving such success. This unrestrained pursuit of material success is what they refer to as "the American Dream"—a societal condition that Merton called anomie. Messner and Rosenfeld state: "In our use of the term 'the American Dream,' we refer to a broad cultural ethos that entails a commitment to the goal of material success, to be pursued by everyone in society, under conditions of open, individual competition."[69]

Four key values are at the foundation of the American Dream: *achievement, individualism, universalism,* and the *fetishism of money* (or "monetary rewards").[70] The achievement orientation ideally is available to all (universalism). It is achieved, however, largely by individuals rather than by groups (individualism), and it is manifest through the visible accumulation of money (monetary rewards). With regard to monetary rewards, Merton observed, "In some large measure, money has been consecrated as a value in itself, over and above its expenditure for articles of consumption or its use for the enhancement of power."[71]

The American Dream of monetary success has been identified as the "defining characteristic of American culture."[72] Because of its preeminence, the American Dream heavily influences the social organization of American society, especially its institutional structure. As a result, the economy comes to dominate all other institutions, including the political system (polity), the family, and education. As such, noneconomic goals, roles, and norms are devalued, and noneconomic institutions are influenced by and must accommodate those that are advanced by the economy. Messner and Rosenfeld offer an example of this: "Education is regarded largely as a means to occupational attainment, which in turn is valued primarily insofar as it promises economic reward. Neither the acquisition of knowledge nor learning for its own sake is highly valued. A revealing illustration of the devaluation of education relative to purely monetary concerns is provided in an interview with a high school student whose grades dropped when she increased her schedule to thirty hours per week at her two after-school jobs. She described her feelings about the value of education this way: 'School's important but so's money. Homework doesn't pay. Teachers say education is your payment, and that just makes me want to puke.' "[73]

Messner and Rosenfeld claim that "the goal of monetary success overwhelms other goals and becomes the principal measuring rod for achievement."[74] Further, the "institutional balance of power" is so centered on the economy that other social institutions are unable to function effectively.[75] They summarize their argument: "Both of the core features of the social organization of the United States—culture and institutional structure—are implicated in the genesis of high levels of crime. At the cultural level, the dominant ethos of the American Dream stimulates criminal motivations while at the same time promoting a weak normative environment (anomie). At the institutional level, the dominance of the economy in the institutional balance of power fosters weak social control."[76]

Critique of Anomie Theory

Anomie theory has intuitive appeal, with its commonsense observations about American society.[77] The heavy emphasis on monetary success in our society is undeniable; at the same time, institutional norms for obtaining success are minimal. The American Dream seems to be an accurate depiction of life in the United States. It is also commonly recognized that not everyone has an equal chance to achieve the American Dream, but this too is consistent with anomie theory. While the intuitive appeal of the theory is its strength, it also has been the target of criticism.[78]

Anomie theory assumes that a single dominant value—monetary success—fills all of American society, and there is widespread consensus regarding this value. Edwin Lemert, however, has argued that American society is more accurately characterized by value diversity and value pluralism.[79] Other researchers and theorists have pointed out that individuals and groups may seek a variety of goals, and that goals may vary depending on age, peer-group association, social-class background, and even by individual preference.[80] Especially relevant to a discussion of juvenile delinquency is the consideration of whether adolescents readily adopt this goal of economic success. Some theorists and researchers claim that adolescents tend to be more concerned with immediate rather than long-term goals. Goals related to economic success, such as educational achievement and occupational status, are seen as secondary to more immediate goals such as popularity, friends, freedom and independence, athletic success, and fun.[81] Goal aspirations may also be closely related to social class. Research by Herbert Hyman, for example, found that lower-class individuals have lower aspirations for success than do middle-class individuals.[82] Even though subsequent research has not always confirmed Hyman's finding, it draws into question whether the goal of monetary success is as universal and strong as anomie theory claims.[83]

The broad and sweeping argument of anomie theory has also been criticized for being virtually untestable. Ruth Kornhauser has led this charge, arguing that anomie is an abstract explanation of crime whose major concepts are difficult to measure and, when they are measured, lack empirical support.[84]

Finally, anomie theory has also been criticized as class-biased. Crime is allegedly concentrated in the lower class due to the lack of opportunities.[85] Messner and Rosenfeld point out that this criticism results from a narrow reading of anomie theory.[86] Merton does indeed include a structural component to the theory, arguing that the lack of opportunity is disproportionately experienced by lower-class individuals. However, Merton's primary theoretical depiction of the imbalance between goals and means applies equally to individuals in other social classes.[87] Anomie explains a variety of crimes, not just lower-class street crime. It can also be applied to crime with a "pecuniary" quality such as white-collar crime.

Nonetheless, Merton contended that crime and delinquency are more likely committed by individuals from the lower class because they, like individuals from other social classes, desire success, but they disproportionately lack opportunities to achieve it. As a result, the lower class should display higher rates of crime and delinquency than other social classes. Research using data from criminal justice agencies reveals consistently that crime and delinquency are concentrated in the lower class.[88] The emergence of self-report studies of delinquent behavior in the 1950s, however, questioned this class distribution of delinquency.[89] Many of these studies showed few differences in levels of delinquent behavior across different socioeconomic statuses. More recent self-report studies that measure frequency and seriousness of delinquent acts reveal that they are indeed committed more frequently by lower-class individuals, and that the offenses of lower-class individuals are more serious than those of youth from other social classes.[90] Nonetheless, as we saw in Chapter 5, research has failed to provide clear, consistent, and conclusive evidence supporting a link between social class and delinquent behavior.[91] One group of researchers has gone so far as to describe this connection as a "myth."[92] Delinquent behavior is not only a lower-class activity—it is engaged in by youth from all social classes. In final analysis, research shows that other factors appear to play a more critical role in causing delinquent behavior than does social class.[93]

Strain Theory

While Merton's theory tries to depict the structural characteristics of society that produce high rates of crime and deviance, it also considers how groups and individuals adapt to those structural characteristics. As we have already seen, this first consideration is referred to as *anomie theory* while the second consideration is sometimes distinguished as *strain theory*.[94] While many criminologists use anomie and strain interchangeably, it is useful to make the distinction between these theories.

Building on anomie theory, strain theory explains how groups and individuals adapt to the condition of anomie in society. Anomie refers to the imbalance or disjunction between goals and institutional means, in which the goal of monetary success is emphasized without clear normative standards for achieving success. In addition, anomie theory states that the acceptable means to success are unavailable to a considerable part of the population.[95] This results in high rates of crime and deviance. Anomie is a structural characteristic of society, not an individual characteristic.[96] Merton proposed that people experience *strain* when the accepted means for achieving success are unclear or unavailable and they must adapt to this societal condition of anomie. While blocked opportunities for success might seem to be the primary source of strain, strain theory centers on an individual's response to anomie in terms of acceptance or rejection of cultural goals and the institutional means to reach those goals.

Merton's Modes of Adaptation

According to Merton, there is not only one possible response to structurally induced strain. Many years ago, psychologist Seymour Halleck observed that "it is difficult to imagine stress which could be adapted to in only one way." Since crime "represents only one of several possible adaptations, the criminologist must grapple with the question of why the criminal solution is 'chosen.' "[97] Merton proposed five possible individual adaptations to the goals–means disjunction of anomie. Three of the adaptations are considered deviant, but the other two usually are not.[98] These adaptations are shown in Expanding Ideas, "Merton's Typology of Adaptations to Anomie." Merton pointed out that a person may shift from one adaptation to another.[99]

EXPANDING IDEAS

Merton's Typology of Adaptations to Anomie

Adaptation	Cultural Goals	Means	
Conformity	+	+	The most common adaptation in all social classes. The individual adopts both cultural goals and institutionalized means.
Innovation	+	−	With the accepted means to success underemphasized or unavailable, some individuals turn to illegitimate means. While this adaptation explains crime by all classes of people, it is most common among the lower class because of disproportionate lack of opportunity.
Ritualism	−	+	An almost compulsive following of the rules. More common among the lower middle class because they have strict patterns of socialization.
Retreatism	−	−	The least common adaptation. Individuals who accept the goals and means, but because goals and means are unattainable they become frustrated, leading to dropping out of society. For example, drug addicts, drunkards, vagrants.
Rebellion	±	±	Individuals who reject cultural goals and means, and then replace them with new ones. This may be a collective adaptation, involving organized struggle for change.

Source: Merton, "Social Structure and Anomie," adapted from page 676.

Conformity is the most common adaptation to anomie. In fact, conformity is the basis for stability and order in society. The existing social structure is supported and reaffirmed, and behavior is controlled to the degree that people accept cultural goals and the means that have been institutionalized to obtain those goals. Conformity represents an acceptance of cultural goals and the institutional means to obtain those goals, regardless of whether those institutional means are available to the individual. According to Merton, conformity is spread widely across all social classes.[100] Most people conform because they accept the cultural goals and the means that have been established to obtain success, regardless of whether these means are available to them or not.

Innovation involves the use of illegitimate means to achieve cultural goals. With the accepted means to success underemphasized or unavailable, some individuals turn to illegitimate means. While this adaptation explains a wide variety of crime by all classes of people, Merton argued that crime is most common among the lower class because of the lack of opportunity afforded its members.

Ritualism involves a rigid compliance with rules, without a clear commitment to the goals. Ritualism often does not result in deviant behavior. In fact, it may produce "over conformity" and the lowest rates of deviance among the five modes of adaptation.[101] A youth who lives within the law but makes little effort in school and has few aspirations displays ritualism in his or her daily life. Merton contended that ritualism is most common among the lower middle class because strict patterns of socialization encourage following the rules of society. The routines of institutional norms provide youth with a sense of security, even without the adoption of success goals.[102]

Retreatism is the least common adaptation. Some individuals initially accept the goals and means, but because the means are unattainable, they eventually drop out of society. Merton says that these frustrated individuals are "aliens" in society. "Not sharing the common frame of orientation," they are "*in* the society but not *of* it."[103] Included in this adaptation are vagrants, chronic drunkards, and drug addicts.

Rebellion depicts individuals who not only reject cultural goals and means, but also replace them with new ones. Merton notes that this adaptation is so different from the others that it must be distinguished and set apart. (His original table actually had a line separating rebellion from the other modes of adaptation.) This form of adaptation "involves an effort to change the existing structure" of goals and means. Rebellion can involve collective adaptation in an organized struggle for change such as occurred during the Vietnam War protests and some segments of the Civil Rights Movement of the 1960s.

Research on Strain

In its most basic form, strain theory argues that "delinquency results when individuals cannot get what they want through legitimate channels."[104] In this way, strain is seen as the motivation for delinquent acts.[105] Though strain is experienced at the individual level, it results from structural factors, especially social class placement and the resulting level of opportunity.[106]

This social psychological character of strain is most frequently depicted as the imbalance or disjunction between *aspirations* (an individual's goals) and *expectations* (an individual's perception of opportunities)—the discrepancy between what people want and what they expect to get.[107] According to strain theory, those youths who adopt goals of economic success, but feel that they are unable to reach those goals because of limited opportunities, are more likely than others to experience emotional strain in the form of frustration or anger, and this, in turn, motivates delinquent acts.[108]

Research testing this basic premise of strain theory has failed to provide much empirical support.[109] One of the most widely cited assessments of strain theory was conducted by Travis Hirschi in his book *Causes of Delinquency*. In advancing his own social bond theory, which stands in contrast to strain theory, Hirschi argued that high aspirations to conventional goals actually constrain delinquency. Delinquent youth, in fact, lack commitment to these goals. Remember that Hirschi identified commitment to conventional goals as one of his four elements of the social bond. In order to extend his control argument in contrast to strain theory, he alleged that the discrepancy between aspirations and expectations does not provide any additional explanation for delinquent behavior beyond that offered by the social bond of commitment. Using measures of educational aspirations and educational expectations, Hirschi found that commitment to educational aspirations decreased the likelihood of delinquent behavior, regardless of the level of perceived expectations. He observed that "discrepancies in educational aspirations and expectations are not important in the causation of delinquency for two reasons: few boys in the sample have aspirations greatly in excess of their expectations; and, *those boys whose educational or occupational aspirations exceed their expectations are no more likely to be delinquent than those boys whose aspirations and expectations are identical.*"[110] Hirschi and others have concluded that strain theory fails to offer explanation beyond that of social bond theory.[111]

Research findings have also provided only limited support to the proposition that strain results when goals for success go unfulfilled because of blocked opportunities.[112] Strain supposedly provides motivation for crime and delinquency and is experienced disproportionately by lower-class individuals.[113] Contrary to the theory's predictions, Ruth Kornhauser's review of the empirical research on strain revealed that delinquents showed both low aspirations and low expectations, indicating that they are not under social psychological strain and have little motivation to engage in delinquency acts.[114]

Margaret Farnworth and Michael Leiber contend that the lack of empirical support for strain theory has more to do with the way in which strain has been measured than with the theory's lack of legitimacy.[115] They argue that Merton's depiction of strain is most appropriately measured as the difference between economic goals (the desire to make lots of money) and educational expectations (how much school one expects to complete).[116] The reliance on educational measures of both aspirations and expectations, as Hirschi used, neglects the emphasis that Merton placed on economic aspirations as the primary goal in American society.[117]

Farnworth and Leiber examined various measures of aspirations and expectations. They found that, consistent with Hirschi's findings, educational aspirations alone are significantly related to delinquency and that the strain of unequal educational aspirations and educational expectations provides no additional explanation.[118] However, they also found that when economic aspirations are used as a measure of goals, instead of educational aspirations, the strain that results from an imbalance between aspirations and expectations is a better predictor of delinquency than are financial aspirations alone.[119] They conclude that "the empirical findings to date are not sufficient to falsify the basic postulates of Merton's theory of strain and deviance."[120]

More recently, Scott Menard has argued that previous research has neglected the social psychological adaptations to strain that are a major part of Merton's theory. By neglecting this, research has failed to consider how the different modes of adaptation are related to involvement in crime, delinquency, and other forms of deviance. Menard analyzed data from the National Youth Survey, a longitudinal data set gathered from a national sample of youth. Just as Merton claimed, Menard found that nearly all youth adopted the goal of economic success, as measured by a question asking the importance of getting a good job or career. Furthermore, there were far more youth who desired success than there were who expected to be successful. This discrepancy between aspirations and expectations indicates a societal condition of anomie.[121]

Also in keeping with Merton's strain theory, Menard found that innovators and retreatists had higher rates of minor and serious offending than did conformists or ritualists. Innovators had the highest rates of offending in early adolescence, while retreatists had higher rates in middle and late adolescence. Throughout the adolescent years, ritualists had the lowest rate of offending.[122] The findings for marijuana and drug use were also consistent with Merton's strain theory, showing that retreatists had the highest rate of drug offending, followed by innovators.[123]

The Strain of Adolescence: General Strain Theory

Even though research has not been overly supportive of strain theory, its major theme remains influential in delinquency theory and research: delinquent acts are motivated by the strain or frustration that results when goals cannot be achieved and go unfulfilled. Revisions of classic strain theory continue to adopt this basic theme but sometimes expand the notion of goals beyond the goal of economic success that Merton emphasized. Some have argued that adolescents pursue a variety of goals in addition to monetary success. Goals such as popularity, friends, freedom and independence, athletic success, and having fun take on great importance during the adolescent years, and these more immediate goals are possibly even more influential than the long-term goal of gaining monetary success.[124]

Robert Agnew's revised strain theory, *general strain theory*, follows a different avenue. Rather than expanding the number and variety of goals that adolescents pursue, Agnew focuses on the central importance of strain as a motivation for delinquency. Crime and delinquency are viewed as adaptations to strain, and Agnew's revised theory identifies several sources of strain beyond that offered by classical strain theory, which points to a single source of strain: culturally prescribed means for achieving success are unclear or unavailable. Drawing on *frustration aggression* and *social learning theories,* Agnew claims that strain also results from an individual's efforts to avoid unpleasant or painful situations.[125] In other words, strain results not only from goal seeking but also from undesirable situations and outcomes.

Agnew's view of strain emphasizes the social psychological aspects of strain, rather than the structural limitation of low social class and lack of opportunity. In particular, he focuses on the strain that results from negative social relationships, "relationships in which others are not treating the individual as he or she would like to be treated."[126] He offers three general types of strain that result from relationships in which others (1) prevent an individual from achieving valued goals, (2) remove or threaten to remove valued stimuli, and (3) present or threaten to present negative stimuli.[127]

Strain Resulting from the Failure to Achieve Valued Goals

The first source of strain includes the traditional emphasis that classical strain theory places on the disjunction between aspirations and expectations as the primary motivation for delinquency. This form of strain, however, has been heavily criticized. Agnew points to social psychological research which shows that "people tend to (a) pursue a variety of goals, (b) place most importance on those goals they are best able to achieve, and (c) exaggerate or distort their actual and expected levels of goal achievement."[128] To illustrate, he refers to his own research which showed that the goals adolescents set for themselves are often exaggerated. "Adolescents doing poorly in school . . . often described themselves as good students who expected to attend college."[129] Agnew concludes that the disjunction between aspirations and expectations is probably not a major source of strain because virtually all adolescents believe they are achieving at least some of their goals, and they often make social psychological adjustments when their goals seem unachievable.[130]

Within this first category of strain, Agnew adds two more sources of strain that emerge in the context of negative social relationships. Both sources of strain draw from social psychological research on justice and equity in interpersonal relations.[131] First, a gap between expectations and actual achievements can lead to feelings of anger, unfairness, and disappointment. Similar to goals, beliefs about what is achievable—expectations—are developed in reference groups. In comparison with others, youths develop a sense of what they can reasonably expect to achieve. Strain results when actual achievement falls short of these expectations, especially when an individual sees that others are successful.[132]

Second, strain can result from an inconsistency between perceptions of "fair outcomes" and actual outcomes.[133] Once again, the context of this source of strain is within social relations, but the emphasis is on the fairness or equity of interactions among people. If interpersonal exchange is viewed as unequal in terms of either what one puts into it or what one gets out of it, the individual may experience "distress." A possible reaction to this distress is deviance.[134]

Strain Resulting from the Loss of Positively Valued Stimuli

A second category of strain involves the experience of "stressful life events" such as the separation or divorce of parents, the loss of a boyfriend or girlfriend, moving, serious illness of a friend, or suspension from school.[135] The stress of adolescence is often associated with the loss of something or someone the youth values. Psychologists have developed numerous inventories of stressful life events and have found that the number of these events experienced by youth is related to delinquent behavior.

Strain Resulting from Negatively Valued Stimuli

The third category of strain relates to another set of stressful life events—those involving negative action by others toward the youth. Experiences such as child abuse, criminal victimization, or negative relations with parents, teachers, or peers are also seen as a source of strain. These "noxious stimuli" may lead to delinquency as the adolescent tries to avoid, compensate, or seek revenge.[136]

Explaining the Strain of Adolescence

Agnew's general strain theory tries to expand and categorize the sources of strain that result from negative relationships with others.[137] These "negative relationships increase the likelihood that individuals will experience anger or frustration."[138] Anger is an especially important emotional reaction because it provides motivation and justification, and it breaks down constraints on delinquent acts. In this way, delinquency can be seen as a "coping response to interpersonal problems."[139]

As in classical strain theory, Agnew acknowledges that strain does not lead all adolescents to crime. Delinquency is only one possible adaptation to strain. A variety of "predispositions" and "constraints" influence the likelihood that a youth will turn to delinquent acts as a solution to strain. Predispositions and constraints include "the adolescent's temperament, problem-solving skills, self-efficacy, self-esteem, level of conventional social support, attributions regarding the causes of strain, level of social control, and association with delinquent peers."[140] In a recent study, Agnew and his colleagues found that youths characterized by personality traits of negative emotionality and low constraint were more likely to react to strain with delinquency.[141]

General strain theory holds that strain can be experienced by anyone, regardless of social class, gender, or race.[142] As a result, the negative pressure or motivation toward delinquency crosses social boundaries. Unlike classical strain theory, general strain theory attempts to explain delinquency generally, not as an adaptation concentrated in the lower class.

Research on General Strain Theory

Agnew's general strain theory receives indirect support from a wide variety of research findings showing that "delinquency is associated with such strains as negative relations with parents and teachers, child abuse, conflict with peers, criminal victimization, neighborhood problems, and a range of stressful life events—like parental divorce, family financial problems, and changing schools. Further, certain studies indicate that these strains increase the likelihood of delinquency by increasing the individual's level of anger and frustration."[143]

More direct tests of the theory have constructed summary measures of strain.[144] Robert Agnew and Helene White, for example, focused on two categories of strain presented in the theory: the loss of valued stimuli and the presence of negative stimuli. Strain was measured through responses to questions about recent negative experiences such as the loss of a close friend through death; the divorce of parents; not getting along with classmates, parents, and teachers; and experience of various types of crime.[145] Adolescents who scored high on this composite measure of strain were far more likely to engage in delinquent acts than those who scored lower. Thus, the experience of strain pressures or motivates youth to engage in delinquency acts. The effects of strain were also found to depend on association with delinquent peers and self-efficacy—factors that general strain theory acknowledges will play a role in producing delinquent outcomes to strain.[146]

Timothy Brezina has recently studied the degree to which delinquent acts are a coping response to strain.[147] His study used data drawn from the second and third waves of the Youth in Transition survey, a nationally representative sample of male public school students. The second wave included 1,886 students in their junior year of high school, and the third wave included 1,799 of those students a year later, at the end of their senior year. Adopting a measure of strain similar to Agnew and White's, Brezina found that the experience of strain was associated with feelings of anger, resentment, anxiety, and depression. He concluded that involvement in delinquent acts enables adolescents to minimize the negative emotional consequences of strained social relationships.[148] As such, "delinquent behavior may possess self-reinforcing qualities—qualities that would promote the pursuit and continuation of delinquent activities."[149] This helps explain the appeal that delinquent acts have for many adolescents and why delinquent behavior is not easily changed.

Strain and Gang Subcultures

Criminologists have used the idea of strain to explain gang delinquency among urban, lower-class males in two important extensions of Merton's strain theory: Albert Cohen's *reaction formation theory* and Richard Cloward and Lloyd Ohlin's *differential opportunity theory*.[150] In both applications, the lack of opportunity and relative deprivation that virtually all lower-class boys experience results in strain and ultimately leads to gang delinquency. However, the key link between strain and gang delinquency is the adoption of subcultural values and norms that encourage and support delinquent acts.[151]

Cohen's Reaction Formation Theory

Albert Cohen observed that much of the delinquent activity in inner-city areas is committed by gang members and that most of these acts are done not for economic gain, but "for the hell of it."[152] He described gang delinquency as "nonutilitarian, malicious, and negativistic"—it appears to serve little purpose and is often hostile, cruel, and contemptuous. Cohen's influential book, *Delinquent Boys: The Culture of the Gang,* is an attempt to account for the character of gang delinquency and to explain the development of gangs and the values and norms they embrace. Cohen's focal point is the delinquent subculture that is "a way of life that has somehow become traditional among certain groups in American society. These groups are the boy's gangs that flourish most conspicuously in the 'delinquent neighborhoods' of our larger American cities."[153]

Like Merton, Cohen claimed that delinquency is ultimately caused by blocked goals. However, he argued that lower-class boys are "not simply concerned with the goal of monetary success. Rather, they want to achieve the broader goal of middle-class status, which involves respect from others as well as financial success."[154] Lower-class boys are disadvantaged in their efforts to achieve status, however, especially status in conventional institutions like schools. Cohen distinguished *ascribed status* from *achieved status*.[155] Ascribed status is tied to the social position of one's family, and achieved status is earned through effort and accomplishment. Inner-city boys typically have low ascribed status because their families are lower-class, and they are at a competitive disadvantage to earn achieved status because their early socialization fails to equip them with the characteristics and skills necessary to be successful in school, the institution in which competition for achieved status first occurs.

According to Cohen, lower-class parents are "easy-going" and "permissive," whereas middle-class parents are "rational, deliberate, and demanding."[156] Lower-class parents also do not model attitudes and behaviors that encourage achievement.[157] Thus, it is Cohen's contention that social class placement structures children's socialization experiences.[158] As a result, lower-class boys fail to learn the values, traits, and skills upon which status in school is judged. Standards such as ambition; responsibility; skills relevant to future economic achievement; deferred gratification; "rationality" in terms of planning; courtesy, manners, and "personability;" and control of physical aggression are rarely taught or stressed through the casual parenting of lower-class parents.[159] Due to such inadequate socialization, lower-class boys inevitably fail in school because teachers impose "middle-class measuring rods."[160] School failure, in turn, breeds **status frustration**—the dominant type of strain lower-class boys experience.

status frustration According to Cohen, the dominant type of strain that lower-class boys experience. It results from school failure brought about by inadequate socialization to the values, traits, and skills that teachers use to judge success in school.

Cohen argued that most people adapt to strain collectively, by joining with others to find solutions.[161] Confronted with the common problem of status frustration, lower-class boys turn to each other to achieve status. Cohen claims that three different adaptations are likely. The *corner boy* accepts the low status ascribed to those from the lower class and disengages from the competitive struggle for status. Instead, he turns to the "sheltering community of like-minded peers"—the corner boy subculture of the lower class.[162] The *college boy* is able to achieve academic success despite the competitive disadvantage that confronts lower-class youth. Relatively few lower-class boys are able to compete in the middle-class arena of education, but the few that are able to master unfamiliar "linguistic, academic, and social skills" achieve status in school and continue on into college.[163] The final adaptation—the *delinquent boy*—is the focus of Cohen's book. Cohen argues that the delinquent subculture begins and is maintained as a solution to this status problem common among lower-class boys.[164]

According to Cohen, lower-class boys often join gangs in order to deal with their collective problem of status frustration experienced in schools. Delinquent gangs offer a social context in which lower-class boys can gain status. "The delinquent subculture . . . is a way of dealing with the problems of adjustment . . . These problems are chiefly status problems: certain children are denied status in the respectable society because they cannot meet the criteria of the respectable status system. The delinquent subculture deals with these problems by providing criteria of status which these children *can* meet."[165]

reaction formation A psychological process of rejecting conventional goals and means for success and substituting alternative goals and means.

In a psychological process called **reaction formation**, lower-class boys in gangs develop alternative values that allow them to experience success and thereby gain status (at least in the eyes of their peers).[166] Delinquent acts easily fulfill the values of the gang subculture. Cohen describes these acts and the delinquent subculture as nonutilitarian, malicious, negativistic, versatile, and hedonistic, and as providing group autonomy.[167] He claims that much of the property crime committed by gang members yields little apparent gain. It is also commonly observed that gang members acquire status, even pleasure, from the harm and trouble they cause others, and that they take pride in reputations of meanness and toughness. Gang delinquency also occurs spontaneously, with little planning, and includes a wide range of illegal activities. Finally, the criminal acts of the gang provide definition, cohesiveness, and autonomy to the delinquent subculture.

Criticism of Cohen's version of strain theory has focused more on its logical completeness than on its adequacy as tested through research.[168] Concern over the theory has been raised primarily by pointing to a number of unresolved questions:

- Do lower-class boys desire middle-class status as achieved through interpersonal respect and economic success?

- Does school failure produce status frustration and low self-esteem? Is this strain enough to explain the development of gang subcultures?

- Do youth seek collective responses to problems?

- Does the gang subculture constitute a wholesale rejection of middle-class culture? Is the gang subculture unique, or do delinquent gangs partially reflect the dominant culture? Do subcultures approve of crime?

- Are the delinquent acts of gangs nonutilitarian, malicious, and negativistic? Does this character of gang delinquency promote group autonomy?

- How is the gang subculture maintained over time?

Cloward and Ohlin's Differential Opportunity Theory

Like Cohen, Richard Cloward and Lloyd Ohlin offer a strain theory that attempts to explain how delinquent gang subcultures arise and persist in lower-class neighborhoods. Clearly building on Merton's strain theory, they state their thesis as follows: "The disparity between what lower-class youth are led to want and what is actually available to them is the source of a major problem of adjustment. Adolescents who form delinquent subcultures . . . have internalized an emphasis upon conventional goals. Faced with limitation of legitimate avenues of access to these goals, and unable to revise their aspirations downward, they experience intense frustrations; the exploration of nonconformist alternatives may be the result."[169]

As Cohen does, Cloward and Ohlin claim that education is central to upward mobility, but there are significant barriers to this *legitimate opportunity* for success.[170] The strain that results from this lack of opportunity is experienced most intensively by adolescent males from urban, lower-class environments. However, just as legitimate opportunities are unequally available, so too are the *illegitimate opportunities* to which youth turn out of frustration. As it points to the existence of both legitimate and illegitimate opportunities for success, the theory is called *differential opportunity theory.*

Cloward and Ohlin claim that lower-class areas are characterized by different types of criminal patterns and traditions. They identify three distinct delinquent subcultures that reflect the type of illegitimate opportunity available in the surrounding community. A *criminal subculture* exists in well-organized neighborhoods where norms all but require criminal involvement and where values support, validate, and rationalize involvement in crime. Criminal role models are readily available, and possibilities for involvement in crime are everywhere. Delinquent gangs that develop within such a tradition of crime are "devoted to theft, extortion, and other illegal means of securing income."[171]

Conflict subcultures predominate in areas that lack criminal traditions. Without criminal patterns to follow and without readily available opportunities for crime, conflict and violence become the primary means of gaining status. Threats and the use of force dominate the activities of these " 'warrior' groups that attract so much attention in the press."[172]

The *retreatist subculture* is culturally and socially detached from the lifestyle and everyday preoccupations of the conventional world.[173] In their place, the retreatist subculture stresses "the continuous pursuit of the 'kick.' "[174] The extensive use of drugs for fun and pleasure is encouraged and expected within the subculture.

Gang delinquency is ultimately an expression of the structure of opportunities, both legitimate and illegitimate. "For delinquency is not, in the final analysis, a property of individuals or even of subcultures; it is a property of the social system in which individuals and groups are enmeshed. The pressures that produce delinquency originate in these structures, as do the forces that shape the content of specialized subcultural adaptations."[175]

Differential opportunity theory became the cornerstone of a delinquency prevention program called Mobilization for Youth, begun in New York City in the early 1960s. While writing *Delinquency and Opportunity*, Cloward and Ohlin helped develop a program for the Henry Street Settlement on the Lower East Side of New York City. As the name implies, Mobilization for Youth was a comprehensive program that attempted to increase legitimate opportunities for lower-class youth through a variety of economic and education reforms. Components included preschool programs, tutoring, in-service teacher training to increase teacher–parent communication and enhance cultural awareness of lower-class communities, vocational training, job placement programs, community organization through neighborhood councils and associations, services to youth and families, and a detached worker program to respond to gangs.

Neither this practical application of differential opportunity theory, nor the theory itself, were subject to thorough empirical testing. As a result, scholarly criticism focused more on the accuracy and logical completeness of the theory's explanation than on the amount of empirical support for the theory or the program. The theoretical argument of differential opportunity theory assumes that blocked educational and economic opportunities make lower-class youth receptive to illegitimate avenues for success—opportunities that are largely illegal. This reflects the common assumption of strain theories that crime and delinquency are committed mainly by lower-class individuals. As described earlier in this chapter and in Chapter 5, research has failed to provide clear and consistent evidence that delinquency is a lower-class activity.[176] Moreover, when blocked educational opportunities are added to the causal explanation, the link between social class and delinquency becomes even more tenuous. School failure is related to delinquent behavior in all social classes, not just the lower class.[177]

Differential opportunity theory also proposes that three types of delinquent subcultures are most common in lower-class areas, depending on the organization of the neighborhood and the availability of illegitimate opportunities for success. The criminal subculture has a tradition of crime in which illegitimate opportunities flourish and are encouraged. The retreatist subculture emphasizes alternative lifestyles and the heavy use of drugs. Violent subcultures lack traditions, values, and norms of any kind. As a result, violent acts and threats of violence are used to gain status. Research has failed to uncover these three distinct subcultures, nor do delinquent gangs specialize in the types of acts suggested by the theory.[178] Instead, delinquent gangs engage in a wide variety of criminal offenses. Moreover, boys in delinquent gangs spend most of their time involved in nondelinquent activities.

■ Summary and Conclusions

Social structure theories attempt to identify and account for the social and societal characteristics that integrate and regulate people's daily lives. Sociologists refer to these organizational features of the social environment as *social structures*. Social structure theories consider societal characteristics such as cultural traditions, institutionalized social relations within the context of families, schools, and employment, and ecological dimensions such as the residential concentration of ethnic groups and social classes and population mobility. When these societal characteristics disrupt social organization, social control breaks down, and crime and delinquency flourish. Particularly important to many social structure theories is the thesis that rapid social change results in the breakdown of primary group controls. This lack of social regulation is called *anomie*.

Shaw and McKay's theory of social disorganization emphasizes three structural characteristics of urban environments that disrupt social organization: *low economic status, ethnic heterogeneity,* and *residential mobility*. Urban areas characterized by these structures typically lack effective social control mechanisms and, as a result, experience high rates of crime and delinquency. These delinquency areas often have strong criminal traditions, or subcultures, in which involvement in illegal activity is a way of life, passed on from adult to youth and youth to youth. Delinquency areas also lack legitimate economic opportunities, resulting in personal frustration, or strain, that can motivate criminal involvement.

Robert Sampson has recently revitalized *social disorganization theory* by advancing his own theory called *collective efficacy*. He argues that structural characteristics related to the social composition of neighborhoods, including concentrated disadvantage, immigrant concentration, and residential stability, are strongly related to rates of violence. The influence of these structural features of neighborhoods, however, depends on the degree to which local residents are interdependent, cohesive, and willing to exercise informal social control. This essential neighborhood characteristic of interconnectedness is called *collective efficacy*.

Delinquency theories that focus on the lack of regulation in society are referred to as *anomie theories*. Robert Merton argued that the *goal* of economic success permeates all of American society, but that the *institutionalized means* or norms to achieve success are neither stressed to the same degree nor equally available to all people. This social structural characteristic of anomie frees people to pursue economic success by whatever means necessary, including crime.

Messner and Rosenfeld's *institutional anomie theory* adds another structural feature to the anomie equation: institutions. They argue that the economy has come to dominate all institutions in the United States because of the ever-increasing emphasis on economic success—"the American Dream." Institutions such as the political system, the family, and education are declining in importance and influence. Domination by the economy prevents these other institutions from functioning effectively. As a result, individuals are poorly socialized and inadequately controlled. Both the cultural and institutional structures of American society generate high levels of crime.

Building on anomie theory, the strain perspective explains how groups and individuals adapt to the condition of anomie in society. Merton contended that individuals experience *strain* when the acceptable means to economic success are unclear or unavailable, and that they must adapt to such strain. He proposed five modes of adaptation as being most common: conformity, ritualism, innovation, retreatism, and rebellion. The last three adaptations normally involve deviant behavior.

Robert Agnew's *general strain theory* identifies additional sources of strain beyond the structural feature of anomie. In particular, he focuses on the strain that results from negative social relationships and efforts to avoid unpleasant or painful situations. As such, he emphasizes the social psychological aspect of strain.

Strain has also been used as the fundamental explanation of gang delinquency among urban, lower-class males. Albert Cohen's version of strain theory is referred to as either *status frustration theory* or *reaction formation theory* because of the central role these two processes play in his argument. Due to inadequate socialization, lower-class boys are poorly equipped to do well in school. As a result, they experience *status frustration,* which leads them to search out collective solutions through involvement in delinquent gangs. The delinquent subculture develops values that stand in sharp contrast to middle-class goals. This process is referred to as *reaction formation.* These alternative values allow lower-class boys to experience success and to gain status, at least in the eyes of their peers. The resulting delinquent gang constitutes a subculture that is nonutilitarian, malicious, negativistic, versatile, hedonistic, and autonomous.

Cloward and Ohlin's strain theory of gang delinquency also points to the importance of blocked educational and economic opportunities in producing delinquent subcultures in lower-class urban areas. When *legitimate opportunities* are unavailable, youth turn to *illegitimate opportunities.* However, like legitimate opportunities, illegitimate opportunities are not always available or accessible. The criminal traditions, values, norms, and roles of lower-class neighborhoods determine the availability of illegitimate opportunities and, in turn, the types of delinquent subcultures that develop. Three types of delinquent gangs are most common in lower-class areas: *criminal subcultures, conflict subcultures,* and *retreatist subcultures.*

THEORIES

- social solidarity
- social disorganization
- collective efficacy
- anomie
- institutional anomie
- strain
- general strain
- reaction formation
- differential opportunity

CRITICAL THINKING QUESTIONS

1. Consider the case of Sidney Blotzman presented at the beginning of this chapter. Explain, in terms of the social structure theories discussed in this chapter, how the characteristics of Sidney's neighborhood may have influenced his involvement in delinquency.

2. Sociologists first proposed social disorganization theory in 1929/1942, and anomie and strain theories in 1938. Discuss the relevance of these theories in contemporary American society. Are these theories as applicable today as when they were first proposed?

3. How accurate is the characterization of American society that anomie and strain theories provide? Are these theories relevant or useful for the explanation of crime in societies with different economic and political structures?

4. Discuss the avenues through which the social structural processes described in social disorganization and anomie theories influence behavior at the individual level.

5. Suppose you were asked to design delinquency prevention efforts based on Sampson, Raudenbush, and Earls' theory of collective efficacy. What would that delinquency prevention program look like?

SUGGESTED READINGS

Bursik, Robert J., Jr., and Harold G. Grasmick. *Neighborhoods and Crime: The Dimensions of Effective Community Control.* New York: Lexington, 1993.

Merton, Robert K. "Social Structure and Anomie." *American Sociological Review* 3 (1938):672–682.

Sampson, Robert J., Steven W. Raudenbush, and Felton Earls. "Neighborhoods and Violent Crime: A Multilevel Study of Collective Efficacy." *Science* 277 (1997):918–924.

Shaw, Clifford R., and Henry D. McKay. *Juvenile Delinquency and Urban Areas: A Study of Rates of Delinquency in Relation to Differential Characteristics of Local Communities in American Cities.* Rev. ed. Chicago: University of Chicago Press, 1969.

GLOSSARY

anomie: According to Merton, a societal emphasis on certain cultural goals without a corresponding emphasis on the institutional means to obtain those goals.

collective efficacy: The willingness of community residents to be involved with each other and to exercise informal social control; collective efficacy influences the degree to which concentrated disadvantage produces crime and delinquency in a community.

cultural goals: Goals that are socially learned and commonly held to be desirable, worthwhile, and meaningful.

institutional means: Accepted avenues for achieving cultural goals.

mechanical solidarity: Social integration and regulation based on social similarity; develops in small, cohesive societies in which people do the same type of work and have the same values and norms.

modes of adaptation: Various possible individual responses to the goals-means disjunction of anomie. Merton identified five possible modes of adaptation: conformity, ritualism, innovation, retreatism, and rebellionism.

organic solidarity: Social integration and regulation based on the complementary nature of social life; develops in modern, complex societies in which interdependence is necessary.

reaction formation: A psychological process of rejecting conventional goals and means for success and substituting alternative goals and means.

social solidarity: Social structures of society that integrate and regulate individuals and groups.

social structures: Organizational features of the social environment, such as neighborhood socioeconomic status, ethnic heterogeneity, residential mobility, cultural traditions, and age and gender distributions.

status frustration: According to Cohen, the dominant type of strain that lower-class boys experience. It results from school failure brought about by inadequate socialization to the values, traits, and skills that teachers use to judge success in school.

REFERENCES

Agnew, Robert. "The Contribution of Social-Psychological Strain Theory to the Explanation of Crime and Delinquency." In *The Legacy of Anomie Theory,* edited by Freda Adler and William S. Laufer, 113–137. New Brunswick, NJ: Transaction, 1995.

———. "Foundation for a General Strain Theory of Crime and Deviance." *Criminology* 30 (1992):47–87.

———. "A General Strain Theory of Community Differences in Crime Rates." *Journal of Research in Crime and Delinquency* 36 (1999):123–155.

———. "Sources of Criminality: Strain and Subcultural Theories." In *Criminology*, 3rd ed., edited by Joseph F. Sheley, 349–371. Belmont, CA: Wadsworth, 2000.

———. "Types of Strain Most Likely to Lead to Crime and Delinquency." *Journal of Research in Crime and Delinquency* 38 (2001):319–361.

Agnew, Robert, Timothy Brezina, John Paul Wright, and Francis T. Cullen. "Strain, Personality Traits, and Delinquency: Extending General Strain Theory." *Criminology* 40 (2002):43–72.

Agnew, Robert, and Helene Raskin White. "An Empirical Test of General Strain Theory." *Criminology* 30 (1992):475–499.

Akers, Ronald, and Christine S. Sellers. *Criminological Theories: Introduction and Evaluation.* 4th ed. Los Angeles: Roxbury, 2004.

Alinsky, Saul. *Reveille for Radicals.* New York: Free Press, 1960.

Bernard, Thomas. "Control Criticisms of Strain Theories: An Assessment of Theoretical and Empirical Adequacy." *Journal of Research in Crime and Delinquency* 21 (1984):353–372.

———. "Testing Structural Strain Theories." *Journal of Research in Crime and Delinquency* 24 (1987):262–280.

———. "Reply to Agnew." *Journal of Research in Crime and Delinquency* 24 (1987):287–290.

Bordua, David J. "Delinquent Subcultures: Sociological Interpretations of Gang Delinquency." *Annals of the American Academy of Political and Social Science* 338 (1961):119–136.

———. "Juvenile Delinquency and 'Anomie': An Attempt at Replication." *Social Problems* 6 (1958):230–238.

Brezina, Timothy. "Adapting to Strain: An Examination of Delinquent Coping Responses." *Criminology* 34 (1996):39–60.

———. "Delinquent Problem-Solving: An Interpretive Framework for Criminological Theory and Research." *Journal of Research in Crime and Delinquency* 37 (2000):3–30.

———. "Teenage Violence Toward Parents as an Adaption to Strain: Evidence from a National Survey of Male Adolescents." *Youth and Society* 30 (1999):416–444.

Burgess, Ernest W. "The Growth of the City." 1925. In *The City,* edited by Robert E. Park, Ernest W. Burgess, and Roderick D. McKenzie, 47–62. Chicago: University of Chicago Press, 1967.

Bursik, Robert J., Jr. "Ecological Stability and the Dynamics of Delinquency." In *Communities and Crime, Crime and Justice: A Review of Research,* vol. 8, edited by Albert J. Reiss, Jr., and Michael Tonry, 35–66. Chicago: University of Chicago Press, 1986.

———. "Social Disorganization and Theories of Crime and Delinquency: Problems and Prospects." *Criminology* 26 (1988):519–551.

Bursik, Robert J., Jr., and Harold G. Grasmick. *Neighborhoods and Crime: The Dimensions of Effective Community Control.* New York: Lexington Books, 1993.

Burton, Velmer S., and Francis T. Cullen. "The Empirical Status of Strain Theory." *Crime and Justice* 15 (1992):1–13.

Burton, Velmer S., Francis T. Cullen, T. David Evans, and R. Gregory Dunaway. "Reconsidering Strain Theory: Operationalization, Rival Theories, and Adult Criminality." *Journal of Quantitative Criminology* 10 (1994):213–239.

Catalano, Richard F., Rolf Loeber, and Kay C. McKinney. "School and Community Interventions to Prevent Serious and Violent Offending." In *Juvenile Justice Bulletin.* Washington, DC: US Department of Justice, 1999.

Chilton, Roland J. "Continuity in Delinquency Area Research: A Comparison of Studies in Baltimore, Detroit, and Indianapolis." *American Sociological Review* 29 (1964):71–83.

Cloward, Richard A., and Lloyd E. Ohlin. *Delinquency and Opportunity: A Theory of Delinquent Gangs.* New York: Free Press, 1960.

Cohen, Albert K. *Delinquent Boys: The Culture of the Gang.* New York: Free Press, 1955.

Coleman, James S. *Foundations of Social Theory.* Cambridge, MA: Harvard University Press, 1990.

———. "Social Capital in the Creation of Human Capital." *American Journal of Sociology* 94 (1988):95–120.

Cullen, Francis T. *Rethinking Crime and Deviance Theory: The Emergence of a Structuring Tradition.* Totowa, NJ: Rowman & Allanheld, 1983.

———. "Were Cloward and Ohlin Strain Theorists? Delinquency and Opportunity Revisited." *Journal of Research in Crime and Delinquency* 25 (1988):214–241.

Cullen, Francis T., and Robert Agnew. *Criminological Theory: Past to Present; Essential Readings.* 2nd ed. Los Angeles: Roxbury, 2003.

Curran, Daniel J., and Claire M. Renzetti. *Theories of Crime.* 2nd ed. Boston: Allyn and Bacon, 2001.

Davidson, Norma R. *Crime and Environment.* New York: St. Martin's Press, 1981.

Durkheim, Emile. *The Division of Labor in Society.* 1893. Translated by George Simpson. New York: Free Press, 1947.

Elliott, Delbert S., and Susan S. Ageton. "Reconciling Race and Class Differences in Self-Reported and Official Estimates of Delinquency." *American Sociological Review* 45 (1980):95–110.

Elliott, Delbert S., Suzanne S. Ageton, and Rachelle J. Cantor. "An Integrated Theoretical Perspective on Delinquent Behavior." *Journal of Research in Crime and Delinquency* 16 (1979):3–17.

Elliott, Delbert S., David Huizinga, and Suzanne S. Ageton. *Explaining Delinquency and Drug Use.* Beverly Hills, CA: Sage, 1985.

Elliott, Delbert S., William Julius Wilson, David Huizinga, Robert J. Sampson, Amanda Elliott, and Bruce Rankin. "Effects of Neighborhood Disadvantage on Adolescent Development." *Journal of Research in Crime and Delinquency* 33 (1996):389–426.

Empey, LaMar T., Mark C. Stafford, and Carter H. Hay. *American Delinquency: Its Meaning and Construction.* 4th ed. Belmont, CA: Wadsworth, 1999.

Esbensen, Finn-Aage, and David Huizinga. "Gangs, Drugs, and Delinquency in a Survey of Urban Youth." *Criminology* 31 (1993):565–589.

Chapter Resources

Farnworth, Margaret, and Michael J. Leiber. "Strain Theory Revisited: Economic Goals, Educational Means, and Delinquency." *American Sociological Review* 54 (1989):263–274.

Farrington, David P., Howard N. Snyder, and Terrance A. Finnegan. "Specialization in Juvenile Court Careers." *Criminology* 26 (1988):461–487.

Finestone, Harold. *Victims of Change: Juvenile Delinquents in American Society.* Westport, CT: Greenwood, 1976.

Gibbons, Don C. *The Criminological Enterprise: Theories and Perspectives.* Englewood Cliffs, NJ: Prentice Hall, 1979.

Greenberg, David. "Delinquency and the Age Structure of Society." *Contemporary Crisis* 1 (1977):66–86.

Hagan, John, Gerd Hefler, Cabriele Classen, Klaus Boehnke, and Hans Merkens. "Subterranean Sources of Subcultural Delinquency Beyond the American Dream." *Criminology* 36 (1998): 309–341.

Halleck, Seymour. *Psychiatry and the Dilemmas of Crime: A Study of Causes, Punishment, and Treatment.* New York: Harper & Row, 1967.

Hay, Carter. "Family Strain, Gender, and Delinquency." *Sociological Perspective* 46 (2003):107–135.

Hinkle, Roscoe C., Jr., and Gisela J. Hinkle. *The Development of Modern Sociology: Its Nature and Growth in the United States.* New York: Random House, 1954.

Hirschi, Travis. *Causes of Delinquency.* Berkeley, CA: University of California Press, 1969.

Hochschild, Jennifer. *Facing Up to the American Dream: Race, Class, and the Soul of the Nation.* Princeton, NJ: Princeton University Press, 1995.

Hoffman, John P., and Alan S. Miller. "A Latent Variable Analysis of General Strain Theory." *Journal of Quantitative Criminology* 14 (1998):83–110.

Hyman, Herbert H. "The Value System of Different Classes: A Social-Psychological Contribution to the Analysis of Stratification." In *Class, Status, and Power,* edited by Reinhard Bendix and Seymour M. Lipset, 426–442. New York: Free Press, 1953.

Jensen, Gary F. "Salvaging Structure Through Strain: A Theoretical and Empirical Critique." In *The Legacy of Anomie Theory,* edited by Freda Adler and William S. Laufer, 139–158. New Brunswick, NJ: Transaction, 1995.

Johnson, Richard E. *Juvenile Delinquency and Its Origins: An Integrated Theoretical Approach.* New York: Cambridge University Press, 1979.

Karp, David A., Gregory P. Stone, and William C. Yoels. *Being Urban: A Social Psychological View of City Life.* Lexington, MA: Heath, 1977.

Kituse, John I., and David C. Dietrick. "Delinquent Boys: A Critique." *American Sociological Review* 24 (1959):208–215.

Kobrin, Solomon. "The Chicago Area Project—A 25 Year Assessment." *Annals of the American Society of Political and Social Science* 322 (1959):20–29.

Kornhauser, Ruth Rosner. *Social Sources of Delinquency: An Appraisal of Analytic Models.* Chicago: University of Chicago Press, 1978.

Kroeber, A. L., and Talcott Parsons. "The Concepts of Culture and Social System." *American Sociological Review* 23 (1958):582–583.

Krohn, Marvin D. "The Web of Conformity: A Network Approach to the Explanation of Delinquent Behavior." *Social Problems* 33 (1986):581–582.

Lander, Bernard. *Toward an Understanding of Juvenile Delinquency.* New York: Columbia University Press, 1954.

Lemert, Edwin M. *Human Deviance, Social Problems, and Social Control.* 2nd ed. Englewood Cliffs, NJ: Prentice Hall, 1972.

Lilly, J. Robert, Francis T. Cullen, and Richard A. Ball. *Criminological Theory: Context and Consequences.* Newbury Park, CA: Sage, 1978.

Liska, Allen. "Aspirations, Expectations, and Delinquency: Stress and Additive Models." *Sociological Quarterly* 12 (1971):99–107.

Lowenkamp, Christopher T., Francis T. Cullen, and Travis C. Pratt. "Replicating Sampson and Groves' Test of Social Disorganization Theory: Revisiting a Criminological Classic." *Journal of Research in Crime and Delinquency* 40 (2003):351–373.

Lundman, Richard. *Prevention and Control of Juvenile Delinquency.* 2nd ed. New York: Oxford, 1993.

Menard, Scott. "A Developmental Test of Mertonian Anomie Theory." *Journal of Research in Crime and Delinquency* 32 (1995):136–174.

Merton, Robert K. "On the Evolving Synthesis of Differential Association and Anomie Theory: A Perspective from the Sociology of Science." *Criminology* 35 (1997):517–525.

———. "Opportunity Structure: The Emergence, Diffusion and Differentiation of a Sociological Concept, 1930s–1950s." In *The Legacy of Anomie Theory* (*Advances in Criminological Theory,* vol. 6), edited by Freda Adler and William S. Laufer, 3–78. New Brunswick, NJ: Transaction, 1995.

———. "Social Structure and Anomie." *American Sociological Review* 3 (1938):672–682.

———. *Social Theory and Social Structure.* 2nd ed. New York: Free Press, 1968.

Messner, Steven F., and Richard Rosenfeld. *Crime and the American Dream.* 2nd ed. Belmont, CA: Wadsworth, 1997.

Moffitt, Terrie E. "Adolescent-Limited and Life Course Persistent Antisocial Behavior: A Developmental Taxonomy." *Psychological Review* 100 (1993):674–701.

Oberschall, Anthony. "The Institutionalization of American Sociology." In *The Establishment of Empirical Sociology,* edited by Anthony Oberschall, 187–251. New York: Harper and Row, 1972.

Park, Robert E., and Ernest W. Burgess. *Introduction to the Science of Sociology.* 2nd ed. Chicago: University of Chicago Press, 1924.

Park, Robert E., Ernest W. Burgess, and Roderick D. McKenzie. *The City.* 1925. Chicago: University of Chicago Press, 1967.

Paternoster, Raymond, and Paul Mazerolle. "General Strain Theory and Delinquency: A Replication and Extension." *Journal of Research in Crime and Delinquency* 31 (1994):235–263.

Rodman, Hyman, and Paul Grams. "Juvenile Delinquency and the Family." In *Juvenile Delinquency and Youth Crime.* The President's Commission on Law Enforcement and Administration of Justice Task Force Report, 188–221. Washington, DC: GPO, 1967.

Roth, Jeffrey A. "Understanding and Preventing Violence." *Research in Brief.* National Institute of Justice. Washington, DC: US Department of Justice, 1994.

Sampson, Robert J., and Dawn Jeglum Bartusch. "Legal Cynicism and (Subcultural?) Tolerance of Deviance: The Neighborhood Context of Racial Differences." *Law and Society Review* 32 (1998): 777–804.

Sampson, Robert J., and W. Byron Groves. "Community Structure and Crime: Testing Social Disorganization Theory." *American Journal of Sociology* 94 (1989):774–802.

Sampson, Robert J., Jeffrey D. Morenoff, and Felton Earls. "Beyond Social Capital: Spatial Dynamics of Collective Efficacy for Children." *American Sociological Review* 64 (1999):633–660.

Sampson, Robert J., and Stephen W. Raudenbush. "Disorder in Urban Neighborhoods—Does It Lead to Crime?" *Research in Brief.* National Institute of Justice. Washington, DC: US Department of Justice, 2001.

———. "Seeing Disorder: Neighborhood Stigma and the Social Construction of 'Broken Windows.' " *Social Psychological Quarterly* 67 (2004):319–342.

———. "Systematic Social Observation of Public Spaces: A New Look at Disorder in Urban Neighborhoods." *American Journal of Sociology* 105 (1999):603–651.

Sampson, Robert J., Steven W. Raudenbush, and Felton Earls. "Neighborhoods and Violent Crime: A Multilevel Study of Collective Efficacy." *Science* 277 (1997):918–924.

Sampson, Robert J., and William Julius Wilson. "Toward a Theory of Race, Crime, and Urban Inequality." In *Crime and Inequality,* edited by John Hagan and Ruth D. Peterson, 37–54. Stanford, CA: Stanford University Press, 1995.

Schuerman, Leo A., and Solomon Kobrin. "Community Careers in Crime." In *Communities and Crime, Crime and Justice: A Review of Research,* vol. 8, edited by Albert J. Reiss, Jr. and Michael Tonry. Chicago: University of Chicago Press, 1986.

Shaw, Clifford R. *The Natural History of a Delinquent Career.* Chicago: University of Chicago Press, 1931.

Shaw, Clifford R., and Henry D. McKay. *Social Factors in Juvenile Delinquency: A Study of the Community, the Family, and the Gang in Relation to Delinquent Behavior.* Report of the National Commission on Law Observance and Enforcement, Causes of Crime, vol. II. Washington, DC: GPO, 1931.

———. *Juvenile Delinquency and Urban Areas: A Study of Rates of Delinquency in Relation to Differential Characteristics of Local Communities in American Cities.* Rev. ed. Chicago: University of Chicago Press, 1969.

Shaw, Clifford R., Frederick M. Zorbaugh, Henry D. McKay, and Leonard S. Cottrell. *Delinquency Areas: A Study of the Geographic Distribution of School Truants, Juvenile Delinquents, and Adult Offenders in Chicago.* Chicago: University of Chicago Press, 1929.

Sherman, Lawrence W., Denise C. Gottfredson, Doris L. MacKenzie, John Eck, Peter Reuter, and Shawn D. Bushway. "Preventing Crime: What Works, What Doesn't, What's Promising." *Research in Brief.* National Institute of Justice. Washington, DC: US Department of Justice, 1998.

Short, James F., Jr. "Introduction to the Revised Edition." In Shaw and McKay, *Juvenile Delinquency and Urban Areas,* xxv–liv.

Chapter Resources

Short, James F., and Fred L. Strodtbeck. *Group Process and Gang Delinquency.* Chicago: University of Chicago Press, 1965.

Snodgrass, Jon. "Clifford R. Shaw and Henry D. McKay: Chicago Sociologists." *The British Journal of Criminology* 16 (1976):1–19.

Srole, Leo. "Social Integration and Certain Corollaries: An Exploratory Study." *American Sociological Review* 21 (1956):709–716.

Stark, Rodney. "Deviant Places: A Theory of the Ecology of Crime." *Criminology* 25 (1987):893–909.

Stinchcombe, Arthur L. *Rebellion in a High School.* Chicago: Quadrangle Books, 1964.

Thomas, William I., and Florian Znaniecki. *The Polish Peasant in Europe and America.* Vol. II. Boston: Gorham Press, 1920.

Thornberry, Terence P., Marvin D. Krohn, Alan J. Lizotte, and Deborah Chard-Wierschem. "The Role of Juvenile Gangs in Facilitating Delinquent Behavior." *Journal of Research in Crime and Delinquency* 30 (1993):55–87.

Tittle, Charles R. "Social Class and Criminal Behavior: A Critique of the Theoretical Foundation." *Social Forces* 62 (1983):334–358.

Tittle, Charles R., and Robert F. Meier. "Specifying the SES/Delinquency Relationship." *Criminology* 28 (1990):271–299.

Tittle, Charles R., Wayne J. Villemez, and Douglas A. Smith. "The Myth of Social Class and Criminality: An Empirical Assessment of the Empirical Evidence." *American Sociological Review* 43 (1978):643–656.

Vold, George B., Thomas J. Bernard, and Jeffrey B. Snipes. *Theoretical Criminology.* 5th ed. New York: Oxford University Press, 2002.

Walderman, Steven, and Karen Springer. "Too Old, Too Fast?" *Newsweek* (November 16, 1992):80–88.

Wiatrowski, Michael D., David B. Griswold, and Mary K. Roberts. "Social Control Theory and Delinquency." *American Sociological Review* 46 (1981):525–541.

Wiatrowski, Michael D., Stephen Hansell, Charles R. Massey, and David L. Wilson. "Curriculum Tracking and Delinquency." *American Sociological Review* 47 (1982):151–160.

Wilson, James Q., and George Kelling. "The Police and Neighborhood Safety: Broken Windows." *Atlantic Monthly* 249 (1982):29–38.

ENDNOTES

1. Shaw, *Natural History,* 13; Shaw et al., *Delinquency Areas*; and Shaw and McKay, *Juvenile Delinquency.*
2. Sampson and Groves, "Community Structure and Crime," 774–775. See also Kornhauser, *Social Sources of Delinquency,* 63–64.
3. In a brief, but important statement in the *American Sociological Review,* A. L. Kroeber and Talcott Parsons tried to make clear the distinction between the concepts of *culture* and *social structure.* Culture, they argued, should be limited to dimensions such as values, beliefs, and knowledge, whereas social structure should refer to relational or interactional dimensions of the social system. Later, Ruth Kornhauser criticized delinquency theory for failing to keep these concepts distinct. She argued that social disorganization theory and anomie theory mingle these concepts and thereby make them indistinguishable and of limited analytic usefulness. While Kornhauser's point is well taken, the concepts simply cannot be neatly separated. They refer to different aspects of the same underlying social phenomenon: mechanisms of social organization. See also Kornhauser, *Social Sources of Delinquency*; and Messner and Rosenfeld, *Crime and the American Dream,* 50.
4. Krohn, "Web of Conformity," 581–582.
5. Durkheim, *Division of Labor.*
6. Ibid., 79.
7. Ibid., 85.
8. Ibid.
9. Karp, Stone, and Yoels, *Being Urban,* 19.
10. Kornhauser, *Social Sources of Delinquency,* 63–64. See also Sampson and Groves, "Community Structure and Crime," 774–775, 780–781.
11. Hinkle and Hinkle, *Development of Modern Sociology.*

Chapter Resources

12. Curran and Renzetti, *Theories of Crime,* 136; Hinkle and Hinkle, *Development of Modern Sociology*; and Oberschall, "Institutionalization of American Sociology."

13. Park and Burgess, *Introduction,* 720. See also Bursik, "Ecological Stability."

14. Park, Burgess, and McKenzie, *The City.*

15. Burgess, "Growth of the City," 47–62.

16. Shaw and McKay, *Juvenile Delinquency,* 315.

17. Stark, "Deviant Places," 893–909.

18. Park and Burgess, *Introduction*; and Bursik, "Ecological Stability."

19. Shaw and McKay, *Juvenile Delinquency,* 10.

20. Ibid., 37.

21. Shaw et al., *Delinquency Areas,* 198–203.

22. Shaw and McKay. *Juvenile Delinquency and Urban Areas* continued work that was commissioned by the National Commission on Law Observance and Enforcement (The Wickersham Commission). An earlier publication provided preliminary findings from this study (Shaw and McKay, *Social Factors*).

23. Bursik, "Ecological Stability."

24. Shaw and McKay, *Juvenile Delinquency,* 315.

25. Sampson and Groves, "Community Structure and Crime," 774–775. See also Kornhauser, *Social Sources of Delinquency,* 63–64; and Robert Bursik, "Social Disorganization," 520.

26. Akers and Sellers, *Criminological Theories,* 162. See also Bursik, "Social Disorganization," 520.

27. Finestone, *Victims of Change,* 88.

28. Bursik, "Social Disorganization," 521. See also Bursik and Grasmick, *Neighborhoods and Crime,* 33; and Kornhauser, *Social Sources of Delinquency,* 75–78.

29. Shaw et al., *Delinquency Areas,* 1.

30. Shaw and McKay, *Social Factors,* 99–100.

31. Bursik, "Social Disorganization," 521.

32. Shaw and McKay, *Social Factors,* 102.

33. Ibid., 117.

34. Kornhauser, *Social Sources of Delinquency.*

35. Ibid., 75–76.

36. Shaw and McKay, *Juvenile Delinquency,* 318.

37. Kornhauser, *Social Sources of Delinquency,* 76.

38. Shaw and McKay, *Juvenile Delinquency,* 315.

39. Lander, *Toward an Understanding*; Bordua, "Juvenile Delinquency and 'Anomie';" Chilton, "Continuity;" and Schuerman and Kobrin, "Community Careers in Crime."

40. Kornhauser, *Social Sources of Delinquency*; and Sampson and Groves, "Community Structure and Crime."

41. See also Bursik, "Social Disorganization."

42. Sampson and Groves, "Community Structure and Crime," 799. See Lowenkamp, Cullen, and Pratt, "Replicating," for a replication of Sampson and Groves findings.

43. Shaw and McKay, *Juvenile Delinquency,* 321.

44. Bursik, "Social Disorganization."

45. Bursik, "Social Disorganization," 519. See also Davidson, *Crime and Environment.*

46. Elliott et al., "Effects of Neighborhood Disadvantage;" Sampson and Groves, "Community Structure and Crime;" Sampson, Morenoff, and Earls, "Beyond Social Capital;" Sampson, Raudenbush, and Earls, "Neighborhoods and Violent Crime;" and Sampson and Wilson, "Toward a Theory."

47. The concept of *social capital* is drawn from the work of James S. Coleman ("Social Capital" and *Foundations of Social Theory*).

48. Sampson and Wilson, "Toward a Theory."

49. Cullen and Agnew, *Criminological Theory,* 104.

50. Sampson, Raudenbush, and Earls, "Neighborhoods and Violent Crime."

51. Recent research by Sampson, Morenoff, and Earls ("Beyond Social Capital") found that residential stability and concentrated advantage, rather than concentrated disadvantage, predict collective efficacy. Further, their findings indicate that the extent to which neighborhood collective efficacy is able to control crime and delinquency is dependent upon the "neighborhood's relative spatial position in the larger city," especially with regard to the collective efficacy of surrounding neighborhoods (657).

52. Cullen and Agnew, *Criminological Theory,* 101.

53. Shaw and McKay, *Juvenile Delinquency,* 180–181.

54. Cullen and Agnew, *Criminological Theory,* 171–174, 198; and Cullen, *Rethinking.* See Bernard ("Testing Structural Strain Theories" and "Reply to Agnew") for a view that emphasizes solely the structural aspect of Merton's theory.

55. Merton, "Social Structure and Anomie."

56. Merton, "On the Evolving Synthesis," 522; this trail of revision goes from Merton, "Social Structure and Anomie," to Merton, "Opportunity Structure."

57. Merton, "Social Structure and Anomie," 672.

58. Merton, "Social Structure and Anomie," 680 and 675, respectively.

59. Ibid., 673.

60. Cullen and Agnew, *Criminological Theory,* 171–174, 198; and Messner and Rosenfeld, *Crime and the American Dream,* 55–56.

61. Merton, "Social Structure and Anomie," 680, emphasis in original.

62. Ibid., 674.

63. Gibbons, *Criminological Enterprise,* 69; and Vold, Bernard, and Snipes, *Theoretical Criminology,* 138.

64. Cullen and Agnew, *Criminological Theory,* 178.

65. Merton, "Social Structure and Anomie," 680–682.

66. Ibid., 681.

67. Lilly, Cullen, and Ball, *Criminological Theory,* 68.

68. Messner and Rosenfeld, *Crime and the American Dream.*

69. Ibid., 5.

70. Ibid., 62–64.

71. Merton, *Social Theory,* 168, cited in Messner and Rosenfeld, *Crime and the American Dream,* 63.

72. Hochschild, *Facing Up,* xi, cited in Messner and Rosenfeld, *Crime and the American Dream,* 62.

73. Messner and Rosenfeld, *Crime and the American Dream,* 70; Walderman and Springer, "Too Old," 80–88.

74. Messner and Rosenfeld, *Crime and the American Dream,* 68.

75. Ibid.

76. Ibid., 76–77.

77. Gibbons, *Criminological Enterprise,* 71.

78. Vold, Bernard, and Snipes, *Theoretical Criminology,* 146.

79. Lemert, *Human Deviance.* See also Agnew, "Sources of Criminality," 353; and Lilly, Cullen, and Ball, *Criminological Theory,* 76.

80. Agnew, "Sources of Criminality," 354; Elliott, Ageton, and Cantor, "Integrated Theoretical Perspective;" and Elliott, Huizinga, and Ageton, *Explaining Delinquency.*

81. Agnew, "Sources of Criminality," 354; Greenberg, "Delinquency;" Elliott, Huizinga, and Ageton, *Explaining Delinquency*; and Moffitt, "Developmental Taxonomy."

82. Hyman, "Value System."

83. Agnew, "Sources of Criminality," 353. See also Elliott, Ageton, and Cantor, "Integrated Theoretical Perspective;" Empey, Stafford, and Hay, *American Delinquency,* 195; and Kornhauser, *Social Sources of Delinquency,* 162–165.

84. Kornhauser, *Social Sources of Delinquency,* 167–180. See Bernard, "Control Criticisms" (359–366), for a very different interpretation of these same studies.

85. Lilly, Cullen, and Ball, *Criminological Theory,* 76; and Messner and Rosenfeld, *Crime and the American Dream,* 55.

86. Messner and Rosenfeld, *Crime and the American Dream,* 55.

87. Kornhauser, *Social Sources of Delinquency,* 47.

88. Tittle, Villemez, and Smith, "Myth of Social Class."

89. Akers and Sellers, *Criminological Theories,* 170–171.

90. Elliott and Ageton, "Reconciling."

91. Akers and Sellers, *Criminological Theories,* 170–171; Tittle, "Social Class;" Tittle and Meier, "Specifying;" and Tittle, Villemez, and Smith, "Myth of Social Class."

92. Tittle, Villemez, and Smith, "Myth of Social Class."

93. Empey, Stafford, and Hay, *American Delinquency,* 194; Hirschi, *Causes of Delinquency*; and Johnson, *Juvenile Delinquency.*

94. Cullen and Agnew, *Criminological Theory,* 171–174, 198; and Cullen, *Rethinking.*

95. Merton, "Social Structure and Anomie," 680.
96. Menard, "Developmental Test," 137; and Bernard, "Testing Structural Strain Theories."
97. Halleck, *Psychiatry,* 223, cited in Cullen, *Rethinking,* 37.
98. Curran and Renzetti, *Theories of Crime,* 115.
99. Merton, "Social Structure and Anomie," 676.
100. Ibid., 677.
101. Merton, "Social Structure and Anomie;" and Menard, "Developmental Test," 138.
102. Merton, *Social Theory,* 151.
103. Merton, "Social Structure and Anomie," 677.
104. Agnew, "Contribution," 113.
105. Cullen, *Rethinking*; and Jensen, "Salvaging Structure Through Strain."
106. Farnworth and Leiber, "Strain Theory Revisited," 263.
107. Agnew, "Sources of Criminality;" Burton and Cullen, "Empirical Status;" Burton et al., "Reconsidering Strain Theory;" Cullen and Agnew, *Criminological Theory*; and Farnworth and Leiber, "Strain Theory Revisited."
108. Agnew, "Contribution," 114.
109. Agnew, "Sources of Criminality;" Burton and Cullen, "Empirical Status;" Burton et al., "Reconsidering Strain Theory;" Hirschi, *Causes of Delinquency*; Kornhauser, *Social Sources of Delinquency*; Jensen, "Salvaging Structure Through Strain;" and Liska, "Aspirations."
110. Hirschi, *Causes of Delinquency,* 172, emphasis in original. See Johnson (*Juvenile Delinquency,* 148–150) and Jensen ("Salvaging Structure Through Strain") for findings consistent with Hirschi's. See Bernard ("Control Criticisms") for a reinterpretation of the theoretical and empirical adequacy of control theory's criticism of strain theory.
111. Hirschi, *Causes of Delinquency*; Liska, "Aspirations;" Kornhauser, *Social Sources of Delinquency*; and Johnson, *Juvenile Delinquency.*
112. Agnew, "Sources of Criminality;" and Burton et al., "Reconsidering Strain Theory."
113. Agnew, "Sources of Criminality," 353; and Burton et al., "Reconsidering Strain Theory."
114. Kornhauser, *Social Sources of Delinquency,* 167–180; see also Agnew, "Sources of Criminality," 352.
115. Farnworth and Leiber, "Strain Theory Revisited," 272.
116. Farnworth and Leiber, "Strain Theory Revisited," 264, 265; see also Bernard, "Control Criticisms."
117. Farnworth and Leiber, "Strain Theory Revisited," 265.
118. Ibid., 271.
119. Ibid., 272.
120. Ibid.
121. Menard, "Developmental Test," 146–147; Farnworth and Leiber, "Strain Theory Revisited;" and Cullen and Agnew, *Criminological Theory,* 173.
122. Menard, "Developmental Test," 166.
123. Ibid.
124. Agnew, "Sources of Criminality," 354; Greenberg, "Delinquency;" Elliott, Huizinga, and Ageton, *Explaining Delinquency*; and Moffitt, "Developmental Taxonomy."
125. Agnew, "Contribution," 115.
126. Agnew, "Foundation," 50; see also 48.
127. Ibid., 50.
128. Agnew, "Contribution," 115.
129. Ibid.
130. Ibid.
131. Agnew, "Foundation," 52–53.
132. Ibid., 52.
133. Ibid., 54–55.
134. Ibid., 54.
135. Ibid., 57.
136. Ibid., 58.
137. Agnew, "Types of Strain."
138. Cullen and Agnew, *Criminological Theory,* 174–175. See also Agnew, "Foundation," 49; Agnew and White, "Empirical Test," 477; and Brezina, "Adapting to Strain," 41.

Chapter Resources

139. Brezina, "Adapting to Strain," 41–42. See also Agnew, "Foundation," 49; Agnew and White, "Empirical Test," 477; and Cullen and Agnew, *Criminological Theory*, 174–175.
140. Agnew and White, "Empirical Test," 477. See also Agnew, "Foundation."
141. Agnew et al., "Extending General Strain Theory."
142. Hay, "Family Strain."
143. Agnew, "Sources of Criminality," 356. See also Hoffman and Miller, "Latent Variable Analysis;" and Paternoster and Mazerolle, "General Strain Theory."
144. Agnew and White, "Empirical Test;" and Paternoster and Mazerolle, "General Strain Theory."
145. Agnew and White, "Empirical Test."
146. See also Paternoster and Mazerolle, "General Strain Theory."
147. Brezina, "Adapting to Strain;" Brezina, "Teenage Violence;" and Brezina, "Delinquent Problem-Solving."
148. Brezina, "Adapting to Strain," 39.
149. Brezina, "Adapting to Strain," 57.
150. Vold, Bernard, and Snipes, *Theoretical Criminology*, 143.
151. Hagan et al., "Subterranean Sources," 311.
152. Cohen, *Delinquent Boys*, 26.
153. Ibid., 13.
154. Cullen and Agnew, *Criminological Theory*, 186. See also Akers and Sellers, *Criminological Theories*, 145–146. Cohen actually referred to "working-class" boys rather than "lower-class" boys. However, the particular group on which he focused was inner-city boys who today would be more commonly classified as lower class.
155. Cohen, *Delinquent Boys*, 84–88.
156. Ibid., 99, 100, 98 respectively.
157. Ibid., 95–97.
158. Rodman and Grams, "Juvenile Delinquency," 192.
159. Cohen, *Delinquent Boys*, 88–91.
160. Ibid., 84, 113–115.
161. Ibid., 49–72, 121–127.
162. Ibid., 129.
163. Ibid., 128.
164. Ibid., 49.
165. Ibid., 121, emphasis in original.
166. Ibid., 122, 132–133.
167. Ibid., 25–32.
168. Gibbons, *Criminological Enterprise*, 94. See also Agnew, "General Strain Theory;" Akers and Sellers, *Criminological Theories*; Bordua, "Delinquent Subcultures;" Kituse and Dietrick, "Delinquent Boys;" and Stinchcombe, *Rebellion*.
169. Cloward and Ohlin, *Delinquency and Opportunity*, 86.
170. Ibid., 97–103.
171. Ibid., 1. See also 22–23, 161–171.
172. Ibid., 20. See also 24–25, 171–178.
173. Ibid., 25.
174. Ibid., 26.
175. Cloward and Ohlin, *Delinquency and Opportunity*, 211. See also Cullen ("Cloward and Ohlin"), who highlights the structural aspect of Cloward and Ohlin's theory of differential opportunity.
176. Tittle, "Social Class;" Tittle and Meier, "Specifying;" and Tittle, Villemez, and Smith, "Myth of Social Class."
177. Empey, Stafford, and Hay, *American Delinquency*, 195; Hirschi, *Causes of Delinquency*; Johnson, *Juvenile Delinquency*; Wiatrowski, Griswold, and Roberts, "Social Control Theory;" and Wiatrowski et al., "Curriculum Tracking and Delinquency."
178. Esbensen and Huizinga, "Gangs;" Farrington, Snyder, and Finnegan, "Specialization;" Short and Strodbeck, *Group Process and Gang Delinquency*; and Thornberry et al., "Role of Juvenile Gangs."

Labeling and Critical Criminologies

Chapter Objectives

After completing this chapter, students should be able to:

- Explain what distinguishes the labeling perspective and critical criminologies from more traditional approaches to delinquency.
- Describe the consequences of formal and informal labeling processes.
- Identify the major themes of critical criminologies.
- Describe specific theories that take a critical approach to the issue of delinquency.
- Identify the defining features of feminist approaches in criminology.
- Describe how feminist perspectives have been applied to the issue of delinquency.
- Understand theories and key terms:
  Theories:
  theory of reintegrative shaming
  integrated structural-Marxist theory

power-control theory
Greenberg's Marxist interpretation of delinquency
Terms:
labeling perspective
critical criminologies
dramatization of evil
primary deviance
secondary deviance
moral crusaders
deviance amplification
reintegrative shaming
stigmatization
interdependency
communitarianism
instrumental Marxism
structural Marxism
patriarchal families
egalitarian families
masculine status anxiety
feminism
liberal feminism
radical feminism
patriarchy
socialist feminism
Marxist feminism

CASE IN POINT   Labeling Rocky "Delinquent"

Robert Sampson and John Laub, in their book *Crime in the Making: Pathways and Turning Points Through Life,* describe a movie about Rocky Sullivan and Jerry Connelly—two childhood friends and companions in crime whose lives follow very different courses. The story told in this movie suggests that it is Rocky's experience of being caught by the police and thrust into the juvenile justice system, as well as Jerry's ability to get away and avoid both the justice system and the label "delinquent," that is responsible for the very different paths of their lives.

The movie was *Angels with Dirty Faces* and starred James Cagney, Pat O'Brien, Humphrey Bogart, Ann Sheridan, and, of course, the Dead

End Kids. James Cagney played the role of Rocky Sullivan, who was by contemporary definitions a high-rate, chronic offender—in other words, a "career criminal." Pat O'Brien played the role of Jerry Connelly, who became a priest in the local neighborhood parish. Both men were childhood friends, committed petty crime together, and were in fact products of the same slum environment. Yet both obviously had very different life experiences with respect to serious and persistent criminal activity. . . .

At the beginning of the film the young boys, James Cagney (Rocky Sullivan) and Pat O'Brien (Jerry Connelly), break into a box car to steal some pens. The police come and both boys run to avoid getting caught. One boy, Jerry, is a little quicker and gets away. Rocky, slower of foot, gets caught. Of course, Rocky does not squeal on his best friend, and he is the one who turns out to be a criminal later in life. The explanation offered in the film is that "reform school made a criminal out of Rocky," "there but for the grace of God go I," and "Rocky was a good kid gone bad." At the end of the film, after Rocky Sullivan is executed for several homicides, Pat O'Brien in the role of Jerry Connelly sadly remarks: "Let's say a prayer for a boy who couldn't run as fast as I could."

Source: Sampson and Laub, *Crime in the Making*, 64, 97.

Over three decades ago, Edwin Schur made the keen observation that "deviance and social control always involve processes of social definition."[1] We often take for granted that "delinquency" is behavior that violates the law. The <u>labeling perspective</u> does not make this common assumption. Instead, labeling theorists maintain that definitions of *what* actions are delinquent and *who* is delinquent result from dynamic social processes. Instead of asking, "Why do they do it?" and attempting to explain individuals' involvement in delinquency, proponents of the labeling perspective ask the question: How and why do certain behaviors and individuals get labeled "deviant," "delinquent," or "criminal?" In Chapter 2, we explored how the legal concept of delinquency first emerged and how it has transformed over time. In this chapter, we examine how the label "delinquent" is imposed legally in the juvenile justice system and socially in the context of relationships. We also consider consequences of the labeling process.

The labeling perspective focuses on social and societal *reactions* to delinquent behavior. The same is true of <u>critical criminologies</u>, or conflict theories, which consider why society defines crime and reacts to crime and criminals the way it does. Critical criminologies explore the structural bases of inequality in society and the way in which inequality and social status influence definitions of crime and criminals as well as societal reactions to specific behaviors— "delinquent," "criminal," and otherwise. We consider several applications of critical approaches to the problem of delinquency, including integrated structural-Marxist theory, power-control theory, and a Marxist interpretation of juvenile delinquency in capitalist societies.[2] We also examine feminist

labeling perspective This perspective focuses on social reactions to delinquency and crime and considers how definitions of what is crime and who is criminal result from dynamic social processes.

critical criminologies Explore the structural bases of inequality in society and the way in which inequality and social status influence definitions of crime and criminals and reactions to behaviors.

approaches to crime and delinquency, which focus on the role gender plays in creating crime and delinquency and in structuring reactions to offending. We distinguish four feminist perspectives—liberal, radical, socialist, and Marxist—and offer examples of each in the field of criminology.

■ The Labeling Perspective

In the simplest terms, the labeling perspective is concerned with how and why behavior is labeled "deviant." The focus is on the *definition* or *social construction* of deviance, rather than on its causes. Labeling theorists do not view deviance as an inherent quality of particular acts. Instead, they see deviance as derived from the social creation of rules that define particular acts as "deviant" and from the application of those rules to particular people. The labeling perspective explicitly recognizes that definitions of deviance are not universal, but rather are unique to particular times, places, and cultures or subcultures.[3] For example, consider the evolution of the concept of delinquency in the United States that we described in Chapter 2.

Labeling theory has its theoretical roots in both symbolic interactionism and the conflict perspective.[4] From *symbolic interactionism,* labeling theorists drew the idea that the imposition of labels (formal or informal) has consequences for those labeled. Through processes of social interaction described by symbolic interactionists, individuals recognize and respond to the labels attached to them, often with behavior that is consistent with those labels. From the *conflict perspective,* labeling theorists drew the idea that definitions of what constitutes deviant behavior are determined by those with sufficient power and resources to make and enforce the rules.[5] We explore these origins of labeling theory in more detail later in this chapter.

In this section, we discuss original statements of labeling theory, which consider how the labels "deviant" or "delinquent" are imposed, and what consequences those labels have for those who bear them. We examine both *formal* labels imposed through the juvenile justice system and *informal* labels imposed in the context of interactions with others such as parents and peers.

Imposing the Label of "Deviant" or "Delinquent"[6]
Formal Societal Reactions to Deviance: Dramatizing Evil

Frank Tannenbaum is typically credited with providing the first statement of the principles of labeling theory, although these principles were not recognized as a formal theory until the 1960s. In *Crime and the Community,* Tannenbaum argues that youth often engage in acts that they consider to be "play, adventure . . . mischief, fun," but that others, including agents of formal social control, consider to be delinquent or "evil."[7] Others begin to react to these actions and to the youths as "bad" or "evil." The process of definition has negative and lasting consequences for youths' self-images and subsequent behavior. In a process that Tannenbaum calls the **dramatization of evil**, "the young delinquent becomes bad because he is defined as bad and because he is not believed if he is good."[8] In other words, the definition or labeling of individuals as "bad" results in a *self-fulfilling prophecy* in which youths become what they are said to be. In a fre-

dramatization of evil A process in which others react to delinquent and mischievous actions of a youth as though these actions and the youth are "evil," and, through these reactions, create a self-fulfilling prophecy in which the youth becomes what he or she is said to be.

quently cited passage, Tannenbaum writes, "The process of making the criminal . . . is a process of tagging, defining, identifying, segregating, describing, emphasizing, making conscious and self-conscious; it becomes a way of stimulating, suggesting, emphasizing, and evoking the very traits that are complained of. . . . The person becomes the thing he is described as being."[9]

Once individuals are formally labeled "delinquent," they are isolated from law-abiding groups and drawn "into companionship with other children similarly defined."[10] In this way, because of the formal labeling process, initial acts of mischief evolve into serious delinquency. Thus, according to Tannenbaum, crime and delinquency are *created* through the dramatization of evil. Tannenbaum's recommendation for dealing with juveniles is "a refusal to dramatize the evil. The less said about it the better."[11]

Primary and Secondary Deviance: Identity and Self-Fulfilling Prophecy

In 1951, Edwin Lemert expanded on Tannenbaum's ideas about labeling and deviance with the concepts of primary and secondary deviance.[12] **Primary deviance** refers to initial acts of deviance, many of which go undetected. Lemert attributes primary deviance to a variety of causes, including cultural, situational, psychological, and even physiological factors.[13] Self-report studies (see Chapter 3) suggest that primary deviance is relatively common. Many individuals engage in delinquency but are never caught and processed by the juvenile justice system. The most important aspect of primary deviance is that it has no significant impact on self-concept or social status. Individuals who engage in primary deviance are able to rationalize their behavior as part of a "socially acceptable role" and continue to view themselves as something other than deviants or delinquents.[14] For example, adolescents might skip school and engage in acts of shoplifting or vandalism, but still not see themselves as "delinquents." Primary deviance is often seen as "a problem of everyday life" and dealt with in the context of established relationships.[15]

primary deviance Initial acts of deviance, many of which go undetected.

In some cases, however, societal reactions to primary deviance include elements of social control, punishment, and stigmatization that change how primary deviants view themselves. According to Lemert, the application of "deviant" or "delinquent" labels through formal responses to primary deviance has significant consequences for social status, self-concept, and subsequent deviance.[16] **Secondary deviance** refers to deviance that occurs in response to problems created by societal reactions to primary deviance. Formal societal reactions to primary deviance cause "deviant" individuals to change their self-concepts and social roles in ways that acknowledge the labels that others have applied to them. Deviance may become a central element of identity for those whose primary deviance has received official sanction. "The secondary deviant . . . is a person whose life and identity are organized around the facts of deviance."[17] In the creation of secondary deviance, the reasons for primary deviance are overshadowed by the importance of the disapproving and isolating societal reactions to that deviance.

secondary deviance Deviance that occurs in response to problems created by societal reactions to primary deviance.

The changes in identity and social roles that occur in response to societal reactions take place gradually. Lemert describes a lengthy process: primary deviance, social penalties, further primary deviance, stronger penalties and reactions, further deviance, formal action by the community that is stigmatizing to the deviant, deviance as a reaction to the penalties applied by the community and the corresponding stigma, and finally "acceptance of deviant social status" and efforts to adjust to that status.[18] In this process, both deviance and social reactions to it are amplified. As a result of societal reactions, individuals come to see themselves as deviant and adjust their behaviors to be consistent with that role.

The fundamental distinction between primary and secondary deviance lies in the acceptance of a deviant identity. Secondary deviance occurs in response to changes in identity, which are based on societal reactions, while primary deviance occurs without effect on identity. Through the process Lemert describes, deviance becomes a self-fulfilling prophecy. Over time, as others respond to acts of primary deviance, individuals come to see themselves as deviants and behave in accord with that label. In other words, deviant or delinquent labels essentially create the very behaviors they were intended to inhibit.

The Emergence of "Labeling Theory"

The works of Tannenbaum and Lemert stood for many years as relatively isolated statements of the consequences of societal reactions to deviance. In the 1960s and early 1970s, an identifiable school of thought emerged in a series of writings that conveyed a unified theme, known as "labeling theory." Central among these writings were Erving Goffman's books *Asylums* (1961) and *Stigma* (1963), John Kitsuse's article on homosexuality (1962), Kai Erikson's essay on the sociology of deviance (1962) and study of *Wayward Puritans* (1966), Howard Becker's book *Outsiders* (1963), and Edwin Schur's article on reactions to deviance (1969) and book *Labeling Deviant Behavior* (1971).[19] The common core of these writings was a focus on how definitions of deviance are applied to individuals and groups, and on the consequences of social control through the application of deviant labels.

Scholars have noted that, during the 1960s, intellectuals were particularly open to the critique of political authority and government power offered by labeling theory.[20] Francis Cullen and Robert Agnew remind us of the historical events: "Recall that during the 1960s and early 1970s, the United States was greeted with revelation after revelation of the government abusing its power—from Civil Rights demonstrators being beaten, to inmates being gunned down at Attica, to students being shot at Kent State, to Viet Nam, to Watergate, and on and on. As trust in the state plummeted—especially on university campuses—a theory that blamed the government for causing more harm than good struck a chord of truth. Labeling theory, of course, did precisely this in arguing that the criminal justice system stigmatized offenders and ultimately trapped them in a criminal career."[21] And so the popularity of the labeling perspective grew.

Theory into Practice, "Juvenile Justice System Reforms Based on the Labeling Perspective," provides a brief description of juvenile justice policies spawned in part by the labeling perspective in the 1960s.

The Creation and Enforcement of Social Rules

One of the most frequently cited works on the social construction of deviance that emerged during the 1960s is Howard Becker's book *Outsiders*. Becker

Juvenile Justice System Reforms Based on the Labeling Perspective

LaMar Empey, Mark Stafford, and Carter Hay describe four reforms of the juvenile justice system in the 1960s and 1970s that were based, at least in part, on the labeling perspective: decriminalization, diversion, due process, and deinstitutionalization.

Decriminalization

"The first reform suggested that status offenses such as running away, defying parents, sexual promiscuity, and truancy should be **decriminalized.**" Labeling theorists believed that responses of the juvenile justice system to minor status offenses produced more harm than good. They proposed that status offenses be dealt with in ways other than legal intervention. In 1967, the President's Commission on Law Enforcement and Administration of Justice concluded that "serious consideration should be given [to] complete elimination from the court's jurisdiction of conduct illegal only for a child" (*Challenge of Crime,* 27).

Diversion

"A second reform—**diversion**—was closely related to decriminalization. It, too, was based on the premise that the evils of children have been overdramatized. To avoid labeling and stigmatization, potential arrestees or court referrals should be diverted from the juvenile justice system into other, less harmful agencies. . . ."

This policy represented an attempt to respond to the problem behaviors of youth, but without the adverse effects of juvenile court processing. Suggestions for accomplishing this included: (1) the creation of community agencies and Youth Services Bureaus that would provide various services, such as counseling, tutoring, and recreational services, to young people—both delinquents and nondelinquents; and (2) the employment of juvenile specialists in police departments, who would decide which juveniles might be diverted from the juvenile court.

Due Process

"The third reform stressed the importance of **due process** in juvenile court proceedings. . . . In [*Kent v. United States*] and subsequent decisions, the Supreme Court concluded that except for jury trials, children should receive most of the [constitutional] protections that adults receive." (See Chapter 2 for a discussion of *Kent v. United States* and the due process revolution in juvenile justice.)

"One of the main themes of labeling theory was evident in the actions of the Supreme Court: the need to limit the discretion of child savers and control their power over young people. Because all delinquent acts could not be decriminalized and some offenders could not be diverted, those who were referred to juvenile court should be protected by constitutional procedures."

Deinstitutionalization

"The fourth reform to which labeling theory contributed was **deinstitutionalization**—the removal of children from detention centers, jails, and reformatories. Like diversion, the goal of deinstitutionalization was to limit the destructive effects of legal processing."

The President's Commission wrote, "Institutions tend to isolate offenders from society, both physically and psychologically, cutting them off from schools, jobs, families, and other supportive influences and increasing the probability that the label of criminal will be indelibly impressed upon them. The goal of reintegration is likely to be furthered much more readily by working with offenders in the community than by incarceration" (*Task Force Report,* 165).

Sources: Empey, Stafford, and Hay, *American Delinquency,* 263–265; President's Commission, *Challenge of Crime;* President's Commission, *Task Force Report.*

concisely describes the process through which deviance is socially constructed: "*Social groups create deviance by making the rules whose infraction constitutes deviance,* and by applying those rules to particular people and labeling them as outsiders. From this point of view, deviance is *not* a quality of the act the person commits, but rather a consequence of the application by others of rules and sanctions to an 'offender.' The deviant is one to whom that label has successfully been applied; deviant behavior is behavior that people so label."[22]

In a similar and often-cited passage, Kai Erikson draws attention to the crucial role of the social audience in the creation of deviance: "Deviance is not a property *inherent in* certain forms of behavior; it is a property *conferred upon* these forms by the audiences which directly or indirectly witness them. Sociologically, then, the critical variable in the study of deviance is the social *audience* rather than the individual *person,* since it is the audience which eventually decides whether or not any given action or actions will become a visible case of deviation."[23] The social audience is important because it determines whether or not the relevant rules should be enforced. Both Erikson and Becker express one of the central tenets of the labeling perspective—that deviance is "in the eye of the beholder."

Becker's work in particular moves beyond the earlier writings of Tannenbaum and Lemert, who focused on societal reactions to deviance. While he is certainly interested in societal reactions, Becker also considers how groups are able to create and enforce the rules that others are selectively called upon to follow and for which they are sanctioned if they do not follow.[24] In other words, as we try to understand why some people are labeled deviant or delinquent, Becker invites us to consider not only societal reactions to their behaviors, but also the creation of rules that direct attention to those behaviors.

In a chapter entitled "Moral Entrepreneurs," Becker describes those who create and enforce rules.[25] He often refers to rule creators as **moral crusaders**, who are attempting to correct some "evil" in the world.[26] Moral crusaders are typically individuals of high social status (like the child savers described in Chapter 2), and this status helps them to convince others of the legitimacy of their position. Thus, according to Becker, it is generally the worldview of influential individuals and groups that is imposed through rules on those who are less powerful.[27] In his book *Outsiders,* Becker describes the creation of the Marihuana Tax Act in 1937, which outlawed marijuana use, as an example of a "moral crusade."

Once created, rules tend to be selectively enforced. Becker writes, "The degree to which an act will be treated as deviant depends . . . on who commits the act and who feels he has been harmed by it. Rules tend to be applied more to some persons than to others."[28] To illustrate his point, Becker uses studies of juvenile delinquency and social class: "Boys from middle-class areas do not get as far in the legal process when they are apprehended as do boys from slum areas. The middle-class boy is less likely, when picked up by the police, to be taken to the station; less likely when taken to the station to be booked; and it is extremely unlikely that he will be convicted and sentenced. This variation occurs even though the original infraction of the rule is the same in the two cases."[29]

Whether or not rules are enforced and individuals are labeled deviant or delinquent depends on many factors other than the actual behavior. These factors may include characteristics of the actor, the victim, the setting, and the observers. For example, the process of enforcing rules and labeling individuals may be influenced by the demeanor of the rule violator toward the "rule enforcer," the rule enforcer's perception that "he must make some show of doing his job in order to justify his position," or the place the act committed holds in the rule enforcer's list of priorities. The labeling process may also be affected by the race, social class, and gender of the rule violator and the victim (see Chapter 5).[30]

Most labeling theorists of the 1960s and 1970s, including Becker, were concerned primarily with the application and consequences of *official* or *formal* labels—those applied through formal agencies of social control, such as the juvenile justice system. In the 1990s, however, scholars began to consider the effects of *informal* labels—those applied in the context of interaction with others, such as parents and teachers.

moral crusaders
Individuals, typically of high social status, who create rules in an attempt to correct some "evil" in the world.

Informal Reactions: Labeling As an Interpersonal Process

In contrast to the emphasis of the traditional labeling perspective on formal reactions to delinquency, Ross Matsueda and his colleagues examine the process of informal labeling that occurs in social interaction.[31] Drawing on symbolic interactionism, Matsueda develops a theory of how self-concept arises and guides behavior, including delinquency. He conceptualizes the self in part as an individual's perceptions of how others view him or her. These perceptions are developed through the process of *role-taking,* which, according to symbolic interactionism, is the mechanism through which individuals influence one another in social interaction. Role-taking "consists of projecting oneself into the role of other persons and appraising, from their standpoint, the situation, oneself in the situation, and possible lines of action."[32] George Herbert Mead, writing 60 years before Matsueda, proposed that role-taking is the key to social control.[33] Based in part on Mead's work, Matsueda and his colleagues argue that, in the process of interaction, role-taking serves as a mechanism of social control, as individuals view themselves and potential lines of action from the standpoint of others and adjust their behavior accordingly.[34]

Matsueda uses the term *reflected appraisals of self* to refer to the view of self that one develops by taking the role of others and appraising oneself from the perspective of those others.[35] He argues that reflected appraisals of self result in part from the actual appraisals made by others. Reflected appraisals are important in the study of delinquency because they are a mechanism of self-control, insofar as individuals tend to behave in accord with them. Reflected appraisals may either constrain or free an individual to particular behaviors, depending on the content of the appraisals. For example, reflected appraisals of self as "bad kids" may free individuals to behave in accordance with that label. If they perceive that others already see them as "bad kids," they have "nothing to lose" by actually behaving that way. According to Matsueda, "Those who see themselves (from the standpoint of others) as persons who engage in delinquent behavior in certain situations are more likely to engage in delinquency."[36]

To test his interactionist theory of delinquency, Matsueda used data from the National Youth Survey (NYS). This dataset includes measures of parents' actual appraisals (or labels) of their children as sociable, distressed, rule violators, and likely to succeed. It also includes children's corresponding reflected appraisals of self (from the standpoint of parents, teachers, and peers). Research in Action, "The Measurement of Parents' Appraisals and Youths' Reflected Appraisals," presents the measures of actual and reflected appraisals in Matsueda's research.

Matsueda specified several hypotheses, which are illustrated in **Figure 13-1.** First, prior delinquent behavior was expected to influence both parents' actual

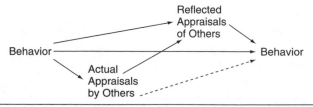

Figure 13-1 Matsueda's model of reflected appraisals and behavior. Matsueda's interactionist model shows that actual appraisals by others and reflected appraisals of self result in part from prior behavior. Actual appraisals by others influence reflected appraisals, which in turn influence subsequent behavior. The dashed path from actual appraisals to subsequent behavior indicates that actual appraisals should have no effect on later behavior, except through their effect on reflected appraisals.

Source: Matsueda, "Reflected Appraisals," 1585. Copyright © 1992, University of Chicago Press. All rights reserved. Reproduced with permission of the publisher.

The Measurement of Parents' Appraisals and Youths' Reflected Appraisals

Ross Matsueda and his colleagues used variables from the National Youth Survey to measure parents' appraisals of their children and youths' reflected appraisals of self. This dataset contains measures of four distinct dimensions of "the self"—as sociable, distressed, rule violator, and likely to succeed.

Parents' Appraisals

Each parent respondent was asked, "I'd like more information about how you see your son or daughter. I will read you a list of words or short phrases. Please listen carefully and tell me . . . how much you agree or disagree with each of the words or phrases as a description of your son or daughter." Five response categories ranged from strongly agree to strongly disagree.

Parent appraisal as sociable:
My son or daughter is well liked.
My son or daughter gets along well with other people.

Parent appraisal as distressed:
My son or daughter is often upset.
My son or daughter has a lot of personal problems.

Parent appraisal as a rule violator:
My son or daughter gets into trouble.
My son or daughter breaks rules.

Parent appraisal as likely to succeed:
My son or daughter is likely to succeed.

Youths' Reflected Appraisals

Each youth respondent was asked, "I'd like to know how your parents, friends, and teachers would describe you. I'll read a list of words or phrases and for each will ask you to tell me how much you think your parents would agree with that description of you. I'll repeat the list twice more, to learn how your friends and your teachers would describe you." Five response categories ranged from strongly agree to strongly disagree.

Reflected appraisal as sociable:
Parents agree I am well liked.
Friends agree I am well liked.
Teachers agree I am well liked.
Parents/friends/teachers agree I get along well with other people.

Reflected appraisal as distressed:
Parents/friends/teachers agree I am often upset.
Parents/friends/teachers agree I have a lot of personal problems.

Reflected appraisal as a rule violator:
Parents/friends/teachers agree I get into trouble.
Parents/friends/teachers agree I break rules.

Reflected appraisal as likely to succeed:
Parents/friends/teachers agree I am likely to succeed.

Sources: Matsueda, "Reflected Appraisals;" Bartusch and Matsueda, "Gender."

appraisals of their children and youths' reflected appraisals of self from the standpoint of significant others. This hypothesis is consistent with the concept of primary deviance setting the labeling process in motion. Second, parents' appraisals of their children were expected to influence youths' reflected appraisals of themselves. Third, youths' reflected appraisals of self (especially as "rule violator") were expected to influence their later involvement in delinquency. Parents' appraisals or labels were expected to influence youths' future delinquency, but only indirectly, through their effects on youths' reflected appraisals of self.[37]

In his research, Matsueda found general support for these hypotheses. Most importantly, parents' appraisals or labels strongly influenced youths' reflected appraisals of self, and youths' reflected appraisals (especially as a rule violator) strongly influenced subsequent delinquent behavior. "Persons who perceive that others view them as one who violates rules, or gets in trouble, engage in more delinquent acts. This supports the major hypothesis of an interactionist theory of delinquency: behavior is strongly influenced by reflected appraisals."[38] However, contrary to his theory, Matsueda also found that parents' appraisals of their children as rule violators *directly* affected youths' delinquent behavior, regardless of youths' perceptions of those appraisals.[39]

This contradictory finding is explained in a later study by Matsueda and his colleague Karen Heimer.[40] In their research, Heimer and Matsueda examine in greater detail the social context of role-taking. Specifically, they consider association with delinquent peers, commitment to particular roles, and anticipation of others' reactions to delinquency. Presumably, through the role-taking process, association with delinquent peers would lead to attitudes favoring delinquency and anticipation of positive reactions by those peers to delinquent behavior. Heimer and Matsueda found that, once association with delinquent peers is taken into account, the direct effect of parental labels on youths' delinquency disappears, and parental labels influence delinquency only indirectly, through reflected appraisals and association with delinquent peers.[41] "This provides support for symbolic interactionism (and labeling theory): Youths who are appraised negatively by parents are likely to commit subsequent delinquent acts, in part because of their perceptions of the appraisal [reflected appraisals] and in part because they are more likely to come into contact with peers who are delinquent."[42]

Matsueda's research highlights the significance of the informal labeling process. In the following section, we examine other research on the effects of informal labels, as well as research on the effects of formal sanctions.

Consequences of Labeling: Stigma or Deterrence?

The labeling perspective postulates that societal reactions to initial acts of deviance may lead to the development of a deviant identity that increases the likelihood of subsequent deviance, or amplifies deviance, among those who have been labeled "deviant" or "delinquent." In other words, the labeling process has a stigmatizing effect that ultimately results in **deviance amplification**. In his seminal book *Stigma*, Erving Goffman describes stigma as "an attribute that is deeply discrediting."[43] He speaks of stigma as "spoiled identity," which can result from being labeled for prior deviant behavior, such as delinquency, or for personal characteristics that others sometimes perceive as deviant, such as mental illness.[44] The stigma associated with a negative label can have powerful and wide-ranging consequences. It may limit an individual's legitimate opportunities (for example, ineligibility for particular jobs because of a criminal record), influence an individual's personal relationships (for example, alienation from parents or teachers as a result of delinquent behavior), and affect an individual's self-concept and self-esteem as he or she accepts the deviant identity associated with the negative label. John Hagan uses the concept of "social embeddedness" to describe how punishment and its accompanying stigma "embed" individuals in relationships and social institutions that make success in the conventional realm (such as employment) less likely and continued deviance more likely.[45]

deviance amplification A process in which societal reactions to initial acts of deviance lead to the development of a deviant identity, which increases the likelihood of subsequent deviance among those who have been labeled "deviant."

The stigmatizing effect of societal reactions that labeling theorists describe is contrary to the deterrent effect proposed by classical theorists (see Chapter 7). Deterrence theory, derived from the classical school of thought, argues that societal reactions to delinquency constitute a form of punishment that deters future delinquency. According to deterrence theory, negative reactions will increase individuals' perceptions of risk associated with future offending, which in turn will deter subsequent delinquency.[46] **Figure 13-2** depicts the competing models offered by the labeling perspective and deterrence theory of the relationship between sanctions or societal reactions and future deviance.

During the past fifteen years, numerous studies have explored the consequences of societal reactions to deviance and delinquency.[47] We briefly review research that has considered the effects of formal and informal reactions.

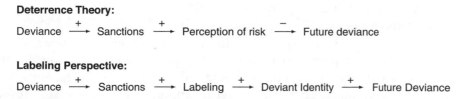

Figure 13-2 Competing models of sanctions and future deviance. Ward and Tittle illustrate the alternative views of sanctions and subsequent deviance offered by deterrence theory and the labeling perspective. Deterrence theory contends that sanctions or societal reactions *decrease* the likelihood of future deviance by increasing the perception of risk associated with deviant behavior. The labeling perspective maintains that societal reactions *increase* the likelihood of future deviance by leading individuals to adopt deviant identities and behave in accord with them. In the figures, plus signs indicate positive relationships and minus signs indicate negative ones.
Source: Based on Ward and Tittle, "Deterrence or Labeling," 44–45.

Formal Sanctions: Stigma or Deterrence?

Initially, labeling theorists tended to focus on the effects of formal "deviant" or "delinquent" labels and the role of formal organizations, such as police or juvenile courts, in creating deviance and delinquency.[48] Much early research on the labeling perspective concentrated on the effects of labeling that occurs in the process of official responses to delinquency, such as arrest or juvenile court processing.[49]

In a 1989 study, Douglas Smith and Patrick Gartin asked a basic labeling-vs-deterrence-theory question: "Does arrest amplify or deter the future criminal activity of those arrested?"[50] The labeling perspective suggests that arrest will increase the likelihood of future offending. Deterrence theory contends that arrest will decrease the likelihood of future offending by increasing the perceived risks associated with delinquency. Smith and Gartin considered two groups of offenders: those who had contact with police but were released without arrest, and those who were arrested. They compared subsequent criminal activity (measured by future police contacts) for the two groups and found that those arrested were less likely than those who were released without arrest to have subsequent contact with police. Smith and Gartin also found that the influence of arrest on future offending differed for "novice" offenders (those with only one or two police contacts) and "experienced" offenders with more established criminal careers. Novice offenders were more likely than experienced offenders to stop their criminal activity as a result of arrest. Smith and Gartin interpreted their results as offering support for deterrence theory.[51]

Later research on the effects of formal sanctions, however, has been more supportive of the labeling perspective.[52] For example, a 1999 study by Spencer De Li examined the effects of legal sanctions (measured as convictions for delinquent offenses) on self-reported delinquency, later delinquency convictions, and achievements in education and work.[53] Results indicated that conviction at an early age (10–13 years old) had particularly strong effects, increasing later delinquency, convictions, and unemployment, and decreasing later educational and occupational achievements. "Overall, convictions seem to have a strong adverse effect on legitimate opportunities for adolescents: Youths who were labeled by the juvenile justice system were more likely to develop antisocial attitudes and unconventional behaviors. They were also more likely to exhibit low educational achievement, low occupational status, and unstable job records."[54] These findings are consistent with the labeling perspective, which suggests that the stigma associated with delinquent labels will have consequences for many aspects of youths' lives.

These results can be interpreted within the developmental framework proposed by Robert Sampson and John Laub.[55] Sampson and Laub consider the stability of delinquency and crime over time and the way in which labeling contributes to that stability. They use the term "cumulative disadvantage" to describe how delinquency disrupts various aspects of life, including family and peer relationships, success in school, and employment opportunities. These disruptions sustain delinquent behavior by severing social bonds and foreclosing legitimate opportunities.

Sampson and Laub incorporate the labeling perspective into their developmental theory of social control, arguing that official sanctions contribute to the process of cumulative disadvantage and play a role in sustaining delinquency. "The theory specifically suggests a 'snowball' effect—that adolescent delinquency and its negative consequences (e.g., arrest, official labeling, incarceration) increasingly 'mortgage' one's future, especially later life chances molded by schooling and employment."[56] The stigma associated with official sanctions prevents individuals from establishing strong ties to conventional lines of action. Thus, Sampson and Laub propose that the effects of official sanctions on future delinquency are *indirect,* operating through the disruption of social relationships and legitimate opportunities.

Sampson and Laub's own research supports this view of official sanctions and subsequent delinquency.[57] Other research is also consistent with the theory that official reactions to delinquency have a cumulative effect and contribute to the stability of delinquent behavior by closing off opportunities and relationships.[58]

Informal Reactions: Stigma or Deterrence?

During the 1990s, researchers began to look beyond official reactions to delinquency and focus on the consequences of informal labels.[59] Earlier, we described studies by Matsueda and his colleagues, who used data from the National Youth Survey (NYS) and found that the process of informal labeling affects youths' subsequent involvement in delinquency.[60] Other criminologists studying the consequences of informal labels have also found support for the labeling perspective.[61] Many of these researchers also used data from the NYS, which contains numerous questions asked of parents about how they view their children (informal labels), and questions asked of youth about how parents, peers, and teachers view them (perceptions of informal labels). See Research in Action, "The Measurement of Parents' Appraisals and Youths' Reflected Appraisals," for a description of these measures.

Like Matsueda and his colleagues, other researchers have examined the effects of youths' perceptions of informal labels on delinquency.[62] Douglas Smith and Robert Brame, for example, used NYS data to examine whether the factors that explain initial involvement in delinquency are the same as those that explain continuation of offending.[63] They found that, consistent with the hypothesis of deviance amplification, youths' perceptions of negative labels by parents and peers contribute to continuity of offending, but not to initiation into delinquency. "Youths who feel that their parents and peers view them negatively are not at increased risk to begin delinquency. However, among those who have offended in the past, this same negative labelling is associated with a significantly higher probability of continued delinquency. This result is exactly what early labelling theorists such as Lemert predict, and it highlights the importance of informal labels and their subjective perception."[64]

Other researchers have explored the effect on delinquency of youths' perceptions of informal labels by teachers. Mike Adams and David Evans used NYS data and found that students' perceptions of negative labeling by teachers, like their perceptions of labeling by parents in other studies, increased youths' subsequent delinquency.[65] When Adams and Evans took into account youths' associations with delinquent peers, though, the direct effect of teacher labeling on youths' delinquent behavior was no longer significant. Instead, the effect was largely mediated by associations with delinquent peers. Adams and Evans interpret this finding as "entirely consistent with Tannenbaum and Becker's descriptions of the labeling process."[66]

Other research by Adams and his colleagues suggests that the effects of informal labeling vary by race.[67] Some studies indicate that formal labeling by the criminal justice system is more detrimental for whites than it is for blacks.[68] But Adams and his associates found that the informal labeling process has a greater effect on delinquency for blacks than it does for whites.[69]

Assessment of the Labeling Perspective

Following its heyday in the 1960s and early 1970s, the labeling perspective began to fall out of favor among many criminologists. Here we briefly note some of the criticisms that have been leveled against the labeling perspective. Space constraints prevent us from elaborating on them.

Perhaps the most damning criticism has been that the labeling perspective is just that—a perspective, rather than a formal theory. Some have argued that the labeling perspective does not rise to the level of a formal theory, because it does not offer testable propositions.[70] (Recall from Chapter 1 that, when evaluating theories, we must consider the extent to which those theories are testable.) Similarly, some have argued that proponents of the labeling perspective have failed to define precisely key terms such as "labeling" and "deviance," which has led to confusion about how to measure key constructs and test the perspective.[71]

Much early empirical research failed to support the labeling perspective. "Whereas they had once embraced the theory without any evidence of its validity, criminologists now rejected labeling theory on the grounds that the perspective 'had no empirical support.' "[72] The quality of that early research has been called into question, however, along with the dismissal of the labeling perspective that it yielded.[73] Because of methodological problems and poor data in some of the early research, scholars cannot legitimately use those studies to reject or support the perspective.[74]

Another criticism of the labeling perspective centers on its avoidance of the issue of causation. According to some critics, the labeling perspective focuses on the continuation of deviance that results from societal reactions, but it does not address the causes of the initial act(s) of deviance, or primary deviance, that set in motion the labeling process.[75] In a related point, critics argue that the labeling perspective views the person who is labeled as *passive* and *acted upon* by society, rather than as *active* in the process of creating deviance. Reid writes, "Most labeling theorists overemphasize the action of society and deemphasize the action of the subject being labeled."[76]

Some have criticized the labeling perspective for its assumption that the labeling process has only negative effects. It is possible that labeling a person "delinquent," rather than increasing involvement in delinquency (labeling effect), might prevent continued offending (deterrent effect). Critics argue that the labeling perspective has not considered factors that might determine whether the labeling process will have positive or negative consequences, nor has it developed potential explanations of different responses to labeling.[77]

Finally, some criminologists observe that the labeling perspective pays little attention to the developmental origins of serious offending.[78] Recall from

Chapter 6 our discussion of developmental patterns of offending and the evolution of childhood problem behaviors into serious delinquency in adolescence and early adulthood. Scholars have argued that labeling theorists should consider more fully the early portion of the life course, when the conduct problems that form the foundation for later serious criminality initially emerge, and the place of societal reactions in the developmental unfolding of problem behaviors.[79]

These critiques of the labeling perspective cannot be ignored. Proponents of this perspective must address these serious and legitimate concerns if they hope to advance and revitalize this approach. Some scholars (for example, John Braithwaite) have begun this work, but much remains to be done.

Although proponents of the labeling perspective have yet to address many of these criticisms, the labeling perspective has generated renewed interest since the late 1980s. As our earlier discussion shows, during the 1990s, research provided some support for the labeling perspective.[80] For example, Smith and Brame report that their research results are consistent with the labeling theory hypothesis of deviance amplification, "despite numerous eulogies of the labelling perspective."[81] In addition, criminologists John Braithwaite and Ross Matsueda have offered innovative contemporary variations of the labeling perspective.[82] Also, Sampson and Laub have recognized the developmental nature of the labeling perspective and incorporated this approach into their influential life-course theory of delinquency (see Chapter 6).[83] Together, these theoretical innovations (and recent research that supports labeling theory) have begun to bring back a perspective that may have been prematurely dismissed. The resurrection of the labeling perspective will likely depend on whether or not its proponents can successfully address the criticisms we noted earlier.

Next, we consider John Braithwaite's theory of reintegrative shaming as an example of recent innovations in labeling theory. Braithwaite uses the labeling perspective, along with traditional theories of crime, to explore the process of shaming that occurs in response to crime and delinquency.[84] The novel aspect of his theory is his distinction between two types of shaming—one that increases subsequent offending through its stigmatizing effects, and another that deters subsequent offending through the reintegration of the offender into the community.

theory of reintegrative shaming Crime can be reduced or controlled through the deterrent effect of shaming, followed by attempts to reintegrate the offender back into the community.

Braithwaite's Theory of Reintegrative Shaming[85]

Braithwaite's <u>theory of reintegrative shaming</u> begins with several traditional theories of crime (control, subcultural, differential association, strain, and labeling theories), and then introduces a new concept—reintegrative shaming— to draw those existing theories together in an integrated framework.[86] Braithwaite defines *shaming* as "all social processes of expressing disapproval which have the intention or effect of invoking remorse in the person being shamed and/or condemnation by others who become aware of the shaming."[87] <u>Reintegrative shaming</u> is "shaming which is followed by efforts to reintegrate the offender back into the community of law-abiding or respectable citizens through words or gestures of forgiveness or ceremonies to decertify the offender as deviant."[88] Thus, reintegrative shaming, according to Braithwaite's theory, reduces or controls crime.

reintegrative shaming "Shaming which is followed by efforts to reintegrate the offender back into the community of law-abiding or respectable citizens through words or gestures of forgiveness or ceremonies to decertify the offender as deviant" (Braithwaite, *Crime, Shame and Reintegration,* 100–101).

Crime prevention through reintegrative shaming is accomplished in several ways.[89] Shaming has a specific deterrent effect on offenders who want to avoid the public humiliation associated with detection of their crimes. Shaming also has a general deterrent effect on others who want to avoid being subjected to the shaming process and therefore choose not to engage in crime. Both of these deterrent effects are greater for persons with strong attachments to others, because these attachments increase the potential social costs of offending.

Reintegrative shaming also prevents crime by leading people to think of crime as "unthinkable." The process of reintegrative shaming affirms and builds commitment to the criminal law, and thus makes future crime (by the offender being shamed and by others) less likely. According to Braithwaite, the element of repentance in reintegrative shaming is the key to this affirmation of law. Finally, shaming is "a participatory form of social control."[90] Citizen participation in the shaming process accomplishes both specific and general deterrence. As Braithwaite writes, "Participation in expressions of abhorrence toward the criminal acts of others is part of what makes crime an abhorrent choice for us ourselves to make."[91]

stigmatization "Disintegrative shaming in which no effort is made to reconcile the offender with the community" (Braithwaite, *Crime, Shame and Reintegration,* 101).

This theory distinguishes reintegrative shaming from **stigmatization**, defined as "disintegrative shaming in which no effort is made to reconcile the offender with the community."[92] Rather than drawing offenders back into the community, stigmatization propels offenders toward deviant subcultures and increases crime by providing opportunities and reinforcements for criminal behavior. Braithwaite derives the concept of stigmatization from labeling theory.

interdependency "The extent to which individuals participate in networks wherein they are dependent on others to achieve valued ends and others are dependent on them" (Braithwaite, *Crime, Shame and Reintegration,* 98, 100). The concept of interdependency is similar to attachment, commitment, and social bonding in control theory.

Braithwaite contends that people vary in susceptibility to shaming based on their level of **interdependency**. The concept of interdependency is similar to attachment, commitment, and social bonding in control theory. Interdependency is "the extent to which individuals participate in networks wherein they are dependent on others to achieve valued ends and others are dependent on them."[93] Those who are most interdependent are most affected by the shaming process. Braithwaite's theory proposes that levels of interdependency are influenced by individual characteristics such as age, gender, marital status, and educational and employment aspirations.[94] **Communitarianism** is the societal-level counterpart to interdependency. According to Braithwaite's theory, shaming is more widespread and effective in communitarian societies (such as Japan, where the culture emphasizes group-relatedness) than in other societies (such as the United States, where the culture is more individualistic).[95] Communitarianism is influenced by urbanization and residential mobility.

communitarianism The societal-level counterpart to interdependency. Shaming is more widespread and effective in communitarian societies than in other societies.

Who Is Labeled, and Based on Whose Rules?

As we noted at the beginning of this chapter, labeling theory is grounded in part in the conflict perspective, which considers how power and resources shape the creation and enforcement of definitions of deviance. Rules and laws that define particular acts as deviant are socially constructed and applied to particular people. In the 1970s, Edwin Schur pointed out that relationships exist between resources and the ability to make or impose rules, and between resources and the ability to resist deviant labels.[96] Howard Becker also described the role of group conflict in the creation of rules. Rather than being universally agreed upon, rules result from a political process characterized by conflict and disagreement.[97]

Conflict characterizes not only the creation of rules, but also the application of those rules and the labeling of individuals based on rule violations. The application of "delinquent" or "deviant" labels is based in part on the status and power of the actor, the victim, and observers. Drawing on the conflict perspective, labeling theorists have argued that "delinquent" or "criminal" labels are disproportionately applied to less powerful members of society, such as racial minorities or those from the lower class.[98]

We turn now to critical criminologies, which share with the labeling perspective certain views of power and conflict. These critical criminologies examine the roles of power and conflict in determining why society defines crime and reacts to crime and criminals in the way that it does. Critical criminologies also place power and conflict within the political and economic structures from which they originate.

■ Critical Criminologies

Though Willem Bonger made a statement on the relationship between capitalism and crime as early as 1916, the ideas that form the basis of critical criminology were quieted by a wave of optimism in the 20-year period following World War II, sometimes referred to as the Great Society period (1950–1970). The unfulfilled expectations of this period, however, turned to economic and social despair in the decades that followed.[99] Beginning in the late 1960s, renewed interest in the structural forces that breed crime led to a wide variety of critical interpretations of crime and delinquency. The social and political unrest that fueled labeling theory also played a role in the rise of critical criminologies.

Before we go any further, we should clarify our use of the term "critical criminologies." Within criminology, several terms have been used to describe the general approach that has been offered as an alternative to traditional or "mainstream" criminology: radical, critical, Marxist, socialist, humanist, and the new criminology. Gregg Barak argues that proponents of this new approach eventually settled on the term "critical" because "it was more inclusive of sympathizers and less offensive to critics," and because it was a "broad enough category" to include the many approaches represented by the "criminological 'left.' "[100] We follow Barak's lead and use the term "critical." We speak of critical *criminologies,* rather than the singular criminology, to indicate that this umbrella term includes a variety of perspectives that share a common core, but are, nonetheless, theoretically distinct.

Major Themes of Critical Criminologies

Because a variety of distinct perspectives are housed under the roof of "critical criminologies," it is difficult to describe the central themes of the critical approach. Francis Cullen and Robert Agnew, however, have distilled this realm of thought down to the following five core themes.[101]

First, "the concepts of inequality and power are integral to any understanding of crime and its control."[102] Critical criminologies focus on the distribution of power in society, and on inequalities among members of different groups, including social classes. Differences in power produce conflict as members of different groups pursue their own interests. Conflict between the "haves" and the "have-nots" is an inherent part of the capitalist economic structure.

Second, crime and delinquency are politically defined. Laws do not reflect social consensus. Instead, political and economic power underlie the definitions of what is crime and who is criminal. "In general, the injurious acts of the poor and powerless are defined as crime, but the injurious acts of the rich and powerful—such as corporations selling defective products or the affluent allowing disadvantaged children to go without health care—are not brought within the reach of the criminal law."[103]

Third, both law and the criminal justice system protect the interests of the most powerful members of society—the capitalist class.[104] Mechanisms of social control, including socialization (which is constrained by social forces), informal controls, law, and the criminal justice system, are necessary to maintain the social order. In addition, in enforcing social order, agents of control sometimes commit crimes themselves, such as police brutality.

Fourth, "capitalism is the root cause of criminal behavior."[105] The capitalist economic system leads powerless members of lower classes, whose needs are ignored, to commit crimes out of economic necessity or resentment and frustration based on the demoralizing conditions of their existence. Capitalism also leads powerful members of upper classes to commit crimes to increase profits, with relatively little chance that these crimes will be punished.

Fifth, "the solution to crime is the creation of a more equitable society."[106] Only by understanding how the inequalities inherent within capitalism create crime can we begin to resolve the crime problem. Criminologists should advocate policies that promote social justice.

With these five themes in mind, we turn to examples of critical approaches. We begin with the early work of Willem Bonger and then describe instrumental and structural Marxist views of crime that emerged in the 1970s and 1980s. We also examine several structural Marxist interpretations of juvenile delinquency, including Mark Colvin and John Pauly's integrated structural-Marxist theory, John Hagan's power-control theory, and David Greenberg's Marxist interpretation of delinquency. We explore these examples thoroughly because the insight and usefulness of critical criminologies become apparent in these applications.

Bonger's View of Capitalism and Crime

Drawing from Karl Marx's bleak view of the social, economic, and political consequences of capitalism, Willem Bonger in 1916 offered the first explanation of how capitalism generates crime.[107] He maintained that capitalism creates a pervasive mental state of "egoism" or ruthless competition, which he contrasted with the altruism of precapitalist societies. By emphasizing profit and the accumulation of wealth, capitalism ignites greed and competition. According to Bonger, crime is a natural outcome of such egoism or competition. He used Marxist ideology to develop a theory to explain both the crimes of the powerless working class and those of the powerful upper classes. *Ideology* is a set of ideas and beliefs that together form a social, political, or moral agenda and that attempt to justify a particular social or political order.[108]

To explain the crimes of the poor and powerless, Bonger turned to the harsh realities of life created for them by the capitalist economic system. The "demoralizing" existence of those in the working class is characterized by disorder of living conditions and family life, a lack of moral training and education to instill altruistic sentiments, the economic necessity of child labor, exposure to criminal influences on the street, and the keen awareness of the existence of those who lack nothing. Bonger argued that the crimes of the working class result from these factors, which lead to the development of "the egoistic side of the human character."[109]

To explain the crimes of the powerful, Bonger considered opportunities for deception and crime and the "class character of the penal law," which favors the economically powerful.[110] Bonger argued that members of the upper classes, in the course of their business dealings, have opportunities to engage in crimes of deception that bring economic gain. In addition, because the law tends to serve the interests of the powerful, and because the penalties for economic crimes by members of the upper classes are relatively light, upper-class criminals have little punishment to fear for their crimes.[111]

Bonger believed that the solution to the problem of crime existed in the transformation of society from capitalism to socialism. According to Bonger, in a socialist state, altruism and concern for the common good of the community would replace the egoism of capitalism, and crime would diminish as a result.[112]

Marxist Views of Crime

In 1974, Richard Quinney proposed an __instrumental Marxist view__ of capitalism and crime.[113] According to this view, "criminal law is an instrument of the state and ruling class to maintain and perpetuate the existing social and economic order."[114] In other words, law and the criminal justice system serve the interests of the powerful and are used to control those who represent a threat to their privileged position. Instrumental Marxists believe crime results from class inequality alone.

These ideas drew criticism from scholars who saw them as too simplistic. Structural Marxists developed a more complex analysis of the role of law and the presence of crime in the capitalist state.[115] According to the __structural Marxist view__, the capitalist ruling class is far less unified and cohesive than instrumental Marxists imply.[116] Rather than seeing law as serving the interests of a cohesive ruling class, structural Marxists see law as a tool that is used to preserve the capitalist system as a whole.[117] For example, some laws, such as those that attempt to ensure safe working conditions for laborers, may actually be costly to capitalists. But in the long run, these laws serve to maintain and protect capitalism by averting crisis and maintaining the relations of production.[118]

Regarding the causes of crime, structural Marxists argue that, as important as economic structures and inequality are, they alone cannot explain the crimes of capitalists or workers.[119] Instead, structural Marxists attempt to situate micro-level explanations of crime, such as theories that focus on family and peer influences, within a broader social, political, and economic context that takes into account class distinctions inherent in the capitalist system.[120] An excellent example of this approach is the work of Mark Colvin and John Pauly, to which we now turn.

instrumental Marxism A perspective on capitalism and crime that states that "criminal law is an instrument of the state and ruling class to maintain and perpetuate the existing social and economic order" (Quinney, *Critique of Legal Order*, 16).

structural Marxism A perspective on capitalism and crime that views law as a tool used to preserve the capitalist system as a whole, rather than to serve the interests of a cohesive ruling class. Some laws may actually be costly to capitalists. But in the long run, laws maintain and protect capitalism by averting crisis in capitalist–worker relations.

integrated structural-Marxist theory Explains delinquency in terms of the structure of social relations grounded in capitalism. Emphasizes control structures within the workplace, family, school, and peer groups, and the way these control structures are shaped by capitalism.

Colvin and Pauly's Integrated Structural-Marxist Theory of Delinquency

In 1983, Colvin and Pauly proposed an <u>integrated structural-Marxist theory</u> of juvenile delinquency. They describe their theory as structural-Marxist insofar as it begins its analysis of delinquency with the structure of social relations grounded in the capitalist economic system.[121] Colvin and Pauly integrate this critical approach with elements from traditional theories of delinquency, including the emphasis of social control theory on family and school and the emphasis of social learning theory on peers.

Workplace Control Structures

Structural Marxists conceptualize classes in terms of relations to the *means of production*—the system by which goods are produced and distributed. For example, the capitalist class owns and controls the means of production, and the working class sells its labor to capitalists. Colvin and Pauly examine mechanisms of control within the workplace, which are necessary because of the typically "antagonistic" relations among classes. They contend that, in a capitalist state, members of different classes are subjected to different levels and types of control within the workplace.[122]

To describe variations in workplace control and compliance structures, Colvin and Pauly rely on Richard Edwards' analysis of three major fractions of the working class.[123] *Fraction I* workers typically labor in low-skill, "dead-end jobs" characterized by a high turnover of employees and little opportunity for advancement, such as nonunion manufacturing, service jobs, and lower-level clerical and sales jobs. The primary mechanism used to control fraction I workers is "simple control" or coercive control, achieved through the application or threat of force in the form of dismissal from the job. This type of coercive control creates a negative, alienated orientation on the part of the worker toward the employer and authority in the workplace.[124]

Fraction II "is composed of organized workers who, through earlier struggles, gained wage and benefit concessions, job protection, and the establishment of industry-wide unions."[125] Examples include auto and steel industries, machine manufacturing, and mining. The primary mechanism used to control fraction II workers is "technical control," which involves the manipulation of material rewards such as pay raises. This form of control relies on workers calculating their own material self-interests. It creates an unstable ideological bond between workers and employers that rests on workers' advancement up the pay ladder and involves limited worker loyalty to the employer.[126]

Fraction III workers hold jobs that require independent initiative and self-motivation, such as middle-level supervisors, foremen, skilled craftsmen (e.g., electricians, plumbers), and salaried professional workers (e.g., accountants, engineers, school teachers). These types of jobs are typically governed in part by formal standards of professional conduct. "Bureaucratic control" is also used to gain compliance from fraction III workers, and involves the manipulation of statuses and symbols to promote commitment to the organization. Fraction III workers often define their identities in terms of their occupation and derive meaning from their work. Bureaucratic control elicits "moral involvement" on the part of workers and uses the possibility of increases in job status, rather than just pay, to achieve compliance. This type of control leads workers to an intense positive orientation toward workplace authority and the organization for which they work.[127]

Family Control Structures

The cornerstone of Colvin and Pauly's theory is their analysis of how these three forms of workplace control and the orientation to authority they produce translate into mechanisms of control within the family. Colvin and Pauly rely on the work of Melvin Kohn, who argues that parents who experience greater external control in the workplace stress (either consciously or unconsciously) conformity to external authority in their child-rearing practices. Parents who experience less external control and have more autonomy in the workplace stress to their children the value of self-control, initiative, and creativity.[128] As Colvin and Pauly write, "An underlying message to the child in both instances is an ideological statement about the world: control of life circumstances and the determinants of one's behavior spring from either external compulsion or internal motivation. The child, in his or her everyday interactions with parents, learns that one acts toward authority either out of fear or calculation of external consequences or out of a sense of internalized respect or commitment. Through the process of parental control over children, a parent's bond to the authority of the workplace is reproduced in the child's initial bond to parental authority."[129]

Just as control structures produce certain ideological orientations toward authority for workers within the workplace, control structures affect orientations toward authority for children within the family. Colvin and Pauly offer these hypotheses about the links between workplace and family controls:

- Fraction I workers tend to use coercive, arbitrary, and inconsistent discipline that is sometimes lax and at other times highly punitive. The result is an "alienated" bond between children and parental authority.

- Fraction II workers tend to use compliance structures that involve parental manipulation of external rewards and children's calculations of self-interest. Children can dependably predict the external consequences of their actions, and form moderate bonds to parental authority.

- Fraction III workers tend to use compliance structures that involve parental manipulation of *symbolic* rewards, and result in children's intense positive bonds to parental authority.[130]

Colvin and Pauly cite research that suggests that family control structures are related to delinquency. For example, studies have shown that erratic and punitive disciplinary practices, as well as weak bonds between parents and children, are positively associated with children's involvement in delinquent behavior.[131]

School Control Structures

According to Colvin and Pauly, schools use various control structures to create a skilled and disciplined workforce, amenable to the compliance required of workers under capitalism.[132] School control structures parallel those in the family and the workplace through several mechanisms. First, children take IQ and aptitude tests that are used to place them in "tracks." Colvin and Pauly argue that these tests may measure motivation and orientation more than they measure innate cognitive ability. Thus, students who earn high IQ scores are likely to be self-motivated—a quality achieved through positive orientation to authority within family compliance structures that involve symbolic rewards. Students who earn low IQ scores are likely to have a more negative orientation to authority and to be placed in "'lower-level' tracks that are more regimented and coercive"—consistent with their experiences of control within the family.[133]

Second, children's orientation to authority, developed initially within family control structures, affects their behavior, which elicits controlling responses from teachers and other school authority figures. "The negatively bonded child, for instance, may give behavioral cues of being (in the teacher's perception) potentially disruptive. These behavioral cues may elicit a labeling process that creates the 'self-fulfilling prophecy' of a disruptive child who elicits still more coercive controls than before."[134]

Finally, financial resources vary among schools located in lower-class, working-class, and middle-class neighborhoods, resulting in tracking *between* schools. "This circumstance creates differences in the availability of rewards and punishments for controlling children in the school setting and often necessitates a greater reliance on coercive measures in lower-class schools."[135]

In sum, school tracking experiences correspond to and prepare students for workplace control structures. "The lower tracks emphasize strict discipline, regimentation, and conformity to external authority. The higher tracks emphasize initiative, creativity, and self-direction."[136]

Colvin and Pauly cite research that indicates a relationship between school control structures and delinquency. For example, studies have shown a negative association between IQ scores and involvement in delinquency. Similarly, research suggests that placement in lower-level educational tracks is positively related to delinquency. Studies also indicate that strong positive bonds to school decrease the likelihood of delinquent behavior.[137]

Peer Associations and Control Structures

Adolescents do not form peer groups randomly. Instead, they are drawn to one another based on common experiences and beliefs. Colvin and Pauly's theory assumes that adolescents with similar orientations or bonds to authority (based on school and family experiences) will be drawn to each other and will form peer groups. Colvin and Pauly offer these specific hypotheses:

- Students in higher educational tracks, which reinforce positive bonds to authority, will form peer groups that offer symbolic rewards, reinforce positive bonds, and "insulate" group members from delinquency.

- Students in intermediate tracks, whose bonds to authority are precarious and based on the calculation of material rewards, will form peer groups centered around the pursuit of external rewards. These peer groups will reinforce occasional involvement in instrumental delinquency intended to achieve material gains, such as theft.

- Students in lower educational tracks, with alienated bonds to authority, will form groups characterized by coercive control relations among peers (because they lack material resources and legitimate opportunities for status achievement needed for other types of control). These peer groups will reinforce serious, patterned delinquency.[138]

According to this theory, control structures within peer groups are based on both control structures within the school and opportunities (legitimate and illegitimate) for material gains and status within neighborhoods.

In summary, Colvin and Pauly argue that workplace control structures shape the forms of control parents use to discipline and reward their children. Children's socialization experiences within the family then influence the forms of control they encounter in school and peer groups. The more coercive the controls children experience in families, schools, and peer groups tend to be, the more negative or alienated their ideological bonds will be, and the more likely they are to engage in serious, patterned delinquency.[139]

Hagan's Power-Control Theory of Delinquency

In the mid-1980s, John Hagan, A. R. Gillis, and John Simpson developed <u>power-control theory</u>, a structural explanation for common delinquency that focuses on the intersection of social class and gender.[140] This theory combines the macro-level concept of *power* derived from relations to the means of production and authority in the workplace with the micro-level concept of *control* established within relationships such as family interactions. Hagan and his colleagues argue that both the presence of power and the absence of control contribute to adolescents' freedom to violate social norms. Their objective is to explain gender differences in delinquency based on gender differences in experiences of power and control. Power-control theory is explicitly intended to explain common, minor forms of delinquency, not serious, violent delinquent behavior.[141]

Like Colvin and Pauly, Hagan and his colleagues begin with the premise that parents' workplace experiences, which are grounded in class, influence parental control and children's socialization experiences within families. "To understand the effects of class position in the workplace on crime and delinquency it is necessary to trace the way that work relations structure family relations."[142] Power-control theory contends that parents who have relatively little authority and who experience greater controls in the workplace (based on class) will exercise greater controls over their children.[143]

Hagan and his colleagues define social class first in terms of relations to the means of production (e.g., employers who are owners and workers who are non-owners), and later in terms of relations to authority in the workplace (e.g., a *command class* that exercises authority and an *obey class* that is subject to the authority of others).[144] They then distinguish ideal types of families (or abstract statements of the essential characteristics of different types of families[145]) based on husbands' and wives' relations to authority. In <u>patriarchal families</u>, family class relations are "unbalanced." Husbands exercise greater authority in the workplace than do wives, who are either unemployed or employed in positions of less authority than those of their husbands. In <u>egalitarian families</u>, family class relations are "balanced." Husbands and wives are employed in positions of similar authority, in either the command class or the obey class.[146]

How, then, do differences in parents' workplace experiences of authority translate into gender differences in delinquency among sons and daughters? Hagan and his colleagues argue that, relative to fathers, mothers assume greater responsibility for the socialization and control of children.[147] In patriarchal

power-control theory A theory of gender differences in common delinquency that examines both power derived from class position and authority in the workplace and control established in family interactions. The theory argues that the presence of power and the absence of control, which differ by class and gender, contribute to adolescents' freedom to engage in delinquency.

patriarchal families According to power-control theory, families in which class relations are unbalanced and husbands exercise greater authority in the workplace than do wives.

egalitarian families According to power-control theory, families in which class relations are balanced and husbands and wives are employed in positions of similar authority.

families, women reproduce gender imbalances in power by exerting greater control over daughters than over sons. In other words, "mothers more than fathers are the instruments of familial controls" and "daughters more than sons are the objects of familial controls."[148] In egalitarian families, on the other hand, parents "redistribute their control efforts so that daughters are subjected to controls more like those imposed on sons. In other words, in egalitarian families, as [wives] gain power relative to husbands, daughters gain freedom relative to sons."[149]

Power-control theory proposes that the amount of control daughters and sons experience within families influences their attitudes toward risk taking.[150] Risk taking is contrary to the emphasis on "passivity" that characterizes patriarchal views of women's and girls' roles. Thus, in patriarchal families, parents (particularly mothers) teach daughters to avoid risk. In egalitarian families, parents encourage both daughters and sons to be more open to taking risks, in part in preparation for the kinds of entrepreneurial activities these children are expected eventually to perform in the workplace.[151]

Delinquency can be regarded as a form of risk taking. Power-control theory predicts that attitudes toward risk taking influence involvement in delinquent behavior, in part through their effect on perceived risks of sanctions for engaging in delinquency. The theory also predicts that both "taste for risk" and the deterrent effect of perceived risks of sanctions will vary by gender as a result of differences in familial controls for daughters and sons. Females are taught to avoid risks, including the risk of legal sanction, more frequently than males are. Power-control theory hypothesizes that "females will be deterred more by the threat of legal sanctions than males and that this effect will be produced more through maternal than paternal controls."[152]

The model of gender and delinquency presented in power-control theory is depicted in **Figure 13-3.** Summarizing their model's predictions, Hagan and his colleagues write, "We use power-control theory to predict that patriarchal families will be characterized by large gender differences in common delinquent behavior while egalitarian families will be characterized by smaller gender dif-

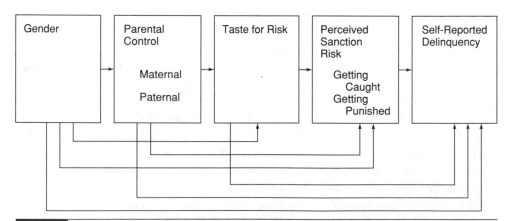

Figure 13-3 Hagan's model of power-control Theory. Hagan, Gillis, and Simpson illustrate their theory with this model of gender and delinquency. They examine the process depicted here within the context of distinct class categories, defined in terms of relations to the means of production and to authority under capitalism.

Source: Hagan, Gillis, and Simpson, "Class Structure," 1157. Copyright © 1985, University of Chicago Press. All rights reserved. Reproduced with permission of the publisher.

ferences in delinquency. In egalitarian families, daughters become more like sons in their involvement in such forms of risk taking as delinquency."[153]

In sum, power-control theory argues that social class and gender together create the conditions of freedom that lead to common delinquency. Both class and gender, by determining "the presence of power and the absence of control," structure the socialization experiences of children that contribute to delinquent behavior.[154]

Hagan and his colleagues, as well as other researchers, have found support for power-control theory.[155] A study by Christopher Uggen, for example, incorporated direct measures of workplace freedom and control for both parents and working youths. Uggen found strong support for Hagan's theory.[156] Other researchers, however, have not found support for power-control theory.[157] In addition, power-control theory has been criticized for its limited definition of patriarchal control, which is reduced to parental supervision, and for its view of female-headed households as equivalent to egalitarian families.[158]

Greenberg's Marxist Interpretation of Delinquency

David Greenberg offers a <u>Marxist interpretation</u> of the disproportionate involvement of adolescents in crime, based on the view that youth occupy a precarious status position in capitalist societies.[159] Greenberg begins by noting that the peak ages for involvement in crime have declined over time as a result of the changing position of juveniles in capitalist industrial societies. He argues that traditional theories of delinquency are unable to explain both this historical shift in the age distribution of crime and the decline in involvement in crime during late adolescence and early adulthood (the "aging out of crime" we discussed in Chapter 5).[160] Greenberg's critical approach accounts for both of these patterns through two primary components. First, he attributes the motivation to engage in delinquency to "the structural position of juveniles in American society." Second, he argues that the willingness to act on this motivation is different across age groups because the costs of engaging in crime vary by age.[161]

Greenberg's Marxist interpretation of delinquency A theory that attributes delinquency to the precarious status position of youth in capitalist societies. Greenberg explains delinquent acts in terms of adolescent needs for financial resources to support social life, status frustrations caused by school experiences, and status anxieties related to occupational prospects and male role expectations.

Adolescent Labor Force Participation and Social Life

Greenberg begins his analysis of delinquency by observing that adolescents in American society share a common status problem. Over time, American teenagers have been increasingly excluded from labor force participation through laws restricting child labor and establishing compulsory education. The result has been twofold. First, changing work and education patterns have resulted in an "age-segregated" society in which peers have become increasingly important for adolescents.[162] Sensitivity to peer expectations, desire for peer approval, and participation in teenage social activities have become significant forces in the lives of adolescents. Participation in teenage social life, however, requires financial resources. This brings us to the second result of the exclusion of adolescents from the labor force. The deterioration of the teenage labor market "has left teenagers less and less capable of financing an increasingly costly social life whose importance is enhanced as the age segregation of society grows."[163]

Greenberg explains delinquent acts of financial gain, such as theft, in terms of this financial problem. "Adolescent theft . . . occurs as a response to the disjunction between the desire to participate in social activities with peers and the absence of legitimate sources of funds needed to finance this participation."[164] As adolescents age, they gain more legitimate opportunities to acquire needed resources (e.g., through work in early adulthood) and the motivation to engage in delinquent acts of financial gain decreases.

School Experiences and Delinquency

How, then, do we explain delinquent acts that provide no financial reward, or "nonutilitarian" delinquency, such as vandalism or violence? Here, Greenberg turns to the status problems of adolescents caused by school experiences. He argues that two features of the school experience contribute to nonutilitarian delinquency: "its denial of student autonomy, and its subjection of some students to the embarrassment of public degradation."[165]

First, schools are filled with rules and restrictions about everything from attendance to acceptable hairstyles and clothing. At a time when adolescents are gaining increased autonomy within the family and elsewhere, schools continue to deny student autonomy. For those who do well in school, the resentment caused by this denial of autonomy tends to be compensated for by rewards associated with success in school. But for those who are less successful academically, the school setting offers only frustration. "It brings no current gratification and no promise of future payoff."[166]

According to Greenberg, class distinctions exist in this process. He argues that "middle and upper class children are supervised more closely than their working class counterparts, and thus come to expect and accept adult authority, while working class youths, who enter an unsupervised street life among peers at an early age, have more autonomy to protect."[167] Also, middle-class adolescents may be more likely to believe that future opportunities are tied to school success, and thus, may accept restrictions on autonomy more readily than will lower-class adolescents.[168]

Second, schools are places where teachers and other personnel evaluate and make distinctions among students and where respect is not universally granted to students. According to Greenberg, those who are "relatively powerless," such as failing students and members of lower social classes, are shown less respect than others. "School personnel continuously communicate their evaluations of students through grades, honor rolls, track positions, privileges, and praise for academic achievement and proper deportment. On occasion, the negative evaluation of students conveyed by the school's ranking systems is supplemented by explicit criticism and denunciation on the part of teachers who act as if the academic performance of failing students could be elevated by telling them they are stupid, or lazy, or both."[169] The experiences of humiliation suffered by some students only increase their frustrations regarding school.

Greenberg argues that nonutilitarian delinquency, such as property destruction and violence, can be understood as a response to the status frustrations associated with school. Through these delinquent acts, adolescents display independence and rejection of authority, restore self-esteem lost to school failure, and preserve moral character by demonstrating a willingness to take risks.[170]

Masculine Status Anxiety

Greenberg contends that school requirements of submission to authority are inconsistent with social expectations about masculinity, but not those about femininity.[171] Thus, the status problems caused by school experiences might help explain gender differences in delinquency in early adolescence. In late adolescence, individuals begin to consider occupational prospects and adult role expectations. In American society, the ability to fulfill traditional male roles centered on work and providing for a family are central to male identity.[172] For those who face structural barriers to fulfilling these roles, such as limited employment prospects based on social class and school failure, the result can be **masculine status anxiety**. Greenberg explains crimes that are exaggerated displays of masculinity, such as rape, homicide, and assault, as attempts to alleviate masculine status anxiety. As Greenberg writes, "In this interpretation a compulsive concern with toughness and masculinity arises . . . as a response to a contradiction between structural economic-political constraints on male status attainment and the cultural expectations for men that permeate American society."[173]

In summary, Greenberg's Marxist interpretation of delinquency is based on the constraints of a capitalist economic system that generally excludes adolescents from workforce participation. This exclusion deprives youth of the monetary resources needed to finance teenage social activities. In addition, the status problems caused by school experiences tend to vary by social class, and they account for involvement in nonutilitarian delinquency. Finally, masculine status anxiety also varies by class and accounts for lower-class involvement in violent crime.

We turn now to a discussion of feminist criminology, which shares with critical criminologies an emphasis on power and inequality within society, a focus on the structural determinants of behavior, and a view that crime and delinquency can be understood in terms of differences in power.

masculine status anxiety
Anxiety about one's ability to fulfill traditional male roles centered on work and providing for a family, caused by structural economic and political constraints on male status attainment.

■ Feminist Criminology

For most of its history, the discipline of criminology ignored the issue of gender. Feminist scholars typically attribute this neglect of gender to one of two causes. First, because crime—especially serious crime—is disproportionately committed by males (see Chapter 5), criminologists tended to view female offending as "tangential to the crime problem" and of little consequence for an understanding of crime and delinquency.[174] Second, criminology was a male-dominated discipline, and male criminologists seemed content to focus on male offenders and to ignore women both as offenders and as victims of crime. Feminist criminologists have lamented not only the historical neglect of women in the discipline of criminology, but also the tangential place historically given to feminist perspectives within theoretical criminology.[175]

Challenges to "male-centered" criminology began in the 1970s, as the women's movement gained momentum and feminist perspectives began to emerge. **Feminism** is not a single theory, but rather "a set of theories about

feminism "A set of theories about women's oppression *and* a set of strategies for social change" (Daly and Chesney-Lind, "Feminism and Criminology," 502). Feminism places gender at the center of attempts to understand human behavior.

EXPANDING IDEAS — Core Elements of Feminist Thought

Kathleen Daly and Meda Chesney-Lind, noted feminist criminologists, describe the complexities involved in defining feminism and suggest that feminist thought is distinguished by five core elements.

1. Gender is not a natural fact but a complex social, historical, and cultural product; it is related to, but not simply derived from, biological sex difference and reproductive capacities.
2. Gender and gender relations order social life and social institutions in fundamental ways.
3. Gender relations and constructs of masculinity and femininity are not symmetrical but are based on an organizing principle of men's superiority and social and political–economic dominance over women.
4. Systems of knowledge reflect men's views of the natural and social world; the production of knowledge is gendered.
5. Women should be at the center of intellectual inquiry, not peripheral, invisible, or appendages to men.

Source: Daly and Chesney-Lind, "Feminism and Criminology," 504.

women's oppression *and* a set of strategies for social change."[176] As criminologist Sally Simpson writes, "Feminism is best understood as both a world view and a social movement."[177] In its various forms, feminism tries to illuminate the origins and mechanisms of gender inequality and to propose strategies for abolishing it. Feminist perspectives place gender at the center of attempts to understand human behavior, including crime and delinquency. Kathleen Daly and Meda Chesney-Lind present five elements of feminist thought that distinguish it from other schools of social thought.[178] Expanding Ideas, "Core Elements of Feminist Thought," lists these five elements.

Feminist perspectives have in common "the recognition of *gender* as a central organizing component of social life," the recognition of imbalances in power and value granted to men and women, and the realization that any understanding of the social world is colored by gender.[179]

Some feminist scholars have challenged not only traditional theoretical perspectives, but also research methods based on the assumption of objectivity. These feminists have rejected the "scientific enterprise" that suggests that we can objectively know anything aside from our own *subjective* experiences of it and the world.[180] They contend that, in order to understand "women's existence," feminist researchers must incorporate their own individual experiences into the research process.[181] Thus, some feminists call upon researchers to employ more subjective methods that consider women's experiences of powerlessness and that address the research question at hand through the lens of gender.

In other words, to understand delinquency and crime by females, researchers must understand the "social worlds" in which females live—social worlds distinct from those of males. Later in this chapter, we offer examples of research that examines female offending in terms of the "social worlds" of females. We turn now to early feminist critiques of criminology, and then consider four feminist perspectives: liberal, radical, socialist, and Marxist feminisms.

Early Feminist Critiques of Criminology and the Criminal Justice System[182]

In the 1970s, Dorie Klein and Carol Smart examined the ways in which scholars in the early twentieth century described and explained female crime.[183] On the rare occasions when scholars paid attention to female offending, they tended to attribute involvement in crime or delinquency to individual or biological factors, rather than to social, political, or economic forces.[184] In the early and middle 1900s, writers who considered female crime typically portrayed female offenders as somehow deviating from the true nature of women. Early explanations for female offending were interwoven with moral judgments about the behaviors of proper and improper women. As Klein writes, the characteristics early writers used to explain female criminality "are of a *physiological* or *psychological* nature and are uniformly based on implicit or explicit assumptions about the *inherent nature of women.* . . . Since criminality is seen as an individual activity, rather than as a condition built into existing structures, the focus is on biological, psychological and social factors that would turn a woman toward criminal activity. To do this, the writers create two distinct classes of women: good women who are 'normal' noncriminals, and bad women who are criminals, thus taking a moral position that often masquerades as a scientific distinction."[185] This focus on individual characteristics diverted attention away from social structural factors, including gender inequality, that might account for female offending.

For over thirty years, Meda Chesney-Lind has offered a feminist critique of the criminal justice system.[186] During the 1970s and 1980s, she drew attention to the way the juvenile justice system "sexualized" female offending—especially female involvement in status offenses.[187] As we noted in Chapter 5, females are more likely to be arrested for status offenses, such as running away, incorrigibility, and curfew violations, than are males, even though males are more likely to report engaging in status offenses.[188] Chesney-Lind documents how police and juvenile courts respond more harshly to female status offenses than to more serious offenses committed by females or to status and other delinquent offenses committed by males.[189] She attributes this "double standard of juvenile justice" to threats to parental authority and female chastity posed by female status offenses. "Like good parents, police and court personnel respond differently to the indiscretions of young men and women. . . . The juvenile justice system is concerned that girls allowed to run wild might be tempted to experiment sexually and thus endanger their marriageability. The court involves itself in the enforcement of adolescent morality and parental authority through the vehicle of status offenses."[190] Chesney-Lind's critique is supported by evidence of juvenile court practices, such as ordering gynecological examinations of females brought before the court, even when those females were referred to juvenile court for offenses that had nothing to do with sexual behavior.[191]

Early feminist critiques of criminology and the criminal justice system also challenged the *separate spheres* assumptions operating in criminological theory and justice practices. "Separate spheres is a set of ideas about the place of men and of women in the social order that emerged in the first quarter of the nineteenth century in the United States, as well as in other countries undergoing capitalist industrialization. This ideology placed men in the public sphere (paid workplace, politics, law) and women in the private sphere (household, family life); it characterized gender relations for white, middle-class, married heterosexual couples."[192] As Daly and Chesney-Lind point out, laws and legal practices were based on separate spheres ideology. During the 1960s and 1970s, feminists began to challenge the assumptions of separate spheres, the oppression of this ideology for women, and the gender-based laws derived from separate spheres ideology.[193] The goal of reform was equality for females and males, and it found expression primarily in liberal feminism.

Liberal Feminism

liberal feminism
Maintains that gender equality can be achieved through (1) changes in socialization experiences and gender roles acquired through socialization, and (2) the provision of equal opportunities for women through the elimination of discriminatory policies and practices and the development of policies that promote equal rights.

Liberal feminism is based on ideals of equal opportunity and freedom of choice for females. Advocates of this perspective view gender inequality as rooted in separate spheres ideology and in attitudes favoring traditional male and female roles, reinforced through discrimination against women in the public sphere (for example, in work, education, and politics).[194] Liberal feminists believe that gender equality can be achieved through (1) changes in socialization experiences and gender roles acquired through socialization, and (2) the provision of equal opportunities for women through the elimination of discriminatory policies and practices, and through the development of policies that promote equal rights.[195] Thus, liberal feminists advocate using laws, such as those prohibiting gender-based discrimination, and the legal system as tools for achieving gender equality.

Within criminology, liberal feminism found its first expressions in the works of Freda Adler and Rita Simon.[196] Both Adler and Simon moved away from notions of female offending as grounded in biology or psychology—ideas that characterized early explanations of female crime. Instead, they focused on social factors, arguing that females' relatively limited involvement in crime was linked to limited opportunities for women in the public sphere. Adler and Simon's view of female offending has been called the "liberation hypothesis" and "emancipation theory." They argued that the liberation ushered in by the women's movement would change gender socialization experiences and opportunities for women, and that both would have an effect on women's involvement in crime.

More specifically, Adler argued that women were becoming more like men—not only through a growing similarity of gender roles and increasing opportunities in the work world, but also in the types of criminal offenses they were committing. Adler saw women embracing their new freedoms, in part by entering the formerly male world of serious and violent crimes.[197] Simon's argument was more complex. She linked increases in female offending primarily to their entry into the paid workforce. Women's increased participation in the labor market was accompanied by increased opportunities for participation in crimes linked to work, such as fraud and embezzlement. Unlike Adler, Simon did not argue that liberation would bring with it increases in violent crimes by women.[198]

Adler and Simon's works are important because they drew attention to the issue of women's crime and proposed that it was a valid topic of inquiry within criminology, but empirical evidence has not supported their hypotheses. As we discussed in Chapter 5, the gender gap in crime has not narrowed in the way or to the extent that Adler and Simon predicted. Female involvement in crime has increased most for minor property offenses, such as larceny–theft. This increase reflects the increasing economic marginalization of women, not liberation or equal opportunities across gender.[199]

Radical Feminism

Radical feminism developed in part in response to criticisms of liberal feminism. Liberal feminists believe that gender equality can be achieved through changes in socialization experiences and the provision of equal opportunities for women through legal reform, such as laws prohibiting discrimination based on gender. Radical feminists see this approach as inadequate because it ignores the structural roots of gender inequality; they take a more critical view of the legal system and other social institutions.

Radical feminists focus on **patriarchy** as the root of women's oppression, and they explore the structurally based imbalance of power between men and women. Patriarchy is "a social structure of men's control over women's labor and sexuality."[200] According to radical feminism, the oppression of women is "fundamental." Male power and privilege are at the center of all inequality, as the relationships between men and women in society influence all other relationships (e.g., those based on class).

Radical feminists focus on male sexuality and sexual aggression as important mechanisms that men use to exert power. Noted radical feminist Catharine MacKinnon contends that male dominance is maintained primarily through the control of female sexuality and reproduction.[201] She argues that compulsory heterosexuality for women and the threat or use of sexual violence (e.g., rape, pornography, sexual abuse, sexual harassment, and wife battering) are the primary mechanisms of male dominance and control of women.[202]

Given their emphasis on the sexual control of women, it is not surprising that radical feminists have focused on females as victims of crime (particularly sexual violence), rather than on females as offenders.[203] Chesney-Lind has also shed light on the ways in which female victimization and female offending are often intertwined. With her colleague, Randall Shelden, Chesney-Lind argues that in order to understand girls' delinquency we must first understand girls' everyday lives.[204] Gender socialization experiences, the sexual double standard (which encourages or at least condones male sexual exploration, but prohibits and punishes female sexual behavior), and female victimization in patriarchal society combine to create a reality for girls that is very different from that of boys.

Chesney-Lind's research shows that many girls who are brought into the juvenile justice system have histories of victimization, especially physical and sexual abuse.[205] "According to a study of girls in juvenile correctional settings . . . a very large proportion of these girls . . . had experienced physical abuse (61.2%), and nearly half said that they had experienced this abuse 11 or more

radical feminism Focuses on patriarchy as the root of women's oppression and explores the structurally based imbalance of power between men and women. Contends that male dominance is maintained in part through the control of female sexuality and reproduction.

patriarchy "A social structure of men's control over women's labor and sexuality" (Daly and Chesney-Lind, "Feminism and Criminology," 511).

times. . . . More than half of these girls (54.3%) had experienced sexual abuse and for most this was not an isolated incident; a third reported that it happened 3 to 10 times and 27.4% reported that it happened 11 times or more."[206] More than 80% of these girls eventually ran away from home—a survival strategy that violates the law and results in juvenile justice system contact for many girls who resort to it. Female runaways are far more likely than male runaways to cite abuse as their reason for leaving home.[207] Once on the street, girls often resort to "survival" offenses such as theft, shoplifting, and prostitution, which increase the risk of contact with the juvenile justice system. Thus, the mechanisms that girls often use to escape victimization are themselves "criminalized."[208] Case in Point, "Barbara's Victimization and Delinquency," offers one girl's story of victimization and offending.

Chesney-Lind calls for greater attention to the relationship between female offending and female victimization. She also maintains that, because of the different life experiences of girls and boys, attempts to explain female delinquency cannot simply begin with traditional theories developed to explain male delinquency—theorists cannot "add gender and stir." Daly and Chesney-Lind call this the "generalizability problem," which concerns the question: Can theories of male offending be generalized and applied to female offending?[209] Because most traditional theories have attempted to explain only the behavior of males, they have not considered the subordination and victimization of girls that often contribute to girls' delinquent behavior. Chesney-Lind writes, "Delinquency theory has all but ignored girls and their problems. As a result, few attempts have been made to understand the meaning of girls' arrest patterns and

CASE IN POINT

Barbara's Victimization and Delinquency

Regina Arnold describes the relationship between abusive home environments and delinquency among young black women. Arnold illustrates this relationship with Barbara's story:

> I was with this foster lady who was very cruel. She was abusive, and I was no more than a maid as a child. That was my purpose. She received welfare for foster kids, that was her purpose. I stayed there 'til I was thirteen, then ran away. I was tired of the physical abuse. I ran to my mother's. The man she lived with sexually abused me, and I ran away again. I was on the street. I got arrested for shoplifting. I was picked up for vagrancy when I was almost fifteen. The judge gave my mother an option to take me home or he would have no other choice but to send me away. She says, "Send her away, I don't want her." So they sent me away. I went to a state home for girls.

Source: Arnold, "Black Women in Prison," 174.

the relationship between these arrests and the very real problems that these arrests mask. . . . Conventional theories of delinquency seem best situated to explain the relative absence of girls from traditional boys' delinquency."[210] But as we noted in Chapter 5, some studies have shown that the processes leading to delinquency are similar for males and females.[211] Yet these studies should not be used to justify ignoring the unique life experiences of females and males. It would be a mistake to ignore the potential of gender-specific theories to illuminate unique causes of offending for females and males, just as it would be a mistake to assume that traditional theories have nothing of value to say about female offending.[212]

In addition to considering the generalizability of traditional theories and the link between female victimization and offending, radical feminist criminologists have also examined the patriarchal biases of the criminal justice system. As Chesney-Lind writes, "Anyone seriously interested in examining women's crime or the subjugation of women . . . must carefully consider the role of the contemporary criminal justice system in the maintenance of modern patriarchy."[213] According to radical feminists, the legal system operates as an instrument of male dominance, as do other major social institutions, such as the economy and the family.

The double standards of the criminal justice system signify how it is used for the perpetuation of male power. Radical feminists point out that the harm of crimes against women, such as pornography and domestic violence, has often been overlooked or downplayed by a system that devotes itself to fighting other types of crime.[214] In addition, in crimes such as rape, female victims are sometimes treated as though they must prove their innocence by justifying, for example, their attire and sexual histories. In an extensive body of research, Chesney-Lind has also revealed the paternalistic attitudes of police and juvenile court judges, who typically have been less concerned with girls' serious delinquent offenses than with their sexual behavior (or even the potential for it, suggested by status offenses such as running away or incorrigibility, which indicate that girls are beyond parental control).[215] Girls are also more likely than boys to be institutionalized for status offenses, and studies reveal that girls are sometimes further victimized while in custody.[216] Radical feminists argue that, because of the patriarchal biases of the legal system, it cannot be used as a means of achieving gender equality, as liberal feminists propose.

Socialist Feminism

Like radical feminism, <u>socialist feminism</u> calls for an understanding of crime and delinquency (by both males and females) within the social context of patriarchy.[217] Socialist feminists also consider the intersecting oppressions of patriarchy and capitalism. They contend that the social organization of society is determined by the *interaction* of class and gender inequalities.[218] According to this perspective, neither class nor gender is given greater weight as the primary basis for oppression. Rather, capitalism and patriarchy are seen as fundamentally intertwined in their influence on the structure of society. "Socialist-feminists attempt a synthesis between two systems of domination, class and patriarchy (male supremacy). Both relations of production and reproduction are structured by capitalist patriarchy."[219]

socialist feminism
Contends that the social organization of society is determined by the *interaction* of class and gender inequalities, and considers the dual and intersecting oppressions of patriarchy and capitalism. Neither class nor gender is given greater weight as the primary basis for oppression.

With the Industrial Revolution in the United States two centuries ago came the development of the nuclear family and a gendered division of labor, both inside and outside the home. Prior to industrialization, the production of goods occurred primarily in the home, and all family members were economically vital to the production and distribution process. The members of a household together formed a "unit of economic production," and the contributions of both women and men were valued economically.[220] The Industrial Revolution brought with it a clear distinction between economic and domestic life. Wage labor was now performed outside the home, primarily by men who assumed the role of economic provider for their families. Women were assigned responsibility for the domestic sphere, where their labor was "invisible" and unpaid, and they became economically dependent on men in a way that they had not been prior to industrialization.[221]

Socialist feminists maintain that patriarchal ideology is fundamentally interwoven with industrial capitalism. Moreover, they argue that laws and the criminal justice system reinforce both patriarchy and class inequality inherent in capitalism. For example, marriage and abortion laws constrain the sexual and reproductive activities of women in ways that support the reproduction of the labor force.[222]

In his book *Capitalism, Patriarchy, and Crime*, James Messerschmidt offers a socialist feminist analysis of crime.[223] He argues that the interaction of patriarchy and capitalism determines the distribution of power in society. In other words, in capitalist societies, power and privilege are defined by both gender and social class. Inequalities based on gender and class create two relatively powerless groups—women and members of the lower classes—and one relatively powerful group—men from the upper classes. Messerschmidt contends that gender and class differences in the seriousness and frequency of crime are based on these group differences in access to power.

Like all forms of behavior, crime requires opportunity. The most powerful members of society have the greatest opportunities for involvement in both law-abiding and criminal behavior. According to Messerschmidt, this explains why, in all social classes, men commit more crimes than women.[224] The types of crime they commit depend on their class standing and corresponding access to legitimate and illegitimate opportunities. For example, people (usually men) in positions of power, such as corporate and government leaders, are able to commit serious crimes of immense harm to society. (Recall the recent scandals at corporations such as Enron and WorldCom, in which the crimes of the corporate elite led to billions of dollars in losses to average citizens.) Men in the lower classes lack opportunities for corporate crime but may commit "street" crimes, such as burglary or robbery, that reflect their disadvantaged class position.[225] Likewise, the types of crime women typically commit, such as larceny–theft and credit card fraud, reflect their exclusion from positions of power and their economic dependence in patriarchal, capitalist society. Even crimes of violence by women—which are relatively rare events—are often perpetrated against their male partners or children, and so are tied to their limited roles as wives and mothers.

Like radical feminists, socialist feminists have also pointed out that the same gender and class inequalities that structure offending also structure victimization. Female offending is often linked to females' experiences as victims of oppression based on gender and class.[226]

Marxist Feminism[227]

Socialist and Marxist feminists share a view of oppression based on both class and gender. But <u>Marxist feminism</u> views class inequality inherent in capitalist societies as the *primary* form of oppression. Gender and race inequalities are seen as *secondary* forms of oppression that derive from and reflect the class inequality of capitalism.[228] This perspective contends that the capitalist system of production determines social relations based on both class and gender. "Although most Marxist feminists examine masculine dominance and sexism in society, they comprehend the roles of men and women in relation to capital, not in relation to a separate system of masculine power and dominance. Women's labor in the home is analyzed not in terms of how it benefits men but, rather, how it provides profits for the capitalist class."[229]

In a Marxist feminist analysis of crime, Sheila Balkan and her colleagues consider how the social position of women is structured and legitimated within capitalist America.[230] As we noted earlier, the development of a gendered division of labor in the United States is linked to industrialization and capitalism. The dominant social position of men is linked to their role as wage earner and status as "breadwinner." The subordinate social position of women is linked to their economic dependence on men, and their roles as wife and mother—roles that provide for the reproduction of the labor force and thus serve capitalist interests. According to Marxist feminism, then, the differential value of men's and women's social positions is fundamentally linked to capitalism. Balkan and her colleagues suggest that women's crime must be understood in terms of how it reflects women's oppressed position within capitalist society. For example, female offenders are often involved in crimes, such as larceny and fraud, that testify to their economic marginalization (see Chapter 5).

Another example of Marxist feminist criminology is Julia and Herman Schwendinger's study of rape.[231] The Schwendingers considered why women who are raped often experience feelings of guilt about the crime perpetrated against them. The Schwendingers argue that the relegation of women to economically dependent roles and the devaluing of women's domestic labor under capitalism contribute to a sexist culture that supports male violence, including rape. They conclude that the sense of inadequacy and diminished self-worth that accompanies women's dependent social status leads women to blame themselves for their own rape victimization.

Marxist feminism Views class inequality inherent in capitalist societies as the *primary* form of oppression, and gender and race inequalities as *secondary* forms of oppression that derive from and reflect class inequality.

■ Summary and Conclusions

The study of juvenile delinquency typically begins with the assumption that delinquency is behavior that violates the law, and then attempts to answer the question, "Why do they do it?" In this chapter, we examined approaches that try to answer a different question: How and why do certain behaviors and individuals get labeled "delinquent" or "criminal?" The labeling perspective and critical criminologies both focus on social and societal reactions to delinquent behavior.

The labeling perspective is concerned with the social construction of deviance and maintains that definitions of what is delinquency and who is delinquent result from dynamic social processes. We described early statements of these processes offered in Tannenbaum's discussion of the dramatization of evil and Lemert's work on primary and secondary deviance. Both argued that, as a result of societal reactions to initial acts of deviance or "mischief," individuals come to see themselves as "bad" and behave according to that label, enacting a self-fulfilling prophecy in which they become what they are said to be. Becker considered the importance of rule creation and enforcement in this process.

In discussing the labeling perspective, we considered both formal and informal reactions to deviance. Matsueda examined how informal labels by significant others, such as parents, peers, and teachers, are perceived and how they influence subsequent behavior, including delinquency. We explored the consequences of both formal and informal labeling. The labeling perspective contends that societal reactions to deviance have a stigmatizing effect that increases the likelihood of future deviance. In contrast, deterrence theory argues that societal reactions constitute a form of punishment that deters future deviance. Results of research on the effects of formal labels have been mixed, but several studies support the labeling perspective. Research on the effects of informal labels, much of which uses National Youth Survey data, also offers support for the labeling perspective and demonstrates the influence of informal labels on delinquency.

We turned next to critical criminologies, which view capitalism as the root cause of crime and delinquency and which explain social responses to behaviors, including crime, in terms of the influence of capitalism. Critical perspectives share a focus on power and inequality in understanding crime and crime control, a view that defining crime and delinquency is a political enterprise that reflects conflict and power imbalances, and a view that law and the criminal justice system protect the interests of the powerful or the capitalist system as a whole. We considered the distinction between instrumental and structural Marxist views of crime and delinquency, and then examined several structural critical approaches to the problem of delinquency.

Colvin and Pauly, Hagan and his colleagues, and Greenberg all offer critical approaches to understanding delinquency that are based on the role of capitalism in the production of delinquent behavior. Colvin and Pauly, in their integrated structural-Marxist theory, and Hagan and his colleagues, in their power-control theory, share an emphasis on the way in which control and authority in the workplace, as determined by social class, shape parental control of children and socialization processes within the family. Integrated structural-Marxist theory also considers class differences in control structures in schools and peer groups and their contribution to delinquency. Power-control theory considers how the translation of workplace control and authority to family control differs for sons and daughters in patriarchal and egalitarian families, and the ability of these gender differences to explain gender differences in delinquency. Greenberg takes a slightly different approach, explaining delinquent behavior in terms of status problems created for adolescents by the structure of work opportunities under capitalism. These status problems include frustrations associated with school experiences and masculine status anxiety regarding work and traditional male roles, both of which vary by social class.

Finally, we explored the critical understanding of crime and delinquency offered in recent decades by feminist criminologists. Feminist approaches share a view that gender must be at the center of attempts to understand human behavior, including crime and delinquency. Early feminist critiques of criminology and the criminal justice system examined (1) how scholars had explained female crime in terms of individual biological and psychological factors, rather than social, political, or economic forces; and (2) the "sexualization" of female delinquency, particularly status offenses, by the juvenile justice system. We considered four feminist approaches—liberal, radical, socialist, and Marxist feminisms—and applications of them to the issue of delinquency and crime. These approaches are distinguished primarily by the following:

- their views of the ability of law and the legal system to accomplish social change and promote gender equality;

- their emphasis on patriarchy and structurally based gender imbalances in power as the root of female oppression; and

- their consideration of the oppressions of capitalism and the ways in which they intersect with the oppressions of patriarchy.

THEORIES

- theory of reintegrative shaming
- integrated structural-Marxist theory
- power-control theory
- Greenberg's Marxist interpretation of delinquency

CRITICAL THINKING QUESTIONS

1. What ideas do the labeling perspective and critical criminologies have in common that distinguish them from other theories discussed in previous chapters?

2. Explain the distinction between primary and secondary deviance.

3. What do critical criminologists mean when they say that crime and delinquency are grounded in the capitalist economic system?

4. Explain the distinctions between the four feminist perspectives described in this chapter: liberal, radical, socialist, and Marxist feminisms.

5. Consider again the story of Rocky Sullivan and Jerry Connelly presented at the beginning of this chapter. In light of what you have read in this chapter, do you believe that the experience of being apprehended, labeled a "delinquent," and formally sanctioned "made a criminal out of Rocky?" Why or why not?

SUGGESTED READINGS

Becker, Howard S. *Outsiders: Studies in the Sociology of Deviance.* New York: Free Press, 1963.

Chesney-Lind, Meda, and Randall G. Shelden. *Girls, Delinquency, and Juvenile Justice* (2nd edition). Belmont, CA: Wadsworth, 1998.

Colvin, Mark, and John Pauly. "A Critique of Criminology: Toward an Integrated Structural-Marxist Theory of Delinquency Production." *American Journal of Sociology* 89 (1983):513–551.

Daly, Kathleen, and Meda Chesney-Lind. "Feminism and Criminology." *Justice Quarterly* 5 (1988): 497–538.

Greenberg, David F. "Delinquency and the Age Structure of Society." In *Crime and Capitalism*, edited by David F. Greenberg, 118–139. Palo Alto, CA: Mayfield, 1981.

Hagan, John. *Structural Criminology.* New Brunswick, NJ: Rutgers University Press, 1989.

Paternoster, Raymond, and Leeann Iovanni. "The Labeling Perspective and Delinquency: An Elaboration of the Theory and an Assessment of the Evidence." *Justice Quarterly* 6 (1989):359–394.

GLOSSARY

communitarianism: The societal-level counterpart to interdependency. Shaming is more widespread and effective in communitarian societies than in other societies.

critical criminologies: Explore the structural bases of inequality in society and the way in which inequality and social status influence definitions of crime and criminals and reactions to behaviors.

deviance amplification: A process in which societal reactions to initial acts of deviance lead to the development of a deviant identity, which increases the likelihood of subsequent deviance among those who have been labeled "deviant."

dramatization of evil: A process in which others react to delinquent and mischievous actions of a youth as though these actions and the youth are "evil," and, through these reactions, create a self-fulfilling prophecy in which the youth becomes what he or she is said to be.

egalitarian families: According to power-control theory, families in which class relations are balanced and husbands and wives are employed in positions of similar authority.

feminism: "A set of theories about women's oppression *and* a set of strategies for social change" (Daly and Chesney-Lind, "Feminism and Criminology," 502). Feminism places gender at the center of attempts to understand human behavior.

Greenberg's Marxist interpretation of delinquency: A theory that attributes delinquency to the precarious status position of youth in capitalist societies. Greenberg explains delinquent acts in terms of adolescent needs for financial resources to support social life, status frustrations caused by school experiences, and status anxieties related to occupational prospects and male role expectations.

instrumental Marxism: A perspective on capitalism and crime that states that "criminal law is an instrument of the state and ruling class to maintain and perpetuate the existing social and economic order" (Quinney, *Critique of Legal Order,* 16).

integrated structural-Marxist theory: Explains delinquency in terms of the structure of social relations grounded in capitalism. Emphasizes control structures within the workplace, family, school, and peer groups, and the way these control structures are shaped by capitalism.

interdependency: "The extent to which individuals participate in networks wherein they are dependent on others to achieve valued ends and others are dependent on them" (Braithwaite, *Crime, Shame and Reintegration,* 98, 100). The concept of interdependency is similar to attachment, commitment, and social bonding in control theory.

labeling perspective: This perspective focuses on social reactions to delinquency and crime and considers how definitions of what is crime and who is criminal result from dynamic social processes.

liberal feminism: Maintains that gender equality can be achieved through (1) changes in socialization experiences and gender roles acquired through socialization, and (2) the provision of equal opportunities for women through the elimination of discriminatory policies and practices and the development of policies that promote equal rights.

Marxist feminism: Views class inequality inherent in capitalist societies as the *primary* form of oppression, and gender and race inequalities as *secondary* forms of oppression that derive from and reflect class inequality.

masculine status anxiety: Anxiety about one's ability to fulfill traditional male roles centered on work and providing for a family, caused by structural economic and political constraints on male status attainment.

moral crusaders: Individuals, typically of high social status, who create rules in an attempt to correct some "evil" in the world.

patriarchal families: According to power-control theory, families in which class relations are unbalanced and husbands exercise greater authority in the workplace than do wives.

patriarchy: "A social structure of men's control over women's labor and sexuality" (Daly and Chesney-Lind, "Feminism and Criminology," 511).

power-control theory: A theory of gender differences in common delinquency that examines both power derived from class position and authority in the workplace and control established in family interactions. The theory argues that the presence of power and the absence of control, which differ by class and gender, contribute to adolescents' freedom to engage in delinquency.

primary deviance: Initial acts of deviance, many of which go undetected.

radical feminism: Focuses on patriarchy as the root of women's oppression and explores the structurally based imbalance of power between men and women. Contends that male dominance is maintained in part through the control of female sexuality and reproduction.

reintegrative shaming: "Shaming which is followed by efforts to reintegrate the offender back into the community of law-abiding or respectable citizens through words or gestures of forgiveness or ceremonies to decertify the offender as deviant" (Braithwaite, *Crime, Shame and Reintegration,* 100–101).

secondary deviance: Deviance that occurs in response to problems created by societal reactions to primary deviance.

socialist feminism: Contends that the social organization of society is determined by the *interaction* of class and gender inequalities, and considers the dual and intersecting oppressions of patriarchy and capitalism. Neither class nor gender is given greater weight as the primary basis for oppression.

stigmatization: "Disintegrative shaming in which no effort is made to reconcile the offender with the community" (Braithwaite, *Crime, Shame and Reintegration,* 101).

structural Marxism: A perspective on capitalism and crime that views law as a tool used to preserve the capitalist system as a whole, rather than to serve the interests of a cohesive ruling class. Some laws may actually be costly to capitalists. But in the long run, laws maintain and protect capitalism by averting crisis in capitalist–worker relations.

theory of reintegrative shaming: Crime can be reduced or controlled through the deterrent effect of shaming, followed by attempts to reintegrate the offender back into the community.

REFERENCES

Adams, Mike S., and T. David Evans. "Teacher Disapproval, Delinquent Peers, and Self-Reported Delinquency: A Longitudinal Test of Labeling Theory." *The Urban Review* 28 (1996):199–211.

Adams, Mike S., James D. Johnson, and T. David Evans. "Racial Differences in Informal Labeling Effects." *Deviant Behavior: An Interdisciplinary Journal* 19 (1998):157–171.

Adams, Mike S., Craig T. Robertson, Phyllis Gray-Ray, and Melvin C. Ray. "Labeling and Delinquency." *Adolescence* 38 (2003):171–186.

Adler, Freda. *Sisters in Crime: The Rise of the New Female Criminal.* New York: McGraw-Hill, 1975.

Andersen, Margaret L., and Howard F. Taylor. *Sociology: Understanding a Diverse Society.* 2nd ed. Belmont, CA: Wadsworth, 2002.

Arnold, Regina. "Black Women in Prison: The Price of Resistance." In *Women of Color in U.S. Society,* edited by Maxine Baca Zinn and Bonnie Thornton Dill, 171–184. Philadelphia: Temple University Press, 1994.

Balkan, Sheila, Ronald J. Berger, and Janet Schmidt. *Crime and Deviance in America: A Critical Approach.* Belmont, CA: Wadsworth, 1980.

Barak, Gregg. "Time for an Integrated Critical Criminology." In *Cutting the Edge: Current Perspectives in Radical/Critical Criminology and Criminal Justice,* edited by Jeffrey Ian Ross, 34–39. Westport, CT: Praeger, 1998.

Bartusch, Dawn Jeglum, and Ross L. Matsueda. "Gender, Reflected Appraisals, and Labeling: A Cross-Group Test of an Interactionist Theory of Delinquency." *Social Forces* 75 (1996):145–177.

Becker, Howard S. *Outsiders: Studies in the Sociology of Deviance.* New York: Free Press, 1963.

Beirne, Piers, and James Messerschmidt. *Criminology.* 3rd ed. Boulder, CO: Westview, 2000.

Belknap, Joanne. "Meda Chesney-Lind: The Mother of Feminist Criminology." *Women and Criminal Justice* 15 (2004):1–23.

Blackwell, Brenda Sims. "Perceived Sanction Threats, Gender and Crime: A Test and Elaboration of Power-Control Theory." *Criminology* 38 (2000):439–488.

Bonger, Willem. *Criminality and Economic Conditions.* Abridged ed. Bloomington, IN: Indiana University Press, 1969.

Braithwaite, John. *Crime, Shame and Reintegration.* Cambridge: Cambridge University Press, 1989.

Canter, Rachelle J. "Sex Differences in Self-Reported Delinquency." *Criminology* 20 (1982):373–393.

Chambliss, William J., and Robert B. Seidman. *Law, Order, and Power.* Reading, MA: Addison-Wesley, 1982.

Chesney-Lind, Meda. "Challenging Girls' Invisibility in Juvenile Court." *The Annals of the American Academy of Political and Social Science* 564 (1999):185–202.

———. *The Female Offender: Girls, Women, and Crime.* Thousand Oaks, CA: Sage, 1997.

———. "Girls' Crime and Woman's Place: Toward a Feminist Model of Female Delinquency." *Crime and Delinquency* 35 (1989):5–29.

———. "Judicial Enforcement of the Female Sex Role: The Family Court and the Female Delinquent." *Issues in Criminology* 8 (1973):51–69.

———. "Judicial Paternalism and the Female Status Offender: Training Women to Know Their Place." *Crime and Delinquency* 23 (1977):121–130.

Chesney-Lind, Meda and Randall G. Shelden. *Girls, Delinquency, and Juvenile Justice.* 2nd ed. Belmont, CA: Wadsworth, 1998.

Colvin, Mark, and John Pauly. "A Critique of Criminology: Toward an Integrated Structural-Marxist Theory of Delinquency Production." *American Journal of Sociology* 89 (1983):513–551.

Cullen, Francis T., and Robert Agnew. *Criminological Theory: Past to Present (Essential Readings).* 2nd ed. Los Angeles: Roxbury, 2003.

Curran, Daniel J., and Claire M. Renzetti. *Theories of Crime.* 2nd ed. Boston: Allyn and Bacon, 2001.

Daly, Kathleen, and Meda Chesney-Lind. "Feminism and Criminology." *Justice Quarterly* 5 (1988):497–538.

De Li, Spencer. "Legal Sanctions and Youths' Status Achievement: A Longitudinal Study." *Justice Quarterly* 16 (1999):377–401.

Edwards, Richard. *Contested Terrain: The Transformation of the Workplace in the Twentieth Century.* New York: Basic, 1979.

Empey, LaMar T., Mark C. Stafford, and Carter H. Hay. *American Delinquency: Its Meaning and Construction.* 4th ed. Belmont, CA: Wadsworth, 1999.

Erikson, Kai T. "Notes on the Sociology of Deviance." *Social Problems* 9 (1962):307–314.

———. *Wayward Puritans: A Study in the Sociology of Deviance.* New York: John Wiley, 1966.

Farrell, Ronald A., and Victoria Lynn Swigert. "A Hierarchical Systems Theory of Deviance." In *Social Deviance,* 3rd ed., edited by Ronald A. Farrell and Victoria L. Swigert, 391–407. Belmont, CA: Wadsworth, 1988.

———. "Prior Offense Record as a Self-Fulfilling Prophecy." *Law and Society Review* 12 (1978):437–453.

Friedrichs, David O. "New Directions in Critical Criminology and White Collar Crime." In *Cutting the Edge: Current Perspectives in Radical/Critical Criminology and Criminal Justice,* edited by Jeffrey Ian Ross, 77–91. Westport, CT: Praeger, 1998.

Gaarder, Emily, and Joanne Belknap. "Tenuous Borders: Girls Transferred to Adult Court." *Criminology* 40 (2002):481–517.

Goffman, Erving. *Asylums.* New York: Anchor, 1961.

———. *Stigma: Notes on the Management of Spoiled Identity.* Englewood Cliffs, NJ: Prentice Hall, 1963.

Gove, Walter R. "The Labelling Perspective: An Overview." In *The Labelling of Deviance,* 2nd ed., edited by Walter R. Gove, 9–33. Beverly Hills, CA: Sage, 1980.

Grasmick, Harold G., John Hagan, Brenda Sims Blackwell, and Bruce J. Arneklev. "Risk Preferences and Patriarchy: Extending Power-Control Theory." *Social Forces* 75 (1996):177–199.

Greenberg, David F. "Delinquency and the Age Structure of Society." In *Crime and Capitalism,* edited by David F. Greenberg, 118–139. Palo Alto, CA: Mayfield Publishing, 1981.

Hagan, John. "The Social Embeddedness of Crime and Unemployment." *Criminology* 31 (1993):465–491.

———. *Structural Criminology.* New Brunswick, NJ: Rutgers University Press, 1989.

Hagan, John, A. R. Gillis, and John Simpson. "Clarifying and Extending Power-Control Theory." *American Journal of Sociology* 95 (1990):1024–1037.

———. "The Class Structure of Gender and Delinquency: Toward a Power-Control Theory of Common Delinquent Behavior." *American Journal of Sociology* 90 (1985):1151–1178.

Hagan, John, and Alberto Palloni. "The Social Reproduction of a Criminal Class in Working-Class London, circa 1950–1980." *American Journal of Sociology* 96 (1990):265–299.

Hagan, John, John Simpson, and A. R. Gillis. "Class in the Household: A Power-Control Theory of Gender and Delinquency." *American Journal of Sociology* 92 (1987):788–816.

Harris, Anthony R. "Imprisonment and the Expected Value of Criminal Choice: A Specification and Test of Aspects of the Labeling Perspective." *American Sociological Review* 40 (1975):71–87.

Heimer, Karen, and Ross L. Matsueda. "Role-Taking, Role Commitment, and Delinquency: A Theory of Differential Social Control." *American Sociological Review* 59 (1994):365–390.

Hirschi, Travis. *Causes of Delinquency.* Berkeley, CA: University of California Press, 1969.

Hirschi, Travis, and Michael J. Hindelang. "Intelligence and Delinquency: A Revisionist Review." *American Sociological Review* 42 (1977):571–587.

Jensen, Gary F. "Delinquency and Adolescent Self-Conceptions: A Study of the Personal Relevance of Infraction." *Social Problems* 20 (1972):84–103.

Jensen, Gary F., and Kevin Thompson. "What's Class Got to Do with It? A Further Examination of Power-Control Theory." *American Journal of Sociology* 95 (1990):1009–1023.

Kaplan, Howard B., and Robert J. Johnson. "Negative Social Sanctions and Juvenile Delinquency: Effects of Labeling in a Model of Deviant Behavior." *Social Science Quarterly* 72 (1991):98–122.

Kitsuse, John I. "Societal Reaction to Deviant Behavior: Problems of Theory and Method." *Social Problems* 9 (1962):247–256.

Klein, Dorie. "The Etiology of Female Crime: A Review of the Literature." *Issues in Criminology* 8 (1973):3–30.

Kohn, Melvin L. *Class and Conformity.* Chicago: University of Chicago Press, 1977.

Lemert, Edwin M. *Human Deviance, Social Problems, and Social Control.* 2nd ed. Englewood Cliffs, NJ: Prentice Hall, 1972.

———. *Social Pathology: A Systematic Approach to the Theory of Sociopathic Behavior.* New York: McGraw-Hill, 1951.

Link, Bruce, and Jo Phelan. "Conceptualizing Stigma." *Annual Review of Sociology* 27 (2001):363–385.

Liu, Xiaoru. "The Conditional Effect of Peer Groups on the Relationship Between Parental Labeling and Youth Delinquency." *Sociological Perspectives* 43 (2000):499–514.

Loeber, Rolf, and Magda Stouthamer-Loeber. "Family Factors as Correlates and Predictors of Juvenile Conduct Problems and Delinquency." In *Crime and Justice: An Annual Review of Research,* vol. 7, edited by Michael Tonry and Norval Morris, 29–149. Chicago: University of Chicago Press, 1986.

Lynch, Michael J., and W. Byron Groves. *A Primer in Radical Criminology.* New York: Harrow and Heston, 1986.

MacDonald, John M., and Meda Chesney-Lind. "Gender Bias and Juvenile Justice Revisited: A Multi-year Analysis." *Crime and Delinquency* 47 (2001):173–195.

Macionis, John J. *Sociology.* 9th ed. Upper Saddle River, NJ: Prentice Hall, 2003.

MacKinnon, Catharine A. "Feminism, Marxism, Method, and the State: An Agenda for Theory." *Signs: Journal of Women in Culture and Society* 7 (1982):515–544.

———. "Feminism, Marxism, Method, and the State: Toward Feminist Jurisprudence." *Signs: Journal of Women in Culture and Society* 8 (1983):635–658.

———. *Feminism Unmodified: Discourses on Life and Law.* Cambridge, MA: Harvard University Press, 1987.

———. *Toward a Feminist Theory of the State.* Cambridge, MA: Harvard University Press, 1989.

Matsueda, Ross L. "Reflected Appraisals, Parental Labeling, and Delinquency: Specifying a Symbolic Interactionist Theory." *American Journal of Sociology* 97 (1992):1577–1611.

Matsueda, Ross L., and Karen Heimer. "A Symbolic Interactionist Theory of Role-Transitions, Role-Commitments, and Delinquency." In *Developmental Theories of Crime and Delinquency,* edited by Terence P. Thornberry, 163–213. New Brunswick, NJ: Transaction, 1997.

McCarthy, Bill, John Hagan, and Todd S. Woodward. "In the Company of Women: Structure and Agency in a Revised Power-Control Theory of Gender and Delinquency." *Criminology* 37 (1999):761–788.

McDermott, M. Joan. "On Moral Enterprises, Pragmatism, and Feminist Criminology." *Crime and Delinquency* 48 (2002):283–299.

Mead, George Herbert. *Mind, Self, and Society.* Chicago: University of Chicago Press, 1934.

Messerschmidt, James W. *Capitalism, Patriarchy and Crime: Toward Socialist Feminist Criminology.* Totowa, NJ: Rowman and Littlefield, 1986.

———. *Masculinities and Crime: Critique and Reconceptualization of Theory.* Lanham, MD: Rowman and Littlefield, 1993.

Mies, Maria. "Towards a Methodology for Feminist Research." In *Theories of Women's Studies,* edited by Gloria Bowles and Renate Duelli Klein, 117–139. Boston: Routledge and Kegan Paul, 1983.

Miller, Walter B. "Ideology and Criminal Justice Policy: Some Current Issues." *Journal of Criminal Law and Criminology* 64 (1973):141–162.

Morash, Merry, and Meda Chesney-Lind. "A Reformulation and Partial Test of the Power Control Theory of Delinquency." *Justice Quarterly* 8 (1991):347–377.

Naffine, Ngaire. *Feminism and Criminology.* Philadelphia: Temple University Press, 1996.

Paternoster, Raymond, and Leeann Iovanni. "The Labeling Perspective and Delinquency: An Elaboration of the Theory and an Assessment of the Evidence." *Justice Quarterly* 6 (1989):359–394.

President's Commission on Law Enforcement and Administration of Justice. *The Challenge of Crime in a Free Society.* Washington, DC: GPO, 1967.

———. *Task Force Report: Juvenile Delinquency and Youth Crime.* Washington, DC: GPO, 1967.

Quinney, Richard. *Class, State, and Crime.* 2nd ed. New York: Longman, 1980.

———. *Critique of Legal Order: Crime Control in a Capitalist Society.* Boston: Little, Brown, 1974.

Raines, Prudence. "Imputations of Deviance: A Retrospective Essay on the Labeling Perspective." *Social Problems* 23 (1975):1–11.

Ray, Melvin C., and William R. Downs. "An Empirical Test of Labeling Theory Using Longitudinal Data." *Journal of Research in Crime and Delinquency* 23 (1986):169–194.

Reid, Sue Titus. *Crime and Criminology.* 9th ed. Boston: McGraw-Hill, 2000.

Sampson, Robert J., and John H. Laub. *Crime in the Making: Pathways and Turning Points Through Life.* Cambridge, MA: Harvard University Press, 1993.

———. "A Life-Course Theory of Cumulative Disadvantage and the Stability of Delinquency." In *Developmental Theories of Crime and Delinquency,* edited by Terence P. Thornberry, 133–161. New Brunswick, NJ: Transaction, 1997.

Schaefer, Walter E., Carol Olexa, and Kenneth Polk. "Programmed for Social Class: Tracking in High School." In *Schools and Delinquency,* edited by Kenneth Polk and Walter E. Schaefer, 33–54. Englewood Cliffs, NJ: Prentice Hall, 1972.

Schur, Edwin M. *Labeling Deviant Behavior: Its Sociological Implications.* New York: Harper and Row, 1971.

———. *Labeling Women Deviant: Gender, Stigma, and Social Control.* Philadelphia: Temple University Press, 1983.

———. "Reactions to Deviance: A Critical Assessment." *American Journal of Sociology* 75 (1969):309–322.

Schwendinger, Julia R., and Herman Schwendinger. *Rape and Inequality.* Beverly Hills: CA: Sage, 1985.

Simon, Rita J. *Women and Crime.* Lexington, MA: Lexington Books, 1975.

Simpson, Sally S. "Feminist Theory, Crime, and Justice." *Criminology* 27 (1989):605–631.

Smart, Carol. *Women, Crime, and Criminology: A Feminist Critique.* Boston: Routledge and Kegan Paul, 1976.

Smith, Douglas A., and Robert Brame. "On the Initiation and Continuation of Delinquency." *Criminology* 32 (1994):607–629.

Smith, Douglas A., and Patrick R. Gartin. "Specifying Specific Deterrence: The Influence of Arrest on Future Criminal Activity." *American Sociological Review* 54 (1989):94–105.

Smith, Douglas A., and Raymond Paternoster. "The Gender Gap in Theories of Deviance: Issues and Evidence." *Journal of Research in Crime and Delinquency* 24 (1987):140–172.

Stanko, Elizabeth Anne. *Intimate Intrusions: Women's Experiences of Male Violence.* Boston: Routledge and Kegan Paul, 1985.

Steffensmeier, Darrell. "National Trends in Female Arrests, 1960–1990: Assessments and Recommendations for Research." *Journal of Quantitative Criminology* 9 (1993):411–440.

Steffensmeier, Darrell, and Emilie Allan. "Gender and Crime: Toward a Gendered Theory of Female Offending." *Annual Review of Sociology* 22 (1996):459–487.

Steffensmeier, Darrell, and Cathy Streifel. "Time-Series Analysis of the Female Percentage of Arrests for Property Crimes, 1960–1985: A Test of Alternative Explanations." *Justice Quarterly* 9 (1992):77–103.

Stewart, Eric A., Ronald L. Simons, Rand D. Conger, and Laura V. Scaramella. "Beyond the Interactional Relationship Between Delinquency and Parenting Practices: The Contribution of Legal Sanctions." *Journal of Research in Crime and Delinquency* 39 (2002):36–59.

Stiles, Beverly L., and Howard B. Kaplan. "Stigma, Deviance, and Negative Social Sanctions." *Social Science Quarterly* 77 (1996):685–696.

Tannenbaum, Frank. *Crime and the Community.* New York: Columbia University Press, 1938.

Teilmann, Katherine S., and Pierre H. Landry, Jr. "Gender Bias in Juvenile Justice." *Journal of Research in Crime and Delinquency* 18 (1981):47–80.

Thorsell, Bernard A., and Lloyd D. Klemke. "The Labeling Process: Reinforcement and Deterrent?" *Law and Society Review* 6 (1972):393–403.

Tittle, Charles R. "Labelling and Crime: An Empirical Evaluation." In *The Labelling of Deviance,* 2nd ed., edited by Walter R. Gove, 241–263. Beverly Hills, CA: Sage, 1980.

Tittle, Charles R., Jason Bratton, and Marc G. Gertz. "A Test of a Micro-Level Application of Shaming Theory." *Social Problems* 50 (2003):592–617.

Triplett, Ruth A., and G. Roger Jarjoura. "Theoretical and Empirical Specification of a Model of Informal Labeling." *Journal of Quantitative Criminology* 10 (1994):241–276.

Uggen, Christopher. "Class, Gender, and Arrest: An Intergenerational Analysis of Workplace Power and Control." *Criminology* 38 (2000):835–862.

———. "Criminology and the Sociology of Deviance." *The Criminologist* 28 (2003):1–5.

Ward, David A., and Charles R. Tittle. "Deterrence or Labeling: The Effects of Informal Sanctions." *Deviant Behavior* 14 (1993):43–64.

Wellford, Charles F. "Labeling Theory and Criminology: An Assessment." *Social Problems* 22 (1975):332–345.

Wellford, Charles F., and Ruth Triplett. "The Future of Labeling Theory: Foundations and Promises." In *New Directions in Criminological Theory* (Advances in Criminological Theory, vol. 4), edited by Freda Adler and William S. Laufer, 1–22. New Brunswick, NJ: Transaction, 1993.

West, Donald J., and David P. Farrington. *Who Becomes Delinquent?* London: Heinemann, 1973.

Westermann, Ted D., and James W. Burfeind. *Crime and Justice in Two Societies: Japan and the United States.* Pacific Grove, CA: Brooks/Cole, 1991.

Zhang, Lening, and Sheldon Zhang. "Reintegrative Shaming and Predatory Delinquency." *Journal of Research in Crime and Delinquency* 41 (2004):433–453.

ENDNOTES

1. Schur, *Labeling Deviant Behavior,* 7.
2. Greenberg, "Age Structure of Society;" Colvin and Pauly, "Critique of Criminology;" Hagan, Gillis, and Simpson, "Class Structure;" Hagan, Simpson, and Gillis, "Class in the Household;" and Hagan, *Structural Criminology.*
3. Schur, *Labeling Deviant Behavior,* 2.
4. Paternoster and Iovanni, "Labeling Perspective and Delinquency," 361–363; and Schur, *Labeling Deviant Behavior,* 8.
5. Paternoster and Iovanni, "Labeling Perspective and Delinquency," 361–363.
6. This section is based in part on the discussion of labeling theory presented in Empey, Stafford and Hay, *American Delinquency,* 257–263.
7. Tannenbaum, *Crime and the Community,* 17.
8. Ibid., 17–18.
9. Ibid., 19–20.
10. Ibid., 20.
11. Ibid., 20.
12. Lemert, *Social Pathology.* See also Lemert, *Human Deviance.*
13. Lemert, *Social Pathology,* 75; and Lemert, *Human Deviance,* 47–48, 62.
14. Lemert, *Social Pathology,* 75–76; and Lemert, *Human Deviance,* 62–63.
15. Lemert, *Human Deviance,* 62–63.
16. Lemert, *Social Pathology,* 75–76; and Lemert, *Human Deviance,* 47–48, 63.
17. Lemert, *Human Deviance,* 63.
18. Lemert, *Social Pathology,* 77.
19. Goffman, *Asylums,* and *Stigma*; Kitsuse, "Societal Reaction;" Erikson, "Notes," and *Wayward Puritans*; Becker, *Outsiders*; and Schur, "Reactions to Deviance," and *Labeling Deviant Behavior.* Cullen and Agnew, *Criminological Theory,* 296; Raines, "Imputations of Deviance," 1; and Lemert, *Human Deviance,* 15, offer lists of the primary works that defined "labeling theory" in the 1960s.
20. Beirne and Messerschmidt, *Criminology,* 178; and Cullen and Agnew, *Criminological Theory,* 298.
21. Cullen and Agnew, *Criminological Theory,* 298.
22. Becker, *Outsiders,* 9. Emphasis in original.
23. Erikson, "Notes," 308. Emphasis in original.
24. Becker, *Outsiders.* Empey, Stafford, and Hay (*American Delinquency,* 261) point out this aspect of Becker's work.
25. Becker, *Outsiders,* 147–163.
26. See also Schur, who considers the issue of "rules, values, and moral crusades" (*Labeling Deviant Behavior,* Chapter 5).
27. Becker, *Outsiders,* 149.
28. Ibid., 12.
29. Ibid., 12–13.
30. Becker, *Outsiders,* 161. See also Matsueda, "Reflected Appraisals," 1588.

31. Matsueda, "Reflected Appraisals;" Heimer and Matsueda, "Role-Taking;" Bartusch and Matsueda, "Gender;" and Matsueda and Heimer, "Symbolic Interactionist Theory."

32. Matsueda, "Reflected Appraisals," 1580.

33. Mead, *Mind, Self, and Society.*

34. Matsueda, "Reflected Appraisals;" Heimer and Matsueda, "Role-Taking;" and Bartusch and Matsueda, "Gender."

35. Matsueda, "Reflected Appraisals," 1584–1587.

36. Ibid., 1582.

37. Ibid., 1596–1597.

38. Ibid., 1601.

39. Matsueda's initial analyses included only males from the National Youth Survey sample. With his colleague, Dawn Jeglum Bartusch, Matsueda explored how well his interactionist theory explained female delinquency (Bartusch and Matsueda, "Gender"). They found a similar pattern of results for males and females. Despite this similarity, Bartusch and Matsueda found that their model of informal labeling and reflected appraisals was able to account for a substantial portion of the gender gap in delinquency. See also Ray and Downs ("Empirical Test"), who examine gender differences in the labeling process, and Schur (*Labeling Women Deviant*), who explores the process through which women are labeled deviant.

40. Heimer and Matsueda, "Role-Taking."

41. Ibid., 378–380.

42. Ibid., 378.

43. Goffman, *Stigma,* 3. See also Link and Phelan, "Conceptualizing Stigma."

44. Goffman, *Stigma.* See also Stiles and Kaplan, "Stigma."

45. Hagan, "Social Embeddedness." See also Uggen, "Criminology."

46. Ward and Tittle, "Deterrence or Labeling."

47. For example, Stewart et al., "Beyond the Interactional Relationship;" Liu, "Conditional Effect;" De Li, "Legal Sanctions;" Adams, Johnson, and Evans, "Racial Differences;" Adams and Evans, "Teacher Disapproval;" Bartusch and Matsueda, "Gender;" Heimer and Matsueda, "Role-Taking;" Smith and Brame, "Initiation and Continuation;" Ward and Tittle, "Deterrence or Labeling;" Matsueda, "Reflected Appraisals;" Kaplan and Johnson, "Negative Social Sanctions;" Hagan and Palloni, "Social Reproduction;" and Smith and Gartin, "Specifying Specific Deterrence."

48. For example, see Schur, *Labeling Deviant Behavior,* 82–99.

49. See Kaplan and Johnson, "Negative Social Sanctions," for a brief review of research on the effects of formal sanctions on subsequent offending.

50. Smith and Gartin, "Specifying Specific Deterrence," 94.

51. Ibid.

52. For example, Hagan and Palloni, "Social Reproduction;" Kaplan and Johnson, "Negative Social Sanctions;" De Li, "Legal Sanctions;" Stewart et al., "Beyond the Interactional Relationship."

53. De Li, "Legal Sanctions."

54. Ibid., 391.

55. Sampson and Laub, "Life-Course Theory." See also Sampson and Laub, *Crime in the Making.*

56. Sampson and Laub, "Life-Course Theory," 147.

57. Ibid.

58. For example, De Li, "Legal Sanctions;" and Stewart et al., "Beyond the Interactional Relationship." See also Hagan and Palloni, "Social Reproduction."

59. Wellford and Triplett, "Future of Labeling Theory;" and Triplett and Jarjoura, "Theoretical and Empirical Specification."

60. Matsueda, "Reflected Appraisals;" Heimer and Matsueda, "Role-Taking;" and Bartusch and Matsueda, "Gender."

61. For example, Adams et al., "Labeling and Delinquency;" Liu, "Conditional Effect;" Adams, Johnson, and Evans, "Racial Differences;" Adams and Evans, "Teacher Disapproval;" Smith and Brame, "Initiation and Continuation;" and Ward and Tittle, "Deterrence or Labeling."

62. Liu, "Conditional Effect;" and Smith and Brame, "Initiation and Continuation."

63. Smith and Brame, "Initiation and Continuation."

64. Ibid., 624.

65. Adams and Evans, "Teacher Disapproval."

66. Ibid., 209.

67. Adams, Johnson, and Evans, "Racial Differences."

68. Jensen, "Delinquency," and Harris, "Imprisonment," cited in Adams, Johnson, and Evans, "Racial Differences," 158–159.

69. Adams, Johnson, and Evans, "Racial Differences."

70. See Tittle, "Labelling and Crime."

71. Tittle, "Labelling and Crime;" Gove, "Labelling Perspective;" Wellford, "Labeling Theory and Criminology." See also Schur's discussion of the criticisms and misunderstandings of the labeling perspective (*Labeling Deviant Behavior*, 13–23).

72. Cullen and Agnew, *Criminological Theory*, 299. Citation omitted.

73. Paternoster and Iovanni, "Labeling Perspective and Delinquency."

74. Reid, *Crime and Criminology*, 175; and Paternoster and Iovanni, "Labeling Perspective and Delinquency."

75. See Schur, *Labeling Deviant Behavior*, 17–23.

76. Reid, *Crime and Criminology*, 175.

77. Thorsell and Klemke, "Labeling Process," cited in Reid, *Crime and Criminology*. See Reid, *Crime and Criminology*, 175; and Paternoster and Iovanni, "Labeling Perspective and Delinquency."

78. Cullen and Agnew, *Criminological Theory*, 301.

79. Ibid.

80. See Cullen and Agnew, *Criminological Theory*, 299.

81. Smith and Brame, "Initiation and Continuation," 624.

82. Braithwaite, *Crime, Shame and Reintegration*; Matsueda, "Reflected Appraisals;" and Heimer and Matsueda, "Role-Taking."

83. Sampson and Laub, *Crime in the Making*, and "Life-Course Theory."

84. Braithwaite, *Crime, Shame and Reintegration*.

85. This discussion of Braithwaite's theory of reintegrative shaming is similar to that provided by Empey, Stafford, and Hay in *American Delinquency* (310–311).

86. Braithwaite, *Crime, Shame and Reintegration*, 5.

87. Ibid., 100.

88. Ibid., 100–101.

89. Ibid., Chapter 5, especially pages 80–83.

90. Ibid., 81.

91. Ibid., 81–82.

92. Ibid., 101.

93. Ibid., 98, 100.

94. In empirical research by Tittle, Bratton, and Gertz ("Test"), Braithwaite's theory received partial support. Zhang and Zhang ("Reintegrative Shaming") also found some support for Braithwaite's theory, but not for the key hypothesis that reintegrative shaming decreases subsequent delinquency.

95. Westermann and Burfeind, *Crime and Justice in Two Societies*.

96. Schur, *Labeling Deviant Behavior*, 148–154.

97. Becker, *Outsiders*.

98. Farrell and Swigert, "Prior Offense Record," and "Hierarchical Systems Theory;" and Becker, *Outsiders*.

99. Cullen and Agnew, *Criminological Theory*, 333.

100. Barak, "Time," 35. See also Friedrichs, who notes that, beginning in the late 1980s, "critical criminology" became the favored term ("New Directions," 80).

101. Cullen and Agnew, *Criminological Theory*, 334–335.

102. Ibid., 334.

103. Ibid.

104. Ibid.

105. Ibid.

106. Ibid., 335.

107. Bonger, *Criminality and Economic Conditions*.

108. Definition derived from Andersen and Taylor, *Sociology* (479–480), and Miller, "Ideology" (142).

109. Bonger, *Criminality and Economic Conditions*, 48.

110. Ibid., 142.

111. Ibid.

112. Ibid.

113. Quinney, *Critique of Legal Order*, and *Class, State, and Crime*.

114. Quinney, *Critique of Legal Order*, 16.

115. *Left realists* also responded critically to the view of crime proposed by some critical criminologists. Left realism is also a critical approach, but it takes a different view of crime than that offered by instrumental Marxists. Curran and Renzetti offer an excellent description of left realism: Left realists believe that "in its zeal to highlight the crimes of the powerful, radical criminology has historically romanticized crimes committed by the poor and working class as crimes of the down-trodden struggling for survival. . . . Most crime, left realists argue, is not a mode of rebellion against the oppression of capitalism, nor is the criminal 'the champion of the underprivileged' whose goal is to redistribute private property. Most offenders, though often poor, are not motivated to commit crimes to secure the necessities of life. Instead, 'they crave luxuries.' . . . Thus, while left realists are correct that the wealthy do commit crimes and that their crimes are often more costly and more injurious than street crime, left realists point out that the poor and working class, as well as people of color, are more likely to be victimized not only by crimes of the powerful, but also by street crime. . . . In short, left realists want radical criminologists to undertake a careful analysis of the causes and consequences of street crime" (*Theories of Crime*, 197–199).

116. Chambliss and Seidman, *Law, Order, and Power*; and Colvin and Pauly, "Critique of Criminology."

117. Curran and Renzetti, *Theories of Crime*, 194.

118. Ibid., 194.

119. Lynch and Groves, *Primer*; and Curran and Renzetti, *Theories of Crime*, 195.

120. Lynch and Groves, *Primer*.

121. Colvin and Pauly, "Critique of Criminology," 513.

122. Ibid.

123. Edwards, *Contested Terrain*.

124. Colvin and Pauly, "Critique of Criminology," 532.

125. Ibid.

126. Ibid., 532–533.

127. Ibid., 534.

128. Kohn, *Class and Conformity*; and Colvin and Pauly, "Critique of Criminology," 534–535.

129. Colvin and Pauly, "Critique of Criminology," 535.

130. Ibid., 536.

131. For example, Hirschi, *Causes of Delinquency*; and West and Farrington, *Who Becomes Delinquent*; Loeber and Stouthamer-Loeber, "Family Factors;" and Sampson and Laub, *Crime in the Making*.

132. Colvin and Pauly, "Critique of Criminology," 537.

133. Ibid.

134. Ibid., 537–538.

135. Ibid., 538.

136. Ibid.

137. For example, Hirschi, *Causes of Delinquency*; Schaefer, Olexa, and Polk, "Programmed for Social Class;" and Hirschi and Hindelang, "Intelligence and Delinquency."

138. Colvin and Pauly, "Critique of Criminology," 539–540.

139. Ibid., 515.

140. Hagan, Gillis, and Simpson, "Class Structure;" and Hagan, Simpson, and Gillis, "Class in the Household."

141. Hagan, Gillis, and Simpson, "Class Structure," 1161–1162.

142. Hagan, Simpson, and Gillis, "Class in the Household," 794.

143. Hagan, Gillis, and Simpson, "Class Structure;" Hagan, Simpson, and Gillis, "Class in the Household;" and Hagan, *Structural Criminology*.

144. Hagan, Gillis, and Simpson, "Class Structure;" and Hagan, Simpson, and Gillis, "Class in the Household."

145. John Macionis defines an ideal type as "an abstract statement of the essential characteristics of any social phenomenon" (*Sociology*, 102). In this sense, "ideal" does not mean good or best. Rather, ideal types are descriptive devices for characterizing and categorizing elements of the social world, such as families.

146. Hagan, Simpson, and Gillis, "Class in the Household." Hagan, Simpson, and Gillis consider female-headed households to be "a special kind of egalitarian family" because they are characterized by "freedom from male domination" and an absence of power imbalances between

parents (793). This categorization of female-headed households has drawn criticism from scholars who object to the equating of "upper-status" egalitarian families with female-headed households that are often characterized by poverty and economic deprivation (Chesney-Lind and Shelden, *Girls,* 120).

147. Hagan, Gillis, and Simpson, "Class Structure," 1156; and Hagan, Simpson, and Gillis, "Class in the Household," 792.
148. Hagan, Gillis, and Simpson, "Class Structure," 1156.
149. Hagan, Simpson, and Gillis, "Class in the Household," 792.
150. Ibid., 793.
151. Ibid.
152. Hagan, Gillis, and Simpson, "Class Structure," 1156.
153. Hagan, Simpson, and Gillis, "Class in the Household," 793.
154. Hagan, Gillis, and Simpson, "Class Structure," 1174.
155. Hagan, Gillis, and Simpson, "Class Structure," and "Clarifying;" Hagan, Simpson, and Gillis, "Class in the Household;" Blackwell, "Perceived Sanction Threats;" and Uggen, "Class, Gender, and Arrest." See also Grasmick et al., "Risk Preferences and Patriarchy;" McCarthy, Hagan, and Woodward, "Company of Women;" and Blackwell, "Perceived Sanction Threats," for revisions and extensions of power-control theory.
156. Uggen, "Class, Gender, and Arrest."
157. For example, Jensen and Thompson, "Further Examination;" and Morash and Chesney-Lind, "Reformulation."
158. Chesney-Lind and Shelden argue that the limited definition of patriarchal control found in power-control theory leads Hagan and his colleagues to argue that "a mother's working outside the home leads to increases in a daughter's delinquency because the daughter finds herself in a more 'egalitarian family'—a family less likely to supervise the female children. . . . This is essentially a not-too-subtle variation on the 'liberation' hypothesis. Now, mother's liberation or employment causes daughter's crime" (Chesney-Lind and Shelden, *Girls,* 120). (See also Chesney-Lind, *Female Offender,* 21–22.) Chesney-Lind and Shelden point out that no evidence supports the idea that girls' delinquency has increased as women's labor force participation has increased.
159. Greenberg, "Age Structure of Society."
160. Ibid., 118–119.
161. Ibid., 121.
162. Ibid., 122.
163. Ibid., 123.
164. Ibid.
165. Ibid., 125.
166. Ibid.
167. Ibid., 126.
168. Greenberg argues that, compared to middle-class youth, upper-class adolescents may also be less concerned with conformity to school requirements because "their inherited class position frees them from the necessity of doing well in school to guarantee their future economic status" ("Age Structure of Society," 126).
169. Greenberg, "Age Structure of Society," 127.
170. Ibid., 128–130.
171. Ibid., 131.
172. Ibid.
173. Ibid.
174. Cullen and Agnew, *Criminological Theory,* 397.
175. For example, Naffine, *Feminism and Criminology.*
176. Daly and Chesney-Lind, "Feminism and Criminology," 502. Emphasis in original. See also Simpson, "Feminist Theory," 606; Curran and Renzetti, *Theories of Crime,* 210–211; and McDermott, "On Moral Enterprises."
177. Simpson, "Feminist Theory," 606.
178. Daly and Chesney-Lind, "Feminism and Criminology," 504.
179. Curran and Renzetti, *Theories of Crime,* 210.
180. See Simpson, "Feminist Theory," 608–609.
181. Miles, "Toward a Methodology," 121, quoted in Simpson, "Feminist Theory," 609.

182. This section is based on Daly and Chesney-Lind's discussion of early feminist critiques ("Feminism and Criminology," 508–510).

183. Klein, "Etiology of Female Crime;" and Smart, *Women, Crime, and Criminology*.

184. Klein, "Etiology of Female Crime," 4.

185. Ibid., 4. Emphasis in original.

186. Chesney-Lind, "Judicial Enforcement," and "Judicial Paternalism," "Girls' Crime," and *Female Offender*; and Chesney-Lind and Shelden, *Girls*. See Belknap, "Mother of Feminist Criminology," for a biography of Chesney-Lind and a discussion of her significant influence on the field of criminology.

187. Chesney-Lind, "Judicial Enforcement," "Judicial Paternalism," and "Girls' Crime."

188. Chesney-Lind, "Judicial Paternalism;" Teilmann and Landry, "Gender Bias;" and Canter, "Sex Differences."

189. Chesney-Lind, "Judicial Paternalism;" and MacDonald and Chesney-Lind, "Multiyear Analysis."

190. Chesney-Lind, "Judicial Paternalism," 122–123.

191. Chesney-Lind, "Judicial Enforcement," 56.

192. Daly and Chesney-Lind, "Feminism and Criminology," 509.

193. Ibid.

194. Simpson, "Feminist Theory," 607.

195. Ibid.

196. Adler, *Sisters in Crime*; and Simon, *Women and Crime*.

197. Adler, *Sisters in Crime*.

198. Simon, *Women and Crime*.

199. See, for example, Steffensmeier and Streifel, "Time-Series Analysis;" Steffensmeier, "National Trends;" and Steffensmeier and Allan, "Gender and Crime."

200. Daly and Chesney-Lind, "Feminism and Criminology," 511.

201. MacKinnon, "Agenda for Theory," "Toward Feminist Jurisprudence," *Feminism Unmodified,* and *Toward a Feminist Theory*. See also Stanko, *Intimate Intrusions*.

202. MacKinnon, *Toward a Feminist Theory*.

203. Curran and Renzetti, *Theories of Crime,* 220.

204. Chesney-Lind and Shelden, *Girls*; and Chesney-Lind, *Female Offender*.

205. Chesney-Lind, *Female Offender,* Chapter 2; and Chesney-Lind and Shelden, *Girls,* Chapter 6. In a recent study of 22 girls incarcerated in a women's prison, Gaarder and Belknap ("Tenuous Borders") also found that these girls reported lives characterized by violence and victimization.

206. Chesney-Lind, *Female Offender,* 26.

207. Chesney-Lind and Shelden, *Girls,* 115.

208. Chesney-Lind, *Female Offender,* and "Challenging Girls' Invisibility."

209. Daly and Chesney-Lind, "Feminism and Criminology."

210. Chesney-Lind, *Female Offender,* 31.

211. See, for example, Smith and Paternoster, "Gender Gap;" and Steffensmeier and Allan, "Gender and Crime."

212. Cullen and Agnew, *Criminological Theory,* 401.

213. Chesney-Lind, *Female Offender,* 4.

214. Daly and Chesney-Lind, "Feminism and Criminology;" and MacKinnon, *Toward a Feminist Theory*.

215. Chesney-Lind, "Judicial Enforcement," "Judicial Paternalism," *Female Offender,* and "Challenging Girls' Invisibility;" Chesney-Lind and Shelden, *Girls*; and MacDonald and Chesney-Lind, "Multiyear Analysis."

216. Chesney-Lind, *Female Offender*; and Chesney-Lind and Shelden, *Girls*.

217. Daly and Chesney-Lind, "Feminism and Criminology," 511.

218. Beirne and Messerschmidt, *Criminology,* 207–208.

219. Simpson, "Feminist Theory," 607.

220. Andersen and Taylor, *Sociology,* 421.

221. Ibid.

222. Curran and Renzetti, *Theories of Crime,* 223–224.

223. Messerschmidt, *Capitalism, Patriarchy and Crime*.

224. Messerschmidt, *Capitalism, Patriarchy and Crime,* and *Masculinities and Crime*.

225. Messerschmidt, *Capitalism, Patriarchy and Crime*.

226. Chesney-Lind, *Female Offender.*
227. This section is based in part on Beirne and Messerschmidt's discussion of Marxist feminism (*Criminology,* 205–206).
228. Simpson, "Feminist Theory," 607; and Curran and Renzetti, *Theories of Crime,* 223.
229. Beirne and Messerschmidt, *Criminology,* 205–206.
230. Balkan, Berger, and Schmidt, *Crime and Deviance.*
231. Schwendinger and Schwendinger, *Rape and Inequality.*

Understanding Delinquency

14

Chapter Objectives

After completing this chapter, students should be able to:

- Identify and explain the key theoretical concepts for explaining delinquency presented in this book.
- Analyze Rick's story in terms of these key theoretical concepts.
- Describe the goals of integrated theories of delinquency and the debate surrounding integrated theories.
- Describe specific integrated theories of delinquent behavior.
- Understand key theories and terms:

  Theories:

  Elliott's integrated theory of delinquent behavior

  Thornberry's interactional theory

  Tittle's control balance theory

  Miethe and Meier's integrated theory of offenders, victims, and situations

Terms:
reciprocal relationships
control balancing

Rick: A "Delinquent Youth"

We begin this chapter with the same case study that opened Chapter 1. Our expectation is that, after reading the previous 13 chapters, the reader will be able to look at this case from a fresh perspective and with greater insight. Perhaps when you read this case at the beginning of the book, you found it and the questions that followed it a bit overwhelming. It is our hope that Rick's story is now less overwhelming, and that you have acquired some tools to use in making sense of it.

The Youth Court adjudicated or judged Rick, a 14-year-old, a "delinquent youth" for motor vehicle theft and placed him on formal probation for six months. He and a good friend took without permission a car that belonged to Rick's father. They were pulled over by the police for driving erratically—a classic case of joyriding.

Rick was already a familiar figure in the juvenile court. When Rick was 12, he was referred to the court for "deviant sex" for an incident in which he was caught engaging in sexual activity with a 14-year-old girl. The juvenile court dealt with this offense "informally." A probation officer met with Rick and his parents to work out an agreement of informal probation that included "conditions" or rules, but no petition in court. Not long after this first offense, Rick was taken into custody by the police for curfew violation and, on a separate occasion, vandalism—he and his good friend had gotten drunk and knocked down numerous mailboxes along a rural road. In both of these instances, Rick was taken to the police station and released to his parents. Even though Rick's first formal appearance in juvenile court was for the auto theft charge, he was already well known to the police and probation departments.

Rick was a very likable kid; he was pleasant and personable. He expressed a great deal of remorse for his delinquent acts and seemed to genuinely desire to change. He had a lot going for him; he was goal-directed, intelligent, and athletic. He interacted well with others, including his parents, teachers, and peers. His best friend, an American Indian boy who lived on a nearby reservation, was the same age as Rick and had many similar personal and social characteristics. Not surprisingly, the boy also had a very similar offense record. In fact, Rick and his friend were often "companions in crime," committing many of their delinquent acts together.

Rick was the adopted son of older parents who loved him greatly and saw much ability and potential in him. They were truly perplexed by the trouble he was in, and they struggled to understand why Rick engaged in delinquent acts and what needed to be done about it. Rick, too, seemed to really care about his parents. He spent a good deal of time with them and apparently enjoyed their company. Because Rick was adopted as an infant, these parents were the people he considered family.

Rick attended school regularly and earned good grades. He was not disruptive in the classroom or elsewhere in the school. In fact, teachers reported that he was a very positive student both in and out of class and that he was academically motivated. He did his homework and handed in assignments on time. He was also actively involved in sports—football, wrestling, and track and field.

Rick's six months of formal probation for auto theft turned into a two-year period as he continued to get involved in delinquent acts. Through regular meetings and enforcement of probation conditions, his probation officer tried to work with Rick to break his pattern of delinquency. Such efforts were to no avail. Rick continued to offend, resulting in an almost routine series of court hearings that led to the extension of his probation supervision period. The continuing pattern of delinquency included a long list of property and status offenses: minor in possession of alcohol, numerous curfew violations, continued vandalism, minor theft (primarily shoplifting), and continued auto theft, usually involving joyrides in his father's car.

Rick's "final" offense was criminal mischief, and it involved extensive destruction of property. Once again, Rick and his best friend "borrowed" his father's car, got drunk, and drove to Edina, an affluent suburb of Minneapolis. For no apparent reason, they parked the car and began to walk along France Avenue, a major road with office buildings along each side. After walking a while, they started throwing small rocks toward buildings, seeing how close they could get. Their range increased quickly and the rocks soon reached their targets, breaking numerous windows. The "fun" turned into thousands of dollars worth of window breakage in a large number of office buildings.

Because of the scale of damage, Rick faced the possibility of being placed in a state training school. As a potential loss of liberty case, Rick was provided with representation by an attorney. This time, the juvenile court's adjudication process followed formal procedures, including involvement of a prosecutor and a defense attorney. In the preliminary hearing, Rick admitted to the petition (statement of charges against him), and the case was continued to a later date for disposition (sentencing). In the meantime, the judge ordered a predisposition report.

The predisposition report is designed to individualize the court's disposition to "fit the offender." The investigation for the report uses multiple sources of information, including information from the arresting officer, parents, school personnel, coaches, employers, friends, relatives, and, most importantly, the offending youth. The predisposition report tries to describe and explain the pattern of delinquency and then offer recommendations for disposition based on the investigation. In Rick's case, the predisposition report attempted to accurately describe and explain his persistent pattern of property and status offending, and it offered a recommendation for disposition. Finding no information to justify otherwise, the probation officer recommended that Rick be committed to the Department of Corrections for placement at the Red Wing State Training School. Depending on one's viewpoint, the state training school represented either a last ditch effort for rehabilitation or a means of punishment through restricted freedom. Either way, Rick was viewed as a chronic juvenile offender, with little hope for reform.

We ask you to consider again the questions we posed in Chapter 1:

- Is involvement in delinquency common among adolescents—that is, are most youths delinquent? Maybe Rick was just an unfortunate kid who got caught. (See Chapter 4.)
- Are Rick's offenses fairly typical of the types of offenses in which youths are involved? (See Chapter 4.)
- Will Rick "grow out" of delinquent behavior? (See Chapters 5 and 6.)
- Is Rick's pattern of offending much the same as those of other delinquent youths? (See Chapter 6.)
- Do most delinquent youths begin with status offenses and then persist and escalate into serious, repetitive offending? (See Chapter 6.)
- Is there a rational component to Rick's delinquency so that punishment by the juvenile court would deter further delinquency? (See Chapter 7.)
- Did the fact that Rick was adopted have anything to do with his involvement in delinquency? Might something about Rick's genetic makeup and his biological family lend some insight into his behavior? (See Chapter 8.)
- What role did Rick's use of alcohol play in his delinquency? (See Chapter 9.)
- Are there family factors that might relate to Rick's involvement in delinquency? (See Chapter 10.)
- Were there aspects of Rick's school experiences that might be related to his delinquency? (See Chapters 10 and 12.)
- What role did Rick's friend play in his delinquent behavior? (See Chapter 11.)
- Did the youth court's formal adjudication of Rick as a "delinquent youth" two years earlier label him and make him more likely to continue in delinquent behavior? (See Chapter 13.)
- Should the juvenile court retain jurisdiction for serious, repeat offenders like Rick? (See Chapter 15.)

- What should the juvenile court try to do with Rick: punish, deter, or rehabilitate him? (See Chapter 15.)
- Should the juvenile court hold Rick less responsible for his acts than an adult because he has not fully matured? (See Chapter 15.)

We will refer to Rick's case frequently in this chapter as we review the nature of offenses, offenders, and offending and the theoretical explanations for delinquency presented in previous chapters.

In this chapter, we summarize the key theoretical concepts discussed in previous chapters and attempt to "put the pieces together." Our emphasis here is on *integration*—the integration of theory and research, and the integration of various elements of different theories.

Our goal in this book has been to draw together the primary theories of delinquent behavior and the relevant empirical research that has tested those theories. In Chapter 1, we described the integral relationship between theory and research. Throughout the book, we have woven together these two facets of the discipline of criminology by presenting major theories of delinquency along with key research findings based on those theories. Theories of delinquency attempt to identify the key causal factors that lead to delinquent behavior and to describe how these factors are interrelated in producing delinquency. The success of theories in accomplishing these goals depends on the outcomes of research designed to assess the validity of theory—the degree to which it accurately and adequately explains delinquent behavior. Research is vital to both testing and developing theory, as it provides empirical observations related to delinquency.

In Chapter 3, we explored various research methods (case studies, ethnography, ecological analysis, and survey analysis) and sources of data ("official data," victimization surveys, and self-report surveys) used to examine the nature and extent of delinquency and to test and develop theories of delinquent behavior. An understanding of delinquency builds on both the explanations offered in theories and the findings revealed in research.

For individuals exploring the nature and causes of delinquent behavior, the task of evaluating theories and developing a real understanding of delinquent behavior is daunting. Rand Conger wrote that the most important task for students of delinquent behavior is to sort through different theories of delinquency to determine "(1) the degree to which they are different or similar; (2) the extent to which their seeming differences are really a result of addressing different questions; (3) which theories or parts thereof can be set aside as empirically refuted; and finally, (4) to what degree those aspects of the different models which appear to have empirical support can be synthesized into a general theory."[1] This is no small challenge, but it is *your* challenge now that you have examined the various theories presented in previous chapters.

We will consider the synthesis of theories about which Conger wrote later in this chapter, when we discuss integrated theories of delinquency. Before turning our attention to integrated approaches, we first summarize the major theories of delinquency presented in this book and what we have learned from key empirical tests of these theories. In doing so, we note how these theories and the research related to them inform our understanding of Rick's story. This review is a logical introduction to integrated theories, because the separate theories presented in previous chapters have provided the building blocks for integrated theoretical approaches.

◼ Key Theoretical Concepts for Explaining Delinquency

Before presenting theoretical explanations for delinquent behavior, we first defined juvenile delinquency by describing how the concept was *socially constructed* (Chapter 2). After tracing the historical origins and recent transformations of this concept, we concluded that delinquency is both a social and a legal construct, subject to changes in society. As such, juvenile delinquency is a status determined both by age and behavior, but this social and legal determination has changed tremendously over time. The original juvenile court was more concerned with the offender than with the offense. It did not matter whether the child was dependent, neglected, or delinquent; all types were within the purview of the child-saving efforts of the juvenile court. Contemporary legal definitions of juvenile delinquency distinguish among dependent and neglected children; status offenders; delinquent youth who violate the criminal code; and serious, violent offenders. Greater emphasis is now given to offender accountability, public safety, and offender competency development.

The Nature of Delinquency

After defining what we meant by delinquency, we attempted to provide readers with an understanding of the nature of delinquent behavior by examining three distinct dimensions of delinquency: offenses, offenders, and patterns of offending.

Extent of Offenses

In Chapter 4, we presented data on the extent of delinquent offenses in contemporary America, to provide readers with an accurate sense of the volume of and recent trends in juvenile offending. Self-report data indicate that a substantial portion of adolescents report involvement in delinquent acts, but that relatively few adolescents report *frequent* or *repetitive* involvement in delinquency. Data also indicate that minor forms of delinquent behavior, such as vandalism, minor theft, and simple assault, are far more common than serious, violent offenses. Data suggest that juvenile property crimes fluctuated somewhat during the 1990s but did not increase substantially. In fact, arrest data even suggest a decline in juvenile property crimes in the late 1990s. Juvenile violent crimes increased in the late 1980s and into the 1990s but have declined since 1995.

Rick's Offenses

Let's think for a moment about Rick's offenses, described at the beginning of this chapter. The data presented in Chapter 4 suggest that involvement in minor delinquency is common among adolescents. Rick's fairly frequent, repeated involvement in delinquency, however, is not typical of adolescent offending. Rick's frequency of offending is unusually high, even among adolescents. Some of the offenses in which Rick engaged are common among adolescent offenders: alcohol use, curfew violations, vandalism, and minor theft. Some of Rick's offenses, though, such as auto theft and extensive destruction of property, are more serious than those of most juvenile offenders.

Social Correlates of Delinquency

In Chapter 5, we asked, "Who are delinquent offenders?" What are the social characteristics that tend to distinguish offenders from non-offenders? We examined age, gender, race, and social class as social correlates of offending. Research shows that age and gender are strongly and consistently related to involvement in crime, which tends to be the domain of young males. The relationships between race and offending, and between social class and offending are less straightforward. But research tends to show that African Americans and those who are economically disadvantaged are more likely than others to be involved in *serious* crime. These relationships, however, do not hold for *minor* forms of offending.

Rick's Age and Gender

Rick's case study tells us nothing about his race or social class. We do know that he was a male who began his delinquent career at age 12, by engaging in sex with another minor. In terms of age and gender, Rick is similar to most delinquent youths. In Chapter 5, we noted that males are more likely than females to engage in delinquency, especially violent acts. We also described the relationship between age and crime. For most offenders, mid-adolescence to early adulthood is the peak period of offending. Though we do not know about Rick's desistance from crime, we know that his initiation into delinquency at age 12 and escalation to fairly frequent involvement in a variety of offenses by about age 16 is consistent with research findings about the age–crime relationship.

Developmental Patterns of Offending

We began Chapter 6 by reviewing research on chronic offenders or career criminals, who are responsible for a disproportionate share of offenses. Compared to the delinquent careers of other offenders, those of chronic offenders are characterized by early ages of initial involvement in delinquency, high frequencies of offending, greater seriousness of offenses, and longer durations of criminal careers. The developmental perspective of the 1990s provides theoretical insight for understanding patterns of involvement in delinquency for both chronic and nonchronic offenders. Developmental theories suggest different patterns of offending based on age of onset of problem behaviors, level of continuity or change in problem behaviors over time, progression of seriousness in offending, generality of deviance or co-occurrence of problem behaviors, and age at desistance from offending. A relatively small proportion of offenders display problem behaviors early in the life course; show stability in antisocial behavior over time, from childhood through adulthood; engage in progressively more serious offenses; display problem behaviors other than delinquency, such as alcohol use and underachievement in school; and continue offending beyond young adulthood. The most common pattern of offending, however, shows offenders committing only one or a limited number of offenses during late adolescence or early adulthood and then desisting in young adulthood without involvement in serious offending.

Rick's Pattern of Offending

Rick's case appears to be atypical in terms of patterns of offending. Rick first came into contact with the juvenile justice system at age 12—somewhat earlier than the typical age of onset of offending, which is in middle to late adolescence.[2] His initial status offense was followed by progressively more serious delinquent acts, including minor theft and motor vehicle theft, and culminating in "criminal mischief" involving extensive destruction of property. His involvement in delinquency was stable over the 4-year period about which we are told. This stability in offending over time and escalation in seriousness of offenses is not characteristic of most offenders, who have brief, non-serious delinquent careers. Indeed, Rick is characterized at the end of the case study as a "chronic juvenile offender." As we noted in Chapter 6, only a very small proportion of all delinquents are considered chronic offenders, based on the frequency, seriousness, and duration of their offending. Rick's pattern of offending from ages 12 to 16 suggests that, as a chronic offender, he is unlikely to "grow out" of crime.

Theoretical Explanations for Delinquency

After laying the necessary foundation by exploring the nature of delinquency in terms of offenses, offenders, and patterns of offending, we turned to theoretical explanations for delinquent behavior.

Classical and Positivist Criminology

We began by considering whether delinquent behavior is chosen or determined (Chapter 7). The *classical school* of thought contends that people, characterized by will (pursuit of self-interest) and hedonism, choose their actions based on rational considerations of gains and losses, pleasure and pain. Contemporary rational choice and deterrence theories are based on this idea and suggest that crime can be deterred through certainty and severity of punishment, which influence individuals' perceptions of risk. Research support for this deterrent effect, however, is modest.

Positivist criminology proposes a profoundly different view of involvement in crime. Using concepts of differentiation and pathology, positivist criminologists view criminals and delinquents as fundamentally different from the average person. The scientific approach advanced by positivist criminology is based on determinism—the assumption that crime and delinquency are caused or determined by identifiable factors, which may be biological, psychological, or social.

Some scholars have suggested that the distinction between will and determinism may not be as great as the debate surrounding these concepts suggests.[3] For example, criminal involvement may be, in part, a matter of choice, but may also be influenced by deterministic forces such as individual temperament and situational characteristics.

Situational Explanations

In Chapter 9, our focus was on the role of situational experiences and routine activities of adolescents in explaining delinquency. Situational explanations of delinquent behavior focus on processes operating at the moment crime occurs. They include characteristics of the current setting that motivate and provide opportunity for delinquent acts (objective content), and the perceptions and interpretations of those involved in the situation (subjective content). According to *routine activities theory*, opportunity for crime involves three key elements: motivated offenders, suitable targets, and the absence of capable guardians. The routine activities of adolescents and young adults provide situational opportunities for delinquency. Youths' daily lives tend to expose them to situations that provide opportunity and reward for deviance because their routine activities involve both time with peers (who may increase the situational potential for deviance) and unstructured, unsupervised leisure activities outside the direct control of authority figures.[4]

Was Rick's Delinquency Chosen or Determined?

How does your understanding of will, determinism, and situational inducements to crime inform your understanding of Rick's involvement in delinquency? Rick was described as an adolescent who "expressed much remorse for his delinquent acts" and "seemed to genuinely desire to change." Yet he continued to commit progressively more serious offenses at a relatively high frequency. In some respects, Rick's delinquent behavior seems more determined than chosen; he seems almost propelled toward delinquency. Formal punishment did not have the expected deterrent effect on Rick. Despite the rational calculus that formal punishment should have provided, Rick continued to offend while on probation. We are given little reason to believe that punishment by the juvenile court would deter further delinquency. Rick's offending appears to be based on something other than rational decision making.

Situational Influences on Rick's Delinquency

Perhaps situational inducements played a role in Rick's decisions to commit delinquent acts. Rick's offenses typically occurred in the company of his friend—his "companion in crime"—after the two had gotten drunk. The theories and research we considered in Chapter 9 suggest that Rick's time spent with a fellow delinquent, in the absence of parental controls (the "situation of company"); their mutual "pursuit of kicks;" and their alcohol use might all provide situational motivations to delinquency, causing Rick to behave in ways that he would not act in the absence of these factors. Research shows that situational inducements are strongest for youths whose commitment to conformity is weakest.[5] We will return to the issue of Rick's commitment to conformity.

Biological and Psychological Approaches

In Chapter 8, we considered biological and psychological explanations for delinquency. Early biological approaches to the study of crime often fell prey to biological determinism, with proponents arguing that offenders and non-offenders could be distinguished by physical characteristics, such as facial features, skull shape, and physique or body type. Although some early biological approaches explored the connection between physical and psychological characteristics, the emphasis was clearly on biology. Contemporary biological approaches, on the other hand, view delinquency as resulting from a combination of biological, psychological, and social causes. This new biosocial approach is based on the fundamental concept of nature–nurture interaction. For example, biosocial development models propose that biological factors, combined with environmental conditions, produce personality traits that are conducive to delinquent behavior.[6]

In Chapter 8, we also explored the ambiguous relationship between IQ and delinquency. On IQ tests, delinquents score, on average, about eight IQ points lower than nondelinquents. Just what IQ tests measure, however, remains an open question. IQ scores are probably most accurately viewed as a measure of academic preparedness and aptitude rather than innate intelligence. Some scholars have suggested that low IQ is indirectly related to delinquency. Low IQ places youth at risk for poor school performance and frustration, which in turn increase the likelihood of delinquent behavior.[7]

Rick's Biological and Psychological Characteristics

In the case we have been considering, we are told that Rick was adopted as an infant. Rick's adoptive parents are presumably law-abiding citizens; we are given no indication otherwise. Given the research we discussed in Chapter 8, we might wonder, then, about the criminal histories of Rick's biological parents. Recent research on genetic influences on crime, or the heritability of crime, has relied primarily on two research designs: adoption and twin studies. Both types of studies suggest some genetic component to antisocial behavior. In Rick's case, we are given no information about his biological parents. But adoption studies suggest that the criminal behaviors of adopted children are more closely related to the criminal behaviors of their biological parents than to those of their adoptive parents.

Rick's case study also notes that he was a "very likable kid," interacted well with others, and was goal-directed and intelligent. Some researchers have suggested that "agreeableness" and "conscientiousness" are the two dimensions of personality most strongly linked to antisocial behavior. Antisocial individuals tend to be low on both dimensions.[8] Joshua Miller and Donald Lynam describe those low in agreeableness as "hostile, self-centered, spiteful, jealous, and indifferent to others," and those low in conscientiousness as tending to "lack ambition, motivation, and perseverance" and to "have difficulty controlling their impulses." These characteristics are not consistent with the description of Rick we are given. Rather than being hostile and indifferent, Rick is "likable" and interacts well with others. He is also academically motivated and appears to try hard to do well in school. Consistent with Miller and Lynam's description, however, Rick appears to be rather impulsive, given some of the offenses he commits. Contrary to what we might expect based on research on IQ and delinquency (which suggests links among low IQ, poor school performance, and delinquent behavior), Rick is "intelligent" and does well in school.

Social Control Theories

Some of the most influential theories in criminology focus on the roles of families and peers in accounting for delinquency. In Chapter 10, we presented control theories, which emphasize the importance of social relationships, especially within the family, in controlling behavior. Social controls that originate from family relationships are called informal social controls. These are characteristics of social relationships that make people conform (for example, parental supervision, emotional attachment, and sensitivity to others' expectations). We considered three social control theories. First, Hirschi's *social bond theory* attempts to explain conformity in terms of four elements of the social bond—attachment, commitment to conventional lines of action, involvement in conventional activities, and belief in the moral validity of law.[9] Second, Sampson and Laub's *life-course theory* examines social bonds over the life course and explores how informal social controls in adulthood involve obligations and restraints that are attached to particular roles, such as marriage and work.[10] Third, Gottfredson and Hirschi's *self-control theory* focuses on controls within individuals and examines how self-control develops through socialization experiences, as well as the consequences of low self-control for behavior, including delinquency.[11]

Rick's Social and Self Controls

We are given the impression that Rick and his parents were quite close. Rick "interacted well" with his parents, as well as with teachers and peers. His parents "loved him greatly," and Rick "seemed to really care about his parents." They spent a "good deal of time" together, and Rick enjoyed the company of his parents. In terms of Hirschi's social bond theory, Rick and his parents clearly shared a strong attachment. Rick was also committed to conventional lines of action and involved in conventional activities. He was "academically motivated," earned good grades, and was not disruptive in school. He was also actively involved in sports. Rick's attachment to others, commitment to success in school, and involvement in conventional activities, both with his parents and through sports, should have led to conforming behavior, according to social bond theory. It also appears, though, that Rick was not subject to much direct control by his parents. He was apparently allowed a good deal of unsupervised time with his best friend, who was his "companion in crime." (Hirschi contended, however, that delinquent acts generally take little time to commit and therefore that direct controls are not as important in inhibiting delinquency as indirect controls.) Finally, it seems that Rick was rather impulsive and lacking in self-control, judging from the types of offenses he committed. So, while indirect, relational controls over Rick seemed relatively strong, direct and self-controls over his behavior apparently were not.

Social Learning Theories

A vast body of research demonstrates the importance of family processes in explaining delinquency. The same is true of peer influences on delinquent behavior, which we discussed in Chapter 11. We considered both learning and control theories of peer influences. Sutherland's *differential association theory* and

Akers' *social learning theory* both point to peer groups as the context in which delinquent behavior is learned.[12] Differential association theory contends that youths become delinquent by learning in the course of social interaction both "definitions favorable to the violation of law" and techniques for committing delinquent acts. According to the theory, associations with others vary in frequency, duration, priority, and intensity, and these four elements influence the learning process.

Social learning theory attempts to state *how* the learning of delinquent behavior occurs. This theory incorporates principles from behavioral learning theory, especially operant conditioning, in which behavior is shaped by rewards and punishments. Like differential association theory, social learning theory employs the concepts of differential association to describe the group context of learning and definitions or beliefs about delinquent acts. Social learning theory also includes the concepts of differential reinforcement, which refers to learning processes that involve rewards and punishments, and imitation, which involves copying behaviors modeled by others.

In sharp contrast to learning theories, control theories argue that peers are largely irrelevant in explaining why individuals become delinquent. According to social bond theory, attachment to peers, like attachment to parents, generates social control and thus conformity. To explain the relationship between delinquent behavior and association with delinquent peers, Hirschi argues that "birds of a feather flock together." In other words, individuals first commit delinquent acts, and then they gravitate toward others whose activities and attitudes are similar to their own. Similarly, self-control theory argues that low self-control increases the tendency to associate with deviant peers, but that delinquent friends are a consequence, not a cause, of delinquent behavior.

Rick's Delinquent Peer

Rick's friend appeared to play an important role in his delinquent career. The two shared similar personal and social characteristics and were "companions in crime," committing many of their delinquent acts together. As a result, they had similar offense records. Rick's delinquent friend may have played a causal role in his delinquent behavior by providing attitudes or definitions favorable to delinquency and by teaching techniques for committing delinquent acts. This situation would be consistent with learning theories. Or Rick and his friend may have been drawn together after the start of their delinquent careers because of their similar attitudes and thrill-seeking interests. This situation would be consistent with control theories. We cannot tell from reading the case study which theoretical interpretation of the delinquency–delinquent-peer relationship is correct in Rick's case.

Social Structure Theories

In Chapter 12, we explored how societal characteristics influence individuals' behavior. We considered three key theories: social disorganization, anomie, and strain. Shaw and McKay's *social disorganization theory* contends that rapid social change is associated with a breakdown in community social control and consequently with high rates of delinquency.[13] Shaw and McKay were interested in how problems associated with urban life, such as rapid population growth due to immigration, residential mobility, poverty, and physical deterioration, affected community cohesion and the ability of communities to exert social control over the behavior of individuals. Shaw and McKay found that the ecological distribution of delinquency in Chicago was closely associated with characteristics of local communities. Delinquency rates were highest in areas characterized by social disorganization.

Anomie and strain theories emphasize cultural goals of success along with the availability of opportunities to achieve success. Anomie theories explain why some societies have high rates of crime. Strain theories explain why some people and groups within society are more likely than others to engage in crime. According to Merton's *anomie theory*, the goal of economic success permeates all of American society, but the institutionalized means or norms for achieving success are neither stressed to the same degree nor equally available to all people.[14] Crime and delinquency are viewed as resulting from the social structural characteristic of anomie, or normlessness, which frees people to pursue economic success by whatever means necessary. *Strain theories* explain how groups or individuals adapt to the condition of anomie in society. Merton argued that individuals experience strain when legitimate means to economic success are unclear or unavailable.[15] He proposed five modes of adaptation to strain—conformity, ritualism, innovation, retreatism, and rebellion—with the last three typically involving deviant behavior. Strain theories have also been used to explain gang delinquency among urban, lower-class males.

Sources of Strain for Rick

Of the theories we considered in Chapter 12, Agnew's *general strain theory* appears most relevant to Rick's case. General strain theory emphasizes the social-psychological aspects of strain and addresses the strain of adolescence as a motivation for delinquency.[16] Agnew proposes three general types of strain: (1) strain resulting from failure to achieve valued goals, such as when expectations about school performance do not match actual achievements; (2) strain resulting from the loss of positively valued stimuli, such as the divorce of parents or suspension from school; and (3) strain resulting from the presence of negatively valued stimuli, such as child abuse or negative relationships with parents or teachers.

In the case study, we are told that Rick was "goal-directed" and did well in school. He attended school regularly, earned good grades, and was not disruptive in school in any way. "Teachers reported that he was a very positive student both in and out of class." It appears that school was not a serious source of strain for Rick. Yet it is possible that, even for good students like Rick, achievements do not live up to expectations about performance. Similarly, Rick's positive relationships with others, including parents, teachers, and peers, appear unlikely to be a source of strain. It may be, however, that Rick's initial brushes with the law themselves constituted a source of strain, if they weakened Rick's relationships with others or caused others to change their expectations or perceptions of Rick, or if they caused Rick to change his own expectations about the achievement of his goals.

Labeling and Critical Criminologies

In Chapter 13, we turned to perspectives that focus on social and societal reactions to delinquency. These approaches recognize that definitions of what actions are delinquent and who is delinquent result from dynamic social processes. We considered two questions addressed by the *labeling perspective:* how the label "delinquent" is imposed legally by the juvenile court, and how this label is imposed socially in the context of relationships. According to the labeling perspective, the delinquent label—whether legally or socially applied—has consequences for future behavior, as individuals tend to behave in accord with labels imposed on them.

In Chapter 13, we also considered critical criminologies and feminist perspectives. Critical approaches challenge traditional explanations for delinquency and offer alternative views of definitions of and societal responses to delinquent behavior. Although a variety of critical approaches exist, they share several key themes: capitalism is at the root of all crime, inequality and power are central to understanding crime, crime and delinquency are politically defined, and law and the criminal justice system protect the interests of the powerful.

We explored in detail three critical interpretations of delinquency: Colvin and Pauly's *integrated structural-Marxist theory,* Hagan's *power-control theory,* and Greenberg's *Marxist interpretation of delinquency.* The first two theories both explore how forms of control within the workplace, which are grounded in the capitalist economic system, translate into mechanisms of control within families. Colvin and Pauly argue that children's socialization experiences within families influence the forms of control they encounter in school and peer groups. The more coercive the controls that children experience in families, schools, and peer groups, the more likely they are to engage in serious delinquency. Hagan's power-control theory contends that both social class and gender determine the presence of power and the absence of control. Together, class and gender structure children's socialization experiences and create conditions of freedom that contribute to delinquent behavior. Greenberg offers a Marxist interpretation of delinquency as a response to the precarious status position of youth in capitalist societies. To explain delinquent acts that offer financial reward, Greenberg describes how the exclusion of adolescents from workforce participation deprives youth of the monetary resources needed to finance teenage social activities. To explain nonutilitarian and violent delinquent acts, Greenberg describes status problems caused by school experiences and masculine status anxiety, both of which vary by social class.

Like critical criminologies, feminist criminology emphasizes power and inequality within society and focuses on the structural determinants of behavior. Feminist criminologists, however, contend that gender must be at the center of attempts to understand human behavior, including crime and delinquency. We examined four distinct feminist approaches: *liberal, radical, socialist,* and *Marxist* feminisms. Feminist scholars understand both offending and victimization in terms of the different life experiences of males and females. Feminist criminologists have also offered critiques of the criminal justice system that illuminate the differential treatment of males and females and the use of the system to perpetuate male power.

Labeling Rick a "Delinquent Youth"

Let's consider Rick's case one last time, in terms of the social and societal responses to delinquency presented in Chapter 13. At age 14, Rick was labeled a "delinquent youth" by the juvenile court. Labeling theorists would argue that this formal label contributed to Rick's continued delinquent career. Once such labels are imposed, individuals tend to take on an identity consistent with the label and behave accordingly, creating a self-fulfilling prophecy. Perhaps the label imposed on Rick by the juvenile court had a stigmatizing effect that changed significant others' perceptions of or reactions to Rick, leaving him with the feeling that he had "nothing left to lose" by continued delinquent behavior.

Rick's first referral to the juvenile court was for "deviant sex" when he was 12 years old. The court dealt with this offense "informally." Research by feminist criminologists shows that the juvenile justice system responds differently to males and females, particularly for status offenses that pose a threat to female chastity. This research suggests that the juvenile court probably would have responded quite differently to Rick's "deviant sex" offense if Rick were female.

■ "Putting the Pieces Together": Integrated Theories of Delinquent Behavior

At this point, you might feel like you're swimming in a sea of potential explanations for delinquency, without a clear sense of your way back to shore. As we stated at the beginning of this chapter, the task of sorting through theoretical explanations for delinquent behavior is a complex one. From the late 1960s to the early 1980s, testing competing theories of delinquency against one another was a popular focus of research. For example, several studies tested competing hypotheses derived from social control and social learning theories.[17] In the late 1970s and throughout the 1980s, a different approach emerged—one that focused on *integrating* existing theories of delinquency. Rather than viewing alternative theoretical explanations as competing with one another, those who favor integrated models focus on combining elements of existing theories that have been supported in empirical research.[18] The goal is to arrive at a more accurate and comprehensive causal perspective on delinquency.

We now briefly describe the debate surrounding integrated approaches, and then present examples of integrated theories.

Debate About Integrated Approaches[19]

Earlier tests of competing theories enabled criminologists in the 1980s to begin to integrate validated elements of theories. These tests provided an accumulated body of empirical and theoretical knowledge that allowed for the development of integrated theoretical models in criminology.[20]

Theoretical integration has intuitive appeal because it helps us organize what we "know" about delinquency in ways that make existing knowledge seem less fragmented. In addition, different theories may explain different aspects of involvement in delinquency, so distinct causal explanations are not necessarily incompatible.[21] For example, integrated theories may combine microsocial and macrosocial levels of explanation (see Chapter 1). Delbert Elliott argues that combining theories that explain different aspects of delinquent behavior would enable criminologists to explain a greater share of delinquency.[22] In other words, integrated theories are potentially "more powerful" than the separate theories used to form them, because integrated models will presumably account for more of the variation in delinquent behavior than will separate theories.[23]

Still, not all scholars have supported integrated approaches. Travis Hirschi responded to the attempt by Delbert Elliott and his colleagues to integrate strain, control, and learning theories with a paper in which the title summarized Hirschi's view: "Separate and Unequal is Better."[24] He argued that the assumptions on which criminological theories are based are often incompatible, so the integration of inherently contradictory theories is impossible. Hirschi maintains that even models that claim to be integrated are sometimes only restatements of one or another of the originating theories. For example, Hirschi argues that the "integrated" model proposed by Elliott and his colleagues basically reverts to differential association theory and ignores the competing tenets of control theory. Rather than integration, Hirschi advocates the careful consideration and testing of individual—not combined—theories.[25]

Terence Thornberry rejects Hirschi's position of complete opposition to integrated approaches. But he expresses concern that the process of integration—of attempting to reconcile potentially contradictory elements of different theories—"increases the risk of watering down the clarity and strength of the [original] theoretical statements."[26]

These criticisms of theoretical integration point out important issues, but they should not be used to dismiss altogether the enterprise of integration. Integrated approaches have much to offer those interested in the causes of delinquent behavior, because they acknowledge that multiple causes of crime and delinquency exist. In developing and testing integrated theories, criminologists must be careful to draw together theoretical elements that are compatible and to explain systematically *how* these elements are related in causing delinquency. The process of integration is not like a recipe in which one combines a little of this and a bit of that.

We turn now to several examples of integrated approaches in the study of crime and delinquency. Our review is not exhaustive.[27] Instead, we have selected four specific theories: Elliott's integrated theory, Thornberry's interactional theory, Tittle's control balance theory, and Miethe and Meier's theory of offenders, victims, and situations. We chose these examples because they represent a classic statement of integrated theory (Elliott), propose reciprocal effects in causal models (Thornberry), include new concepts in addition to the traditional theoretical elements they combine (Tittle), or incorporate broader social contexts (Miethe and Meier). We should also note that several of the theories we discussed in previous chapters are themselves integrated theories of delinquency. For example, Gottfredson and Hirschi's general theory of crime, Messner and Rosenfeld's institutional anomie theory, Agnew's general strain theory, Colvin and Pauly's integrated structural-Marxist theory, and Braithwaite's theory of reintegrative shaming could also be included in this discussion of integrated approaches.

Elliott's Integrated Theory of Delinquent Behavior

Delbert Elliott, along with his colleagues Suzanne Ageton and Rachelle Canter, developed an <u>integrated theory</u> to explain delinquency and drug use that weaves together elements of strain, social control, and social learning theories.[28] This integrated theory is straightforward insofar as it does not introduce any new concepts or variables; instead, it simply combines elements of existing theories.

Elliott and his colleagues first combined strain and social control theories. The strength of one's bonds to the conventional social order is central to Elliott's integrated theory. This theory postulates that three factors affect the strength of conventional bonding: early socialization experiences; social disorganization at home (such as marital discord or divorce) and in the community (such as high rates of residential mobility or economic disadvantage); and strain produced by limited opportunities to achieve conventional goals.[29] Elliott and his colleagues combine strain and social control theories by arguing that limited opportunities to achieve conventional goals are a source of strain only for those who are committed to conventional goals. For those weakly bonded to the conventional social order, limited opportunities should have little effect in the process leading to delinquency.[30]

Next, Elliott and his colleagues integrated social learning theory into their strain–control perspective. Consistent with differential association theory, Elliott's integrated theory maintains that "delinquent behavior, like conforming behavior, presupposes a pattern of social relationships through which motives, rationalizations, techniques, and rewards can be learned and maintained."[31] This learning occurs primarily in delinquent peer groups that reinforce and reward delinquent behavior and are thus essential to the development of delinquent patterns of behavior. Elliott and his colleagues tie the social learning process to conventional bonding in this way: "Those committed to conventional goals, although they may have been exposed to and learned some delinquent behaviors, should not establish patterns of such behavior unless (1) their ties to the conventional social order are neutralized through some attenuating experiences and (2) they are participating in a social context in which delinquent behavior is rewarded. In social-learning terms, they may have acquired or learned delinquent behavior patterns, but the actual performance and maintenance of such behavior are contingent on attenuation of their commitment to conventional norms and their participation in a social context supportive of delinquent acts."[32]

Elliott and his colleagues contend that the process leading to delinquent behavior typically involves weak bonds to the conventional social order and strong ties to peers who reward and reinforce delinquency. They also hold open the possibility that strain may lead directly to delinquent behavior, if commitment to unattainable goals is sufficiently strong.[33] Similarly, weak conventional bonding may lead directly to delinquency, though delinquent behavior patterns cannot be *sustained* without a peer group that supports delinquency.[34]

Thornberry's Interactional Theory[35]

Terence Thornberry's <u>interactional theory</u> combines elements primarily of social control and social learning theories, and also of strain and culture conflict theories.[36] Like Elliott's integrated theory, Thornberry's interactional theory

Elliott's integrated theory of delinquent behavior An integrated approach to explain delinquency and drug use that combines elements of strain, social control, and social learning theories.

Thornberry's interactional theory Combines elements of social control, social learning, strain, and culture conflict theories. Thornberry's theory focuses on the processes and settings of social interaction through which delinquency is learned, performed, and reinforced. This theory allows for reciprocal relationships among the variables of interest.

begins with the social-control-theory premise that weakening of bonds to the conventional social order contributes to delinquent behavior. Interactional theory diverges from previous theories, however, in its focus on the processes and settings of *social interaction* through which delinquent behavior is learned, performed, and reinforced.

Thornberry's interactional model includes six concepts: attachment to parents, commitment to school, belief in conventional values, association with delinquent peers, formation of delinquent values, and delinquent behavior.[37] The first three are derived from social control theory and represent bonds to conventional society. The next two are derived primarily from social learning theory and represent the "interactive settings that reinforce delinquency." **Figure 14-1** shows the proposed relationships among these variables in early adolescence.

The distinguishing features of interactional theory are its allowance for <u>reciprocal relationships</u> among the variables of interest and its specification of delinquent behavior as both an outcome of social processes and a causal factor influencing those social processes over time. For example, interactional theory moves beyond the criminological debate about which comes first—association with delinquent peers or delinquent behavior—by proposing a reciprocal relationship between peer association and behavior. Not only do people behave like those with whom they associate (association influences behavior), but they also associate with others whose behaviors are like their own (behavior influences association).

Interactional theory contends that reciprocal relationships exist among associations with delinquent peers, delinquent values, and delinquent behavior. "Interactional theory sees these three concepts as embedded in a causal loop, each reinforcing the others over time. Regardless of where the individual enters the loop, the following obtains: delinquency increases associations with delinquent peers and delinquent values; delinquent values increase delinquent behavior and associations with delinquent peers; and associations with delinquent peers increase delinquent behavior and delinquent values."[38] See Figure 14-1.

The next logical question is: What factors lead some individuals to enter that causal loop? To answer that question, Thornberry turns to concepts of social control theory. Attachment to parents, commitment to school, and belief in conventional values are viewed as the key mechanisms through which adolescents are bound to conventional society. According to interactional theory, weakening of these conventional bonds explains how some individuals enter the causal loop described above.

In Thornberry's model, the relationships among attachment to parents, commitment to school, and belief in conventional values are also reciprocal. These variables interact with each other over time during the developmental process.[39] For example, attachment to parents, who are likely to reinforce conventional values, increases adolescents' belief in conventional values and commitment to school. School success, which generally derives from commitment, in turn, reinforces attachment to parents, who value that success. Similarly, reciprocal relationships exist between the bonding variables (at-

reciprocal relationships
Relationships in which variables are both causes and effects and exert mutual influences on each other (Vogt, *Dictionary*, 190). For example, interactional theory proposes a reciprocal relationship between delinquency and delinquent peers: having delinquent peers causes one to engage in delinquency, and involvement in delinquency causes one to associate with others who are delinquent.

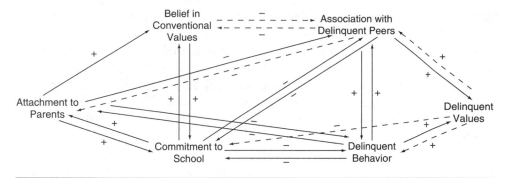

Figure 14-1 Thornberry's model of delinquency in early adolescence. Thornberry's interactional theory proposes reciprocal relationships among the concepts of interest. Solid lines represent stronger effects; dashed lines represent weaker effects. Plus and minus symbols indicate positive and negative causal relationships.
Source: Thornberry, "Toward an Interactional Theory," 871. Copyright © 1987, American Society of Criminology. All rights reserved. Reproduced with permission of the publisher.

tachment, commitment, and belief) and both delinquent behavior and association with delinquent peers. Weak bonds to conventional society lead to delinquent behavior and association with delinquent peers. Delinquency and association with other delinquent youth, in turn, further weaken bonds to the conventional world.[40]

Finally, interactional theory takes into account the developmental process by proposing separate, but similar, models for early, middle, and late adolescence. The differences among these models represent changes in relationships and attitudes associated with aging through adolescence. For example, the major difference between the models for early and middle adolescence concerns the importance of attachment to parents as a causal factor. The early adolescence model shows attachment strongly affecting all other variables in the model except delinquent values. But in middle adolescence, attachment to parents exerts weaker effects, indicating a less influential role of parents at a time when youth are increasingly affected by school and peer relationships.

Tittle's Control Balance Theory

In 1995, Charles Tittle presented <u>control balance theory</u>, a complex integrated theory that draws together elements of social control, differential association, anomie, labeling, deterrence, Marxian conflict, and routine activities theories.[41] With this theory, Tittle tries to account not only for law-violating behaviors, but for all forms of deviance, defined as "any behavior that the majority of a given group regards as unacceptable or that typically evokes a collective response of a negative type."[42]

At the center of Tittle's theory is the concept of <u>control balancing</u>—balancing the control that might be gained from deviant behavior against the control that might be exercised against a person for that behavior.[43] The central premise of control balance theory is that "the amount of control to which an individual is subject, relative to the amount of control he or she can exercise, determines the probability of deviance occurring as well as the type of deviance

Tittle's control balance theory An integrated theory that combines elements of social control, differential association, anomie, labeling, deterrence, Marxian conflict, and routine activities theories. This theory introduces the concept of *control balance*, contending that the probability and type of deviance are determined by "the amount of control to which an individual is subject, relative to the amount of control he or she can exercise" (Tittle, *Control Balance*, 135).

control balancing The "balancing of the control you might gain from deviant behavior against the control that will likely be directed back at you" (Tittle, "Control Balance," 316).

likely to occur."[44] The key variable in this theory is one's *control ratio,* which is "the extent to which an individual can potentially exercise control over circumstances impinging on him, relative to the potential control that can be exercised by external entities and conditions against the individual."[45] According to the theory, control imbalances—either control deficits or control surpluses—and a desire for autonomy produce motivations toward deviance. Deviant behavior is a mechanism for escaping control deficits or extending control surpluses.[46]

Control balance theory does not propose, however, that control imbalances necessarily lead to deviance. In addition to motivation, or "predisposition" toward deviance, other factors must also be present, including situational provocation, opportunity, and lack of constraint.[47] Situational provocations, such as verbal insults or challenges to one's authority, alert individuals to control imbalances.[48] For individuals to respond in a deviant way to motivations and provocations, the opportunity to commit deviance must exist.[49] In addition, for deviance to occur, the constraint imposed by external controls on individuals must be less than the control they will achieve by engaging in deviance. While serious offenses are likely to produce the greatest increases in one's sense of control, they are also likely to activate the strongest external controls. This level of constraint and its effect on one's control ratio also factor into the decision to engage in deviance.[50]

Control balance theory allows for the possibility that, even when motivation, situational provocations, opportunity, and lack of constraint exist, deviance may not occur. Other conditions, called *contingencies,* may intervene in the control balancing process.[51] Tittle defines a contingency as "any aspect of an individual; social relationships; organizational structures; or the physical environment that influences how completely or strongly the control balancing process operates."[52] Personal contingencies include perceptual accuracy, habits, moral commitments, personality, and ability to commit deviance. Tittle highlights "subcultural affiliations" as a key organizational contingency. Situational contingencies concern the opportunity for deviance, risk of counter control, and intensity and frequency of provocation toward deviance. For example, a person may be motivated toward deviance, be provoked by situational factors, have the opportunity to commit deviance, and perceive limited counter controlling responses. He may still abstain from deviance because he is morally opposed to the deviant act or lacks the ability to carry it out, or because he belongs to a peer group that disapproves of deviance as a means to assert control. These factors are contingencies that intervene to prevent deviance when the control balancing process would otherwise make deviance appear likely.

Finally, control balance theory is able to account for different types of deviance. Tittle's theory proposes that deviance takes two distinct forms: *repressive deviance* and *autonomous deviance.*[53] Repressive deviance includes acts of "submission" (e.g., physical abuse and sexual degradation), "defiance" (e.g., curfew violations, vandalism, and political protests), and "predation" (e.g., theft, rape, homicide, robbery, and assault).[54] According to control balance theory, individuals characterized by control deficits—those who can exercise *less* control over impinging circumstances than the potential control that can be exercised against them—are most likely to resort to repressive forms of deviance.

Autonomous deviance includes acts of "exploitation" (e.g., corporate price-fixing, and influence peddling by political figures), "plunder" (e.g., oppressive taxation of the poor to sustain the wealth of corrupt leaders, and pollution by oil companies which then increase oil prices to pay for cleanup efforts), and "decadence" (e.g., sexual exploitation of children, and humiliation of others for entertainment purposes).[55] According to control balance theory, individuals characterized by control surpluses—those who can exercise *more* control over impinging circumstances than the potential control that can be exercised against them—are most likely to engage in autonomous forms of deviance.

Miethe and Meier's Integrated Theory of Offenders, Victims, and Situations

Terance Miethe and Robert Meier go beyond an integration of risk factors for delinquency or factors that motivate an individual's involvement in delinquent behavior. Instead, they attempt to provide an <u>integrated theory of offenders, victims, and situations</u>.[56] Traditionally, criminologists have tried to understand delinquency by studying criminals and their social worlds. A variety of perspectives, ranging from differential association and social control theories to anomie and social disorganization theories, take this approach.

Since the mid-1970s, however, criminologists have begun to recognize, by examining victimization data, that the risks of criminal victimization vary based on demographic characteristics. Recall from Chapter 5 that violent victimization risk is greatest for those who are young, male, African American, and economically disadvantaged. Relatively recently, criminologists have developed "systematic theories of victimization," such as "routine activity" and "criminal opportunity" approaches.[57] The basic premise of these perspectives is that the routine activities and lifestyles of individuals contribute to criminals' perceptions of opportunities for crime and their selection of "suitable targets" or victims of crime.[58]

In trying to explain crime, Miethe and Meier consider not only offenders and victims, but also the situational contexts in which crimes occur. They point to the obvious fact that every crime occurs within a particular *social context*—"a micro-environment that has physical and social dimensions."[59] The importance of social context becomes apparent when we recognize that crime is more likely in some environments than others.[60] For example, the recent work of Robert Sampson and his colleagues suggests that urban neighborhoods characterized by concentrated economic disadvantage, residential mobility, and population heterogeneity are especially favorable contexts for crime.[61]

Miethe and Meier ambitiously draw together these three elements—offenders, victims, and social contexts—in their integrated theory. They recognize that a potential offender's motivation to commit crime is not sufficient to account for crime. Instead, the activities and lifestyles of potential victims create opportunities for crime that enable an offender to act on his or her criminal motivations. Both of these factors—offender motivation and criminal opportunities created by the actions of potential victims—operate within a particular social context. As Miethe and Meier write, "The integration of theories of criminality and theories of victimization recognizes that crime involves a two-step process: (1) the

Miethe and Meier's integrated theory of offenders, victims, and situations Considers a potential offender's motivation to commit crime; the activities and lifestyles of potential victims that create opportunities for crime; and the particular social context in which crime occurs, including physical location, offender–victim relationship, and behavioral setting.

decision to engage in crime (crime commission) and (2) the selection of a particular source for victimization (target-selection). . . . Criminal intentions without the availability of a suitable crime target and a facilitating context will go unfulfilled."[62]

Figure 14-2 shows Miethe and Meier's model for the integration of offenders, victims, and social contexts to explain criminal events. Sources of offender motivation in this model are compatible with several theories of criminality, including social learning, social control, and strain theories. These sources include "economic disadvantage, weak social bonds, pro-crime values, psychological or biological attributes, generalized needs (e.g., money, sex, friendship, excitement), and the availability of non-criminal alternatives."[63] The victim characteristics that provide criminal opportunities are those suggested by routine activity and criminal opportunity theories. These include "proximity to offenders, exposure to high-crime situations, target attractiveness, and the absence of guardianship."[64] The social context of crime involves three elements: (1) a physical location, including amount of physical space in a setting, lighting, "pace" or "rhythm," and history of criminal activity; (2) the interpersonal relationship between offender and victim; and (3) a behavioral setting that "establishes the activities of the victim at the time of the offense."[65]

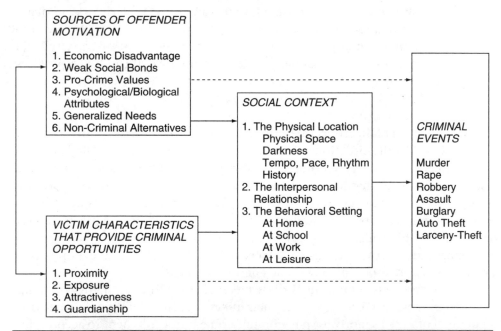

Figure 14-2 Miethe and Meier's model of criminal events. Miethe and Meier present this model depicting their theoretical integration of offenders, victims, and social contexts to explain criminal events. Solid arrows represent direct causal relationships; dashed arrows represent "direct relationships that are of a smaller absolute magnitude, because the primary influence of a variable is being transmitted through another variable" (in this case, social context); and the double-headed arrow represents correlation or association between variables that is not causal in nature.

Source: Miethe and Meier, *Crime and Its Social Context*, 63–65. Copyright © 1994, State University of New York Press. All rights reserved. Reproduced with permission of the publisher.

This model depicts a particular causal ordering of variables. Offender motivation and victim characteristics that provide criminal opportunities are related, though not causally. Both of these factors may lead directly to criminal events (as shown by the dashed arrows), but their primary effects on criminal events operate through the social context in which crimes occur (as shown by the three solid arrows).[66] In this model, social context is the central factor in understanding crime. Crime may occur when only one of these factors—offender motivation, victim characteristics that provide criminal opportunities, or a social context conducive to crime—is present. The combination of all three factors, though, substantially increases the likelihood of crime.[67]

■ Summary and Conclusions

In this chapter, we have summarized the theories and research presented in previous chapters. In an attempt to put a "human face" on theoretical explanations for delinquency, we have considered Rick's case from the various perspectives discussed in this book.

We have also discussed integrated theories in this chapter, to encourage the reader to think beyond distinct theories and to begin to draw together the building blocks provided in this text. We presented four examples of integrated theories: Elliott's integrated theory of delinquent behavior, Thornberry's interactional theory, Tittle's control balance theory, and Miethe and Meier's integrated theory of offenders, victims, and situations.

Elliott's integrated theory is a fairly straightforward union of strain, social control, and social learning theories. Essentially, Elliott and his colleagues argue that the process leading to delinquency involves weak bonds to the conventional social order and strong ties to peers who reward and reinforce delinquent behavior. Limited opportunities to achieve conventional goals are viewed as a source of strain for those who are committed to those goals. Thornberry's approach is similar to Elliott's insofar as it draws together the same theories. But Thornberry's theory is unique in its focus on reciprocal relationships and the processes and settings of social interaction. Reciprocal relationships involve variables that mutually influence each other. In other words, each variable is both a cause and an effect of the other. Interactional theory views delinquency as both an outcome of social processes and a causal factor influencing those social processes over time.

Tittle's control balance theory introduces new concepts, in addition to the traditional theoretical elements it combines. This theory is a complex integration of elements of social control, differential association, anomie, labeling, deterrence, Marxian conflict, and routine activities theories. Tittle introduces the concept of control balancing and states that the central premise of control balance theory is that "the amount of control to which an individual is subject, relative to the amount of control he or she can exercise, determines the probability of deviance occurring as well as the type of deviance likely to occur."[68] According to the theory, control deficits (in which individuals can exercise *less* control over impinging circumstances than the potential control that can be exercised against them) are associated with repressive forms of deviance. Control surpluses (in which individuals can exercise *more* control over impinging circumstances than the potential control that can be exercised against them) are associated with autonomous forms of deviance.

Finally, we presented Miethe and Meier's integrated theory of offenders, victims, and situations. To account for criminal events, this theory considers a potential offender's motivation to commit crime; the activities and lifestyles of potential victims that create opportunities for crime; and the particular social context in which crime occurs, including physical location, offender–victim relationship, and behavioral setting. In Miethe and Meier's model, the primary effects of offender motivation and victim characteristics on criminal events operate through the social context in which crime occurs.

THEORIES

- Elliott's integrated theory of delinquent behavior
- Thornberry's interactional theory
- Tittle's control balance theory
- Miethe and Meier's integrated theory of offenders, victims, and situations

CRITICAL THINKING QUESTIONS

1. Which theories of delinquency presented in this book seem *most* plausible to you? In other words, which theories do you think best explain delinquency? Why? Which theories seem *least* plausible? Why?

2. Think of an individual you know who has engaged in delinquency. Consider his or her delinquent behavior from the various perspectives presented in this book, as we considered Rick's delinquency in this chapter.

3. Do you think theoretical integration in the study of crime and delinquency is a strategy worth pursuing? Why or why not? Is true theoretical integration an achievable goal? In other words, can alternative theories be *fully* integrated? Why or why not?

SUGGESTED READINGS

Elliott, Delbert S., Suzanne S. Ageton, and Rachelle J. Canter. "An Integrated Theoretical Perspective on Delinquent Behavior." *Journal of Research in Crime and Delinquency* 16 (1979):3–27.

Hirschi, Travis. "Separate and Unequal Is Better." *Journal of Research in Crime and Delinquency* 16 (1979):34–38.

Messner, Steven F., Marvin D. Krohn, and Allen E. Liska, eds. *Theoretical Integration in the Study of Deviance and Crime: Problems and Prospects.* Albany, NY: State University of New York Press, 1989.

Miethe, Terance D., and Robert F. Meier. *Crime and Its Social Context: Toward an Integrated Theory of Offenders, Victims, and Situations.* Albany, NY: State University of New York Press, 1994.

Thornberry, Terence P. "Toward an Interactional Theory of Delinquency." *Criminology* 25 (1987):863–891.

Tittle, Charles R. *Control Balance: Toward a General Theory of Deviance.* Boulder, CO: Westview Press, 1995.

GLOSSARY

control balancing: The "balancing of the control you might gain from deviant behavior against the control that will likely be directed back at you" (Tittle, "Control Balance," 316).

Elliott's integrated theory of delinquent behavior: An integrated approach to explain delinquency and drug use that combines elements of strain, social control, and social learning theories.

Miethe and Meier's integrated theory of offenders, victims, and situations: Considers a potential offender's motivation to commit crime; the activities and lifestyles of potential victims that create opportunities for crime; and the particular social context in which crime occurs, including physical location, offender–victim relationship, and behavioral setting.

reciprocal relationships: Relationships in which variables are both causes and effects and exert mutual influences on each other (Vogt, *Dictionary*, 190). For example, interactional theory proposes a reciprocal relationship between delinquency and delinquent peers: having delinquent peers causes one to engage in delinquency, and involvement in delinquency causes one to associate with others who are delinquent.

Thornberry's interactional theory: Combines elements of social control, social learning, strain, and culture conflict theories. Thornberry's theory focuses on the processes and settings of social interaction through which delinquency is learned, performed, and reinforced. This theory allows for reciprocal relationships among the variables of interest.

Tittle's control balance theory: An integrated theory that combines elements of social control, differential association, anomie, labeling, deterrence, Marxian conflict, and routine activities theories. This theory introduces the concept of *control balance,* contending that the probability and type of deviance are determined by "the amount of control to which an individual is subject, relative to the amount of control he or she can exercise" (Tittle, *Control Balance,* 135).

REFERENCES

Agnew, Robert. "An Integrated Theory of the Adolescent Peak in Offending." *Youth and Society* 34 (2003):263–299.

———. "The Contribution of Social-Psychological Strain Theory to the Explanation of Crime and Delinquency." In *The Legacy of Anomie Theory* (Advances in Criminological Theory, vol. 6), edited by Freda Adler and William S. Laufer, 113–137. New Brunswick, NJ: Transaction, 1995.

———. "Foundation for a General Strain Theory of Crime and Deviance." *Criminology* 30 (1992):47–87.

———. "A Revised Strain Theory of Delinquency." *Social Forces* 64 (1985):151–167.

Beirne, Piers. *Inventing Criminology: Essays on the Rise of Homo Criminalis.* Albany, NY: State University of New York Press, 1993.

———. "Inventing Criminology: The 'Science of Man' in Cesare Beccaria's *Dei Delitti e Delle Pene* (1764)." *Criminology* 29 (1991):777–820.

Braithwaite, John. *Crime, Shame and Reintegration.* Cambridge: Cambridge University Press, 1989.

Briar, Scott, and Irving Piliavin. "Delinquency, Situational Inducements, and Commitment to Conformity." *Social Problems* 13 (1965):35–45.

Burgess, Robert L., and Ronald L. Akers. "A Differential Association–Reinforcement Theory of Criminal Behavior." *Social Problems* 14 (1966):128–147.

Canter, Rachelle J. "Sex Differences in Self-Reported Delinquency." *Criminology* 20 (1982):373–393.

Catalano, Richard F., and J. David Hawkins. "The Social Development Model: A Theory of Antisocial Behavior." In *Delinquency and Crime: Current Theories,* edited by J. David Hawkins, 149–197. Cambridge: Cambridge University Press, 1996.

Chesney-Lind, Meda. "Girls' Crime and Women's Place: Toward a Feminist Model of Female Delinquency." *Crime and Delinquency* 35 (1989):5–29.

———. "Judicial Paternalism and the Female Status Offender: Training Women to Know Their Place." *Crime and Delinquency* 23 (1977):121–130.

Chesney-Lind, Meda, and Randall G. Shelden. *Girls, Delinquency, and Juvenile Justice.* 2nd ed. Belmont, CA: Wadsworth, 1997.

Cohen, Lawrence E., and Marcus Felson. "Social Change and Crime Rate Trends: A Routine Activity Approach." *American Sociological Review* 44 (1979):588–608.

Colvin, Mark, Francis T. Cullen, and Thomas VanderVen. "Coercion, Social Support, and Crime: An Emerging Theoretical Consensus." *Criminology* 40 (2002):19–42.

Colvin, Mark, and John Pauly. "A Critique of Criminology: Toward an Integrated Structural-Marxist Theory of Delinquency Production." *American Journal of Sociology* 89 (1983):513–551.

Conger, Rand D. "Social Control and Social Learning Models of Delinquent Behavior: A Synthesis." *Criminology* 14 (1976):17–40.

Curry, Theodore R., and Alex Piquero. "Control Ratios and Defiant Acts of Deviance: Assessing Additive and Conditional Effects with Constraints and Impulsivity." *Sociological Perspectives* 46 (2003):397–415.

Drass, Kriss A., and Terance D. Miethe. "Qualitative Comparative Analysis and the Study of Crime Events." In *The Process and Structure of Crime: Criminal Events and Crime Analysis,* Advances in

Criminological Theory, vol. 9, edited by Robert F. Meier, Leslie W. Kennedy, and Vincent F. Sacco, 125–140. New Brunswick, NJ: Transaction, 2001.

Elliott, Delbert S. "The Assumption that Theories Can Be Combined with Increased Explanatory Power: Theoretical Integrations." In *Theoretical Methods in Criminology,* edited by Robert F. Meier, 123–149. Beverly Hills, CA: Sage, 1985.

Elliott, Delbert S., Suzanne S. Ageton, and Rachelle J. Canter. "An Integrated Theoretical Perspective on Delinquent Behavior." *Journal of Research in Crime and Delinquency* 16 (1979):3–27.

Elliott, Delbert S., David Huizinga, and Suzanne S. Ageton. *Explaining Delinquency and Drug Use.* Beverly Hills, CA: Sage, 1985.

Gottfredson, Michael, and Travis Hirschi. *A General Theory of Crime.* Stanford, CA: Stanford University Press, 1990.

———. "The Positive Tradition." In *Positive Criminology,* edited by Michael Gottfredson and Travis Hirschi, 9–22. Newbury Park, CA: Sage, 1987.

Greenberg, David F. "Delinquency and the Age Structure of Society." *Contemporary Crises: Crime, Law and Social Policy* 1 (1977):189–223.

Hepburn, John R. "Testing Alternative Models of Delinquency Causation." *Journal of Criminal Law and Criminology* 67 (1976):450–460.

Hickman, Matthew, and Alex Piquero. "Exploring the Relationships between Gender, Control Balance, and Deviance." *Deviant Behavior* 22 (2001):323–351.

Hickman, Matthew, Alex Piquero, Brian Lawton, and Jack Greene. "Applying Tittle's Control Balance Theory to Police Deviance." *Policing* 24 (2001):497–519.

Hindelang, Michael J., Travis Hirschi, and Joseph G. Weis. *Measuring Delinquency.* Beverly Hills, CA: Sage, 1981.

Hirschi, Travis. *Causes of Delinquency.* Berkeley, CA: University of California Press, 1969.

———. "Exploring Alternatives to Integrated Theory." In *Theoretical Integration in the Study of Deviance and Crime: Problems and Prospects,* edited by Steven F. Messner, Marvin D. Krohn, and Allen E. Liska, 37–49. Albany, NY: State University of New York Press, 1989.

———. "Separate and Unequal Is Better." *Journal of Research in Crime and Delinquency* 16 (1979):34–38.

Hirschi, Travis, and Michael J. Hindelang. "Intelligence and Delinquency: A Revisionist Review." *American Sociological Review* 42 (1977):571–587.

Jensen, Gary F. "Parents, Peers, and Delinquent Action: A Test of the Differential Association Perspective." *American Journal of Sociology* 78 (1972):562–575.

Johnson, Richard E. *Juvenile Delinquency and Its Origins: An Integrated Theoretical Approach.* Cambridge: Cambridge University Press, 1979.

Matsueda, Ross L. "Testing Control Theory and Differential Association: A Causal Modeling Approach." *American Sociological Review* 47 (1982):489–504.

Merton, Robert. "Social Structure and Anomie." *American Sociological Review* 3 (1938):672–682.

Messner, Steven F., Marvin D. Krohn, and Allen E. Liska, eds. *Theoretical Integration in the Study of Deviance and Crime: Problems and Prospects.* Albany, NY: State University of New York Press, 1989.

Miethe, Terance D., and Robert F. Meier. *Crime and Its Social Context: Toward an Integrated Theory of Offenders, Victims, and Situations.* Albany, NY: State University of New York Press, 1994.

Miller, Joshua D., and Donald Lynam. "Structural Models of Personality and Their Relation to Antisocial Behavior: A Meta-Analytic Review." *Criminology* 39 (2001):765–798.

Osgood, D. Wayne, Janet K. Wilson, Patrick M. O'Malley, Jerald G. Bachman, and Lloyd D. Johnston. "Routine Activities and Individual Deviant Behavior." *American Sociological Review* 61 (1996):635–655.

Pearson, Frank S., and Neil Alan Weiner. "Toward an Integration of Criminological Theories." *Journal of Criminal Law and Criminology* 76 (1985):116–150.

Piquero, Alex, and Matthew Hickman. "Extending Tittle's Control Balance Theory to Account for Victimization." *Criminal Justice and Behavior* 30 (2003):282–301.

Sampson, Robert J., and John H. Laub. *Crime in the Making: Pathways and Turning Points Through Life.* Cambridge, MA: Harvard University Press, 1993.

Sampson, Robert J., Stephen W. Raudenbush, and Felton Earls. "Neighborhoods and Violent Crime: A Multilevel Study of Collective Efficacy." *Science* 277 (1997):918–924.

Shaw, Clifford R., and Henry D. McKay. *Juvenile Delinquency and Urban Areas: A Study of Rates of Delinquency in Relation to Differential Characteristics of Local Communities in American Cities.* Rev. ed. Chicago: University of Chicago Press, 1969.

Chapter Resources

Sutherland, Edwin H. *Principles of Criminology.* 4th ed. Philadelphia: Lippincott, 1947.

Teilmann, Katherine S., and Pierre H. Landry, Jr. "Gender Bias in Juvenile Justice." *Journal of Research in Crime and Delinquency* 18 (1981):47–80.

Thornberry, Terence P. "Empirical Support for Interactional Theory: A Review of the Literature." In *Delinquency and Crime: Current Theories,* edited by J. David Hawkins, 198–235. Cambridge: Cambridge University Press, 1996.

———. "Reflections on the Advantages and Disadvantages of Theoretical Integration." In *Theoretical Integration in the Study of Deviance and Crime: Problems and Prospects,* edited by Steven F. Messner, Marvin D. Krohn, and Allen E. Liska, 51–60. Albany, NY: State University of New York Press, 1989.

———. "Toward an Interactional Theory of Delinquency." *Criminology* 25 (1987):863–891.

Thornberry, Terence P., Alan J. Lizotte, Marvin D. Krohn, Margaret Farnworth, and Sung Joon Jang. "Delinquent Peers, Beliefs, and Delinquent Behavior: A Longitudinal Test of Interactional Theory." *Criminology* 32 (1994):47–83.

Tittle, Charles R. "Continuing the Discussion of Control Balance." *Theoretical Criminology* 3 (1999):344–352.

———. "Control Balance." In *Explaining Criminals and Crime: Essays in Contemporary Criminological Theory,* edited by Raymond Paternoster and Ronet Bachman, 315–334. Los Angeles: Roxbury, 2001.

———. *Control Balance: Toward a General Theory of Deviance.* Boulder, CO: Westview Press, 1995.

Vogt, W. Paul. *Dictionary of Statistics and Methodology: A Nontechnical Guide for the Social Sciences.* Newbury Park, CA: Sage, 1993.

Vold, George B., Thomas J. Bernard, and Jeffrey B. Snipes. *Theoretical Criminology.* 5th ed. New York: Oxford University Press, 2002.

Weis, Joseph G., and J. David Hawkins. *Preventing Delinquency.* Washington, DC: GPO, 1981.

ENDNOTES

1. Conger, "Social Control," 17–18.
2. Rick's case study tells us that his first contact with the juvenile justice system was at age 12. It may be, however, that Rick actually began involvement in delinquency at an earlier age, but that his earlier delinquent acts escaped notice by police.
3. Beirne, "Inventing Criminology," and *Inventing Criminology*; and Gottfredson and Hirschi, "Positive Tradition."
4. Osgood et al., "Routine Activities."
5. Briar and Piliavin, "Delinquency."
6. See Miller and Lynam, "Structural Models."
7. Hirschi and Hindelang, "Intelligence and Delinquency."
8. Miller and Lynam, "Structural Models."
9. Hirschi, *Causes of Delinquency.*
10. Sampson and Laub, *Crime in the Making.*
11. Gottfredson and Hirschi, *General Theory of Crime.*
12. Sutherland, *Principles of Criminology*; and Burgess and Akers, "Differential Association–Reinforcement Theory."
13. Shaw and McKay, *Juvenile Delinquency.*
14. Merton, "Social Structure and Anomie."
15. Ibid.
16. Agnew, "Revised Strain Theory," "Foundation," and "Contribution."
17. For example, Hirschi, *Causes of Delinquency*; Jensen, "Parents;" Hepburn, "Testing Alternative Models;" and Matsueda, "Testing."
18. Elliott, Ageton, and Canter, "Integrated Theoretical Perspective," 20.
19. Messner, Krohn, and Liska (*Theoretical Integration*) present a collection of thoughtful essays about the problems and prospects regarding theoretical integration in the study of crime and delinquency.
20. Elliott, Ageton, and Canter, "Integrated Theoretical Perspective."
21. Elliott, "Assumption."

22. Ibid.

23. Vold, Bernard, and Snipes, *Theoretical Criminology*, 301. See also Elliott, "Assumption;" and Elliott, Huizinga, and Ageton, *Explaining Delinquency*.

24. Hirschi, "Separate and Unequal."

25. Hirschi, "Separate and Unequal;" and Hirschi, "Exploring Alternatives."

26. Thornberry, "Reflections," 56.

27. For example, we could have also discussed Johnson, *Juvenile Delinquency*; Weis and Hawkins, *Preventing Delinquency*; Colvin and Pauly, "Critique of Criminology;" Pearson and Weiner, "Toward an Integration;" Braithwaite, *Crime, Shame and Reintegration*; Catalano and Hawkins, "Social Development Model;" Colvin, Cullen, and VanderVen, "Coercion;" and Agnew, "Integrated Theory."

28. Elliott, Ageton, and Canter, "Integrated Theoretical Perspective;" and Elliott, Huizinga, and Ageton, *Explaining Delinquency*.

29. Elliott, Ageton, and Canter, "Integrated Theoretical Perspective," 9–11; and Elliott, Huizinga, and Ageton, *Explaining Delinquency*, Chapter 4.

30. Elliott, Ageton, and Canter, "Integrated Theoretical Perspective," 9.

31. Ibid., 13.

32. Ibid., 14.

33. Ibid.

34. Ibid., 14–15.

35. Thornberry argues that theoretical *elaboration* extends existing theory by "using available theoretical perspectives and empirical findings to provide a more accurate model of the causes of delinquency. In the process of elaboration, there is no requirement to resolve disputes among other theories—for example, their different assumptions about the origins of deviance; all that is required is that the propositions of the [new] model . . . be consistent with one another" ("Toward an Interactional Theory," 865). Genuine theoretical *integration,* on the other hand, requires reconciliation of conflicting elements of different theories (Thornberry, "Reflections," 59–60). We include Thornberry's interactional theory in our discussion of integrated theories, but we note that Thornberry himself considers interactional theory to be an *elaboration*, rather than an *integration,* of key criminological theories.

36. Thornberry, "Toward an Interactional Theory." See also Thornberry et al., "Delinquent Peers;" and Thornberry, "Empirical Support."

37. Thornberry, "Toward an Interactional Theory," 866.

38. Ibid., 873.

39. Ibid., 875.

40. Ibid., 876.

41. Tittle, *Control Balance*. See also Tittle, "Continuing the Discussion," and "Control Balance." In several articles, Alex Piquero, Matthew Hickman, and their colleagues have tested, extended, and applied control balance theory (see Hickman and Piquero, "Exploring the Relationships;" Hickman et al., "Applying;" Piquero and Hickman, "Extending;" and Curry and Piquero, "Control Ratios").

42. Tittle, *Control Balance,* 124. See also Tittle, "Control Balance," 330–331.

43. Tittle, "Control Balance," 316.

44. Tittle, *Control Balance,* 135.

45. Tittle, "Control Balance," 317.

46. Tittle, *Control Balance,* 142.

47. Tittle, *Control Balance,* 142; and Tittle, "Control Balance," 317.

48. Tittle, *Control Balance,* 162–166.

49. Tittle, *Control Balance,* 169.

50. Tittle, *Control Balance,* 167–168; and Tittle, "Control Balance," 321–322.

51. Tittle, *Control Balance,* Chapter 8; and Tittle, "Control Balance," 328–330.

52. Tittle, *Control Balance,* 201.

53. Tittle, "Control Balance," 325.

54. Ibid., 327–328.

55. Ibid., 326–327.

56. Miethe and Meier, *Crime*. See also Drass and Miethe, "Qualitative Comparative Analysis."

57. Miethe and Meier, *Crime*, 2–3.

58. Cohen and Felson, "Social Change."
59. Miethe and Meier, *Crime,* 3.
60. Ibid.
61. Sampson, Raudenbush, and Earls, "Neighborhoods and Violent Crime."
62. Miethe and Meier, *Crime,* 61.
63. Ibid., 64.
64. Ibid., 66.
65. Ibid.
66. Ibid., 64.
67. Ibid.
68. Tittle, *Control Balance,* 135.

Responding to Juvenile Delinquency

Contemporary Juvenile Justice

15

Chapter Objectives

After completing this chapter, students should be able to:

- Describe the structure of juvenile justice systems in the United States.
- Provide examples of how discretion and diversion are exercised throughout the juvenile justice process.
- Discuss the different roles that police take in their encounters with juveniles.
- Define due process of law and describe how due process affects juvenile law enforcement.
- Identify and describe the juvenile court process, including both informal and formal procedures.
- Inventory the full range of dispositional options that are authorized under state statutory law.

- Describe the three major areas of juvenile corrections: probation, community-based corrections, and residential placements.
- Understand key terms:
  discretion
  diversion
  problem behavior
  risk factor
  protective role
  taking into custody
  crime-fighting role
  collaborative role
  law enforcer role
  search and seizure
  arrest
  custodial interrogation
  referral
  intake
  petition
  informal disposition
  detention
  adjudication
  predisposition report
  formal disposition
  commitment
  probation conditions
  probation supervision
  revocation
  residential placement
  group homes
  custodial institutions
  aftercare
  parole

CASE IN POINT

The Juvenile Court's Dispositional Order for Rick

The case of Rick is real; however, his name has been changed to maintain confidentiality. Nor can Rick's final dispositional order be disclosed. While the following dispositional order is hypothetical, with fictitious names and places, it is similar in structure and content to the one that was handed down in Rick's case.

```
1    JEFFREY H. HOUSTON
2    District Judge
3    Fifth Judicial District
4    Hastings County Courthouse
5    Belfry, Minnesota
6
7         FIFTH JUDICIAL DISTRICT COURT, HASTINGS COUNTY
8
9    IN THE MATTER OF:      )              Cause No. DJ 96 - 58
10   RICHARD A. SMITH      )              ORDER OF COMMITMENT
11
12   A Youth Under the Age of 18.
13       On the 1st day of October 2004, Richard Smith, a youth of this
14   County, 17 years of age, was brought before me pursuant to a
15   Petition of Gross Criminal Mischief previously filed. The youth
16   appeared with his parents, Bernard and Mary Smith, and his
17   attorney, Michael Langley. The Court fully advised the youth and his
18   parents of all legal rights accorded by the laws of the State of
19   Minnesota. Also present were Deputy County Attorney Jane Kline,
20   and Juvenile Probation Officer James Burfeind.
21       Upon being questioned by the Court concerning the allegations in
22   the petition, the youth admitted the allegations to make a finding
23   of delinquency in addition to a finding that the youth had violated
24   his probation. After reviewing the predispositional report, and it
25   having been established that reasonable efforts have been made to
26   prevent or eliminate removal of the youth from the home, it is
27   determined that removal of the youth from the home is necessary at
28   this time because continuation in the home would be contrary to the
29   welfare of the youth. The Court entered the following Order:
30       IT IS HEREBY ORDERED that Richard Smith is found to be a Serious
31   Juvenile Offender, having previously been found to be a Delinquency
32   Youth.
33       IT IS HEREBY ORDERED that Richard Smith is committed to the
34   Minnesota State Department of Correction for commitment to Red
35   Wing State Training School.
```

1. **1.** The youth is committed to the Department of Corrections for an indefinite period of time for placement at Red Wing State Training School.
2. **2.** The youth will follow and abide by the rules of Red Wing State Training School.
3. **3.** The youth will attend school regularly, he will follow all the rules of the school, and he will maintain passing grades.
4. **4.** The youth will participate in treatment programs available at Red Wing State Training School.
5. **5.** The youth will submit to random drug testing and UA's upon request of staff at Red Wing State Training School or the Probation Officer.
6. **6.** The youth will submit to a search of his person or living area at the request of staff at Red Wing State Training School or the Probation Officer.
7. **7.** The Department of Corrections will notify the Judge two weeks in advance of the youth's release from Red Wing State Training School.
8. **8.** Upon release, the youth will report to his Probation Officer and be placed on Intensive Supervision Probation for a ninety-day period, followed by regular probation supervision.
9. **9.** The youth is liable for payment of restitution in the amount to be determined by the Court, to be paid in full by the time the youth turns 18 years of age. The Court will retain jurisdiction over the youth should he fail to pay restitution.
10. **10.** The youth will not be in possession or consume any alcohol or illicit drugs.

DONE IN OPEN COURT this 1st day of October, 2004.
   DATED this 28th day of October, 2004.

Jeffrey H. Houston
DISTRICT JUDGE

Back in Chapter 2 we saw that the "second revolution" resulted in major redefinition of the legal concept of juvenile delinquency and redirection of the philosophy, authority, and process of juvenile justice systems across the United States.[1] In general, the *parens patriae* authority of the original juvenile court was reduced substantially and the **rehabilitative ideal** was seriously questioned. Detention, formal adjudication, and more punitive disposition, including incarceration, are now used commonly for serious juvenile offenders, whereas informal procedures and diversion, especially those involving community-based alternatives, are employed for status and nonviolent juvenile offenders.

In his critique of contemporary juvenile justice, Barry Feld examined four areas of change: (1) greater emphasis of procedural due process with minimal impact; (2) the shift in legal philosophy from the rehabilitative ideal to punishment and criminal responsibility, especially for serious, violent juvenile offend-

rehabilitative ideal The traditional legal philosophy of the juvenile court which emphasizes assessment of the youth and individualized treatment, rather than determination of guilt and punishment.

ers; (3) reforms to divert, deinstitutionalize, and decriminalize status offenders (the "soft end" of the juvenile court), thereby sending them to a "hidden," private sector treatment system of social control, administered by mental health and chemical dependency service providers; and (4) the transfer of serious juvenile offenders (the "hard end" of the juvenile court) to criminal court jurisdiction.[2] As a result of these sweeping changes, the structure and process of contemporary juvenile justice systems have come to resemble those of adult criminal justice systems. Nonetheless, elements of *parens patriae* authority and the rehabilitative ideal continue to survive, preserving some level of distinction for juvenile justice systems.

This chapter describes contemporary juvenile justice in terms of its structure and process—how the different elements of juvenile justice are organized and how they operate. In a formal sense, the juvenile justice system is comprised of police, courts, and corrections. But as we will see, a substantial amount of juvenile delinquency is dealt with informally, sometimes by agencies not usually considered a part of the "system." We begin by describing how juvenile justice is really not a "system" at all, but is highly decentralized and fragmented among a wide variety of both public and private agencies. Moreover, each element of juvenile justice exercises tremendous discretion and relies on diversion to operate efficiently. After clarifying these important characteristics of contemporary juvenile justice, we turn to prevention strategies, as they have become increasingly important since the second revolution. Then, we examine both informal and formal procedures of contemporary juvenile justice systems: police, courts, and corrections.

◼ Structure of Juvenile Justice Systems: Decentralized and Fragmented

It may surprise you to learn that there is not a single juvenile justice system in the United States. Rather, each state has a separate juvenile justice system, and they are "systems" in only a limited sense. The various parts of these systems—police, courts, and corrections—are all operated by different levels and branches of government, and sometimes even by private organizations. For example, courts with jurisdiction over juvenile matters are a judicial entity, and they operate at the federal, state, and local levels, but there is not a single, standard structure to juvenile courts. In most states, juvenile courts operate within district courts (courts of original trial jurisdiction). However, some lower courts, like city and municipal courts, have special or limited jurisdiction in juvenile cases. Juvenile traffic violations, for example, are often handled by these lower courts. The federal court system also deals with juvenile matters, usually at the District Court level, but the number of juvenile cases dealt with by federal courts is minuscule compared to state juvenile courts. Thus, the juvenile justice system is said to be *decentralized,* operating at every level of government.

The juvenile justice system is also *fragmented* among the different branches of government. Each branch of government—legislative, judicial, and executive—plays an important role in juvenile justice. For example, laws are the foundation of juvenile justice systems. Under the United States Constitution, such laws are the responsibility of legislative bodies at all levels—local, state, and federal. The structure, jurisdiction, and authority of juvenile justice systems are primarily determined by state legislatures. For instance, the law that created the first juvenile court in Chicago was enacted by the Illinois Legislature in 1899. Similarly, contemporary juvenile courts are legally authorized by state youth court acts. State and federal legislatures also pass codes that regulate people's actions (criminal codes) and establish procedures to be followed (procedural codes) when someone breaks the law. Thus, legislatures play a central role in juvenile justice, even though they do not actually operate juvenile justice agencies.

Table 15-1 shows the structural elements of juvenile justice in terms of their level and branch of government. As you can see, the various parts of juvenile justice systems are spread broadly among the different levels and branches of government, making them decentralized and fragmented. Let's look at a few of these components of juvenile justice systems. Beyond statutory law, legislative units also allocate funds for the different components of juvenile justice systems. Courts, including juvenile courts, are a judicial responsibility. Juvenile courts provide both informal and formal procedures to deal with cases of delinquency that are referred to them. The many facets of probation are often a judicial responsibility, operated by juvenile courts. Appellate courts at both the state and federal levels have tremendous influence on juvenile justice process, establishing due process of law in juvenile matters. The executive branch of government is responsible for law enforcement, the operation of detention centers, prosecution, departments of corrections, and sometimes probation and parole. Probation and parole are perhaps the most structurally varied components of juvenile justice. Juvenile probation is often a judicial function, but in some states it is operated by the department of corrections. In addition, parole is sometimes administratively attached to probation, in which case both operate under the same governmental unit—either the courts or the department of correction. In other

Table 15-1	Juvenile Justice Structure: Decentralized and Fragmented		
	Legislative	**Judicial**	**Executive**
Local	Ordinances Funding	Courts of limited jurisdiction	Law enforcement Detention centers Prosecution Corrections
State	Statutory law (criminal codes, youth court acts, and procedural codes) Funding	Juvenile courts Probation Appellate courts	Law enforcement Detention centers Prosecution Corrections Probation
Federal	Statutory law (criminal code, procedural code) Funding	District courts Probation Appellate courts	Law enforcement Detention centers Prosecution Corrections

states, probation is a part of the juvenile court and parole is a part of the department of correction.

■ Discretion and Diversion: Here, There, and Everywhere

The structure and process of juvenile justice vary considerably from state to state and from community to community. Nevertheless, some common elements and procedures can be observed in most juvenile justice systems.[3] Two features are key to the juvenile justice process in literally every jurisdiction throughout the country: discretion and diversion.

Well-known legal scholar Roscoe Pound defined **discretion** as "authority conferred by law to act in certain conditions or situations in accordance with an official's or an official agency's considered judgment and conscience."[4] While discretion is usually discussed with regard to decision making by the police, virtually all decisions in the juvenile justice process carry with them wide-ranging discretion. We will continually refer to this authority and latitude in decision making as we describe the structure and process of juvenile justice.

A second prevalent feature of the juvenile justice process is diversion. **Diversion** is the tendency of juvenile justice systems to deal with juvenile matters informally, without formal processing and adjudication, by referring cases to special programs and agencies inside or outside the juvenile justice system. Police officers commonly divert some cases to special programs within the department or to a variety of community resources such as family counseling and drug treatment. Probation officers, when screening cases that have been referred to the juvenile court, have legal authority to decide whether cases should be dealt with formally or informally. Informal handing may involve a diversionary referral to some community treatment option. Prosecutors can choose not to petition a case and instead to provide for some informal handling of a case, often involving a referral to community resources.

discretion "Authority conferred by law to act in certain conditions or situations in accordance with an official's or an official agency's considered judgment and conscience" (Pound, "Discretion, Dispensation, and Mitigation," 926).

diversion The tendency to deal with juvenile matters informally, without formal processing and adjudication, by referring cases to agencies outside the juvenile justice system.

■ Delinquency Prevention

While juvenile justice systems are the primary mechanism by which communities respond to delinquent acts, delinquency prevention programs represent a proactive approach to juvenile crime, attempting to prevent it before it occurs. In the late 1960s, the President's Commission on Law Enforcement and the Administration of Justice issued a series of important reports. One of them, the *Task Force Report on Juvenile Delinquency and Youth Crime*, concluded: "In the last analysis, the most promising and so the most important method of dealing with crime is by preventing it—by ameliorating the conditions of life that drive people to commit crime and that undermine the restraining rules and institutions erected by society against antisocial conduct."[5] Similarly, in 1973, the National Advisory Commission on Criminal Justice Standards and Goals recommended that "first priority . . . be given to preventing juvenile delinquency, to minimizing the involvement of young offenders in the juvenile and criminal justice system, and to reintegrating delinquent and young offenders into the community."[6] Delinquency prevention became a national priority through the Juvenile Justice and Delinquency Prevention (JJDP) Act of 1974. This Act also created the Office of Juvenile Justice and Delinquency Prevention (OJJDP), whose charge was to assist states and local communities to develop policies, practices, and programs directed at delinquency prevention. The resulting prevention perspective stands in sharp contrast to formal processes of juvenile justice.

The Prevention Perspective: Adolescents At Risk

As a developmental stage, adolescence is a period of maturation, change, and unrest. The word adolescence is derived from the Latin verb *adolescere,* meaning to grow up or come to maturity.[7] In her book *Adolescence at Risk,* Joy Dryfoos offers seven critical tasks of adolescence that are drawn from the psychological literature on adolescent development: (1) the search for self-definition; (2) the search for a personal set of values; (3) the acquisition of competencies necessary for adult roles, such as problem solving and decision making; (4) the acquisition of skills for social interaction with parents, peers, and others; (5) the achievement of emotional independence from parents; (6) the ability to negotiate between the pressure to achieve and the acceptance of peers; and (7) experimentation with a wide variety of behaviors, attitudes, and activities.[8]

The developmental tasks of adolescence take place during a time in life that is often characterized by "storm and stress."[9] According to G. Stanley Hall, an early psychologist, the turmoil of adolescence is caused by hormonal changes and the accompanying growth spurts and mood swing. These changes, he said, are at the root of adolescent misbehavior, including delinquency. Contemporary research, however, indicates that the majority of teenagers weather the challenges of the adolescent years without developing significant social, emotional, or behavioral difficulties.[10] Nonetheless, adolescence is inherently a period of *dependency and risk* because of the pivotal importance of adolescent development and the tremendous influence that the social environment has on such development. Psychologist Richard Jessor describes successful adolescent development as "the accomplishment of normal developmental tasks, the fulfillment of expected social roles, the acquisition of essential skills, the achievement of a sense of adequacy and competence, and the appropriate preparation for transition to the next stage in the life trajectory—young adulthood."[11] Thus, adolescence involves a striving for independence and a dependence on others.

During the adolescent years, involvement in **problem behaviors** compromises successful development.[12] As described more fully in Chapter 6, research suggests that juvenile delinquency is a component of a larger group of problem behaviors that tend to occur together—behaviors such as drug and alcohol use, mental health problems, behavior problems and underachievement in school, and precocious and risky sexual behavior. While many delinquent youth are "multiple problem youth," the origins and development of these problems are distinct.[13] Youth who display problem behaviors early in life are more likely to develop persistent and serious patterns of problem behaviors.[14] In particular, children (age twelve and younger) who engage in delinquent behavior are two to three times more likely to become chronic offenders than youths who first engage in delinquency in their adolescent years.[15]

problem behaviors
Behaviors that compromise successful adolescent adjustment, such as drug and alcohol use, mental health problems, underachievement in school, precocious and risky sexual behavior, and delinquency.

Delinquency prevention programs seek to reduce the likelihood of such problem behaviors, since problem behaviors compromise successful adolescent development. Research has identified consistently a number of <u>risk factors</u> that increase the likelihood of problems behaviors and, ultimately, negative developmental outcomes. The OJJDP Study Group on Serious and Violent Juvenile Offenders and the Study Group on Very Young Offenders reviewed this literature and concluded that a limited number of empirically verified risk factors can be identified, as listed in Research In Action, "Risk Factors: Causes and Correlates of Delinquency."[16] These risk factors are grouped into five categories: individual, family, peer, school, and neighborhood and community factors. Contemporary approaches to delinquency prevention argue that these risk factors must be first identified through research and then addressed through a broad range of prevention programs at the family, school, and community levels.[17]

> **risk factor** Any individual trait, social influence, or environmental condition that leads to greater likelihood of problem behaviors and ultimately negative developmental outcomes during the adolescent years.

RESEARCH IN ACTION

Risk Factors: Causes and Correlates of Delinquency

Individual Factors

- verbal deficits (affecting listening, reading, problem solving, speech, writing, and memory)
- inattentiveness and impulsiveness
- difficult temperament/limited agreeableness (disagreeable, oppositional, defiant, rebellious)
- limited conscientiousness
- low intelligence and academic failure
- limited behavioral inhibition (self-control)
- risk-taking/sensation seeking
- early onset of problem behaviors
- aggression
- social withdrawal
- attitudes favorable to problem behaviors

Family Factors

- poor family management and direct controls (monitoring and supervision, standards for behavior, recognizing and responding to problem behaviors)
- lax, harsh, and inconsistent discipline
- child maltreatment (especially neglect)
- low levels of parental involvement
- weak family attachment and indirect controls
- parental attitude favorable to deviance
- parental criminality, substance abuse, psychopathology
- family and marital conflict
- family disruption, including divorce and separation
- residential mobility

School and Academic Factors

- early and persistent classroom disruption and antisocial behaviors
- poor academic performance
- weak social bonds to school (attachment, commitment, and involvement)
- limited academic aspirations
- truancy and dropping out
- frequent school transition
- school with high rate of delinquency
- low academic quality of school
- low parent and community involvement in school

Peer-Related Factors

- association with delinquent peers (frequency, duration, priority, and intensity)
- peer rejection
- delinquent attitudes and behaviors are modeled, imitated, and reinforced in a delinquent peer group
- delinquent values and attitudes
- learned techniques for committing crimes
- gang membership

Neighborhood and Community Factors

- concentrated disadvantage (geographic concentration, social isolation, joblessness)
- social disorganization (poverty, cultural heterogeneity, and residential mobility)
- ineffective social control (collective efficacy)
- delinquent and criminal subculture (tradition of crime, socialization, availability of drugs and access to weapons)

Source: Hawkins et al., "Predictors of Youth Violence;" Loeber, Farrington, and Petechuk, "Child Delinquency;" Miller and Lynam, "Structural Models of Personality;" Moffitt, "Developmental Taxonomy;" Wasserman et al., "Risk and Protective Factors."

Delinquency Prevention Programs

"Risk-focused prevention is based upon the simple premise that to prevent a problem [behavior] from occurring, we need to identify the factors that increase the risk of that problem developing and then find ways to reduce the risk."[18] Nonetheless, the task of identifying and reducing risk factors is not simple or easy. Only in the last fifteen years have delinquency prevention efforts been based on empirical findings of the causes and correlates of delinquent behavior and evaluation research of program effectiveness.[19]

Dryfoos identifies eleven key characteristics of effective prevention programs.[20] These features involve coordinated attention to the different risk factors and the different contexts for service delivery—families, schools, peer groups, neighborhoods, mental and public health, child welfare, and juvenile justice.

1. **Intensive individualized attention.** Successful programs focus on the individual needs of the youth and individualize intervention accordingly.

2. **Communitywide, multiagency collaboration.** Community programs and services need to work together collaboratively.

3. **Early identification and intervention.** Successful programs begin at the earliest point in the development of problem behaviors by offering early assessment and intervention.

4. **Locus in schools.** Because of the important link between academic achievement and problem behaviors, successful prevention programs are often located in schools.

5. **Administration of school programs by agencies outside of schools.** While successful programs are centered in the schools, they are administered by nonschool community agencies.

6. **Location of programs outside of schools.** Successful programs are not limited to schools. Consistent with the need for community-wide, multi-agency collaboration, prevention programs also involve nonschool, community intervention.

7. **Arrangements for training.** Staff training and supervision is central to success.

8. **Social skills training.** Successful programs include social skills training for youth, including such skills as the development and maintenance of relationships, responses to peer pressure, and conflict resolution.

9. **Engagement of peers in interventions.** Successful programs effectively mobilize peer groups to generate change.

10. **Involvement of parents.** Parents too have an important influence on youth. Successful programs involve parents, educating them, training them, and allowing them to participate in programs.

11. **Link to the world of work.** Successful prevention programs include the opportunity for work experiences in the labor force.[21]

Beyond these program features, effective delinquency prevention depends on program implementation—how the program is put in place, how it operates, and how it is administered. Even programs that have been evaluated as effective through research must be implemented properly to produce the desired effects. The Center for the Study and Prevention of Violence at the University of Colorado–Boulder has recently conducted an evaluation of implementation quality, looking at aspects of program implementation rather than program features. Nine model programs at 147 sites were part of the study.[22] Questionnaires were administered at each site every four months, over a two-year period. The study revealed six key aspects of program implementation.

1. **Effective organization:** Successful programs depend on strong administrative support for the implementing staff; agency stability; a shared vision of program goals; and interagency links, especially with other programs involved in clients' treatment plans.

2. **Qualified staff:** Program success is fostered by staff who support the program and who are motivated to implement the program day in and

day out. Program staff must also be skilled, experienced, and have the necessary credentials to carry out the program. They must also be given the time necessary to implement the program, especially with a new program. Paid staff have been found to be more effective than volunteers.

3. **Program champion(s):** Every successful program needs a person who champions the program—who motivates, innovates, guides, and fosters program delivery.

4. **Program integration:** Success is most likely when a program is integrated into a larger organizational structure, in which the prevention program supports and augments the host agency's goals and objectives. Bullying prevention, for example, is most effectively implemented within a school context in which student safety and security is taught as a part of the academic curriculum.

5. **Training and technical assistance:** "A strong, proactive package of training and technical assistance builds confidence and can help agencies overcome and even avoid many implementation barriers."[23]

6. **Implementation fidelity:** Sometimes referred to as integrity, fidelity has to do with the degree to which the program is actually delivered as it was designed. Successful programs tend to be those that closely follow the goals and methods of the program's design.

Over the course of this book, you have read about a variety of delinquency prevention programs through "Theory into Practice" applications. Here we offer several brief descriptions of prevention programs that have been evaluated as effective. These are often referred to as "model programs," and there is a great deal of agreement in identifying these prevention programs as effective.[24] The OJJDP Web site is a rich source of descriptions and evaluation research on delinquency prevention programs (see Links, "Office of Juvenile Justice and Delinquency Prevention").

Links

Office of Juvenile Justice and Delinquency Prevention

The Web site of the Office of Juvenile Justice and Delinquency Prevention (OJJDP) is a rich source for resources on matters related to juvenile delinquency. In the context of the present discussion, the OJJDP Web site offers resource materials on risk factors and delinquency prevention. On the sidebar, choose "Topics." "Delinquency Prevention" is then one of the topics offered. When you click on this, numerous resources related to funding, programs, and publications are easily accessible. Find the link to this Web site at:

http://criminaljustice.jbpub.com/burfeind

Prenatal and Infancy Home Visits by Nurses

Designed for low-income, first-time single mothers, the nurse visitation program targets three risks factors associated with the development of early antisocial behavior in children: adverse maternal health-related behaviors during pregnancy, child abuse and neglect, and troubled maternal life course. Public health nurses with small caseloads make prenatal and infancy home visits every one to two weeks and continue home visitation until the child is two years old. Nurses provide support and instruction on prenatal health, infancy caregiving, personal health, child development, parenting, pursuing education, and career development. Extensive evaluation shows that the program reduced rates of child abuse and neglect by helping mothers learn effective parenting skills and personal controls. Mothers receiving nurse home visits had fewer months on welfare, fewer behavioral problems, and lower arrest and conviction rates than mothers who did not participate in the program. Adolescents whose mothers participated in the nurse visitation program were less likely to run away, be arrested, or be convicted of a crime. As compared to the children of nonparticipants, these adolescents smoked fewer cigarettes, consumed less alcohol, and had fewer behavioral problems related to alcohol and drug use. By the time a child reached age fifteen, cost savings were estimated as four times the original investment as a result of reductions in crime, welfare and health care costs, and taxes paid by working parents.[25]

The Incredible Years Series: Family and Teacher Training

The Incredible Years Parent, Teacher, and Child Training Series is a comprehensive set of curricula that provide parent training, teacher training, and child training. The lessons are designed to promote social competence and to prevent and respond to conduct problems in young children. The program targets children, ages 2 to 8, who exhibit or are at risk for conduct problems. "Trained facilitators use interactive presentations, videotape modeling, and role playing techniques to encourage group discussion, problem solving, and sharing of ideas."[26] The parent training component comprises three series: BASIC, ADVANCE, and SCHOOL. The BASIC series is the core of the program, teaching parents how to engage in interactive play, reinforcement skills, nonphysical discipline techniques, logical and natural consequences, and problem-solving strategies. The teacher training series is directed at enhancing classroom management skills. This includes reinforcing prosocial behavior, encouraging cooperation with peers and teachers, teaching discipline techniques, and teaching anger management and problem-solving skills. The child training series is known as the Dina Dinosaur curriculum. The curriculum is written for children in the preschool and primary grades and tries to promote awareness of emotions, empathy with others, the making and keeping of friends, anger management, interpersonal problem solving, obedience of school rules, and success in school. Evaluation research on the Incredible Years Series has shown it to improve parenting interactions, to reduce aggression in participating families, and to improve classroom management by teachers.[27]

Promoting Alternative THinking Strategies (PATHS): Classroom-Based Prevention

PATHS is a school-based prevention curriculum taught by teachers in kindergarten through fifth grade. The curriculum is designed to promote social and emotional competence and to decrease risk factors for delinquency. Lessons are taught in twenty-minute sessions, three times per week, and they cover the topics of self-control, emotional understanding, self-esteem, relationships, and interpersonal problem-solving skills. "Lessons are sequenced according to increasing developmental difficulty and include activities such as dialoguing, role-playing, storytelling, modeling by teachers and peers, and social and self-reinforcement. Among other lessons, youth are taught to identify and label their feelings; express, understand, and regulate their emotions; understand the difference between feelings and behaviors; control impulses; and read and interpret social cues. Youth are given activities and strategies to use inside and outside the classroom, and parents receive program materials to reinforce behaviors at home."[28] Studies have compared classrooms that are taught the PATHS curriculum with several different types of matched control groups. Compared to students in control groups, students who participated in the PATHS program were significantly better at recognizing and understanding emotions, understanding social problems, and thinking of effective alternative solutions, and they less frequently showed aggressive behaviors. Teachers reported improvement in student's self-control, emotional understanding, ability to tolerate frustration, and use of conflict resolution strategies.[29]

Big Brothers Big Sisters of America: Mentoring

Founded in the early 20th century, Big Brothers Big Sisters of America (BBBSA) attempts to provide significant persons in the lives of at-risk youth. Volunteer mentors are carefully screened, trained, and matched with an appropriate youth, age 6 to 18. "A mentor meets with his or her youth partner at least three times a month for 3 to 5 hours, participating in activities that enhance communication skills, develop relationship skills, and support positive decision making. Such activities are determined by the interests of the child and the volunteer and could include taking walks, attending school activities or sporting events, playing catch, visiting the library, or just sharing thoughts and ideas about life."[30] The mentoring relationship provides a significant relationship and a positive role model in the life of an at-risk youth, thereby promoting prosocial socialization. "An 18-month study of eight BBBSA affiliates found that when compared with a control group on a waiting list for a match, youth in the mentoring program were 46 percent less likely to start using drugs, 27 percent less likely to start drinking, and 32 percent less likely to hit someone. Mentored youth skipped half as many days of school as control youth, had better attitudes toward and performance in school, and had improved peer and family relationships."[31]

Minnesota Delinquents Under 10 Program

The Delinquents Under 10 Program in Hennepin County, Minnesota involves coordinated case planning and intervention among several county departments, including Children and Family Services, Economic Assistance, Community Health, and the County Attorney's Office. A screening team reviews police reports and then determines appropriate early intervention for children deemed to be at the highest risk for future delinquency. A "wraparound network" is developed for each targeted child, including: "a community-based organization to conduct indepth assessments, improve behavior and school attendance, and provide extracurricular activities; an integrated service delivery team made up of county staff who coordinate service delivery and help children and family members access services; a critical support person or mentor; and a corporate sponsor that funds extracurricular activities."[32]

As these delinquency prevention programs illustrate, prevention programs operate in a variety of service sectors: mental health, public health, education, child welfare, and juvenile justice. Most are based in the local community, sometimes using community volunteers. Even prevention programs that operate in the juvenile justice service sector rely on community resources and services. Regardless of where they occur, prevention programs seek to minimize youth involvement in the juvenile justice system and attempt to integrate youth into the community.[33]

■ Police Encounters with Juveniles[34]

Juveniles constitute a unique population that poses a special set of problems for the police.[35] First, police have frequent contact with juveniles. In 2003, police made an estimated 2.2 million arrests of juveniles nationwide, but this number represents only a fraction of all police–juvenile encounters, many of which never result in arrest.[36] Nonetheless, juveniles account for a significant portion of the crime problem. Second, law enforcement with juveniles includes a broad array of offenses, ranging from status offenses to violent crime. Third, police commonly view juvenile delinquency as minor crime, and "taking a youth into custody" is often considered not a "real arrest."[37] Fourth, police are often expected to take an approach with juveniles that involves prevention, protection, and rehabilitation, while at the same time they are expected to enforce the law.[38] A report on police work with juveniles captures this dilemma: "crime prevention can be viewed as 'social work,' a role that police often see as taking time away from what they consider to be their primary role—the apprehension of criminals."[39] Fifth, juveniles must be processed through a justice system that is different from the adult system in philosophy, jurisdiction, structure, and process. This is unfamiliar terrain for some officers. Moreover, police have little support for a system that they view as pandering—one that does not hold youths accountable for their actions. Some police officers have trouble accepting all of this, and as a result they have little interest in working with juveniles.[40] The feeling is apparently mutual, as a long track record of research indicates that most high school students are lukewarm in their attitudes toward the police. "Young people consistently express more negative attitudes toward the police than older people."[41]

Most police departments of at least moderate size have specialized juvenile units to deal with juveniles and delinquency.[42] Criminologist Larry Gaines describes the reason why juvenile units within police departments are so common.

> *Although most of a department's contact with juveniles is by patrol officers, juvenile officers generally process youth who are placed in*

custody and coordinate programs designed to reduce juvenile delinquency. Departments developed juvenile units because often juvenile cases are complex and many patrol officers do not remain familiar with how juvenile cases are processed. Juvenile courts have implemented specific rules to protect the juvenile, and these rules are quite different from criminal procedures. The juvenile unit maintains all files relating to juvenile cases and works with other officers in presenting their cases in juvenile court. This unit is generally housed in the criminal investigation or detective division within the department.[43]

The Police Role with Juveniles

Police role refers to the expectations associated with the position of police officer. As Samuel Walker and Charles Katz point out, "There is significant controversy over the proper police role toward juveniles. Some people favor a strict law enforcement role, emphasizing the arrest of offenders. Others prefer a crime prevention role, arguing that the police should emphasize helping young people who are at risk with advice, counseling, and alternatives to arrest."[44] Police officers sometimes struggle with the apparently inconsistent expectations of their work, believing that their primary duty is to fight crime by apprehending criminals, while at the same time they must also assist others and provide services to the public. Such role conflict can make police work difficult to manage for some individual officers and for departments as a whole. This inconsistency in role expectations seems especially pertinent to the policing of juveniles. In addition, uncertain and undefined departmental policies with regard to juveniles create further confusion for police officers.

Parens Patriae Policing: The Original Protective Role

protective role The original role of police in dealing with juveniles, in which police were expected to practice "kindly discipline" in place of parents—to function in *loco parentis*. As an extension of the juvenile court, police were to deal with juveniles in an informal, parent-like fashion, acting in the "best interests of the child."

Historically, police have been viewed as an extension of the juvenile court. In fact, a substantial portion of the staff of the first juvenile court in Chicago was police officers, paid by the Chicago Police Department (see Chapter 2, Case in Point, "Personnel of the Original Chicago Juvenile Court"). In this way, the original "juvenile court provided its own policing machinery and removed many distinctions between the enforcement and adjudication of laws."[45] While these officers were assigned to assist probation officers in supervision, their authority was derived from the court's *parens patriae* doctrine, and their role was consistent with the rehabilitative ideal that came to dominate juvenile justice. As such, police were expected to practice "kindly discipline" in the place of parents.[46] As an extension of the juvenile court, police were to deal with juveniles in an informal, parent-like fashion, acting in the "best interests of the child."[47] Recall from Chapter 2 that the early juvenile justice system was deliberately established to operate as a child-welfare approach, not as a judicial system.[48] This **protective role** was adopted generally by the police, giving them broad discretion in dealing with juveniles. In fact, the legal distinction between "taking into custody" and adult arrest was created to signify the protective approach that police have taken traditionally with juveniles. Under authority of statutory law, **taking into custody** involves taking control of a youth, much like in an arrest, but for protective purposes; this is sometimes referred to as *protective custody*. If the police

taking into custody The statutory authority given to police officers to take control of a juvenile, either physically or by the youth's voluntary submission, in an effort to separate that youth from his or her surroundings. Taking into custody is used when the juvenile has broken the law or is in need of protection.

officer believes that the youth is in need of protection or has violated the law, the police officer may take control of the youth and remove the youth from his or her surroundings. This can be accomplished by physical control or by voluntary submission of the youth. The youth may be placed in the back of the squad car, told to sit in a chair in the waiting area of the police station, or placed into a holding cell. Taking into custody may or may not involve placement in a detention facility.

Despite the clear connection that police had with the early juvenile court, police departments created few special programs designed to implement this protective role. In the first half of the twentieth century, "the two most common programs were Officer Friendly programs and police athletic leagues (PALS). The Officer Friendly programs consisted of officers visiting schools and making presentations about how the police operated. They were primarily intended to develop better relations with students. The PAL programs, which still exist, consist of involving youth, especially those from disadvantaged families, in sports programs. The police organize and sponsor baseball, basketball, and football programs. Both programs were designed to mentor youth, provide them with positive experiences, and keep them from wandering the streets and becoming involved in delinquency."[49]

The degree to which individual police officers and police departments adopted this protective role is a matter of debate. Criminologist Larry Gaines contends that throughout the first three quarters of the last century, "police officers generally saw their role as detaining and processing juvenile suspects."[50] Regardless, the protective origins of the police role established broad discretion for police in dealing with juveniles, and such broad discretion was exercised within an incredibly wide range of police activities related to juveniles, from prevention education and truancy reduction in school to investigation and arrest of criminal activity. Sociologists refer to this broad range of activities and expectation associated with a given position as *role diversity*. Role diversity aptly characterizes juvenile policing.

Professional Policing: The Crime-Fighting Role

Beginning in the early 1920s, during an historical period of policing known as the Reform Era, a strong and concerted movement emerged that significantly influenced juvenile policing, both in terms of its structure and role.[51] Referred to as *professional policing,* this movement produced at least two important innovations in policing. First, because of long-standing corruption and lack of centralized control in law enforcement, bureaucratic structure and authority was implemented in local law enforcement across the United States.[52] Police organization and management became more hierarchical, involving a chain and unit of command. Bureaucratic policing also resulted in specialization, in which different units performed special tasks simultaneously. Besides the creation of patrol, detective, and vice divisions, juvenile units were put in place in most medium and large police departments. Second, the protective role of policing juveniles was to a great degree replaced with the **crime-fighting role**. Enforcement of laws and crime control became the primary interest of local police. The relationship between the police and public was redefined, with the police being portrayed as the "thin blue line" that protected the lives and property of a passive

crime-fighting role The police role associated with professional policing that emphasizes enforcement of laws and crime control. The primary strategies for crime fighting involve motorized preventive patrol, rapid response, and reactive investigation.

and dependent public. Police began to view themselves as the professional experts on crime—the ones most qualified to fight a war on crime. While the protective role of the police was not out of line with the notion of the "thin blue line," those youth who formerly were viewed as dependent and at risk, requiring police protection, now began to be viewed as juvenile suspects who were most likely to violate the law.

Professional policing relied on three innovations to fight crime: motorized preventive patrol, the doctrine of rapid response to calls for emergency service, and reactive investigation of crime.[53] These innovations transformed the activities of police from *proactive* law enforcement to *reactive* law enforcement. The image of professional policing that emerged was that of a well-organized, tough-minded, highly trained crime fighter. Thus "professional policing" became synonymous with bureaucratic structure and authority within police departments and with the crime-fighting role.

Beginning in the mid-1960s, the due process revolution in juvenile justice introduced a variety of procedural requirements in law enforcement that can be viewed as either a limitation on crime fighting or as an enhancement of professional policing. Some police officers believed that due process requirements limited their abilities to gain evidence and statements, making arrest less likely; other officers, especially supervising officers, believed that procedural requirements actually enhanced professional policing by requiring officers to know the law and to enforce it in a standardized and objective fashion. Regardless of the point of view, the introduction of due process into juvenile policing substantially reduced the protective role of police, while emphasizing the importance and quality of arrest. We'll look further at due process in juvenile policing later.

Community-Oriented Policing: The Collaborative Role

Beginning as a "quiet revolution" in the early 1980s, community-oriented policing (COP) signifies a dramatic change in the philosophy, structure, and authority of law enforcement, especially as it relates to juveniles.[54] In contemporary law enforcement, community-oriented policing stands in sharp contrast to the crime-fighter role that persists in the popular "get tough" approach (which will be discussed in the next section).

Beginning in the 1960s, research, public sentiment, and changing views among the police themselves began to raise questions about professional policing. At least four developments spurred the growth of community-oriented policing.[55] First, mounting research evidence in the 1960s and 1970s showed that the strategies of professional policing were largely unsuccessful. Drawn under scrutiny were professional policing's key strategies of motorized patrol, rapid response, and reactive investigation.[56] Second, administrators and researchers became much more aware of role diversity in law enforcement. Despite the popularity of the crime-fighting role, it became increasingly obvious that the police role was far more diverse and included aspects of protection, public safety, order maintenance, and service. Third, the preoccupation with bureaucratic organization and the crime-fighting role seriously failed to fulfill its promises of managerial efficiency and crime control. Fourth, a long history of problematic relationships between the police and the public was worsened by the crime-fighting role when its strategies failed to control the crime problem.

In response, community-oriented policing was born. Jerome Skolnick and David Bayley attempt to capture the essence of this approach: "Thinking police professionals have had to develop some new ideas. The key reformulation has been that police and the communities they are policing must try to become co-producers of crime prevention. Roughly speaking, this concept of co-production, of increased cooperation between police and the community, is what has taken hold as 'Community Policing.' "[57] Community-oriented policing is made up of four basic elements: (1) community crime prevention, (2) reorientation of patrol activities to emphasize service and public order, (3) increased accountability to the public, and (4) decentralization of command through the creation of neighborhood police centers.[58] Under community-oriented policing, it is assumed that the police cannot fight crime effectively without the public's cooperation and involvement. As a result, the crime-fighting role began to be viewed as inadequate to prevent and control crime. In contrast, the **collaborative role** emphasizes that police are to engage the public in order to cooperatively prevent crime, maintain public order, and solve community problems.

To varying degrees, most police departments have incorporated the idea of community-oriented policing into their police strategies, which has affected how they deal with juveniles and juvenile crime.[59] While the community-oriented policing role entails police efforts to foster relationships with neighborhood youth in order to prevent crime and maintain order, the role has been implemented most extensively through school-based education programs, directed at preventing drug use and gang involvement. Theory into Practice, "D.A.R.E.," and "Gang Resistance Education and Training (G.R.E.A.T.)" provide brief descriptions of D.A.R.E. and G.R.E.A.T., both of which use juvenile officers working in schools in an educational capacity.

collaborative role The police and the community working together to prevent crime, maintain public order, and solve community problems.

THEORY INTO PRACTICE   D.A.R.E.

School-based drug prevention programs have become increasingly popular. While a large number of programs have emerged, none is more popular than Project D.A.R.E.: Drug Abuse Resistance Education. Created in 1983 by the Los Angeles Police Department in collaboration with the Los Angeles Unified School District, D.A.R.E. uses specially trained, uniformed police officers to teach a drug prevention curriculum to fifth grade students. D.A.R.E. has been implemented in more than 75% of all local school districts nationwide. The chief goal of D.A.R.E. is to provide students with the knowledge and decision-making skills necessary to resist social pressures to use drugs. While drug use prevention is the primary focus of the program, the curriculum is much more broadly conceived, covering topics such as knowledge and skills related to peer pressure, decision making using the D.A.R.E. Decision-Making Model, assertiveness training, and media influences. The curriculum is divided into nine lessons, with an additional "culmination" session. Topics include tobacco, alcohol, and illicit drug use, peer pressure, assertiveness skills, decision making, and refusal skills. The D.A.R.E. curriculum has been revised at lease twice since its inception, and new programs are being developed targeting middle school and high school students.

Despite its widespread adoption, evaluation research of D.A.R.E.'s effectiveness has produced mostly negative results.[60] However, evaluation of any drug education program such as D.A.R.E. is exceedingly complex. For example, research usually includes multiple outcome, or criterion measures, including knowledge about drugs, attitudes about drug use, social skills, self-esteem, attitude toward police, and drug use.

David Carter has argued that drug education programs such as D.A.R.E. can be a central component of community-oriented policing. He identifies a number of ways in which D.A.R.E. complements community-oriented policing, including the following:

- D.A.R.E. "humanizes" the police: that is, young people can begin to relate to officers as people, rather than in terms of uniforms or parts of an institution.
- D.A.R.E. permits student to see police officers in a helping role, not just in an enforcement role.
- D.A.R.E. opens lines of communication between law enforcement and youth.
- D.A.R.E. serves as a source of feedback to the police department to better communicate the fears and concerns of youth, thus helping the police develop problem-solving efforts that extend beyond drugs. . . .
- D.A.R.E. opens lines of communication between the school district and the police department to deal with a wide range of issues such as violence in the schools, drug abuse, and any other problem about which there is mutual concern.
- D.A.R.E. shows the police in a different light to many adults within the community: parents, teachers, school staff, administrators, and school board members. . . .

Source: Carter, "Community Policing and D.A.R.E." 5–6.

 **D.A.R.E.**

Drug Abuse Resistance Education (D.A.R.E.) is operated and managed by D.A.R.E. America, a national nonprofit organization. The link to the D.A.R.E. Web site is provided at:

http://criminaljustice.jbpub.com/burfeind

Gang Resistance Education and Training (G.R.E.A.T.)

G.R.E.A.T. is a school-based education program originally developed through a combined effort of the Bureau of Alcohol, Tobacco, and Firearms (ATF) and the Phoenix Police Department. In 2004, the overall program administration was transferred from ATF to the Office of Justice Programs, Bureau of Justice Assistance. G.R.E.A.T.'s violence prevention curriculum is designed to allow middle school students to develop beliefs and practice behaviors that will help them avoid gang pressure and youth violence. The nine-hour curriculum is taught in school by uniformed officers and includes thirteen sessions. Lessons include information about gangs and their connection to drugs and violence; education about peer pressure; and skill training on communication, active listening, empathy, and conflict resolution.

G.R.E.A.T. has been evaluated through a five-year longitudinal study conducted by Finn-Aage Esbensen and D. Wayne Osgood. The evaluation showed modest positive effects on adolescents' attitudes and delinquency risk factors, including peer group associations, attitudes about gangs and law enforcement, and risk-seeking behaviors, but no effects on involvement in gangs and actual delinquent behaviors.

Sources: Esbensen, "Evaluating G.R.E.A.T.;" Esbensen et al., "How Great is G.R.E.A.T.?"; Esbensen and Osgood, "Gang Resistance Education;" Esbensen and Osgood, "National Evaluation."

G.R.E.A.T.

Find the link to the Gang Resistance Education and Training (G.R.E.A.T.) Web site at:

http://criminaljustice.jbpub.com/burfeind

In recent years, school-based, community-oriented policing has been implemented most extensively through the school resource officer (SRO) program. Officers are assigned to schools (usually junior high and high schools) and provide a number of services to the school, including crime-related education in the classroom, crime prevention by walking the halls and outside school building, and investigation of crimes on or near campus. The underlying strategy is to develop relationships with students and school personnel in an effort to prevent crime and to respond to crime committed by students. "One of the important functions of an SRO program is for the police and schools to advise each other about criminal behavior involving students. . . . Historically, the police and schools have not communicated, which has led to a number of problems when students commit crimes on campus. SRO programs not only link the police with schools, they allow juvenile units to collect better information about delinquents who may be violent, possess weapons, or deal in drugs."[61] The school resource officer is primarily a collaborator between police, the schools, families, and the community. Thus, SRO programs provide crime prevention, order maintenance, and problem-solving collaboration with students, families, and school staff. As such, the SRO program is a fairly full reflection of the COP role.

The Office of Community Oriented Policing Services (COPS), an office within the US Department of Justice, is the primary federal agency promoting community-oriented policing in local law enforcement agencies. COPS provides major funding for school resource officers through its COPS in Schools Program. COPS has awarded "almost $748 million to more than 3,000 law enforcement agencies to fund over 6,500 school resource officers (SRO) through the COPS in Schools (CIS) Program."[62] In a time of tight budgets in many police departments, this program is one way in which some have been able to add officers and to implement community-oriented policing in schools.

Getting Tough: The Law-Enforcer Role

Beginning in the early 1980s, the "get tough" approach to crime introduced a <u>**law-enforcer role**</u> for police, even as community-oriented policing was coming onto the field at the same time. Greater attention was given to public safety and holding juvenile offenders accountable. As we saw in Chapter 2, contemporary trends in balanced and restorative justice provided justification and resources for an approach to policing that emphasized swift and certain law enforcement. The Juvenile Accountability Incentive Block Grants Program (JAIBG), first enacted in 1997, provided a rich coffer of federal funds to local and state juvenile justice agencies, including police departments.[63] "The underlying premise of juvenile accountability programming is that young people who violate the law should be held accountable for their offenses through the swift, consistent application of sanctions that are proportionate to the offenses—both as a matter of basic justice and as a way to combat delinquency and improve the quality of life in the nation's communities."[64] Several of the 16 "purpose areas" in the Juvenile Accountability Block Grants (JABG) program (which replaced JAIBG in 2003) specifically relate to law enforcement and provide grants to police departments: training for law enforcement personnel with respect to preventing and controlling juvenile crime; information sharing between juvenile and criminal justice systems, schools, and social service agencies; school safety efforts involving police; and the hiring of construction, staffing, and training detention and correctional personnel to improve facility practices and programming (detention facilities are usually operated by sheriff's departments).[65]

Contemporary juvenile policing is increasingly characterized by the law-enforcer role, emphasizing vigorous enforcement of laws, both criminal laws and status offense laws. Reflecting this, the legal distinction between taking into custody and arrest in statutory law has been reduced in some states, authorizing police to arrest juveniles just as adults.

law-enforcer role
Associated with the "get tough" approach to crime and delinquency, the law-enforcer role emphasizes public safety and offender accountability and involves vigorous enforcement of laws through arrest and referral to the juvenile court.

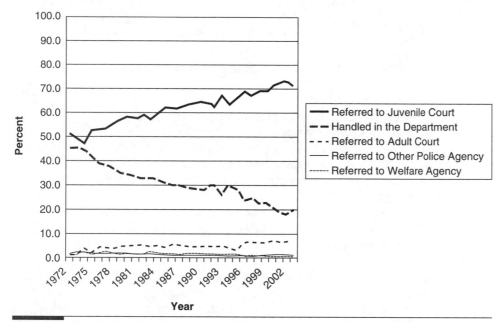

Figure 15-1 Method of handling juveniles taken into custody. Over the last thirty years, local police have shown an increasing tendency to deal formally with juveniles taken into custody through "referral to juvenile court" or "referral to adult criminal court," and a decreasing tendency to deal informally with juveniles taken into custody by handling them within the department and then releasing them—a practice commonly referred to as "warn and release."
Source: Maguire and Pastore, eds. *Sourcebook of Criminal Justice Statistics,* Table 4.26.

Perhaps the clearest example of this contemporary law-enforcer role in juvenile policing lies in the increasing tendency of police to deal formally with juveniles taken into custody through a "referral to juvenile court" or "referral to adult criminal court." Correspondingly, police are also far less likely to deal informally with juveniles taken into custody through what is known as "warn and release." **Figure 15-1** shows that in the early 1970s, almost half of all juveniles taken into custody were handled informally within police departments and then released. In recent years, however, this has been true for only about one fifth of all juveniles that the police have taken into custody. In contrast, almost three fourths of all juveniles taken into custody are now referred to juvenile courts, and a small but increasing portion are referred to adult criminal court. These trends in handling juveniles taken into custody are a clear indication of a shift in law enforcement efforts toward greater accountability and public safety.

The contemporary law enforcer role is also evident in the area of drug and gang control efforts. Most police departments of moderate to large size have developed specialized units to deal with drugs and gangs. Some of these gang and drug units are inter-agency organizations, made up of local police together with state and federal law enforcement. Theory into Practice, "Gang Suppression and Intervention," provides a brief description of suppression-based approaches to gangs and youth violence.

Discretion and Diversion

Regardless of changes in the police role toward juveniles, it is local police (municipal and county) that have most frequent contact with juveniles. State and

THEORY INTO **PRACTICE** **Gang Suppression and Intervention**

"Law enforcement agencies have pursued an increasingly sophisticated suppression approach to youth gangs, including surveillance, stakeouts, aggressive patrol and arrest, followup investigations, intelligence gathering, coupled with some prevention and community relations activities. Police have created complex data and information systems and improved coordination among law enforcement."[66]

Suppression-based strategies target high-incidence areas and deploy the same officers to those areas for extended periods of time. Effective suppression is based on gathering and organizing intelligence information on youth gangs and their members. Police officers must be experienced and specifically trained to recognize gang problems in particular parts of the city. These officers also ensure that judges are aware of gang affiliations of defendants before sentencing. These efforts result in large numbers of gang members being imprisoned.[67]

"Some police departments have developed community-oriented strategies, with considerable attention to community collaboration, social intervention, and even opportunity enhancement. Some police officers assigned to the gang problem have directly provided counseling, job development and referral, and tutoring, and engaged in extensive community relations and development of activities."[68]

You may also want to study several excellent sources on the connection between youth gangs, drugs, and violence, and police responses to gangs:

- Howell, James C., and Scott H. Decker. "The Youth Gangs, Drugs, and Violence Connection." In *Juvenile Justice Bulletin*. Washington, DC: Office of Juvenile Justice and Delinquency Prevention, 1999.
- Spergel et al., "Gang Suppression and Intervention: Problem and Response: Research Summary." In *OJJDP Summary*. Washington, DC: Office of Juvenile Justice and Delinquency Prevention, 1994.
- Decker, Scott H., ed. *Policing Gangs and Youth Violence*. Belmont, CA: Wadsworth, 2003.
- Bureau of Justice Assistance. *Addressing Community Gang Problems: A Model for Problem Solving*. Washington, DC: Bureau of Justice Statistics, 1997.
- Katz, Charles M., and Vincent J. Webb. *Police Response to Gangs: A Multi-Site Study*. 2003. Washington, DC: National Institute of Justice, 2003.

The problem of gangs and criminal violence has also been confronted through the courts. The California Street Terrorism Enforcement and Prevention Act (STEP), signed into law in 1998, provided for the use of criminal prosecution, civil liability, and asset forfeiture to combat the gang problem.[69]

Sources: Spergel et al., "Gang Suppression and Intervention;" Worrall, "California Street Terrorism."

federal law enforcement officers have comparatively few confrontations with youth. Local patrol officers are typically the first line of contact with juveniles in everyday encounters and with juveniles suspected of crime. State statutory laws provide authority to police officers to stop and talk to individuals and to investigate suspicious situations and crime incidents. These laws also give police officers arrest authority and authority to take a youth into custody. The application of this statutory law enforcement authority gives police officers broad, wide-ranging discretion.

A large proportion of police contacts with juveniles never result in a citation, custody, arrest, or formal referral to the juvenile court. Instead, a majority of juveniles are dealt with informally or are referred to agencies other than the juvenile court. In his classic study, David Bordua found that in 1964, the Detroit police had 106,000 "encounters" with juveniles.[70] Only 9% (9,445) of these were classified as "official contacts" by the police, resulting in formal enforcement action. Bordua concluded that a majority of police–juvenile contacts are dealt with informally and never result in official action such as issuing a ticket or citation, taking the youth into custody, or referring the youth to juvenile court.[71]

As we saw in Figure 15-1, more recent data show a decreasing tendency for the police to deal informally with juveniles that are taken into custody. Instead, referral to juvenile court has become far more common. However, it should be noted that these data report only those juveniles who were "taken into custody." Those contacts that were dealt with by the police in a more informal manner are not reported. We are left to speculate about the proportion of police contacts that do not result in a youth being taken into custody. If Bordua's findings are still relevant to contemporary policing, a substantial portion of all police encounters with juveniles are dealt with informally, rather than through formal referral to the juvenile court.

Informal handling of juveniles by the police involves a number of diversionary approaches intended to keep youth out of the juvenile justice system, including the following:

- Warn and release at the scene—a "street adjustment."
- Take the youth into custody, but release the youth to the parents at the police station—a "station house adjustment."
- Refer the case to a juvenile officer or juvenile unit within the department for further investigation or for involvement in an intervention program operated by the police department. Such referral may or may not involve taking the youth into custody.
- Refer the youth to community-based services, such as a youth services bureau, school, or church. Like in-house referrals, community-based referral may or may not involve taking the youth into custody.
- Refer the case to a child protective service agency for services, family intervention, or dependency and neglect investigation.[72]

Taking a youth into custody and arrest may involve the use of secure detention; it is a discretionary decision. Police are usually the ones to initiate the initial detention decision. In most states, the decision for detention requires that the officer inform or request authorization from the probation department and provide a written report that specifies why the officer believes the youth must be held in detention. After initial authorization, statutory law typically requires that a detention hearing be held within a set time period, often within 24 hours.[73] The detention decision is based on the consideration of whether the youth poses a risk to him- or herself or others and whether detention is needed to ensure the youth's appearance at upcoming court hearings.[74]

Police discretion involves much broader authority than just the decisions whether to take a youth into custody or to make an arrest and whether to detain the youth. Patrol officers make decisions regarding where, when, and how intensely to patrol; whether to engage in a high-speed pursuit; whether to stop, question, or frisk a suspect; and how much force to use in implementing an arrest. Detectives and juvenile officers make decisions related to criminal investigation, including how vigorously to investigate the case; whom to talk to; and whether to conduct surveillance, seek a warrant, or to involve other police investigative agencies. Additionally, police administrators have discretion in the style of law enforcement that they encourage among their officers, the type of crimes they emphasize for enforcement, and the degree to which they emphasize supervision and accountability of officers.

Factors Influencing the Use of Discretion

Police officers consciously or unconsciously consider a number of factors when deciding how to handle juveniles suspected of crime. Research reveals several key factors that influence police discretion.

- **Seriousness of the offense.** Perhaps the most obvious factor influencing police discretion is the seriousness of the offense. More serious offenses are more likely to lead to formal handling, including taking into custody, arrest, and referral to the juvenile court.[75] Police use of discretion decreases in direct proportion to the seriousness of the crime.[76] However, since a majority of all police–juvenile contacts relate to minor offenses in which the probability of arrest is very low, police discretion is extensive in juvenile matters.[77]

- **Previous contact with the police and prior arrest.** Research documents consistently that previous contact with the police, as well as prior arrest record, influence the chance of arrest. Youths with more frequent and serious prior offenses are more likely to be arrested.[78]

- **Attitude and demeanor of the juvenile.** The attitude and demeanor of the youth are especially important in police encounters with juveniles. One study found that 53% of the youths who were cooperative with the police were warned and released, and only 4% were arrested. In contrast, only 5% of the uncooperative youths were warned and released, and 67% were arrested.[79]

- **Social characteristics of the juvenile.** Race, gender, and social class appear to enter into the mix of factors considered by police officers in making discretionary decisions. However, these factors cannot be studied in isolation from other factors that influence discretion. The effect that any one of these factors has on police decision making depends on other factors—collectively, all of these factors shape the officer's decision. For example, while some studies find that race significantly influences discretion, other studies indicate that race must be considered in conjunction with legal factors, such as seriousness of the offense and prior record.[80] Thus, the influence of juveniles' social characteristics on discretion is not simple and straightforward; rather, these characteristics are interrelated with a variety of other factors.

- **Source and attitude of the complainant.** In their classic study, Donald Black and Albert Reiss found that over three fourths of all police–juvenile contacts came about from citizen complaints, rather than by officer initiation. The preferences of the complainant influenced the likelihood of arrest: if a citizen initiated the complaint, remained involved, and desired the juvenile to be arrested, the likelihood of arrest increased in contrast to situations when an officer initiated contact.[81]

- **Community characteristics.** Various community characteristics appear to affect police discretion. For example, research by Robert Samson points to an "ecological bias" in police control of juveniles: police are more likely to arrest juveniles in lower-class areas, in contrast to middle and upper-class areas, even when offense seriousness is taken into consideration.[82]

    The availability of community alternatives to arrest is of special importance regarding police discretion with juveniles. The more diversionary alternatives are available in the community, the less likely the officer is to take the youth into custody or arrest. Officers are more likely to arrest when they think that nothing else can be done.[83]

- **Organization, policies, customs, and styles of the police department.** Police departments vary extensively in their organizational structure.[84] Departments that are highly bureaucratic, emphasizing a hierarchical authority structure with extensive rules and regulation, are more likely to be separated and isolated from the communities they serve.[85] In contrast, departments that adopt community-oriented policing stress the importance of police–community cooperation in preventing and controlling crime and delinquency. Community-oriented policing is usually decentralized, with the use of neighborhood police stations or precincts.

    James Q. Wilson has argued that a department's organizational structure and law-enforcement style reflect the political and socioeconomic character of the local community.[86] In his study of two police agencies in an "Eastern City" and a "Western City," he found substantial differences in the attitude and approach of juvenile officers toward youth. The police department in the Western City was organized bureaucratically and emphasized "professional policing," whereas the Eastern City police department was less formally organized and far more informal in its law enforcement practices with juveniles. Juvenile officers in the Eastern City tended to take a moralistic perspective toward youth, viewing juvenile delinquency as a product of weak personal or family morality. Western City juvenile officers were less moralistic and more therapeutic in their views of delinquent youth. Interestingly, the youth officers in the Western City had more contact with youth and made more extensive use of arrest, followed by more frequent referral to the youth court.

Police departments also vary in terms of their formal policies, put in place by the administration. Standard operating procedural manuals lay down policies that officers are expected to follow. While these manuals establish a particular style of enforcement, a department's style of enforcement is also affected by informal traditions and customs of police officers. Such informal expectations are extended through peer pressure, exerted interpersonally among police officers. Important here is the custom of enforcement with regard to juveniles: different departments have different customs regarding juvenile policing.

Due Process in Law Enforcement

Though police discretion in juvenile matters is broad and extensive, it is not unbridled. During the 1960s and 1970s, the due process revolution in juvenile justice addressed several key law enforcement methods. Drawn into question was whether constitutionally protected due process rights apply to juveniles in a system and process that is intentionally protective and rehabilitative in nature, operating in "the best interests of the child." Also of concern is whether the due process revolution, as advanced by state and federal appellate courts, is binding on local law enforcement practices with juveniles. Three methods of law enforcement have been addressed most extensively by state and federal appellate court decisions: (1) search and seizure, (2) custodial interrogation, and (3) arrest. The case decisions that make up the due process revolution established procedural laws that attempt to extend basic rights and freedoms advanced in state and federal constitutions.

Search and Seizure

The Fourth Amendment to the US Constitution provides a right to privacy to all citizens. More specifically, it protects citizens from unauthorized **search and seizure**:

search and seizure The Fourth Amendment protection against unreasonable search and seizure by the police, requiring a search warrant, which is issued by the courts and based on probable cause.

> *The right of the people to be secure in their persons, houses, papers, and effects, against unreasonable searches and seizures, shall not be violated, and no Warrants shall be issued, but upon probable cause, supported by Oath or affirmation, and particularly describing the place to be searched, and the persons or things to be seized.*

The outcome of police violations of unreasonable search and seizure is a case law doctrine called the *exclusionary rule,* which demands that illegally obtained evidence be ruled inadmissible in a court of law. With several significant exceptions, the Fourth Amendment prohibits search and seizure without a valid search warrant. A search warrant must be issued by a court, based on probable cause—the reasoned conclusion, drawn from knowledge and facts, that a particular person committed a particular crime.[87] Without a search warrant, a search is considered "unreasonable." The exceptions to this Constitutional right are few, but they are significant. The Supreme Court has ruled that police, under special circumstances, may stop a person, ask questions, and conduct a limited search. At the time of arrest, the police may also carry out a search of the immediate area. In addition, the police may conduct a warrantless search when a person voluntarily consents to the search. Lastly, an automobile search without a warrant is generally legal when there is probable cause to believe that the vehicle contains criminal evidence.

In *State v. Lowry* (1967), the Supreme Court considered whether Fourth Amendment protections against unreasonable search and seizure apply to juveniles. The Court first acknowledged that the juvenile justice system is distinct in purpose, in that it attempts to protect and reform youth who are dependent and at risk. Does a system that pursues "the best interests of the child" have to extend due process rights? In particular, do juveniles have Fourth Amendment rights? Here's how the court answered these questions: "Is it not more outrageous for the police to treat children more harshly than adult offenders, especially when such is in violation of due process and fair treatment? Can a court countenance a system, where, as here, an adult may suppress evidence with the usual effect of having the charges dropped for lack of proof, and on the other hand a juvenile be institutionalized—lose the most sacred possession a human being has, his freedom—for 'rehabilitative' purposes because the Fourth Amendment right is unavailable to him?"[88] Thus, the Court's answer was a resounding yes—juveniles have the right against unreasonable search and seizure.

A juvenile's right against unreasonable search and seizure is conditioned in the context of school, however, where school safety takes priority. In the 1985 landmark decision, *New Jersey v. T.L.O.*, the US Supreme Court ruled that school officials can search students and their possessions if the students are suspected of violating school rules. In this case, a 14-year-old high school freshman was observed smoking in a school lavatory by a teacher. When confronted by the assistant vice-principal, she denied that she had been smoking and claimed that she did not smoke at all. The assistant vice-principal opened her purse and found a pack of cigarettes and a package of cigarette rolling papers, commonly used to smoke marijuana. Upon more thorough search of the purse, he found some marijuana, a pipe, a list of students who owed the young woman money, and two letters implicating her in marijuana dealing. Based on the need for schools to provide a safe and secure learning environment, this case grants school officials search and seizure authority beyond that of the police. In addition, evidence seized by school officials can be turned over to the police for use in their own criminal investigation.[89]

Arrest

Most states do not have specific statutory provisions distinguishing the arrest of juveniles from the arrest of adults. Even when states maintain the legal distinction between "taking into custody" and arrest, the protective purpose of taking a youth into custody has diminished in recent years. In practice, taking into custody has come to mean almost the same as arrest. In fact, the due process rights applied to a youth taken into custody are very similar to those applied to juveniles who are arrested (see Case in Point, "Statutory Authorization for Taking a Youth into Custody").

As we have seen, the police have very broad arrest authority because of the protective origins of juvenile policing and because of the extended range of

Statutory Authorization for Taking a Youth into Custody

41-5-331. Rights of youth taken into custody—questioning—waiver of rights.

(1) When a youth is taken into custody for questioning upon a matter that could result in a petition alleging that the youth is either a delinquent youth or a youth in need of intervention, the following requirements must be met:

(a) The youth must be advised of the youth's right against self-incrimination and the youth's right to counsel.

(b) The investigating officer, probation officer, or person assigned to give notice shall immediately notify the parents, guardian, or legal custodian of the youth that the youth has been taken into custody, the reasons for taking the youth into custody, and where the youth is being held. If the parents, guardian, or legal custodian cannot be found through diligent efforts, a close relative or friend chosen by the youth must be notified. . .

Source: Montana Code Annotated 2005.

arrest The deprivation of an individual's freedom by a person of authority on the grounds that there is probable cause to believe that that individual has committed a criminal offense.

juvenile offenses, particularly status offenses. Authority for arrest is granted in state statutory law (see Case in Point, "Statutory Authority for Arrest"). **Arrest** involves the deprivation of an individual's freedom by a person of authority on the grounds that there is probable cause to believe that that individual has committed a criminal offense.[90] Arrest is purposive: to protect the public and to initiate a legal response to the suspected crime. The legality of arrest lies in the Fourth Amendment's "right against unreasonable . . . seizures." The seizure that is specified refers to both property and people; therefore, the requirement of a warrant applies to arrest as well as to searches. However, just as there are exceptions to the constitutional requirement for search warrants, so too are there exceptions to the requirement that arrests be accomplished through the use of warrants. Two general exceptions have been extended to this constitutional requirement for warranted arrests.[91] First, for misdemeanor offenses, arrest without a warrant is permissible when the offense is committed in the "officer's presence," or when the officer can "sense" the offense—smell it, feel it, or hear it. Second, for felony offenses, arrest without a warrant is permissible when based on probable cause—the knowledge and facts that lead an officer to conclude that a crime has been committed and that the suspect committed it.

Use of force is the primary legal issue arising from the law enforcement practice of arrest. Under Common Law tradition, most state statutes authorize that "all necessary and reasonable force may be used in making an arrest, but the per-

CASE IN POINT

Statutory Authority for Arrest

> **46-6-104. Method of arrest.**
>
> **(1)** An arrest is made by an actual restraint of the person to be arrested or by the person's submission to the custody of the person making the arrest.
>
> **(2)** All necessary and reasonable force may be used in making an arrest, but the person arrested may not be subject to any greater restraint than is necessary to hold or detain that person.
>
> **(3)** All necessary and reasonable force may be used to effect an entry into any building or property or part thereof to make an authorized arrest.
>
> *Source: Montana Code Annotated 2005.*

son arrested may not be subjected to any greater restraint than is necessary to hold or detain that person" (see Case in Point, "Statutory Authority for Arrest").[92] The defining Supreme Court case regarding use of deadly force involved a juvenile—a fifteen-year-old youth from Tennessee named Edward Garner. *Tennessee v. Garner* (1985) is a tragic case dealing with an officer's hard-pressed, spur-of-the-moment interpretation of "all necessary and reasonable force."[93] The Supreme Court ruled that use of deadly force against an apparently unarmed and nondangerous fleeing felon is an illegal seizure under the Fourth Amendment.[94] Since the *Garner* decision, four key factors have come to be emphasized in the use of deadly force: (1) a felony has been or is being committed, (2) the person to be arrested committed the crime, (3) the officer believes that there is substantial risk that the suspect poses harm to others (public safety), and (4) the use of deadly force poses no danger to others.[95]

Custodial Interrogation

<u>Custodial interrogation</u> has to do with police questioning of suspects while they are under arrest or when their freedom is restricted in a significant way. Two Constitutional Amendments are drawn into question with the practice of custodial interrogation: the Fifth Amendment's right against self-incrimination and the Sixth Amendment's right to counsel. These two fundamental rights were linked together for the first time in *Miranda v. Arizona* (1966), in which the Supreme Court ruled that in order to ensure the right against self-incrimination, it is necessary for the accused to have counsel.[96] The often-repeated Miranda Warning requires police to notify persons taken into custody that they have a right to remain silent and that any statements they make can and will be used against them; and that they have a right to counsel, and that if they cannot afford counsel, it will be provided at public expense. It should be noted, however,

custodial interrogation
Police questioning of suspects when the custody involves arrest or the significant deprivation of freedom and the questioning may be incriminating.

that Ernesto Miranda was 23 years old when he was charged with kidnapping and rape. Do these same rights apply to juveniles?

One year after the Miranda decision, the Supreme Court extended due process rights to juvenile court proceedings in the case *In re Gault* (1967).[97] However, the Court intentionally declined to address whether the right to counsel and the right against self-incrimination is required in pre-adjudication stages of juvenile justice, including police procedures with juveniles. Two earlier Supreme Court cases ruled that statements made by juveniles, when under custodial interrogation, must be considered in terms of the *totality of circumstances*—a variety of considerations including the absence of parents and lawyers, the age of the suspect, the length of questioning, and the approach used by officers.[98] In *Haley v. Ohio* (1948), five hours of apparently coercive interrogation, without parents or attorney present, led the Supreme Court to conclude, "The age of the petitioner, the hours when he was grilled, the duration of his quizzing, the fact that he had no friend or counsel to advise him, the callous attitude of the police toward his rights combine to convince us that this was a confession wrung from a child by means which the law should not sanction. Neither man nor child can be allowed to stand condemned by methods which flout constitutional requirements of due process of law."[99]

After *Miranda,* the totality-of-circumstances standard was enhanced by the requirement that juveniles be given the Miranda warning by the police when questioning may extract incriminating statements. The critical question that follows is whether a juvenile is able to knowingly, intelligently, and voluntarily waive the rights stated in the Miranda warning. In most circumstances, the waiver of Miranda can be made by a juvenile only with the presence of parents or lawyers. In *California v. Prysock* (1981), the US Supreme Court addressed the adequacy of the Miranda warning when it was paraphrased by a police officer to allow the youth to better understand the expressed rights. The Court ruled that even though the Miranda warning was not given verbatim, its meaning was plain and easily understandable for a juvenile. Thus, an officer's attempt to make Miranda understandable for youth is permissible law enforcement practice.

◼ Juvenile Court Processes: Formal and Informal

Police encounters with juveniles involve much discretion and rely heavily on diversion, and it is often said that the juvenile courts would collapse from the sheer number of cases referred by the police. Even so, juvenile courts across the United States handle some 1.6 million delinquency cases each year. Juvenile courts, too, rely heavily on discretion and diversion—many cases are diverted or handled informally by juvenile courts. The juvenile court process involves a series of discretionary decisions, including intake, petitioning, transfer to adult court, adjudication, and disposition.

Referral to Juvenile Court

The juvenile court process is initiated by a **referral** to the juvenile court. Referrals are written documents that request or result in court consideration of a particular juvenile matter. Referrals take various forms: police make referrals

referral Written documents, usually from the police, parents, or schools, that request or result in juvenile court consideration of a particular matter involving a juvenile.

through incident reports, citations, and tickets; and parents, schools, and victims sign complaints. Almost 85% of all referrals nationwide come from the police, almost always from local law enforcement.[100] The remaining referrals come from probation officers, parents, schools, and victims. Police, however, refer comparatively few of the status offense cases dealt with formally by the juvenile court. Only 10% of the juvenile courts' truancy cases and 11% of the ungovernability cases were referred by the police, whereas 40% of the runaway and 92% of the liquor law violations were referred by the police.[101] Status offense cases are more commonly referred by other agencies such as schools and social welfare agencies. In most other respects, the processing of status offense cases parallels that of delinquency [102]

Referrals: Extent, Trends, Age, Gender, Race

Table 15-2 shows that in 2000, more than 1.6 million delinquency cases were referred to juvenile courts across the United States. This represents a 43% increase from 1985 to 2000 in the number of referrals to juvenile courts. When differences in the size of the juvenile population across these years is taken into consideration by calculating an annual case rate per 1,000 juveniles, the resulting rates of referrals increased by 43% between 1985 and 1996, and then declined by 14% through 2000.[103] The increase in number of juvenile court cases was greatest for drug law violations (164%), but person offenses and public order offenses also increased by 107% and 106% respectively. While the number of property offense cases declined by 3%, from 1985 to 2000, property offenses made up more than 40% of the delinquency cases handled by juvenile courts. In order to understand the type of delinquency cases handled by juvenile courts, it must be noted that property offenses make up the largest offense category in the

Table 15-2 Juvenile Court Delinquency Caseload, 2000						
Most serious offense	Number of cases	Percent change 1985–2000	Percent of total delinquency cases	Percent detained	Percent petitioned	Percent adjudicated
TOTAL DELINQUENCY	1,633,300	43%	100%	20%	58%	66%
Person offenses	375,600	107	23	23	60	63
Property offenses	668,600	−3	41	16	54	67
Drug law violations	194,200	164	12	23	61	68
Public order offenses	395,000	106	23	23	59	68

Violent offenses: homicide, rape, robbery, aggravated assault, simple assault, other violent sex offenses, other person offenses.

Property offenses: burglary, larceny–theft, motor vehicle theft, arson, vandalism, trespassing, stolen property, other property offenses.

Public order offenses: obstruction of justice, disorderly conduct, weapons offenses, liquor law violation, nonviolent sex offenses, other public order offenses.

Sources: Puzzanchera et al., *Juvenile Court Statistics 2000,* 6, 7, 31, 39; Harms, "Detention in Delinquency Cases," 1 (percent detained report 1999 data).

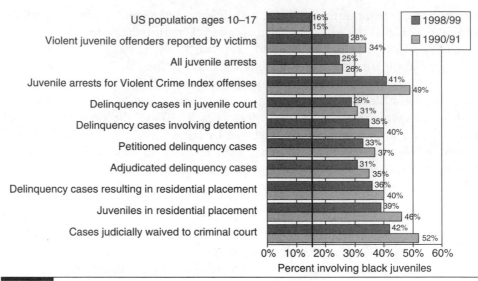

Figure 15-2 Black juveniles are overrepresented at all stages of the juvenile justice system compared with their proportion in the population.
Source: Sickmund, "Juveniles in Corrections," 12.

caseload of juvenile courts: 41% property offenses, 23% person offenses, 23% public order offenses, and 12% drug violations.[104]

Three-fourths (75%) of all delinquency referrals to juvenile courts involve males. The increase in delinquency cases from 1985 to 2000, though, was greater for females than it was for males (a 58% increase for females versus a 14% increase for males).[105] Also noteworthy is that over two thirds (68%) of all delinquency referrals to juvenile courts involve white youth, compared to one fourth (28%) that involve blacks. However, a disproportion of delinquent cases involve black youth, given their proportion of the juvenile population (15%). The percent increase in delinquency cases from 1985 to 2000 was greater for blacks (61%) than it was for whites (36%).[106] As will become evident in the discussion that follows, the problem of overrepresentation of black juveniles exists in all stages of the juvenile justice system (see **Figure 15-2**). Referral to juvenile court is also progressively more common across the adolescent years, with 13-year-olds having a case rate of 41 per 1,000 juveniles and 17-year-olds having a case rate of 112 per 1,000 juveniles.[107]

Intake and Petition

intake The screening of cases referred to the juvenile court, usually by the probation department. Intake screening results in a determination of how the case will be handled—formally, informally, or by dismissal.

<u>Intake</u> is the initial screening of cases referred to the juvenile court that determines how they will be handled. Juvenile court intake is a discretionary decision, normally the responsibility of the juvenile probation department, and less typically, of the prosecutor's office. Four options are most common:

- Case dismissal
- Diversion for nonjudicial handling, usually through community-based services
- Informal processing by the probation department
- Formal processing by the juvenile court, called *adjudication*.[108]

In most jurisdictions, intake screening involves three key determinations. The first is a legal determination of probable cause: are there legally admissible facts and information to warrant formal processing of the case in juvenile court? Second, do the facts of the case allow the juvenile court to have jurisdiction? This is especially important because contemporary juvenile courts do not always have original jurisdiction in juvenile matters. Recall that two recent trends limit the original jurisdiction of many juvenile courts: statutory exclusion of certain offenses and offenders and concurrent jurisdiction, allowing prosecutors discretion in filing certain types of cases in either juvenile or criminal court. Third, is formal processing by the juvenile court in the best interests of the child? While this determination was the overarching concern of the traditional juvenile court, the contemporary emphasis on offender accountability and public safety has introduced a second question that follows immediately after the first: Is formal processing by the juvenile court in the best interests of the community?

In some jurisdictions, intake screening is based on a preliminary inquiry, conducted by a probation officer under authority of statutory law. A preliminary inquiry collects pertinent information about the alleged offense and the youth. Information about the youth is sometimes referred to as a "youth assessment" and includes chemical dependency evaluation, educational assessment, and determination of mental health and family service needs. Thus, the preliminary inquiry provides information relevant for determining probable cause, juvenile court jurisdiction, and the "best interests" of the youth and community.

Petition: Extent, Trends, Age, Gender, Race

If intake screening determines that the case should be dealt with formally by the juvenile court, a **petition** is written by the prosecutor's office to initiate the formal juvenile justice process. The petition is a legal document that specifies the facts of the case, the charges, and the basic identifying information on the youth and his or her parents. **Figure 15-3** indicates that in 2000, 58% of all delinquency cases referred to juvenile court were petitioned, and 42% were not petitioned. These numbers represent an increasing tendency for referred delinquency cases to be dealt with formally by juvenile courts. Ten years earlier, in 1990, half of all delinquency cases were processed formally through petition.[109] Compared to cases that are not petitioned into juvenile court, formally processed cases "tend to involve more serious offenses, older juveniles, and juveniles with longer court histories."[110] It is also the case that males are more likely than females and blacks are more likely than members of other races to have their cases petitioned.[111]

Juvenile Court Statistics provides information about status offenses only for cases that are petitioned. Therefore, we have no indication of how many or what types of status offense cases are referred to juvenile courts or what portion is petitioned. Instead, *Juvenile Court Statistics* reports on the case processing of four major categories of petitioned status offenses: running away, truancy, ungovernability, and underage liquor law violations (e.g., minor in possession of alcohol and underage drinking). In 1997, 158,500 status offense cases were petitioned into juvenile courts nationwide. Of these cases, 15% involved running away, 26% truancy, 13% ungovernability, and 26% liquor law violations; the remaining 20% were categorized as miscellaneous.[112]

petition A legal document filed in juvenile court that specifies the facts of the case, the charge(s), and the basic identifying information of the youth and his or her parents.

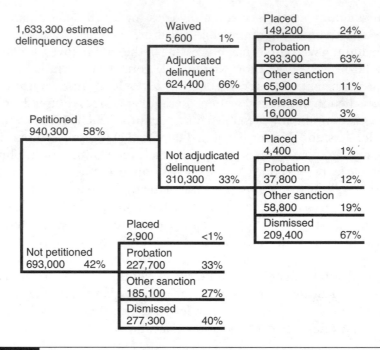

Figure 15-3 Juvenile court processing of delinquency cases, 2000. While a majority of cases (58%) referred to juvenile courts are petitioned, a sizeable proportion (42%) are not petitioned and are either dismissed or provided some form of informal disposition. Of the cases that are petitioned, two thirds (66%) result in adjudication.
Source: Puzzanchera et al., *Juvenile Court Statistics 2000,* 52.

informal disposition A juvenile court's response and handling of a case that has been referred but not adjudicated. Informal disposition may occur in cases that have not been petitioned and in cases that have been petitioned but not adjudicated. A variety of informal dispositional options are available to juvenile courts, including referral to another agency for services (e.g., chemical dependency treatment or counseling), community service, restitution, and informal probation through the probation department.

Delinquency cases that are not petitioned are either dismissed (as are over one third of the cases that are not petitioned) or dealt with informally by the juvenile court. Even though these cases are not formally petitioned, almost two thirds receive some form of **informal disposition** by the juvenile court. Informal disposition is usually based on the youth's admission of a delinquent act or acts. A variety of disposition options are then possible, including referral to another agency for services (e.g., chemical dependency treatment or counseling), community service, restitution, and informal probation through the probation department.[113] These informal dispositions usually involve a *consent adjustment without petition*, a voluntary agreement between the youth and his or her parents that is approved by a juvenile court judge. The written agreement is given legal authority under statutory law and, in terms of informal probation, the consent adjustment involves rules or conditions that the youth is expected to follow, under supervision of a probation officer.

Detention

detention The secure confinement of a youth under court authority, pending court action, disposition, or placement.

One of the most crucial decisions at intake is whether to detain the youth pending further court action. In most states, initial authorization for **detention** is provided by probation officers in response to police officers' request for use of secure detention. Police officers are usually required to provide a written report that describes why they believe the youth must be held in detention. Continued detention requires court authorization, usually through a judicial hearing within a statutorily specified period of time, commonly 24 hours. In cases involving detention, the detention hearing is the first court action. As specified in statutory law, the detention hearing must establish probable cause that the youth is a delinquent youth or status offender. The detention hearing follows procedural rules, also set forth in statutory law. The juvenile court's detention decision is sometimes informed by the preliminary inquiry, done by the probation officer at intake. The basic decision is whether continued detention is necessary "to protect the community, to ensure a juvenile's appearance at subsequent court hearings, to secure the juvenile's own safety, or for the purpose of evaluating the juvenile."[114]

Statutory authorization for detention was tested in the US Supreme Court case *Schall v. Martin* (1984).[115] In this case, Martin was arrested for first-degree robbery, second-degree assault, and criminal possession of a weapon.[116] At issue was whether the statutory provisions for preventive detention provided adequate due process of law under the Fourteenth Amendment of the US Constitution. The statutory provisions in question here were under the New York Family Court Act, which allowed for preventive detention of juveniles charged with delinquent acts, pending completion of juvenile court adjudication. The Supreme Court ruled that the use of preventive detention was constitutional because the Family Court Act provided specific procedures that must be followed for a juvenile to be detained.[117]

Detention: Extent, Trends, Age, Gender, Race

Most delinquency cases referred to the juvenile court do not involve detention. In 2000, 329,800 delinquency cases were detained between referral and disposition—20% of the delinquency cases processed by juvenile courts (see Table 15-2).[118] The number of delinquency cases involving detention increased by 41% between 1985 and 2000. This increase was largest for drug cases (139%) and person offense cases (100%), whereas property offense cases (−10%) declined during this period.[119] However, the increase in the number of delinquency cases involving detention during this time period is due to the increasing number of cases handled by juvenile courts across the nation.[120] Property offenses (33%) account for the largest portion of delinquency cases involving detention, as compared to person (28%), public order (28%), and drug (11%) offenses.[121] Males are more likely than females to be detained for all types of delinquencies.[122] However, females charged with person offenses showed the most dramatic increase in detention population.[123] Detention was used more often for older juveniles than for younger juveniles.[124] Of serious concern is the disproportionate confinement of black youth: "Although black youth made up 28% of all delinquency cases processed in 2000, they were involved in 35% of detained cases. This overrepresentation was greatest for drug offenses: blacks accounted for 22% of all drug cases processed but 37% of drug cases detained."[125] In addition, the percentage of delinquency cases involving detention is higher for blacks than it is for whites or other races, and this is found across all types of offenses. The problem of *disproportionate minority confinement* has been clearly identified and documented for over 30 years, but the problem persists.

A much lower proportion of petitioned status offenses involve detention: 17% of the runaway offenses, 3% of the truancy offenses, 10% of the ungovernability offenses, and 7% of the liquor law offenses. For most status offense categories, males were more likely to be detained than females, and blacks and members of other non-white races were more likely to be detained than whites. The percentage of petitioned status offense cases involving detention did not vary for adolescents in different age categories.[126]

Disproportionate Minority Confinement

The Juvenile Justice and Delinquency Prevention (JJDP) Act of 1974 was directed at problematic detention practices across the United States. As we discussed in Chapter 2, the Act sought to do the following: (1) remove status offenders from detention or secure confinement, (2) promote sight and sound separation of juveniles from adults while in detention, (3) remove juveniles from adult jail facilities, and (4) reduce the number of minorities in secure facilities. The Office of Juvenile Justice and Delinquency Prevention was established to promote these initiatives. Amendments to the JJDP Act in 1988 and 1992 required states participating in the Formula Grants Program to address disproportionate minority confinement.[127] In recent years, OJJDP has attempted to help states deal with the problem of disproportionate minority confinement through technical assistance, consultation, and evaluation grants.

Evidence indicates that federal initiatives through OJJDP have influenced state and local juvenile detention practices in each of the areas addressed in the JJDP Act: (1) the number of status offenders held in secure detention or secure corrections has been reduced substantially; (2) new and remodeled state and local detention facilities provide greater levels of sight and sound separation of juveniles from adults while in detention; (3) since the Act's passage, there has been a significant decline in the number of juveniles held in adult jails; and (4) the number of minorities in secure detention facilities has declined.[128] Despite significant progress, compliance has not been fully achieved, and race effects in the processing of youth are still found in the majority of studies, indicating the "existence of disparities and potential biases in juvenile justice processing" (see Figure 15-3).[129]

Transfer to Criminal Court

Intake screening may also result in the decision to transfer juvenile cases to adult criminal court. As we discussed in Chapter 2, several transfer provisions have become increasingly popular, including judicial waiver, concurrent jurisdiction, and statutory exclusion.[130] These statutory mechanisms allow juvenile court judges, prosecutors, and the legislature to designate the types of offenders and offenses that are beyond the jurisdiction of the juvenile court. *Judicial waiver* is the oldest transfer provision, providing statutory authority for juvenile court judges to waive the court's original jurisdiction and transfer cases to criminal court. While judicial waiver provisions vary from state to state, the basic rationale is that certain types of offense and offenders, especially violent ones, are beyond the scope of the juvenile court, and juvenile court judges should be given authority to transfer these cases to adult court. *Concurrent jurisdiction* gives statutory authority to prosecutors to file certain types of offenses in either juvenile or adult court. Some states, for example, allow prosecutors, at their discretion, to file felony offenses directly in adult criminal courts. Under *statutory exclusion*, state statutes exclude certain juvenile offenders and offenses from juvenile court jurisdiction, and these cases originate in criminal, rather than juvenile court. State statutes providing for statutory exclusion usually set age and offense limits for excluded offenses.

Judicial Waiver: Extent, Trends, Age, Gender, Race

Though all states have expanded their statutory provisions for transferring juvenile cases to adult court, the actual number of juvenile cases transferred is relatively small.[131] Judicial waiver remains the most common transfer provision, and it is the only one for which data are available from the *Juvenile Court Statistics*.[132] In 2000, juvenile courts waived less than 1% of all petitioned delinquency cases (see Figure 15-3). The number of juvenile cases waived judicially to adult court peaked in 1994 with 12,100 cases and has since declined to less than 6,000 in 2000. Judicial waiver is often thought to target serious violent offenses; however, the proportion of person offenses among waived cases peaked in 1995 with 47%, and then dropped to 40% in 2000. Property offenses were 54% of the judicially waived cases in 1985, but they were 36% of cases waived in 2000. Drug offenses (14%) and public order offenses (10%) round out the remaining judicially waived cases in 2000. Judicial waiver is more common for older juveniles than for younger juveniles, for males than for females, and for black youth than for youth of other races.[133] "Among both white juveniles and black juveniles, the number of delinquency cases judicially waived to criminal court peaked in the mid-1990s and then declined."[134]

Adjudication

When formal court intervention is deemed necessary, cases proceed through a series of hearings, with varying names, that serve particular purposes. The adjudication process is initiated by the petition. The first adjudication hearing often is referred to as a *probable cause hearing*. This hearing makes an initial determination about whether sufficient information and evidence exist to substantiate that a youth was likely involved in activity that violated state criminal or status offense law. As already mentioned, determinations of continued detention are also made by judges in probable cause hearings. An *arraignment* normally follows at a separate time in order to advise the youth of his or her rights and obligations, to read the petition, and to allow the youth to admit or deny the petition. The vast majority of juveniles admit the petition at arraignment. However, if the youth denies the petition, a *fact-finding* or *adjudicatory hearing* follows. This hearing comes closest to an adversarial trial in adult court, but the youth has fewer due process rights than an adult would. The outcome of this fact-finding process is a legal determination by the judge about whether the juvenile committed the offense(s). This is sometimes referred to as a *finding of fact* or a *finding of delinquency*. If the youth admitted to or was found to have committed the offense(s), the judge then decides if the youth should be **adjudicated** a *delinquent youth* or adjudicated a *status offender* (or some equivalent term). Both terms designate a legal status that results from the juvenile court's determination that the youth has in fact violated the criminal law (delinquent youth)

adjudication A case dealt with formally by the juvenile court goes through a series of hearings, involving determination of probable cause, arraignment, and fact finding. The purpose of this adjudicatory process is to determine whether the youth is responsible for the offense(s) charged in the petition, and if the youth should be legally declared a "delinquent youth" or a "status offender" by the juvenile court.

or status offense law (status offender) and is in need of the juvenile court's assistance. Recall from Chapter 2 that most state statutes now refer to status offenders as a youth in need of supervision (YINS), a minor in need of supervision (MINS), a youth in need of intervention (YINI), or some equivalent term.

Adjudication: Extent, Trends, Age, Gender, Race

Figure 15-3 indicates that two thirds of the petitioned delinquency cases result in adjudication (624,400 out of 940,300 petitioned delinquency cases), while one third are not adjudicated (310,300). Even when a youth is *not* adjudicated delinquent, the court can impose various dispositions, including out-of-home placement, probation, and other options such as restitution and community service. The largest portion of delinquency cases that are not adjudicated are dismissed by the juvenile court (67%). The remaining nonadjudication cases result in probation (12%), placement (1%), or some other disposition (19%). The various dispositions of nonadjudicated cases are often based on a *consent decree with petition*—an agreement between the juvenile court judge, parents, youth, and probation officer.

Across different delinquency offense categories, youth were more likely to be adjudicated delinquent in 2000 than in 1985. Younger juveniles (15 years old or younger) were more likely than older juveniles to be adjudicated delinquent. Also, petitioned delinquency cases involving males were more likely to be adjudicated delinquent than those involving females, while cases involving black juveniles were less likely to be adjudicated delinquent than were cases involving juveniles of other races.[135]

Similarly, most petitioned status offense cases result in adjudication. About two-thirds of the truancy, ungovernability, and liquor law violations resulted in adjudication, whereas about half of the runaway cases were adjudicated. While the likelihood of adjudication was about the same for males and females, and black and white youth, adjudication was more likely for younger status offenders, as compared to older status offenders.[136]

Disposition

predisposition report A court-ordered report written by a probation officer that provides information relevant for court disposition in an effort to make disposition individualized and rehabilitative. The report usually includes three major sections (offense, social history, and summary and recommendation) that furnish information on the adjudicated youth's prior record, current offense, and social background. The probation officer also makes a recommendation for disposition.

Once a youth is adjudicated a delinquent youth or a status offender, the court decides the most appropriate sanction—referred to as disposition. Disposition is similar to sentencing in adult courts. Traditionally, disposition has been one of the key decision points that considers "the best interests of the child." Disposition is imposed at a *dispositional hearing,* which is separated in time from the various adjudication hearings. After adjudication but before the dispositional hearing, a **predisposition report** is usually written by a probation officer.[137] Authorized by court order and statutory law, the report provides an evaluation of the youth and his or her background and social environment. Even with contemporary initiatives for punishment and accountability, the predisposition report attempts to provide assessment information to the judge so that disposition can be individualized and directed at rehabilitation. The predisposition report is commonly organized with three major sections: the offense section, a social history, and a summary and recommendation. The offense section covers four main areas: (1) an official version of the offense, usually provided by the police report;

(2) commentary regarding the juvenile's version of the offense; (3) prior record, including previous arrests, petitions, adjudications, and dispositions; and (4) a victim impact statement (required by many states). The social history section provides observations about the juvenile and his or her family background, educational experiences and achievements, friends, employment, and neighborhood context. The final section, the summary and recommendation, is the shortest, but perhaps most important. Sometimes it is the only section that the judge reads. Two areas of concise discussion are included. First, an evaluative summary highlights and summarizes the key findings of the assessment. Second, the probation officer offers a recommendation regarding disposition that is followed by the judge in most cases. In addition to the predisposition report, various supplemental assessments can be ordered by the court, including psychological, educational, chemical dependency, and family assessments. The predisposition report (as well as any other relevant information) is directed at making disposition individualized and seeks to promote rehabilitation.

<u>Formal disposition</u> is distinct from adjudication in purpose and process. The dispositional hearing is not a fact-finding or adversarial process; rather it maintains an individualized and rehabilitative focus. Rules of evidence do not strictly apply, nor are due process rights fully extended. As the subject areas of the predisposition report suggest, non-legal factors become the center of attention in the dispositional hearing. The youth's attitude, family background, school experiences, and friends all have an important influence on disposition. Disposition attempts to fulfill multiple, seemingly inconsistent, goals: public safety, nurturance, deterrence, accountability, and rehabilitation. Disposition results in a court order prescribing sanctions, and the order usually involves more than one disposition. For example, probation is commonly included with restitution and community service. Under state statutory law, a full range of dispositions is possible, including the following:

formal disposition The juvenile court's response to a youth's admission of a petition or the court's finding that a youth committed an offense(s). Among the dispositional options are probation, out-of-home placement, restitution, and community service.

- probation
- foster or group home placement
- placement in a residential treatment center
- restitution
- community service
- counseling for the youth, parents, or guardians
- further medical or psychological evaluation of the youth, parents, or guardian
- services that the court may designate, provided by parents or guardians
- commitment to a mental health facility
- home arrest
- confiscation of the youth's driver's license
- fine
- payment of court costs
- payment of victim counseling

- deferred imposition of sentence
- mediation
- placement in a secure facility
- placement in a youth assessment center[138]

A disposition involving out-of home-placement usually involves a court order of **commitment**. Commitment is the juvenile court's legal authorization for out-of-home placement of a youth who has been adjudicated a delinquent youth. Legal custody is transferred to "the state"—some state agency such as the juvenile court or department of corrections.

commitment The juvenile court's legal authorization for out-of-home placement for a youth who has been adjudicated a delinquent youth, involving a transfer of legal custody to "the state"—some state agency such as the juvenile court or department of corrections.

Disposition: Extent, Trends, Age, Gender, Race

Probation was the most restrictive disposition given in almost two thirds of the cases adjudicated in 2000 (63% or 393,300 cases; see Figure 15-3). Out-of-home placement was used in one quarter (24%) of the adjudicated cases, resulting in residential placement at a training school, drug treatment center, boot camp, private treatment center, or group home.[139] Other dispositions (11% of the adjudicated cases) include drug court, nonresidential community programs, restitution, and community service. Three percent of the adjudicated cases were released without disposition. Once adjudicated, juveniles age 15 or younger were more likely than older juveniles to be placed on probation. In contrast, adjudicated cases involving older juveniles were more likely to result in out-of-home placement than were those cases involving younger juveniles. Adjudicated cases involving females were more likely to be placed on probation than were those cases involving males, whereas delinquency cases involving males were more likely to result in out-of-home placement. The disposition of probation was used at about the same rate for all racial groups.[140] "Once adjudicated delinquent, cases involving black youth were more likely to result in out-of-home placement than were cases involving white youth or youth of other races."[141]

Probation is also the most common dispositional order in adjudicated status offense cases. Juvenile court data from 1985 to 2000 show that probation is the most restrictive disposition in 57% of the runaway cases, 78% of the truancy cases, 64% of the ungovernability cases, and 56% of the liquor law violations. Across the different types of status offenses, there are observable racial differences in the use of probation as a disposition, with black youth and youth of other races receiving probation more frequently than white youth. Probation is also more common for females adjudicated of runaway (as compared to males), older adolescents (16 years or older) adjudicated of truancy, and younger adolescents (15 years or younger) adjudicated of liquor law violations.[142]

Out-of-home placement is the next most common disposition for youths adjudicated of status offenses, ranging from 8% of the liquor law adjudications to 27% of all runaway and ungovernability adjudications.[143] Adjudicated status offenders 15 years of age or younger are more likely than older adolescents to receive a disposition of out-of-home placement. In addition, black youths who have been adjudicated as runaways or as liquor law violators are more likely than youth of other races to be subject to out-of-home placement.[144]

■ Juvenile Corrections

Juvenile court disposition is the gateway to corrections because it legally establishes how the court desires to deal with a case. Regardless of whether a case is adjudicated or not, juvenile court disposition seeks correctional options that are individualized, rehabilitative, and least restrictive. Reflecting the rehabilitative ideal, juvenile court disposition has traditionally involved a wide range of correctional options. Here we structure these correctional options in three arenas: probation, community-based corrections, and residential placement. Before describing these various forms of juvenile corrections, we first need to briefly describe how the second revolution has changed juvenile corrections.

Juvenile Corrections and the Second Revolution

In Chapter 2, we described three areas of change that made up the "second revolution" in juvenile justice: (1) the due process revolution; (2) the enactment of the JJDP Act of 1974; and (3) "getting tough"—contemporary initiatives for punishment and accountability. Common among these three areas of change was a questioning of the rehabilitative ideal, which was the foundation of the original juvenile court. Rehabilitation was called into question by a number of studies in the 1970s that took inventory of the effectiveness of juvenile and adult corrections.[145] Summarizing a large body of evaluation research, these studies yielded limited evidence of rehabilitative effectiveness of correctional programs for both juveniles and adults. Even though several successful programs were uncovered, the general conclusion drawn was that "nothing works." While considerable debate ensued, the "nothing works" view toward rehabilitation was widely adopted by the public and criminal justice professionals. The questioning of rehabilitation was much stronger with regard to adult corrections than it was for juvenile corrections; nonetheless, the contemporary "get tough" approach, advocating for punishment and accountability in juvenile justice, was fueled by this questioning of the rehabilitative ideal.

The second revolution introduced a number of additional goals beyond rehabilitation into juvenile corrections. No longer was it "assumed that matters of treatment and reform of the offender are the only questions worthy of serious attention."[146] Instead, aspects of public safety, deterrence, punishment (sometimes legally referred to as "consequences"), accountability, nurture (promoting mental and physical development), competency development, and due process of law have displaced rehabilitation as the first and foremost goal of juvenile corrections.

Probation

Probation is unquestionably the most frequently imposed disposition in juvenile courts (see Figure 15-3). It is used informally in cases that are not petitioned and in cases that are not adjudicated, and it is the most common formal disposition for adjudicated cases. "In 1999, probation supervision was the most severe disposition in 40 percent (677,000) of all delinquency cases. The number of cases placed on probation grew 44 percent between 1990 and 1999."[147]

Probation has its origins in the innovative work of John Augustus, a Boston shoemaker, who in the mid-1800s convinced a judge to release juvenile and adult offenders to him with the promise that he would provide moral guidance to them: he would establish certain expectations and rules for them to follow, and he would supervise their daily activities and progress. Thus, probation, as conceived by Augustus, involved the conditional release of an offender into the community, under supervision. Today, conditions and supervision are still the core of probation.

Probation Conditions

Probation conditions are court-imposed rules that juveniles must obey in order to live in the community and avoid confinement. In this way, probation involves the conditional release into the community. Probation conditions are written into a legal document. In some juvenile court jurisdictions, probation conditions are a part of the court order of disposition; other juvenile courts use a standardized supervision agreement.

Probation conditions are of two types: *general conditions* that are common for everyone on probation and *special* or *individualized conditions*. General conditions involve common rules of probation such as to abide by all state laws, to attend school regularly, and to obey parents. Special conditions are individualized to the youth and are often based on needs identified in the predisposition report. Urinalyses, participation in various treatment programs, counseling, and restitution are examples of special conditions. A major reason for the extensive use of probation is that probation conditions serve multiple purposes—they provide public safety through supervision, they render punishment through restriction of freedom, they deter future crime, and they promote offender accountability and rehabilitation. As such, probation has adapted to the changing philosophies of the juvenile justice system.

Juvenile courts have extensive discretion when imposing conditions of probation.[148] While an incredibly wide range of probation conditions have been authorized under appellate court review, the basic standard is that they be *reasonable* and *relevant*. Probation conditions must be clear and not excessive (reasonable), and they must be related to the offense, the prevention of future delinquency, or to public safety (relevant).[149] For example, is a probation condition restricting freedom of association constitutional? A Florida appellate court case called *In the Interest of D. S. and J. V. Minor Children* (1965) dealt with two juveniles who were found guilty of simple battery and placed on probation.[150] One of the conditions of their probation was not to associate with gang members, under the rationale that this would prevent further delinquency. Their appeal alleged that such restriction of freedom of association was unconstitutional. The appellate court ruled that this condition was proper.[151] While the standard of reasonable and relevant cannot be defined with precision, a wide range of conditions are used in juvenile probation cases.

Probation Supervision

Probation supervision, too, serves multiple purposes. The approach to supervision taken by probation officers determines the relative emphasis given to offender rehabilitation or enforcement of probation rules. It is also the case that

probation conditions
Court-imposed rules that are a central part of the disposition of probation. Juveniles placed on probation by the court must obey these conditions in order to live in the community and avoid confinement. In this way, probation involves conditional release into the community.

EMPHASIS ON CONTROL

		Low	High
EMPHASIS ON REHABILITATION	**Low**	*Service Broker*	*Rule Enforcer*
	High	*Therapeutic Caseworker*	*Moral Reformer*

Figure 15-4 A model of probation supervision.
Source: Based on Jordan and Sasfy, *National Impact Program Evaluation,* 29.

styles of supervision have varied over the history of juvenile probation since it was first implemented with the original juvenile court in Chicago in 1899. <u>Probation supervision</u>, then, refers to the assistance and monitoring of probationers by probation officers.

A descriptive typology of probation supervision can be constructed based on two important dimensions: rehabilitation and control, as shown in **Figure 15-4.**[152] *Rehabilitation* refers to the degree to which probation officers emphasize reform and the role that they take in bringing about offender change. *Control* has to do with the level of monitoring and surveillance by probation officers and how closely they enforce probation conditions. The relative emphasis on these two important aspects of supervision influences probation officers' views and approaches to (1) probation conditions as a vehicle for change, (2) enforcement of conditions through monitoring and surveillance, and (3) readiness to revoke probation.[153]

The *moral-reformer* style of supervision is the original approach used in probation. Like many others of his day, John Augustus viewed crime and delinquency as a moral problem, requiring moral guidance. Augustus worked closely with the offenders released to him, getting involved in many aspects of their lives—family, work, church, recreation—and providing assistance. He believed that with proper guidance, offenders could be morally reformed. Following this tradition, early juvenile probation officers often used a moral-reformer style of supervision. Active involvement in the daily lives of juvenile offenders was meant to promote rehabilitation and high levels of control.

The professionalization of the juvenile court beginning in the 1920s was associated with the increasing use of full-time, highly trained probation officers rather than volunteers. State statutes began to require probation officers to have a college degree. Consistent with the child savers' view that the juvenile court should be more of a welfare system than a judicial system, professional probation officers emphasized a *therapeutic-casework* supervision style.[154] Following a "casework approach," probation officers engaged in case assessment, in which they evaluated the needs of their "clients" and developed rehabilitative goals based on such case assessment. Probation officers also provided direct services—they engaged in therapeutic counseling, including individual, family, and peer counseling.

Beginning in the 1960s, growing consideration was given to the community's role in generating and responding to crime and delinquency. In this

probation supervision
Assistance and monitoring of probationers by probation officers. The approach to supervision taken by probation officers determines the relative emphasis given to offender rehabilitation or enforcement of probation rules.

context, a community-based approach to probation supervision developed. Rather than having probation officers engage in therapeutic casework, the *service-broker* supervision style emphasized the need to use resources in the community, including educational programs, job training, job placement, community mental health, certified counselors, and drug treatment. It was thought that by using community resources, the offender would be better able to be reintegrated back into the community. With this approach to supervision, probation officers become brokers of community services and resources.

The final supervision style to emerge was that of the *law enforcer*. Associated with the "get tough" approach to crime and delinquency during the 1980s and 1990s, the law-enforcer style of supervision emphasized restriction of freedom, enforcement of probation conditions, monitoring and surveillance, and revocation of probation for rule violation. Seeking to provide offender accountability and public safety, this contemporary approach to supervision is characterized by the credo: "trail 'em, nail 'em, jail 'em." The law-enforcer style of supervision seeks punishment and retribution through restriction of freedom, and accountability and deterrence through enforcement of probation conditions.

Regardless of supervision style, probation supervision is accomplished primarily through three basic techniques: interviews with probationers and others, urinalysis, and records checks. These supervision techniques are used in office visits, field visits, and collateral contacts. Probation officers conduct interviews in their offices and at probationers' homes, schools, and places of work. Most commonly, interviews are simply informal talks with the offenders, their parents, teachers, counselors, and employers. Depending on supervision style, interviews with probationers may be therapeutic, involving counseling techniques, or they may be directed at information gathering to bring about offender accountability and enforcement of conditions. Urinalysis is most commonly conducted randomly during office visits, but it may also be done during field visits at home, school, or work. Probation officers also routinely check records that are kept by various agencies including police reports, detention intakes, school attendance records, and grades. Records information like this is normally gathered through collateral contacts with people and agencies that have relevance to the probation case. These collateral contacts supplement the probation officers' knowledge about probationers' daily activities and attitudes.[155]

Revocation

revocation The legal termination of probation by the court that granted probation, based on the violation of the probation agreement/order.

Violations of probation conditions result in a highly discretionary decision called **revocation**. Revocation is the legal termination of probation by the court, based on the violation of probation. Violations are of two types: law violations and technical violations. *Law violations* involve the commission of a new crime, in violation of statutory law. *Technical violations* are infractions of probation conditions that do not involve violations of law, such as the failure to report to a probation officer, curfew violations, and association with criminals and disreputable characters.[156] Probation conditions are court-imposed, and consequently the decision to revoke probation also rests with the court that imposed the conditions. Nonetheless, it is the probation officer who initiates the revocation process with a revocation report to the court. The probation officer has extensive discretion in deciding whether to instigate a revocation report. The probation officer has authority to deal with the violation informally, or formally through a revocation report that instigates a revocation hearing. Judges also have considerable discretion in whether to revoke probation or not. The revocation hearing follows due process of law, with certain rights that have been extended through appellate court review.[157] Upon finding that probation has been violated, the judge may continue the youth on probation, perhaps with new conditions, or may revoke the probation with a suspended sentence imposed or carried out.

Intensive Probation Supervision

An innovation in probation supervision, called intensive probation supervision (ISP), attempts to provide public safety and offender accountability through intensive monitoring and surveillance of the probationer. As an alternative to secure confinement, ISP is highly structured in terms of conditions and supervision standards. ISP officers have greatly reduced case loads that allow for much higher levels of supervision. ISP commonly involves the use of electronic monitoring devices (at least in initial phases), a specified number of contacts per week, mandatory school and work checks, frequent urinalysis, and a required number of community service hours. While these standards are intended to provide public safety, they are also designed to promote competency development by monitoring performance in school and work. In this way, ISP attempts to accomplish all three goals of balanced and restorative justice: offender accountability, public safety, and competency development.

Community-Based Corrections

The rise of community-based corrections in the 1960s and 1970s coincided with the transformation of the juvenile justice system during the second revolution. This was no accident. As described in Chapter 2, the President's Commission on Law Enforcement and Administration of Justice, established in 1965, recommended that nonviolent offenders be handled in the community rather than the juvenile courts. The logic of this recommendation stems from the second revolution's serious questioning of the rehabilitative ideal. In the words of the Commission's report: "the great hopes originally held for the juvenile court have not been fulfilled. It has not succeeded significantly in rehabilitating delinquent youth, in reducing or even stemming the tide of delinquency, or in bringing justice and compassion to the child offender."[158] In addition, the Commission emphasized the importance of dealing with juvenile delinquency in the context in which it develops—the local community: ". . . Crime and delinquency are symptoms of failures and disorganization of the community. . . . The task of corrections, therefore, includes building or rebuilding social ties, obtaining employment and education, securing in the larger sense a place for the offender in the routine functioning of society. This requires not only efforts directed toward changing the individual offender, which have been almost the exclusive focus of rehabilitation, but also mobilization and change of the community and its institutions."[159] Instead of rehabilitation, the major goal of community-based corrections is *reintegration*—correctional efforts designed to keep juvenile offenders in the community and to help them participate in community life.

Beginning in the second half of the 1970s, federal financial support through the OJJDP led to the development of community-based programs that utilized community resources. Community-based programs tend to be diversionary, attempting to deal with delinquency in the community rather than in the juvenile justice system. As we have already stressed, diversion occurs throughout the juvenile justice system. With regard to juvenile corrections, diversion refers to the use of community-based programs that provide needed services to the youth in the local community, regardless of at what point the case is diverted. Community-based programs also attempt to deinstitutionalize juvenile corrections, keeping delinquent youths in the local community, as opposed to placing them in correctional institutions.

A wide variety of community-based juvenile correctional programs were developed during this time period. Chief among them were youth service bureaus that provided a wide variety of services to youth and families, including individual, peer, and family counseling; crisis hotlines; drop-in centers; job placement services; educational tutoring; and recreation programs.[160] Also developed were foster care programs; community service programs; and various residential programs, including group homes that emphasize treatment and education, drug treatment centers, and halfway houses. Though it originated much earlier than community-based corrections, probation is conceptually consistent with these community-based approaches because it provides conditional release into the community with supervision. Since their inception, community-based correctional programs have attempted to promote delinquency prevention, diversion, and community reintegration as an alternative to placement in secure, custodial correctional institutions.

Balanced and Restorative Justice

Contemporary correctional efforts in balanced and restorative justice are often community-based, because they attempt to actively involve the community in delinquency prevention and intervention. This popular approach is typically nonresidential and attempts to actively involve community members and use community resources. As described in Chapter 2, the balanced and restorative approach to juvenile justice is directed at three fundamental goals: offender accountability, public safety, and competency development. The balanced and restorative approach to juvenile corrections attempts to accomplish these goals through what Gordon Bazemore and Mark Umbreit call "restorative conferencing": "a range of strategies to bring the victim, offender, and other members of the community together in a nonadversarial, community-based process."[161] Bazemore and Umbreit identify and describe four restorative conferencing models.[162]

1. *Victim–offender mediation:* With the assistance of a trained mediator, victims voluntarily meet with offenders in a safe and structured setting. The victim is allowed to tell the offender about the crime's physical, emotional, and/or financial impact and to ask the offender questions about the crime. Together they work out a restitution plan for the offender.[163]

2. "*Community reparative boards* are composed of small groups of citizens who have received intensive training and conduct public, face-to-face meetings with offenders who have been ordered by the court to participate. During the meeting, board members discuss with the offender the

nature and seriousness of the offense. The board develops agreements that sanction offenders, monitor compliance, and submit reports to the court."[164]

3. *Family group conferencing model:* "To initiate a family group conference, a trained facilitator contacts the victim and the offender to explain the process and invite them to participate. The victim and offender are asked to identify key members of their support system who will also be asked to attend. Participation is voluntary. The conference usually begins with the offender describing the incident, after which the other participants describe how the incident has affected their lives." The process is designed to make an offender confront his or her crime and the impact it has had on others, including victims, families, and friends. Victims can ask questions and express their feelings. "After the discussion, the facilitator asks the victim what he or she wants the outcome of the conference to be. All participants can contribute suggestions and how the harm can be repaired, and an agreement is reached that outlines what is expected of the offender."[165]

4. *Circle sentencing* is "based on traditional practices of aboriginal peoples in Canada and the native Americans in the United States. . . . Circle sentencing includes participation of the victim, offender, both of their families and friends, personnel from the justice agency, police department and relevant social service agencies, and interested community members who together develop a sentencing plan that addresses the concerns of all the parties. After an offender petitions to participate in a circle, a healing circle is formed for the victim, a healing circle is formed for the offender, and sentencing circle is formed to reach consensus on a reparation plan, and a follow-up circle takes responsibility for monitoring the enforcement of the agreement."[166]

Residential Placement

The most restrictive disposition available to juvenile courts is placement in secure, custodial institutions—most commonly, state training schools. There is tremendous variation in juvenile **residential placement** facilities, especially in terms of administrative control (private or public operation), size, physical setting, level of custody, length of placement, and the degree to which they pursue rehabilitation. As we saw in Chapter 2, the use of an institutional setting for juvenile corrections dates back to the New York House of Refuge, established in 1824. Large, highly structured correctional institutions have been the hallmark of juvenile corrections in the United States since then. Contemporary juvenile corrections include a wide range of residential facilities for out-of-home placement, including psychiatric hospitals, residential treatment centers, custodial institutions, foster homes, and group homes.

A one-day count in 1999 revealed that there were nearly 109,000 juveniles placed in private and public residential facilities as a result of being charged with or adjudicated for a delinquent offense.[167] **Table 15-3** shows that most (96%) of these youths were placed for delinquent offenses, rather than status offenses. Most of the residential placements (74%) involved a commitment. Commitment

residential placement Court-authorized or court-ordered out-of-home placement of a youth in a group living facility. Residential placement facilities vary greatly in terms of administrative control, size, physical setting, level of custody, length of placement, and degree of emphasis on rehabilitation.

Table 15-3	Juvenile Offenders in Residential Placement, 1999					
	All facilities		**Public facilities**		**Private facilities**	
	Number	Percent	Number	Percent	Number	Percent
Total offenders	108,931	100%	77,158	100%	31,599	100
Delinquency	104,237	96	75,537	98	28,536	90
Person	38,005	35	28,056	36	9,897	31
Property	31,817	29	22,725	29	9,051	29
Drug	9,882	9	6,819	9	3,054	10
Public order	10,487	10	7,380	10	3,087	10
Technical violation	14,046	13	10,557	14	3,447	11
Status offense	4,694	4	1,623	2	3,063	10

Source: Sickmund, "Juveniles in Corrections," 6.

group homes Small, private, nonsecure residential facilities that emphasize rehabilitation. Group homes are located in neighborhoods and use community resources and services in their attempt to integrate youth into the local community.

custodial institutions Closed and secure residential facilities for delinquent youth, providing long-term custody.

is the juvenile court's legal authorization for out-of-home placement of a youth who has been adjudicated a delinquent youth or a status offender. Two thirds (68%) of the committed youths were placed in public facilities and one third (31%) were placed in private facilities, even though there are far more private than public residential facilities (1,794 and 1,136, respectively). The number of youths committed for residential placement increased by 51% from 1991 to 1999.[168] During the same time period, the number of residential placements for status offenses declined by 32%.[169] Minority youth accounted for 66% of the juveniles committed to public residential facilities across the nation, almost twice their proportion of the juvenile population (34%).[170]

Most private residential facilities for juveniles are small and nonsecure, emphasizing rehabilitation.[171] While no accurate count exists, **group homes** are undoubtedly the most common type of private residential facility. As such, group homes are better classified as community-based corrections, because they are usually located in neighborhoods and use community resources and services in their attempt to integrate youth into the local community. In contrast, about two thirds of the youth in public institutions were placed in "closed and secure" facilities, providing "long-term" custody.[172] Most often, **custodial institutions** take the form of state training schools. However, state training schools vary considerably, from campus-like grounds with cottages to large, secure buildings much like adult prisons. Some state training schools are very large, housing hundreds of youths.[173] With the contemporary "get tough" approaches to juvenile corrections, many states have remodeled some of their state training schools or built new high-security buildings, to serve as juvenile prisons. While contemporary versions of state training schools tend to emphasize custody over rehabilitation, they also commonly offer programs for competency development, especially education and job training. They typically operate under a school-day schedule and incorporate behavior modification programs in an effort to promote behavior change and to teach living skills.

Other residential facilities for juveniles feature specific rehabilitative purposes. While psychiatric hospitals and residential drug treatment centers are usually private institutions, they are sometimes used as a dispositional placement by juvenile courts. Contemporary "get tough" efforts in juvenile corrections have led to the development of boot camp programs in many states, which provide relatively short-term placement in a military-style environment. A regimented schedule stresses discipline, physical training, work, and education. More general in purpose are large residential child-care facilities for juveniles such as boys and girls "homes," "ranches," and "towns." These residential centers can be quite large, but they are usually broken down into smaller units identified as cottages or houses, where house parents or a small staff attempts to provide a family-like environment. Thus, residential placement facilities vary considerably in administrative control (public versus private), size, physical setting, level of custody, length of placement, and the degree to which they pursue rehabilitation.

One of the biggest issues for residential placement is **aftercare**—the assistance, services, and programs that follow residential placement upon a resident's release. Traditionally, youth released from residential placement are placed on **parole**. Parole involves conditions and supervision, just like probation. While essentially similar, probation and parole are distinguished in relation to residential placement: probation is used before placement and parole afterwards. In many states, parole supervision is provided by officers who are both probation and parole officers.

In today's "get tough" approach, new aftercare programs have been developed, including the intensive aftercare program (IAP). The IAP model is designed to reduce recidivism among high-risk parolees by combining intensive supervision with treatment, as a bridge between residential placement and reentry into the community. Five principles guide the structured reentry process:

1. Prepare youth for progressively increased responsibility and freedom in the community.

2. Facilitate youth–community interaction and involvement.

3. Work with the offender and targeted community support systems (e.g., schools, family) on qualities needed for constructive interaction and for the youth's successful community adjustment.

4. Develop new resources and supports where needed.

5. Monitor and test the youth and the community on their ability to deal with each other productively.[174]

IAP attempts to develop individualized parole reentry plans by careful case selection, assessment, and classification. In essence, it is a balanced and restorative approach to aftercare, in that it provides intensive supervision and treatment services, balanced incentives, and graduated consequences.

aftercare The assistance, services, and programs that follow residential placement upon a resident's release.

parole The conditions and supervision provided after release from residential placement, intended to smooth the transition back into the community.

■ Summary and Conclusions

The term "juvenile justice system" is grossly inaccurate, a misnomer. There is not a single juvenile justice system; rather, each state has a separate juvenile justice system and these are "systems" in only a limited sense. The various parts of these systems—police, courts, and corrections—are all operated by different levels and branches of government, sometimes even by private organizations. As a result, juvenile justice is highly decentralized and fragmented. In addition, many juvenile justice cases are diverted from "the system" or dealt with informally with only limited due process of law. As a result, "justice" is highly discretionary in juvenile justice systems across the United States.

Delinquency prevention programs represent a proactive approach to juvenile crime, attempting to keep it from occurring. Such programs not only try to prevent delinquency, but also try to prevent involvement in juvenile justice systems. Research has established that juvenile delinquency is a component of a larger group of problem behaviors that tend to occur together, including drug and alcohol use, mental health problems, behavior problems and underachievement in school, and precocious and risky sexual behavior. "Many youth who are seriously delinquent also experience difficulty in other areas of life."[175] Research In Action, "Risk Factors: Causes and Correlates of Delinquency," listed a number of risk factors that have been found to be related to these problem behaviors. Contemporary approaches to delinquency prevention argue that these risk factors must first be identified through research and then addressed through a broad range of prevention programs at the individual, family, school, and community levels.[176] We considered several delinquency prevention programs that have been evaluated as successful and identified as "model programs."

Police encounters are usually a juvenile's entry point into juvenile justice systems. As is true throughout the juvenile justice process, many cases are dealt with informally through diversion, such as warning and release at the scene (a "street adjustment") and referral to various community resources and services. Formal processing of juveniles suspected of crime involves a police referral to the juvenile court. Over the last thirty years, formal juvenile court referrals by the police have become increasingly common in comparison to informal police processing. In recent years, almost three fourths of all juveniles taken into custody are referred to juvenile courts.

Police have extensive discretion in dealing with juvenile matters. We described a number of factors that have an influence on discretion, such as seriousness of the offense; previous police contact; prior arrest; attitude and demeanor of the juvenile; social characteristics of the juvenile; and organization, policies, and customs of the police department. The arrest of a juvenile in some states is referred to as "taking into custody" in order to convey the protective role of juvenile policing. The role of police has evolved over time to include both a collaborative role, which involves the public in a collective effort to prevent crime and maintain public order, and a law-enforcer role, which emphasizes arrest and formal referral to the juvenile court. While police discretion in juvenile matters is broad and extensive, it is not unbridled. During the 1960s and 1970s, the due process revolution in juvenile justice addressed several key law-enforcement methods related to juveniles. The case decisions that make up the due process revolution established procedural law that attempts to extend basic rights and freedoms advanced in state and federal constitutions, especially with regard to search and seizure, custodial interrogation, and arrest.

Juvenile courts, too, rely heavily on discretion and diversion—many cases are diverted or handled informally. The juvenile court process involves a series of discretionary decisions, including intake, petitioning, transfer to adult court, adjudication, and disposition. Intake by a probation officer refers to the initial screening of a case referred to the juvenile court. If secure detention is requested by the police, it must be initially authorized by a probation officer as a part of intake. Continued detention, however, requires court authorization, usually through a judicial hearing within a statutorily specified period of time. While most delinquency cases do not involve secure detention, it is used disproportionately in those cases that involve black youth. The problem of disproportionate minority confinement has been clearly identified and documented for over 30 years, but the problem persists.

More than two thirds of the cases referred to juvenile courts are dealt with formally through petition or transfer to adult court. Several transfer provisions have become increasingly popular, including judicial waiver, concurrent jurisdiction, and statutory exclusion. Even when cases are not formally petitioned, they can receive some form of informal disposition by juvenile courts. Informal disposition is usually based on the youth's admission of delinquency. A variety of disposition options are then possible, including diversionary referral to another agency for services (e.g., chemical dependency treatment or counseling), community service, restitution, and informal probation through the probation department.[177] These informal dispositions usually involve a consent adjustment without petition, a voluntary agreement between the youth and his or her parents that is approved by a juvenile court judge. Less than one percent of all petitioned delinquency cases are waived to adult court.

When formal court intervention is deemed necessary, cases proceed through a series of hearings that have varying names and that serve particular purposes. The probable cause hearing makes an initial determination about whether sufficient information and evidence exist to substantiate that a youth was likely involved in activity that violated state criminal or status offense law. In addition, probable cause hearings determine whether continued detention is necessary. Arraignments involve the advising of rights, reading of the petition, and the youth's admission or denial of the petition. The adjudicatory hearing is a fact-finding process with due process rights that results in a judicial determination of whether the youth has in fact violated criminal law (delinquent youth) or status offense law (status offender) and whether the youth is in need of the juvenile court's assistance and supervision. If a youth is adjudicated a delinquent youth or a status offender, the court decides the most appropriate sanction at a dispositional hearing. The dispositional hearing is usually separated in time and purpose from the adjudicatory procedures that precede it. A predisposition report, written by a probation officer, informs disposition through evaluation of the youth and his or her background and social environment. A full range of dispositions are normally authorized under state statute, including probation, out-of-home placement, restitution, community service, counseling, fines, and mediation.

Juvenile court disposition is the gateway to corrections, because it legally establishes how the court desires to deal with a case. Regardless of whether a case is adjudicated or not, court disposition seeks correctional options that are individualized, rehabilitative and the least restrictive option. Probation is the most frequently imposed disposition in juvenile courts. Probation is usually combined with other dispositions, such as restitution or community service, and it involves a set of court-imposed rules, called conditions, and supervision by a probation officer. Community-based corrections emphasize the community's role in causing and responding to delinquency and crime. Community corrections often utilize community resources such as counseling services, tutoring and educational programs, and job training and placement. Residential placement is the most restrictive disposition available to juvenile courts. There is tremendous variation in residential placement facilities, especially in terms of administrative control, size, physical setting, level of custody, length of placement, and degree of emphasis on rehabilitation. Group homes are typically small, private treatment facilities in residential neighborhoods. In contrast, custodial institutions, such as state training schools, are usually much larger and far more oriented toward security rather than rehabilitation.

Juvenile justice in the United States is incredibly diverse in philosophy, structure, and process. There is not a single juvenile justice system. Rather, each state has a separate juvenile justice system, and the various parts of these systems—police, courts, and corrections—are all operated by different levels and branches of government, sometimes even by private organizations. Each part of the "system" has a different purpose and area of responsibility. It is no wonder that each has its own juvenile justice philosophy and process.

CRITICAL THINKING QUESTIONS

1. In what ways are juvenile justice systems decentralized and fragmented among the different levels and branches of government?
2. Describe how discretion is used throughout the juvenile justice process.
3. Describe how diversion is used throughout the juvenile justice process.
4. What does the term "adolescents at risk" mean and how is it related to delinquency prevention efforts?
5. Identify and briefly describe the various roles that police adopt in juvenile law enforcement.
6. How does due process influence law enforcement with juveniles?
7. Outline the formal juvenile court process.
8. Provide an inventory of juvenile court dispositions.
9. What are probation conditions? What does it mean that probation conditions must be "reasonable and relevant"?
10. What is "community"-related about community-based corrections?
11. Describe how residential placement facilities vary with regard to administrative control, size, physical setting, level of custody, length of placement, and degree of emphasis on rehabilitation.
12. What is commitment? Who has legal authority to impose it? When is it used? What does it entail?

SUGGESTED READINGS

Sickmund, Melissa. "Juveniles in Court." *Juvenile Offenders and Victims: National Report Series.* Washington, DC: Office of Juvenile Justice and Delinquency Prevention, 2003.

———. "Juveniles in Corrections." *Juvenile Offenders and Victims: National Report Series.* Washington, DC: Office of Juvenile Justice and Delinquency Prevention, 2004.

Snyder, Howard N. "Law Enforcement and Juvenile Crime." *Juvenile Offenders and Victims: National Report Series.* Washington, DC: Office of Juvenile Justice and Delinquency Prevention, 2001.

Feld, Barry C. *Bad Kids: Race and the Transformation of the Juvenile Court.* New York: Oxford University Press, 1999.

Howell, James C. *Preventing and Reducing Juvenile Delinquency: A Comprehensive Framework.* Thousand Oaks, CA: Sage, 2003.

GLOSSARY

adjudication: A case dealt with formally by the juvenile court goes through a series of hearings, involving determination of probable cause, arraignment, and fact finding. The purpose of this adjudicatory process is to determine whether the youth is responsible for the offense(s) charged in the petition, and if the youth should be legally declared a "delinquent youth" or a "status offender" by the juvenile court.

aftercare: The assistance, services, and programs that follow residential placement upon a resident's release.

arrest: The deprivation of an individual's freedom by a person of authority on the grounds that there is probable cause to believe that that individual has committed a criminal offense.

collaborative role: The police and the community working together to prevent crime, maintain public order, and solve community problems.

commitment: The juvenile court's legal authorization for out-of-home placement for a youth who has been adjudicated a delinquent youth, involving a transfer of legal custody to "the state"—some state agency such as the juvenile court or department of corrections.

crime-fighting role: The police role associated with professional policing that emphasizes enforcement of laws and crime control. The primary strategies for crime fighting involve motorized preventive patrol, rapid response, and reactive investigation.

custodial institutions: Closed and secure residential facilities for delinquent youth, providing long-term custody.

custodial interrogation: Police questioning of suspects when the custody involves arrest or the significant deprivation of freedom and the questioning may be incriminating.

detention: The secure confinement of a youth under court authority, pending court action, disposition, or placement.

discretion: "Authority conferred by law to act in certain conditions or situations in accordance with an official's or an official agency's considered judgment and conscience" (Pound, "Discretion, Dispensation, and Mitigation," 926).

diversion: The tendency to deal with juvenile matters informally, without formal processing and adjudication, by referring cases to agencies outside the juvenile justice system.

formal disposition: The juvenile court's response to a youth's admission of a petition or the court's finding that a youth committed an offense(s). Among the dispositional options are probation, out-of-home placement, restitution, and community service.

group homes: Small, private, nonsecure residential facilities that emphasize rehabilitation. Group homes are located in neighborhoods and use community resources and services in their attempt to integrate youth into the local community.

informal disposition: A juvenile court's response and handling of a case that has been referred but not adjudicated. Informal disposition may occur in cases that have not been petitioned and in cases that have been petitioned but not adjudicated. A variety of informal dispositional options are available to juvenile courts, including referral to another agency for services (e.g., chemical dependency treatment or counseling), community service, restitution, and informal probation through the probation department.

intake: The screening of cases referred to the juvenile court, usually by the probation department. Intake screening results in a determination of how the case will be handled—formally, informally, or by dismissal.

law-enforcer role: Associated with the "get tough" approach to crime and delinquency, the law-enforcer role emphasizes public safety and offender accountability and involves vigorous enforcement of laws through arrest and referral to the juvenile court.

parole: The conditions and supervision provided after release from residential placement, intended to smooth the transition back into the community.

petition: A legal document filed in juvenile court that specifies the facts of the case, the charge(s), and the basic identifying information of the youth and his or her parents.

predisposition report: A court-ordered report written by a probation officer that provides information relevant for court disposition in an effort to make disposition individualized and rehabilitative. The report usually includes three major sections (offense, social history, and summary and recommendation) that furnish information on the adjudicated youth's prior record, current offense, and social background. The probation officer also makes a recommendation for disposition.

probation conditions: Court-imposed rules that are a central part of the disposition of probation. Juveniles placed on probation by the court must obey these conditions in order to live in the community and avoid confinement. In this way, probation involves conditional release into the community.

probation supervision: Assistance and monitoring of probationers by probation officers. The approach to supervision taken by probation officers determines the relative emphasis given to offender rehabilitation or enforcement of probation rules.

problem behaviors: Behaviors that compromise successful adolescent adjustment, such as drug and alcohol use, mental health problems, underachievement in school, precocious and risky sexual behavior, and delinquency.

protective role: The original role of police in dealing with juveniles, in which police were expected to practice "kindly discipline" in place of parents—to function in *loco parentis*. As an extension of the juvenile court, police were to deal with juveniles in an informal, parent-like fashion, acting in the "best interests of the child."

referral: Written documents, usually from the police, parents, or schools, that request or result in juvenile court consideration of a particular matter involving a juvenile.

rehabilitative ideal: The traditional legal philosophy of the juvenile court, which emphasizes assessment of the youth and individualized treatment, rather than determination of guilt and punishment.

residential placement: Court-authorized or court-ordered out-of-home placement of a youth in a group living facility. Residential placement facilities vary greatly in terms of administrative control, size, physical setting, level of custody, length of placement, and degree of emphasis on rehabilitation.

revocation: The legal termination of probation by the court that granted probation, based on the violation of the probation agreement/order.

risk factor: Any individual trait, social influence, or environmental condition that leads to greater likelihood of problem behaviors and ultimately negative developmental outcomes during the adolescent years.

search and seizure: The Fourth Amendment protection against unreasonable search and seizure by the police, requiring a search warrant, which is issued by the courts and based on probable cause.

taking into custody: The statutory authority given to police officers to take control of a juvenile, either physically or by the youth's voluntary submission, in an effort to separate that youth from his or her surroundings. Taking into custody is used when the juvenile has broken the law or is in need of protection.

REFERENCES

Albert, Rodney L. "Juvenile Accountability Incentive Block Grants Programs." *OJJDP Fact Sheet* (#76). Washington, DC: Office of Juvenile Justice and Delinquency Prevention, 1998.

Allen, Francis. "Legal Values and the Rehabilitative Ideal." *Journal of Law, Criminology, and Police Science* 50 (1959):226–232.

Allen-Hagen, Barbara. "Public Juvenile Facilities: Children in Custody 1989." Washington, DC: Office of Juvenile Justice and Delinquency Prevention, 1991.

American Friends Service Committee. *Struggle for Justice.* New York: Hill and Wang, 1971.

Andrews, Chyrl, and Lynn Marble. "Changes to OJJDP's Juvenile Accountability Program." *Juvenile Justice Bulletin.* Washington, DC: Office of Juvenile Justice and Delinquency Prevention, 2003.

Bartollas, Clemens, and Stuart H. Miller. *Juvenile Justice in America.* 4th ed. Upper Saddle River, NJ: Prentice Hall, 2005.

Bazemore, Gordon, and Mark Umbreit. *Balanced and Restorative Justice for Juveniles: A Framework for Juvenile Justice in the 21st Century.* Washington, DC: Office of Juvenile Justice and Delinquency Prevention, 1997.

Becker, Harold K., Michael W. Agopian, and Sandy Yeh. "Impact Evaluation of Drug Abuse Resistance Education (D.A.R.E.)." *Journal of Drug Education* 22 (1992):283–291.

Belbott, Barbara A. "Restorative Justice." In *Encyclopedia of Juvenile Justice,* edited by Marilyn D. McShane and Frank P. Williams, 322–325. Thousand Oaks, CA: Sage, 2003.

Beyer, Marty. "Best Practices in Juvenile Accountability: An Overview." *Juvenile Accountability Incentive Block Grants Program Bulletin.* Washington, DC: Office of Juvenile Justice and Delinquency Prevention, 2003.

Bishop, Donna, and Charles Frazier. "Gender Bias in Juvenile Justice Processing: Implication of the JJDP." *Journal of Criminal Law and Criminology* 82 (1992):1162–1186.

———. "The Influence of Race in Juvenile Justice Processing." *Journal of Research in Crime and Delinquency* 25 (1988):242–261.

Black, Donald J., and Albert J. Reiss, Jr. "Police Control of Juveniles." *American Sociological Review* 35 (1970):63–77.

Bordua, David J. "Recent Trends: Deviant Behavior and Social Control." *Annals of the American Academy of Political and Social Science* 359 (1967):36–43.

Borrero, Michael. "The Widening Mistrust Between Youth and Police." *Families in Society: The Journal of Contemporary Human Services* 82 (2001):399–408.

Botvin, Gilbert J., Eli Baker, Linda Dusenbury, Elizabeth M. Botvin, and Tracy Diaz. "Long-Term Follow-Up Results of a Randomized Drug Abuse Prevention Trial in a White Middle-Class Population." *JAMA* 273 (1995):1106–1112.

Burns, Barbara J., James C. Howell, Janet K. Wiig, Leena K. Augimeri, Brendan C. Welsh, Rolf Loeber, and David Petechuk. "Treatment, Services, and Intervention Programs for Child Delinquency." Washington, DC: Office of Juvenile Justice and Delinquency Prevention, 2003.

Cicourel, Aaron. *The Social Organization of Juvenile Justice*. New York: Wiley, 1968.

Cohen, L., and J. Kluegel. "Determinants of Juvenile Court Disposition: Ascription and Achievement in Two Metropolitan Courts." *American Sociological Review* 43 (1978):162–176.

Coolbaugh, Kathleen, and Cynthia J. Hansel. "The Comprehensive Strategy: Lessons Learned From Pilot Sites." Washington, DC: Office of Juvenile Justice and Delinquency Prevention, 2000.

Cromwell, Paul F., Rolando V. del Carmen, and Leanne Fiftal Alarid. *Community-Based Corrections*. 5th ed. Belmont, CA: Wadsworth, 2002.

Czajkoski, Eugene H. "Exposing the Quasi-Judicial Role of Probation Officers." *Federal Probation* 37 (1973):9–13.

Danegger, Anna E., Carole E. Cohen, Cheryl D. Hayes, and Gwen A. Holden. *Juvenile Accountability Incentive Block Grants: Strategic Planning Guide*. Washington, DC: Office of Juvenile Justice and Delinquency Prevention, 1999.

Dannefer, Dale, and Russel Schutt. "Race and Juvenile Justice Processing in Police and Court Agencies." *American Journal of Sociology* 87 (1982):1113–1132.

Decker, Scott H. "Citizen Attitudes Toward the Police: A Review of Past Findings and Suggestions for Future Policy." *Journal of Police Science and Administration* 9 (1981):80–87.

———. "Increasing School Safety Through Juvenile Accountability Programs." *Juvenile Accountability Incentive Block Grants Program Bulletin*. Washington, DC: Office of Juvenile Justice and Delinquency Prevention, 2000.

Dryfoos, Joy G. *Adolescents at Risk: Prevalence and Prevention*. New York: Oxford University Press, 1990.

Elliott, Delbert S., David Huizinga, and Scott Menard. *Multiple Problem Youth: Delinquency, Substance Use, and Mental Health Problems*. New York: Springer-Verlag, 1989.

Ennett, Susan, D. P. Rosenbaum, Robert L. Flewelling, G. S. Bieler, Christopher L. Ringwalt, and S. L. Bailey. "Long-Term Evaluation of Drug Abuse Resistance Education." *Addictive Behavior* 19 (1994):113–125.

Ennett, Susan, Nancy S. Tobler, Christopher L. Ringwalt, and Robert L. Flewelling. "How Effective Is Drug Abuse Resistance Education? A Meta-Analysis of Project D.A.R.E. Outcome Evaluations." *American Journal of Public Health* 84 (1994):1394–1401.

Esbensen, Finn-Aage. "Evaluating G.R.E.A.T.: A School-based Gang Prevention Program." *NIJ Research for Policy*. Washington, DC: National Institute of Justice, 2004.

Esbensen, Finn-Aage, and D. Wayne Osgood. "Gang Resistance Education and Training (G.R.E.A.T.): Results from the National Evaluation." *Journal of Research in Crime and Delinquency* 36 (1999):194–225.

———. "National Evaluation of G.R.E.A.T." *Research in Brief*. Washington, DC: National Institute of Justice, 1997.

Esbensen, Finn-Aage, D. Wayne Osgood, Terrance J. Taylor, Dana Peterson, and Adrienne Freng. "How Great is G.R.E.A.T.? Results from a Longitudinal Quasi-Experimental Design." *Criminology and Public Policy* 1 (2001):87–118.

Fagan, Jeffrey, Ellen Slaughter, and Eliot Hartstone. "Blind Justice? The Impact of Race on the Juvenile Justice Process." *Crime and Delinquency* 33 (1987):224–258.

Feld, Barry C. *Bad Kids: Race and the Transformation of the Juvenile Court*. New York: Oxford University Press, 1999.

Ferdico, John N. *Criminal Procedure for the Criminal Justice Professional*. 5th ed. Minneapolis/St. Paul: West, 1993.

Ferdinand, Theodore N., and Elmer C. Luchtenhand. "Inner-City Youths, the Police, the Juvenile Court and Justice." *Social Problems* 17 (1970):510–527.

Fogelson, Robert M. *Big City Police*. Cambridge, MA: Harvard University Press, 1977.

Gaines, Larry K. "Police Responses to Delinquency." In *Encyclopedia of Juvenile Justice*, edited by Marilyn D. McShane and Frank P. Williams, 286–290. Thousand Oaks, CA: Sage, 2003.

Goldman, Nathan. *The Differential Selection of Juvenile Offenders for Court Appearance.* Washington, DC: National Council on Crime and Delinquency, 1963.

Goldstein, Herman. *Problem-Oriented Policing.* New York: McGraw-Hill, 1990.

Hall, G. Stanley. *Adolescence: Its Psychology and its Relations to Physiology, Anthropology, Sociology, Sex, Crime, Religion and Education.* New York: Appleton, 1904.

Harmon, Michele Alicia. "Reducing the Risk of Drug Involvement Among Early Adolescents: An Evaluation of Drug Abuse Resistance Education (D.A.R.E.)." *Evaluation Research* 17 (1993):221–239.

Harms, Paul. "Detention in Delinquency Cases, 1990–1999." Washington, DC: Office of Juvenile Justice and Delinquency Prevention, 2003.

Hawkins, J. David, Todd I. Herrenkohl, David P. Farrington, Devon Brewer, Richard F. Catalano, Tracy W. Harachi, and Lynn Cothern. "Predictors of Youth Violence." Washington, DC: Office of Juvenile Justice and Delinquency Prevention, 2000.

Hellum, Frank. "Juvenile Justice: The Second Revolution." *Crime and Delinquency* 25 (1979):299–317.

Hemmens, Craig, Benjamin Steiner, and David Mueller. *Significant Cases in Juvenile Justice.* Los Angeles: Roxbury, 2004.

Hickman, Matthew J., and Brian A. Reaves. "Community Policing in Local Police Departments, 1997 and 1999." In *Bureau of Justice Statistics A Special Report.* Washington, DC: Bureau of Justice Statistics, 2001.

Howell, James C. *Juvenile Justice and Youth Violence.* Thousand Oaks, CA: Sage, 1997.

———. *Preventing and Reducing Juvenile Delinquency: A Comprehensive Framework.* Thousand Oaks, CA: Sage, 2003.

Hsia, Heidi M., George S. Bridges, and Rosalie McHale. *Disproportionate Minority Confinement 2002 Update.* Washington, DC: Office of Juvenile Justice and Delinquency Prevention, 2004.

Huizinga, David, Rolf Loeber, Terence P. Thornberry, and Lynn Cothern. "Co-Occurrence of Delinquency and Other Problem Behaviors." *Juvenile Justice Bulletin.* Washington, DC: Office of Juvenile Justice and Delinquency Prevention, 2000.

Hurst, Yolander G., and James Frank. "How Kids View Cops: The Nature of Juvenile Attitudes Toward the Police." *Journal of Criminal Justice* 28 (2000):189–202.

Jessor, Richard. "Risk Behavior in Adolescence: A Psychosocial Framework for Understanding and Action." In *Adolescents at Risk: Medical and Social Perspectives,* edited by D. Rogers and E. Ginzberg, 374–389. Boulder, CO: Westview, 1996.

Jordon, Frank, and Joseph M. Sasfy. *National Impact Program Evaluation: A Review of Selected Issues and Research Findings Related to Probation and Parole.* Washington, DC: Mitre Corporation, 1974.

Kelling, George L. "Police and Communities: The Quiet Revolution." In *Perspective on Policing #1.* Washington, DC: US Department of Justice, 1988.

Kelling, George L., and Mark H. Moore. "The Evolving Strategy of Policing." In *Perspective on Policing #4.* Washington, DC: US Department of Justice, 1988.

Klinger, David. "Demeanor of Crime? Why 'Hostile' Citizens are More Likely to be Arrested." *Criminology* 32 (1994):475–493.

Klockers, Carl B., Jr. "A Theory of Probation Supervision." *Journal of Criminal Law, Criminology, and Police Science* 63 (1972):550–557.

Krisberg, Barry, and James Austin. *The Children of Ishmael: Critical Perspective on Juvenile Justice.* Palo Alto, CA: Mayfield, 1978.

Lattimore, Pamela, Christy Visher, and Richard Linster. "Predicting Rearrest for Violence Among Serious Youthful Offenders." *Journal of Research in Crime and Delinquency* 32 (1995):54–83.

Lehman, Joseph D., J. David Hawkins, and Richard F. Catalano. *Corrections Today* (August 1994): 92–100.

Leiber, Michael J., Mahesh K. Nalla, and Margaret Farnworth. "Explaining Juveniles' Attitudes Toward the Police." *Justice Quarterly* 15 (1998):151–171.

Lerman, Paul. *Community Treatment and Control.* Chicago: University of Chicago Press, 1975.

Lerner, Richard M., and Nancy L. Galambos. "Adolescent Development: Challenges and Opportunities for Research, Programs, and Policies." *Annual Review of Psychology* 49 (1998):413–446.

Lewett, Allan E. "Centralization of City Police in Nineteenth Century United States." PhD diss., University of Michigan, 1975.

Lipton, Douglas, Robert Martinson, and Judith Wilks. "The Effectiveness of Correctional Treatment." New York: Praeger, 1975.

Loeber, Rolf, David Farrington, and David Petechuk. "Child Delinquency: Early Intervention and Prevention." Washington, DC: Office of Juvenile Justice and Delinquency Prevention, 2003.

Lundman, Richard. "Demeanor and Arrest: Additional Evidence from Previously Unpublished Data." *Journal of Research in Crime and Delinquency* 33 (1996):349–353.

—. "Demeanor of Crime? The Midwest City Police–Citizen Encounters Study." *Criminology* 32 (1994):631–653.

Lundman, Richard J., Richard E. Sykes, and John P. Clark. "Police Control of Juveniles: A Replication." *Journal of Research in Crime and Delinquency* 15 (1978):74–91.

Lynam, Donald R., Richard Milich, and Richard Zimmerman. "Project D.A.R.E.: No Effects at a 10-Year Follow Up." *Journal of Consulting and Clinical Psychology* 67 (1999):590–593.

Maguire, Kathleen and Ann L.Pastore, eds. *Sourcebook of Criminal Justice Statistics* [Online]. Available at: http://www.albany.edu/sourcebook.

Marans, Steven, and Miriam Berkman. "Child Development—Community Policing: Partnership in a Climate of Violence." In *Juvenile Justice Bulletin*. Washington, DC: Office of Juvenile Justice and Delinquency Prevention, 1997.

Mihalic, Sharon, Katherine Irwin, Delbert Elliott, Abigail Fagan, and Diane Hansen. "Blueprints for Violence Prevention." Washington, DC: Office of Juvenile Justice and Delinquency Prevention, 2001.

Mihalic, Sharon, Katherine Irwin, Abigail Fagan, Diane Ballard, and Delbert Elliott. "Successful Program Implementation: Lessons from Blueprints." Washington, DC: Office of Juvenile Justice and Delinquency Prevention, 2004.

Miller, Joshua D., and Donald Lynam. "Structural Models of Personality and Their Relation to Antisocial Behavior: A Meta-Analytic Review." *Criminology* 39 (2001):765–798.

Moffitt, Terrie E. "Adolescence-Limited and Life-Course-Persistent Antisocial Behavior: A Developmental Taxonomy." *Psychological Review* 100 (1993):674–701.

Moore, Mark H., Robert C. Trojanowicz, and George L. Kelling. "Crime and the Police." In *Perspective on Policing #2*. Washington, DC: US Department of Justice, 1988.

National Advisory Commission on Criminal Justice Standards and Goals. *Task Force Report on Corrections*. Washington, DC: GPO, 1973.

Office of Community Oriented Policing Services. "COPS in Schools: The COPS Commitment to School Safety." *COPS Fact Sheet*. Washington, DC: Office of Community Oriented Policing Services.

Ohlin, Lloyd E., Herman Piven, and D. M. Pappenfort. "Major Dilemmas of the Social Worker in Probation and Parole." *National Probation and Parole Association Journal* 2 (1956):21–25.

Olds, David, Peggy Hill, and Elissa Rumsey. "Prenatal and Early Childhood Nurse Home Visitation." Washington, DC: Office of Juvenile Justice and Delinquency Prevention, 1998.

Piliavin, Irving, and Scott Briar. "Police Encounters with Juveniles." *American Journal of Sociology* 70 (1964):206–214.

Platt, Anthony. *The Child Savers: The Invention of Delinquency*. 2nd ed. Chicago: University of Chicago Press, 1977.

Pope, Carl E., Rick Lovell, and Heidi M. Hsia. "Disproportionate Minority Confinement: A Review of the Research Literature From 1989 Through 2001." Washington, DC: Office of Juvenile Justice and Delinquency Prevention, 2002.

Pound, Roscoe. "Discretion, Dispensation, and Mitigation: The Problem of the Individual Special Case." *New York University Law Review* 35 (1960):925–937.

President's Commission on Law Enforcement and Administration of Justice. *The Challenge of Crime in a Free Society*. Washington, DC: GPO, 1967.

—. *Task Force Report: Corrections*. Washington, DC: GPO, 1967.

—. *Task Force Report on Juvenile Delinquency and Youth Crime*. Washington, DC: GPO, 1967.

Puzzanchera, Charles M. "Juvenile Delinquency Probation Caseload, 1990–1999." Washington, DC: Office of Juvenile Justice and Delinquency Prevention, 2003.

Puzzanchera, Charles, Anne L. Stahl, Terrence A. Finnegan, Howard Snyder, Rowen S. Poole, and Nancy Tierney. *Juvenile Court Statistics 1997*. Washington, DC: Office of Juvenile Justice and Delinquency Prevention, 2000.

Puzzanchera, Charles, Anne L. Stahl, Terrence A. Finnegan, Nancy Tierney, and Howard Snyder. *Juvenile Court Statistics 2000*. Washington, DC: Office of Juvenile Justice and Delinquency Prevention, 2004.

—. *Juvenile Court Statistics 1999*. Washington, DC: Office of Juvenile Justice and Delinquency Prevention, 2003.

—. *Juvenile Court Statistics 1998*. Washington, DC: Office of Juvenile Justice and Delinquency Prevention, 2003.

Quinn, James F. *Corrections: A Concise Introduction*. 2nd edition. Prospect Heights, IL: Waveland Press, 2003.

Chapter Resources

Rainville, Gerard A., and Steven K. Smith. "Juvenile Felony Defendants in Criminal Courts." Washington, DC: Bureau of Justice Statistics, 2003.

Riksheim, Eric, and Steven Chermak. "Causes of Police Behavior Revisited." *Journal of Criminal Justice* 21 (1993):353–382.

Ringwalt, C. S., S. T. Ennett, and K. Holt. "An Outcome Evaluation of D.A.R.E. (Drug Abuse Resistance Education)." *Health Education Research Theory and Practice* 6 (1991):327–337.

Rosenbaum, Dennis P., R. L. Flewelling, S. L. Baieler, G. L. Ringwalt, and D. L. Wilkinson. "Cops in the Classroom: A Longitudinal Evaluation of Drug Abuse Resistance Education (D.A.R.E.)." *Journal of Research in Crime and Delinquency* 3 (1994):3–31.

Rosenbaum, Dennis P., and Gail S. Hanson. "Assessing the Effects of School-Based Drug Education: A Six-Year Multilevel Analysis of D.A.R.E." *Journal of Research in Crime and Delinquency* 35 (1998):381–412.

Sampson, Robert. "Effects of Socioeconomic Context of Official Reaction to Juvenile Delinquency." *American Sociological Review* 51 (1986):876–885.

Schlossman, Steven. *Love and the American Delinquency: The Theory and Practice of "Progressive" Juvenile Justice.* Chicago: University of Chicago Press, 1977.

Sealock, Miriam D., and Sally S. Simpson. "Unraveling Bias in Arrest Decisions: The Role of Juvenile Offender Type-Scripts." *Justice Quarterly* 15 (1998):427–457.

Seiter, Richard P., and Angela D. West. "Supervision Styles in Probation and Parole: An Analysis of Activities." *Journal of Offender Rehabilitation* 38 (2003):57–75.

Sellin, Thorsten, and Marvin Wolfgang. *The Measurement of Delinquency.* New York: Wiley, 1964.

Sherman, Lawrence W., Denise C. Gottfredson, Doris L. Mackenzie, John Eck, Peter Reuter, and Shawn D. Bushway. "Preventing Crime: What Works, What Doesn't, What's Promising." Washington, DC: National Institute of Justice, 1998.

Sickmund, Melissa. "Juveniles in Corrections." In *Juvenile Offenders and Victims: National Report Series.* Washington, DC: Office of Juvenile Justice and Delinquency Prevention, 2004.

———. "Juveniles in Court." In *Juvenile Offenders and Victims: National Report Series.* Washington, DC: Office of Juvenile Justice and Delinquency Prevention, 2003.

Skolnick, Jerome H., and David H. Bayley. "Community Policing: Issues and Practices Around the World." In *Perspective on Policing.* Washington, DC: US Department of Justice, 1988.

Smith, Douglas. "The Organizational Context of Legal Control." *Criminology* 22 (1984):19–38.

Smith, Douglas, and Jody Klein. "Police Control of Interpersonal Disputes." *Social Problems* 31 (1984):468–481.

Smith, Douglas, and Christy Visher. "Street-Level Justice: Situational Determinants of Police Arrest Decision." *Social Problems* 31 (1981):468–481.

Smith, Douglas, Christy A. Visher, and Laura A. Davidson. "Equity and Discretionary Justice: The Influence of Race on Police Arrest Decisions." *Journal of Criminal Law and Criminology* 75 (1984):234–249.

Snyder, Howard N. "Juvenile Arrests 2002." In *Juvenile Justice Bulletin.* Washington, DC: Office of Juvenile Justice and Delinquency Prevention, 2004.

———. "Law Enforcement and Juvenile Crime." In *Juvenile Offenders and Victims: National Report Series.* Washington, DC: Office of Juvenile Justice and Delinquency Prevention, 2001.

Snyder, Howard N., and Melissa Sickmund. *Juvenile Offenders and Victims: 1999 National Report.* Washington, DC: Office of Juvenile Justice and Delinquency Prevention, 1999.

Spergel, Irving, Ron Chance, Kenneth Ehrensaft, Thomas Regulus, Candice Kane, Robert Laseter, Alba Alexander, and Sandra Oh. "Gang Suppression and Intervention: Community Models. Research Summary." Washington, DC: Office of Juvenile Justice and Delinquency Prevention, 1994.

Steinberg, Laurence, and Amanda Sheffield Morris. "Adolescent Development." *Annual Review of Psychology* 52 (2001):23–45.

Taylor, Terrance J., K. B. Turner, Finn-Aage Esbensen, and Thomas L. Winfree, Jr. "Coppin' an Attitude: Attitudinal Differences Among Juveniles Toward Police." *Journal of Criminal Justice* 29 (2001):295–305.

Terry, Robert M. "Discrimination in the Handling of Juvenile Offenders by Social Control Agencies." *Journal of Research in Crime and Delinquency* 4 (1967):218–230.

Thornberry, Terence P. "Race, Socioeconomic States, and Sentencing in the Juvenile Justice System." *Journal of Criminal Law and Criminology* 70 (1979):164–171.

Thornberry, Terence P., David Huizinga, and Rolf Loeber. "The Causes and Correlates Studies: Findings and Policy Implications." *Juvenile Justice* 9, no. 1 (2004):3–19.

Walker, Samuel, and Charles M. Katz. *Police in America: An Introduction.* 5th ed. Boston: McGraw-Hill, 2005.

Wasserman, Gail A., Kate Keenan, Richard E. Tremblay, John D. Cole, Todd I. Herrenkohl, Rolf Loeber, and David Petechuk. "Risk and Protective Factors of Child Delinquency." Washington, DC: Office of Juvenile Justice and Delinquency Prevention, 2003.

Wasserman, Gail A., Laurie S. Miller, and Lynn Cothern. "Prevention of Serious and Violent Juvenile Offending." Washington, DC: Office of Juvenile Justice and Delinquency Prevention, 2000.

Webster-Statton, Carolyn. "The Incredible Years Training Series." Washington, DC: Office of Juvenile Justice and Delinquency Prevention, 2000.

West, Steven L., and Keir K. O'Neal. "Project D.A.R.E. Outcome Effectiveness Revisited." *American Journal of Public Health* 94 (2004):1027–1029.

Westermann, Ted D., and James W. Burfeind. *Crime and Justice in Two Societies: Japan and the United States.* Pacific Grove, CA: Brooks/Cole, 1991.

White, Helen Raskin. "Early Problem Behavior and Later Drug Problems." *Journal of Research in Crime and Delinquency* 29 (1992):412–429.

Wiebush, Richard G., Betsie McNulty, and Thao Le. "Implementation of the Intensive Community-Based Aftercare Program." Washington, DC: Office of Juvenile Justice and Delinquency Prevention, 2000.

Wilson, James Q. "The Police and the Delinquent in Two Cities." In *Controlling Delinquents,* edited by Stanton Wheeler, 9–30. New York: Wiley, 1968.

———. *Varieties of Police Behavior.* Cambridge, MA: Harvard University Press, 1968.

Wilson, John J., and James C. Howell. *Comprehensive Strategy for Serious, Violent, and Chronic Juvenile Offenders: Program Summary.* Washington, DC: Office of Juvenile Justice and Delinquency Prevention, 1993.

Worrall, John. "California Street Terrorism Enforcement and Prevention Act." In *Encyclopedia of Juvenile Justice,* edited by Marilyn D. McShane and Frank P. Williams, 39–40. Thousand Oaks, CA: Sage, 1993.

ENDNOTES

1. Hellum, "Juvenile Justice."
2. Feld, *Bad Kids.*
3. Snyder and Sickmund, *Juvenile Offenders and Victims,* 97.
4. Pound, "Discretion, Dispensation, and Mitigation," 926.
5. President's Commission, *Youth Crime,* 41.
6. Howell, *Juvenile Justice,* 18.
7. Oxford English Dictionary: http://dictionary.oed.com
8. Dryfoos, *Adolescents at Risk,* 25.
9. Hall, *Adolescence.*
10. Lerner and Galambos, "Adolescent Development;" and Steinberg and Morris, "Adolescent Development."
11. Jessor, "Risk Behavior in Adolescence."
12. Jessor, "Risk Behavior in Adolescence;" Lerner and Galambos, "Adolescent Development;" and Steinberg and Morris, "Adolescent Development."
13. Elliott, Huizinga, and Menard. *Multiple Problem Youth*; Huizinga et al., "Co-occurrence," 1; Thornberry, Huizinga, and Loeber, "Causes and Correlates Studies;" and White, "Early Problem Behavior," 414.
14. Lerner and Galambos, "Adolescent Development;" and Loeber, Farrington, and Petechuk, "Child Delinquency."
15. Burns et al., "Treatment," 1–2.
16. Hawkins et al., "Predictors of Youth Violence;" and Wasserman et al., "Risk and Protective Factors."
17. Coolbaugh and Hansel, "Comprehensive Strategy," 3.
18. Lehman, Hawkins, and Catalano, *Corrections Today,* 94.
19. Burns et al., "Treatment," 3; Howell, *Preventing and Reducing;* and Thornberry, Huizinga, and Loeber, "The Causes and Correlates Studies."

20. Dryfoos, *Adolescents at Risk,* 228–233; Lerner and Galambos, "Adolescent Development," 436; and Burns et al., "Treatment," 3–6.

21. Lerner and Galambos, "Adolescent Development," 436, based on Dryfoos, *Adolescents at Risk,* 228–233.

22. Mihalic et al., "Successful Program Implementation," 3–9.

23. Ibid., 7.

24. Burns et al., "Treatment;" Coolbaugh and Hansel, "Comprehensive Strategy;" Mihalic et al., "Blueprints for Violence Prevention;" Mihalic et al., "Successful Program Implementation;" Sherman et al., "Preventing Crime;" Wasserman, Miller, and Cothern, "Prevention of Offending;" and Wilson and Howell, *Comprehensive Strategy.*

25. Mihalic et al., "Blueprints for Violence Prevention," 4–5; and Olds, Hill, and Rumsey, "Prenatal."

26. Mihalic et al., "Blueprints for Violence Prevention," 5–6.

27. Mihalic et al., "Blueprints for Violence Prevention," 5–6; and Webster-Statton, "Incredible Years."

28. Mihalic et al., "Blueprints for Violence Prevention," 6.

29. Ibid., 6–7.

30. Ibid., 8.

31. Ibid.

32. Burns et al., "Treatment," 9.

33. National Advisory Commission, *Task Force Report*; and Howell, *Juvenile Justice,* 18.

34. This section title is drawn from Piliavin and Briar's classic study of the same name.

35. Walker and Katz, *Police in America,* 253; see also Bartollas and Miller, *Juvenile Justice in America,* 112; and Gaines, "Police Responses to Delinquency," 286.

36. Snyder, "Juvenile Arrests 2003," 1.

37. Bartollas and Miller, *Juvenile Justice in America,* 112.

38. Gaines, "Police Responses to Delinquency," 286.

39. Cited in Walker and Katz, *Police in America,* 253.

40. Bartollas and Miller, *Juvenile Justice in America,* 112.

41. Walker and Katz, *Police in America,* 253. See Borrero, "Widening Mistrust;" Decker, "Citizen Attitudes;" Hurst and Frank, "How Kids View Cops;" Leiber, Nalla, and Farnworth, "Explaining Juveniles' Attitudes;" and Taylor et al., "Coppin' an Attitude."

42. Gaines, "Police Responses to Delinquency," 286; and Snyder and Sickmund, *Juvenile Offenders and Victims,* 139.

43. Gaines, "Police Responses to Delinquency," 286–287.

44. Walker and Katz, *Police in America,* 253.

45. Platt, *Child Savers,* 139–140.

46. Feld, *Bad Kids*; Platt, *Child Savers*; and Schlossman, *Love.*

47. Feld, *Bad Kids,* 66; and Platt, *Child Savers,* 141.

48. Feld, *Bad Kids,* 66.

49. Gaines, "Police Responses to Delinquency," 290.

50. Ibid.

51. Kelling and Moore, "Evolving Strategy of Policing;" Fogelson, *Big City Police*; and Westermann and Burfeind, *Crime and Justice,* 68–70, 154.

52. Lewett, "Centralization."

53. Moore, Trojanowicz, and Kelling, "Crime and the Police."

54. Kelling, "Police and Communities."

55. Goldstein, *Problem-Oriented Policing.*

56. Moore, Trojanowicz, and Kelling, "Crime and the Police."

57. Skolnick and Bayley, "Community Policing," 2.

58. Ibid.

59. Gaines, "Police Responses to Delinquency," 289; and Hickman and Reaves, "Community Policing."

60. Becker, Agopian, and Yeh, "Impact Evaluation;" Botvin et al., "Long-Term Follow-Up Results;" Ennett et al., "Long-Term Evaluation;" Ennett et al., "Meta-Analysis;" Harmon, "Reducing the Risk;" Lynam, Milich, and Zimmerman, "Project D.A.R.E.;" Ringwalt, Ennett, and Holt, "Outcome Evaluation;" Rosenbaum et al., "Cops in the Classroom;" Rosenbaum and Hanson, "Assessing the Effects;" and West and O'Neal, "Project D.A.R.E."

61. Gaines, "Police Responses to Delinquency," 287–289.

62. Office of Community Oriented Policing Services, "COPS in Schools," 1. SROs are also identified as a mechanism to increase school safety by increasing juvenile accountability (Decker, "Increasing School Safety," 8).

63. Andrews and Marble, "Changes;" Danegger et al., *Strategic Planning Guide*; and Albert, "Juvenile Accountability."

64. Andrews and Marble, "Changes," 1.

65. Andrews and Marble, "Changes," 2; Beyer, "Best Practices;" and Decker, "Increasing School Safety," 8.

66. Spergel et al., "Gang Suppression and Intervention," 8.

67. Ibid., 13.

68. Ibid., 8.

69. Worrall, "California Street Terrorism."

70. Bordua, "Recent Trends."

71. See also Goldman, *Differential Selection*; and Piliavin and Briar, "Police Encounters with Juveniles."

72. Krisberg and Austin, *Children of Ishmael*, 83.

73. Sickmund, "Juveniles in Court," 18.

74. Ibid., 18.

75. Black and Reiss, "Police Control of Juveniles;" Lundman, Sykes, and Clark, "Police Control of Juveniles;" Piliavin and Briar, "Police Encounters with Juveniles;" and Sellin and Wolfgang, *Measurement of Delinquency*.

76. Gaines, "Police Responses to Delinquency," 289.

77. Black and Reiss, "Police Control of Juveniles," 63; see also Lundman, Sykes, and Clark, "Police Control of Juveniles;" Piliavin and Briar, "Police Encounters with Juveniles;" and Terry, "Discrimination."

78. Cicourel, *Social Organization*; Cohen and Kluegel, "Determinants;" Lattimore, Visher, and Linster, "Predicting Rearrest;" Piliavin and Briar, "Police Encounters with Juveniles;" Sellin and Wolfgang, *Measurement of Delinquency*; and Terry, "Discrimination."

79. Piliavin and Briar, "Police Encounters with Juveniles." See also Cicourel, *Social Organization*; Goldman, *Differential Selection*; Klinger, "Demeanor of Crime;" Lundman, "Demeanor of Crime," and "Demeanor and Arrest;" and Sellin and Wolfgang, *Measurement of Delinquency*.

80. Black and Reiss, "Police Control of Juveniles;" Bishop and Frazier, "Influence of Race," and "Gender Bias;" Cohen and Kluegel, "Determinants;" Dannefer and Schutt, "Race;" Ferdinand and Luchtenhand, "Inner-City Youths;" Fagan, Slaughter, and Hartstone, "Blind Justice;" Lundman, "Demeanor and Arrest;" Riksheim and Chermak, "Causes;" Sealock and Simpson, "Unraveling Bias;" Smith and Visher, "Street-Level Justice;" Terry, "Discrimination;" and Thornberry, "Race."

81. Black and Reiss, "Police Control of Juveniles;" Goldman, *Differential Selection*; and Terry, "Discrimination."

82. Sampson, "Effects of Socioeconomic Context;" see also Smith, Visher, and Davidson, "Equity and Discretionary Justice."

83. Smith and Klein, "Police Control."

84. Wilson, "Two Cities," and *Varieties of Police Behavior*.

85. Smith, "Organizational Context."

86. Wilson, "Two Cities," and *Varieties of Police Behavior*.

87. *Brinegar v. U.S.*, 338 U.S. 160 (1948).

88. *State v. Lowry*, 230 A. 2d 907 (1967).

89. Hemmens, Steiner, and Mueller, *Significant Cases*, 26.

90. Ferdico, *Criminal Procedure*, 108–111.

91. Ibid., 121–126.

92. Montana Code Annotated 46-6-14. Method of arrest.

93. *Tennessee v. Garner*, 471 U.S. 1 (1985).

94. Ferdico, *Criminal Procedure*, 135.

95. Ibid., 136.

96. *Miranda v. Arizona*, 384 U.S. 436 (1966).

97. *In re Gault*, 387 U.S. 1 (1967).

98. *Haley v. Ohio*, 332 U.S. 596 (1948); *Gallegos v. Colorado*, 370 U.S. 49 (1962).

99. *Haley v. Ohio*, 332 U.S. 596 (1948).

100. Puzzanchera et al., *Juvenile Court Statistics 2000*, 24; and Sickmund, "Juveniles in Court," 2, 12.

101. Puzzanchera et al., *Juvenile Court Statistics 2000,* 65; and Sickmund, "Juveniles in Court," 30.

102. Sickmund, "Juveniles in Court," 4.

103. Puzzanchera et al., *Juvenile Court Statistics 2000,* 8.

104. Puzzanchera et al., *Juvenile Court Statistics 2000,* 6.

105. Puzzanchera et al., *Juvenile Court Statistics 2000,* 14.

106. Puzzanchera et al., *Juvenile Court Statistics 2000,* 18–19; and Puzzanchera et al., *Juvenile Court Statistics 1999,* 15.

107. Puzzanchera et al., *Juvenile Court Statistics 2000,* 9.

108. Puzzanchera et al., *Juvenile Court Statistics 2000,* 2; and Snyder and Sickmund, *Juvenile Offenders and Victims,* 97.

109. Puzzanchera et al., *Juvenile Court Statistics 2000,* 26; and Sickmund, "Juveniles in Court," 20.

110. Sickmund, "Juveniles in Court," 20.

111. Puzzanchera et al., *Juvenile Court Statistics 2000,* 32–33; and Sickmund, "Juveniles in Court," 21.

112. Puzzanchera et al., *Juvenile Court Statistics 1997,* 38. 1997 was the last year for which *Juvenile Court Statistics* provided estimates of the trends and volume of petitioned status offense case, due to variation in data collection by agencies processing status offenders (Puzzanchera et al., *Juvenile Court Statistics 1998,* 4).

113. Sickmund, "Juveniles in Court," 24.

114. Puzzanchera et al., *Juvenile Court Statistics 2000,* 22.

115. *Schall v. Martin* 104 U.S. 2403 (1984).

116. Hemmens, Steiner, and Mueller, *Significant Cases,* 58–59.

117. Ibid.

118. Harms, "Detention in Delinquency Cases," 1; and Puzzanchera et al., *Juvenile Court Statistics 2000,* 26.

119. Puzzanchera et al., *Juvenile Court Statistics 2000,* 26.

120. Harms, "Detention in Delinquency Cases," 1.

121. Puzzanchera et al., *Juvenile Court Statistics 2000,* 26.

122. Harms, "Detention in Delinquency Cases," 1.

123. Ibid.

124. Puzzanchera et al., *Juvenile Court Statistics 2000,* 28.

125. Puzzanchera et al., *Juvenile Court Statistics 2000,* 27.

126. Puzzanchera et al., *Juvenile Court Statistics 2000,* 68–69.

127. Hsia, Bridges, and McHale, *Disproportionate Minority Confinement,* 1.

128. Hsia, Bridges, and McHale, *Disproportionate Minority Confinement*; Pope, Lovell, and Hsia, *Disproportionate Minority Confinement*; Puzzanchera et al., *Juvenile Court Statistics 2000*; and Sickmund, "Juveniles in Corrections."

129. Pope, Lovell, and Hsia, *Disproportionate Minority Confinement,* 5.

130. Sickmund, "Juveniles in Court," 6–10.

131. Ibid., 6–10, 26–28.

132. Sickmund, "Juveniles in Court," 7. Rainville and Smith, in "Juvenile Felony Defendants," examine juvenile felony defendants in criminal courts. Our concern here is with juvenile court processes.

133. Puzzanchera et al., *Juvenile Court Statistics 2000,* 34–38; and Sickmund, "Juveniles in Court," 6–10, 26–28.

134. Puzzanchera et al., *Juvenile Court Statistics 2000,* 38.

135. Puzzanchera et al., *Juvenile Court Statistics 2000,* 39–43.

136. Puzzanchera et al., *Juvenile Court Statistics 2000,* 70.

137. Sickmund, "Juveniles in Court," 3.

138. Montana Code Annotated 41-5-1512 and 1513.

139. Sickmund, "Juveniles in Court," 3.

140. Puzzanchera et al., *Juvenile Court Statistics 2000,* 56–59.

141. Puzzanchera et al., *Juvenile Court Statistics 2000,* 56.

142. Puzzanchera et al., *Juvenile Court Statistics 2000,* 71.

143. Puzzanchera et al., *Juvenile Court Statistics 2000,* 71.

144. Puzzanchera et al., *Juvenile Court Statistics 2000,* 71.

145. Lipton, Martinson, and Wilks, "Effectiveness of Correctional Treatment." See also American Friends Service Committee, *Struggle for Justice;* and Lerman, *Community Treatment and Control.*

Chapter Resources

146. Allen, "Legal Values," 226.
147. Puzzanchera, "Juvenile Delinquency Probation Caseload," 1.
148. Cromwell, del Carmen, and Alarid, *Community-Based Corrections,* 71.
149. Ibid., 75–81.
150. *In the Interest of D. S. and J. V., Minor Children,* 652 So.2D 892 (Fla. Appl. 1995).
151. Hemmens, Steiner, and Mueller, *Significant Cases,* 124–125.
152. Jordon and Sasfy, *National Impact Program Evaluation*; Klockers, "Theory of Probation Supervision;" Ohlin, Piven, and Pappenfort, "Major Dilemmas;" and Seiter and West, "Supervision Styles."
153. Czajkoski, "Exposing;" and Klockers, "Theory of Probation Supervision."
154. Feld, *Bad Kids,* 66.
155. Quinn, *Corrections,* 84–87.
156. Cromwell, del Carmen, and Alarid, *Community-Based Corrections,* 138.
157. Ibid., 144–150.
158. President's Commission, *Challenge of Crime,* 80.
159. Ibid., 7.
160. Ibid., 83.
161. Belbott, "Restorative Justice," 324.
162. Bazemore and Umbreit, *Balanced and Restorative Justice.*
163. Belbott, "Restorative Justice," 324.
164. Ibid.
165. Ibid.
166. Ibid., 325.
167. Sickmund, "Juveniles in Corrections," 3.
168. Ibid., 4.
169. Ibid., 5.
170. Ibid., 10.
171. Allen-Hagen, "Public Juvenile Facilities."
172. Allen-Hagen, "Public Juvenile Facilities." See also Sickmund, "Juveniles in Corrections," 16.
173. Allen-Hagen, "Public Juvenile Facilities."
174. Wiebush, McNulty, and Le, "Implementation," 2.
175. Huizinga et al., "Co-Occurrence," 1.
176. Coolbaugh and Hansel, "Comprehensive Strategy," 3.
177. Sickmund, "Juveniles in Court," 24.

Author Index

Subject Index